Methods in Cell Biology

VOLUME 44
Drosophila melanogaster:
Practical Uses in Cell and Molecular Biology

Series Editors

Leslie Wilson
Department of Biological Sciences
University of California, Santa Barbara
Santa Barbara, California

Paul Matsudaira
Whitehead Institute for Biomedical Research and
Department of Biology
Massachusetts Institute of Technology
Cambridge, Massachusetts

Methods in Cell Biology

Prepared under the Auspices of the American Society for Cell Biology

VOLUME 44

Drosophila melanogaster:
Practical Uses in Cell and Molecular Biology

Edited by

Lawrence S. B. Goldstein

Howard Hughes Medical Institute
Division of Cellular and Molecular Medicine
Department of Pharmacology
School of Medicine
University of California, San Diego
La Jolla, California

Eric A. Fyrberg

Department of Biology
Johns Hopkins University
Baltimore, Maryland

ACADEMIC PRESS

San Diego   New York   Boston   London   Sydney   Tokyo   Toronto

Cover photograph (paperback edition only): Twin-spot clones of ovarian follicle cells in a stage 9 follicle.   For more details see Chapter 34, Figure 3, Part B.

This book is printed on acid-free paper. ∞

Academic Press, Inc.
A Division of Harcourt Brace & Company
525 B Street, Suite 1900, San Diego, California 92101-4495

United Kingdom Edition published by
Academic Press Limited
24-28 Oval Road, London NW1 7DX

International Standard Serial Number: 0091-679X

International Standard Book Number: 0-12-287070-0 (comb)
International Standard Book Number: 0-12-564145-1 (case)

PRINTED IN THE UNITED STATES OF AMERICA
94 95 96 97 98 99 EB 9 8 7 6 5 4 3 2 1

CONTENTS

PART III Biochemical Preparation Methods

25. Immunofluorescence Analysis of the Cytoskeleton during
Oogenesis and Early Embryogenesis

William E. Theurkauf

26. High-Resolution Microscopic Methods for the Analysis of Cellular
Movements in *Drosophila* Embryos

*Daniel P. Kiehart, Ruth A. Montague, Wayne L. Rickoll,
Graham H. Thomas, and Donald Foard*

27. The Use of Photoactivatable Reagents for the Study of Cell
Lineage in *Drosophila* Embryogenesis

Charles H. Girdham and Patrick H. O'Farrell

PART V   Analysis of Gene Expression *in Situ*

CONTRIBUTORS

Numbers in parentheses indicate the pages on which the authors' contributions begin.

Andrew J. Andres (565), Howard Hughes Medical Institute, University of Utah, Salt Lake City, Utah 84112

Deborah J. Andrew (353), Department of Cell Biology and Anatomy, Johns Hopkins University School of Medicine, Baltimore, Maryland 21205

Peter B. Becker (207), Gene Expression Programme, European Molecular Biology Laboratory, 69117 Heidelberg, Germany

Anke E. L. Beermann (715), Department of Molecular and Cellular Biology, Harvard University, Cambridge, Massachusetts 02138

Sanford I. Bernstein (237), Department of Biology, and Molecular Biology Institute, San Diego State University, San Diego, California 92182

Ann Beyer (613), Department of Microbiology, University of Virginia Health Sciences Center, Charlottesville, Virginia 22908

Allan J. Bieber (683), Department of Biological Sciences, Purdue University, West Lafayette, Indiana 47907

Paul M. Bingham (599), Department of Biochemistry and Cell Biology, State University of New York at Stony Brook, Stony Brook, New York 11794

Silvia Bonaccorsi (371), Istituto Pasteur, Fondazione Cenci-Bolognetti, and Centro di Genetica Evoluzionistica del CNR, Dipartimento di Genetica e Biologia Molecolare, Universitá di Roma "La Sapienza," 00185 Rome, Italy

Andrea H. Brand (635), Wellcome/CRC Institute, and Department of Genetics, University of Cambridge, Cambridge CB2 1QR, United Kingdom

Lucy Cherbas (161), Department of Biology, Indiana University, Bloomington, Indiana 47405

Peter Cherbas (161), Department of Biology, Indiana University, Bloomington, Indiana 47405

Lynn Cooley (545), Department of Genetics, Yale University School of Medicine, New Haven, Connecticut 06510

Susan Cumberledge (143), Department of Biochemistry, Stanford University School of Medicine, Stanford, California 94305

Gideon Dreyfuss (191), Howard Hughes Medical Institute, and Department of Biochemistry and Biophysics, University of Pennsylvania School of Medicine, Philadelphia, Pennsylvania 19104

Bruce Edgar (697), Fred Hutchinson Cancer Research Center, Seattle, Washington 98104

Sarah C. R. Elgin (99, 185), Department of Biology, Washington University, St. Louis, Missouri 63130

J. H. Fessler (303), Molecular Biology Institute, and Department of Biology, University of California, Los Angeles, Los Angeles, California 90024

L. I. Fessler (303), Molecular Biology Institute, and Department of Biology, University of California, Los Angeles, Los Angeles, California 90024

Donald Foard (507), Department of Cell Biology, Duke University Medical Center, Durham, North Carolina 27710

Eric Fyrberg (237), Department of Biology, Johns Hopkins University, Baltimore, Maryland 21218

Maurizio Gatti (371), Istituto Pasteur, Fondazione Cenci-Bolognetti, and Centro di Genetica Evoluzionistica del CNR, Dipartimento di Genetica e Biologia Molecolare, Universitá di Roma "La Sapienza," 00185 Rome, Italy

Charles H. Girdham (533), Department of Biochemistry and Biophysics, University of California, San Francisco, San Francisco, California 94143

Michael L. Goldberg (33), Section of Genetics and Development, Cornell University, Ithaca, New York 14853-2703

Lawrence S. B. Goldstein (3), Howard Hughes Medical Institute, Division of Cellular and Molecular Medicine, Department of Pharmacology, School of Medicine, University of California, San Diego, La Jolla, California 92093-0683

Shermali Gunawardena (393), Departments of Anatomy and Molecular and Cellular Biology, and Graduate Program in Genetics, University of Arizona College of Medicine, Tucson, Arizona 85724

Bruce A. Hamilton (81), Whitehead Institute for Biomedical Research, Cambridge, Massachusetts 02142

Stephen D. Harrison (655), Department of Molecular and Cell Biology, University of California, Berkeley, Berkeley, California 94720

Michael Hortsch (289), Department of Anatomy and Cell Biology, University of Michigan, Ann Arbor, Michigan 48109

Daniel G. Jay (715), Department of Molecular and Cellular Biology, Harvard University, Cambridge, Massachusetts 02138

James T. Kadonaga (225), Department of Biology, and Center for Molecular Genetics, University of California, San Diego, La Jolla, California 92093-0347

Rohinton T. Kamakaka (225), Department of Biology, University of California, San Diego, La Jolla, California 92093-0347

Douglas R. Kellogg (259), Department of Physiology, University of California, San Francisco, San Francisco, California 94143

Daniel P. Kiehart (507), Department of Cell Biology, Duke University Medical Center, Durham, North Carolina 27710

Joseph Kramer (599), Department of Biochemistry and Cell Biology, State University of New York at Stony Brook, Stony Brook, New York 11794

Mark Krasnow (143), Department of Biochemistry, Stanford University School of Medicine, Stanford, California 94305

Ruth Lehmann (575), Whitehead Institute for Biomedical Research, Howard Hughes Medical Institute, Massachusetts Institute of Technology, Cambridge, Massachusetts 02142

Anthony P. Mahowald (129), Department of Molecular Genetics and Cell Biology, University of Chicago, Chicago, Illinois 60637

Armen S. Manoukian (635), Department of Genetics, Harvard Medical School, Boston, Massachusetts 02115

Kathleen A. Matthews (13), Department of Biology, Indiana University, Bloomington, Indiana 47405

Erika L. Matunis (191), Rockefeller University, New York, New York 10021

Michael J. Matunis (191), Laboratory of Cell Biology, Rockefeller University, New York, New York 10021

Kent L. McDonald (411), Electron Microscopy Laboratory, University of California, Berkeley, Berkeley, California 94720

Kathryn G. Miller (259), Department of Biology, Washington University, St. Louis, Missouri 63130

Ruth A. Montague (507), Department of Cell Biology, Duke University Medical Center, Durham, North Carolina 27710

Robert Moss (161), Department of Biology, Wofford College, Spartanburg, South Carolina 29301

Jeanette E. Natzle (109), Section of Molecular and Cellular Biology, University of California, Davis, Davis, California 95616

R. E. Nelson (303), Molecular Biology Institute, and Department of Biology, University of California, Los Angeles, Los Angeles, California 90024

Patrick H. O'Farrell (533), Department of Biochemistry and Biophysics, University of California, San Francisco, San Francisco, California 94143

Yvonne Osheim (613), Department of Microbiology, University of Virginia Health Sciences Center, Charlottesville, Virginia 22908

Mary-Lou Pardue (333), Department of Biology, Massachusetts Institute of Technology, Cambridge, Massachusetts 02139

Nipam H. Patel (445), Department of Embryology, Carnegie Institution of Washington, Baltimore, Maryland 21210

Norbert Perrimon (635), Department of Genetics, and Howard Hughes Medical Institute, Harvard Medical School, Boston, Massachusetts 02115

Sergio Pimpinelli (371), Istituto Pasteur, Fondazione Cenci-Bolognetti, and Centro di Genetica Evoluzionistica del CNR, Dipartimento di Genetica e Biologia Molecolare, Università di Roma "La Sapienza," 00185 Rome, Italy

Wayne L. Rickoll (507), Department of Biology, University of Puget Sound, Tacoma, Washington 98416

Mary Rykowski (393), Departments of Anatomy and Molecular and Cellular Biology, and Graduate Program in Genetics, University of Arizona College of Medicine, Tucson, Arizona 85724

William M. Saxton (279), Department of Biology, Indiana University, Bloomington, Indiana 47405

Gerold Schubiger (697), Department of Zoology, University of Washington, Seattle, Washington 98195

Matthew P. Scott (353), Howard Hughes Medical Institute, Department of Developmental Biology and Genetics, Stanford University School of Medicine, Stanford, California 94305

Christopher D. Shaffer (99, 185), Department of Biology, Washington University, St. Louis, Missouri 63130

Martha Sikes (613), Department of Microbiology, University of Virginia Health Sciences Center, Charlottesville, Virginia 22908

Diethard Tautz (575), Zoologisches Institut, Universität München, D-80333 München, Germany

William E. Theurkauf (489), Department of Biochemistry and Cell Biology, State University of New York at Stony Brook, Stony Brook, New York 11794

Graham H. Thomas (507), Department of Biology, and Department of Biochemistry and Molecular Biology, Pennsylvania State University, University Park, Pennsylvania 16802

Carl S. Thummel (565), Howard Hughes Medical Institute, University of Utah, Salt Lake City, Utah 84112

Toshio Tsukiyama (207), Laboratory of Biochemistry, National Cancer Institute, National Institutes of Health, Bethesda, Maryland 20892

Esther Verheyen (545), Department of Genetics, Yale University School of Medicine, New Haven, Connecticut 06510

Gwendolyn D. Vesenka (109), Section of Molecular and Cellular Biology, University of California, Davis, Davis, California 95616

K. VijayRaghavan (237), National Centre for Biological Sciences, Tata Institute of Fundamental Research, Indian Institute of Science Campus, Bangalore 560012, India

Mariana F. Wolfner (33), Section of Genetics and Development, Cornell University, Ithaca, New York 14853

Carl Wu (207), Laboratory of Biochemistry, National Cancer Institute, National Institutes of Health, Bethesda, Maryland 20892

Joann M. Wuller (99, 185), Department of Biology, Washington University, St. Louis, Missouri 63130

Tian Xu (655), Boyer Center for Molecular Medicine, Department of Genetics, Yale University School of Medicine, New Haven, Connecticut 06536

Zuzana Zachar (599), Department of Biochemistry and Cell Biology, State University of New York at Stony Brook, Stony Brook, New York 11794

Kai Zinn (81), Department of Biology, California Institute of Technology, Pasadena, California 91125

PREFACE: WHY THE FLY?

During the 1950s and 1960s the prominence of *Drosophila* research waned as rapid developments in molecular biology focused attention on simple bacterial and phage systems. Many felt that *Drosophila* research was obsolete. However, in the early 1970s the little fly came back into favor, in part because molecular biologists began to study the more complex phenomena unique to metazoa, in part because the small genome and large chromosomes of *Drosophila* facilitated isolation of mutationally identified genes, and in part because of the fast, compact, and easily manipulated lifestyle of the fly. Perhaps most important has been the realization by neurobiologists, behavioral biologists, biochemists, developmental biologists, and cell biologists that the path to understanding most biological phenomena runs through the forest of genetic analysis. Such resurgences in the use of experimental organisms or systems are rare, and this one underscores the importance of decades of genetic analysis and the value of genetic tractability.

The publication of *Molecular Biology of the Cell* (Alberts *et al.*) in 1983 ushered molecular cell biology into the center of biological research. It encouraged the biochemist and behaviorist alike to put their research into a cellular context. This maturation of perspectives is nowhere more evident than in *Drosophila* research, wherein increasingly purposeful mutant collections have been subjected to sophisticated cell biological analyses. This general approach has helped to relate numerous molecular perturbations to cellular defects, and more are discovered daily. Consequently, we now understand in cellular and molecular detail many previously obscure and tantalizing processes, for example, how the coordinates of the embryo are laid down and how individual photoreceptors of the compound eye adopt particular fates. No other metazoan is likely to be as intimately understood by biologists, because none offers the ability to investigate such diverse cellular and molecular behaviors. We envision that molecular genetic dissection of *Drosophila* cell biology will continue to yield valuable insights for many more years.

Drosophila melanogaster: Practical Uses in Cell and Molecular Biology is a compendium of technical chapters designed both to provide state-of-the-art methods to the broad community of cell biologists and to put molecular and cell biological studies of flies into perspective. We thought a good deal about whether such special attention was warranted, and ultimately concluded that there is a genuine need for this information. On one hand, we hope that this book makes the more baroque aspects of genetic nomenclature and procedure accessible to cell biologists. On the other hand, these chapters contain a wealth of information that will be of use to beginning as well as advanced *Drosophila* workers. Finally, because the chapters were written within a year of their publication, the information is, with only rare exceptions, state-of-the-art. Hence, we feel that the book will be a useful general laboratory reference for many years.

In choosing which areas to cover, we were somewhat confounded by the breakdown of the boundaries between cell biology, developmental biology, neurobiology, and molecular biology. Thus, in order to avoid an unwieldy volume, choices in coverage had to be made. Therefore, we focused on including chapters that detail methods most useful to the biologist studying cellular organization, structure, and interaction, or the cellular basis of development in *Drosophila melanogaster*. We did not include chapters covering methods that can be simply transplanted from other systems (e.g., immunoprecipitations or kinase assays), or chapters covering systems where genetic analyses have revealed processes (e.g., signal transduction) for which no unique cellular or biochemical methods have yet to be developed. Finally, we did not include specialized neurobiological or neurophysiological methods, because from our perspective these methods and approaches remain somewhat outside of the main line of cellular analysis. Some scientists might find our choices biased or even idiosyncratic, and perhaps they are. However, our most fervent hope is that this volume will be useful and will serve as a stepping-stone to ever greater levels of understanding for those who study the cell biology of *Drosophila*.

We wish to express our gratitude to each of the contributors who so generously offered their insights, experiences, and hard-won methodological improvements. Were it not for these authors this book would never have been completed. We are also grateful to Phyllis B. Moses and the Academic Press staff, who put together this volume in record time, and to Melanie Field and Samantha Gibba, who devoted considerable time and effort to the completion of this project.

Lawrence S. B. Goldstein
Eric A. Fyrberg

PART I

Basic Genetic Methodologies

During the 90 years in which *Drosophila* has been used for biological studies, a staggering amount of useful information has been generated. Moreover, many of the genetic methods of most potential use to the cell biologist are so far removed from simple Mendelian procedures as to require substantive specialized knowledge. Although this volume cannot substitute for years of laboratory work and experience, the four discussions that follow go a long way toward making genetic studies of *Drosophila* available to the nonexpert. These chapters also provide an overview of more detailed chapters later in this volume, and so should help readers to understand these more advanced works.

Goldstein starts the discussion by describing the most commonly used sources of information (Chapter 1). Most of the cited technical and intellectual references are useful and familiar, but one, *The Development of Drosophila melanogaster,* should be consulted even by seasoned investigators contemplating or engaged in detailed studies of fly biology. Noteworthy also is the update of Lindsley and Grell (now Lindsley and Zimm). Significant and potentially even more valuable, are two electronic databases, *FlyBase* and *Drosophila Information Newsletter,* as these are being continuously updated and will undoubtedly remain more useful than printed compendia. Access to this material is straightforward using personal computers.

Matthews provides a comprehensive description of *Drosophila* husbandry and handling, as well as a guide to nomenclature (Chapter 2). She has the responsibility for keeping the world's largest stock collection, has experienced nearly every problem, and generously discusses solutions. For the beginner, here is a well-written chapter that conveys everything you need to know about *Drosophila* stocks and their manipulation. For veteran cell biologists who think that emc^2 has something to do with Einstein, this is an opportunity to learn how mutants and complex chromosomes and genotypes are named. Even the experienced *Drosophila* worker should resist the temptation to skip over this chapter because there are so many useful suggestions for improving the accuracy and efficiency of stock handling.

Wolfner and Goldberg have put together an equally comprehensive treatise (Chapter 3), but the emphasis is on strategies for using *Drosophila* genetics to improve our understanding of cell biological problems. Wolfner and Goldberg illustrate how to go from genes to mutants and vice versa and how to isolate mutants that perturb either a common process or a particular phase of development. Also discussed are analytical methods that collectively will extract the most information from a given set of mutant defects. Again, we consider this essential reading for novices and we trust that more seasoned investigators will find useful information here as well.

In the final chapter (Chapter 4), Hamilton and Zinn discuss strategies for directed mutagenesis of genes already identified molecularly. They address a significant problem facing *Drosophila* geneticists, which is that it remains difficult to efficiently target mutagenesis to a particular gene. Occasionally, when one can screen for loss of an assayable product or loss of an antibody epitope, this is not a problem. However, if the gene in question has no links to specific reagents, a more general assay is required. Since P-elements tend to jump locally, and also to delete chromosomal regions flanking them, mobilization of one or more P-transposons located close to a gene in question is proving to be a satisfactory, if not necessarily rapid, means by which to induce mutations. It is clear, however, that further modifications of this basic protocol will continue to streamline directed mutagenesis, as will continued work on homologous recombination.

CHAPTER 1

Sources of Information about the Fly: Where to Look It Up

Lawrence S. B. Goldstein

Howard Hughes Medical Institute
Division of Cellular and Molecular Medicine
Department of Pharmacology
School of Medicine
University of California, San Diego
La Jolla, California 92093

I. Introduction

One of the most valuable assets in working with *Drosophila melanogaster* is the large body of literature and technical information dating back to the early days of this century. While valuable, this body of knowledge is simultaneously daunting to the investigator who is new to working with *D. melanogaster*. It is the purpose of this chapter to briefly summarize the most relevant sources of information and how to access them.

II. Technical References

Using *D. melanogaster* in research requires an understanding of how to grow and maintain the organism as well as knowledge of the genetic system. Fortunately, as described below, several excellent resources exist and are readily available.

A. *Drosophila: A Practical Approach*

This slender volume (Roberts, 1986) is quite useful for basic care and feeding of the fly. While it contains an eclectic mix of information, most of it is quite useful and is written by an excellent group of authors. It includes sections covering theory and practice of *Drosophila* care and feeding, mutagenesis of various sorts, molecular biology, transformation, *in situ* hybridization, embryo analysis, etc. The chapters are well laid out and include numerous detailed protocols directly suitable for use in the laboratory.

B. *Drosophila: A Laboratory Handbook*

This ample volume (Ashburner, 1989a) and its companion, *Drosophila: A Laboratory Manual* (Ashburner, 1989b), represent an integrated and impressive compendium of information about the genetics and biology of Drosophila. The handbook contains considerable useful information about such topics as genetic nomenclature, genome composition and organization, development, virgin collection, chromosome rearrangements and their behavior, and evolution. It is a fascinating and useful reference volume containing the most important information, as well as a collection of useful and interesting trivia. By and large, it represents Michael Ashburner's distillation of the vast *Drosophila* literature and his conversations with numerous *Drosophila* workers at meetings and during his travels. The companion volume, *Drosophila: A Laboratory Manual*, consists of 137 useful laboratory protocols referred to in *Drosophila: A Laboratory Handbook*. Together, these two volumes present a very useful set of methodological approaches and procedures for basic *Drosophila* manipulation.

C. *The Genome of Drosophila melanogaster*

An organism with the rich genetic resources of *Drosophila* needs a bible/ dictionary of the various genetic elements. This resource has recently been upgraded with the release of *The Genome of Drosophila melanogaster*, a.k.a.

"The Red Book" (Lindsley and Zimm, 1992a,b). Contained herein is an alpha-betized list of all mutants, alleles, and loci current up to the end of 1989. In addition, there are extensive sections on chromosomal rearrangements (inversions, translocations, duplications, and deficiencies as well as special chromosomes including balancers and compound chromosomes). The book is extensively referenced and provides an invaluable guide to the previous literature. Of special use are richly detailed genetic and chromosomal maps. Thus, one can use this book to figure out what the strange inscriptions on stock tags mean, what mutants and genes are known to map near one's own gene or region of interest, or what other genes with similar phenotypes are out there. The full text of this book is also accessible electronically in *FlyBase* (as is updated information; see below). No fly lab or serious investigator studying *Drosophila* can function without at least one copy of this volume.

D. *Drosophila Information Service*

Drosophila Information Service (*DIS*) is a periodical devoted exclusively to the technical aspects of *Drosophila* biology. It is published as material warrants. Much valuable technical information and genetic information appeared in older versions; its unreviewed nature means that one should test methods, etc. before relying on them. Whether this volume will ultimately be supplanted by electronic versions (see *Drosophila Information Newsletter* (*DIN*) below) remains to be seen. *DIS* can be ordered by contacting the address below.[1]

III. Intellectual References

Many excellent volumes concerning the developmental biology, cell biology, and neurobiology of *Drosophila* have been published. Of these, a few are worthy of particular attention and are described below.

A. *Biology of Drosophila*

First published over 40 years ago and reprinted in 1965, this volume (Demerec, 1965) remains an outstanding source of information about the basic development and anatomy of *Drosophila*. The book is edited by Demerec and has contributions from many leading *Drosophila* workers of the 1950s. All stages of the life cycle are covered and many useful anatomical diagrams are presented. The only problem is finding a copy, as it has been out of print for many years. Nonetheless, many libraries have copies and it is still useful when one is trying

---

[1] James N. Thompson, Jr., Department of Zoology, 730 Van Vleet Oval, University of Oklahoma, Norman, OK, 73019. Fax: 405-325-7560. Internet: THOMPSON@AARDVARK.UCS.UOK-NOR.EDU. Bitnet: THOMPSON@UOKUCSVX.

to figure out the identity of a piece of the larva or adult just encountered in a dissection.

B. *The Genetics and Biology of Drosophila*

This excellent series consists of three volumes. Volume 1 (Ashburner and Novitski, 1976a,b,c) covers many aspects of the genetic system and genetic phenomenology. Topics include care and feeding of the fly, mass culture, meiosis, mutagenesis, etc. Volume 2 (Ashburner and Wright, 1978a,b,c,d) covers various aspects of *Drosophila* development, cell biology, and neurobiology. Volume 3 (Ashburner *et al.*, 1981, 1982, 1983a,b, 1984, 1985), covers the population genetics of *Drosophila*. Volumes 1 and 2 are probably of the most use to cell biologists (although Volume 3 provides useful evolutionary perspective and interest). While Volumes 1 and 2 are somewhat out of date and not as complete as more recent and more specialized works, they nonetheless contain much useful information and are essential reference sources for users of *Drosophila*. Their comparatively great expense means that most smaller labs may not have a complete set, but most libraries should.

C. *The Development of Drosophila melanogaster*

This outstanding text (Bate and Martinez Arias, 1993a,b) covers all aspects of the development of the fly. Written by a superb collection of authors, this text is beautifully illustrated and reviews what is currently known about all of the major organ systems, tissues, and embryonic stages (e.g., oogenesis, pattern formation, embryonic cell cycles, CNS development, etc.). The two volumes are also accompanied by a useful short companion volume (Hartenstein, 1993) showing in pictorial form all of the major organ systems at different stages of development. The only drawback of these volumes, perhaps, is cost, but what price quality?

D. Other References of Note

While not a complete list, a few other books are noteworthy for the cell biologist seeking to use *Drosophila* in his or her research. Early development of the embryo and nervous system is covered in extensive detail in a monograph by Hartenstein and Campos-Ortega (1985). This book is one of the most complete anatomical references to the *Drosophila* embryo. While expensive, access to this volume is well worthwhile if one is dealing with embryonic structures or problems. Finally, for those wishing access to older works in the *Drosophila* literature (stretching back over 70 years), there is the *Bibliography of Drosophila* series edited by I. H. Herskowitz. These volumes are available from most university libraries.

IV. Electronic References

For the computer literate (or near literate) investigator, there are excellent databases available by computer through the Internet. This technology and the attendant databases are evolving rapidly, so the following is only a ''snapshot'' in time of what is likely to grow very rapidly in capacity and capability. In the discussion that follows, a basic understanding of the Internet and how to connect to it is assumed. For those wishing to learn more about the Internet and its capabilities, an excellent resource is *The Whole Internet User's Guide and Catalog* (Krol, 1992). This book covers many of the methods described below such as ftp and Gopher as well as providing information about how to obtain these utilities.

A. *Drosophila Information Newsletter*

DIN is an electronic newsletter covering various aspects of *Drosophila* technology and genetics. It is distributed quarterly by e-mail; back issues are available in *FlyBase* (see below) and are also published in *DIS*. A full description taken directly from *Drosophila Information Newsletter* (*DIN*) is reprinted here (with thanks to and permission from K. Matthews and C. Thummel).

''The Drosophila Information Newsletter has been established with the hope of providing a timely forum for informal communication among Drosophila workers. The Newsletter will be published quarterly and distributed electronically, free of charge. We will try to strike a balance between maximizing the useful information included and keeping the format short; priority will be given to genetic and technical information. Brevity is essential. If a more lengthy communication is felt to be of value, the material should be summarized and an address made available for interested individuals to request more information. Submitted material will be edited for brevity and arranged into each issue. Research reports, lengthy items that cannot be effectively summarized, and material that requires illustration for clarity should be sent directly to Jim Thompson (THOMPSON@AARDVARK.UCS.UOKNOR.EDU) for publication in DIS. Back issues of DIN are available from *Flybase* in the directory flybase–news or in News/ when accessing *Flybase* with Gopher. Material appearing in the Newsletter may be cited unless specifically noted otherwise.

''Material for publication should be submitted by e-mail. Figures and photographs cannot be accepted at present. Send technical notes to Carl Thummel and all other material to Kathy Matthews. The e-mail format does not allow special characters to be included in the text. Both superscripts and subscripts have been enclosed in square brackets; the difference should be obvious by context. Boldface, italics, underlining, etc. cannot be retained. Please keep this in mind when preparing submissions. To maintain the original format when printing DIN, use Courier 10cpi font on a standard 8.5'' × 11'' page with 1'' margins.

"Drosophila Information Newsletter is a trial effort that will only succeed if a broad segment of the community participates. If you have information that would be useful to your colleagues, please take the time to pass it along."

The editors of DIN are:

Carl Thummel
Department of Human Genetics
Eccles Institute
University of Utah
Salt Lake City, UT 84112
CTHUMMEL@HMBGMAIL.MED.
   UTAH.EDU

Kathy Matthews
Department of Biology
Indiana University
Bloomington, IN 47405
MATTHEWK@INDIANA.EDU
MATTHEWK@INDIANA.BITNET

"To add your name to the Newsletter distribution list, send one of the following E-mail messages from the account at which you wish to receive DIN.
Via Bitnet—To: LISTSERV@IUBVM
        Subject:
        Message: SUB DIS-L Your real name
Via Internet To: LISTSERV@IUBVM.UCS.INDIANA.EDU
        Subject:
        Message: SUB DIS-L Your real name
LISTSERV will extract your user name and node from the E-mail header and add you to the list. Use your Internet address if you have one. You will receive confirmation by E-mail if you have successfully signed on to the list. If you are on the list and do not wish to receive DIN, or you want to remove a soon-to-be-defunct address, replace SUB in the above message with UNS. The SUB command can also be used to correct spelling errors in your real name; the new entry will simply replace the old as long as it was sent from the same USERID@NODE address."

While this is a new form of technical information distribution in the *Drosophila* community, it is very convenient and the expectation is that it will continue and expand.

B. *FlyBase*

FlyBase is a new electronic database replete with the most current published information about *Drosophila*. It is easy to access and the information is quite up-to-date. As would be expected, the organizing principle is based on the genetic map of *Drosophila,* but substantial information pertaining to the cell, developmental, molecular and evolutionary biology of *Drosophila* is contained in *FlyBase* as well. *Flybase* is in its initial developmental stages and currently is electronically accessible only through Internet protocols. In the future, other interfaces will be available, and it is recommended that the user check the current information manuals in *FlyBase* to understand the available options. Currently, accessing *FlyBase* is straightforward and can be done either by anonymous ftp to FTP.BIO.INDIANA.EDU or by using Gopher (available by anonymous ftp from BOOMBOX.MICRO.UMN.EDU in pub/gopher). If you access *Flybase* via ftp, you will be operating in a UNIX-like interface; however, if you type "get Readme" (be careful to use exactly this case since this system is case sensitive), a text file containing instructions on how to get around in the database will download to your local system.

The best method for accessing *Flybase* is probably to use Gopher. Gopher provides a more easily managed interface and allows one to search the most current view of the various databases online interactively without having to download them to your local system. For example, the most up-to-date information on genes and alleles is currently contained in a set of files called genes94, which can be easily searched through the Gopher interface. Using Gopher, *Flybase* can be reached by any of several routes but the most straightforward is by starting at the University of Minnesota Gopher Server and going through "Other Gopher Services" to USA, then Indiana, and then IUBIO Archive. An added benefit of using IUBIO archive in this way is that GenBank access is also available.

For further information, I have reproduced here "About *Flybase*" from *Flybase* itself (with permission of William Gelbart on behalf of the *Flybase* consortium).

"About Flybase
20 January 1994

"Flybase is a database of genetic and molecular data for *Drosophila*. It contains all the information in the 'Red Book' (D. L. Lindsley and G. G. Zimm, The genome of *Drosophila melanogaster,* Academic Press, 1992) merged with more recently published information. You can access FlyBase data by using the Internet Gopher or anonymous ftp to the IUBio server ftp.bio.indiana.edu at Indiana University. To use anonymous ftp, type:

ftp ftp.bio.indiana.edu

and type the user name anonymous when it asks for your login name. All of the Flybase files are in a directory called flybase, and gopher clients for a variety of computers are in a directory called util/gopher. If you do not have Internet access we suggest you talk to your local computer experts. FlyBase will periodically publish extracts of its data as special issues of the Drosophila Information Service (DIS).

"FlyBase includes the following things:

A complete merge of the corrected GENES information from Lindsley and Zimm with the Ashburner files. It includes over 1500 new genes not described in Lindsley and Zimm.

An updated list of chromosome aberrations.

A unified bibliography on *Drosophila* with over 58,000 records.

Stock lists of the Bloomington, Bowling Green, Umea, and Species stock centers and from some individual labs.

A genetic map of *Drosophila*.

A list of *Drosophila* genes sorted by function.

Lists of clones, including the European cosmids and US P1 bacteriophage, sorted by cytological location.

A unified directory of *Drosophila* workers, with their mail and email addresses, phone and fax numbers.

Associated databases, e.g., Amero's cytological features database, Baechli's list of species names in the family Drosophilidae, and P element lists from the Berkeley Genome Center, among others.

News about FlyBase and an archive of the bionet.drosophila news group.

"To use FlyBase effectively, you should read the FlyBase User Manual (flybase/docs/User-manual.text). For further information, read our Reference Manual (flybase/docs/Reference-manual.text for plain text or flybase/docs/Reference-manual.ps for Postscript printers).

"You can contact FlyBase by email to:

flybase@morgan.harvard.edu

or by regular mail to:

FlyBase, Biological Laboratories
Harvard University
16 Divinity Avenue
Cambridge, MA 02138

FlyBase welcomes comments and suggestions from the community.

"FlyBase is supported by grants from the U.S. National Institutes of Health (Washington) and Medical Research Council (London). The project is carried out by a consortium of Drosophila researchers and computer scientists at Harvard University, University of Cambridge (UK), Indiana University, and UCLA. A complete list of consortium members is available in the file flybase/docs/flybasemembers.doc.

"FlyBase is an electronic publication that is copyrighted 1993 by the Genetics Society of America. Copying in whole or part for commercial uses requires written consent. Copying for non-commercial, scientific uses is permitted. Other copyrights pertain to portions of FlyBase from other sources; see flybase/docs/Copyright-GSA-1993.doc."

V. Final Remarks

Good information is essential to the fruitful use of *D. melanogaster* in cell biological and other research projects. As the above should make clear, many valuable resources exist, and are continually being developed, that should significantly drop the activation energy of using *Drosophila* by new researchers.

Acknowledgments

My thanks to Melanie Field and Samantha Gibba for help in preparing this chapter and to Bill Gelbart and the rest of the FlyBase consortium for their advice about FlyBase capabilities.

References

Ashburner, M. (1989a). "*Drosophila:* A Laboratory Handbook" New York: Cold Spring Harbor Laboratory Press.

Ashburner, M. (1989b). "*Drosophila:* A Laboratory Manual" New York: Cold Spring Harbor Laboratory Press.

Ashburner, M., and Novitski, E. (1976a). "The Genetics and Biology of *Drosophila*," Vol. 1a, pp. 1–486. New York: Academic Press.

Ashburner, M., and Novitski, E. (1976b). "The Genetics and Biology of *Drosophila*," Vol. 1b, pp. 487–954. New York: Academic Press.

Ashburner, M., and Novitski, E. (1976c). "The Genetics and Biology of *Drosophila*," Vol. 1c, pp. 955–1427, New York: Academic Press.

Ashburner, M., and Wright, T. R. F. (1978a). "The Genetics and Biology of *Drosophila*," Vol. 2a, New York: Academic Press.

Ashburner, M., and Wright, T. R. F. (1978b). "The Genetics and Biology of *Drosophila*," Vol. 2b, New York: Academic Press.

Ashburner, M., and Wright, T. R. F. (1978c). "The Genetics and Biology of *Drosophila*," Vol. 2c, New York: Academic Press.

Ashburner, M., and Wright, T. R. F. (1978d). "The Genetics and Biology of *Drosophila*," Vol. 2d, New York: Academic Press.

Ashburner, M., Carson, H. L., and Thompson, J. N. (1981). "The Genetics and Biology of *Drosophila*," Vol. 3a, New York: Academic Press.

Ashburner, M., Carson, H. L., and Thompson, J. N. (1982). "The Genetics and Biology of *Drosophila*," Vol. 3b, New York: Academic Press.

Ashburner, M., Carson, H. L., and Thompson, J. N. (1983a). "The Genetics and Biology of *Drosophila*," Vol. 3c, New York: Academic Press.

Ashburner, M., Carson, H. L., and Thompson, J. N. (1983b). "The Genetics and Biology of *Drosophila*," Vol. 3d, New York: Academic Press.

Ashburner, M., Carson, H. L., and Thompson, J. N. (1984). "The Genetics and Biology of *Drosophila*," Vol. 3e, New York: Academic Press.

Bate, M., and Martinez Arias, A. (1993a). "The Development of *Drosophila melanogaster*," Vol. 1, pp. 1–746. New York: Cold Spring Harbor Laboratory Press.

Bate, M., and Martinez Arias, A. (1993b). "The Development of *Drosophila melanogaster*," Vol. II, pp. 747–1558. New York: Cold Spring Harbor Laboratory Press.

Campos-Ortega, J. A., and Hartenstein, V. (1985). "The Embryonic Development of Drosophila melanogaster." Germany: Springer-Verlag.

Demerec, M. (1965). "Biology of Drosophila." New York/London: Hafner Publishing.

Hartenstein, V. (1993). "Atlas of Drosophila Development," New York: Cold Spring Harbor Laboratory Press.

Herskowitz, I. H. (1953). Bibliography on the genetics of *Drosophila*, Part II, "Commonwealth Bureau of Animal Breeding and Genetics." England: Farnham Royal.

Herskowitz, I. H. (1958). Bibliography on the genetics of *Drosophila*, Part III, Indiana: Indiana University Publications.

Herskowitz, I. H. (1963). Bibliography on the genetics of *Drosophila*, Part IV, New York: McGraw–Hill.

Herskowitz, I. H. (1969). Bibliography on the genetics of *Drosophila*, Part V, New York: Macmillan.

Herskowitz, I. H. (1974). Bibliography on the genetics of *Drosophila*, Part VI, New York: Collier–Macmillan.

Herskowitz, I. H. (1974–80). Bibliography on the genetics of *Drosophila*, Part VII, *Drosophila Information Service*. 1974. **51,** 159–193; 1977, **52,** 186–226; 1978, **53,** 219–244; 1980, **55,** 218–262; 1980, **56,** 198–256.

Herskowitz, I. H. (1982–83). Bibliography on the genetics of *Drosophila*, Part VIII, *Drosophila Information Service* 1982, **58,** 227–270; 1983, **59,** 162–256.

Krol, E. (1992). The Whole Internet User's Guide and Catalog. Sebastopol, California: O'Reilly and Associates, Inc.

Lindsley, D. L., and Zimm, G. G. (1992a). The Genome of *Drosophila melanogaster*. San Diego: Academic Press.

Lindsley, D. L., and Zimm, G. G. (1992b). The Genome of *Drosophila melanogaster*. Polytene maps in a separate eight-map packet. San Diego: Academic Press.

Roberts, D. B. (1986). Drosophila—A Practical Approach. Oxford: IRL Press.

Note added in proof

FlyBase (and many other useful electronic databases) is also easily accessed through World Wide Wes (WWW); the Uniform Resource Locator (URL) for *FlyBase* is: gopher://ftp.bio.indiana.edu/hh/Flybase/

An excellent piece of client software is Mosaic, which can be obtained from ftp.ncsa.uiuc.edu or nic.switch.ch or some other "mirror" site. Mosaic clients are available for Macintosh, Unix X Windows Systems, and Microsoft Windows for the PC.

CHAPTER 2

Care and Feeding
of *Drosophila melanogaster**

Kathleen A. Matthews

Department of Biology
Indiana University
Bloomington, Indiana 47405

I. Introduction
II. Tools of the Trade
    A. Microscope and Light Source
    B. Anesthetizer
    C. Fly Pusher
    D. Aspirator
    E. Pounding Pad
    F. Morgue
III. Room and Board
    A. Bottles and Vials
    B. Media
    C. Plugs
    D. Trays
IV. Culturing *Drosophila*
    A. Stockkeeping
    B. Culture Contaminants
    C. Experimental Populations
V. Reading the Runes—Nomenclature
    A. Genes
    B. Aberrations
VI. Resources
    A. Bottles, Vials, and Plugs
    B. CO_2 Plates
    C. Etherizer
    D. Half-Spear Needle for Fly Pusher
    E. Media Pump and Filling Unit
    F. Media—Premixed

* Keep your stocks clean, your genotypes complete, and your aspirator to yourself.

METHODS IN CELL BIOLOGY, VOL. 44

I. Introduction

The ease and economy of laboratory culture of *Drosophila melanogaster* are not the most interesting qualities of this exceptional fly, but the newcomer will find them invaluable. A working colony can be established and maintained with a minimum amount of equipment, space, and effort, making it practical for almost any laboratory to work with *Drosophila*. This chapter is a basic guide to fly husbandry intended for the neophyte. Most of the topics covered here are discussed in greater depth elsewhere, most notably in Ashburner's excellent epic *Drosophila: A Laboratory Handbook* (1989). My intention is to cover only the fundamentals of *Drosophila* culture and provide some simple solutions to common problems.

II. Tools of the Trade

Basic fly handling equipment includes a binocular microscope with a good light source, an etherizer or a CO_2 plate for anesthetization, a fly pusher, an aspirator, a pounding pad, and a morgue. For most purposes flies can be kept at room temperature, but one or two constant temperature rooms, preferably humidified, or incubators are generally useful and are necessary for some techniques.

A. Microscope and Light Source

Choose a binocular or trinocular (for photos) microscope with good quality optics, easy access to the magnification changer, and a smooth, accessible focusing mechanism. $10\times$ ocular lenses with a zoom objective providing a magnification range of 8 to $40\times$ provide a good basic set up. A scope with a longer working distance is easier to position comfortably. Ideally, you should be able to look through the eyepieces without bending your back, and your forearms should rest comfortably on the bench top without elevating your shoulders.

A fiber optic light source with dual light guides provides the most adaptable illumination. The intensity, focus, and direction of the light are all adjustable, and the long guides allow you to place the box itself out of the way of your microscope. Smaller units with fluorescent of halogen light sources, although less flexible, are also fine for routine fly work.

B. Anesthetizer

Ether and CO_2 are the fly anesthetics of choice. CO_2 requires more setup and maintenance than ether, but it is easier on all except newly eclosed flies and has the advantage of being nonexplosive. If you choose ether, you will need an etherizer and a sorting plate. A sorting plate should be rigid, with a smooth, opaque, light-colored surface, and washable and fit comfortably in the hand. Small plastic etherizers, shown in Fig. 1, are available commercially. These are serviceable, but can be somewhat awkward to use. If you like ether but do not like these student anesthetizers, canvass your fly colleagues for their favorite etherizer design and copy one that suits you.

A CO_2 pad serves as both anesthetizer and sorting plate. A shallow box-shaped plate accepts a gas line at the back or side and has a firm but porous top that allows gas to continually flow across the surface of the plate, keeping the flies immobile until they are removed. Commercial plates are available (see Resources), but most are homemade. A good pad should be stable, provide an even flow of gas across the surface, fit comfortably in the hand, and be cleanable. A functional design, as shown in Fig. 1, is an open-faced plexiglass box about $14 \times 7 \times 2.5$ cm with a 5- to 8-mm i.d. rigid tube entering at the back of the box and extending about 2 cm into the interior. The cavity is filled with a dry

Fig. 1 Tools for manipulation of *Drosophila*. Left to right, cultures in 8-dram glass vials and a half-pint glass bottle with raw cotton plugs, etherizer, CO_2 plate with two kinds of gas concentrators, and yeast shaker with bottle top covering perforated shaker lid, all sitting on a sponge rubber pounding pad. Foreground, spear and brush combination fly pusher. Background, aspirator.

sponge cut to size; the sponge supports the top while allowing an even flow of gas. The plate can be topped in several ways. The easiest is a sheet of high-density but uniformly porous polyethylene (see Resources) that is cut to size and attached to the plate with tape (clear, heavy-duty packing tape, 3.2 ml thick, works well) or silicone cement. A more cushioned top (my preference) can be made from a layer of heavy-gauge nylon mesh (5.1 ends per inch), two layers of cheesecloth, and a layer of fine-gauge polypropylene mesh cloth (99.6 ends per inch), all attached as a unit with tape (as previously). Use an easily detachable fitting to connect the gas supply line to the plate so you can direct gas into culture containers when necessary. If static is a problem, moisten the sponge (through the gas inlet tube) with water or wipe the surface of the plate with antistatic solution (sold for computer screens). Adjust the flow of gas to the pad such that flies, once anesthetized, are motionless and have minimal extension of wings and legs (excess gas causes hyperextension of wings and legs, making the flies difficult to manipulate). This can be done at the regulator or by an inline valve or adjustable clamp.

It takes more gas to render flies senseless than to keep them in that state. The gas flow can be adjusted for each operation but this is time consuming. One option is to knock the flies out while still in the culture container by filling the inverted vessel with gas (if you do this routinely, split your gas supply line and attach one line to a cannula that can be easily passed along the side of the plug without releasing flies, as suggested by Ashburner (1989)). Alternatively, a simple tool, shown in Fig. 1, can be used to increase the local CO_2 concentration at the surface of the pad. A 65-mm powder funnel mounted in a No. 8 rubber stopper placed on the surface of the pad will accept bottles, vials, and 60-mm petri dishes. Place the funnel at the center of the pad, tap the flies away from the top of a bottle or vial, unplug, and invert into the funnel. Shake/tap the flies into the neck of the funnel. When the flies are immobile, remove the bottle or vial and lift the funnel from the plate, shaking or brushing out any flies that might be caught in the neck.

A more elegant device, also shown in Fig. 1, can be constructed from a student etherizer, a glass or clear plastic tube, and a piece of nylon or plastic mesh. Remove the wadding from the etherizer and cut the chamber off just below the ether application arm (cut the arm off too, sealing the opening with silicone cement). Cover one end of the clear tube with mesh and insert the other end into the neck of the etherizer. Bang flies from culture containers into this apparatus and then hold the mesh end of the tube against the surface of the CO_2 plate until the flies are immobilized. Disengage the tube from the funnel top and pour the flies out on the plate.

C. Fly Pusher

Anesthetized flies can be manipulated with brushes, probes, or an aspirator. Each has its advantages. I prefer a metal tool to manipulate flies on the pad and a brush to flip them into vials or sweep them into bottles. My favorite fly pushing tool has a small steel blade on one end and a fairly large sable brush on the other (see Fig. 1). The probe end is a half-spear dissecting "needle" with an acute angle on the blade edge. The probe is mounted in a pin vise such that it extends 2.5–3 cm from the vise cap. The neck of the blade is bent forward about 20°. The pin vise is inserted about 8 cm into a 9- to 10-cm section of flexible tubing, and a No. 4 or No. 5 sable brush is mounted in the other end of the tubing. The tool fits comfortably in the hand and, if not dropped on the blade, will survive many years of heavy use.

D. Aspirator

An aspirator allows flies to be handled without anesthetization and is a speedy way to do "pair" matings (two from the female pile, one from the male pile, 20,000×, and your screen is done); you provide the suction and propulsion. A simple aspirator consists of a rigid but thin chamber with a porous barrier at the base connected by flexible tubing to a mouth piece (see Fig. 1). The chamber should be long enough to reach to the surface of the food in a vial, and the tubing should be long enough to drape around your neck while in use. The chamber can be made from a variety of objects. If you are adept at working glass, a piece of narrow-walled glass tubing about 4–5 mm I.D., with one end tapered, makes a good versatile chamber. A 1-ml plastic serological pipet works well for many purposes, although the tip is too narrow to draw large numbers of flies into the chamber at once. Nylon or polypropylene mesh cloth makes a stable and washable barrier to keep the flies in the chamber and out of your mouth, and the butt end of a tuberculin syringe makes a good mouthpiece.

E. Pounding Pad

Pounding on your bench top with impunity is one of the perks of fly work. A shock-absorbing foam pad reduces the pleasure very little and saves wear and tear on your equipment and your colleagues' nerves. The best pad I have used is a half-inch-thick 8 × 14-inch piece of sponge rubber (Fig. 1), intended as a kneeling pad, but I have not found a current supplier for these. A computer mouse pad or other foam pad designed for office equipment makes an acceptable substitute, although these are noticeably less shock absorbent.

F. Morgue

Anesthetized flies are most conveniently discarded directly into a morgue. Any stable vessel with a wide mouth or a funnel in the neck can make a serviceable morgue. Partially fill the container with heavy mineral oil, motor oil, or 70% ethanol. Flies that are to be discarded can be dumped directly from your sorting or CO_2 plate into the morgue (avoid dipping the end of your plate into the muck).

=========   ## III. Room and Board

A. Bottles and Vials

The classic *Drosophila* culture chamber is a glass half-pint milk bottle. These bottles are now manufactured specifically for *Drosophila* research. Eight-dram glass vials are also in wide use, as are a variety of plastic bottles and vials (see Resources). Population cages for growing larger numbers of flies are described elsewhere in this volume. In general, glass is more economical if you use large numbers of bottles and vials, will continue to do so for some years, and have access to a glassware washing facility (autoclave, electric bottle brushes, or a commercial dishwasher). Plastic is probably more economical if you use smaller quantities and do not have to pay directly for disposal costs or if you would have to bear the full cost of establishing and staffing a glassware washing facility. If you choose plastic, stick with the fully transparent varieties. Semiopaque walls make it difficult to check for contaminants, monitor the progress of a culture, or examine phenotypes.

B. Media

The choice of medium depends on the amount of food needed and the equipment and personnel available. Premixed dry ingredients are available from Carolina Biological Supply, Sigma, and Phillip Harris (see Resources). The agar-free varieties are prepared by distributing the premix into bottles or vials and adding water. The Carolina and Phillip Harris mixes do not require stirring. The Sigma product, which is closer to cooked medium in consistency, must be stirred; it can be mixed in larger batches and poured into bottles, but its viscosity makes if difficult to pour into vials. An agar-based premixed medium that requires boiling and the addition of mold-inhibitor is also available from Sigma. This medium is essentially the Caltech recipe developed by E. B. Lewis and can be poured or pumped as with any cooked medium. Without volume discounts, these premixed media cost about $2–$2.25 per liter.

A cooked medium is usually the most cost-effective approach for all but the small-scale user because it can be prepared in large volume and rapidly dispensed with a pump. A variety of formulas are in use. Choose one based on use, cost, your routine, and the equipment available. Ashburner (1989) provides an assortment of recipes, including the widely used Caltech food. The semidefined medium described by Backhaus *et al.* (1984) is the best food I have used, judged by the productivity of cultures, the size of larvae grown on it, and its excellent storage properties, but it is expensive (about $2/liter) and typical cultures are so fecund that it is difficult to avoid overinoculation with mass-transfer stockkeeping. The formula we use (see Resources) is simple and inexpensive (about 60 cents per liter), pumps smoothly, and keeps well at room temperature for about a week after cooking.

C. Plugs

A variety of materials make acceptable stoppers for *Drosophila* glassware and plasticware. Disposable disk-shaped paper caps are available for glass and plastic bottles. Foam plugs are sold for both vials and bottles, and rayon balls and rayon coil are available for vials. Plugs of any size can be made from raw or pharmaceutical coil cotton, but the latter is considerably more expensive and more difficult to handle. Rayon balls are more expensive than cotton (about 80 cents per 100 vials vs 20 cents for raw cotton) but are quicker to insert—100 vials can be plugged with rayon balls in well under 2 min vs about 6 min for cotton. In my opinion, raw cotton makes a superior plug for both bottles and vials. The same material can be used for any size vessel, plugs can be made to fit snugly, and the cotton fibers retain their elasticity through many rounds of compression. However, it must be bought in large volume (500-lb bales) and it is the most labor-intensive plugging method.

D. Trays

Trays are needed to hold fresh media, stocks and crosses, and, if you use glassware, discarded cultures. If you use disposable culture vessels, any kind of tray with sides tall enough to keep bottles and vials in but short enough to allow easy access to the contents will serve all the necessary functions. Trays with straight instead of sloping sides make the most efficient use of space. Lightweight but sturdy plastic trays with grooves that accept dividers are particularly nice (see Resources).

Glassware handling is minimized by the use of autoclavable trays for fresh food and discards. A sheet metal shop can make autoclavable trays of any size from galvanized steel. The bottom of the trays should be perforated to allow water to drain and the sides should be at least an inch shorter than the intended contents. The same trays can be used for storing flies, so take into account the size of shelving and incubators you will be using as well as the size of your autoclave and storage shelves in the kitchen area. Keep in mind that glassware is heavy, especially when filled with food. Do not make the trays so large that they are difficult to lift when filled.

IV. Culturing *Drosophila*

A. Stockkeeping

1. Mechanics

Most stocks can be successfully cultured by periodic mass transfer of adults to fresh food. Bottles or vials are tapped on the pounding pad to shake flies away from the plug, the plug is rapidly removed, and the old culture is inverted over a fresh bottle or vial. Flies are tapped into the new vessel, or some are shaken back into the old one, as necessary, and the two are rapidly separated and replugged. Good tossing technique combined with plugs that are easily removed and replaced results in very few escapees. You will learn from experience which stocks require a medium or large inoculation of adults and which do better with only a few.

The frequency with which new subcultures need to be established depends on the health and fecundity of the genotype, the temperature at which it is raised, and the density of the cultures. Temperature has a large effect on the rate of *Drosophila* development. Generation time (from egg to adult) is approximately 7 days at 29°C, 9 days at 25°C, 11 days at 22°C, and 19 days at 18°C. For most purposes, stocks are maintained by live culture, transferring adults to fresh medium every few generations. Stocks kept at room temperature should be transferred to fresh food every 20 to 30 days. Mites and mold are more likely to be a problem in older stocks, so it is good policy to set 30 days as an absolute upper limit for room-temperature stocks. This period can be extended by keeping stocks at a lower temperature. 18°C buys more time than 22°C, for example, but a significant number of genotypes fail to thrive at 18°C, and mold can be a serious problem. It is wise to keep a room-temperature backup of stocks to be maintained at low temperature for the first two or three transfers in case the stocks do poorly. If the quality of your fly food is unreliable, keep at least two cultures for each stock, staggered to assure the use of different batches of medium (at least until you find a new cook).

Cryopreservation of ovaries (see Ashburner, 1989) or embryos (Cole *et al.*, 1993; Steponkus and Caldwell, 1993) is a viable alternative to continuous culture for some purposes. Genotypes that are unstable due to reversion, breakdown, or accumulation of modifiers, especially those with nonvisible phenotypes that are time consuming to select, are good candidates for freezing. Also, if you have a large number of stocks that will not be in use but must be kept for many years, it might be cost-effective to maintain these as frozen stocks. However, for most routine stockkeeping purposes, live culture remains the simplest route.

Identify stocks with tags showing the complete genotype of the stock, sans shorthand. Writing the genotype on the vial or bottle at each transfer invites transcription errors and takes longer than moving a tag. Do not use a stock center stock number or other potentially cryptic symbol as the only identifier of a stock. Stocks are often kept for many years and what is obvious to you now may be meaningless even to you in a few years and is easily misinterpreted by someone inheriting your stocks. Unless you are careful to maintain complete stock data elsewhere, record all relevant information on the tag.

2. Balanced Stocks and Balancers

Mutations that are homozygous viable and fertile are most easily kept as homozygous stocks. Lethal or sterile mutations must be maintained in a heterozygous state. A balanced stock (Fig. 2) is one that regenerates the same set of

A. Balanced Lethal Stock

$$Df(3R)by10, red\ e/TM3,\ ri\ p^p\ sep\ l(3)89Aa\ Sb\ bx^{34e}\ e$$

$$Df(3R)by10/TM3\ ♀\ \ x\ \ Df(3R)by10/TM3\ ♂$$

↓

$Df(3R)by10/Df(3R)by10$ ♀, ♂	$Df(3R)by10/TM3$ ♀, ♂	$TM3/TM3$ ♀, ♂
lethal	viable, fertile	lethal

B. Balanced Sterile Stock

$$dsx^D/TM2,\ emc^2\ Ubx^{130}\ e^s\ \&\ ru\ h\ st\ \beta Tub85D^D\ ss\ e^s/TM2,\ emc^2\ Ubx^{130}\ e^s$$

$$dsx^D/TM2\ ♂\ \ x\ \ \beta Tub85D^D/TM2\ ♀$$

↓

$dsx^D/TM2$ intersex ♀	$dsx^D/TM2$ ♂	$\beta Tub85D^D/TM2$ ♀	$\beta Tub85D^D/TM2$ ♂
sterile	fertile	fertile	sterile

C. Balanced Sex-linked Lethal Stock

$$l(1)Eb^3/FM6,\ y^{31d}\ sc^8\ dm\ B$$

$$l(1)Eb^3/FM6\ ♀\ \ x\ \ FM6/Y\ ♂$$

↓

$FM6/FM6$ ♀	$l(3)Eb^3/FM6$ ♀	$FM6/Y$ ♂	$l(3)Eb^3/Y$ ♂
sterile	viable, fertile	viable, fertile	lethal

Fig. 2 The self-maintaining property of balanced genotypes is illustrated for three kinds of balanced stocks, an autosomal lethal, an autosomal dominant sterile, and an X-linked lethal. The genotype as it might appear in a stock list is shown first in each case. The genotypes of fertile adults, followed by the fate of each genotype produced by matings among adults of the stock, are shown. (A) The stock is balanced because only one genotype is viable. *Df(3R)by10* is a deficiency that removes a section of chromosome 3 containing approximately 21 genes. (B) Two genotypes are present in the stock (indicated here by an ampersand in the stock genotype), but only one sex of each genotype is fertile, assuring that the same two genotypes are regenerated each generation. dsx^D is a dominant mutation in a gene that regulates somatic sexual differentiation; $\beta Tub85D^D$ is a dominant mutation in the testis-specific β-tubulin gene. (C) X-linked lethals are typically passaged through females. The *dm* allele on *FM6* is responsible for the sterility of *FM6* homozygous females.

heterozygotes each generation so that the stock can be maintained by mass transfer of adults instead of by mating specific genotypes each generation. A simple, balanced lethal stock carries different recessive lethals on each of the two homologs, allowing only heterozygotes to survive. Dominant male or female steriles (*Ms* or *Fs*) can be maintained in stock without selection by double balancing—one *Ms*, one *Fs*, and one recessive lethal. *Fs/lethal* male and *Ms/ lethal* females will be the only fertile genotypes present in the stock each generation.

In most cases, balanced lethal schemes work only if one of the lethal chromosomes is itself a balancer chromosome. Recombination between lethal (or sterile) mutations on different homologs can produce one homolog with both mutations and one wild-type homolog. The wild-type chromosome will rapidly predominate or become fixed in the stock. Balancers are structurally rearranged chromosomes that prevent recombination between homologs in females (meiotic recombination is absent in *D. melanogaster* males and in the tiny fourth chromosome in females). This is accomplished in part by reducing recombination directly and in part by preventing transmission of recombinant chromosomes. The most commonly used balancers carry overlapping sets of inversions and prevent recombination throughout most of the length of the chromosome. Some special-purpose balancers work well only for specific regions of a chromosome. Suppression of recombination is less effective when balancers for two or more heterologs are present in a stock.

B. Culture Contaminants

Drosophila is largely pestilence-free, but mites, fungi, and bacteria can be problems in laboratory cultures. It is good practice to clean your bench top and fly-pushing equipment regularly. This is particularly important if a problem is evident. Clean the bench top and all equipment that comes into contact with potentially contaminated stocks with 70% ethanol or soap and water after use. Sharing pounding pads, CO_2 pads, fly pushers, and sorting plates can aid in the spread of contaminants. If sharing is unavoidable, the need for cleanliness should be understood by all and enforced.

1. Mites

"I have mites" is not an admission you want to have to make to your fly colleagues (but you must make it, if true). The most dangerous species are egg predators, but even those that simply feed on the medium can outcompete weak genotypes and compromise experimental observations. Frequent stock transfer, tight plugs, and zero mite tolerance by all fly workers in a building are the best defenses. In general, cultures grown at 24–25°C should never be kept for more than 30 days. If mites are known to be a problem in your lab or building, cultures should be checked and discarded after 18 to 20 days. Lining stock trays with benzyl benzoate-treated cheesecloth (soak cloth in 10% benzyl benzoate in 95% ethanol and air dry; replace the cloths every 6 months) may help prevent infestation. However, some kinds of plastic are dissolved by benzyl benzoate, so test first if you use plastic vials or trays (the cloth will fuse with the plastic within 24 hr) and protect paper items such as stock tags from direct contact with the cloths.

To prevent the importation of mites from outside sources, all stocks new to your lab should be quarantined for at least two generations. Never open a foreign bottle or vial at your fly bench (or your neighbor's) without first inspecting the culture for mites. Using a microscope, examine the surface of the medium and the walls of the container, especially around pupae or pupal cases. If no mites are evident, replace foam or paper stoppers with tight cotton plugs and isolate cultures in a quarantine tray. As an added precaution, cultures can be wrapped in the mite cloths described above. Keep the original bottle or vial for about 20 days, even though you have established fresh cultures, rechecking for mites every 5 to 10 days. I check the new cultures too, just to be safe, but I have never found mites in a subculture when the parental vial was mite free.

Any culture found to contain mites should be autoclaved immediately if it can be replaced from a mite-free source. If replacement is not possible, use one of the methods described in Ashburner (1989) to disinfect the culture, such as daily transfer of adults for about a week, using only the final transfer to establish a new and hopefully mite-free culture. Keep infected cultures wrapped in mite cloths until they have been mite free for three generations.

2. Fungi and Bacteria

If mold is a problem in isolated cultures, it can usually be eliminated by daily transfer of adults for 7–10 days. Visually inspect cultures from the later transfers for hyphae (look around the pupal cases) and use one that appears to be free of fungal growth for further subculture. In extreme cases this process may need to be repeated for an additional generation. If fungal contamination is a widespread problem, be sure that fungal inhibitor (*p*-hydroxy-benzoic acid methyl ester) is being added to the medium after it is cooked (boiling destroys the inhibitor), add a small amount of live baker's yeast to every culture (the yeast tends to inhibit the growth of unwanted fungi), and check for sources of infection in the lab, such as incubator fan housings. Clean any contaminated or suspicious areas with disinfectant.

A variety of bacterial contaminants can occur in fly cultures. The most common problems are caused by mucus-producing bacteria. Although not directly toxic, larvae, and to some extent adults, become trapped in the heavy layer of mucus that coats the surface of the food. Large numbers of larvae overcome the effects of the bacteria in a healthy stock, but weak stocks or pair matings can be seriously compromised. A widespread bacterial problem may indicate that the pH of your medium is too high; try lowering the pH to about

5. Individual stocks can be treated with antibiotics for one generation. A quick approach that often works: add 100 μl of penicillin–streptomycin solution (10,000 u/ml and 10,000 μg/ml, respectively) to the surface of the medium in a vial and allow it be absorbed. Add a small amount of yeast and transfer flies to the treated medium. Discard adults before progeny eclose; subculture progeny on untreated medium. Other antibiotics may be tried if the contaminant proves to be resistant to penicillin and streptomycin.

Alternatively, clean cultures can be established from embryos dechorionated with 5.25% sodium hypochlorite (liquid household bleach, full strength). A convenient method is to transfer eggs to a bleach-soaked wedge of filter paper, wick away bleach after chorions have dissolved (3–5 min), wash eggs several times with water, and transfer to a fresh piece of filter paper (small enough to sit on the surface of the medium in a vial) moistened with water. Place the filter paper with eggs into a fresh vial of food and place a larger strip of filter paper along the wall of the vial. Wet this strip of paper to maintain high humidity in the vial until the eggs hatch.

C. Experimental Populations

1. Matings

While mass transfer of adults works well for most stockkeeping purposes, it often results in overcrowded cultures. Overcrowding can affect the outcome of crosses and experimental procedures. Development time is slowed, different genotypes may be disproportionately affected by competition for food and pupation sites, and many pupae and adults will drown in the soup of larvae and liquified medium. The best yield of healthy adults is obtained from cultures established with an optimum number of animals. Expect 50–100 adults from a vigorous 8-dram vial culture and 300–600 from a comparable half-pint bottle culture. For most genotypes, the optimum number of females will range from 1 to 3 per vial and 5 to 20 per bottle. Set up a few test bottles or vials to determine this number empirically for the genotypes involved; control the age of the food, the age of the females, and the number of days the females are left in the vials. One or two males are usually sufficient to rapidly inseminate several females, but some genotypes will require an equal or excess number of males to ensure rapid mating. If necessary, the effects of overcrowding can be reduced midculture by adding baker's yeast (dispense from a salt shaker as in Fig. 1) and a tissue (such as a Kimwipe) to provide additional nutrition and pupation sites, respectively, for the excess larvae. Extremely crowded cultures are best dealt with by distributing larvae (scoop them out with a spatula) to several fresh bottles or vials.

If you cannot distinguish parent from progeny by phenotype, parents should be discarded before the progeny begin to hatch. Experimental crosses maintained at 25°C should be discarded after 18 days to prevent recovery of second-generation progeny. An effective schedule is to establish crosses on Day 0 (start on a Friday if you want to begin virgin collection on a Monday), discard adults and add yeast and papers (optional) on Day 7, and collect virgins or score progeny on Days 10 through 18.

Some mutant phenotypes are affected by temperature or genetic background. Before setting up a large-scale effort such as a screen, make a test cross of the relevant genotypes under the conditions to be used and confirm that all phenotypes are scorable. It is also prudent to assure the absence of "background" lethals in a stock to be used for mutagenesis by isogenizing a chromosome for use in a screen. To isogenize a *ri e* chromosome, for example, cross to an appropriate balancer stock, recover 10–20 progeny heterozygous for *ri e* and the balancer, backcross them individually to the balancer strain, cross sibling *ri e*/balancer progeny, and then recover *ri e* homozygotes from one of the lines and establish a stock. Only lines carrying lethal-free chromosomes will produce homozygotes among the progeny of the sib matings.

2. Virgins

Most experimental schemes require virgin females. *Drosophila melanogaster* adults do not mate for about 10 hr after eclosion, allowing virgins to be collected within 8–12 hr after the culture has been cleared of adults. The timing of this virgin window is genotype dependent. Collect females within 8 hr to be safe or determine empirically for a given strain how long you can wait and still recover virgins. Ashburner (1989) describes a variety of environmental and genetic tricks to facilitate the collection of large numbers of virgin females.

For many mating schemes, virginity is desirable for sake of efficiency, but not essential because the progeny of nonvirgin females can be distinguished phenotypically from the progeny of interest. If your scheme requires virginity, hold females for 3–4 days and check for larvae in the holding vial before using the females in matings. Do not overcrowd females in holding vials—50 or so in an 8-dram vial, fewer if you doubt their virginity (you will have to discard all of the females in the holding vial if any have mated). The peak of female fertility is genotype dependent, but on average it is best to use females between 4 and 10 days old.

V. Reading the Runes—Nomenclature

The rules of *Drosophila* nomenclature are well defined and reasonably straightforward. Once you have learned the arcanum, even complex genotypes are easily deciphered if properly described. However, the symbolism is extensive and a complete explanation is beyond the scope of this work. The conventions for naming loci, alleles, and chromosome aberrations are described in detail in the introductions to relevant sections of Lindsley and Zimm (1992).

The basics presented here will serve as an introduction to *Drosophila* nomenclature and will allow you to interpret a variety of genotypes that you are likely to encounter, assuming shorthand has been kept to a minimum. See Lindsley and Zimm (1992) for a more complete set of rules and for more thorough explanations of the conventions described here.

A. Genes

Each gene has both a unique name and a unique gene symbol that is usually shorter than the name and contains no spaces, allowing genotypes to be described in an unambiguous and manageable way. Gene symbols are always italicized; I have followed Lindsley and Zimm (1992) here and not italicized gene names, although for clarity both the name and symbol are often italicized. In general, genes are named in one of three ways.

Genes may be named according to a mutant phenotype of the gene (generally the phenotype of the first mutant allele identified), e.g., white (*w*), Shaker (*Sh*), cubitus interruptus (*ci*). The name and symbol are capitalized if the phenotype of the mutant allele for which the gene was named is dominant, i.e., the mutant phenotype is expressed in animals heterozygous for the mutant allele and a wild-type allele. However, be aware that many nominally "dominant" genes have recessive alleles and many "recessive" genes have dominant alleles.

Genes may be named according to a category of phenotypic effect, such as suppressor, enhancer, minute, lethal, or sterile, along with identifying information relevant to the class (the name of the gene that is suppressed or enhanced or the chromosomal location of minutes, lethals, and steriles), for example, suppressor of forked (*su(f)*), Enhancer of Star (*E(S)*), Minute (1)15D (*M(1)15D*), lethal (3)85Ea (*l(3)85Ea*), or male sterile (2)1 (*ms(2)1*).

When the product of a gene is known, the gene is typically named according to the product encoded, with a chromosomal location or series number if part of a multigene family, for example, α-tuburlin (3)67C (*aTub67C*), superoxide dismutase (*Sod*), or transfer RNA arginine (*tRNA-Arg1*). Superscripts identify individual mutant alleles of a gene: w^a, $l(2)40fg^1$, $Antp^{LC}$. A + superscript indicates a wild-type allele of the gene, and no superscript implies allele *1* (in theory). A + in place of a gene symbol indicates that the chromosome or the complete genotype, depending on the context, is wild type.

B. Aberrations

Chromosome aberrations are named according to the type of rearrangement, the chromosome or chromosomes involved, and an identifying symbol. The types of aberrations and their abbreviations are deficiency (Df), duplication (Dp), inversion (In), transposition (Tp), translocation (T), compound (C), ring (R), free arm (F), levosynaptic element (LS), and dextrosynaptic element (DS). These are written as type(chromosome)identifier. The identifier may or may

not convey information about the rearrangement. For example, *Df(3R)by10* is the name of a deficiency in the right arm of the third chromosome; in this case, the identifier reflects the inclusion of the blistery (*by*) gene within the deficiency and the *10* distinguishes it from others in a series. Superscripts, which define unique alleles, are not used with symbols of genes deleted by deficiencies. *Df(3R)by10, Df(3R)by62,* and *Df(3R)by77* represent three unique deficiencies, but not unique alleles of *by*—the gene is equally absent in all three aberrations. *T(2;3)ap^{Xa}* refers to a translocation between chromosomes 2 and 3; here the translocation is named for the mutant allele of the *apterous* gene that results from one of the translocation breakpoints. *Tp(1;3)O4* names a three-break event that resulted in the insertion of a piece of chromosome 1 into chromosome 3. In this case, the identifier, *O4,* is arbitrary, formed from the name of the person who recovered the aberration and a series number.

Balancers, as described previously, are an important class of aberration, and one for which shorthand is commonly used. The most popular balancers exist in a variety of marker combinations, all with at least one dominant visible marker. Ideally, each balancer name would define both a specific set of inversions (described in Lindsley and Zimm (1992)) and a specific set of markers. For example, the second chromosome balancer *SM6a* (second multiple 6a) carries two pericentric inversions, *In(2LR)O* and *In(2LR)SM1,* and the markers *al^2 Cy dp^{lvl} cn^{2P} sp^2*. *SM6b* has an additional dominant mutation, *Roi*. In most cases, however, the balancer name defines only the inversions, and genotypes often include only the balancer name. Dominant markers can be identified by inspection if they are not specified in the genotype. If it is important to know which recessive markers are present on a balancer, you can make test crosses with marked stocks or simply replace the balancer with one of known constitution.

A properly assembled genotype represents all mutant components of the stock in the order *1;Y;2;3;4*. Within a chromosome, aberrations precede gene symbols. A comma and space separate aberrations from gene symbols, and genes are listed in the left–right order of the unrearranged chromosome. Gene symbols are separated by a space. Commas should be placed between genes only if it is necessary to indicate that the written order does not correspond to the true standard order of the genes, for example, if the genotype has been rearranged to allow electronic sorting according to a given component of the genotype. Homologs are separated by a solidus(/) and heterologs are separated by a semicolon. Homozygous chromosomes are defined only once: *cn bw* implies *cn bw/cn bw,* and + implies +/+. For example:

cv; sp; th. The stock is homozygous for three recessive mutations, crossveinless on chromosome 1, speck on 2, and thread on 3. In theory, these alleles are *cv^1, sp^1,* and *th^1*, since no superscript is designated, but in practice this assumption will not always hold. If the specific allele is important for your purposes, verify the relevant characteristics before investing a large amount of effort in the stock. This advice applies to all aspects of a stock's genotype.

In(1)dl-49+B^{M1}, sc v^{Of}. This stock is homozygous for two inversions on the X, delta-49 and Bar of Muller, and two recessive mutations, scute and vermillion.

Df(3L)emc5, red/TM2, emc^2 p^p Ubx e^s. The stock is heterozygous for a deficiency on the left arm of chromosome 3 that includes the extra macrochaetae locus. The deficiency chromosome also carries a mutation in the red gene. The stock will express the recessive *emc* phenotype as well as the dominant *Ubx* phenotype because the balancer carries a mutant allele of *emc* in addition to the standard *TM2* markers pink, Ultrabithorax, and ebony.

T(2;3)CyOTM6/pr cn Adh^{ufs}; mwh ry^{506} e. A translocation is superimposed on two balancer chromosomes, *CyO* and *TM6*. The normal sequence homologs carry mutations in the second chromosome genes purple, cinnabar, and alcohol dehydrogenase and the third chromosome genes multiple wing hair, rosy, and ebony.

The rules for designating autosomal homologs cannot be strictly applied to sex chromosomes. Sometimes the genotypes of both sexes are explicitly defined, using the form *X/X* x *X/Y*, or *X/X* & *X/Y*. More often a condensed notation is used and it is left to the user to apply the rules of segregation and sex determination to identify the genotype of each sex. For example, compound first, or attached X, chromosomes are commonly used to create balanced stocks of X-linked female sterile mutations. In a stock of the female sterile mutation diminutive, the genotypes of males and females are *dm/Y* and *C(1)DX/Y*, respectively, but the stock genotype is usually written as *dm/C(1)DX*. The latter seems to imply a stock of triplo-X flies, but triplo-X metafemales have extremely low viability and survivors are sterile. The only interpretation consistent with the biology is that females carry a maternally inherited compound X, males carry a paternally inherited *dm* X, and both sexes carry a wild-type Y chromosome inherited from the opposite sex.

Attempts to standardize transposon nomenclature are currently underway. Transposons are usually indicated by a symbol for the source of the mobile ends such as *P* (P-element ends) or *H* (hobo ends) and brackets of some kind ([ ], { }, and < > are all in use) surrounding the selectable marker or some other component of the transposon that was most relevant to the worker at the time. Although transposon genotypes that describe all components of the construct can become quite long, the potential for extreme complexity in this class of genetic objects makes it imperative that an unambiguous nomenclature be used, for example, *y w^{1118} P{ry$^+$, Hsp70:FLP=hsFLP}1; TM3, ry$^-$/TM6B, ry$^-$*. This stock carries mutant alleles of yellow and white on the X as well as a P-element transposon that carries a wild-type allele of the rosy gene and a chimeric gene containing the promoter of heat-shock protein 70 and the coding sequences from the yeast FLP gene. (Anticipating the completion of transposon nomenclature revision, the symbol for this specific construct is likely to become *hsFLP*, and the identifier for this particular insertion of the *hsFLP* construct is likely to be

1). Both balancers carry unnamed loss-of-function rosy mutations in addition to other, unspecified markers.

VI. Resources

The following list of suppliers is by no means complete. Many of these items are undoubtedly available from other companies, and in many cases similar products will be just as useful as the ones I have described. Prices for some items are included just to give you an idea of the cost; they are catalog prices with no bulk or institutional discounts, and some are almost certainly out-of-date.

A. Bottles, Vials, and Plugs

1. Half-pint glass milk bottles (~$50 per case of 48), paper caps ($12 per case of 500), and foam plugs are available from Sun Brokers, Inc., Wilmington, NC 28402.
2. Eight-dram glass vials ($38.75/144 from Baxter) are available from Sun Brokers, Inc. and Baxter Diagnostics Inc., Scientific Products Division, McGaw Park, IL 60085.
3. Plastic vials and foam plugs are available from Sun Brokers, Inc., Applied Scientific, South San Francisco, CA 94080; Carolina Biological Supply, Burlington, NC 27215; and Sarstedt Inc., Newton, NC 28658.
4. Rayon rope and balls are available from Applied Scientific and a variety of medical supplies providers.
5. Raw cotton, Cotton Queen short staple cotton, ginned, is available from A. Lassberg & Co., Austin, TX 78767. 500-lb bale, $330 plus $132 freight to Bloomington. Autoclave for 1 hr in brown paper grocery bags before using.

B. CO_2 Plates

1. Plexiglass boxes, $11.5 \times 7 \times 3$ cm with quick-release gas inlet and perforated plastic support for top (you provide final covering), are available from Jordan Scientific Co., Bloomington, IN 47401.
2. Hard CO_2 plate tops, high-density polyethylene, 35 um pore size, $\frac{1}{16}$ inch thick, 18×18-inch sheet, $30, are available from Bolab, Lake Havasu City, AZ 86405.
3. Soft CO_2 plate tops, polypropylene mesh cloth, mesh count 99.6/inch, 42 inch wide and nylon mesh cloth, mesh count 5.1/inch, 42 inch wide, are available from Tetko, Inc., Elmsford, NY 10523.

C. Etherizer

This item is available from Carolina Biological Supply.

D. Half-Spear Needle for Fly Pusher

I obtained this item from Clay Adams, a division of Becton, Dickinson & Co., Parsippany, NJ 07054 (Product No. 6121-000-000). I recently spoke with eight people at various divisions of Becton, Dickinson & Co. about the availability of these blades; none were unable to confirm or deny the continued manufacture of this product. Carolina Biological sells a dissecting tool with the same description, but I do not know if it has the same angle on the blade, which is a critical factor.

E. Media Pump and Filling Unit

National Instrument Company, Inc., 4119 Fordleigh Road, Baltimore, MD 21215. We use a Filamatic AB-5 (~$2000) with a FKS-60 filling unit (~$1200).

F. Media—Premixed

Instant *Drosophila* medium is available from Carolina Biological, Sigma Chemical Company, P.O. Box 14508, St. Louis, MO 63178, and Phillip Harris Ltd., Oldmixon Crescent, Weston-super-Mare, Avon BS24 9BJ, United Kingdom. Sigma also sells an agar-based premix that requires cooking.

G. Media Recipes

Bloomington Food (soft agar recipe for general use); makes about 2.75 liters.

Water	2.66 liters (we use tap water)
Agar	16 g
Cornmeal	163 g (Quaker, yellow)
Inactive dry yeast	33 g
Molasses, unsulfured	200 ml
10% *p*-hydroxy-benzoic acid	38 ml
Methyl ester in 95% ethanol	

Stir the agar into about 80% of the water. Bring to a boil and boil until agar in dissolved. In the meantime, dissolve the yeast in most of the remaining water (save some for rinsing) and then stir in the cornmeal (press all the lumps out). When the agar is dissolved, add the molasses to the agar/water mix and then stir in the yeast/cornmeal/water mix. Use the remaining water to rinse the dregs into the cooking pot. Return to a boil and simmer for 5 to 10 min. Allow to cool for 5 min or more and then carefully stir in *p*-hydroxy-benzoic acid methyl ester; dispense.

The amount of agar should be determined empirically for your brand of agar. We buy 100-lb barrels from Moorhead & Co., Inc., Van Nuys, CA 91406 ($1295). Yeast is from SAF Products Corporation, 310 Grain Exchange Building, P.O. Box 15066, Minneapolis, MN 55414 ($70 per 50 lbs). Grandma's (Dole) brand molasses works well (do not use the cheap stuff).

Hard agar medium for egg collection plates (makes about 60 6 × 2-cm plates).

Water	2.25 liters
Agar	88 g
Light corn syrup	350 ml
10% *p*-hydroxy-benzoic acid	38 ml
Methylester in 95% ethanol	

Mix water, agar, and syrup. Boil until agar is dissolved. Cool for 5 min or more and then stir in *p*-hydroxy-benzoic acid methylester. Dispense into petri dishes. When wrapped, these keep at 4°C for at least 6 weeks.

H. Miscellaneous

Available from a variety of laboratory chemical suppliers, including Sigma Chemical are benzyl benzoate; *p*-hydroxy-benzoic acid methyl ester (also known as tegosept, nipagin, methyl *p*-hydroxybenzoate, methyl paraben, and methyl parasept), and penicillin–streptomycin solution.

I. Plastic Trays

Expensive but sturdy and durable plastic trays and dividers are available in a variety of sizes from Healthmark Industries Co., MI 48080. $2\frac{1}{4} \times 10\frac{7}{8} \times 17\frac{3}{8}$-inch trays are about $14 each.

VII. Summary

The information provided here should allow you to begin working with *Drosophila*. Mine your colleagues for alternative approaches, improvements, and refinements and develop your own. If you find a new and better way to do any aspect of fly work, take the time to share it with your colleagues through bionet.drosophila or DIN.

References

Ashburner, M. (1989). "*Drosophila:* A Laboratory Handbook." Cold Spring Harbor, NY: Cold Spring Harbor Laboratory Press.

Backhaus, B., Sulkowski, E., and Schlote, F. W. (1984). A semi-synthetic, general-purpose medium for *D. melanogaster*. *Drosophila Information Service* **60,** 210.

Cole, K. W., Schreuders, P. D., Mahowald, A. P., and Mazur, P. (1993). Procedure for the permeabilization and cryobiological preservation of Drosophila embryos, Version 1.1. Oak Ridge National Laboratory Technical Manual TM-12394.

Lindsley, D. L., and Zimm, G. G. (1992). "The Genome of *Drosophila melanogaster*." San Diego: Academic Press.

Steponkus, P. L., and Caldwell, S. (1993). A procedure for the cryopreservation of *Drosophila melanogaster* embryos. Handout distributed at a workshop on *Drosophila* cryobiology at the 1993 *Drosophila* Conference, San Diego, CA. Available from the authors by request.

CHAPTER 3

Harnessing the Power of *Drosophila* Genetics

Mariana F. Wolfner and Michael L. Goldberg

Section of Genetics and Development
Cornell University
Ithaca, New York 14853

I. Introduction

In this chapter, we present an overview of the strategies and methods of *Drosophila* genetics that we believe will be most generally useful to the cell biologist. Given this topic's very broad theoretical scope, we simply outline approaches and direct the reader to other articles in which the methods are discussed or presented in detail. Because of the enormous literature on *Drosophila,* we can cite only a tiny subset of relevant articles; in many cases the numerous original articles are cited in the books we reference, such as Lindsley and Zimm (1992) or Ashburner (1989), both of which we recommend to the aspiring *Drosophila* geneticist. We apologize to those whose articles were not cited in our limited, and perhaps eclectic, list of references.

To present the most generalized discussion, we have further restricted our attention to mutations in genes required for viability or fertility (Ashburner, 1989; Lindsley and Zimm, 1992). Although visible (e.g., eye color, wing shape, bristle form, or sexual characteristics), behavioral (e.g., defects in learning (e.g., Davis and Dauwalder, 1991), or mating behavior (e.g., Hall, 1979)) phenotypes can provide crucial clues to important processes in cell biology, or can be exploited in designing screens for new mutations in a gene or pathway of interest, the relationship between a given visible phenotype and the underlying affected process is not always straightforward. For example, it is not intuitively obvious that a mutation in the *Glued* gene, which encodes a protein homologous to dynactin (a component of the molecular motor involving cytoplasmic dynein), would result in a fusion of the ommatidia (facets) in the eye (Holzbaur *et al.,* 1991; Gill *et al.,* 1991).

A. Setting the Stage—General Strategic Considerations

In undertaking a genetic analysis that will identify and interrelate the components of a given biological process, we strive to catalog as many relevant genes as possible and to obtain for each gene (1) one, and hopefully several, mutant alleles, (2) a detailed phenotypic analysis of each mutant, (3) cloned DNA representing the gene, (4) a sequence of the gene and its predicted product, (5) antibody to the gene product, (6) an understanding of the biochemical functions performed by the gene product, and (7) identification of other molecules with which a particular gene product may interact.

These objectives can be accessed through either of two major entry points. In the "mutation first" approach, collections of mutants can be screened for those with a desired general class of phenotype (e.g., lethality or sterility), which can then be analyzed in further detail. Once a mutation of interest has been identified, the challenge is to clone the corresponding gene. Alternatively, in the "gene first" strategy, knowledge of the gene product through a DNA sequence or antibody permits the cloning of its gene. The subsequent challenge is to obtain mutations in that gene.

Each of these approaches has unique advantages and disadvantages. The "mutation first" plan offers the opportunity to uncover novel gene products that would otherwise remain unknown, but this very novelty can make subsequent investigations into biochemical function more difficult. The "gene first" approach ultimately allows one to determine the *in vivo* importance of a protein whose biochemical function is already known to some extent. However, directed gene knock-outs are not yet routine in *Drosophila,* and the alternative methods used to generate mutations in a given gene are not always successful (see following). It also happens with some frequency that mutations in a gene are not associated with an informative phenotype, for example, in cases of redundant gene function. As an example, neither mutations in *D-abl* nor in *fasciclin I* show aberrant neural connections, but the double mutant displays abnormal axon guidance (Elkins *et al.,* 1990). This suggests that the two genes likely perform important but compensatory roles in this process. As a very rough guideline, most studies suggest that mutations in only about one-third of the ca. 15,000 genes in the *Drosophila* genome cause lethality or sterility (Merriam *et al.,* 1991).

II. How Do We Get Mutants from Cloned Genes or Clones from Mutated Genes?

A. The "Mutation First" Strategy

In this section, we first discuss how to obtain a collection of mutagenized chromosomes and then outline how to clone a gene associated with an interesting mutant phenotype. We defer consideration of how to screen for desired phenotypes to Section III.

1. How Do We Induce Mutations?

Examining large numbers of flies for a rare mutation is very labor intensive; screens of over 100,000 flies are rarely attempted. Thus, the efficiency of a mutagen is a major consideration, although this needs to be weighed in terms of the desirability of the sorts of mutations produced. Below, we discuss the various types of mutagenesis that are routinely employed, along with considerations that may govern the choice of a particular method. In many cases, investigators induce mutations on wild-type chromosomes so as to avoid any decrease in viability resulting from the use of a multiply marked chromosome. However, inducing the mutation on a marked chromosome can be subsequently advantageous in identifying homozygotes. When screens are done, one should be careful to identify only independent mutational events. Thus, ony one mutant per vial or per mutagenized parent is saved for further study.

a. Mutants by Phone

From the point of view of any geneticist, the most efficient mutageneses are those that have been done by someone else. *Drosophila* stock centers (see Chapters 1 and 2) maintain a large number of mutagenized lines, particularly those induced by mobilization of single P-elements (see following). In addition, a number of regions of the *Drosophila* genome have been screened to approximate saturation for lethal or sterile mutations (e.g., Judd *et al.*, 1972; see Lindsley and Zimm, 1992). These mutant lines are generally held by the investigators who isolated them, with the result that in some cases, mutants have unfortunately been lost over the years.

b. Ionizing Radiation

X-rays and gamma rays tend to cause chromosomal rearrangements. These can be of considerable value in gene cloning. About one-half of these rearrangements are cytologically detectable (multilocus deletions, inversions, or translocations), while a substantial fraction of the remainder are small intragenic deletions (Ward and Alexander, 1957; Pastink *et al.*, 1987,1988). However, use of ionizing radiation has two relative disadvantages. First, irradiation of developing sperm or oocytes causes mortality of a large number of F1 zygotes. Oocytes are so sensitive to irradiation that doses sufficient for effective mutagenesis cannot be used. To irradiate sperm without excessive mortality, males are usually subjected to doses of no more than 4000 roentgens; under these conditions, mutations of a typical locus would be induced in only 1 in every 5000 chromosomes (reviewed in Ashburner, 1989). Second, the investigator needs to be very cautious in interpreting abnormalities associated with radiation-induced mutations. Chromosomal rearrangements can simultaneously affect many genes and thus can give a composite phenotype that may be difficult to interpret.

c. Chemicals

Because of its effectiveness and low toxicity, the most widely used chemical mutagen is the alkylating agent ethylmethane sulfonate (EMS). The majority of EMS-induced alleles are single-base changes, consistent with its known chemical action in causing GC to AT transitions. The most generally used dose of EMS (25 mM fed to males) is approximately 5–10 times more efficient at inducing mutations than 4000 rad of X-rays (Alderson, 1965; Huang and Baker, 1976). However, at this dose, chromosomes would receive multiple "hits" (about three recessive lethal mutations per chromosome), so the correspondence between a phenotype and a particular mutation must subsequently be established. As this can involve considerable additional effort, lower EMS doses are desirable for some purposes, such as large-scale induction of random lethal or sterile mutations.

Other chemical agents have also been used for *Drosophila* mutagenesis, including *N*-ethyl-*N*-nitrosourea (ENU), triethylmelamine (TEM), diepoxybu-

tane (DEB), and formaldehyde (reviewed in Ashburner, 1989). ENU appears
to be relatively specific for the production of point mutations, while TEM,
DEB, and formaldehyde seem to be more effective for inducing deletions. There
is variation in the experiences of investigators concerning the type of lesions
induced with these reagents. For example, most DEB-induced mutations obtained by Crosby and Meyerowitz (1986) and by Reardon *et al.* (1987) were
either apparent point mutations or small deletions, while most DEB-induced
mutations found by Olsen and Green (1982) or Goldberg *et al.* (1989) were
multilocus deletions or other large-scale chromosome rearrangements.

A simple, generic screen for autosomal lethal or sterile mutations induced
with irradiation or chemical mutagens is shown in Fig. 1A. In addition, Ashburner (1989) catalogs several strategies for obtaining X-linked mutations. Xu
and Rubin (1993) developed a strategy that allows faster, easier screens for
recessive mutations, without the need for generations of breeding of stocks of
individual mutagenized chromosomes prior to screening. By inducing mitotic
recombination via the FLP-FRT system (see Section IVB and Chapter 34) one
can generate, in animals heterozygous for a mutagenized chromosome, regions
or tissues homozygous for the mutagenized chromosome. If the homozygous
areas are of the desired phenotype, the mutagenized chromosome can subsequently be placed into stock. This system can be applied even if the mutation
causes organismal lethality or sterility, but it does require that the mutation
produce an externally observable phenotype. Although this method may not
be applicable for all situations, it allows a very large number of mutagenized
chromosomes to be screened rapidly. Thus, it would be wise for the investigator
to consider this method before embarking on a mutagenesis.

d. Transposon Mutagenesis

The most important technical innovation in *Drosophila* mutagenesis in the
past decade has been the method for inducing mutations with single transpositions, or "hops," of marked P-elements (Cooley *et al.*, 1988). A marked P-
element is introduced by germline transformation into a genome otherwise
devoid of *P*-elements. In its most modern form, this P-element lacks sequences
encoding the transposase that would cause it to move on its own (thus it is
stable once integrated). Instead, the P-element contains (1) sequences encoding
a dominant visible marker, usually a wild-type *white*$^+$ or *rosy*$^+$ eye-color gene,
to allow the investigator to detect animals containing the P-element and to
screen for hops (Rubin and Spradling, 1982; Klemenz *et al.*, 1987); (2) bacterial
plasmid sequences including a selectable marker like ampicillin resistance to
facilitate cloning of *Drosophila* sequences adjacent to the insertion site of the
P-element (Steller and Pirrotta, 1985); and (3) (sometimes) *lacZ* sequences fused
to a weak but general promoter, to detect the presence of enhancers with a
developmental expression pattern of interest ("enhancer traps"; Bier *et al.*,
1989; Bellen *et al.*, 1989; Wilson *et al.*, 1989).

Once inserted in the genome, the P-element is mobilized by introducing a

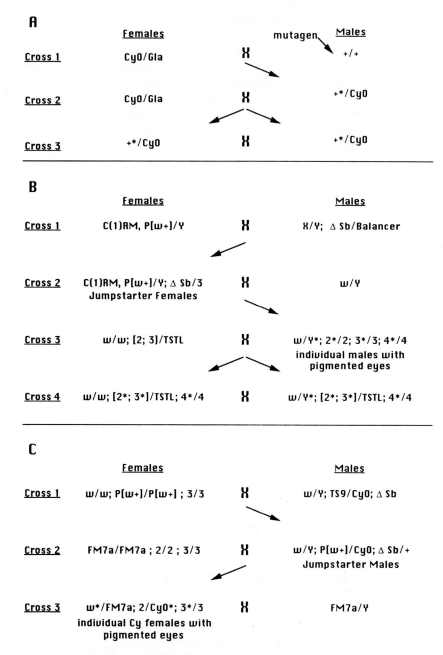

A

	Females		mutagen	**Males**
Cross 1	CyO/Gla	X		+/+
Cross 2	CyO/Gla	X		+*/CyO
Cross 3	+*/CyO	X		+*/CyO

B

	Females		**Males**
Cross 1	C(1)RM, P[w+]/Y	X	X/Y; Δ Sb/Balancer
Cross 2	C(1)RM, P[w+]/Y; Δ Sb/3 Jumpstarter Females	X	w/Y
Cross 3	w/w; [2; 3]/TSTL	X	w/Y*; 2*/2; 3*/3; 4*/4 individual males with pigmented eyes
Cross 4	w/w; [2*; 3*]/TSTL; 4*/4	X	w/Y*; [2*; 3*]/TSTL; 4*/4

C

	Females		**Males**
Cross 1	w/w; P[w+]/P[w+] ; 3/3	X	w/Y; TS9/CyO; Δ Sb
Cross 2	FM7a/FM7a ; 2/2 ; 3/3	X	w/Y; P[w+]/CyO; Δ Sb/+ Jumpstarter Males
Cross 3	w*/FM7a; 2/CyO*; 3*/3 individual Cy females with pigmented eyes	X	FM7a/Y

Fig. 1 Mutagenesis schemes. (A) Generic mutagenesis of the second chromosome with irradiation or chemical mutagens (from Ashburner, 1989). Mutagenized males are mated with females heterozygous for two different second chromosome balancers (*CyO* (with *Curly* wings) and *Glazed* (*Gla*, with *Glazed* eyes). Individual male progeny with *Curly* wings and a mutagenized chromosome (+*) are mated with *CyO/Gla* females. Male and female progeny of this second cross are interbred (Cross 3). The mutagenized second chromosome contains a lethal mutation if all of the progeny in Cross 3 have *Curly* wings, and the stock can be maintained through +*/*CyO* siblings. The

source of transposase into the germline. This is usually done by crossing flies carrying the marked P-element with others carrying a transposase gene called Δ2-3 (Robertson *et al.*, 1988). Among the progeny are "jumpstarter" flies carrying both the marked P-element and the transposase source. Hops of the marked P-element are identified in progeny of the jumpstarter fly in which the marker no longer segregates with the chromosome on which the P-element was originally located.

To obtain P-element hops onto the autosomes, one usually begins with marked P-elements on an X or attached X chromosome (attached-X flies, with or without

---

presence of a sterile mutation on the second chromosome is indicated if the *Curly* progeny, but not the non-*Curly* progeny, in Cross 3 of the appropriate sex are fertile. This same scheme can be applied to chromosome 3, using the chromosome 3 balancers *TM3* and *TM6* in place of *CyO* and *Gla*. (B) P-element mutagenesis of the autosomes. Only one of several possible schemes is depicted. For simplicity, wild-type chromosomes are not always listed, where they are, they are listed as 2 for chromosome 2, etc. Here, a female with an attached X chromosome (C(1)RM) containing a P-element marked with the w^+ gene (P[w +]) is mated to a male containing a source of P-element transposase (Δ2,3; abbreviated here as Δ) on a third chromosome that also has the dominant marker *Stubble bristles* (*Sb*). [The transposase-bearing chromosome is maintained over a third chromosome balancer like TM6B, which does not have *Sb*. In addition, subsequent recombination between Δ and *Sb* can be suppressed if both are contained on a multiply inverted balancer chromosome such as TMS (see Gatti and Goldberg, 1991).] The female progeny of this cross with *Sb* bristles are "jumpstarter" females, in whose germline the marked P-elements will be mobilized. Individual jumpstarter females are crossed to white-eyed males (Cross 2). If P[w⁺] has moved to the Y chromosome or any of the autosomes (chromosomes that could contain a novel P-element jump are symbolized as Y*, 2*, 3*, or 4*), male progeny with pigmented eyes will appear among the progeny of this cross. Individual pigmented-eye males that do not have the Δ *Sb* chromosome (so their P-elements can no longer hop) are now mated in Cross 3 to white-eyed females bearing balancer chromosomes for the second and/or third chromosomes (not shown) or a translocation between second and third chromosome balancers [such as *TSTL14* (abbreviated here as TSTL), which contains both *Tubby* and *Curly* dominant markers; see Gatti and Goldberg, 1991]. Male and female progeny, from each line, that have pigmented eyes and display the balancer marker(s) are intercrossed to generate a stock. Rare P-element insertions on the Y chromosome are identified if the only pigmented progeny of Cross 3 are males. Rare insertions on the fourth chromosome are identified by crossing females of the same genotype used in Cross 4 to white-eyed males (not shown), looking for pigmented progeny that have also received *TSTL*. (C) P-element mutagenesis of the X chromosome, roughly based upon Bellen *et al.* (1989). All X chromosomes contain a *white* mutation, so that pigmentation indicates presence of the P[w⁺] transposon. Jumpstarter males are made by crossing females homozygous for a P[w +] insertion on the second chromosome with males containing the *TMS* balancer containing the Δ source of transposase and the *Sb* marker (see B). These males also carry the *CyO* second chromosome balancer; TMS and *CyO* move as a unit into progeny of this male because they are placed over *TS9*, a translocation between chromosomes 2 and 3 (see Gatti and Goldberg, 1991). Individual jumpstarter males, containing P[w⁺], Δ (on TMS), and *CyO* are mated with females homozygous for the X chromosome balancer *FM7a* [marked with yellow body (*y*), apricot eyes (w^a), and Bar eyes (*B*)]. Individual female progeny with darkly pigmented eyes and *Curly* wings are selected; these must have a novel P-element insertion because normally P[w⁺] segregates from *CyO*. Flies carrying a newly induced X-linked lethal mutation are identified if the progeny of Cross 3 do not include non-*FM7a* w*/Y males with darkly pigmented, non-*Bar* eyes.

a Y chromosome, are fertile females). Because of the characteristics of sex-linked inheritance, marker-bearing progeny of the ''unexpected'' sex are likely to contain an autosomal jump (see Fig. 1B for an example). These flies are then used individually to establish a stock carrying the autosome with the inserted P-element, initially by crossing in balancers for the second and third chromosomes, or a translocation between second and third chromosome balancers that will be inherited as a unit (as in Fig. 1B). Males and females heterozygous for the autosomal P-element insertion and the appropriate balancer are then intercrossed, so that their progeny can be evaluated for phenotype and the stock can be maintained (see following).

A slightly different crossing strategy is used to obtain P-element hops onto the X chromosome (Fig. 1C). Here, it is necessary to start with a marked P-element on an autosome and to look for individual progeny in which the marker fails to segregate with the chromosome on which it was originally found. For female-sterile or visible phenotypes, X-linkage of the new P hops can be easily determined because only the female progeny of a male with a new insertion of the P-element on the X chromosome will show the marker phenotype. This is not possible for lethal or male-sterile mutations. Instead, in these cases one expects that the only male progeny of a female heterozygous for the P-element-bearing chromosome and an X chromosome balancer to survive or to be fertile are those carrying the balancer chromosome. Because the number of genes on the autosomes is approximately four times the number of X-linked genes, few investigators have to date attempted systematic P-element mutagenesis of the X chromosome. This should therefore be considered a potentially rewarding source of novel P-element-induced mutations.

The strengths of the P-element insertion mutagenesis strategy are the relative ease of cloning genes altered by the insertion of the marked P-element, the possibility of detecting genes with a particular expression pattern of interest (using enhancer traps), and the assurance that a large fraction of mutagenized chromosomes will have only a single novel P-element insertion. However, this type of mutagenesis is comparatively inefficient. Because P-element hops can occur during the mitotic expansion of the germline, only one hop from each jumpstarter parent should be isolated to ensure independence; this requires significant input in labor. Usually about 10–15% of new P-element insertions cause lethality (reviewed in Gatti and Goldberg, 1991), suggesting that a mutation of a specific locus would be obtained only once in 50,000 progeny (assuming there are 5000 essential loci). P-element mutagenesis may be particularly inefficient for certain genes, as insertions of P-elements are not random (Kidwell, 1987). Some genes are clearly hot spots for P-element insertions, while others are almost never targeted. A final potential disadvantage is that P-element insertions usually occur in regulatory regions or introns (Searles et al., 1986), altering the amount, but not the nature, of gene products. Thus, potentially interesting missense mutations would not be uncovered by P-element mutagenesis.

Other transposons such as *hobo* can also be used as mutagens (Blackman *et al.*, 1987,1989; Smith and Corces, 1991; Simonova *et al.*, 1992), although this has not yet been widely employed. Presumably, the site specificity for insertion of other transposons may be different from that of P-elements. Thus, one might be able to identify a novel group of mutants in this way or to obtain mutations in a gene that has not successfully been tagged with a P-element.

e. Natural Populations

Natural populations can be a rich source of variants in genes of interest (e.g., Sandler *et al.*, 1968), including transposon insertions (Green, 1988; Clark *et al.*, 1994), although the transposons are unmarked. These mutant sources are not work-free, though, as it is necessary to collect sufficient flies in the field and to set up crosses to isolate and then test individual chromosomes.

2. How Do We Clone an Interesting Gene Identified by a Mutation?

a. Mutations Made by Chemical or Irradiation Mutagenesis

Cloning genes known only through the existence of chemical- or irradiation-induced alleles may be difficult (Pirrotta, 1986). A prerequisite is to establish the chromosomal position of the gene. First, the segregation pattern of the mutation is noted relative to that of markers on the balancer chromosomes (Lindsley and Zimm, 1992); the mutation will segregate away from the balancer for its chromosome. Mutations on the X or Y chromosome will show sex-specific inheritance. (For many mutagenesis screens, such as that shown in Fig. 1A, the chromosome on which the mutant resides is already known because the crossing scheme requires segregation of a mutation from a particular balancer). Once one knows the chromosome on which the mutation resides, the mutation must be mapped by genetic crosses relative to known markers. If there are already preexisting deletions ("deficiencies", Df) of the relevant chromosomal region, flies can be generated that are heterozygous for the mutation and each of the various deletions. If such "hemizygous" flies display the phenotype, the mutation lies within the region missing from the deletion chromosome. In addition to localizing more precisely the cytological position of the mutation, this analysis rules out that the phenotype is due to extraneous mutations outside the region "uncovered" by the deletion. This precise cytological mapping gives the investigator a tool with which to search *Drosophila* databases (Chapter 1) to see whether the mutation is in a previously identified gene. Deletions are available for many regions of the *Drosophila* genome (Lindsley and Zimm, 1992). For other regions, deletions can be generated by "segmental aneuploidy," which is described in detail by Lindsley *et al.* (1972) and by Stewart and Merriam (1973) (see also Lindsley and Zimm, 1992). If no deletion is available in the chromosomal region of interest, it is desirable to make a deletion with irradiation, looking for loss of a closely linked marker. Mapping the position of a mutation relative to other chromosomal rearrangements (inversions (In), translocation (T), or duplications (Dp)) is also possible (Lindsley and Zimm, 1992), although more difficult and thus generally less widely used.

Once the mutation has been localized on the cytogenetic map, one may discover in a database search that DNA from that chromosomal region has already been cloned, perhaps during the course of one of the *Drosophila* genome projects (Kafatos *et al.*, 1991; Hartl *et al.*, 1992). Using this material, chromosomal walks (Bender *et al.*, 1983) in the appropriate chromosome interval can be begun or extended. The position of the gene of interest within the walk can often be narrowed by molecular localization of pertinent rearrangement breakpoints (e.g., Gunaratne *et al.*, 1986). Within the appropriate region, transcripts with the expected expression pattern (e.g., an ovarian transcript for a gene with female-sterile mutations; Lin and Wolfner, 1989) may help to pinpoint candidates for the mutated gene, as might the existence of restriction fragment length polymorphisms associated with mutant alleles. Ultimate proof requires the rescue of the mutant phenotype by germline transformation (Rubin and Spradling, 1982; Karess, 1985; Spradling, 1986; Ashburner, 1989) with wildtype genomic DNA encoding the candidate transcript (e.g., Gunaratne *et al.*, 1986) or with a cDNA for the transcript under the control of an appropriate promoter (e.g., Lopez *et al.*, 1994).

b. Mutations Made by Transposon Mutagenesis

Once an interesting mutation is identified, one must prove that the P-element insertion actually caused the mutation. This is generally accomplished by using transposase to excise the P-element and demonstrating that the excision (detected by loss of the marker carried on the P-element) causes reversion of the mutation (e.g., Cooley *et al.*, 1992). Only a subset of excisions can do this: reversion usually requires precise excision, which may account for 10% or fewer of all losses of P-element markers; most excisions are imprecise (Daniels *et al.*, 1985). Occasionally the inserted P-element proves impossible to mobilize. In that case, one can allow the mutant chromosome to recombine freely with a normal chromosome (Note: In *Drosophila*, meiotic recombination occurs only in females!), looking in the progeny for tight linkage of the P-element marker and the mutant phenotype. Alternatively, one can determine the cytological position of the P-element by *in situ* hybridization (e.g., see Ashburner, 1989) and test whether the mutant phenotype is seen in an animal carrying the mutant chromosome and a chromosome deleted for the region containing the P-element. If the location of the P-element insert is known, the investigator can also search *Drosophila* databases for mutations in that chromosomal region and determine whether the P insertion is allelic to a previously identified gene.

Even if the location of the gene has not been determined, the gene can be cloned by plasmid rescue (if the P-element contains bacterial plasmid sequences; Steller and Pirrotta, 1985; Cooley *et al.*, 1988), inverse PCR strategies using primers related to the P-element ends (Ochman *et al.*, 1988), or by making a genomic library from a stock containing the P-element mutation and screening the library with P-element probes (e.g., Searles *et al.*, 1986). As above, ultimate proof that the cloned gene is the one responsible for the mutant phenotype requires transformation rescue.

B. The "Gene First" Strategy

The gene first strategy can be used if the investigator has some knowledge of the gene or gene product of interest. This can include a homologous gene sequence, the amino acid sequence of at least part of a protein, or an antibody recognizing a polypeptide epitope. Using these tools, the investigator clones the gene of interest by low-stringency hybridizations of heterologous probes or appropriate oligonucleotide mixtures to *Drosophila* genomic libraries (e.g., Zheng *et al.*, 1991), by a PCR-based strategy (e.g., Endow and Hatsumi, 1991), or by antibody screening of *Drosophila* cDNA expression libraries (e.g., Byers *et al.*, 1987). Once the gene is cloned, its position is established by *in situ* hybridization to wild-type polytene chromosomes; the position can be further narrowed by examining hybridization to chromosomes carrying rearrangements in the relevant chromosomal region.

1. How Do We Get Mutants in the "Gene First" Strategy?

Methods to screen for mutation in a gene without prior knowledge of their phenotype have been developed. However, none of these gene first strategies are guaranteed to find a mutation in the gene—they are all based upon sifting through a collection of random genetic alterations. This problem cannot be overcome until a system allowing directed gene knockouts by recombination becomes routine in *Drosophila*. Such a system has been reported for cases in which one has a P-element insert very near the gene of interest (Gloor *et al.*, 1991). In future years this system may gain more general applicability.

a. Mutants by Phone

Once the chromosomal position of the gene is known, the databases (see Chapter 1) and the literature may reveal mutations in the region that could potentially disrupt the gene. Homozygotes for such mutations, if they can be obtained, can be examined by Northern or Western blot analyses to see if they alter the production or size of the gene's transcript or protein product. Alternatively, transformation rescue with the cloned gene can prove that the mutation disrupts the gene of interest. Finally, the mutant alleles can be sequenced to determine whether they disrupt the putative open reading frame (e.g., Weinzierl *et al.*, 1987).

b. Screening Mutants for Protein Loss

If a previously existing mutation in the gene is unavailable, one can screen newly made collections of mutagenized chromosomes directly for ones that fail to make the gene product (e.g., VanVactor *et al.*, 1988). This strategy requires the production of animals identifiably homozygous for the mutation or (better) hemizygous for the mutation over a deletion for the corresponding locus. Hemizygous or homozygous animals carrying an individual mutagenized chromosome are ground in a well of a microtiter dish. The resultant extract is spotted in an array onto media suitable for screening with antibodies or for enzyme activity. Wells in which little or no gene product is detected are identified; these presumably contain flies mutant in the gene of interest. The mutation is then placed into stock from the mutant/balancer sibs of the tested fly.

c. Screening for Nearby P-element Inserts

Another strategy for obtaining mutants in a cloned gene is to find insertions of transposons within or adjacent to the gene. Except when a telephonic or computer requisition (via Flybase) provides a line of flies carrying such a nearby P-element, a screen must be done to mobilize P-elements and to identify inserts in or near the gene by molecular means (see Chapter 4). A substantial increase in frequency of useful hops can be obtained if one begins with a P-element relatively close to the gene of interest (Zhang and Spradling, 1993). However, if there is no available P-element, one can mount a general screen.

Lines of flies resulting from individual P-element hops are established. DNA is extracted from pools of flies representing 50–100 lines. The DNA is subjected to PCR using one primer derived from the transposon end and another primer within the gene (Fig. 2) and examined for the generation of a band indicating the presence of a P-element in the vicinity of the gene (Ballinger and Benzer, 1989; Kaiser and Goodwin, 1990; Clark et al., 1994). Alternatively, pools of plasmids derived from the insertion sites by plasmid rescue are hybridized to Southern blots of the clone of interest (Hamilton et al., 1991) to identify a pool containing a P-element in or near the gene.

Positive pools are subdivided and rescreened to identify the single line that contains the desired P-element. Because the animals from which the DNA was extracted are dead, the insert is maintained in stock via their siblings or progeny. Clark et al. (1994) have noted that natural populations, which contain many P-elements (~20/genome), are an excellent source material for this strategy because the chance that a P-element will be found adjacent to a particular gene is much higher than in chromosomes mutagenized with single P hops.

It is much more efficient to screen simultaneously for insertions adjacent to several genes of potential interest. Clark et al. (1994) have estimated that if P-element insertion were random (unfortunately not a true assumption), the chance of getting a P-element insertion within 2 kb of any particular gene among 1000 naturally derived chromosomes would be 0.21, but the chance of obtaining at least one P insertion near any of 10 target genes in the same sample would be 0.90. The lion's share of the work is in maintaining mutagenized stocks and preparing DNA from them; the marginal effort involved in performing additional PCR reactions or hybridizations (for more genes) is relatively small.

These strategies may pick up P-element insertions within or adjacent to genes.

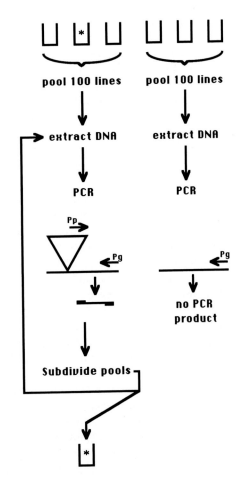

Fig. 2 Scheme for detecting P-element inserts in or near a gene using PCR. This scheme, modified from Ballinger and Benzer (1989), begins with the setting up of crosses each containing one or a few flies with new P-insertions (see Fig. 1 for generation of insertions). In most vials, the flies will not contain an insert in the region of interest. The rare vial containing an insert is designated by *. Flies from 100 vials are pooled and their DNA is extracted. The DNA is subjected to PCR using two primers, one in the P-element (Pp) and one in the gene of interest (Pg). A PCR product will result only when the P-element has inserted in the correct orientation near the gene of interest. If a PCR product is detected, the pool is subdivided, and the extraction and PCR are repeated on the smaller pools. This process is repeated until the investigator has identified the single vial whose flies contain the P-element insert. The rounds of rescreening take place within a single fly generation. Several variations of this method, involving differences in the detection of the insert or in the screening schedule, have been developed by Kaiser and Goodwin (1990), Hamilton *et al.* (1991), and Clark *et al.* (1994).

Although the latter might not themselves alter the expression of a gene, small deletions or other aberrations of the gene can subsequently be generated as imprecise excisions of the P-element (Engels, 1989; Engels *et al.*, 1990; Cooley *et al.*, 1990). Because these excisions could potentially affect additional genes, it is essential to prove that the mutant phenotype can be completely rescued by germline transformation with cloned DNA from the gene of interest.

III. How Do We Determine Which Stages of the *Drosophila* Life Cycle Are Affected by a Mutation?

To define the role of a particular gene product in cells and developing tissues, we begin with a simple sifting procedure to determine the life stage at which the effects of mutations in a gene are first apparent. In many cases, mutations are on unmarked chromosomes maintained in stocks over a balancer chromosome (mutant/Balancer); the balancer chromosome carries a dominant visible marker, a recessive lethal mutation, and a crossover-suppressing rearrangement. Thus, to obtain mutant homozygotes, one mates males and females from the stock and examines their progeny. If all adult progeny of this mating show the phenotype specified by the balancer chromosome, one concludes that the mutation is lethal. If adult progeny that lack the balancer chromosome are obtained, the mutation is viable; the mutant progeny can be examined for fertility and visible and behavioral phenotypes.

An alternative procedure to identify adults homozygous for the mutation is possible if the mutation was induced on a chromosome containing a recessive or dosage-sensitive marker. For example, one can distinguish animals homozygous or heterozygous for mutations caused by insertions of a P-element that carries the hs-w^+ marker ($white^+$ driven by a heat-shock promoter (Klemenz *et al.*, 1987)), because eye color darkens with increasing copies of hs-w^+.

Ashburner (1989) correctly points out that phenotypic analysis of animals homozygous for a particular mutagenized chromosome may be very dangerous, because of the possibility that the chromosome might harbor more than one mutation. To make certain that one is examining the consequences of mutation in a single gene, it is thus better to look at heteroallelic combinations of the mutant chromosome with another carrying a deletion for the region or combinations of two independent, noncomplementing mutations in the gene. These strategies are unfortunately impractical for some uses, particularly for large-scale screens for lethal or sterile mutations that exhibit particular phenotypes. The problem of multiple hits on a chromosome can be mitigated to some degree by using low dosages of mutagens, by performing single P hop mutagenesis, and by mutagenizing chromosomes previously known to be free of lethal or sterile mutations (i.e., by starting the stock to be mutagenized from a single wild-type chromosome). Ultimately, one must prove that the phenotype is caused by a particular single mutation. This can be done either by using heteroallelic combinations as suggested by Ashburner, by recombination experiments to separate out potential multiple hits, by demonstrating simultaneous reversion of the phenotype and loss of the P element marker as described previously, or by rescuing the phenotype with transformed pieces of DNA.

In the following sections, we discuss how to categorize further lethal or sterile mutants identified by the above criteria. More involved phenotypic studies are considered in Section IV.

A. If Your Mutant Is Lethal

For an essential gene, the first order of priority is to determine when mutant animals die, as this will dictate what types of detailed phenotypic analysis could subsequently be performed. A rough first step is to examine a culture to see whether such mutant homozygous animals are represented at normal frequencies at all life stages or whether they only survive to a certain point in development. An important precondition necessary to make this assessment is that one be able to identify homozygotes at any time in development. Morphological features allow one to distinguish a series of stages in embryonic and pupal development; larval development does not involve such dramatic changes and therefore is less well subdivided, except for the three larval instars and for some behavioral changes at the very end of the last instar.

1. Mutants That Die during Embryonic Stages

Because there is little gene transcription immediately postfertilization, early embryogenesis proceeds with gene products mostly supplied to the egg by the mother (Edgar and Schubiger, 1986). If a protein is essential, a mutant zygote will survive until the store of maternal product (supplied by its heterozygous mother) is depleted. This often happens soon after cellularization; indeed, certain maternal mRNA molecules are degraded just prior to this time (Edgar and O'Farrell, 1989,1990).

To determine whether a mutant arrests during embryonic stages, embryos produced from a brief egg lay by heterozygotes are examined visually (Wieschaus and Nusslein-Volhard, 1986). Adult flies heterozygous for the mutation are crossed in plastic beakers with holes punched in them, which are inverted over apple juice agar plates. After a short period of egg laying (typically 30 min, but this can be adjusted depending on the nature of the mutant and the fecundity of the cross), the parents in their beaker are removed. The embryos on the agar plates are either prepared for observation (following) or aged as appropriate before observation.

A drop of voltalef oil placed over a developing embryo allows one to observe the embryo's development under the compound or dissecting scope with transmitted light. The oil does not interfere with development. By comparing the appearance of the embryo to the defined stages of embryogenesis (Foe and

Alberts, 1983; Wieschaus and Nusslein-Volhard, 1986; Foe, 1989; Foe *et al.*, 1993), one can determine when defects are first seen in the mutant. Videomicroscopy of living embryos (under oil) can provide important information on major developmental or cytoskeletal events such as cell or cytoplasmic movements (technique described in Wieschaus and Nusslein-Volhard, 1986; see also Chapter 26).

In the cross between animals heterozygous for an embryonic lethal mutant over a balancer chromosome, one would expect 25% of the fertilized eggs to be homozygous mutant and another 25% to be homozygous for the balancer (Fig. 3A). Thus, if approximately one-quarter of the resultant embryos do not hatch, and display a consistent arrested phenotype (unlike that of the balancer homozygotes, if those are embryo lethal), then by inference, these represent the dying mutant homozygotes. In some cases, it may be possible to distinguish the mutant embryos directly (Table I). Mutants can be identified *post mortem* (after fixing and staining) by generating animals heterozygous for the mutant chromosome and a chromosome containing a *lacZ* gene fusion that is expressed in embryos (Meise and Janning, 1993). When such flies are interbred, homozygous mutant progeny will not stain blue with X-Gal, whereas flies containing the *lacZ* fusion will. One can also determine *post mortem* whether an individual animal was mutant by performing PCR on its DNA, using probes that give different bands for the mutant and the balancer chromosomes. Because these techniques can be used only on fixed embryos, one cannot follow the development of a mutant embryo once identified.

2. Mutants That Arrest at Postembryonic Stages

Death can occur at any time in larval or pupal development for a variety of reasons. In some cases, mutant animals die as larvae if there was sufficient maternal product to carry the embryo beyond hatching, but the homozygote runs out of product as a larva. Alternatively, some proteins may be required for events specific to larval or pupal stages, e.g., eclosion from the pupal case (Cavener and MacIntyre, 1983). The larval/pupal boundary is a particularly sensitive period for many mutations. Much of larval life simply involves cell growth; cell division is required only in tissues such as the imaginal discs that eventually will form adult structures. Thus, at least some mutants in genes necessary for cell proliferation die at the larval/pupal boundary or during the pupal period, as their imaginal discs contain too few cells to support adult development (Gatti and Baker, 1989; Gatti and Goldberg, 1991).

At postembryonic stages, homozygotes can be recognized in stocks containing balancers with dominant markers that exhibit a phenotype during these stages (Table I). For the autosomes, such balancer markers include *Tubby (Tb)* on the third chromosome, which results in short, squat larvae and pupae (for this and all other markers, we refer the reader to Lindsley and Zimm (1992), which gives much information as well as relevant references), and *Black cells*

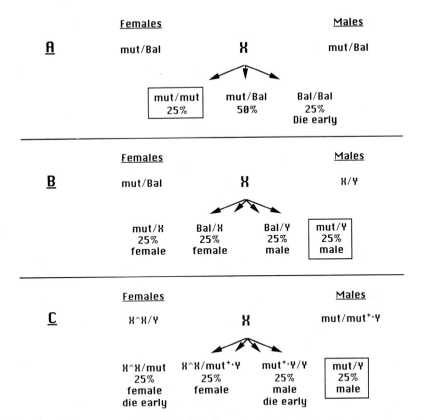

Fig. 3 Identifying mutant animals for phenotypic analysis. In A–C, "mut" refers to a recessive lethal mutation, and "Bal" is a balancer for the chromosome on which the mutation is found. (A) If the mutation is on an autosome, stocks are normally maintained by interbreeding mutant/Balancer flies. Twenty-five percent of the zygotes obtained from this cross are homozygous for the mutation. These can be discriminated from the other progeny if the balancer chromosome contains a dominant marker visible at or before the time of developmental arrest (see text and Table I). (B) If the mutation is X-linked, 25% of the zygotes created by crossing mutant/Balancer females with any males (X/Y) will be hemizygous mutant males. These are identified by lack of the dominant phenotype associated with the balancer and by the appearance of male-specific structures or gene products. (C) In some instances, Y chromosomes are available that contain specific translocated pieces of a wild-type X chromosome (mut$^+$·Y), allowing mutant/mut$^+$·Y to survive. This is particularly advantageous if the developmental arrest is post cellularization. When such males are crossed to attached X/Y females ($X^\wedge XY$), the only male progeny that survive past cellularization of the embryo are the desired hemizygous mutants.

(Bc) on the second chromosome, which results in black cells throughout the body, or the y^+ marker (such as that transposed into some versions of the TM3 balancer) in a *y (yellow)* background. In addition, both the second and third chromosomes can be "marked" with *Tb,* using the balancer translocation *TSTL14,* synthesized by J. Casal and P. Ripoll (see Gatti and Goldberg, 1991).

Table I
Some Useful Markers On the Major Chromosomes

	Gene name	Map position	Useful for[a]	Cell marker?[b] (vs whole animal)	Comments[c]
			X chromosome		
y	yellow	0.0	A: cuticle (bristles, some alleles); L/P: mouth parts	✔	y^+ has been translocated to autosomes and can thus also be used to mark them
w	white	1.5	A: eyes, testes; L: malpighian tubules	✔	
cho	chocolate	5.5	A: eyes L. malpighian tubules		
svb	shaven baby	9.8	E: cuticle	✔	
Sxl	Sex lethal	19.2	Postblastoderm	✔	Postmortem stain
sn	singed	21.0	A: bristle (shape)	✔	
f	forked	56.7	A: bristle and hair (shape)	✔	
B	Bar	57.0	A: eye shape		Dominant, on many balancers
mal	maroonlike	64.8	Eyes and internal organs	✔	Adehyde oxidase, pyridoxal oxidase, xanthine dehydrogenase affected
fog	folded gastrulation	65	E		
bb	bobbed (rDNA)	66	E: nucleoli	✔	Postmortem stain
			Chromosome 2		
Cy	curly	6.1	A: wing position		Dominant, on many balancers
Ki	Kinked	47.6	A: bristle and hair (shape)	✔	Dominant
Bl	Bristle	54.8	A: bristle and hair (shape)	✔	Dominant
stw	straw	55.1	A: bristle (color)	✔	
pwn	pawn	55.4	A: bristle and hair (shape)	✔	
Bc	Black cells	80.6	L/P/A		Dominant
sdp	sandpaper	83	A: abdominal hairs (color, pattern)	✔	
Sdh	Succinate dehydro-genase	89	Internal organs	✔	Postmortem stain
bw	brown	104.5	A: eyes, testes L: malpighian tubules	✔	In st, cn, or v background
Gla	Glazed		A: eye texture		Dominant, on balancers
			Chromosome 3		
mwh	multiple wing hairs	0.3	A: hair number and pattern	✔	
jv	javelin	19.2	A: bristle and hair shape		
flr	flare	38.8	A: bristle and hair shape	✔	
st	scarlet	44.0	A: eyes L: malpighian tubules	✔	In bw background
trc	tricorner	46	A: wing, notum hair pattern	✔	

continues

Table I (*Continued*)

Sb	*Stubble*	58.2	A: bristle shape	✔	Dominant, on some balancers
Ubx	*Ultrabithorax*	58.8	A: haltere shape		Dominant, on some balancers
e	*ebony*	70.7	A: body color		
Tb	*Tubby*	90.6	L,P: body shape		Dominant, on TSTL and TM6B balancers
Bsb	*Blunt short bristles*	100.6	A: bristle (shape)	✔	Dominant
			Other		
lacZ			(From enhancer traps; depends on line but E, L/P, A markers available)	✔	Postmortem stain
GFP	(green fluorescent protein)		All stages?	✔	
epitope tags				✔	Postmortem stain

Note. Reference for GFP is Chalfie *et al.* (1994). At present there are no reports of any limits to the use of this marker at particular stages. Reference for epitope tags is Xu and Rubin (1993).

[a] E, embryo; L, larva; P, pupa; A, adult.

[b] ✔, can be used as cell-autonomous marker in clones

[c] *Gla* is in a rearrangement and thus has no known map position

TSTL14 is a translocation between the second chromosome balancer *SM5* (carrying the dominant *Curly (Cy)* wing mutation) and the third chromosome balancer *TM6B* (marked with *Tb*). This translocation can simultaneously balance both the second and third chromosomes. For example, in a genetic screen to identify lethals on the major autosomes, one searches for mutant/*TSTL14* stocks with an absence of adult flies without curly wings. Whether there are any non-*Tubby* larvae or pupae observed in the stock allows rough determination of the time of arrest.

For recessive lethal mutations on the X chromosome, it is generally easiest to maintain a stock in the form mutant/Balancer females crossed with Balancer/Y males (Fig. 3B). If the mutant-bearing chromosome also carries a *white*$^+$ (w^+) or *yellow*$^+$ (y^+) allele, whereas the balancer carries a mutant *white* or *yellow* allele (for example, the *white*^{a} marker on the balancer *FM7a*), or vice versa, it is possible to distinguish dying male larvae or pupae hemizygous for the mutation of interest. This is because (1) male larvae can be distinguished from female larvae based upon the relative size of their gonads, and male and female pupae can be differentiated on the basis of the presence or absence of the sex-comb on the foreleg of the pupae in the pupal case (Bodenstein, 1950), and (2) the w^+ and the y^+ alleles produce distinguishable phenotypes in larvae and pupae. w^+ animals have yellow Malpighian tubules in larvae and pupae, and dark eyes in later pupal stages, whereas *white* mutant animals have colorless Malpighian tubules and lighter eye color in late pupae. y^+ animals have black mouthhooks in larval stages and black bristles in pupal stages, whereas *y* mutants have brown mouthhooks and bristles.

For some X-linked recessive lethal mutations, the stock can be maintained in an alternative way (Judd *et al.*, 1972; see Fig. 3C) if the mutation is within a region that has been translocated or otherwise inserted into another chromosome; there are a number of Y chromosomes carrying such insertions of X chromosome regions (Lindsley and Zimm, 1992). As an example, a lethal mutation near the *white* gene can be maintained in stock over a $w^+ \cdot Y$ chromosome, which contains the wild-type copy of the region surrounding the *white* gene. The stock is maintained as mutant/$w^+ \cdot Y$ males $\times X^\wedge X /w^+ \cdot Y$ females, where $X^\wedge X$ represents an attached X chromosome. All male larvae or pupae that can be obtained from a subsequent cross between mutant/$w^+ \cdot Y$ males and $X^\wedge X /Y$ females would be hemizygous for the mutation.

3. What Does the Time of Arrest Indicate?

In its simplest interpretation, the time of arrest reflects the first time in development at which the gene product is necessary. However, one should also consider that establishing a time of arrest does not necessarily prove that the gene product of interest is required precisely at the time the animal died. The actual defect might be in an earlier process, whose consequence is not obvious until the time of death. Thus, earlier stages of development should also be examined for possible abnormalities. Also, it should be remembered that arrest at one stage will prevent detection of important effects of the mutation later in development. One method to deal with this issue is to isolate additional, perhaps weaker alleles of the gene that can develop past the earlier block. A second possibility is to generate clones of homozygous mutant tissue in an otherwise heterozygous animal as described under Section IV.

B. If Your Mutant Is Sterile

If adults homozygous for a mutation-bearing chromosome are obtained, one determines whether the mutants are sterile by mating homozygous mutant animals with fertile individuals of the opposite sex and checking for the appearance of progeny. Sterility often results from defects in gametogenesis, which is exquisitely sensitive to perturbation, presumably because of the highly regulated series of events that must occur and the high levels of gene products required.

1. Female-Sterile Mutations

Many female-sterile mutants simply do not produce or lay eggs because of defects in oogenesis or problems in the development of the genitalia or in mating. These mutants are easily recognized by the relative absence of eggs deposited on the agar surface. A normal mated female lays 30–60 eggs per day for the first few days after mating; a female who does not mate lays fewer than 10 eggs per day (Kubli, 1992; Kalb *et al.*, 1993). Direct observation can establish whether the mutant is defective in mating behaviors. Dissection and examination of reproductive organs can detect abnormalities in oogenesis or in genital development (Mahowald and Kambysellis, 1980; Spradling, 1993). If oogenesis is blocked, the defect can be further narrowed by determining which oogenic stages or ovarian gene products are absent in the mutant (see also Chapter 28).

Other female-sterile mutants lay eggs that do not develop into adults. A subset of these mutants lay eggs that cannot complete meiosis or cannot be fertilized. These can be identified by, respectively, DAPI staining to detect the four meiotic products (Doane, 1960; Foe *et al.*, 1993) or staining with an antibody against a sperm tail antigen (Karr, 1991). Many other female-sterile mutants lay eggs that have completed meiosis and been fertilized, but still do not develop; these mutants are called *maternal effect lethals* (MELs). In cases of no development or very early arrest in development, the eggs often remain white in appearance under incident light (as do normal eggs, but of course those hatch after a day). The cytoplasm of these eggs appears grainy or forms a ball within the egg membranes (Wieschaus and Nusslein-Volhard, 1986). In contrast, when development proceeds beyond the earliest stages but halts prior to embryo hatching, the eggs can appear brown, and the embryos inside show signs of development. In some cases (*strict* MELs), no progeny are produced even if the eggs are fertilized with wild-type sperm. In general, strict MELs encode products required to construct a functional egg or products that are deposited in the egg and are required for the earliest events subsequent to fertilization. Because the zygotic genome is essentially not transcribed until the blastoderm stage (Edgar and Schubiger, 1986), these very early events cannot be influenced by the genetic make up of the sperm. Other MELs can be rescued by wild-type sperm. This implies that the activity of the corresponding gene is not needed until after the zygotic genome begins to be transcribed or that mutational disruption of some very early activity can be reversed by subsequent wild-type gene function or by something donated by the sperm. The defects in the embryos laid by MEL females can be categorized using the methods summarized above for embryonic lethals.

2. Male–Sterile Mutations

Male sterility can result from abnormalities in behavior, in genital development, in gametogenesis, or in the ability of the sperm to donate to the egg items necessary for zygotic development. All but the last of these defects can generally be recognized by direct observation of mating behaviors (Hall, 1979) or by dissection and examination of reproductive organs (Lindsley and Tokuyasu, 1980; Fuller, 1993). Male-sterile mutations that affect spermatogenesis can be categorized in terms of the first stage in spermatogenesis that is abnormal, such as spermatogonial mitosis, meiosis, sperm maturation, or motility. Among mutants that produce motile sperm, there is a very small, but highly interesting class whose sperm can fertilize eggs, but cannot support their further development (e.g., Baker, 1975; Fuyama, 1986); these mutations are called *paternal effect lethals*. These mutations could result in defects in the formation or function of the male pronucleus during the first mitotic divisions in the zygote.

C. Caution: Different Mutations in the Same Gene Can Have Very Different Effects—Get Several Alleles!

It is not uncommon for a gene to have alleles with disparate effects, particularly when the gene product is utilized by many different cell types at various stages of development. For example, some mutations in *deep orange (dor)* result in lethality, others in male sterility, others in maternal effect female sterility, and others simply in eye-color changes (Perrimon *et al.*, 1986; Lindsley and Zimm, 1992). Different mutant alleles may alter cell type- or developmental-stage-specific expression of the gene (e.g., Weinzierl *et al.*, 1987); alternatively, certain biological processes may be more differentially sensitive to particular alterations in the amount or primary structure of a protein (e.g., Cooley *et al.*, 1992). Thus, to gain a comprehensive understanding of the role of a particular gene product, it is critical to obtain and analyze as wide a range of alleles as possible.

To obtain new alleles of a gene of interest, one can look through preexisting collections of mutations for those that map in the chromosomal region of the original mutation and that give a phenotype of any sort when heterozygous with the original mutant. New mutageneses, preferably with a different mutagen from that which induced the allele already in hand, can be attempted to find noncomplementing novel mutations. For mutations caused by the insertion of a marked P-element, one can generate imprecise excisions of the P-element to obtain some that delete part of the gene, resulting in a stronger allele.

The most straightforward evaluation of the importance of a particular gene product to the organism is by the analysis of the phenotype associated with a null, or complete loss-of-function, mutation. Null mutations often, but not always, cause the most severe phenotype. Unambiguous proof that a mutation is null is the demonstration that the mutant fails to make the gene's mRNA or protein product. If the requisite molecular reagents are unavailable, or if the above experiment is unsuccessful (as would be the case for a null mutant that produced a nonfunctional product of normal length), mutations can nonetheless be provisionally classified as nulls by a classical genetic test (Muller, 1932). If the phenotype of a homozygote is identical to that of a mutant hemizygote, the mutation is likely to be null. Further support would come from the observation that the phenotype of any heterozygote between the putative null and another allele at the locus is identical to that of the other allele over the deletion. Of course, these genetic tests are limited by the sensitivity at which one can actually evaluate potentially small phenotypic differences.

Although the importance of null mutations is uncontestable, they may still provide only a very limited picture of the role of the gene. Nulls can in theory identify the first process affected by lack of a protein, but this might obscure the importance of the protein for other processes that occur later in development. Individuals mutant for leaky (also called hypomorphic or partially functional) alleles may develop further, allowing examination of later processes that are sensitive to aberrations in the gene product. In addition, leaky mutations provide a more sensitive background than null alleles in which to search for, or observe the effects of, intergenic interactions (Section V).

Two additional classes of mutant alleles may be extremely valuable, although they can be difficult to obtain. Temperature-sensitive (ts) alleles of a gene can be used to determine when the gene's function is of importance (e.g., Belote *et al.*, 1985a; Belote and Baker, 1987; Chapman and Wolfner, 1988; Section 4). They can also be utilized to vary the amount of functional gene product by raising ts mutants at a series of intermediate temperatures (Belote and Baker, 1982). However, since mutagenesis experiments in *Drosophila* are usually so time- and labor-intensive, few screens for relatively rare ts alleles have been successful. Nonetheless, it is easy and important to test any new allele of a gene of interest for phenotypic differences at 16 and 29°C relative to normal room temperature. A final class of potentially useful mutants are dominant mutations. These mutations can result from, for example, inappropriate or elevated expression of a gene product or expression of a product with constitutive or elevated activity (e.g., Schneuwly and Gehring, 1985; Frischer *et al.*, 1986; Schneider *et al.*, 1991). Dominant mutants can be useful in double-mutant studies (e.g., Anderson *et al.*, 1985; Section 5). Although most dominant mutations are associated with visible phenotypes, some mutations causing conditional dominant lethality (e.g., *Sex lethal* (*Sxl*; Cline, 1978) or *Dominant ts lethals* (*DTS*; Suzuki and Procunier, 1969; Holden and Suzuki, 1973)) are known. In addition, a group of dominant female-sterile mutations that has been of great value in making homozygous mutant germline clones has been discovered (e.g., Busson *et al.*, 1983; Perrimon, 1984; see Section 4).

IV. How Do We Analyze Mutants?

To undertake more detailed phenotypic analysis of a mutant, it is often worthwhile to mark the mutant chromosome to identify directly mutant homozygote animals or cells. This can be accomplished by recombining appropriate recessive markers onto the mutant chromosome. This procedure will be aided if the map position of the mutation is known (from previous recombination experiments, mapping relative to deletions, or from the results of *in situ* hybridizations; see preceding). Alternatively, if properly designed, the process of crossing in appropriate markers will itself yield the approximate map position. Some recessive markers of particular utility that can distinguish homozygous larvae, pupae, and/or adults include (Table I; Lindsley and Zimm, 1992, and references therein) on the X chromosome, *w* and *y*; on the second chromosome, *cinnabar (cn)*, *brown (bw)*, and the combination *cn bw* (which gives white Malpighian tubules and eyes); and on the third chromosome, *scarlet (st)* (which in combination with *bw* also gives white Malpighian tubules and eyes) and *ebony (e)*. The X-linked marker *folded gastrulation* (*fog;* Zusman and Wieschaus, 1985) can be used to identify homozygous or hemizygous late embryos. Other markers for more specialized experiments are discussed later in this section.

A. Detailed Morphological Analysis

1. Embryonic Lethals

As described above, embryonic death can occur among the progeny of females homozygous for MEL mutations (due to defects in maternal contributions) or as hemizygous or homozygous mutant progeny of a heterozygous female (due to defective zygotic product). To examine the nature of these defects, fixed embryos can be stained with DAPI to follow nuclear distributions or division patterns (Foe and Alberts, 1983; Foe, 1989), and simultaneously with fluorescent antibodies against molecules such as cytoskeletal components, or molecules characteristic of particular cells or structures (e.g., Karr and Alberts, 1986; also see Chapter 25). Many important events that occur near the surface of the embryo can easily be examined by standard immunofluorescence. Where appropriate, further details can be obtained by electron microscopy (e.g., Stafstrom and Staehelin, 1984) of the mutant embryos. Processes that occur in the embryo's interior are best examined either in sectioned material or by confocal microscopy. Techniques allowing one to follow in real time by sequential confocal micrographs the distribution of certain molecules (DNA, tubulin, and actin) in living, developing embryos that have been microinjected with appropriate fluorescent tags have been developed (Kellogg *et al.*, 1988; Minden *et al.*, 1989).

If embryos survive to the point of cellularization but die soon thereafter, one can check whether cell movement and cell specification processes are disrupted. For example, preparations of the cuticle can easily be made and can be examined for development of anterior/posterior or dorsal/ventral polarities and for the correct development of segments (Wieschaus and Nusslein-Volhard, 1986). This can demonstrate in a simple way whether positional information is being assessed and responded to correctly.

2. Larval or Pupal Lethals

One first searches for differences in morphology between the mutant and wild type; normal morphologies are well-illustrated in Bodenstein (1950) and Bainbridge and Bownes (1981), and specific systems are discussed and illustrated in further detail in the books edited by Ashburner and Wright (1978–1980) and by Bate and Martinez-Arias (1993). One aid in navigating the morphology of the animal is to employ well-characterized markers for particular tissues. A ready source for such markers is the many existing enhancer trap lines that express β-galactosidase in tissue-specific patterns (e.g., Bier *et al.,* 1989; Bellen *et al.,* 1989; Wilson *et al.,* 1989). In addition, it is almost always helpful to focus particular attention on those tissues in which the gene is known to be expressed (if such information is available).

Morphological defects can give clues about the consequences of the mutation; for example, problems in cell proliferation are often indicated by a reduced size of the imaginal discs (Shearn and Garen, 1974; Szabad and Bryant, 1982; Gatti and Baker, 1989). Because it is difficult to perform high-resolution cytology of nuclei in imaginal discs, confirmation that a small-disc mutant causes defects in mitosis requires examination of other tissues. Whole-mounted or squashed larval brains (see Chapter 21) can clearly reveal mitotic characteristics; polytene chromosomes of the salivary glands will be smaller if there is a defect in their replication, and meiotic problems will be observable in stained testis material (reviewed in Gatti and Goldberg, 1991). It is also possible to examine certain behavioral characteristics of larvae (such as phototaxis; Lilly and Carlson, 1990); if unusual behaviors are found, this may indicate the existence of subtle alterations in the nervous system.

Detailed morphological analysis of specific tissues in pupae is much more difficult than in larvae, because pupae are harder to dissect and because there is substantial histolysis that may mask or mimic abnormal degeneration. However, the imaginal discs, which are essential to metamorphosis, can be isolated by dissection and grown in culture (Fristrom *et al.,* 1973) or injected into larvae and examined after metamorphosis (Bodenstein, 1950) to see if they eventually form the appropriate adult structures.

3. Mutants That Survive to Adulthood

As described previously, one of the most informative processes to examine is gametogenesis, because of its ordered stages and stringent demands for products. Similarly, the eye, wing blade, bristles, and hairs are also very sensitive indicators of abnormalities because of their precise patterns of development (Ready *et al.,* 1976; Bryant, 1975; Hartenstein and Posakony, 1989). Examination of virtually any structure may prove rewarding in the investigation of various problems in cell biology, but as discussed previously, the relationship between particular phenotypes and biological processes is not always clear at the outset.

B. When and Where Does the Gene Act, and How Important Is Its Function in Different Cell Types?

The approaches outlined here provide overlapping data that are essential to a full understanding of gene function in the context of a multicellular organism.

1. Use of Molecular Probes

If a clone of the gene is in hand, its use as a probe for *in situ* hybridizations to RNA or for Northern blots can determine when and where the gene is transcribed (see Chapter 30). Similarly, if antibodies against the gene product are available, the gene product can be followed by staining Western blots or preparations of whole mounts or sections. It is critical to note that the expression pattern of a gene must include the times and places in which it acts, but is not limited to them. It is possible, for example, that a transcript or gene product is present but is unable to function under certain circumstances. For example, the *transformer (tra)* gene is transcribed in males and females, but its transcript is productively spliced only in the latter (Boggs *et al.*, 1987); *Mst87F* is transcribed in the spermatocyte, but translated only later in spermatogenesis (Kuhn *et al.*, 1988). In addition, even though a gene product is made at a particular time or place, absence of the function of this protein at these times or locations may be without phenotypic consequence (e.g., *tra2* in the male soma; Mattox *et al.*, 1990; Amrein *et al.*, 1990). Finally, the action of a gene product at one time may have only detectable consequences much later in development (e.g., Chapman and Wolfner, 1988, for *tra2*). Despite these caveats, definition of the expression pattern of a gene is an excellent first step to interpreting phenotypes and understanding function.

2. Genetic Mosaics

a. Theoretical Considerations

If a mutation in a gene causes lethality, one cannot examine the role of the gene product in a particular tissue if the animal has died at an earlier developmental stage. Even if the mutant animal is still alive, there may be degeneration or lack of development of some tissues because of indirect problems. To overcome these difficulties, one can generate cells that are homozygous for a recessive lethal mutation within an animal that is otherwise either wild type or heterozygous for the mutation. These mutant homozygous cells and their clonal descendants can then be examined for a phenotype. If homozygous "clones" are never obtained, the gene is likely to be essential for the life of individual cells. On the other hand, a gene might be essential for the growth of the animal (e.g., for the development of a particular essential tissue) but not for the viability of individual cells. In that case, clones homozygous for the mutation should be detectable in the heterozygous animal, except in the tissue in which the normal allele is essential. In the following paragraphs, we briefly discuss some methods for generating and analyzing clones homozygous for a recessive lethal mutation (for details see Chapters 33 and 34). To begin, we mention a few applications and limitations common to all of these techniques:

First, the production of clones requires that cell division occur subsequent to the generation of a cell homozygous for the mutation. This limits the application of clonal analysis to tissues or cells that are still capable of division.

Moreover, the later in development that a clone is produced, the smaller it will be because it can undergo fewer subsequent divisions. This can be a problem in detecting the clone, although it can be partially overcome by the *Minute* technique described below.

Second, clones must be marked so that they can be identified. Usually, this is accomplished by using a cell autonomous (see following) recessive marker on the same chromosome arm as the mutation of interest. Thus, when a cell is made homozygous or hemizygous for the mutation, it will display the marker phenotype as well. A number of externally visible markers are available (Table I), although their number, position, and expression patterns still limit the technique to some extent. External markers include those that affect body color, bristle morphology or color (obviously limited to regions with bristles or hairs), and eye color (limited to the eye). Another possibility is to use a dominant marker (see Table I) on the homologous chromosome (i.e., the one that does not contain the mutation of interest) in the original heterozygote, so that clones homozygous for the mutation are recognized by the lack of the dominant phenotype. One innovation of considerable utility is the use of insertions of y^+ into various sites in the genome, so that it can be used in essence as a dominant marker (in a y background) to identify clones of mutations located on any chromosome; this can also be done with w^+ insertions.

Markers to identify clones of homozygous recessive cells within the interior of the animal are also available (Table I). These include *Drosophila* enzymes and the expression of β-galactosidase from ubiquitous (Lis *et al.*, 1981) or enhancer trap promoters (Bier *et al.*, 1989; Bellen *et al.*, 1989; Wilson *et al.*, 1989; Meise and Janning, 1993). In this latter case, the *lacZ* fusion gene must be on the chromosome in the heterozygote that carries the wild-type allele of the gene in question; thus, the background of heterozygous cells will stain for β-galactosidase, whereas a homozygous mutant clone will not. A related innovation is the use of genes that express epitope tags not normally found in *Drosophila*, allowing clones to be identified due to their lack of staining against antibody to the epitope tag (Chapter 34). Finally, a marker of particular utility for cell transplantation studies, as well as in mitotic recombination, is the enzyme succinate dehydrogenase (*Sdh;* Lawrence, 1981). This mitochondrial enzyme, encoded on chromosome 2, is present in all cell types. When Sdh^+ cells are injected into an sdh^{ts} host, the progeny of injected cells can be detected by staining for Sdh activity at restrictive temperature.

Third, these studies can determine whether expression of a gene is important only for the cell in which it is expressed or whether the gene product can influence other cells. If the gene product is nondiffusible and is located in the cell interior, the direct effects of the mutation are usually restricted only to cells homozygous for the mutation. For such a cell autonomous mutation, regions displaying a mutant phenotype will exactly coincide on a cell by cell basis with regions homozygous for the linked, recessive cell autonomous marker employed. Thus, clones mutant in a gene for a transcription factor such as *doublesex (dsx)* (Coschigano and Wensink, 1993) are unaffected by whether the surrounding cells are dsx^+ or dsx^- (Baker and Ridge, 1980). Genes whose products receive signals from other cells, such as *sevenless (sev)*, whose product is a receptor tyrosine kinase (Basler and Hafen, 1988), will also in general behave in a cell autonomous fashion (Harris *et al.*, 1976). On the other hand, when a gene encodes a product that can diffuse to, or otherwise signal or influence, adjacent cells mutations in this gene will usually be nonautonomous. Either the phenotype of a homozygous mutant clone ''spreads'' to surrounding non-mutant cells or, conversely, the homozygous clone shows a nonmutant phenotype because of interactions with surrounding normal cells or their products. As an example of the first case, clones of cells mutant in *bride-of-sevenless (boss)*, a gene that encodes an anchored ligand for the product of the *sev* receptor, result in abnormal development of adjacent cells of normal genotype (Reinke and Zipursky, 1988; Hart *et al.*, 1990). In the converse sort of a nonautonomous effect, in some clones of the *nullo* mutant, cells near the periphery of the clones appear wild-type, while those farther away from normal cells look increasingly mutant (Simpson and Wieschaus, 1990). The researcher should be aware that although establishing autonomy or nonautonomy can help formulate hypotheses for gene function or gene product location, these hypotheses can be proven only with supporting molecular data.

Fourth, mosaics can sometimes be used to determine when the gene of interest has acted sufficiently to allow normal development (e.g., Baker and Ridge, 1980). To do this, one makes clones, at different times in development, of cells homozygous for the null mutation. If a clone shows the mutant phenotype, the function of the gene was still required for normal development after the time at which the clone was made. If the clone's phenotype is normal, function of the gene must not be necessary after the time at which the clone was induced. By the time the clone was made, sufficient gene product had accumulated to allow normal activity of a process or pathway or the process was already complete. One must use caution in interpreting these results because of a phenomenon known as perdurance, in which a gene product continues to have an effect after the gene itself is lost or inactivated. For example, if the gene encoded a very stable protein, removal of the gene might be without phenotypic consequence if sufficient residual protein were present.

b. How Does One Make Genetic Mosaics?

An excellent description of classical techniques to make genetic mosaics is given by Ashburner (1989). Here, we briefly summarize the essence of strategies for clonal analysis with particular emphasis on more recently developed methods.

Chromosome Loss. If cells can proliferate without a particular chromosome, mitotic clones can be generated from which that chromosome has been lost. Cells with only one X chromosome (XO somatic cells are male in phenotype) can survive so loss of an X chromosome from an XX female heterozygous for an X-linked mutation of interest can generate hemizygous mutant, male cells. There are several techniques available to destabilize X chromosomes during mitotic segregation at a sufficiently high frequency to generate enough XX// XO mosaics for further analysis. One method is to create a female heterozygous for the mutation-bearing X chromosome and for an unstable ring X chromosome (reviewed in Hall *et al.*, 1976; Ashburner, 1989). This latter chromosome tends to get lost in the first few nuclear divisions after fertilization. Since the early division planes are oriented randomly, a collection of such gynandromorphs or gynanders (female//male mosaics) should contain animals in which the cells of interest are hemizygous for the mutation. Since the chromosome is lost in the very first divisions, the clones are large (with many descendants); however, in the case of lethals, only those gynanders with small mutant patches may survive. An X chromosome can also be lost at high frequency in embryos fathered by male flies mutant at the *paternal loss* (*pal;* Baker, 1975) or the *Horka (Hor)* (Szabad *et al.*, 1994) genes or in embryos that are the progeny of females homozygous for *claret-non-disjunctional* (*ca^{nd;}* (e.g., Davis, 1969)) or *mitotic loss-inducer* (*mit;* (Gelbart, 1974)). These mutations can also be used to generate cell-viable clones with only a single copy of the small fourth chromosome; cells with only a single copy of chromosome 2 or 3 are inviable. In addition to the detection of clones via cell markers (previously), clones in gynanders can also be distinguished by sexual phenotype (or presumably by *Sxl* expression; Bopp *et al.*, 1991)) and, in embryos, by nucleolar size in certain marked configurations (Zusman and Wieschaus, 1987).

Mitotic Recombination. In a heterozygote, mitotic recombination in the interval between the centromere and the gene in question can produce homozygous mutant cells (Fig. 4A) in an animal otherwise heterozygous for the mutation. Irradiation of embryos or larvae with X-rays can be used to induce mitotic recombination (reviewed in Ashburner, 1989). A newer, more efficient, method to induce mitotic clones is the use of site-specific recombination such as that exhibited by the FLP-FRT system (Fig. 4b; Golic and Lindquist, 1989; Golic, 1991; also see Chapter 34). Expression of the FLP recombinase gene is made inducible by placing it under the control of a heat-shock promoter (hs-FLP). When synthesized, the FLP recombinase catalyzes recombination between its FRT target sites located in the same position on both homologous chromosomes. This will result in the generation of homozygous mutant clones if the FRT site is located between the gene of interest and the centromere. This technique requires a certain amount of genetic manipulation (you must construct a chromosome containing the mutation and an FRT site and then make animals heterozygous for this chromosome and a chromosome containing the FRT site at the same location, while at the same time adding an hs-FLP gene). However, Xu and Rubin (1993) have described strains for these purposes that contain epitope-tagged genes that can mark mutant clones, potentially making this effort extremely rewarding.

In a few cases, clones induced late in development are unable to grow large

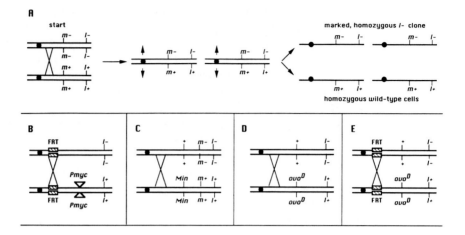

Fig. 4 Generating marked, homozygous mutant clones by mitotic crossing-over. (A) A generic scheme. An animal is constructed that is heterozygous for one chromosome arm bearing both a recessive mutation of interest (here, a recessive lethal mutation l^-) and a recessive cell marker (m^-; see Table I) and a homologous chromosome carrying dominant alleles of both genes (l^+ and m^+, respectively). If mitotic crossing-over is induced by X-rays between the marker and the centromere (·), cells that display the recessive marker and that are homozygous for the mutation of interest can be produced. (Here, only the orientation of homologous chromosomes on the mitotic spindle that would produce such cells is depicted; an equally likely orientation would yield $m^-\ l^-/$ $m^+\ l^+$ cells displaying the wild-type pheonotype associated with both loci.) For B–E, only the starting configuration is shown. (B) Mitotic recombination catalyzed by *FLP* recombinase (Xu and Rubin, 1993). Clones can be generated at high efficiency if both homologous chromosomes contain near their centromere an FRT site recognized by the *FLP* recombinase. When the *FLP* recombinase is expressed in such an animal, mitotic crossing-over will occur at high frequency due to recombination between the FRT sites. Absence of a nuclear marker (*Pmyc*, a fusion between part of the human *c-myc* gene and part of the *Drosophila* P-element transposase) identifies homozygous mutant clones. (C) Synthesis of larger clones using the *Minute* technique. After mitotic recombination, resultant homozygous, marked mutant cells will be *Minute*$^+$ (designated here as *Min*$^+$) and will outgrow other cells. (D) Generation of homozygous mutant germline clones. Here, the marker is the wild-type allele of a gene recessive to a dominant female-sterile mutation that prevents ovarian development (like *ovo*^{D}; see text for details). The only functional germ cells in a female with this genotype will be homozygous for the mutation. (E) Using *FLP* recombinase to make homozygous mutant germline clones (Chou and Perrimon, 1992). Germline clones are generated at high efficiency by utilizing *ovo*^{D} or other dominant female sterile mutations as shown in D and the *FLP*–FRT system described in B.

enough to be recognized. In such situations, one can exploit *Minute* mutations, which are dominant mutations that interfere with the growth of cells and are also recessive lethals (Lindsley and Zimm, 1992; e.g., Garcia-Bellido *et al.*, 1973). Mitotic crossovers in flies that carry your mutation on one homolog and a *Minute* mutation in a nearby position on the other homolog will generate cells homozygous for your mutation that are simultaneously *Minute*$^+$ (Fig. 4C). These mutant but *Minute*$^+$ cells will outgrow the surrounding *Minute* cells and thus will make a larger clone than would normally be obtained from such a late induction (as long as the mutation is not itself a cell lethal).

Injection of Cells or Nuclei. Another method to get clones is to inject mutant tissues, cells, or nuclei carrying a detectable marker into wild-type animals lacking that marker, or vice versa. The injection of tissues such as imaginal discs is most widely used to test whether the the development of the tissue requires (or is influenced by) exogenous factors (e.g., Beadle and Ephrussi, 1937). The injected tissue can be traced by any genetic marker differentiating it from the recipient. For other purposes, such as cell lineage studies, one can inject cells (or nuclei in precellular stages) taken from various sites in a donor into various sites in a recipient. As discussed previously, the enzyme Sdh, or, in theory, the expression of *lacZ* enhancer traps or epitope-tagged constructs, is very useful to trace the fates of injected cells (e.g., Lawrence and Johnston, 1986), but the animal must be sacrificed for the analysis.

Germline Clones. There are two important reasons for making clones in the germline. (1) For male- or female-sterile mutations that disrupt gametogenesis, one can determine whether the gene product is necessary in the germline or in nongermline supporting tissues. (2) For mutations lethal to the whole organism, one can establish whether the gene also has a maternal function that is necessary for embryogenesis or for development of the subsequent embryo. The strategies used to make mitotic clones in general can also be employed, with some modification, to make germline clones.

Dominant female-sterile mutations that disrupt oogenesis can assist in mitotic recombination experiments to generate homozygous mutant germline clones. Mitotic recombination induced in a mutant/*Dom-FS* heterozygous female will result in germline stem cells that are simultaneously homozygous for the mutation in question and *Dom-FS*$^+$ (Fig. 4d). As long as the mutation of interest does not itself block the earliest stages of oogenesis, this means that the only eggs developing within, or laid by, the heterozygous female will be mutant. The phenotype of these eggs or of the embryos that are formed from them allows an assessment of the importance of the gene product to the female germline. Several dominant female-sterile mutations can be used for these studies, but the most commonly used is the *ovo*^{D} mutation, which prevents oogenesis (Lindsley and Zimm, 1992; e.g., Perrimon, 1984). Although the *ovo* gene itself is X-linked, stocks containing the *ovo*^{D} mutation on any chromosome arm are available (Chou *et al.*, 1993; Mével-Ninio *et al.*, 1994). Because the induction of mitotic recombination in the germline with X-rays is relatively inefficient, Chou and Perrimon (1992) have designed an extension of the hs-FLP-FRT system particularly suited to the generation of germline clones with *ovo*^{D} (Fig. 4e).

One can also transplant the germline of one animal into another. To do this, pole cells (precursors of the germ cells) are taken from one embryo and injected into embryos that do not have functional germlines because they are heterozygous for *ovo*^{D} (if one is concerned only with a female-sterile mutation) or are the progeny of mothers homozygous for certain mutations in the *oskar* (*osk*) gene (Lehman and Nusslein-Volhard, 1986). As long as the transfers are within the same sex, the introduced pole cells will form germ cells in the soma of the recipient (VanDeusen, 1976). One can introduce normal pole cells into a sterile mutant embryo or mutant pole cells into a normal embryo to ask whether the sterility (whether male or female) correlates with mutant germline or soma. This procedure has been used elegantly to study the somatic vs germline contributions to dorsal-ventral polarity (e.g., Schüpbach, 1987) and to sex determination in the germline (Marsh and Wieschaus, 1978; Schüpbach, 1982).

3. Temperature-Sensitive Mutations

If you are lucky enough to have a temperature-sensitive allele of the gene you are studying, you can essentially turn on or off the mutant protein by shifting the temperature (e.g., Belote *et al.*, 1985a; Belote and Baker, 1987; Chapman and Wolfner, 1988). Shifting at different times in development may allow determination of the times at which the gene product acts: When does a shift to the restrictive temperature include a mutant phenotype? When can a shift from the restrictive temperature prevent the mutant phenotype? However, one should be aware that some ts alleles may also exhibit perdurance, if the thermolabile step is the synthesis of an otherwise stable product or if the product is stable once it is assembled into a complex.

V. Using Mutants to Explore Processes and Pathways

While mutations are useful in identifying and assessing the importance of individual components of a process *in vivo, Drosophila* genetics can also provide further insights into the process as a whole. Once a mutant of interest is in hand, new types of genetic screens can be used to identify additional components of the same process. Once several such genes are available, genetics can provide information about the way in which their various products interact. However, *Drosophila* still has limitations relative to unicellular systems in which large screens (10^6 or greater) can be mounted: obtaining one or two interacting genes is in itself a tour de force in *Drosophila*.

A. Screening for Suppressors or Enhancers of a Phenotype

If two proteins interact with each other either directly or indirectly, the phenotype associated with a mutation in one of the corresponding genes may be modified by mutations in the other gene. For example, if two proteins are in physical contact, an alteration in one protein that prevents the interaction might be compensated by a mutational change in the other protein, leading to phenotype suppression. On the other hand, if a mutation results in reduced protein activity, a mutation reducing the activity of a second gene product involved in the same biochemical pathway may lower the throughput of the pathway even more, resulting in phenotype enhancement. These examples illustrate only a few of the ways in which screens for genetic modifiers of a phenotype (both suppressors and enhancers) might identify genes that encode proteins involved in the same biological process.

One can devise procedures to select either dominant or recessive modifiers. However, almost all investigators look for dominant modifiers because the large additional effort required to establish lines homozygous for recessive modifiers is daunting. Theoretically, screens for dominant suppressors of mutations causing lethality or sterility should be extremely powerful in that en masse matings would allow the examination of large numbers of newly mutagenized chromosomes to see if they result in the appearance of viable adults (Su; lethal/lethal) or viable progeny (of Su; sterile/sterile flies). However, in practice only a few such screens have been successfully employed (e.g., Gertler *et al.*, 1990). One of the major difficulties is that tremendous changes may be required to correct the lethal or sterile effects of a mutation. It may therefore be more prudent (although more effort-intensive) to search for suppressors that have less dramatic effects. One could for example look for suppressors that would change the lethal phase of a lethal mutation from an embryonic or larval to a pupal stage (Gertler *et al.*, 1990). Alternatively, one might look for suppression of a "sensitized" phenotype, such as that encountered when an animal with a temperature-sensitive allele is raised at semipermissive conditions (Simon *et al.*, 1991).

As this last example shows, choice of an allele is critical in the design of screens for genetic modifiers. Suppression of a null allele of a gene would require a complete bypass of the gene's function (often improbable except in situations of epistasis (see following), whereas suppressors of alleles associated with insertion of a transposable element may reflect alterations in gene expression rather than of gene function (Spana *et al.*, 1988; Harrison *et al.*, 1989). In general, screening for suppression of missense mutations probably offers the best chances for success, although it is rarely apparent at the outset which particular missense allele might work best. In addition, suppression of a missense mutation but not a null mutation of a gene provides some circumstantial evidence of direct interaction between the products of the two loci.

Even fewer attempts to look for dominant enhancers of lethal or sterile mutations have been made as these would only make the animals more dead or sterile. One of the few success stories is the case of mutations in the gene *Disabled* (*Dab*), which act as dominant enhancers of mutations in the *Drosophila* Abelson tyrosine kinase gene (*abl*) (Gertler *et al.*, 1989). Animals homozygous for *abl* mutations die in the late pupal stages, whereas $Dab^-\ abl^-/Dab^+\ abl^-$ animals die as embryos or larvae; this early lethality can be demonstrated to be absolutely dependent upon the original mutation in *abl*. This example again suggests that finding a phenotype sensitive to small changes would normally be a precondition for conducting a screen for genetic modifiers. If an enhancer is found, the underlying molecular mechanism for the genetic interaction may not be completely straightforward; a theoretical basis for interpreting enhancement has recently been formulated by Guarente (1993).

B. Screening for Second-Site Noncomplementation

A slightly different kind of screen for genes making interacting proteins or proteins otherwise involved in the same process is based on the following concept. If recessive mutations occur in two genes whose products interact, it is conceivable that a double heterozygote for the two mutations might show a phenotype, whereas heterozygotes for either mutation alone would be wild-type. For example, suppose the two gene products make a heterodimeric protein. A heterozygote for a null allele of one gene would produce only 50% of the normal gene product. If the mutation of the other gene produced a protein that was nonfunctional but could nevertheless interact with the first protein, then the intracellular concentration of the functional dimer might be only 25% of wild-type. This is one of the several potential scenarios that could result in an aberrant phenotype in a double heterozygote.

In accordance with this theory, mutations in a major α-tubulin gene act as such "second-site non-complementors" of a mutation in the gene encoding a testis-specific isoform of β-tubulin (β2t) (Hays *et al.*, 1989). However, other cases of second-site noncomplementation are less simple to interpret. Mutations in the *haywire* gene also fail to complement null mutations in β2t (Regan and Fuller, 1988, 1990). The *haywire* gene encodes a protein of the helicase super-family homologous to a human polypeptide associated with xeroderma pigmentosum B (Mounkes *et al.*, 1992). The relationship, if any, between a helicase apparently involved in DNA repair and β-tubulin remains uncertain; we know of no data indicating a direct interaction between β-tubulin and the *haywire* product. Instead, it seems more likely that the *haywire* mutation causes nonspecific defects in transcription or translation that further decrease the amount of functional β-tubulin underlying the failure of these mutations to complement each other.

Although failure to complement a null mutation might be the result of very indirect effects, the second-site noncomplementation technique might yield a narrower group of true interactors if an effect is seen only between particular nonnull allele of *both* genes in question; of course one needs multiple, molecularly characterized alleles of both genes for this approach. Second-site noncomplementation has proven of value in a different context: "sensitizing" a system to allow screens for genetic modifiers as discussed above. For example, double heterozygotes for mutations in *transformer* (*tra*) and *transformer-2* (*tra-2*) are wild-type in appearance. In this sensitized genotype, heterozygosity for recessive mutations in other sex determination genes results in a dominant intersexual phenotype (Belote *et al.*, 1985b).

C. Biochemical Screens for Interacting Proteins

Since genetic approaches to find additional proteins involved in a biological process do not always succeed, we briefly enumerate a few biochemical procedures that can also identify interacting proteins; ideally the genetic and biochemical routes would be applied in concert. The molecular strategies include (a) screening cDNA expression libraries with a labeled protein (Blanar and Rutter, 1992; Kaelin, Jr., *et al.*, 1992); (b) use of the two-hybrid system in yeast (Fields and Song, 1989; Zervos *et al.*, 1993); and (c) immunoaffinity chromatography or immunoprecipitation of complexes from cell extracts (e.g., Kellogg and Alberts, 1992). Techniques a and b yield cDNA sequences that encode proteins that putatively interact with the "bait" protein; the existence of a direct interaction between the proteins must subsequently be verified by other methods. Technique c purifies proteins from interacting complexes; the corresponding gene can then be obtained by making an antibody to the interacting protein and screening expression libraries or by characterizing the N-terminus of the protein by microsequencing and then probing a recombinant library with appropriate oligonucleotides. In either of these cases, the investigator must then obtain mutants in the gene (as outlined previously in Section 11B) for further analysis.

D. Looking at Double Mutants to Explore Gene Interactions

Examination of the phenotype of double mutants (here, animals homozygous for recessive null mutations in two different genes), sometimes called epistasis analysis, can often give useful information about the roles of the corresponding gene products. Epistasis analysis allows one to determine, within limitations, whether the products of the two genes act sequentially in a single pathway, whether they function in two independent pathways, or whether their activity is interdependent, such as genes encoding two subunits of a multisubunit enzyme (see Hereford and Hartwell, 1974, for discussion). In general, one must use null mutations for epistasis analysis; leaky mutations introduce such complexity that the results can be impossible to interpret. In addition, interpretable results are obtained only if the phenotype associated with each of the two mutations individually is distinct. Even when these conditions are met, additional information is often needed to decide between several options that may still be consistent with the findings (Hereford and Hartwell, 1974; Avery and Wasserman, 1992).

The most informative circumstance occurs if the phenotype of the double mutant is exactly the same as one of the mutants. For example, if an $a^- b^-$ double mutant has the phenotype of the null mutation in gene A rather than that of the null mutation in gene B (i.e., a^- is epistatic to b^-), one can hypothesize that the two genes act in a dependent pathway. Epistasis of a^- to b^- can result, however, from two very different situations (Fig. 5). In the "biochemical

a⁻ is epistatic to b⁻ :

Biochemical Pathway: A and B catalyze steps in a pathway

a⁻ phenotype is 1

b⁻ phenotype is 2

a⁻b⁻ phenotype is 1

Regulatory Pathway: B mediates the choice between two activities of A

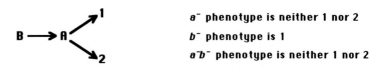

a⁻ phenotype is neither 1 nor 2

b⁻ phenotype is 1

a⁻b⁻ phenotype is neither 1 nor 2

Fig. 5 Two models of gene action that could result from the observation that a^- is epistatic to b^-. In the biochemical pathway model, the epistasis of a^- implies that the A gene product must catalyze a step prior to that catalyzed by the B gene product. A single mutant's phenotype results from the accumulation of the metabolite it was unable to use (1 for a^- and 2 for b^-). If the double mutant lacks A activity, it will accumulate product 1, regardless of whether it could utilize product 2; hence the phenotype will be like that of a^-. This sort of epistasis is seen in the pathway for synthesis of eye pigments (e.g., reviewed in Phillips and Forrest, 1980). In the "regulatory pathway model," the epistasis of a^- implies that the B gene product regulates the activity of the A gene product, causing it to catalyze outcome 1 rather than outcome 2. The a^- mutant produces no functional A protein and thus is unable to direct either outcome 1 or outcome 2. The b^- mutant is unable to direct A to produce outcome 2, and thus its phenotype is constitutive outcome 1. If the double mutant lacks A activity, it will be unable to direct either outcome 1 or outcome 2, hence resulting in an a^- phenotype. This sort of epistasis is seen in the pathway for sex determination, where a is the *doublesex* gene and b is the *tra* gene and outcome 1 is maleness (Baker and Ridge, 1980; reviewed in Belote, 1992).

pathway" model, activity of protein A yields the (direct or indirect) substrate of protein B; in other words, activity of B is dependent on prior activity of A. In the absence of functional A enzyme, the pathway is blocked at the earlier step, and the phenotype of the double mutant is that of a^-. For example, a fly with a null mutation at the *white* gene will always have white eyes regardless of its genotype at other eye color genes, apparently because pigment precursors are not transported into *white⁻* pigment-producing cells (Montell *et al.*, 1992). In the "regulatory pathway" model, protein B regulates or acts through protein A. In this case, in the absence of functional A protein, any regulation by protein B is irrevelant and the phenotype is again that of a^-. A good example of this type of epistasis is in the hierarchy of genes controlling sexual differentiation. Since the role of the *transformer* (*tra*) protein is to splice the *doublesex* (*dsx*)

RNA into a female-specifying mode (e.g., Nagoshi *et al.*, 1988; reviewed in Belote, 1992), the function or lack of function of *tra* is irrevelant in the absence of any functional *dsx* protein. Thus, a *tra dsx* double mutant has the *dsx* phenotype (Baker and Ridge, 1980).

In other cases, the double mutant phenotype is more severe than, or at least different from, that associated with either single mutant. A likely explanation for this result is that the two genes participate in independent pathways (e.g., a *cinnabar brown* double mutant has white eyes due to the simultaneous loss of two pigment classes, whereas flies homozygous for either mutant alone have pigmented eyes (reviewed in Phillips and Forrest, 1980)) or in otherwise parallel pathways (for example, if the genes have overlapping function). Thus, simultaneous loss of both pathways can have more severe phenotypic consequences than loss of one alone (see Hereford and Hartwell (1974), Guarente (1993), and Thomas (1993) for further discussion).

In a few cases, the phenotype of the double null mutant may appear closer to wild type than that of either mutant alone. Formally, one of the mutations can thus be regarded as a recessive suppressor of the other. For further discussion, we refer the reader back to Section VA, calling particular attention to the potential pitfalls attendant upon the suppression of null alleles.

Rarely, double mutant combinations using alleles other than null alleles—particularly gain-of-function dominant mutations—have been very valuable in ordering gene action. For example, a dominant gain-of-function allele of *Toll* (*Toll*^{D}) that presumably results in consitutive activity of the *Toll* gene product causes embryos to be ventralized. This phenotype is epistatic to some but not all recessive dorsalizing mutations (Anderson *et al.*, 1985). *Toll*^{D} is epistatic to mutations in genes such as *snake* (that is, the *snake Toll*^{D} double mutant looks like *Toll*^{D}). On this basis, the *snake* gene is hypothesized to act upstream of *Toll* to regulate its activity. On the other hand, epistasis data suggest that *Toll* acts through genes such as *dorsal* because the *dorsal Toll*^{D} double mutant resembles *dorsal* alone. These hypotheses on the order of gene action have now been strongly supported by molecular analyses of the dorsal-ventral pathway (for reviews see Govind and Steward, 1991; Anderson *et al.*, 1992).

E. Do Mutations Disrupt the Distribution or Synthesis of Other Gene Products?

Much interesting information concerning potential gene interactions can be obtained by determining whether particular mutations affect the localization or synthesis of other gene products. This analysis is simple and is gaining increasing use as more antibodies and gene probes are becoming available. An example of the use of this analysis concerns two genes, *1(1)zw10* and *rough deal*, which have very similar effects on chromosome segregation (Karess and Glover, 1989; Williams *et al.*, 1992). The *1(1)zw10* gene product is distributed in a cell-cycle-dependent manner to various spindle-associated structures. In cells mutant for *rough deal*, the *1(1)zw10* gene product is unable to achieve its normal localization during mitosis, even though normal amounts of the *l(1)zw10* protein are present. This indicates that *rough deal* function is required for proper distribution of the *l(1)zw10* protein (Williams and Goldberg, 1994). This type of analysis has also been used extensively to determine the order of segmentation gene activities (see Ingham and Martinez-Arias, 1992; St. Johnston and Nusslein-Volhard, 1992; Lawrence, 1992, for reviews).

F. Ectopic or Overexpression of a Protein

Once a gene has been cloned, one can engineer flies that synthesize wild-type or altered proteins at inappropriate times, in inappropriate tissues, or at unusually high or low levels. These flies can be used to address questions such as (1) Does one see phenotypic consequences of novel expression patterns or levels?; if so, is this effect dependent upon particular alleles of other, interacting genes? (2) What parts of the protein are essential for various aspects of its function? In the background of a null mutation in the gene, one can assess whether transgenes with site-directed lesions can rescue the mutant phenotype. (3) What happens when expression of the gene is limited in particular tissues? In a few cases, it has been possible to express antisense RNA at high levels in certain tissues to abolish translation of the corresponding mRNA.

As indicated by this very incomplete list of questions, the uses of *in vitro*-modified transgenes are limited only by the imagination of the investigator. To achieve the desired control of gene expression, a range of promoters or inducible expression systems suitable for use in the intact organism is available. These are treated in detail in Chapter 33; here, we briefly summarize some of the techniques and constructs that will prove most widely profitable for experiments involving ectopic expression or overexpression.

A simple method to increase gene expression over a limited range is to augment the number of copies of the gene. For example, an animal homozygous for a gene in its normal position and simultaneously homozygous for two wild-type transgenes would be expected to produce three times the normal amount of the protein. To obtain even higher levels of expression, one can fuse a cDNA to a manipulable promoter such as that for the hsp70 gene. This promoter can be induced about 100- to 200-fold at almost any stage of development (e.g., Lis *et al.*, 1981). However, since the *hsp70* promoter has a basal level of expression in animals that are not heat shocked, further control of the system is sometimes needed, for example, when the cDNA encodes a deleterious molecule. A recent innovation that addresses this issue is to clone the cDNA of interest under the control of a UAS element responsive to the yeast transcription factor Gal4 (Brand and Perrimon, 1993). This construct has no basal activity in normal *Drosophila*. Flies carrying this construct are then crossed with flies carrying a transgene in which the *Gal4* gene is under control of the *hsp70* promoter. Heat induction of progeny results in the production of Gal4 protein and the consequent induction of the cDNA of interest.

It is also possible to target expression of a gene to a given tissue. In some cases, one can induce heat-shock only in localized regions of an animal carrying an *hsp70* fusion of the type described above (Monsma *et al.*, 1988). Another option is to use a promoter expressed specifically in a tissue (or sex) of choice. For example, a cDNA for the *fs(1)Ya* protein fused to an ovary-expressing promoter (*hsp26*, which also allows heat-inducibility; e.g., Glaser *et al.*, 1986) rescues the maternal-effect lethality of *fs(1)Ya* mutations (Lopez *et al.*, 1994). In this same vein, a tissue- or stage-specific promoter can be used to drive expression of Gal4 in a modification of the system described in the previous paragraph.

Finally, because of the large size of syncytial *Drosophila* blastoderm embryos, microinjection can be used to introduce various reagents into particular positions within the embryo. For example, sense or antisense RNA (e.g., Anderson and Nusslein-Volhard, 1984; Rosenberg *et al.*, 1985), or protein of the gene of interest (e.g. Kaufman and Rio, 1991), or cytoplasm can be injected into various positions in mutant (or wild-type) embryos. This technique has been used, among other things, to assess the localization and function of molecules essential for establishing the embryonic axes during early embryogenesis.

VI. Conclusion

The great length of this article reflects the enormous amount of ingenuity and effort expended by *Drosophila* researchers over the last century in developing genetic tools applicable to a wide range of biological questions. As we have tried to show, these techniques are extremely powerful, particularly for the identification of genes involved in a process and for determining interrelationships and pathways involving these genes. Although *Drosophila* genetics is neither a panacea nor an end in itself, the cell biologist will be amply rewarded by an acquaintance with, and use of, its potentials.

Acknowledgments

We thank Drs. R. Nöthiger and C. Aquadro for their critical reading of the manuscript and for their helpful suggestions and A. Clark for the statistics on screening for P-element hops. Research in our labs is supported by grants from NIH, NSF, and HFSP; M.F.W. also gratefully acknowledges the support of a Faculty Research Award from the American Cancer Society.

References

Alderson, T. (1965). Chemically induced delayed germinal mutation in *Drosophila*. *Nature* (*London*) **207,** 164–167.

Amrein, H., Maniatis, T., and Nöthiger, R. (1990). Alternatively spliced transcripts of the sex-determining gene *tra-2* of *Drosophila* encode functional proteins of different size. *EMBO J.* **9,** 3619–3630.

Anderson, K. V., and Nusslein-Volhard, C. (1984). Information for the dorsal-ventral pattern of the *Drosophila* embryo is stored as maternal mRNA. *Nature* **311,** 223–227.

Anderson, K. C., Jurgens, G., and Nusslein-Volhard, C. (1985). Establishment of dorsal-ventral polarity in the *Drosophila* embryo: Genetic studies on the role of the *Toll* gene product. *Cell* **42,** 779–790.

Anderson, K. C., Schneider, D. S., Morisato, D., Jin, Y., and Ferguson, E. L. (1992). Extracellular morphogens in *Drosophila* embryonic dorsal-ventral patterning. *Cold Spring Harbor Symp. Quant. Biol.* **57,** 409–417.

Ashburner, M. (1989). "Drosophila: A Laboratory Handbook." Cold Spring Harbor, NY: Cold Spring Harbor Laboratory Press.

Ashburner, M. A., and Wright, T. R. F. (1978–1980). "The Genetics and Biology of *Drosophila*." New York: Academic Press.

Avery, L., and Wasserman, S. (1992). Ordering gene function: the interpretation of epistasis in regulatory hierarchies. *Trend Genet.* **8,** 312–316.

Bainbridge, S. P., and Bownes, M. (1981). Staging the metamorphosis of *Drosophila melanogaster*. *J. Embryol. Exp. Morphol.* **66,** 57–80.

Baker, B. (1975). *Paternal loss (pal)*: A meiotic mutant in *Drosophila melanogaster* causing the loss of paternal chromosomes. *Genetics* **80,** 267–296.

Baker, B. S., and Ridge, K. A. (1980). Sex and the single cell. I. On the action of major loci affecting sex determination in *Drosophila melanogaster*. *Genetics* **94,** 383–423.

Ballinger, D. G., and Benzer, S. (1989). Targeted gene mutations in *Drosophila*. *Proc. Natl. Acad. Sci. U.S.A.* **86,** 9402–9406.

Basler, K., and Hafen, E. (1988). Control of photoreceptor cell fate by the sevenless protein requires a functional tyrosine kinase domain. *Cell* **54,** 299–312.

Bate, M., and Martinez-Arias, M. (1993). "The Development of *Drosophila melanogaster*." New York: Cold Spring Harbor Press.

Beadle, G. W., and Ephrussi, B. (1937). Diffusible substances and their interrelations. *Genetics* **22,** 76–80.

Bellen, H. J., O'Kane, C. J., Wilson, C., Grossniklaus, U., Pearson, R. K., and Gehring, W. J. (1989). P-element-mediated enhancer detection: A versatile method to study development in *Drosophila*. *Genes Dev.* **3,** 1288–1300.

Belote, J. M. (1992). Sex determination in *Drosophila melanogaster:* From the X:A ratio to *doublesex*. *Semin. Dev. Biol.* **3,** 319–330.

Belote, J. M., and Baker, B. S. (1982). Sex determination in *Drosophila melanogaster:* Analysis of *transformer-2,* a sex-transforming locus. *Proc. Natl. Acad. Sci. U.S.A.* **79,** 1568–1572.

Belote, J. M., and Baker, B. S. (1987). Sexual behavior: Its genetic control during development and adulthood in *Drosophila melanogaster*. *Proc. Natl. Acad. Sci. U.S.A.* **84,** 8026–8030.

Belote, J. M., Handler, A. M., Wolfner, M. F., Livak, K. J., and Baker, B. S. (1985a). Sex-specific regulation of yolk protein gene expression in *Drosophila*. *Cell* **40,** 339–348.

Belote, J. M., McKeown, M. B., Andrew, D. J., Scott, T. N., Wolfner, M. F., and Baker, B. S. (1985b). Control of sexual differentiation in *Drosophila melanogaster*. *Cold Spring Harbor Symp. Quant. Biol.* **50,** 605–614.

Bender, W., Akam, M., Karch, F., Beachy, P. A., Peifer, M., Spierer, P., Lewis, E. B., and Hogness, D. S. (1983). Molecular genetics of the *Bithorax* complex in *Drosophila melanogaster*. *Science* **221,** 23–29.

Bier, E., Vaessin, H., Shepherd, S., Lee, K., McCall, K., Barbel, S., Ackerman, L., Carretto, R., Uemura, T., Grell, E., Jan, L. Y., and Jan, Y. N. (1989). Searching for pattern and mutation in the *Drosophila* genome with a *P-lacZ* vector. *Genes Dev.* **3,** 1273–1287.

Blackman, R. K., Grimaila, R., Koehler, M. M. D., and Gelbart, W. M. (1987). Mobilization of hobo elements residing within the *decapentaplegic* gene complex: Suggestion of a new hybrid dysgenesis system in *Drosophila melanogaster*. *Cell* **49,** 497–506.

Blackman, R., Koehler, M. M. D., Grimaila, R., and Gelbart, W. M. (1989). Identification of a fully functional hobo transposable element and its use for germ-line transformation of *Drosophila*. *EMBO J.* **8**, 211–218.

Blanar, M. A., and Rutter, W. J. (1992). Interaction cloning: Identification of a helix-loop-helix protein that interacts with c-Fos. *Science* **256**, 1014–1018.

Bodenstein, D. (1950). The postembryonic development of *Drosophila*. *In* ''Biology of *Drosophila*'' (M. Demerec, ed.), pp. 275–367. New York: Wiley.

Boggs, R. T., Gregor, P., Idriss, S., Belote, J. M., and McKeown, M. (1987). Regulation of sexual differentiation in *D. melanogaster* via alternative splicing of RNA from the *transformer* gene. *Cell* **40**, 739–747.

Bopp, D., Bell, L. R., Cline, T. W., and Schedl, P. (1991). Developmental distribution of female-specific Sex-lethal proteins in *Drosophila melanogaster*. *Genes Dev.* **5**, 403–415.

Brand, A. H., and Perrimon, N. (1993). Targeted gene expression as a means of altering cell fates and generating dominant phenotypes. *Development* **118**, 401–415.

Bryant, P. J. (1975). Pattern formation in the imaginal wing disk of *Drosophila melanogaster:* Fate map, regeneration, and duplication. *J. Exp. Zool.* **193**, 49–78.

Busson, D., Gans, M., Komitopoulou, K., and Masson, M. (1983). Genetic analysis of three dominant females sterile mutations located on the X chromosome of *Drosophila melanogaster*. *Genetics* **105**, 309–325.

Byers, T. J., Dubreuil, R., Branton, D., Kiehart, D. P., and Goldstein, L. S. B. (1987). *Drosophila* spectrin. II. Conserved features of the alpha-subunit are revealed by analysis of cDNA clones and fusion proteins. *J. Cell Biol.* **105**, 2103–2110.

Cavener, D. R., and MacIntyre, R. J. (1983). Biphasic expression and function of glucose dehydrogenase in *Drosophila melanogaster*. *Proc. Natl. Acad. Sci. U.S.A.* **80**, 6286–6288.

Chalfie, M., Tu, Y., Euskirchen, G., Ward, W. W., and Prasher, D. C. (1994). Green fluorescent protein as a marker for gene expression. *Science* **230**, 802–805.

Chapman, K. B., and Wolfner, M. F. (1988). Determination of male-specific gene expression in *Drosophila* accessory glands. *Dev. Biol.* **126**, 195–202.

Chou, T.-B., and Perrimon, N. (1992). Use of a yeast site-specific recombinase to produce female germline chimeras in *Drosophila*. *Genetics* **131**, 643–653.

Chou, T.-B., Noll, E., and Perrimon, N. (1993). Autosomal *P[ovo^Dl]* dominant female-sterile insertions in *Drosophila* and their use in generating germline chimeras. *Development* **119**, 1359–1369.

Clark, A. G., Silveira, S., Meyers, W., and Langley, C. H. (1994). Nature Screen, an efficient method for screening natural populations in *Drosophila* for targeted P elements. *Proc. Natl. Acad. Sci. U.S.A.* **91**, 719–722.

Cline, T. W. (1978). Two closely linked mutations in *Drosophila melanogaster* that are lethal to opposite sexes and interact with *daughterless*. *Genetics* **90**, 683–698.

Cooley, L., Kelley, R., and Spradling, A. C. (1988). Insertional mutagenesis of the *Drosophila* genome with single P elements. *Science* **239**, 1121–1128.

Cooley, L., Thompson, D., and Spradling, A. C. (1990). Constructing deletions with defined endpoints in *Drosophila*. *Proc. Natl. Acad. Sci. U.S.A.* **87**, 3170–3173.

Cooley, L., Verheyen, E., and Ayers, K. (1992). *chickadee* encodes a profilin required for intercellular cytoplasmic transport during *Drosophila* oogenesis. *Cell* **69**, 173–184.

Coschigano, K. T., and Wensink, P. C. (1993). Sex-specific transcriptional regulation by the male and female doublesex proteins of *Drosophila*. *Genes Dev.* **7**, 42–54.

Crosby, M. A., and Meyerowitz, E. M. (1986). Lethal mutations flanking the 68C glue gene cluster on chromosome 3 of *Drosophila melanogaster*. *Genetics* **112**, 785–802.

Daniels, S. B., McCarron, M., Love, C., and Chovnick, A. (1985). Dysgenesis-induced instability of *rosy* locus transformation in *Drosophila melanogaster:* Analysis of excision events and the selective recovery of control element deletions. *Genetics* **109**, 95–117.

Davis, D. G. (1969). Chromosome behavior under the influence of *claret-nondisjunctional* in *Drosophila melanogaster*. *Genetics* **61**, 577–594.

Davis, R. L., and Dauwalder, B. (1991). The *Drosophila dunce* locus: Learning and memory genes in the fly. *Trends Genet.* **7**, 224–229.

Doane, W. W. (1960). Completion of meiosis in uninseminated eggs of *Drosophila melanogaster*. *Science* **132**, 677–678.

Edgar, B. A., and O'Farrell, P. H. (1989). Genetic control of cell division patterns in the *Drosophila* embryo. *Cell* **57**, 177–187.

Edgar, B. A., and O'Farrell, P. H. (1990). The three postblastoderm cell cycles of *Drosophila* embryogenesis are regulated in G2 by *string*. *Cell* **62**, 469–480.

Edgar, B. A., and Schubiger, G. (1986). Parameters controlling transcriptional activation during early *Drosophila* development. *Cell* **44**, 871–877.

Elkins, T., Zinn, K., McAllister, L., Hoffman, F. M., and Goodman, C. S. (1990). Genetic analysis of *Drosophila* neural cell adhesion molecules: Interaction of fasciclin I and Abelson tyrosine kinase mutations. *Cell* **60**, 565–576.

Endow, S. A., and Hatsumi, M. (1991). A multimember kinesin gene family in *Drosophila*. *Proc. Natl. Acad. Sci. U.S.A.* **88**, 4424–4427.

Engels, W. R. (1989). P-elements in *Drosophila melanogaster*. *In* "Mobile DNA" (D. E. Berg and M. M. Howe, eds.), Washington, D.C.: American Society for Microbiology.

Engels, W. R., Johnson-Schlitz, D. M., Eggleston, W. B., and Sved, J. (1990). High-frequency P element loss in *Drosophila* is homolog dependent. *Cell* **62**, 515–526.

Fields, S., and Song, O.K. (1989). A novel genetic system to detect protein–protein interactions. *Nature (London)* **340**, 245–246.

Foe, V. E., and Alberts, B. M. (1983). Studies of nuclear and cytoplasmic behavior during the five mitotic cycles that precede gastrulation in *Drosophila* embryogenesis. *J. Cell Sci.* **61**, 31–70.

Foe, V. E. (1989). Mitotic domains reveal early commitment of cells in *Drosophila* embryos. *Development* **107**, 1–22.

Foe, V., Odell, G., and Edgar, B. A. (1993). Mitosis and morphogenesis in the *Drosophila* embryo: Point and counterpoint. *In* "The Development of *Drosophila melanogaster*" (M. Bate and M. Martinez-Arias, eds.), pp. 149–300. New York: Cold Spring Harbor Press.

Frischer, L. E., Hagen, F. S., and Garber, R. L. (1986). An inversion that disrupts the *antennapedia* gene causes abnormal structure and localization of RNA. *Cell* **47**, 1017–1024.

Fristrom, J. W., Logan, W. R., and Murphy, C. (1973). The synthetic and minimal culture requirements for evagination of imaginal disks. *Dev. Biol.* **33**, 441–456.

Fuller, M. T. (1993). Spermatogenesis. *In* "The Development of *Drosophila melanogaster*" (M. Bate and M. Martinez-Arias, eds.), pp. 71–147. New York: Cold Spring Harbor Press.

Fuyama, Y. (1986). Genetics of parthenogenesis in *Drosophila melanogaster*. II. Characterization of a gynogenetically reproducing strain. *Genetics* **114**, 495–510.

Garcia-Bellido, A., Ripoll, P., and Morata, G. (1973). Developmental compartmentalization of the wing disc of *Drosophila*. *Nature New Biol. (London)* **245**, 251–253.

Gatti, M., and Baker, B. S. (1989). Genes controlling essential cell-cycle functions in *Drosophila melanogaster*. *Genes Dev.* **3**, 438–453.

Gatti, M., and Goldberg, M. L. (1991). Mutations affecting cell division in *Drosophila*. *Methods Cell Biol.* **35**, 543–586.

Gelbart, W. M. (1974). A new mutant controlling mitotic chromosome distinction in *Drosophila melanogaster*. *Genetics* **76**, 51–63.

Gertler, F. B., Bennett, R. L., Clark, M. J., and Hoffmann, F. M. (1989). *Drosophila abl* tyrosine kinase in embryonic CNS axons: A role in axonogenesis is revealed through dosage-sensitive interactions with *disabled*. *Cell* **58**, 103–113.

Gertler, F. B., Doctor, J. S., and Hoffmann, F. M. (1990). Genetic suppression of mutations in the *Drosophila abl* proto-oncogene homolog. *Science* **248**, 857–860.

Gill, S. R., Schroer, T. A., Szilak, I., Steuer, E. R., Sheetz, M. P., and Cleveland, D. W. (1991). Dynactin—A conserved, ubiquitously expressed component of an activator of vesicle motility mediated by cytoplasmic dynein. *J. Cell Biol.* **115**, 1639–1650.

Glaser, R. L., Wolfner, M. F., and Lis, J. T. (1986). Spatial and temporal pattern of *hsp26* expression during normal development. *EMBO J.* **5,** 747–754.

Gloor, G. B., Nassif, N. A., Johnson-Schlitz, D. M., Preston, C. R., and Engels, W. R. (1991). Targeted gene replacement in *Drosophila* via P element-induced gap repair. *Science* **253,** 1110–1117.

Goldberg, M. L., Colvin, R. A., and Mellin, A. F. (1989). The *Drosophila zeste* locus is nonessential. *Genetics* **123,** 145–155.

Golic, K. G. (1991). Site-specific recombination between homologous chromosomes in *Drosophila*. *Science* **252,** 958–961.

Golic, K. G., and Lindquist, S. (1989). The FLP recombinase of yeast catalyzes site-specific recombination in the *Drosophila* genome. *Cell* **59,** 499–510.

Govind, S., and Steward, R. (1991). Dorsoventral pattern formation in *Drosophila* signal transduction and nuclear targeting. *Trends Genet.* **7,** 119–125.

Green, M. M. (1988). Mobile DNA elements and spontaneous gene mutation. *Banbury Rep.* **30,** 41–50.

Guarente, L. (1993). Synthetic enhancement in gene interaction: A genetic tool comes of age. *Trends Genet.* **9,** 362–366.

Gunaratne, P. H., Mansukhani, A., Liapari, S. E., Liou, H.-C., Martindale, D. W., and Goldberg, M. L. (1986). Molecular cloning, germline transformation and transcriptional analysis of the *zeste* locus of *Drosophila melanogaster*. *Proc. Natl. Acad. Sci. U.S.A.* **83,** 701–705.

Hall, J. C. (1979). Control of male reproductive behavior by the central nervous system of *Drosophila:* Dissection of a courtship pathway by genetic mosaics. *Genetics* **92,** 437–457.

Hall, J. C., Gelbart, W. M., and Kankel, D. R. (1976). Mosaic systems. *In* "The Genetics and Biology of *Drosophila*" (M. Ashburner and E. Novitski, eds.), Vol. 1a, pp. 265–314. New York: Academic Press.

Hamilton, B. A., Palazzolo, M. J., Chang, J. H., Raghavan, K. V., Mayeda, C. A., Whitney, M. A., and Meyerowitz, E. M. (1991). Large scale screen for transposon insertions into cloned loci by plasmid rescue of flanking DNA. *Proc. Natl. Acad. Sci. U.S.A.* **88,** 2731–2735.

Harris, W., Stark, W. S., and Walker, J. A. (1976). Genetic dissection of the photoreceptor system in the compound eye of *Drosophila melanogaster*. *J. Physiol.* **256,** 415–439.

Harrison, D. A., Geyer, P. K., Spana, C., and Corces, V. G. (1989). The gypsy retrotransposon of *Drosophila melanogaster:* Mechanisms of mutagenesis and interaction with the *suppressor of Hairy wing* locus. *Dev. Genet.* **10,** 239–248.

Hart, A. C., Kramer, H., VanVactor, D. L., Paidhungat, M., and Zipursky, S. L. (1990). Induction of cell fate in the *Drosophila* retina: The bride of sevenless protein is predicted to contain a large extracellular domain and seven transmembrane segments. *Genes Dev.* **4,** 1835–1847.

Hartenstein, V., and Posakony, J. W. (1989). Development of adult sensilla on the wing and notum of *Drosophila melanogaster*. *Development* **107,** 389–405.

Hartl, D. L., Ajioka, J. W., Cai, H., Lohe, A. R., Lozovskaya, E. R., Smoller, D. A., and Duncan, I. W. (1992). Toward a *Drosophila* genomic map. *Trends Genet.* **8,** 70–75.

Hays, T. S., Deuring, R., Robertson, B., Prout, M., and Fuller, M. T. (1989). Interacting proteins identified by genetic interactions: A missense mutation in alpha-tubulin fails to complement alleles of the testis-specific β-tubulin gene of *Drosophila melanogaster*. *Mol. Cell. Biol.* **9,** 875–884.

Hereford, L. M., and Hartwell, L. H. (1974). Sequential gene function in the initiation of *Saccharomyces cerevisiae* DNA synthesis. *J. Mol. Biol.* **84,** 445–461.

Holden, J., and Suzuki, D. T. (1973). Temperature-sensitive mutations in *Drosophila melanogaster*. XII. The genetic and developmental characteristics of dominant lethals on chromosome 3. *Genetics* **73,** 445–458.

Holzbaur, E. L. F., Hammarback, J. A., Paschal, B. M., Kravit, N. G., Pfister, K. K., and Vallee, R. B. (1991). Homology of a 150K cytoplasmic dynein-associated polypeptide with the *Drosophila* gene *Glued*. *Nature (London)* **351,** 579–583.

Huang, S. L., and Baker, B. S. (1976). The mutability of the *Minute* loci of *Drosophila melanogaster* with ethyl methanesulfonate. *Mutat. Res.* **34,** 407–414.

Ingham, P. W., and Martinez-Arias, A. (1992). Boundaries and fields in early embryos. *Cell* **68,** 221–235.

Judd, B. H., Shen, M. W., and Kaufman, T. C. (1972). The anatomy and function of a segment of a chromosome of *Drosophila melanogaster*. *Genetics* **71,** 139–156.

Kaelin, W. G., Jr., Krek, W., Sellers, W. R., DeCaprio, J. A., Ajchenbaum, F., Fuchs, C. S., Chittenden, T., Li, Y., Farnham, P. J., Blanar, M. A., Livingston, D. M., and Flemington, E. K. (1992). Expression cloning of a cDNA encoding a retinoblastoma-binding protein with E2F-like properties. *Cell* **70,** 351–364.

Kafatos, F. C., Louis, C., Savakis, C., Glover, D. M., Ashburner, M., Link, A. J., Siden-Kiamos, I., and Saunders, R. D. C. (1991). Integrated maps of the *Drosophila* genome: Progress and prospects. *Trends Genet.* **7,** 155–161.

Kaiser, K., and Goodwin, S. F. (1990). Site-selected transposon mutagenesis of *Drosophila. Proc. Natl. Acad. Sci. U.S.A.* **87,** 1686–1690.

Kalb, J. M., DiBenedetto, A. J., and Wolfner, M. F. (1993). Probing the function of *Drosophila melanogaster* accessory glands by directed cell ablation. *Proc. Natl. Acad. Sci. U.S.A.* **90,** 8093–8097.

Karess, R. E. (1985). P element mediated germ line transformation of *Drosophila*. In: "DNA Cloning: A Practical Approach" (D. M. Glover, ed.), pp. 121–141. Oxford: IRL Press.

Karess, R. E., and Glover, D. M. (1989). *rough deal:* A gene required for proper mitotic segregation in *Drosophila. J. Cell Biol.* **109,** 2951–2961.

Karr, T. L. (1991). Intracellular sperm/egg interactions in *Drosophila:* A three-dimensional structural analysis of a paternal product in the developing egg. *Mech. Dev.* **34,** 101–112.

Karr, T. L., and Alberts, B. M. (1986). Organization of the cytoskeleton in early *Drosophila* embryos. *J. Cell Biol.* **102,** 1494–1509.

Kaufman, P. D., and Rio, D. C. (1991). Germline transformation of *Drosophila melanogaster* by purified P element transposase. *Nucleic Acids Res.* **19,** 6336.

Kellogg, D. R., and Alberts, B. M. (1992). Purification of a multiprotein complex containing centrosomal proteins from the *Drosophila* embryo by chromatography with low-affinity polyclonal antibodies. *Mol. Biol. Cell* **3,** 1–11.

Kellogg, D. R., Mitchison, T. J., and Alberts, B. M. (1988). Behavior of microtubules and actin filaments in living *Drosophila* embryos. *Development* **103,** 675–686.

Kidwell, M. G. (1987). A survey of success rates using P element mutagenesis in *Drosophila melanogaster. Nature* (*London*) **253,** 755–776.

Klemenz, R., Weber, U., and Gehring, W. J. (1987). The *white* gene as a marker in a new P-element vector for gene transfer in *Drosophila. Nucleic Acids Res.* **15,** 3947–3959.

Kubli, E. (1992). The sex peptide. *Bioessays* **14,** 779–784.

Kuhn, R., Schaefer, U., and Schaefer, M. (1988). Cis-acting regions sufficient for spermatocyte-specific transcriptional and spermatid-specific translational control of the *Drosophila melanogaster* gene *mst(3)gl-9. EMBO J.* **7,** 447–454.

Lawrence, P. A. (1981). A general cell marker for clonal analysis of *Drosophila* development. *J. Embryol. Exp. Morphol.* **64,** 321–332.

Lawrence, P. A. (1992). "The Making of a Fly: The Genetics of Animal Design." Oxford: Blackwell Scientific Publishers.

Lawrence, P. A., and Johnston, P. (1986). The muscle pattern of a segment of *Drosophila* may be determined by neurons and not by contributing myoblasts. *Cell* **45,** 505–513.

Lehmann, R., and Nusslein-Volhard, C. (1986). Abdominal segmentation, pole cell formation, and embryonic polarity require the localized activity of *oskar,* a maternal gene in *Drosophila. Cell* **47,** 141–152.

Lilly, M., and Carlson, J. (1990). *Smellbind:* A gene required for *Drosophila* olfaction. *Genetics* **124,** 293–302.

Lin, H., and Wolfner, M. F. (1989). Cloning and analysis of *fs(1)Ya*, a maternal effect gene required for the initiation of *Drosophila* embryogenesis. *Mol. Gen. Genet.* **215**, 257–265.

Lindsley, D. L., Sandler, L., Baker, B. S., Carpenter, A. T. C., Denell, R. E., Hall, J. C., Jacobs, P. A., Miklos, G. L. G., Davis, B. K., Gethmann, R. C., Hardy, R. W., Hessler, A., Miller, S. M., Nozawa, H., Parry, D. M., and Gould-Somero, M. (1972). Segmental aneuploidy and the genetic gross structure of the *Drosophila* genome. *Genetics* **71**, 157–184.

Lindsley, D. L., and Tokuyasu, K. T. (1980). Spermatogenesis. *In* "The Genetics and Biology of *Drosophila*" (M. Ashburner and T. R. F. Wright, eds.), pp. 225–294. New York: Academic Press.

Lindsley, D. L., and Zimm, G. (1992). "The Genome of *Drosophila melanogaster*." San Diego: Academic Press.

Lis, J. T., Neckameyer, W., Dubensky, R., and Costlow, M. (1981). Cloning and characterization of nine heat-shock-induced mRNAs of *Drosophila melanogaster*. *Gene* **15**, 67–80.

Lopez, J., Song, K., Hirshfeld, A., Lin, H., and Wolfner, M. F. (1994). The *fs(1)Ya* protein of *Drosophila* is a component of the lamina of cleavage nuclei, pronuclei and polar bodies in early embryos. *Dev. Biol.*, **163**, 202–211.

Mahowald, A. P., and Kambysellis, M. P. (1980). Oogenesis. *In* "The Genetics and Biology of *Drosophila*" (M. Ashburner and T. R. F. Wright, eds.), pp. 141–224. New York: Academic Press.

Marsh, J. L., and Wieschaus, E. (1978). Is sex determination in germ line and soma controlled by separate genetic mechanisms? *Nature (London)* **272**, 249–251.

Mattox, W., Palmer, M. J., and Baker, B. S. (1990). Alternative splicing of the sex determination gene *transformer-2* is sex-specific in the germline but not in the soma. *Genes Dev.* **4**, 789–805.

Meise, M., and Janning, W. (1993). Cell lineage of larval and imaginal thoracic anlagen cells of *Drosophila melanogaster* as revealed by single-cell transplantations. *Development* **118**, 1107–1121.

Merriam, J., Ashburner, M., Hartl, D. L., and Kafatos, F. C. (1991). Toward cloning and mapping the genome of *Drosophila*. *Science* **254**, 221–225.

Mével-Ninio, M., Guénal, I., and Limbourg-Buchon, B. (1994). Production of dominant female sterility in *Drosophila melanogaster* by insertion of the *ovo*^{Dl} allele on autosomes: Use of transformed strains to generate germline mosaics. *Mech. Dev.* **45**, 155–162.

Minden, J. S., Agard, D. A., Sedat, J. W., and Alberts, B. M. (1989). Direct cell lineage analysis in *Drosophila melanogaster* by time lapse three-dimensional optical microscopy of living embryos. *J. Cell Biol.* **109**, 505–516.

Monsma, S. A., Ard, R., Lis, J. T., and Wolfner, M. F. (1988). Localized heat-shock induction in *Drosophila melanogaster*. *J. Exp. Zool.* **247**, 279–284.

Montell, I., Rasmuson, A., Rasmuson, B., and Holmgren, P. (1992). Uptake and incorporation in pteridines of externally supplied GTP in normal and pigment-deficient eyes of *Drosophila melanogaster*. *Biochem. Genet.* **30**, 61–75.

Mounkes, L. C., Jones, R. S., Liang, B.-C., Gelbart, W., and Fuller, M. T. (1992). A *Drosophila* model for xeroderma pigmentosum and Cockayne's syndrome: *haywire* encodes the fly homolog of *ERCC3*, a human excision repair gene. *Cell* **71**, 925–937.

Muller, H. J. (1932). Further studies on the nature and causes of gene mutations. *Proc. VI Intl. Congr. Genet.* **1**, 213–255.

Nagoshi, R. N., McKeown, M., Burtis, K. C., Belote, J. M., and Baker, B. S. (1988). The control of alternative splicing at genes regulating sexual differentiation in *D. melanogaster*. *Cell* **53**, 229–236.

Ochman, H., Gerber, A. S., and Hartl, D. L. (1988). Genetic applications of an inverse polymerase chain reaction. *Genetics* **120**, 621–623.

Olsen, O.-A., and Green, M. M. (1982). The mutagenic effects of diepoxybutane in wild-type and mutagen-sensitive mutants of *Drosophila melanogaster*. *Mutat. Res.* **92**, 107–115.

Pastink, A., Schalet, A. P., Vreeken, C., Parádi, E., and Eeken, J. C. J. (1987). The nature of radiation-induced mutations at the *white* locus of *Drosophila melanogaster. Mutat. Res.* **177,** 101–115.

Pastink, A., Vreeken, C., Schalet, A. P., and Eeken, J. C. J. (1988). DNA sequence analysis of X-ray induced deletions at the *white* locus of *Drosophila melanogaster. Mutat. Res.* **207,** 23–28.

Perrimon, N. (1984). Clonal analysis of dominant female-sterile germline-dependent mutations in *Drosophila melanogaster. Genetics* **108,** 927–939.

Perrimon, N., Mohler, D., Engstrom, L., and Mahowald, A. P. (1986). X-linked female-sterile loci in *Drosophila melanogaster. Genetics* **113,** 695–712.

Phillips, J. P., and Forrest, H. S. (1980). Ommochromomes and pteridines. *In* "Genetics and Biology of *Drosophila*" (M. Ashburner and T. R. F. Wright, eds.), Vol. 2, pp. 541–623. New York: Academic Press.

Pirrotta, V. (1986). Cloning *Drosophila* genes. *In* "*Drosophila:* A Practical Approach" (D. B. Roberts, ed.), pp. 83–110. Oxford: IRL Press.

Ready, D. F., Hanson, T. E., and Benzer, S. (1976). Development of the *Drosophila* retina, a neurocrystalline lattice. *Dev. Biol.* **53,** 217–240.

Reardon, J. T., Liljestrand-Golden, C. A., Dusenbery, R. L., and Smith, P. D. (1987). Molecular analysis of diepoxybutane-induced mutations at the *rosy* locus of *Drosophila melanogaster. Genetics* **115,** 323–331.

Regan, C. L., and Fuller, M. T. (1988). Interacting genes that affect microtubule function: The *nc2* allele of the *haywire* locus fails to complement mutations in the testes-specific β-tubulin gene of *Drosophila. Genes Dev.* **2,** 82–92.

Regan, C. L., and Fuller, M. T. (1990). Interacting genes that affect microtubule function in *Drosophila melanogaster:* Two classes of mutation revert the failure to complement between hay^{nc2} and mutations in tubulin genes. *Genetics* **125,** 77–90.

Reinke, R., and Zipursky, S. L. (1988). Cell–cell interaction in the *Drosophila* retina: The *bride-of-sevenless* gene is required in photoreceptor cell 8 for R7 cell development. *Cell* **55,** 321–330.

Robertson, H. M., Preston, C. R., Phillis, R. W., Johnson-Schlitz, D., Benz, W. K., and Engels, W. R. (1988). A stable genomic source of P element transposase in *Drosophila melanogaster. Genetics* **118,** 461–470.

Rosenberg, U. B., Preiss, A., Seifert, E., Jäckle, H., and Knipple, D. C. (1985). Production of phenocopies by *Kruppel* antisense RNA injection into *Drosophila* embryos. *Nature (London)* **313,** 703–706.

Rubin, G. M., and Spradling, A. C. (1982). Transformation of *Drosophila* with transposable elements. *Science* **218,** 348–353.

Sandler, L., Lindsley, D. L., Nicoletti, B., and Trippa, G. (1968). Mutants affecting meiosis in natural populations of *Drosophila melanogaster. Genetics* **80,** 525–558.

Schneider, D. S., Hudson, K. L., Lin, T. Y., and Anderson, K. V. (1991). Dominant and recessive mutations define functional domains of Toll, a transmembrane protein required for dorsal-ventral polarity in the *Drosophila* embryo. *Genes Dev.* **5,** 797–807.

Schneuwly, S., and Gehring, W. J. (1985). Homeotic transformation of thorax into head: Developmental analysis of a new *Antennapedia* allele in *Drosophila melanogaster. Dev. Biol.* **108,** 377–386.

Schüpbach, T. (1982). Autosomal mutations that interfere with sex determination in somatic cells of *Drosophila* have no direct effect on the germline. *Dev. Biol.* **89,** 117–127.

Schüpbach, T. (1987). Germ line and soma cooperate during oogenesis to establish the dorsoventral pattern of egg shell and embryo in *Drosophila melanogaster. Cell* **49,** 699–708.

Searles, L. L., Greenleaf, A. L., Kemp, W. E., and Voelker, R. A. (1986). Sites of P element insertion and structures of P element deletions in the 5′ region of *Drosophila melanogaster RpII215. Mol. Cell. Biol.* **6,** 3312–3319.

Shearn, A., and Garen, A. (1974). Genetic control of imaginal disk development in *Drosophila. Proc. Natl. Acad. Sci. U.S.A.* **71,** 1393–1397.

Simon, M. A., Bowtell, D. D. L., Dodson, G. S., Laverty, T. R., and Rubin, G. M. (1991). Ras1 and a putative guanine nucleotide exchange factor perform crucial steps is signalling by the sevenless protein tyrosine kinase. *Cell* **67,** 701–716.

Simonova, O. B., Petruk, S. F., and Gerasimova, T. I. (1992). Participation of the *hobo* elements in the transposition events in a long-term instability system in *Drosophila melanogaster. Genetika* **28,** 73–79.

Simpson, L., and Wieschaus, E. (1990). Zygotic activity of the *nullo* locus is required to stabilize the actin-myosin network during cellularization in *Drosophila. Development* **110,** 851–864.

Smith, P. A., and Corces, V. G. (1991). *Drosophila* transposable elements: Mechanisms of mutagenesis and interactions with the host genome. *In* "Advances in Genetics" (J. G. Scandalios and T. R. F. Wright, eds.), pp. 229–300. San Diego: Academic Press.

Spana, C., Harrison, D. A., and Corces, V. G. (1988). The *Drosophila melanogaster* suppressor of Hairy wing protein binds to specific sequences of the gypsy retrotransposon. *Genes Dev.* **2,** 1414–1423.

Spradling, A. C. (1986). P element-mediated transformation. *In* "*Drosophila:* A Practical Approach" (D. B. Roberts, ed.), pp. 175–197. Oxford: IRL Press.

Spradling, A. C. (1993). Developmental genetics of oogenesis. *In* "The Development of *Drosophila melanogaster*" (M. Bate and M. Martinez-Arias, eds.), pp. 1–70. New York: Cold Spring Harbor Press.

Stafstrom, J. P., and Staehelin, L. A. (1984). Dynamics of the nuclear envelope and of nuclear pore complexes during mitosis in the *Drosophila* embryo. *Eur. J. Cell Biol.* **34,** 179–189.

Steller, H., and Pirrotta, V. (1985). A transposable P vector that confers selectable G418 resistance to *Drosophila* embryos. *EMBO J.* **4,** 167–171.

Stewart, B., and Merriam, J. R. (1973). Segmental aneuploidy of the X chromosome. *Drosophila Information Service* **50,** 167–170.

St. Johnston, D., and Nusslein-Volhard, C. (1992). The origin of pattern and polarity in the *Drosophila* embryo. *Cell* **68,** 201–219.

Suzuki, D. T., and Procunier, D. (1969). Temperature-sensitive mutations in *D. melanogaster.* III. Dominant lethals and semilethals on chromosome 2. *Proc. Natl. Acad. Sci. U.S.A.* **62,** 369–376.

Szabad, J., and Bryant, P. J. (1982). The mode of action of "discless" mutations in *Drosophila melanogaster. Dev. Biol.* **93,** 240–256.

Szabad, J., Mathe, E., and Puro, J. (1994). *Horka,* a dominant mutation of *Drosophila,* induces nondisjunction and, through paternal-effect, chromosome loss and genetic mosaics. *Genetics,* in press.

Thomas, J. H. (1993). Thinking about genetic redundancy. *Trends Genet.* **9,** 395–399.

VanDeusen, E. B. (1976). Sex determination in the germ line of *Drosophila melanogaster. J. Embryol. Exp. Morphol.* **37,** 173–185.

VanVactor, D., Krantz, D. E., Reinke, R., and Zipursky, S. L. (1988). Analysis of mutants in chaoptin A photoreceptor cell-specific glycoprotein in *Drosophila* reveals its role in cellular morphogenesis. *Cell* **52,** 281–290.

Ward, C. L., and Alexander, M. L. (1957). Cytological analysis of X-ray induced mutations at eight specific loci in the third chromosome of *Drosophila melanogaster. Genetics* **42,** 42–54.

Weinzierl, R., Axton, J. M., Ghysen, A., and Akam, M. (1987). *Ultrabithorax* mutations in constant and variable regions of the protein coding sequence. *Genes Dev.* **1,** 386–397.

Wieschaus, E., and Nusslein-Volhard, C. (1986). Looking at embryos. *In* "*Drosophila,* A Practical Approach" (D. B. Roberts, ed.), pp. 199–227. Oxford: IRL Press.

Williams, B. C., Karr, T. L., Montgomery, J. M., and Goldberg, M. L. (1992). The *Drosophila l(1)zw10* gene product, required for accurate chromosome segregation, is redistributed at anaphase onset. *J. Cell Biol.* **118,** 759–773.

Williams, B. C., and Goldberg, M. L. (1994). Determinants of *Drosophila zw10* protein localization and function. *J. Cell Sci.* **107,** 785–798.

Wilson, C., Pearson, R. K., Bellen, H. J., O'Kane, C. J., Grossniklaus, U., and Gehring, W. J. (1989). P-element-mediated enhancer detection: An efficient method for isolating and characterizing developmentally regulated genes in *Drosophila*. *Genes Dev.* **3,** 1301–1313.

Xu, T., and Rubin, G. (1993). Analysis of genetic mosaics in developing and adult *Drosophila* tissues. *Development* **117,** 1223–1237.

Zervos, A. S., Gyuris, J., and Brent, R. (1993). Mxi1, a protein that specifically interacts with max to bind myc-max recognition sites. *Cell* **72,** 223–232.

Zhang, P., and Spradling, A. C. (1993). Efficient and dispersed local P element transposition from *Drosophila* females. *Genetics* **133,** 361–373.

Zheng, Y., Jung, M. K., and Oakley, B. R. (1991). Gamma-tubulin is present in *Drosophila melanogaster* and *Homo sapiens* and is associated with the centrosome. *Cell* **65,** 817–823.

Zusman, S. B., and Wieschaus, E. F. (1985). Requirements for zygotic gene activity during gastrulation in *Drosophila melanogaster*. *Dev. Biol.* **111,** 359–371.

Zusman, S. B., and Wieschaus, E. (1987). A cell marker system and mosaic patterns during early embryonic development in *Drosophila melanogaster*. *Genetics* **115,** 725–726.

CHAPTER 4

From Clone to Mutant Gene

Bruce A. Hamilton* and Kai Zinn†

*Whitehead Institute for Biomedical Research
Cambridge, Massachusetts 02412

†Department of Biology
California Institute of Technology
Pasadena, California 91125

I. Introduction

Targeted mutagenesis has become a second essential paradigm in molecular genetics for correlating mutant phenotypes with molecular properties. The ability to produce null mutations in cloned genes has been invaluable in the yeast and mouse genetic systems. In *Drosophila*, phenotypic screening for mutations has remained the primary means of genetic analysis because methods for targeted mutagenesis have until recently been rather limited. However, many newly identified fly genes are isolated on the basis of sequence similarities (Michelson *et al.*, 1990; Tian *et al.*, 1991), bulk expression pattern (Levy *et al.*, 1982; Palazzolo *et al.*, 1989), or biochemical properties of their products (Goodrich *et al.*, 1993). Even in the most genetically tractable organisms, large numbers of transcripts for which mutations are not known to exist have been identified (Oliver *et al.*, 1992; Palazolo *et al.*, 1989; Sulston *et al.*, 1992). Isolating a null mutation in a molecularly defined gene can be immediately useful in at least three ways: for direct assessment of any nonredundant functions provided by the gene, as a host background in which to test transgene constructs for restoration or alteration of function, and as a starting stock for phenotypic suppressor and enhancer screens to identify interacting genes. In this chapter, we briefly review several methods for isolating mutations in cloned *Drosophila* genes and provide detailed protocols for generating P-element-induced mutations at targeted loci. In these protocols, local transposition of P-elements is used to increase the probability of insertion into the site of interest. We also describe flanking sequence rescue screens to identify lines bearing insertions into the gene of interest from collections of P-element transposants.

A. Screens Based on Chemical or Radiation Mutagenesis

In those cases in which the phenotype of a desired mutation can be reasonably guessed, standard phenotypic screens of mutagenized chromosomes are often feasible (e.g., Saxton *et al.*, 1991). In this approach, mutagenized chromosomes are crossed into a marked stock carrying a deficiency that spans the cytological interval of interest. The progeny of this cross are assayed for a predicted phenotype (usually lethality of individuals heterozygous for a mutagenized chromosome and a deficiency chromosome). In this way, genes outside of the region defined by the deficiency can be ignored and only genes within the defined interval that can be mutagenized to the predicted phenotype are isolated. This method is limited primarily by the ability to predict a mutant phenotype, since for many genes a null mutation will not confer lethality. In addition, for some regions of the genome suitable deficiency stocks may not be available.

Another approach that has been used successfully is screening for loss of an epitope by antibody binding (Katz *et al.*, 1988; Van Vactor *et al.*, 1988). An antibody raised against the gene product is used to assay crude protein extracts from animals carrying the mutagenized chromosome, either as homozygotes or as heterozygotes over a deficiency. This approach does not require a prediction of phenotype, except that the mutation must not be lethal prior to the time when readily detectable levels of the gene product accumulate in unaffected animals. This tactic does require a good antibody for each gene targeted and a separate assay for each chromosome screened. Dolph *et al.* (1993) recently used this method to isolate three mutations in the *arrestin-2* gene among 20,800 third chromosomes screened and two *arrestin-1* alleles among 15,481 second chromosomes screened.

B. Screens Based on Transposon Mutagenesis

Several properties of P-elements (in addition to being nontoxic) make them attractive as mutagenic agents for targeted genetic screens. Recombinant P-element derivatives whose mobility is controlled by mating to a source of transposase and whose transposition is selectable by phenotype (either a change in eye color or G418 resistance) are widely available, making the generation of single or multiple hit transposants straightforward. Since P-element transpositions tag the inserted site with a known DNA sequence, insertions into the target region can be detected at the DNA level without a prediction of phenotype. P-element insertions isolated in or near a gene of interest can be used as substrates to generate new alleles at very high efficiencies. For example, remobilization of a P-element often generates flanking sequence deletions by imprecise excision (Daniels *et al.*, 1985; Salz *et al.*, 1987; Voelker *et al.*, 1984). These can extend into an adjacent target gene to produce a null mutation.

Remobilization also generates local reinsertions that can often be selected by scoring the dominant marker on the transposon. This local transposition property is especially useful in light of the large number of enhancer detector P-elements that have been carefully mapped and collected. Spradling and co-workers have reported insertion rate enhancements for linked sites up to 100 kb away from a starting P-element (Tower *et al.*, 1993) on a minichromosome derived from the X. We have likewise observed several instances of local transposition that cross an entire lettered division on ordinary autosomes (Hamilton, 1993). These findings suggest that saturation of P-mutable sites within a region might be obtained efficiently using a local transposition strategy.

Several detection assays have been used for targeted mutagenesis experiments with P-elements. These include genomic Southern blot hybridization to detect RFLPs, PCR assays to detect the juxtaposition of P-element sequences to known sequences in a gene of interest, and flanking sequence rescue techniques in which genomic sequences adjacent to P-insertion sites are isolated and screened by hybridization to target DNA. Each of these designs has inherent strengths and weaknesses, which we outline below.

Genomic Southern blot hybridization can be an effective screen for insertions generated by local transposition. The high targeting efficiencies achieved in some local transposition experiments and the definitiveness and relative simplicity of the assay argue that this is a reasonable approach for screening up to a few hundred lines. Furthermore, rehybridizing the same filter membranes with transposon-derived probes can provide useful information about the success of the transposition strategy, such as the true rate of transposition, the existence and dominance of insertion hot spots among recovered lines, and the fate of the initial insertion site in local transposition experiments. Our lab has isolated PlacW insertions in or near two receptor-linked protein tyrosine phosphatase genes using this type of screen (B. A. Hamilton, A. Ho, and K. Zinn, in preparation: C. Desai and K. Zinn, unpublished experiments). Two insertions adjacent to the 3' end of *DPTP99A* were isolated from 132 transposants from a PlacW element at 99B, and an insertion into the *DPTP69D* locus was identified from about 300 transposants from an element at 69D.

PCR strategies have been designed to detect novel fragments amplified between a P-element terminal repeat primer and one or more gene-specific primers (Ballinger and Benzer, 1989; Kaiser and Goodwin, 1990). This scheme has the advantages of being rapid, assaying large numbers of mutagenized chromosomes in a single experiment, and allowing the use of naturally occurring defective P-elements such as are found in the Birmingham and Harwich strains. This kind of screen has also been adapted to local transposition experiments in the mutagenesis of the *Drosophila* synaptotagmin gene (Littleton *et al.,* 1993). The limitations of PCR-based screens are the size of the target (effectively about 2 kb per gene-specific primer) and the potential for false positives if the primer sequences and PCR conditions are not carefully controlled.

Flanking sequence rescue screens present a third alternative for P-element insert detection. The essence of this method is to isolate genomic DNA fragments that flank the transposon insertion sites and hybridize them against the targeted sequences. In the original version of this screen, a P-element that carries a miniplasmid in one end (PlacW) is used as the mutagen and sequences that flank insertion sites are rescued by digesting transposant genomic DNAs with *Eco*RI or *Sac*II, cyclizing the fragments with DNA ligase and transforming the circles into *Escherichia coli.* Circles capable of transforming *E. coli* to drug resistance carry both the plasmid end of the transposon and the flanking genomic restriction fragment. The resulting plasmid library is then used as a hybridization probe against segments of the targeted genes. This method was used to isolate insertions near or within several genes represented by cDNA clones in an array of hybridization targets (Hamilton *et al.,* 1991) and to isolate insertions adjacent to *DPTP99A,* using an array of genomic DNA clones (Hamilton, 1993; B. A. Hamilton, A. Ho, and K. Zinn, in preparation). The strengths of this approach are that little molecular characterization of the targeted gene is required before the screen is begun and that relatively large pools of transposants (up to 200 or more) can be represented in each probe. Furthermore, each probe can be hybridized against a large array of targets representing the entire genomic region spanning each targeted gene. The advantage of screening large targets is that they allow detection of insertions into introns and nonessential flanking regions, and these insertions can be used to generate null alleles at very high frequencies by local transposition or imprecise excision of the element. Recently, Dalby and Goldstein have used inverse PCR (Ochman *et al.,* 1988) on genomic DNA to generate flanking sequence probes in a successful screen of similar design (personal communication, 1993). Because it does not require a transformation step, this adaptation streamlines probe preparation and broadens the variety of useful starting transposons.

II. Overview

A. Steps in Targeted P-Element Screens Using Flanking Sequence Rescue Hybridization

1. Design the experiment and select the materials, including one or more transposon stocks, a source of transposase, and target clones. Sources for these materials and some considerations in their use are summarized in Table 1. The version of flanking rescue described here requires the use of a transposon that contains a plasmid in one end, such as PlacW.
2. Set up matings to generate transpositions, select transposants, and establish lines (Fig. 1).
3. Prepare blots of clones representing the mutagenesis targets.
4. Prepare DNA from pools of mated transposants.
5. Rescue the transposon-flanking sequences by transformation into *E. coli* (Fig. 2).
6. Hybridize labeled rescue plasmids to blots.
7. Identify mutant lines from positive pools.

B. Planning a Targeted Mutagenesis Screen

For local transposition experiments, it is wise to gather several P-element single insert lines as potential starting stocks. As a rule of thumb, one-half to one lettered division of the salivary gland chromosome map from the target gene is a reasonable proximity from which to start. Not all P-element derivatives will transpose with the same efficiency, and the distribution of second insertion sites may vary both with the type of element and the starting location. Furthermore, local hot spots may act as sinks for insertions in some regions. Beginning with transposons at distinct sites near the gene of interest may therefore offer an advantage in some cases and having several alternatives already passed through your quarantine procedures would be worthwhile. Transposons marked with partially complementing w^+ minigenes, such as PlacW (Bier *et al.*, 1989), are the most useful for scoring increases in transposon copy number caused by local transpositions. Most PlacW single-insert lines have yellow or orange eyes, and increases in PlacW copy number produce easily visible eye-color changes. Many mapped single-insert lines of this type are available through stock centers and individual researchers. In designing the crossing scheme, it is also worth bearing in mind that the distribution of inserted sites is more random in female germlines than in male (Zhang and Spradling, 1993).

Table I
Starting Materials

Type of material	Description	Availability	Advantages	Disadvantages
Transposons	Attached X chromosome that carries eight PlacW elements	Bloomington Stock Center	X-to-autosome transpositions, targeting several sites at once	Low targeting efficiency for any one locus
	Single-insert enhancer trap lines	Bloomington Stock Center; individual labs	Local transpositions	Starting site remains linked
Transposase source	P[ry]Δ2-3(99B)	Many sources	Stable	May recombine away from chromosomal markers
	TMS, *Sb* P[ry]Δ2-3(99B)	Bloomington Stock Center	Stable, not recombinogenic	
Balancer stocks		Many sources	Select stocks that are compatible with markers on the chosen P-elements	
Target clones	cDNA	Listed in *Drosophila Information Service (DIS)*, Vol. 72, pp. 180–183; updated in *Drosophila Information Network (DIN)*	Detects insertions into the gene	May not detect insertions into introns or control sequences
	Lambda and cosmid genomic clones	Listed in DIS, Vol. 72, pp. 180–183; updated in DIN	Span entire gene with a short walk, represent as discrete fragments on a gel blot	
	P1 genomic clones	Genome Systems	May span an entire gene with a single clone	

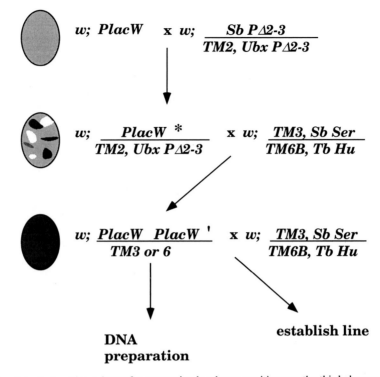

Fig. 1 A typical mating scheme for generating local transpositions on the third chromosome. A line bearing a single w^+ PlacW element (light eye color) is mated to a source of P transposase. Somatic transpositions in the progeny are evident as a mosaic eye. Selecting F2 animals for eye color darker than that of nonmobilized controls (w; PlacW/ +) greatly enriches for lines carrying multiple transposons. The majority of these are likely to be local transposants that retain the original insertion site intact.

Several alternatives are also possible for target clones; we recommend a combination of cDNA and genomic clones to span the entire gene. Full-length cDNA clones are useful targets because any insertions detected by them are likely to disrupt the function of the gene. However, insertions into introns or flanking control regions may also be mutagenic. Insertions into these sites can be detected by genomic clones that span the length of the targeted gene. Insertions of this kind may also be useful for generating deletion alleles by imprecise excision in a subsequent mobilization. This approach may be especially useful for genes that are refractory to direct insertion by P-elements.

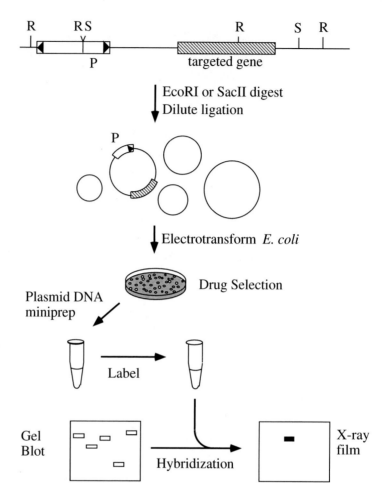

Fig. 2 Transposon insertions can be screened by plasmid rescue of flanking genomic DNA. Pools of DNA are digested with *Eco*RI (R) or *Sac*II (S) to release the plasmid carried in the transposon (P) along with a flanking genomic fragment. DNA fragments are ligated in dilute solution to form circles and transformed into *E. coli*. Plasmid DNA from drug-resistant colonies is labeled by random priming and used as a hybridization probe against Southern blots of cloned fragments from the genes of interest. Modified from Hamilton *et al.* (1991).

III. Protocols

A. Protocol I: Generating Local Transposition Lines

Most dysgenic P-element mating schemes use the stable transposase source P[ry$^+$]Δ2-3(99B), also called PΔ2-3 (Robertson *et al.*, 1988). A typical mating scheme for local transposition on the third chromosome in a female germline is illustrated in Fig. 1. For dysgeneses in female germlines, the balancer chromosome TM2, *Ubx* PΔ2-3 has been used; however, it tends to be a weak source of transposase. Recombinations between PΔ2-3 and visible markers on noninverted chromosomes are only slightly problematic as they are usually evident by eye-color mosaicism and can be discarded. A recent alternative chromosome is TMS, *Sb* PΔ2-3. Mutagenesis on other chromosomes or in male germlines follows essentially the same design, except for the choice of balancers. The steps of a typical mating protocol are illustrated in Fig. 1 and enumerated here.

1. Mate the stock carrying the starting insertion to a PΔ2-3 stock *en masse*. This can be done in either direction with respect to sexes. Ten to 15 virgin females to 3 or 4 males in a half-pint bottle, transferred every 3 days for a total of four bottles, provides a good density of offspring without overcrowding.

2. Collect dysgenic progeny. These should be evident by mosaic patches of eye color caused by the somatic activity of PΔ2-3. For stocks in which the transposition rate is low, mosaicism in the eyes of F1 progeny may be a good indicator of which animals will produce frequent transpositions in the germline.

3. Mate single dysgenic F1 animals to an appropriate balancer stock, such as *w; TM3/TM6*, in a small vial. Selecting one F2 transposant from each vial guards against repeated isolations of the same insertion event. At the same time, set up several control vials, mating the original transposon strain to the balancer stock. Progeny from this vial will be controls for scoring transposants in the next generation. Use two or three flies from the balancer stock for each vial, as single-pair matings often have a high failure rate. If the transposition rate is low, including two to three dysgenic flies per vial may be more efficient. Although dysgenesis in female germlines is thought to produce a more dispersed pattern of insertion sites, it may be of use to set up matings from a limited number of males as well.

4. Select transposants by scoring for darker eye color. Taking only one transposant from each vial ensures that each line analyzed represents an independent transposition. These flies will be enriched for lines that carry multiple inserts. Most of these lines will carry the original plus one new insertion, although single, apparently new insertions are relatively common and lines carrying mutliple new insertions are occasionally seen as well. Comparing the F2 dysgenesis offspring with control offspring of the same age is important, as eye color deepens with age. Males are preferable at this point because they can be mated to several balancer females for a prodigious stock vial before being used in DNA preparation and because eye-color changes are often more difficult to score in younger (virgin) flies.

5. Mate the selected transposants to the balancer stock, again in vials. Once larvae appear in the vial, remove the original transposants for DNA preparation.

B. Protocol II: DNA Preparation

This protocol is based on the method described by Bender *et al.* (1983) and is intended to provide consistent quality and yield among many samples.

1. Collect 20 F2 flies per 1.7-ml microcentrifuge tube and freeze on dry ice. Each tube constitutes a subpool in the following screening protocol. It is essential to faithfully record which fly lines are collected in each tube, so that mutant lines can be identified after screening.

2. Remove a tube from dry ice and add 100 μl grinding buffer (5% sucrose, 80 mM NaCl, 100 mM Tris, pH 8.5, 0.5% SDS, 50 mM EDTA). Grind the flies quickly but carefully with a small baked glass rod. Rinse the end of the rod into the tube with 100 μl grinding buffer. Return the tube to dry ice until all samples have been processed.

3. Incubate the homogenate at 70°C for 30 min.

4. Add 35 μl 8 M KOAc. Incubate on ice for 30 min. Pellet the precipitate by spinning for 10 min in a microcentrifuge. Decant the supernatant into a clean, prelabeled tube.

5. Add 150 μl isopropanol. Mix well by inverting the tube several times. Let stand at room temperature for 5 min. Centrifuge for 10 min to pellet the DNA. The pellet should be visible, but may be dispersed along the side of the tube. Rinse the pellet once with about 0.5 ml 70% EtOH and a brief spin in the centrifuge. Aspirate the supernatant. Dry the pellet briefly by leaving the tube inverted on a paper towel or in a finger rack.

6. Resuspend the pellet in 100 μl TE + RNase A (approximately 0.1 mg/ml). Dissolution of the pellet is aided by incubation at 55–60°C for 1–3 hr.

C. Protocol III: Plasmid Rescue of Flanking Sequences

1. Restriction digests release the plasmid and flanking genomic DNA on the same fragment. Pool 50 μl from each of 5 to 10 DNA preparations (100 to 200 fly lines). Digest each of four 30-μl aliquots of DNA (approximately 3 μg each) with 20 units of the chosen restriction enzyme for 1 hr in a 40-μl reaction. Heat inactivate the enzyme at 70°C for 15 min.

2. The fragments are then ligated to form closed circles. To each digest, add 120 μl 5× ligase buffer (Sambrook *et al.*, 1989) and 440 μl sterile distilled water. Mix well. Add 1 μl T4 DNA ligase (400 U/μl; New England Biolabs) and mix by inverting the tube several times. Incubate at room temperature for 4 to 16 hr to cyclize the fragments.

4. Precipitate the DNA circles by adding 12 μl 8 M LiCl and 600 μl isopropanol to each tube. Incubate on ice for 20 min. Recover the DNA by centrifugation for 10 min. Aspirate the supernatant. Wash the pellet once with 0.5 ml 70% EtOH and centrifuge for 5 min. Aspirate the supernatant. Dry the pellet *in*

vacuo. Resuspend the pellet in 8 µl sterile distilled water. Centrifuge 10 min to pellet any remaining insoluble material.

4. Transformation is accomplished by electroporation. Add 4 µl resuspended DNA to 40 µl electroporation-competent LE392 cells (Dower *et al.,* 1988) on ice. Incubate 5 to 10 min on ice. Electroporate (we use a Bio-Rad Gene Pulser, following the manufacturer's recommended conditions for *E. coli*) and immediately take up the cells with 1 ml SOC medium into a 15-ml culture tube. Shake the cells 200–220 rpm at 37°C for 45 min. Spread 250 µl of the culture onto each of four LB agar plates containing 100 µg/ml carbenicillin. Incubate overnight or until colonies are 1 to 2 mm in diameter. The yield should be at least two to three colonies per fly line used.

5. Plasmid DNA is prepared by the rapid boiling method (Holmes and Quigley, 1981). Resuspend the colonies from all four plates in 2–3 ml LB with a sterile pipet. Split the suspension into two 1.7-ml microcentrifuge tubes and pellet the cells with a brief centrifuge spin. Resuspend each pellet in 0.6 ml STET (8% sucrose, 5% Triton X-100, 50 mM EDTA, 50 mM Tris, pH 8.0) with a vortex mixer. Incubate with 35 µl lysozyme (10 mg/ml) for 0.5 to 2 min. Boil for 2 to 2.5 min. Centrifuge for 10 min. Scoop out the viscous pellet with a toothpick and discard it. Precipitate DNA with 0.5 ml 75% isopropanol, 2.5 M NH$_4$OAc. Spin down the precipitate and wash the pellet with 0.5 ml 70% EtOH. Dry the pellet *in vacuo* and resuspend in 100 µl TE, 50 µg/ml RNase A.

6. To estimate the DNA concentration and complexity of recovered plasmids, compare restriction digests of each sample to known standards by electrophoresis through an agarose gel containing ethidium bromide.

D. Protocol IV: Hybridization Screening

1. Preincubate the clone blot filters that represent the targeted loci with hybridization buffer and agitation at 42°C while preparing the probe. Our hybridization buffer is 50% formamide, 5x SSPE, 1% SDS, 1x Denhardt's solution, and 100 µg/ml denatured salmon testes DNA. The buffer should be heated above 80°C for 5 min before use to denature the DNA.

2. Label approximately 100 ng of restriction-digested plasmid DNA by random priming method (Feinberg and Vogelstein, 1983). High-quality kits are available from several commercial sources. Follow manufacturer recommendations in their use. Dilute 100 ng digested plasmid DNA with sterile distilled water so that the final reaction volume will be 75 µl. Boil for 2 min to denature the DNA and immediately chill it on dry ice. Most of the reaction components can be added while thawing the denatured DNA between fingers. Add 8 µl [α^{32}P]dCTP (80 µCi at 3000 Ci/mmole) to a reaction mix that otherwise lacks dCTP before adding the polymerase.

3. Purify the probe through a Sephadex G-50 spin column (Sambrook *et al.,*

1989). Add hybridization buffer to approximately 1 ml. Determine the specific activity of the probe by Cerenkov or scintillation counting.

4. Denature the probe by heating to 80–100°C for 4 min.

5. Add the probe to the preincubated filters to a final concentration of $1-3 \times 10^6$ cpm/ml and incubate at 42°C with agitation for 18–36 hr.

6. Wash the filters with three changes of $0.1 \times$ SSPE, 0.3% SDS prewarmed to 50°C.

7. Wrap the filters between single layers of SaranWrap and expose to film between two intensifying screens (the order is screen, film, filter, screen) at −70°C or below. Signals are often evident after 24 hr, although signals from 200-line pool probes have required as long as 1 week with a single intensifying screen.

8. Identify lines that carry the relevant insertions. Once a pool is identified as containing a line that has a candidate mutation in a gene of interest, each DNA preparation from that pool should be digested, Southern blotted, and hybridized to a probe from the appropriate target clone. This identifies a subpool of 20 lines. Genomic DNA from each line in this subpool is prepared and used to make a Southern blot. Probing this blot with the identified target clone fragment indentifies the line.

9. Depending on the characteristics of the identified insertion lines, generating derivative deletion alleles by imprecise excision of the P-element may be desirable. To do this, the insertion is remobilized using essentially the same crossing scheme by which it was produced. The principle differences are that in excising the element, one scores for a reduction in eye color and molecular screening is done by genomic Southern blot hybridization of individual lines to probes flanking the insertion sites.

IV. Conclusion

We have used the protocols presented here routinely and with success. We have isolated insertions at several target sites from relatively small numbers of examined transposant lines (Hamilton, 1993; Hamilton *et al.*, 1991). We have not yet observed any artifacts that obscure the results of this screening procedure. We note, however, that modifications to these protocols could streamline this approach. In particular, the inverse PCR probe adaptation of Dalby and Goldstein may save substantial effort when the target sites are represented by large, contiguous clones. Enhanced chemiluminescence detection may offer additional time savings over conventional autoradiography.

The availability of large collections of mapped single-insert enhancer detector P-element strains makes targeted mutagenesis of almost any locus in *Drosophila* feasible without the need of specialized reagents. Although targeting efficiencies will vary by target site and the availability of nearby elements, local transposition strategies, combined with the ability to generate flanking deficiencies by imprecise excision, offer a powerful genetic approach for cell biologists.

Acknowledgments

We thank M. J. Palazzolo, K. VijayRaghavan, and E. M. Meyerowitz for their encouragement and contributions to the development of these procedures. We thank C. Desai, B. Dalby, and L. S. B. Goldstein for discussing their results prior to publication. Our work cited in this manuscript was supported by NIH Grant NS28182, Basil O'Connor Starter Scholar Research Award 5-816 from the March of Dimes Birth Defects Foundation, a Pew Scholars Award, and a McKnight Scholars Award to K.Z. B.A.H. was supported in part by National Research Service Award 5 T32 HG00021-02 from the National Center for Human Genome Research during the development of these protocols and is currently supported by a fellowship from the Helen Hay Whitney Foundation.

References

Ballinger, D. G., and Benzer, S. (1989). Targeted gene mutations in Drosophila. *Proc. Natl. Acad. Sci. U.S.A.* **86,** 9402–9406.

Bender, W., Spierer, P., and Hogness, D. S. (1983). Chromosomal walking and jumping to isolate DNA from the Ace and rosy loci and the bithorax complex in Drosophila melanogaster. *J. Mol. Biol.* **168,** 17–33.

Bier, E., *et al.* (1989). Searching for pattern and mutation in the Drosophila genomes with a P-*lacZ* vector. *Genes Dev.* **3,** 1273–1287.

Daniels, S. B., McCarron, M., Love, C., and Chovnick, A. (1985). Dysgenesis-induced instability of rosy locus transformation in Drosophila melanogaster: analysis of excision events and the selective recovery of control element deletions. *Genetics* **109,** 95–117.

Dolph, P. J., Ranganathan, R., Colley, N. J., Hardy, R. W., Socolich, M., and Zuker, C. S. (1993). Arrestin function in inactivation of G protein-coupled receptor rhodopsin in vivo. *Science* **260,** 1910–1916.

Dower, W. J., Miller, J. F., and Ragsdale, C. W. (1988). High efficiency transformation of E. coli by high voltage electroporation. *Nucleic Acids Res.* **16,** 6127–6145.

Feinberg, A. P., and Vogelstein, B. (1983). A technique for radiolabeling DNA restriction endonuclease fragments to high specific activity. *Anal. Biochem.* **132,** 6–13.

Goodrich, J. A., Hoey, T., Thut, C. J., Admon, A., and Tjian, R. (1993). Drosophila TAFII40 interacts with both a VP16 activation domain and the basal transcription factor TFIIB. *Cell* **75,** 519–530.

Hamilton, B. A. (1993). Assessing molecular function in the Drosophila nervous system: A reverse genetic approach. Ph.D. thesis, California Institute of Technology, Pasadena, CA.

Hamilton, B. *et al.* (1991). Large scale screen for transposon insertions into cloned genes. *Proc. Natl. Acad. Sci. U.S.A.* **88,** 2731–2735.

Holmes, D. S., and Quigley, M. (1981). A rapid boiling method for the preparation of bacterial plasmids. *Anal. Biochem.* **114,** 193–197.

Kaiser, K., and Goodwin, S. F. (1990). "Site-selected" transposon mutagenesis of Drosophila. *Proc. Natl. Acad. Sci. U.S.A.* **87,** 1686–1690.

Katz, F., Moats, W., and Jan, Y. N. (1988). A carbohydrate epitope expressed uniquely on the cell surface of Drosophila neurons is altered in the mutant *nac* (neurally altered carbohydrate). *EMBO J.* **7,** 3471–3477.

Levy, L. S., Ganguly, R., Ganguly, N., and Manning, J. E. (1982). The selection, expression and organization of a set of head-specific genes in *Drosophila. Dev. Biol.* **94,** 451–464.

Littleton, J. T., Stern, M., Schulze, K., Perin, M., and Bellen, H. J. (1993). Mutational analysis of Drosophila synaptotagmin demonstrates its essential role in Ca2 + -activated neurotransmitter release. *Cell* **74**, 1125–1134.

Michelson, A. M., Abmayr, S. M., Bate, M., Arias, A. M., and Maniatis, T. (1990). Expression of a MyoD family member prefigures muscle pattern in Drosophila embryos. *Genes Dev.* **4**, 2086–2097.

Ochman, H., Gerber, A. S., and Hartl, D. L. (1988). Genetic applications of an inverse polymerase chain reaction. *Genetics* **120**, 621–623.

Oliver, S. G., *et al.* (1992). The complete DNA sequence of yeast chromosome III. *Nature* **357**, 38–46.

Palazzolo, M. J., *et al.* (1989). Use of a new strategy to isolate and characterize 436 Drosophila cDNA clones corresponding to RNAs detected in adult heads but not in early embryos. *Neuron* **3**, 527–539.

Robertson, H. M., *et al.* (1988). A stable genomic source of P element transposase in *Drosophila melanogaster*. *Genetics* **118**, 461–470.

Salz, H. K., Cline, T. W., and Schedl, P. (1987). Functional changes associated with structural alterations induced by mobilization of a P element inserted in the *Sex-lethal* gene of Drosophila. *Genetics* **117**, 221–231.

Sambrook, J., Fritsch, E. F., and Maniatis, T. (1989). ''Molecular Cloning: A Laboratory Manual,'' 2nd ed. Cold Spring Harbor, NY: Cold Spring Harbor Laboratory.

Saxton, W. M., Hicks, J., Goldstein, L. S., and Raff, E. C. (1991). Kinesin heavy chain is essential for viability and neuromuscular functions in Drosophila, but mutants show no defects in mitosis. *Cell* **64**, 1093–1102.

Sulston, J., *et al.* (1992). The C. elegans genome sequencing project: A beginning. *Nature* **356**, 37–41.

Tian, S.-S., Tsoulfas, P., and Zinn, K. (1991). Three receptor-linked protein-tyrosine phosphatases are selectively expressed on central nervous system axons in the Drosophila embryo. *Cell* **67**, 675–685.

Tower, J., Karpen, G. H., Craig, N., and Spradling, A. C. (1993). Preferential transposition of Drosophila P elements to nearby chromosomal sites. *Genetics* **133**, 347–359.

Van Vactor, D., Jr., Krantz, D. E., Reinke, R., and Zipursky, S. L. (1988). Analysis of mutants in chaoptin, a photoreceptor cell-specific glycoprotein in Drosophila, reveals its role in cellular morphogenesis. *Cell* **52**, 281–290.

Voelker, R. A., Greenleaf, A. L., Gyurkovics, H., Wisely, G. B., Huang, S.-M., and Searles, L. L. (1984). Frequent imprecise excision among reversions of a P element-caused lethal mutations in Drosophila. *Genetics* **107**, 279–294.

Zhang, P., and Spradling, A. C. (1993). Efficient and dispersed local P element transposition from Drosophila females. *Genetics* **133**, 361–373.

PART II

Isolation and Culture Methods

As starting material for biochemical isolations, or for analytical work, one frequently requires precisely staged embryos, particular organs and tissues, or highly purified cell types. Generally, the preparation of *Drosophila* cells and tissues is neither as straightforward nor as sophisticated as for mammals, because flies are small and because invertebrate culture methods are not as well developed as their vertebrate counterparts. Nonetheless, the insights offered by mutant fly strains and by comparative studies of larval and imaginal tissues have consistently warranted the extra effort. The good news, as related in the five chapters that follow, is that large-scale *Drosophila* rearing methods, as well as cell and organ culture methods and cell fractionation methods, work well and are readily reproducible in most labs.

Shaffer *et al.* begin with a discussion of mass rearing methods (Chapter 5). Their laboratory has refined the systems originally devised by earlier workers to the point where 100 g of embryos can be harvested daily from a colony of 400,000 flies. In addition to providing instructions for colony maintenance, media making, and equipment construction, the authors describe a simple method for rigorously quantifying the developmental age of particular embryo preparations. By following these simple instructions, it is possible to harvest sufficient quantities of precisely staged embryos for biochemical isolation protocols or large-scale organ culture.

Natzle and Vesenka next describe imaginal tissue isolation and culture (Chapter 6). As the authors point out, *in vitro* manipulations of imaginal tissue, already having played a prominent role in understanding how ecdysone affects alterations of gene activity as well as morphogenesis, now represent the means by which to delineate cellular behaviors and interactions that lay down adult tissue pattern. Natzle and Vesenka describe methods for large- and small-scale preparation of highly purified imaginal discs and also provide recipes for Robb's rich synthetic medium, a defined medium that will support imaginal disc morphogenesis and epithelial differentiation. The authors end with a discussion of the further improvements of culture conditions required to foster imaginal disc growth and peripheral nervous system development *in vitro*.

Mahowald focuses on the isolation of three very important cell types: pole cells, embryonic neuroblasts, and defined stages of oocytes (chapter 7). He begins by describing a mass rearing method utilizing liquid medium and killed yeast. Use of this medium for raising embryos from which cells will be cultured reduces both labor and the chances of contaminating cell cultures with live yeast. Next, the uses of isopycnic gradients and centrifugal elutriation for cell fractionation are described. Finally, Mahowald ends with instructions on how to prepare large or small numbers of staged *Drosophila* oocytes. The importance of this latter protocol is heightened by the fact that oocytes have proven to lend themselves well to study of a variety of cellular transport and localization phenomena (e.g., Chapters 25 and 28).

Cumberledge and Krasnow provide a provocative description of the whole animal cell sorting technique for purification of particular cell types (Chapter 8). One problem inherent to fractionation of embryonic cells is that they have few distinguishing characteristics prior to their overt differentiation. These investigators circumvent this problem by sorting cell populations according to their expression of *lacZ* reporter genes. This general method is broadly useful because so many enhancer trap lines expressing *lacZ* in different cell types already exist. Remarkably, sorted cells can be prepared within a few hours of embryo dissociation. A major drawback of the method is that the number of cells that can be prepared is small, limiting its applicability, with rare exceptions, to cytological procedures. However, the value of this method for further illuminating the cell biology of early embryos is great indeed, especially if reporter proteins not requiring the problematic FDG substrate can be developed.

Finally, Cherbas *et al.* present a compendium of methods for maintaining, cloning, and transforming *Drosophila* cell lines (Chapter 9). Most of the extant lines probably derive from hemocytes and hence do not represent the biological diversity of *Drosophila*. Nevertheless, cell lines transfected with particular *Drosophila* genes, for example, *Notch* and *Delta,* have been instrumental in furthering our understanding of the roles of particular macromolecules in adhesion, cell–cell signaling, and hormone reception. Opportunities along these lines will markedly increase as more antibodies recognizing proteins encoded by particular cloned genes are manufactured. Cherbas *et al.* provide detailed discussions of the suitabilities and transfection frequencies of particular lines and provide a wealth of knowledge regarding the usage of particular transformation protocols, vectors, and selection systems.

CHAPTER 5

Raising Large Quantities of *Drosophila* for Biochemical Experiments

Christopher D. Shaffer, Joann M. Wuller, and Sarah C. R. Elgin

Department of Biology
Washington University
St. Louis, Missouri 63130

I. Introduction

Many experiments using the techniques of biochemistry and molecular biology require large amounts of experimental material. *Drosophila* is a suitable organism for such efforts; it is relatively easy to grow adequate quantities of embryos, larvae, and adults. Mass cultures of *Drosophila* have been used in the preparation of a variety of cellular components, including RNA, proteins, DNA, and nuclei. In this chapter, we describe the system we use to rear and maintain a population of about 400,000 flies capable of producing 100 g of embryos a day.

II. The *Drosophila* Colony

The minimal requirements for growth of the flies are simple and can be easily met. However, to obtain maximum yield of embryos, the following environmental conditions should be maintained: humidity of 70%, temperature of 25°C, and a constant 24-hr light–dark cycle. Constant conditions are critical if a strict schedule of inoculation, feeding, and collection is to be maintained. For added convenience, one may use a reverse light cycle to take advantage of the burst in egg-laying at dusk (a 50–100% increase in the number of embryos/hour for the first dark hours). Twenty-four-hour cycles of 12–14 hr light, 12–10 hr dark have been successfully used; the 12:12-hr cycle appears to be more readily entrained (Rensing and Hardeland, 1967). The maintenance of stringent environmental conditions requires some mechanical sophistication, such as a specially constructed room containing temperature and humidity servomechanisms (variation should be no more than ±0.3°C, ±2% humidity), even air flow distribution, and a light-regulating timing unit. Such a control chamber is initially expensive; however, close control will result in at least a doubling of productivity (in grams of embryos/hour/fly), which is highly significant in view of labor costs.

The following system was developed by D. S. Hogness, W. J. Peacock, and L. Prestidge at Stanford University Medical School Department of Biochemistry and has been used in our lab for over 20 years (Elgin and Miller, 1978). Embryos are harvested from feeding trays placed in population cages holding 50,000 adults. The adult flies are raised over a 22-day cycle, with cycles overlapping so that when one adult population has become less productive, another is ready to take its place. More closely overlapping cycles may also be used.

Day 1: Inoculation of containers with embryos (population A)

Day 5: Feed larvae (A) at third instar

Day 12: Transfer flies (A) to cages

Day 15: Inoculation of containers with embryos (population B)

Day 19: Feed larvae (B) at third instar

Day 22: Dispose of adults (A)

Day 26: Transfer flies (B) to cages

Egg laying generally peaks on the third or fourth day after transfer of the adults and declines linearly at about 5% per day thereafter. A population of 400,000 (eight cages) will produce 30–120 g embryos/day, depending on the age of the adults and strain used.

A. Feeding of Flies—Harvesting of Embryos

The adult flies are fed yeast food (see recipes below). A single strip of yeast paste is spread on a grape plate, and the whole plate is placed inside the fly cage. Although the grape plate is probably fed on lightly, its predominant purpose is to stimulate egg laying on its surface as a consequence of the attractive aroma and high surface humidity. The amount of yeast to be spread on the plate is determined by the amount of time the plate is to be left in the cage. For 6- to 12-hr collections, we find that a large strip of yeast about 3 cm wide and 0.5 cm deep down the center of the long axis of a grape plate is sufficient. If desirable, the yeast can be omitted for a short-time collection ($\leq$ 2 hr). However, the rate of egg production can be severely compromised if the flies remain unfed for longer periods of time. Also, if short collection times are used, it is advisable to avoid using grape plates directly from the refrigerator. Instead, allow the plates to warm up to room temperature to increase the yield of embryos.

At each collection one unties the curtain and shakes vigorously to get all flies into the cage; grape plates can then be inserted and/or removed. To remove grape plates, pick them up, turn vertically, and bang sharply on the bottom of the cage to knock off the adults. Bring the plate into the curtain tube and blow on the plate to remove any remaining adults. After the plate is removed, shake any adults in the curtain tube back into the cage and retie the curtain. (It should be noted that this technique creates a considerable amount of airborne allergen; people susceptible to hay fever, etc. will frequently become allergic to *Drosophila*. We recommend the use of 3M's 8500 Comfort Mask and gloves for all persons making transfers and collecting grape plates.)

The feeding and collection schedule used will depend on the stage desired. For collections to isolate nuclei, we use a schedule with three collection times a day: 9 a.m., 3 p.m., and 9 p.m. The overnight (9 p.m.–9 a.m.) collection is allowed to mature for 6 hr (9 a.m.–3 p.m.); it is then harvested and stored frozen at $-80°C$. The morning (9 a.m.–3 p.m.) and evening collections (3 p.m.–9 p.m.) are aged for 6 and 12 hr, respectively, to give more precisely staged embryos (6–12 hr and 12–18 hr). If very large amounts of fresh rather than frozen embryos are desired, one can cover and store the grape plates at 4°C until sufficient embryos have been recovered. It has been reported that embryos can be stored for up to 3 days at 4°C (Biggin and Tjian, 1988).

When the embryos are ready to harvest, they are washed off the plates using saline solution (7 g NaCl/liter) and collected in a Teflon-coated pan. A nylon paint brush and a squirt bottle containing the saline solution are useful in this procedure. Any remaining lumps of yeast and embryos are resuspended using the paint brush. For ease in completing this and the subsequent washing steps, the yeast food originally placed on the plate is usually left behind and the embryos are collected by brushing the area not covered by yeast. When all the plates have been cleaned, the saline solution containing embryos and contaminating adults is differentially filtered using a plexiglass filter holder (see Fig. 1). A screen of 630-mesh Nitex, held within the plexiglass lid that fits atop the upper cylinder, is used to trap the adults. The 116-mesh Nitex screen below (Nitex No. 3-160/53) is held in place between the two Plexiglass cylinders. The 116-mesh Nitex is cut to fit the cylinders and trimmed to weigh exactly 1 g to

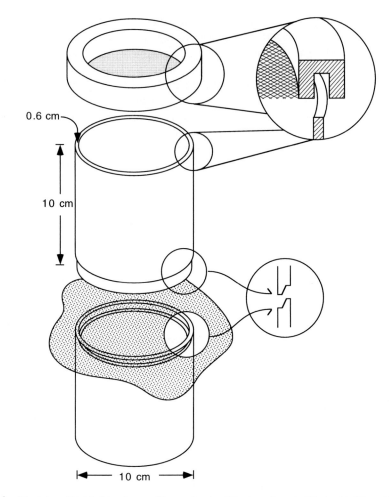

0.6 cm

10 cm

10 cm

Fig. 1 Plexiglass filter holder for washing embryos. The top piece is filled with Nitex H630 to trap adults and designed to fit down inside the top cylinder. The lower filter, Nitex H116, will trap embryos and is held in place by the interlocking cylinders.

assist in determining the weight of the collected embryos. The embryos are washed thoroughly with water at room temperature (or colder) and finally washed with saline solution. After the final wash, the cylinders are separated and the mesh is folded with the embryos inside. The mesh/embryo packet is then placed between several layers of paper towels, and the entire stack is gently pressed to remove wash solution. The packet is then weighed to determine the amount of embryos collected.

The washed embryos can be used immediately for experimental purposes or to continue the population; alternatively, they can be frozen at $-80°C$ for future experiments. If the embryos are to be frozen, they are removed from the Nitex, transferred to foil, wrapped tightly, and frozen in flat packages of no more than 40 g. The frozen embryos appear to be the equivalent of fresh embryos for the preparation of nuclei for several months; subsequently, some desiccation appears to occur and the yield of nuclei/gram embryos decreases. Note that the harvesting of embryos subjects them to anaerobic conditions; if the embryos are spread on a grape plate and allowed to recover, they will show induction of the heat–shock response at about 10% of maximum inducible levels. This can be avoided by freezing the harvested embryos immediately in liquid nitrogen without allowing any opportunity to recover from anaerobic conditions.

If it is important to know the accuracy of the timing in the collection the embryos, a 2-hr collection can be stained (0–2 hr embryos) with Feulgen (method of W. J. Peacock and L. Prestidge, personal communication). Embryos are dechorionated with 50% Clorox, washed in saline, and fixed for 1 hr in 3:1:5 methanol–glacial acetic acid–toluene (two changes). They are then rehydrated by successive equilibrium in 100, 95, 70, 50, 30, and 10% ethanol and water, for a minimum of 5 min in each. The last four solutions contain 0.1% Triton X-100. The DNA in the embryos is then hydrolyzed by immersing the embryos in 1 N HCl plus 0.1% Triton X-100 at 60°C for 8 min. Subsequently, the embryos are washed with water and stained in Feulgen stain plus 0.1% Triton X-100 for 40–60 min. Following staining, early syncytial embryos will be white, blastula embryos will be pink, and cellular embryos will be deep purple, corresponding to the DNA content. In a typical assessment of a 2-hr collection in which 2580 embryos were stained and counted, 2306, or 89.4%, were white (preblastula), 225, or 9.9%, were pink (blastula), and 19, or 0.7%, were purple (cellular).

[Feulgen stain preparation (W. J. Peacock, personal communication). Bring 200 ml distilled water to a boil and pour into a beaker containing 1 g basic Fuchsin. Stir for 5 min. Cool to 50°C and then filter (Whatman No. 1 paper). Add 30 ml 1 N HCl and 3 g $K_2S_2O_5$; stir ca. 1 min. Pour into capped, light-tight flask and let stand ca. 24 hr at room temperature. Add 2 g activated charcoal, stir a few minutes, and filter. Store in refrigerator in dark. The stain should be made fresh every 2 weeks.]

B. Inoculation for Continuing Culture

Inoculation of growth chambers for the next generation of adult flies is carried out as follows. The embryos collected on the Nitex filter (approximately 23 g or scaled up proportionately) are added to a 250-ml graduated cylinder containing 20 ml of a 0.7% NaCl–0.1% Triton X-100 solution. The cylinder is filled to 100 ml, covered tightly, and inverted several times. The embryos are allowed to settle at 1 g for 10 min (to about the 50 ml point). Note that the *Drosophila* embryos are sticky and that labware, pipets, and any other items should be rinsed just before use with a small amount of NaCl–Triton X solution to keep the embryos from sticking. After settling, the top solution is removed with an aspirator or poured off. The cylinder is now filled to 230 ml with 0.7% NaCl–13% sucrose solution and inverted to achieve mixing. Sucrose increases the viscosity of the solution so that it is easier to obtain an even suspension of the embryos. The suspension is poured into a large beaker. The embryos are maintained in even suspension by magnetic stirring; stir just fast enough to keep the embryos suspended in the solution. Two milliliters of embryo suspension is added to a larval container using a wide-mouthed transfer pipet and is distributed evenly over the surface. Finally, a small amount of dry Fleischmann's yeast is sprinkled over the surface and the containers are closed and placed in the environmentally controlled room. The embryo concentration in the solution can be determined by counting the embryos from 0.5-ml aliquots of a 10-fold dilution on black filter paper using a dissecting microscope. The concentration is adjusted to 3000 eggs/milliliter. After a few trial sessions, the counting step may be omitted and a simple weight of embryos per milliliter NaCl solution can be used.

On Day 5 (assuming the inoculation to be Day 1), the larvae are fed a few milliliters of the larval food solution. The larval food solution is placed with a squeeze bottle on the bottom of each container around the edge where the food meets the sidewall of the container. Under proper conditions, the adults emerge on Day 11 or 12, depending on the age of the embryos used for inoculation, and are transferred to the population cages. Generally, 2500–4000 flies (2.5 to 4.0 g depending on strain and rearing conditions) will hatch in each container. It should be noted that these procedures are designed to minimize the human labor input; the flies are crowded in both the larval and adult cages. However, increasing the number of embryos per tub any further results in decreased yield and causes developmental delays of the overcrowded animals.

When being cultured, *Drosophila* can have several competitors, generally mites or mold. To avoid these problems, several precautions are necessary. As an aid in slowing down the growth of mold, propionic acid and tegosept are added to most media. In addition, to help cut down the number of spores as well as guard against mites, all equipment (including adult cages) is washed in 50% Clorox. It is also possible to autoclave many of the media preparations if necessary.

C. Transfer of Flies

The flies are transferred to the population cages as follows. The larval growth container in which the adults have emerged is held on its side (to prevent immobilized adults from becoming stuck in the food) and CO_2 is introduced with tubing through the Nitex screening in the lid. The adults are pooled together in a clean dry tub with a screened lid but without food. Adults are collected from several larval growth containers until approximately 50 g of flies (50,000 flies) have been accumulated. The anesthetized adults are then dispersed on the bottom of the population cage, and a grape plate containing a strip of yeast food is also placed in the cage.

D. Cages

1. Larval Containers

The plastic containers used to grow the larvae have been of several types, but generally are about 1 liter in volume and of approximately equal height and diameter, with a larvae- and fly-tight lid (see Fig. 2). Critical parameters are (1) a configuration such that proper humidity is maintained; (2) sufficient surface area for larvae to crawl up the walls and pupate; and (3) a proper surface such that the food adheres firmly to the bottom of the container and the larvae will crawl up and pupate on the sides of the container. It is most convenient to use disposable plastic containers, minimizing labor costs. Louisiana Plastics 32SRL-N 32-ounce natural plastic containers with LR32P lids work well. The lids must be modified by cutting a 7- to 8-cm hole and filling the hole with a 64-mesh Nitex screen inset (Nitex high capacity No. 3-64/65). The holes are easily made

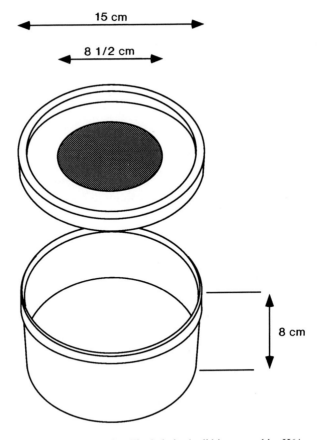

Fig. 2 Plastic larval culture tubs. The hole in the lid is covered by H64 mesh Nitex.

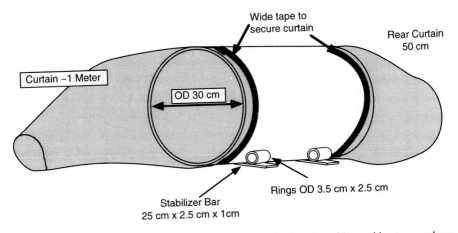

Fig. 3 Population cages. The dacron curtains are held in place by wide masking tape and are loosely knotted to seal the cage.

by using a drill press and the Nitex can be "melted" into the top with an appropriate sealing iron (e.g., Super Sealer, Clamco Corp., Cleveland, Ohio). After each use, the lids are frozen or soaked in 50% Clorox, washed, and reused.

2. Population Cages

The population/embryo collection cages are made from 40-cm sections of 30-cm O.D. (12 inch) standard cast leucite or other acrylic tubing 0.6 cm (1/4 inch) thick (see Fig. 3). This seamless design makes for easy cleaning and low breakage. The cylinder is mounted horizontally on plastic supports. (Note that a flat sheet of plastic can be formed into a tube; this is frequently cheaper.) After curtain end pieces are slipped on the tube, the curtain edges are sealed securely with masking tape. The curtains are dacron, ca. 70 × 70 threads/inch (Dupont). The back curtain can be a single flat piece; the front curtain is a 750-cm to 1-m-long cylinder, closed with a knot.

III. Media Recipes

1. *Cornmeal Media* (22 liters, for 80–100 larval containers, for eight cages with ca. 50,000 flies each)

Water	22,000 ml
Agar	250 g
D(+)-glucose (Sigma No. G-8270)	1313 g
Sucrose (table sugar grade)	655 g

Cornmeal	2273 g
Yeast 500 + vitamins	398 g
Acid Mix A (see below)	253 ml
Tegosept (see below)	150 ml

Add cold water to large kettle or other large pot appropriate for combining all ingredients. Slowly add dry ingredients to cold water, mixing to avoid lumping. Stir constantly, bring mixture to a boil, and cook until mixture is thickened and homogeneous. A motorized stirrer is convenient for mixing. Cool to 60°C. Add Acid Mix A and Tegosept; the final pH should be 4.2–4.3. Dispense into tubs; avoid getting cornmeal media on sides of tub. Allow each container to cool completely before use.

    2. *Tegosept*

            p-hydroxybenzoic acid methyl ester   50 g

The *p*-hydroxybenzoic acid methyl ester (Sigma No. H-5501) is dissolved in 500 ml of 95% ethanol.

    3. *Acid Mix A*

100% propionic acid	418 ml
85% phosphoric acid	41.5 ml

Add the propionic and phosphoric acid to 500 ml water; adjust volume to 1 liter with water.

    4. *Larval Food* (for 80–100 containers)

Water (cold)	604 ml
Propionic acid	1.4 ml
Active dry yeast	270 g

Combine water and acid; slowly add yeast while stirring constantly. A consistency of maple syrup is desirable. Use immediately by squirting a ring of the yeast larval food around the tub where the food meets the sidewall. A hand-held plastic squirt bottle with a short tip is convenient for this procedure.

    5. *Grape Plates* (5 liters, ca. 70 plates)

Water	11,000 ml
Grape juice, 48-oz cans	six cans
Agar	450 g
D(+)-glucose (Sigma No. G-8270)	1160 g
Sucrose (table sugar grade)	580 g
Yeast 500 + vitamins	360 g
Acid Mix A	224 ml
1.25 *N* NaOH	440 ml

Combine cold water and the six cans of grape juice in a large kettle. Slowly add dry ingredients to cold liquids, mixing to avoid lumping. Continue to stir and bring mixture to a boil; simmer for 10 min, scraping the sides of the kettle occasionally. Cool to 60°C. Add Acid Mix A and NaOH solution. Dispense ca. 200 ml of solution into 8 × 10 × $\frac{3}{8}$-inch styrofoam meat-packing trays (e.g., Amoco Stock No. FT8SW; it is important to use the "webbed" foam plates for ease in handling). Allow trays to cool, cover with SaranWrap, and store in the refrigerator in small plastic trash bags.

6. *Yeast Food for Adults*

Active dry yeast	~300 g
Water (cold)	450 ml
Propionic acid	2.72 ml

Combine water and acid; stir constantly while slowly adding yeast and mix thoroughly. Loosely cover the container and store in refrigerator. The container used should be large enough to allow some expansion of the live yeast culture. Note: A consistency of smooth peanut butter is desirable; add more yeast if necessary.

IV. Materials Used

1. Disposable plastic containers and styrofoam meat packing trays are available from local packaging supply houses.

2. Nitex is a nylon monofilament bolting cloth. The number indicates the opening size in micrometers. Available from Tetko, Inc. P.O. Box 346 Lancaster, NY 14086.

3. Yeast 500 + vitamins is available from SPI Nutritionals 3300 Hyland Avenue Costa Mesa, CA 92626.

4. Active dry yeast (Fleischmann's) can be purchased in cases of 12 2-lb bags from any grocery supply company.

5. Agar. We have found that it is not necessary to use a high-quality grade of agar (e.g., Bacto agar). We obtain our agar in 25 kilo lots from Colony Imports 226 7th Street, Suite 103 Garden City, NY 11530.

The list of suppliers and equipment is given for the convenience of others working in the field; no warranty or endorsement is to be construed.

References

Biggin, M. D., and Tjian, R. (1988). Transcription factors that activate the *Ultrabithorax* promoter in developmentally staged extracts. *Cell* **53**, 699–711.

Elgin, S. C. R., and Miller, D. W. (1978). Mass rearing of flies and mass production and harvesting of embryos. *In* "The Genetics and Biology of *Drosophila*" (M. Ashburner and T. R. F. Wright, eds.), Vol. 2a, pp. 112–121. Academic Press: New York.

Rensing, L., and Hardeland, R. (1967). Zur wirkung der Circadianen Rhythmik auf die Entwicklung von *Drosophila*. *J. Insect Physiol.* **13**, 1547–1568.

CHAPTER 6

Isolation and Organ Culture of Imaginal Tissues

Jeanette E. Natzle and Gwendolyn D. Vesenka[1]

Section of Molecular and Cellular Biology
University of California, Davis
Davis, California 95616

[1] Current address: Department of Zoology and Genetics, Iowa State University, Ames, IA 50011.

I. Perspectives and Applications

A. Advantages and Disadvantages of Organ Culture

Many developmental events are accessible via genetic approaches but cannot be experimentally manipulated at the biochemical or molecular level *in vivo*. Organ mass isolation and culture procedures can provide access to sufficient material for biochemical and molecular analyses and allow perturbation of developmental processes with agents such as hormones, growth factors, and inhibitors. Tissues removed directly from an animal and analyzed after short-term culture may be more appropriate for some analyses than are cell lines that have been maintained in culture for many years.

On the other hand, isolated tissues may respond in unpredictable ways to *in vitro* culture. When feasible, it is important to verify that events observed *in vitro* mirror *in vivo* responses. Some aspects of normal organ development (e.g., growth of intact imaginal discs) may not occur in culture because the conditions do not replicate the internal milieu maintained in the hemolymph. It is also prudent to remember that mass-isolated tissues typically represent a variable population with respect to developmental status and physiology, reflecting the fact that isolation procedures start with fairly large quantities of animals that cannot be precisely synchronized.

B. Research Applications

Organ culture of mass-isolated larval and imaginal tissues has been particularly important for analysis of the hormonally regulated events associated with metamorphosis. Detailed studies of the effects of the insect steroid hormone 20-hydroxyecdysone (20-HE) on puffing patterns of larval salivary gland polytene chromosomes cultured *in vitro* provided the basis for models of steroid hormone activation of gene transcription that have served as a paradigm for vertebrate hormone response models (Ashburner *et al.*, 1974). These early studies have recently been extended to the molecular level (Segraves and Hogness, 1990; Guay and Guild, 1991; Karim and Thummel, 1991; Andres *et al.*, 1993). Current techniques for analysis of gene expression in salivary glands are discussed in Chapter 29.

Historically, the first organ culture experiments with imaginal tissues used the adult female abdomen or the metamorphosing larva as the culture dish and medium. These experiments used serial transplantation and culture of intact or fragmented imaginal discs to investigate problems such as transdetermination (reviewed by Hadorn, 1978), competence of discs to undergo metamorphosis, and regulation of growth and pattern formation (reviewed by Bryant, 1978). More recently, short-term *in vitro* cultures of imaginal discs have facilitated analysis of mechanisms of epithelial morphogenesis and regulation of the 20-HE response in imaginal tissues at metamorphosis (DiBello *et al.*, 1991; Apple and Fristrom, 1991; Natzle, 1993; Appel *et al.*, 1993). Imaginal discs retain the ability to undergo morphogenesis (appendage eversion) and differentiation (secretion of pupal cuticle) *in vitro* in response to appropriate concentrations of 20-HE during short-term culture (Fristrom *et al.*, 1973; Milner and Sang, 1974; Eugene *et al.*, 1979). Elaboration of an adult cuticle and differentiation of sensory structures such as bristles and hairs have been observed in small-scale long-term axenic cultures (e.g., Mandaron, 1973; Giangrande *et al.*, 1993; reviewed by Martin and Schneider, 1978).

II. Mass Isolation of Imaginal Discs

A. Larval Culture

Imaginal discs are mass isolated from large-scale cultures of well-fed late third instar larvae. The larval culture scheme presented here produces larvae that span a range of ages, due to the use of a relatively long egg-collection period to obtain sufficient material for mass isolation in a single preparation. For more precise timing, one could use shorter egg collections from a larger initial fly population or hand dissect tissues from carefully timed individuals for small-scale experiments (see Section IIC following and also Chapter 29).

1. Protocol

Eggs are collected on standard *Drosophila* medium in styrofoam meat trays (6 × 11 × 1 inch) with a lengthwise stripe of thin yeast paste (1 vol dry yeast:2 vol water) in the center of each tray. Trays (including yeast stripe) should be aged at least 1 day to dry the food surface. A population cage (12 × 14 × 18 inch) stocked with healthy adults from approximately 40 bottles is provided with feeder trays for at least 1–2 days before initiating egg collections. Change the feeder trays every 12 hr so that females remain well-fed to avoid retention of unlaid eggs. Cover with a dark cloth and keep it in a relatively quiet area to encourage egg laying. Replace the feeder trays with 2–3 trays for two consecutive 12-hr egg collections (Set 1 and Set 2). When trays are removed from the cage, place each one into a Tupperware bread box (5 × 6 × 14 inch, with a polycloth-lined ventilation hole cut in top cover).

Larvae are aged at 25°C and fed with yeast paste according to the schedule in Table I. Thin (1 vol yeast:3 vol water, consistency of light syrup), medium (2 vol yeast:3 vol water, consistency of thin milk shake), or thick (6 vol yeast:7 vol water, consistency of thick milk shake) yeast paste is poured onto the surface of food to cover the brownish areas that consist of concentrations of feeding larvae. The collections in Set 1 should be moved to 18°C approximately 24–36 hr prior to harvest to slow development slightly and improve the age synchrony of the two sets of larvae.

Table I
Schedule for Growth and Harvest of Larvae for Imaginal Disc Isolation

Collection Day	Set 1 Age (hr)	Set 2 Age (hr)	Feed
Saturday 8 p.m.–Sunday 8 a.m.	0–12	—	—
Sunday 8 a.m.–8 p.m.	12–24	0–12	—
Monday 8 a.m.	24–36	12–24	Monday p.m. light yeast (24–36 hr)
Tuesday 8 a.m.	48–60	36–48	Tuesday p.m. medium yeast (48–60 hr)
Wednesday 8 a.m.	72–84	60–72	Wednesday p.m. heavy yeast (72–84 hr)
Thursday 8 a.m.	96–108[a]	84–96	—
Friday 8 a.m.	120–132[a]	108–120	Harvest

[a] Place Set 1 boxes at 18°C to retard development. Developmental age of Set 1 and Set 2 will be more equivalent at the time of harvest.

Larvae are harvested on the morning of the fifth day after collections are done; at this time most of the larvae should be crawling on the walls of the boxes, with a small number (less than 10–15%) having reached pupariation. Larvae are washed from the walls of the boxes and the surface of the food tray with a stream of cool water into a 4-liter beaker. Larvae that are deep in the food and not easily dislodged during this washing should be discarded. Larvae are thoroughly washed by 3–4 cycles of rinsing and gravity settling. Remove heavy food particles by "floating" larvae in a 1-liter beaker by stirring in NaCl or sucrose until larvae float. Cleaned larvae are removed to fresh water. Drain water and measure approximate larval volume (typically 150–250 ml for this collection and culture scheme). Chill larvae on ice until used for tissue preparation.

B. Large-Scale Imaginal Disc Preparation

This procedure has been adapted from the standard disc isolation protocol (Eugene et al., 1979), which was designed for very large quantities of tissue, to use smaller quantities of larval starting material. The reader is referred to the detailed descriptions in the original source for reagent volumes, centrifugation conditions, and other aspects of the protocol designed for very large-scale preparations. In brief, mass isolation involves grinding a suspension of larvae to release imaginal discs from their attachments to the larval epidermis, filtration to remove larval carcasses, gravity settling to separate dense (e.g., imaginal discs) from less dense (e.g., fat body) tissues, density gradient separation of tissues on a Ficoll step gradient, and differential adhesion of disc tissue to glass surfaces (Fig. 1). It is important that buffers, equipment, and samples be kept at 4°C throughout the procedure to minimize tyrosinase activity released during

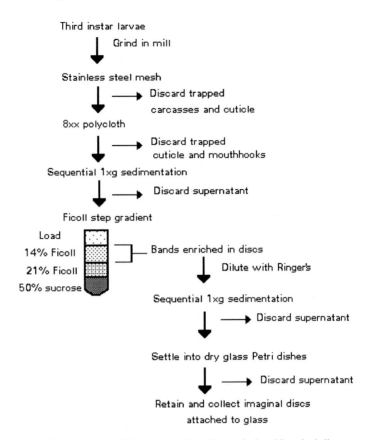

Third instar larvae
Grind in mill

Stainless steel mesh
Discard trapped carcasses and cuticle

8xx polycloth
Discard trapped cuticle and mouthhooks

Sequential 1xg sedimentation
Discard supernatant

Ficoll step gradient
Load
14% Ficoll
21% Ficoll
50% sucrose
Bands enriched in discs
Dilute with Ringer's

Sequential 1xg sedimentation
Discard supernatant

Settle into dry glass Petri dishes
Discard supernatant

Retain and collect imaginal discs attached to glass

Fig. 1 Method for preparation of mass-isolated imaginal discs.

larval grinding, which causes darkening and clumping of tissue. The procedure can be carried out at normal room temperature by prechilling the grinder apparatus at −20°C, packing collection beakers in ice, and using ice-cold reagents throughout.

1. Protocol

Prepare sufficient Ringer's (4–6 liters), O-medium (4–6 liters), and Ficoll gradient solutions (see recipes Section IID) and prechill at 4°C. Prechill grinder blades and feed assembly at −20°C. Presoak all centrifuge tubes and transfer pipets in Ringer's to minimize tissue adherence. Set up a pressurized delivery system for O-medium/Ringer's to be used during tissue grinding and disc preparation (see Section IID).

Larvae suspended in several hundred milliliters of O-medium are passed through a general-purpose grinding mill (Section IID) with rotating plates tightened empirically so that approximately 90% of larvae are broken open. The grindate passes through a stainless-steel mesh sieve (No. 30, 0.6 mm opening), and the flow-through is collected in a 4-liter beaker packed in ice. If the grinder plates are not tight enough, the material collected on the screen can be rinsed back into the grinder for one additional pass. The grinder and the screen material are rinsed with a jet of pressurized O-medium to release any trapped imaginal discs. The larval and imaginal tissues suspended in the beaker are poured through an 8xx polycloth (No. 8xx multifilament polyester cloth; Western Sign Supplies, Inc., Oakland CA 94607) screen into another chilled 4-liter beaker, starting from an elevated position to provide extra force during pouring. Material that collects on the screen is rinsed with a jet of O-medium to free trapped discs and to bring the volume in the beaker to 2 liters.

The imaginal discs are separated from less dense tissue (e.g., fat body, trachea) by a series of gravity settling steps until the supernatant becomes quite clear. (1) Agitate top of liquid to free floating discs and let material settle by gravity on ice for 20 min. Reduce volume to 400 ml by aspiration from the surface. Transfer to 1-liter beaker on ice. (2) Bring volume to 800 ml with O-medium. Settle 10–15 min. Aspirate to 200 ml volume. Transfer to 600-ml beaker. (3) Bring volume to 500 ml with O-medium. Settle 10 min. Aspirate to 100 ml volume. Transfer to 250 ml beaker. (4) Bring volume to 200 ml with O-medium. Settle 10 min. Aspirate to 50 ml. (5) Bring volume to 200 ml with O-medium. Settle 10 min. Aspirate to 30 ml.

The layer of settled tissue, which includes imaginal discs and other larval organs, is fractionated by density on a Ficoll step gradient. Transfer the final settled tissue into two prewet plastic 15-ml centrifuge tubes and spin in a clinical centrifuge for 1 min ($77g$, 800 rpm). Aspirate the supernatant from loose pellet and add 7–8 ml 14% Ficoll gradient solution (see Section IID) to each tube. Mix the contents gently but thoroughly with plastic transfer pipets. Prepare 6–8 Ficoll step gradients (use of more gradients gives cleaner final preparation) as follows in prewet Ultraclear centrifuge tubes ($1 \times 3\frac{1}{2}$ in) or other plastic tubes of similar proportions: 16 ml 50% sucrose pad, 8 ml 21% Ficoll in O-medium, 10 ml 14% Ficoll in O-medium, 3–4 ml tissue suspension. Spin the gradients 8 min in a swinging bucket rotor at $800g$ (e.g., 2250 rpm for Beckman JS13.1). Discard the material at the top of the gradient. Collect the material at the load/14% interface and the 14/21% interface with a plastic transfer pipet into a 250-ml beaker containing Ringer's (see Section IID). These bands will be enriched in discs but will still contain contaminating larval organs (primarily salivary glands, gut, brain tissue, some malphigian tubules). The 21/50% interface and the pellet contain muscle, malphigian tubules, gut, and proventriculus. Carry out several cycles of gravity settling in 250-ml beakers of Ringer's on ice, aspirating the material that does not settle in 10 min.

If further purification is necessary (monitor purity with a dissecting microscope), a final enrichment for imaginal discs is based on preferential adhesion of discs to glass surfaces relative to contaminating tissues. Suspend the discs in 100 ml of Ringer's and pour into four to five dry glass petri dishes. Settle 3–5 min, then gently pour off solution and unattached tissue into a collection beaker. Quickly wash attached imaginal discs into a different beaker using a gentle stream of cold Ringer's from a squeeze bottle. Inspect the collected discs and discarded material and repeat with either fraction if the discs are not sufficiently pure or if the discarded material still contains a significant proportion of imaginal discs. The yield of discs can be estimated by centrifugation of collected discs in a prewet graduated conical plastic centrifuge tube at 55g (700 rpm) for 2 min. An estimate of yield is extrapolated based on 1 ml = 850,000 discs (Eugene *et al.*, 1979). Discs can be used for biochemical preparations at this point or can be cultured *in vitro* as described below (Section III). With this version of the disc preparation protocol, we have typically obtained 100,000–200,000 discs per 100 ml starting larvae. Note that recovery of some disc types such as genital discs is disproportionately low using this protocol and that eye/antennal discs are frequently separated and rarely found as a single unit.

C. Preparation of Hand-Dissected Imaginal Discs

This approach allows more precise assessment of larval age (see Chapter 29) or genetic background, but can provide only limited amounts of material. For immunocytological studies, analysis of morphology by microscopy, and some *in situ* hybridization applications, this material would be sufficient and might be preferable in quality relative to mass-isolated discs. We have been able to isolate sufficient mRNA from as few as 20–25 imaginal discs to allow filter hybridization analysis using sensitive probes, but it is difficult to obtain sufficient material for extensive comparisons of temporal gene expression or varied culture conditions. By using axenically reared larvae (David, 1962), these small-scale cultures could be maintained for a longer time period than is practical for mass-isolated disc cultures.

1. Protocol

Larvae washed well in Ringer's to remove adhering media are severed at approximately one-third of body length using fine forceps (Dumont Biologie No. 5). The anterior third is inverted by supporting the body opening with the forcep tips and using another forceps (or a blunt insect pin) to push the larval mouth parts through the opening to invert this section of the body. Other tissue is dissected away using forceps and/or tungsten needles, and the discs are freed from their connections to the body wall and to the CNS. Discs are gently transferred to a "basket" using a fine glass or plastic capillary pipet, and then they are rinsed, cultured, or otherwise processed. The basket consists of a tubular section cut from a plastic microfuge tube (0.5 or 1.5 ml), which has been heated briefly on one cut end and quickly sealed onto a base of fine-textured Nitex filter. These baskets fit into the wells of 24-well plates and can be easily moved from well to well to facilitate transfer of relatively small numbers of discs between solutions without loss of material or drying that would damage disc structure.

D. Equipment and Solutions

1. Special Equipment for Imaginal Disc Preparation

Grinding Mill. Model No. 4E, Quaker City Mill, Straub Co. (Fisher Cat. No. 08-415). Grinder plates (Fine, No. 4B) must be replaced after a period of use when they no longer efficiently disrupt larvae.

Pressurized container for Ringers and O-medium. Carboy equipped with a filling/venting closure (e.g., Nalgene 2162-0830). Firmly clamp (e.g., worm-drive clamp) inlet opening in cap to air line and clamp a flexible tubing to the outlet opening. Operate the outlet tubing with a pinch clamp. Do not overpressurize the container.

2. Solutions

a. Drosophila Ringer's

130 mM NaCl, 5 mM KCl, 1.5 mM CaCl$_2$, 2 mM Na$_2$HPO$_4$, 0.37 mM KH$_2$PO$_4$. For 6 liters, mix 45 g NaCl, 9.3 ml 3 M KCl, 2.3 ml 4 M CaCl$_2$, 0.3 g Na$_2$HPO$_4$ (anhydrous), and 0.3 g KH$_2$PO$_4$.

b. O-medium

25 mM disodium glycerophosphate, 10 mM KH$_2$PO$_4$, 30 mM KCl, 10 mM MgCl$_2$, 3 mM CaCl$_2$, 162 mM sucrose. For 6 liters, mix 59.3 ml 3 M KCl, 60 ml 1 M KH$_2$PO$_4$, 25 ml 2.4 M MgCl$_2$, 47.3 g α-glycerophosphate, and 330 g sucrose. Adjust pH to 7.2 and volume to 6 liters and add slowly while stirring 4.5 ml 4 M CaCl$_2$ and 6 ml PenStrep (Sigma No. P-0781).

c. Ficoll gradient solutions

A stock of 28% (w/v) Ficoll is stored frozen in aliquots. To make 28% (w/v), mix 50 g Ficoll (Type 400; 400,000 mole wt) with 40 ml O-medium (room temperature). When dissolved, add 40 ml additional O-medium. Store at 4°C overnight until solution clears. Warm the Ficoll solution to room temperature and add 50 ml of room-temperature O-medium. Mix thoroughly but gently by pouring between two beakers. Density of this solution should be 1.12 g/cc. Dilute this stock with O-medium to give desired concentration as needed for gradient preparation.

III. *In Vitro* Culture of Imaginal Discs

A. General Considerations

In vitro culture techniques can be adapted to a variety of different experimental purposes ranging from very small-scale cultures of only a few tissues (Johnson and Milner, 1990) to culture of millions of discs on a large scale (Eugene *et al.*, 1979). In all cases, it is essential to utilize a culture medium that allows physiological or developmental responses that are as close as possible to those of the analogous tissue *in vivo*. When feasible, a chemically defined culture medium (i.e., one that does not contain unknown biochemical mixtures such as serum) allows more precise control of exposure of the tissue to developmental or physiological regulators under study. A rich synthetic medium (Robb's medium; Robb, 1969) supports essentially normal imaginal disc morphogenesis and epithelial differentiation (Fig. 2). For extended culture periods, the beneficial effects of elevated protein concentration in the medium that would be obtained with addition of serum are accomplished by supplementation of the

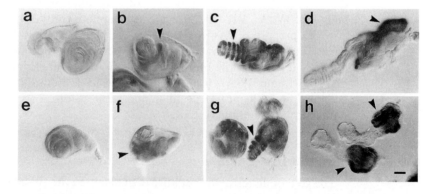

Fig. 2 Development of leg imaginal discs *in vivo* and during *in vitro* culture. Morphogenesis of the disc epithelium is similar for discs that develop *in vivo* (a–d) and for mass-isolated discs that are cultured in Robb's medium with 1 μg/ml (2 μ*M*) 20-HE (e–h). Leg discs that developed *in vivo* were assayed at late third instar (a), white prepupa (b), 3.5-hr pupa (c), and 5.5-hr pupa (d). Cultures of mass-isolated discs were sampled at the time of isolation (e), 4 hr +20-HE (f), 8 hr +20-HE (g), and 18 hr +20-HE (h). The distribution of a 20-HE-inducible transcript, *IMP-E1* (Natzle *et al.*, 1986,1988; Natzle, 1993), was assayed in the same samples by whole-mount *in situ* hybridization to digoxygenin-labeled probes visualized with an alkaline–phosphatase-coupled antibody. The transcript is abundant in the distal portion of the disc during appendage elongation and shifts to the proximal disc region later in both populations of imaginal discs (arrowheads). Cultured imaginal discs are slightly smaller than comparable discs that developed *in vivo* because of the absence of continued growth in the cultured discs following their isolation. Photographs are oriented with distal to the left and proximal to the right. Scale bar, 50 μm.

118

medium with purified bovine serum albumin (BSA Fraction V). This supplement increases the time period over which imaginal discs can be productively maintained in culture without affecting their developmental responses to cues such as addition of 20-HE.

B. Culture Medium

A detailed recipe for preparation of the stock components of Robb's medium can be found in Eugene *et al.* (1979). The components of this medium are summarized here (Table II), with the amounts necessary to make the appropriate stock volume for 1 liter of complete medium included. Larger quantities of each component can be prepared, aliquoted, and stored at −20°C for later use in media preparation. Glass-distilled water should be used throughout. The media stock solutions are assembled as follows to prevent component precipitation during assembly. (1) Add 50 ml 0.1 *N* HCl to 62 mg L-cystine in a small beaker. Heat to boiling to dissolve L-cystine. (2) Add the dissolved L-cystine to 2 mg cholesterol in 1-liter beaker. While the L-cystine solution is still hot, add stock solution II (alternatively, the precipitate that will form in frozen stock II can be dissolved before adding to the cystine/HCl mixture by incubating the stock bottle in a boiling water bath). (3) Add other stock solutions sequentially (Ib, IV, V, VII–XVII) while stirring constantly. (4) If media with BSA is to be prepared, dissolve 10 g BSA in about 200 ml H$_2$O. Filter through nonsterile millipore filter apparatus (0.45 μm). Combine filtered BSA with media mixture. (5) Adjust pH of solution with 1 *N* NaOH (to pH 7.4 with BSA; to pH 7.1 without BSA). (6) Adjust volume to 1 liter. Filter through 0.45-μm millipore filter, using sterile filter apparatus. Store frozen (−20°C) in sterile bottles in 100-ml aliquots.

C. Large–Scale Imaginal Disc Cultures

The goal of imaginal disc culture is to maintain the tissue in a physiologically unstressed state to allow access to large quantities of synchronously developing tissue that can be experimentally manipulated or sampled during the culture period. During imaginal disc culture, it is important to prevent the local build-up of metabolic byproducts and to keep the tissue well-oxygenated. We have found that the following culture parameters typically maintain healthy disc cultures and do not result in the induction of stress-related genes (J. Natzle, unpublished).

1. Protocol

Discs are cultured in 125-ml polycarbonate screw-cap flasks (Nalgene No. 4108-0125) containing 20–25 ml Robb's culture medium or in 50-ml flasks (Nalgene No. 4108-0050) containing 5–10 ml medium. This volume of medium is designed to provide maximum surface area for oxygen exchange. Following isolation, imaginal discs are suspended in a small volume of Robb's medium at a known concentration of discs and distributed among flasks using a plastic transfer pipet marked with a volume calibration. Discs should be cultured at a density less than 5×10^3/ml; lower densities are optimal. Small aliquots of discs are removed for small-scale culture (Section IIID) under standard incubation conditions to monitor the condition of the discs used in the large-scale cultures.

Disc cultures are incubated in a shaking water bath or other temperature-controlled incubator at 25°C with constant gentle shaking to enhance aeration and mixing of the medium. After desired culture period, discs and medium are poured into plastic beakers in which the disc tissue settles to a layer and the medium volume is reduced by aspiration. Morphogenetic response (e.g., degree of disc eversion) is evaluated by microscopic inspection according to the standard eversion scale reported in Chihara et al. (1972). Discs are harvested by transfer to plastic centrifuge tubes for several cycles of gravity settling and washing with cold Ringer's to remove the Robb's medium. Discs are pelleted after the final wash by 3 min centrifugation in a clinical centrifuge ($77g$, 800 rpm). If not used immediately for molecular or immunocytochemical analyses (see Section IV), the disc pellet is quick-frozen in liquid nitrogen and stored at −80°C.

D. Small–Scale Imaginal Disc Cultures

For small quantities of imaginal discs, cultures are carried out in small plastic wells (15 mm diameter) in 1 ml of culture medium. Twenty-four-well tissue culture plates (e.g., Falcon No. 3047) are convenient. These small cultures are routinely used to monitor disc condition and morphogenetic response of aliquots of large-scale preparations. They are also satisfactory for incubation of small quantities of tissue obtained by hand dissection. For small amounts of hand-dissected material, the discs can be left in plastic baskets for culture (see Section IIC). Fewer than 100 discs should be included in each well. It is important to minimize evaporation by enclosing the small cultures in a humidified container.

E. Hormonal Treatment and Media Supplements

1. Minimizing Bacterial and Fungal Contamination

Because the larvae used for mass- isolation of tissues are not grown axenically, it is nearly impossible to avoid some contamination of cultures by bacteria, yeast, or other fungi. The inclusion of penicillin/streptomycin in Robb's medium reduces most bacterial contamination and may be sufficient for short cultures (e.g., less than 12 hr), although we have found the addition of neomycin sulfate (final concentration 10 μM) to be essential for longer cultures. Many antifungal

Jeanette E. Natzle and Gwendolyn D. Vesenka

Table II
Robb's Tissue Culture Medium Components

Stock	Concentration[a]	Vol/liter Robbs[b]	Components	Gm/vol[c] stock	Final concentration[d] mole/liter
Ia	No stock	—	Cholesterol	0.002	Saturated
Ib	50×	20 ml	L-Alanine	0.014	1.6×10^{-4}
			L-Asparagine hydrate	0.030	2.0×10^{-4}
			L-Aspartic acid	0.021	1.6×10^{-4}
			L-Proline	0.037	3.2×10^{-4}
			L-Serine	0.017	1.6×10^{-4}
			Glycine	0.012	1.6×10^{-4}
			β-Alanine	0.009	1.0×10^{-4}
II	25×	40 ml	L-Arginine (free base)	0.348	2.0×10^{-3}
			L-Glutamine	0.585	4.0×10^{-3}
			L-Histidine (free base)	0.062	4.0×10^{-4}
			L-Isoleucine	0.084	6.4×10^{-4}
			L-Leucine	0.084	6.4×10^{-4}
			L-Lysine HCl	0.146	8.0×10^{-4}
			L-Methionine	0.030	2.0×10^{-4}
			L-Phenylalanine	0.066	4.0×10^{-4}
			L-Threonine	0.088	7.4×10^{-4}
			L-Tryptophan	0.016	8.0×10^{-5}
			L-Tyrosine	0.058	3.2×10^{-4}
			L-Valine	0.075	6.4×10^{-4}
III	No stock		L-Cystine HCl	0.062 in 50 ml 0.1 N at 100°C	2.6×10^{-4}
IV	1000×	1 ml	Phenol red	1.14 mg	3.3×10^{-6}
V	50×	20 ml	Fumaric acid, Na salt	0.174	1.5×10^{-3}
			α-Ketoglutaric acid, Na salt	0.278	1.9×10^{-3}
			Malic acid	0.577	4.3×10^{-3}
			Pyruvic acid, Na salt	0.088	8.0×10^{-4}
			Succinic acid, Na salt	1.22	4.5×10^{-3}
VI	(Adjust pH at end to 7.1 or 7.4 with BSA in medium, using 0.1–1 N NaOH)				
VII	40×	25 ml	Biotin (0.01 ml of 0.006%)	6.0 μg	2.5×10^{-8}
			Choline chloride	0.014	1.0×10^{-4}
			Folic acid	0.8 mg	2.0×10^{-6}
			Myo(i)Inositol	0.141	7.8×10^{-4}
			Nicotinamide	0.8 mg	6.6×10^{-6}
			d-Ca pantothenate	0.002	3.3×10^{-6}
			p-aminobenzoic acid	0.8 mg	6.0×10^{-6}
			Pyridoxal HCl	0.9 mg	4.6×10^{-6}
			Riboflavin (0.1 ml of 0.0075%)	75 μg	2.0×10^{-7}
			Thiamine HCl	0.6 mg	2.6×10^{-6}
VIII	10×	100 ml	Glucose	1.8	1.0×10^{-2}
			Sucrose	30.1	8.8×10^{-2}

(continues)

Table II (*continued*)

Stock	Concentration[a]	Vol/liter Robbs[b]	Components	Gm/vol[c] stock	Final concentration[d] mole/liter
IX	100×	10 ml	$FeSO_4 \cdot 7H_2O$	0.7 mg	2.4×10^{-6}
X	20×	50 ml	$MgSO_4 \cdot 7H_2O$	0.296	1.2×10^{-3}
			$MgCl_2 \cdot 6H_2O$	0.244	1.2×10^{-3}
XI	10×	100 ml	$CaCl_2 \cdot 2H_2O$	0.176	1.2×10^{-3}
XII	100×	10 ml	Vitamin B12	108 μg	8.0×10^{-7}
			Putrescine di-HCl (0.01 ml of 0.01%)	98 μg	8.0×10^{-7}
XIII	5×	200 ml	NaCl	1.46	2.5×10^{-2}
			KCl	2.83	3.8×10^{-2}
			$Na_2HPO_4 \cdot 7H_2O$	0.536	2.0×10^{-3}
			KH_2PO_2	0.05	3.7×10^{-4}
XIV	100×	10 ml	Lipoic (Thioctic) acid	0.002	8.0×10^{-6}
			Thymidine	0.6 mg	2.4×10^{-6}
			$CuSO_4 \cdot 5H_2O$	20 μg	8.0×10^{-8}
XV	1000×	1 ml	$ZnSO_4 \cdot 7H_2O$	0.7 mg	2.4×10^{-6}
XVI	100×	10 ml	Penicillin	100 units/ml	
			Streptomycin (prepared stock, Gibco)	100 μg/ml	
XVII	10,000×	0.01 ml	Linoleic acid	67 μg	2.4×10^{-7}
			(Stock is 0.67 mg (75 μl) in 100 ml absolute ethanol; fresh monthly)		

[a] Concentration of the stock solution relative to the final concentrations in Robb's medium.

[b] This volume of stock solution should be added per liter of Robb's medium prepared. These are convenient volumes to prepare as aliquots for storage of components.

[c] This amount is dissolved in the listed volume to make an aliquot of a stock solution. This quantity per single stock solution aliquot will be used to prepare 1 liter of Robb's medium. These amounts and volumes would be increased proportionately to prepare multiple stock aliquots for −20°C storage.

[d] Final molar concentration of components in Robb's medium after additions to make 1 liter.

agents cannot be used without adverse effects on imaginal disc cells. Nystatin (at 2 units/ml final concentration) has proven somewhat effective for reducing some mold contaminants without affecting disc metabolism, but high concentrations are toxic. If it is necessary to carry out long-term cultures (18–24 hr or longer), transferring discs into fresh medium containing necessary supplements halfway through the culture period is recommended to minimize contamination.

2. 20-Hydroxyecdysone Induction

Imaginal discs exposed to appropriate concentrations of the insect steroid hormone 20-HE during *in vitro* culture undergo morphogenesis (eversion into extended appendage) and differentiation (secretion of pupal cuticle). 20-HE is prepared as a 1 mg ml stock in 95% ethanol and concentration is determined by absorbance at 242 nm ($\varepsilon=12,400$; Horn, 1971). Appropriate volumes are added directly to Robb's culture medium during disc culture. An equivalent volume of the 95% ethanol solvent should be added to uninduced control cultures. Imaginal discs prepared by the mass-isolation procedure vary somewhat in age and some discs are no longer hormonally naive, having been exposed to 20-HE *in vivo* prior to initiation of the disc isolation procedure. To allow dissociation of endogenous 20-HE/receptor complexes, a 4- to 6-hr preculture in the absence of 20-HE can be included. It should be noted that the variation in hormonal exposure *in situ* may affect the transcriptional response in cultured tissues *in vitro* in unpredictable ways (Andres *et al.*, 1993; Hurban and Thummel, 1993), and results should be compared to an *in vivo* developmental profile if possible.

Imaginal disc development can be controlled by experimental manipulation of 20-HE levels in culture. For standard analyses of morphogenetic and molecular responses, 20-HE is added to imaginal disc cultures at 1 μg/ml final concentration (approx. 2 μM), approximating recent estimates of the physiological level of 20-HE in larval hemolymph at pupariation (Bainbridge and Bownes, 1988). To induce differentiation (e.g., cuticle secretion) in addition to morphogenesis, it is necessary to apply 20-HE as a pulse, mimicking the natural variations in hormone titer (Fristrom *et al.*, 1982). Pupal cuticle deposition is effectively induced by a 6-hr incubation with 20-HE (1 μg/ml), transfer of the discs into hormone-free medium by three to four gravity settlings in large volumes of fresh Robb's, and continued incubation in fresh medium without 20-HE for 12 hr (Fristrom *et al.*, 1982; Apple and Fristrom, 1991). A second hormonal pulse at lower concentrations (0.1 μg/ml between 14 and 18 hr) is necessary to induce expression of the complete set of pupal cuticle protein genes (Doctor *et al.*, 1985). In contrast, epithelial morphogenesis and the associated cellular and transcriptional changes can be induced in the continuous presence of 20-HE at concentrations ranging from 0.1 to 10 μg/ml (Fristrom and Yund, 1975; Natzle, 1993), affording the opportunity to utilize culture conditions that effectively separate the molecular events associated with morphogenesis from some of those associated with epidermal differentiation.

3. Analysis of Primary Hormone Responses Using Protein Synthesis Inhibitors

Induction of transcription following hormonal stimulus in the absence of protein synthesis has been taken as one criterion for distinguishing between a direct (primary) hormone response and an indirect (secondary) response (Ashburner, 1974; Thummel *et al.*, 1990; Natzle, 1993). In addition, the response to 20-HE in the absence of protein synthesis also distinguishes between two classes of primary response loci: (1) those regulatory loci such as *E74* and *E75* that are superinduced due to lack of synthesis of a hormone-dependent negative regulator (Ashburner *et al.*, 1974; Thummel *et al.*, 1990) and (2) nonregulatory loci such as *IMP-E1* that are induced to equivalent levels in the presence and absence of protein synthesis inhibitors (Natzle, 1993). The most commonly used protein synthesis inhibitor for these studies has been cycloheximide; inhibition of amino acid incorporation is 95–99% complete at 10 μg/ml, yet inhibition of protein synthesis can be reversed (recovery to 60–90% uninhibited levels) by several washes to place the cultured tissue into fresh, drug-free medium (Ashburner, 1974; Natzle, 1993). Toxicity is always a problem during extended incubations with protein synthesis inhibitors and must be considered when planning experiments. In addition, some discrepancies between induction patterns *in vivo* and results obtained after culture plus cycloheximide and 20-HE *in vitro* have been observed (Hurban and Thummel, 1993).

Protein synthesis can be periodically monitored in large-scale cultures by addition of radiolabeled amino acids to 1-ml aliquots of cultured tissue (e.g., 30 min labeling with 50 μCi [^{3}H]leucine, 163 Ci/mmol). Labeled tissue aliquots are pelleted in 1.5-ml microfuge tubes and washed several times with Ringer's. Tissue pellets are resuspended in 200 μl homogenization buffer (100 mM NaCl, 10 mM Tris, pH 7.5, 1 mM EDTA, 1% SDS), heated 5 min to 50°C, and dispersed well with a vortex mixer and homogenization with a Kontes pellet pestle. Insoluble material is pelleted and duplicate 90-μl aliquots of supernatant are diluted into 900 μl H$_2$O and TCA-precipitated 15 min on ice by the addition of 0.5 ml 50% TCA. Precipitates are trapped on GF/A filters, washed extensively with cold 5% TCA, rinsed with cold 95% ethanol, and counted in a liquid scintillation counter.

IV. Analysis of Biosynthetic Activity

A. Protein Synthesis

To radiolabel proteins in organ culture to high specific activity, it is necessary to prepare culture media missing the amino acid(s) corresponding to the desired radiolabeled amino acid(s) (e.g., Robb's minus leucine or methionine). Typical conditions successfully used for metabolic labeling of protein components in imaginal disc cultures include the following: pupal cuticle proteins (Doctor *et al.*, 1985)—10 μCi/ml of ^{3}H amino acid mixture (e.g., ICN No. 20063, 194 mCi/mg); total proteins and membrane proteins (Rickoll and Fristrom, 1983); and extracellular matrix (Birr *et al.*, 1990)—10 μCi/ml[^{35}S]methionine. Protocols for isolation and electrophoretic analysis of pupal cuticle proteins, membrane proteins, and extracellular matrix components of metabolically labeled imaginal discs are described in the respective references listed.

B. RNA Synthesis

Mass-isolated organ cultures facilitate isolation and analysis of relatively large quantities of tissue-specific mRNA. A variety of RNA isolation protocols would produce good quality RNA from mass-isolated imaginal discs, which can be handled much like a comparable pellet of cultured cells. For large-scale cultures, we have had good success with a modification (Natzle *et al.,* 1988) of the Chirgwin protocol (Chirgwin *et al.,* 1979) in which RNA from tissue dissolved in guanidinium thiocyanate is pelleted through a CsCl cushion. With this protocol, it is important to check each lot of guanidinium thiocyanate for RNA-degrading activity prior to use. For small and moderate cultures we use a simpler phenol extraction protocol. Pelleted tissue is homogenized in a 1 : 1 emulsion of phenol : isolation buffer (20 mM Hepes, pH 7.5, 100 mM NaCl, 10 mM EDTA, 0.5% SDS treated with 0.1% diethylpyrocarbonate and autoclaved) at 65°C using a Kontes pellet pestle in a 1.5-ml microfuge tube for very small samples and a Dounce homogenizer for larger samples. The sample is mixed extensively on a vortex mixer after adding chloroform ($\frac{1}{2}$ aqueous volume), and the aqueous phase is recovered following brief centrifugation. The aqueous phase is reextracted sequentially with an equal volume of phenol/chloroform (1 : 1) and then with chloroform. RNA is selectively precipitated by chilling the sample at −20°C for at least 3 hr after adding an equal volume of DEP-treated 8 M LiCl, 5 mM EDTA (Sambrook *et al.,* 1989). The RNA pellet (10,000 g, 30 min) is washed with cold 70% ethanol. Total RNA yields using the guanidinium thiocyanate method average 60 μg/10^4 discs (starting with 5×10^4–1×10^6 discs) and slightly higher (140 μg/10^4 discs) starting with smaller amounts of tissue. Yields from the phenol method are approximately 50% of the guanidinium thiocyanate method. It is also possible to fractionate homogenates of imaginal discs to purify specific RNA populations, such as membrane-bound polysomes (Natzle *et al.,* 1986). The RNA distribution in mass-isolated and hand-dissected cultured imaginal discs can also be analyzed by whole-mount *in situ* hybridization as described in detail in other chapters. The distribution of hormone-inducible transcripts in cultured discs parallels that observed for discs that develop *in vivo* (Fig. 2; Vesenka and Natzle, in preparation).

V. Isolation of Other Larval Organs

Other larval tissues can be mass-isolated from late third instar larvae, although few tissue types have been as extensively utilized as imaginal discs for *in vitro* culture. The disc isolation protocol described here is optimized for imaginal disc recovery and results in significant damage to some of the larger tissues such as salivary glands and gut. Andres and Thummel (Chapter 29) report procedures for culture of unfractionated larval tissues and salivary glands. A protocol for isolation and fractionation of larval fat body, malphigian tubules, discs, gut, and salivary glands after organ extrusion using precisely spaced metal rollers has been reported (Wolfner and Kemp, 1988). Alternative protocols for isolation of larval salivary glands (Boyd, 1978) and for isolation of fat body and imaginal discs (Roberts, 1986) have been described. During *in vitro* culture of any of these organs, it is important to consider the ability of the tissue to metabolize any required media additions. For instance, while imaginal discs do not metabolize ecdysteroids (Chihara *et al.*, 1972), other organs such as fat body, malphigian tubules, and gut can do so (Hoffmann, 1986), introducing additional variables into the experimental design.

VI. Summary and Prospects

Mass isolation and culture of organs provides experimental access to large quantities of synchronously developing tissue. This technology has been particularly valuable for analysis of hormone-dependent events during tissue morphogenesis and differentiation of imaginal discs because these processes do not depend on continued cell proliferation during the culture period. *In vitro* disc culture has recently been applied to analysis of development of glial cells of the peripheral nervous system (Giangrande *et al.*, 1993), indicating the potential for application of disc culture for analysis of PNS development. Imaginal disc pattern formation is another research area that could potentially benefit from improvements in organ culture procedures. Establishment of positional information during larval disc development can currently be investigated only by analysis of larval disc development *in vivo* (e.g., Williams *et al.*, 1993), by inference from adult appendage morphology (e.g., Held, 1993), or by the classic techniques of disc transplantation and culture in adult and/or larval hosts (for recent applications see Simcox *et al.*, 1991; Cohen *et al.*, 1993). Growth and proliferation of cells in intact imaginal discs will not occur *in vitro*, although media that allow proliferation of cells at cut edges of discs in culture (Davis and Shearn, 1977) and that permit establishment of short- and long-term imaginal disc cell lines have been devised (Wyss, 1982; Ui *et al.*, 1987; Currie *et al.*, 1988). Development of culture media/conditions that support imaginal disc growth and pattern formation *in vitro* would provide additional experimental access to these intriguing events of disc development prior to metamorphosis.

Acknowledgments

This work was supported by a grant to J.E.N. from the National Science Foundation. The authors gratefully acknowledge the invaluable help of O. Eugene and J. W. Fristrom (UC Berkeley) in providing patient instruction in the intricacies of imaginal disc preparation.

References

Andres, A., Fletcher, J., and Thummel, C. (1993). Molecular analysis of the initiation of insect metamorphosis: A comparative study of *Drosophila* ecdysteroid-regulated transcription. *Dev. Biol.* **160**, 388–404.

Appel, L. F., Prout, M., Abu-Shumays, R., Hammonds, A., Garbe, J. C., Fristrom, D., Fristrom, J. (1993). The *Drosophila* Stubble-stubbloid gene encodes an apparent transmembrane serine protease required for epithelial morphogenesis. *Proc. Natl. Acad. Sci. U.S.A.* **90**, 4937–4941.

Apple, R. T., and Fristrom, J. W. (1991). 20-Hydroxyecdysone is required for and negatively regulates transcription of *Drosophila* pupal cuticle protein genes. *Dev. Biol.* **146,** 569–582.

Ashburner, M. (1974). Sequential gene activation by ecdysone in polytene chromosomes of *Drosophila melanogaster*. II. The effects of inhibitors of protein synthesis. *Dev. Biol.* **39,** 141–157.

Ashburner, M., Chihara, C., Meltzer, P., and Richards, G. (1974). Temporal control of puffing activity in polytene chromosomes. *Cold Spring Harbor Symp. Quant. Biol.* **38,** 655–662.

Bainbridge, S. P., and Bownes, M. C. (1988). Ecdysteroid titers during *Drosophila* metamorphosis. *Insect Biochem.* **18,** 185–197.

Birr, C. A., Fristrom, D., King, D. S., and Fristrom, J. W. (1990). Ecdysone-dependent proteolysis of an apical surface glycoprotein may play a role in imaginal disc morphogenesis in *Drosophila*. *Development* **110,** 239–248.

Boyd, J. B. (1978). Mass isolation of salivary glands from larvae of *D. hydei, In* ''The Genetics and Biology of Drosophila'' (M. Ashburner and T. R. F. Wright, eds.), Vol. 2A, pp. 127–135. New York: Academic Press.

Bryant, P. J. (1978). Pattern formation in imaginal discs. *In* ''The Genetics and Biology of Drosophila'' (M. Ashburner and T. R. F. Wright, eds.). Vol. 2C, pp. 230–336. New York: Academic Press.

Chihara, C., Petri, W., Fristrom, J. W., and King, D. (1972). The assay of ecdysones and juvenile hormones on *Drosophila* imaginal discs *in vitro*. *J. Insect Physiol.* **18,** 1115–1123.

Chirgwin, J. M., Przybyla, A. E., MacDonald, R. J., and Rutter, W. J. (1979). Isolation of biologically active ribonucleic acid from sources enriched in ribonuclease. *Biochemistry* **24,** 5294–5299.

Cohen, B., Simcox, A. A., and Cohen, S. M. (1993). Allocation of the thoracic imaginal primordia in the *Drosophila* embryo. *Development* **117,** 597–608.

Currie, D. A., Milner, M. J., and Evans, C. W. (1988). The growth and differentiation *in vitro* of leg and wing imaginal disc cells from *Drosophila melanogaster*. *Development* **102,** 805–814.

David, J. (1962). A new medium for rearing *Drosophila* in axenic conditions. *Drosophila Inf. Service* **36,** 128.

Davis, K. T., and Shearn, A. (1977). In vitro growth of imaginal disks from *Drosophila melanogaster*. *Science* **196,** 438–440.

DiBello, P. R., Withers, D. A., Bayer, C. A., Fristrom, J. W., and Guild, G. M. (1991). The *Drosophila Broad-Complex* encodes a family of related zinc finger-containing proteins. *Genetics* **129,** 385–397.

Doctor, J., Fristrom, D., and Fristrom, J. W. (1985). The pupal cuticle of *Drosophila:* Biphasic synthesis of pupal cuticle proteins *in vivo* and *in vitro* in response to 20-hydroxyecdysone. *J. Cell Biol.* **101,** 189–200.

Eugene, O. E., Yund, M. A., and Fristrom, J. W. (1979). Preparative isolation and short-term organ culture of imaginal discs of *Drosophila melanogaster*. *Tissue Cult. Assoc. Manual* **5,** 1055–1062.

Fristrom, J. W., Logan, W. R., and Murphy, C. (1973). The synthetic and minimal culture requirements for evagination of imaginal discs of *Drosophila melanogaster in vitro*. *Dev. Biol.* **33,** 441–456.

Fristrom, J. W., and Yund, M. A. (1975). Uptake and binding of β-ecdysone in imaginal discs of *Drosophila melanogaster*. *Dev. Genet.* **10,** 198–209.

Fristrom, J. W., Doctor, J., Fristrom, D., Logan, W. R., and Silvert, D. J. (1982). The formation of the pupal cuticle by *Drosophila* imaginal discs *in vitro*. *Dev. Biol.* **91,** 337–350.

Giangrande, A., Murray, M. A., and Palka, J. (1993). Development and organization of glial cells in the peripheral nervous system of *Drosophila melanogaster*. *Development* **117,** 895–904.

Guay, P. S., and Guild, G. M. (1991). The ecdysone-induced puffing cascade in *Drosophila* salivary glands, a *Broad-complex* early gene regulates intermolt and late gene transcription. *Genetics* **129,** 169–175.

Hadorn, E. (1978). Transdetermination. *In* ''The Genetics and Biology of Drosophila'' (M. Ashburner and T. R. F. Wright, eds.), Vol. 2C, pp. 556–617. New York: Academic Press.

Held, L. I. (1993). Segment-polarity mutations cause stripes of defects along a leg segment in *Drosophila*. *Dev. Biol.* **157**, 240–250.

Hoffmann, J. A. (1986). Ten years of ecdysone workshops: Retrospect and perspectives. *Insect Biochem.* **16**, 1–9.

Horn, D. H. S. (1971). The ecdysones. *In* "Naturally Occurring Insecticides" (M. Jacobson and D. G. Crosby, eds.), pp. 333–459. New York: Dekker.

Hurban, P., and Thummel, C. (1993). Isolation and characterization of fifteen ecdysone-inducible *Drosophila* genes reveal unexpected complexities in ecdysone regulation. *Mol. Cell Biol.* **13**, 7101–7111.

Johnson, S. A., and Milner, M. J. (1990). Cuticle secretion in *Drosophila* wing imaginal discs *in vitro:* Parameters of exposure to 20-hydroxyecdysone. *Int. J. Dev. Biol.* **34**, 299–307.

Karim, F. D., and Thummel, C. S. (1991). Ecdysone coordinates the timing and amounts of E74A and E74B transcription in *Drosophila*. *Genes Dev.* **5**, 1067–1079.

Mandaron, P. (1973). Effects of α-ecdysone, β-ecdysone and inokosterone on the *in vitro* evagination of *Drosophila* leg discs and the subsequent differentiation of imaginal integumentary structures. *Dev. Biol.* **22**, 298–320.

Martin, P., and Schneider, I. (1978). *Drosophila* organ culture. *In* "The Genetics and Biology of Drosophila" (M. Ashburner and T. R. F. Wright, eds.). Vol. 2A, pp. 266–316. New York: Academic Press.

Milner, M., and Sang, J. H. (1974). Relative activities of alpha-ecdysone and beta-ecdysone for the differentiation *in vitro* of *Drosophila melanogaster* imaginal discs. *Cell* **3**, 141–143.

Natzle, J. E. (1993). Temporal regulation of *Drosophila* imaginal disc morphogenesis: A hierarchy of primary and secondary 20-hydroxyecdysone-responsive loci. *Dev. Biol.* **155**, 516–532.

Natzle, J. E., Hammonds, A. S., and Fristrom, J. W. (1986). Isolation of genes active during hormone-induced morphogenesis in *Drosophila* imaginal discs. *J. Biol. Chem.* **261**, 5575–5583.

Natzle, J. E., Fristrom, D. K., and Fristrom, J. W. (1988). Genes expressed during imaginal disc morphogenesis: *IMP-E1*, a gene associated with epithelial cell rearrangement. *Dev. Biol.* **129**, 428–438.

Rickoll, W. L., and Fristrom, J. W. (1983). The effects of 20-hydroxyecdysone on the metabolic labeling of membrane proteins in *Drosophila* imaginal discs. *Dev. Biol.* **95**, 275–287.

Robb, J. (1969). Maintenance of imaginal discs of *Drosophila melanogaster* in chemically defined media. *J. Cell Biol.* **41**, 876–884.

Roberts, D. B. (1986). Basic *Drosophila* care and techniques. *In* "*Drosophila*, a Practical Approach" (D. B. Roberts, ed.), pp. 1–38. Oxford: IRL Press Limited.

Sambrook, J., Fritsch, E. F., and Maniatis, T. (1989). *In* "Molecular Cloning: A Laboratory Manual", 2nd ed. Cold Spring Harbor, NY: Cold Spring Harbor Laboratory Press.

Segraves, W., and Hogness, D. S. (1990). The E75 ecdysone-inducible gene responsible for the 75B early puff in *Drosophila* encodes two new members of the steroid receptor superfamily. *Genes Dev.* **4**, 204–219.

Simcox, A. A., Hersperger, E., Shearn, A., Whittle, J. R. S., and Cohen, S. M. (1991). Establishment of imaginal discs and histoblast nests in *Drosophila*. *Mech. Dev.* **34**, 11–20.

Thummel, C. S., Burtis, K. C., and Hogness, D. S. (1990). Spatial and temporal patterns of E74 transcription during *Drosophila* development. *Cell* **61**, 101–111.

Ui, K., Ueda, R., and Miyake, T. (1987). Cell lines from imaginal discs of *Drosophila melanogaster*. *In Vitro Cell. Dev. Biol.* **23**, 707–711.

Williams, J. A., Paddock, S. W., and Carroll, S. B. (1993). Pattern formation in a secondary field: A hierarchy of regulatory genes subdivides the developing *Drosophila* wing disc into discrete subregions. *Development* **117**, 571–584.

Wolfner, M. F., and Kemp, D. J. (1988). A method for mass-isolating ecdysterone-inducible tissues of *D. melanogaster*. *Drosophila Inf. Serv.* **67**, 104–106.

Wyss, C. (1982). Ecdysterone, insulin, and fly extract needed for the proliferation of normal *Drosophila* cells in defined medium. *Exp. Cell Res.* **139**, 297–307.

CHAPTER 7

Mass Isolation of Fly Tissues

Anthony P. Mahowald

Department of Molecular Genetics and Cell Biology
University of Chicago
Chicago, Illinois 60637

I. Introduction

Recent advances in the identification of genes responsible for specific cellular processes in *Drosophila* have heightened the need for the biochemical characterization of their products *in vivo*. Although individual proteins can be obtained in biochemical quantities through recombinant DNA technology, the analysis of function in the cell requires the ability to isolate native cellular structures and protein complexes. The small size of *Drosophila*, relative to organs and tissues of many vertebrates, continues to be a pervading obstacle in the use of

Drosophila. However, excellent tissue culture lines are available, and, as I will point out in this chapter, techniques for isolation of biochemical quantities of *Drosophila* cells have been developed. As others join the effort, additional methodologies are certain to be devised.

One of the principal limitations in the use of *Drosophila* has been the difficulty in obtaining sufficient quantities of cellular material from an organism that weighs only 1.5 mg and an egg weighing approximately 0.01 mg. Although many protocols are available for mass rearing of flies, the methods described here were developed because of the need for having large quantities of embryos. Using this protocol, we have typically produced 300 gm of viable eggs daily and a kilogram of adult flies weekly.

II. Method for Rearing Flies on Liquid Medium

Most procedures for mass rearing of flies presuppose the availability of a media preparation room with autoclaves, food pumps, etc. To maintain population cages of flies producing many grams of eggs and other stages of *Drosophila* development, we developed a protocol that eliminated the requirement for the typical cooked fly food. This approach (Travaglini and Tartof, 1972) utilized a liquid yeast broth, poured onto an absorbant cotton matrix, on which 1 ml of eggs were pipetted. By using dead yeast, populations of flies could be maintained with sufficiently low levels of contaminating yeast so that sterile cell cultures could be readily established later.

A. Population Cages

A single 4 × 8-ft (0.25 inch thickness) sheet of plexiglas is sufficient to produce three cages 18 × 18 × 12 inches in size. A 12 × 12-inch opening covered with Nytex netting on the top and an 8-inch circular hole on one side to which a sleeve is attached for entry are the only special adaptations required. An inbred strain of Oregon R flies (Oregon R-P2), originally developed by Jack Schultz at the Fox Chase Institute for Cancer Research, has become especially useful in cages because this strain readily deposits its eggs as they mature. Flies raised from 3 ml of eggs (see below) are seeded into each cage and are fed daily (or more frequently).

B. Food for Growth of Larvae to Pupal Stage

A liquid broth is prepared by mixing 180 g of brewer's (e.g., Red Star) yeast and 90 g of commercial sugar to 630 ml of acid solution (0.06% H_3PO_4, 0.4% propionic acid). Three-hundred milliliters of the broth is decanted into a 9 × 9 × 3-inch plastic box whose bottom has been covered with a sheet of absorbant cotton. One milliliter of embryos collected overnight from the prior week's cages are pipetted onto the wet cotton. We routinely suspend embryos (with chorions) in 70% ethanol in 15 ml graduated centrifuge tubes for pipetting because embryos settle reproducibly in alcohol. We notice no loss of viability by using this step. The cover of the plastic box has a 3-inch square hole covered on the inside surface with Nitex mesh. These boxes should be incubated at a set temperature in a well-ventilated room to ensure eclosion at a predetermined time. One day prior to eclosion, the cotton pads are removed from the boxes and placed in the population cages (on top of absorbent paper to control moisture). The flies will eclose in the cage.

C. Feeding Population Cages

For the first 18 hr after eclosion, very little food is required because flies do not begin to feed appreciably until their larval fat body is consumed. However, before the beginning of Day 2, three 4×8-inch plastic meat trays with a thick layer of 2% agar (colored with molasses) are covered with dead yeast paste (recipe below) and placed in the cages. The trays have to be replaced daily with a copious supply of yeast food. After most of the flies have emerged from the pupal cases, the cotton is given a few firm shakes and removed through the sleeve, and the food trays are placed on the floor of the cage. If the walls and floor of the cage are not allowed to get moist, all of the eggs will be laid on the agar trays, allowing easy harvesting of embryos.

The Oregon R-P2 strain lays eggs optimally only during the first week. Because of the simplicity of this rearing operation, we routinely regenerate the population cages every week (with flies eclosing Friday afternoon and reaching their peak laying period on Monday). Adult flies are routinely collected either by transporting the cages to a 4°C room or by using CO_2 to anesthetize the flies. From 10 cages, containing nearly a million flies, 200 to 300 g of embryos can be collected in the morning, and with 2-hr staged collections, 20 to 50 g of specifically aged embryos can be readily harvested.

D. Composition of Dead Yeast Paste

Approximately 100 g of dried yeast is mixed with a 1-lb cake of brewer's yeast and the mixture is placed in a microwaveable dish and microwaved on high for 3 min. Stir until completely mixed. If the mixture is too thin, add more dry yeast. Cook 3.5 min more on high. Refrigerate mixture after cooling. The texture of the yeast mixture should allow the yeast to be readily spread on the agar egg-laying trays.

III. Methods for Isolating Specific Cells

The availability of large quantities of embryos has led to the development of techniques to isolate specific cell types and to study their development in primary cultures. Specific cell types have been isolated in relative purity by

routine cell-separation procedures (isolation of myoblasts is described in Fyrberg *et al.*, 1977; the use of affinity purification is described in Chapter 8). By
using dead yeast paste during rearing and sterile procedures during cell separation, sterile cultures can be established for all of these cell types.

A. Collection of Embryos

Embryos are washed off the trays with regular tap water and collected through
a series of steel meshes designed to separate dead flies, wings, legs, and other
debris from the embryos (we use 841- , 420- , and 125-μm meshes, which can
be obtained from any major scientific supply house). After the embryos are
washed with tap water several times, they are transferred to a Buchner funnel
lined with a 70-μm Nitex mesh, attached to house vacuum or an aspirator, and
washed again with cold 0.02% Triton X-100 followed by cold water and then
cold 70% ethanol. After the ethanol is drawn off, bleach freshly diluted 1/1
with 95% ethanol is added for 2.5 min. The bleach is removed by suction and
the dechorionated embryos are washed repeatedly with cold 70% ethanol. The
embryos are then transferred to a beaker by inverting the mesh and washing
the embryos into the beaker with a stream of cold 70% ethanol. After the
embryos have settled, fresh cold ethanol is added and drawn off and 70%
ethanol is added again. These washes remove any contaminating structures not
removed by the original straining through wire meshes.

All of the ethanol is then removed and replaced with 0.1% benzalkonium
chloride in 70% ethanol and allowed to stand for 15 min for sterilization. During
this time the beaker should be transferred to a laminar flow hood and the
embryos transferred to 50-ml conical tubes. Decant sterilizing solution and
wash embryos three times in cold Chan and Gehring buffer (see Section VI).

B. Isolation of Pole Cells (cf. Allis *et al.*, 1977, 1979; Fig. 1)

(Our early work used embryos derived from population cages fed with live
yeast, which resulted in yeast contamination of the pole cell cultures. By using
dead yeast paste for food in the population cages, sterile cultures can be obtained.)

Pole cells (precursors to primordial germ cells) form precociously at approximately 1.5 hr of embryonic development (at 25°C), whereas the remainder of
the embryo does not cellularize until 3.5 hr. Thus, embryos that are 2.5 ±
1 hr old should have only pole cells. However, since a post-gastrula-stage
embryo will contribute more than 20 times the number of cells found in the
younger embryos, a lack of synchrony in the initial embryo collections will
make subsequent isolation of pole cells difficult. If the embryo collections are
carefully done, the following protocol will produce significantly enriched pole
cell cultures.

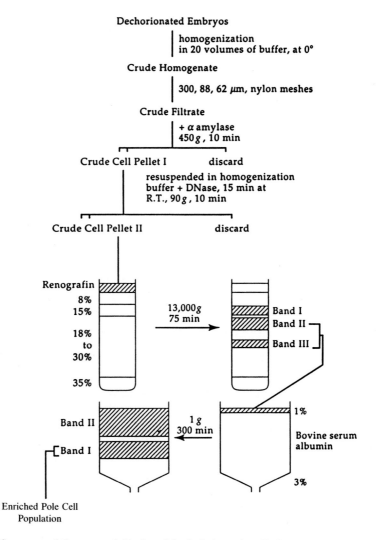

Fig. 1 Summary of the protocol developed for isolating pole cells from precellular blastoderm stage embryos (adapted from Allis *et al.*, 1977, and Ashburner, 1989). The use of α-amylase to prevent clumping in the initial homogenate is assumed to be due to the excessive amount of glycogen in the early embryo. This addition is not critical. The use of DNase I, however, is essential because free DNA forms a gel-like mass in the Renografin solutions. The final critical step is the slow dilution of the Renografin fractions with Chan and Gehring buffer. We use a slow rotary shaker to agitate the receiving flask, which is immersed in ice.

1. Five milliliters of embryos are resuspended by stirring and transferred to a 40-ml sterile Dounce, prechilled on ice. Approximately 5 vol of cold sterile Chan and Gehring buffered saline is added and embryos are carefully homogenized with at least 10 strokes of a loose-fitting (Type A) Dounce pestle at 4°C. The homogenate is filtered through a sterile 35-μm Nitex mesh into a sterile beaker, 10 μg of a α-amylase is added to filtrate (to reduce aggregation due to glycogen), and a crude cell pellet is obtained by centrifugation at 450g for 10 min. The supernatant is drawn off and the pellet resuspended in 40 ml of Chan and Gehring saline and incubated at 25°C for 15 min with 850 units/ml of DNase I (any inexpensive grade is fine). After centrifugation, the pellet is gently resuspended in Chan and Gehring saline and recentrifuged. This last step is repeated twice to remove debris.

2. Pole cells can be fractionated on isopycnic Renografin-60 (Squibb) gradients. All solutions are made up in Chan and Gehring buffer to which is added 850 units/ml of DNase I. (The DNase addition is required to break up the DNA released from blastema nuclei or broken cells.) The cell pellet is resuspended in 2 ml of buffer and layered over freshly prepared gradients containing 5 ml of 35% Renografin, 25 ml of a 30–18% linear Renografin gradient, 5 ml of 15% Renografin, and 5 ml of 8% Renografin. The interface between the cells and the 8% Renografin is lightly mixed, and the tubes are spun in a swinging bucket rotor for 75 min at 13,000g at 4°C. Following centrifugation, fractions are collected from the top and their refractive index is measured.

3. Cells are recovered from the hypertonic Renografin fractions by slowly diluting the fractions with 10 vol of Chan and Gehring buffer during a 45- to 60-min period at 4°C, while continuously agitating the cells on a rotary shaker. (This step is critical: If dilution is too rapid, cells will lyse because of the rapid change in tonicity.) After dilution, cells are collected by centrifugation. Pole cells concentrate at a density of 1.120 mg/ml. They can be distinguished from other embryonic cells by their low content of neutral lipid droplets, detected by using either dark-field microscopy of living cells or oil red O staining of cell spreads fixed with formaldehyde vapor.

4. Depending upon how many post-gastrula-stage embryos are present in the initial embryo collection, the final pole cell concentration can represent between 20 and 70% of the cells. Further enrichment can be obtained by velocity sedimentation, using the staput method of Lam et al. (1970) (cf. Allis et al., 1977).

Pole cells are found at a greater density in the Renografin gradient because of their relatively smaller size compared to most post-gastrula cells, resulting in a greater density. The contaminating cells with similar density can be separated from pole cells by sedimentation velocity due to the fact that pole cells lack the abundant neutral lipid droplets characteristic of other embryonic *Drosophila* cells and consequently settle more rapidly. By using embryos grown on sterile yeast, sterile solutions, and sterile procedures, pole cell cultures free of contaminating yeast can be established. Pole cells in culture will complete their final round of DNA synthesis and then arrest, thus repeating what occurs in the intact embryos; cultured pole cells continue both RNA and protein syntheses (Allis *et al.*, 1979).

C. Isolation of Embryonic Neuroblasts

Neuroblasts are a class of distinctly larger cells in 4.5- to 6.5-hr embryos. Centrifugal elutriation is especially powerful for separating cells that differ in total mass because of larger cell volume (e.g., Furst *et al.*, 1981). We typically use embryos aged to 4.5 ± 0.5 hr (gastrula stage) to prepare neuroblast cultures. Cells are collected according to the procedure described under Section IIIA, except that neither amylase nor DNase I have to be added to embryos after gastrulation. Up to 10^9 cells in a total of 20 vol of Chan and Gehring medium can be used in the procedure utilizing the elutriator rotor (Furst and Mahowald, 1985a,b).

1. Preparation of Elutriator Rotor

After setting up the rotor according to the manufacturer's directions, a sterile loading chamber is attached between the pressure gauge and the inlet tube of the separation chamber (see Fig. 2). While bypassing the loading chamber, the system is filled initially with water to check for leaks and remove bubbles. The centrifuge is set at 12°C (0 compression) at 2600 rpm with the brake on and the timer on hold. Start the rotor while watching the pressure gauge. Purge system of air bubbles by pinching and releasing the outlet tubing and increasing and decreasing the pressure (but never more than 10 psi). This should jar all bubbles free. The fluid pressure should be as low as 1–2 psi. After bubbles are removed and the system is stable with the pressure as low as possible, carefully add 50 ml of 50% H_2O_2 to the water (to make a 5% solution), and run the pump and rotor for 10–15 min to sterilize the system.

At this point, begin working in a laminar flow hood. Turn off the pump and rotor. Place the uptake tube into a sterile flask with cold water. Then attach the outlet tube to a sterile filling bell connected to a 250- to 500-ml flask. The system is then restarted and the peroxide flushed out with water, using about 1 liter of ice cold sterile water during a 20- to 30-min period. After the system is rinsed, the rotor and pump off are turned off and, after both openings are flamed, the intake tube is attached to a new 2-liter flask containing 1 to 1.5 liters of Chan and Gehring medium. The rotor and pump are turned back on and the system is filled with buffer. At this point, the loading chamber is connected to the system and filled with buffer. Again, all bubbles have to be shaken loose, and the valves isolating the loading chamber are then closed. The chamber and large flask are placed on ice. The system is now ready for the cell suspension.

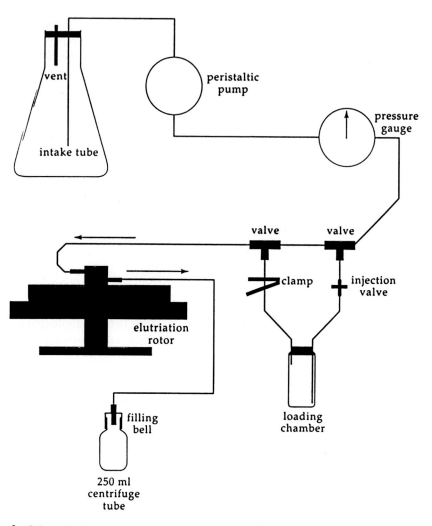

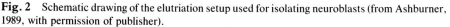

Fig. 2 Schematic drawing of the elutriation setup used for isolating neuroblasts (from Ashburner, 1989, with permission of publisher).

2. Elutriation of Gastrula-Stage Cells

a. Loading of Elutriator Rotor

Approximately 10^9 cells in 20 ml of Chan and Gehring medium are drawn into a 30-cc syringe through a 22-gauge needle (the shear will break up clumps of cells, but if done too forcefully cells will also be broken). The rotor is started and the pump set for its initial speed (about 12.5 ml/min). The scissor clamp is removed from the outlet line from the loading chamber and the cell suspension is injected into the loading chamber through the injection valve. After loading, the injection valve is closed and the syringe removed. Now the flow to the loading chamber is begun by opening the valve, and the cells are loaded into the rotor. The filling bell is placed on the first of the 250-ml centrifuge tubes and 200 ml is collected (keep this and subsequent collection tubes on ice). For the first 125 ml, rock loading chamber to prevent cells from clumping and allow full delivery to the elutriation chamber.

b. Collection of Cells

After the first fraction is collected, the loading chamber is bypassed and five additional fractions are collected while increasing the pump speed by 2.5 ml/min for each fraction (i.e., 200 ml is collected with the pump set for 15 ml/min and 125 ml at speeds of 17.5, 20.0, 22.5, and 25.0 ml/min). When transferring the collecting bell to a new tube, the bell must be flamed to maintain sterility. Neural cells will be in the last three fractions.

3. Culture of Cells

Cells are pelleted in a swinging bucket at $750g$ for 10 min at 10°C. The cells are resuspended in modified Schneider's medium (MSM) (see Section VI), counted, and then repelleted before being resuspended in MSM at a density of 10^5 cells/ml. The cells are incubated in 35-mm petri dishes in a humidified incubator at 26°C. Neural cells adhere well to glass and contaminating cells can be removed by washing dishes 2 hr after plating.

4. General Observations

We have succeeded in obtaining 5×10^7 to 10^8 neuroblasts from a 10^9 starting cell population. Although the initial pooled fractions are enriched in neuroblasts (approximately 50% neuroblasts), further enrichment can be achieved by removing loosely adhering cells after the first 2 hr of culture or by reelutriating the combined fractions. Neuroblasts will complete their normal cycle of cell divisions in culture (Furst and Mahowald, 1985a,b). If individual neuroblasts are plated in dispersed culture, each cell will form a small clone of neurons and glia (Fredieu and Mahowald, 1989) that display in culture a normal pattern of cell differentiation (cf. Huff et al., 1989). Sufficient quantities of RNA and protein can be obtained from these cultures to identify neuroblast specific transcripts (cf. Perkins et al., 1990; Noll et al., 1993).

IV. Isolation of Specific Stages of Oogenesis

A. Introduction

Recent advances have increased the importance of oogenesis for the study of a variety of cellular and developmental questions. Detailed analyses of actin-mediated contractile systems, tyrosine kinase phosphorylation cascades, RNA subcellular localization, and DNA replication patterns have been carried out with genetic and molecular approaches (cf. Spradling, 1993, for a review). Because of mass rearing techniques, it is possible to acquire biochemical amounts of specific stages of oogenesis for biochemical and cellular studies. Except for the procedures utilizing dissection to obtain relatively pure fractions, no detailed analyses to determine whether the isolation procedures have caused injury to the tissues have been carried out on isolated ovarian stages. Hence, I will first describe the relatively easy dissections that provide sufficient quantities of late stages for many biochemical and molecular uses. I will then give two methods for isolating mature eggs from whole flies by homogenization approaches. Large quantities of mature eggs can readily be isolated, but it is relatively difficult to accumulate the early stages in sufficient quantities for use in biochemical studies.

B. Isolation of Late Stages of Oogenesis by Dissection

Oogenesis in *Drosophila melanogaster* is not synchronous as it is in the muscoid flies, mosquitoes, and many other diptera. Consequently, there is no simple procedure to obtain large quantities of specific stages of oogenesis. However, by careful control of feeding cycles and mating, a relatively synchronous wave of oogenesis can be achieved so that large quantities of specific stages can be obtained.

1. Collection of Virgin Females

With the procedures described under Section II for mass rearing of flies, it is relatively easy to collect thousands of virgin flies due to the simple fact that females emerge slightly earlier than males. Small portions of the cotton pads from boxes seeded earlier with embryos are placed into bottles, and flies are collected on the first day of eclosion. At first, nearly all of the flies will be females, thus providing copious supplies of unmated flies.

2. Feeding Schedule to Obtain Specific Stages of Oogenesis

Virgin females will feed but not lay eggs until at least 4 days old unless mated (Wyman, 1979). Hence, virgin flies are fed with a rich yeast paste in bottles for 36 hr after eclosion and then transferred to unyeasted bottles containing ordinary fly food. Without additional yeast, the flies will produce two or more mature oocytes per ovariole, resulting in approximately 100 mature eggs per fly. It is relatively easy to dissect more than a thousand mature oocytes from these flies in less than an hour. If these flies are now mated and placed on yeasted food, they will rapidly lay these held eggs, and a new wave of vitellogenic stages will begin in the ovaries. By selecting appropriate times after mating and refeeding, it is possible to enrich for vitellogenic stages. These can be collected by dissection or by sieving methods (see below).

3. Efficient Method for Obtaining Late Stages of Oogenesis by Dissection (Mahowald, 1980)

Rapidly laying flies have extended abdomens because of the presence of many vitellogenic stages in each ovariole. These late stages can be readily obtained by immersing anesthetized flies in saline and puncturing the posterior abdomen with watchmakers forceps to break open the posterior portion of the ovary. The regular contractions of the ovarian sheath result in a steady emission of oocytes through the hole in the abdomen. Each oocyte from the ovariole is released separately from the preceding oocyte, thus eliminating the need to separate stages of oogenesis. This procedure works very well for any stage from Stage 10 or later. By selecting the time after refeeding, the number of eggs at the desired stage of development can be increased. The final purification of stages is done manually under the dissecting microscope. In practice, we have isolated between 200 and 500 oocytes of stages 10 to 14) in 2 hr (cf. Spradling and Mahowald, 1979).

C. Isolation of Mature Oocytes by Fractionation

Two approaches, both of which provide gram quantities of mature eggs, have been developed.

1. Fractionation with a Laboratory Mill (Petri et al., 1978; cf. Ashburner, 1989)

Well-fed flies are collected by cooling to 0–4°C; they are then immersed in 5 vol of saline and ground in a laboratory mill. The brei is collected and passed through the mill a second time and then collected in a 2-liter beaker in which it is rapidly stirred to liberate follicles from oocytes. The mixture is passed through a 400-μm mesh to trap large pieces, and the filtrate is collected in a 4-liter beaker and allowed to settle. Ovarian follicles settle faster than most contaminants, most of which can be withdrawn by suction. This procedure is then repeated in a tall 1-liter Berzelius beaker in which the follicles are allowed to settle for 5 min before the supernatant is drawn off. This procedure is again repeated, this time allowing only 3 min for the follicles to settle before drawing off the upper fluid. This step is repeated until the supernatant is clear. This procedure can provide as many as 10^5 mixed age follicles from 10^4 females.

Mature eggs can be readily separated from this mixed follicle preparation by allowing the mature eggs to stick to clean glass surfaces. The preparation is poured on acetone-cleaned large petri dishes and allowed to settle for 0.5 min; the supernatant can be poured onto another clean petri dish, and the mature eggs allowed to adhere again. After 0.5 min, the nonadhering follicles can be washed away with a soft stream of saline, and subsequently the mature eggs can be recovered with a strong stream. Mature eggs can be collected on a 150-μm mesh on a filter flask connected to a vacuum. $3–6 \times 10^4$ mature eggs can be rapidly collected by this procedure.

2. Preparation of Mature Eggs by Homogenization with a Waring Blender
(Mahowald *et al.*, 1983)

Well-fed, 4- to 6-day old virgin flies are transferred to 5 vol of cold Chan and
Gehring medium in a precooled Waring blender. The flies are homogenized by
10-sec pulses of the blender with a rheostat setting of 30 V. At this speed, the
abdomens are broken open and the ovaries disrupted, but few eggs are broken.
(If a slight red color develops, the rheostat setting is too high, resulting in the
breakage of pigment cells in the eyes.) The homogenate is filtered through a
297-μm mesh, and the fraction retained on the mesh is run through the blender
a second time. After filtering again, the process is repeated twice more, each
time pooling the material that flows through the mesh. Mature eggs are allowed
to settle from the pooled filtrate and the supernatant drawn off. A further
purification can be achieved by the method of sticking to glass, mentioned
under Section IV.1 above.

V. Concluding Observations

Although *Drosophila* adults are relatively small, the potential to rear them
in large quantities and to obtain relatively synchronous cultures makes many
biochemical procedures possible. We have developed techniques to isolate
specific cell types from embryos, and others have developed analogous protocols to isolate myoblasts (Fyrberg *et al.*,1977). Similarly, protocols have been
developed to obtain gram quantities of pure preparations of salivary glands and
fat body (Zweidler and Cohen, 1971) and imaginal discs (Eugene and Fristrom,
1978). Because of the ability to rear flies sterilely, it should be possible to
extend both the cell types and tissues available in biochemical quantities for
detailed analysis.

VI. Solutions

Chan and Gehring Saline
NaCl	55 mM
KCl	40 mM
CaCl$_2$	5 mM
Sucrose	67.5 mM
Glucose	40 mM
Tricine	10 mM
BSA	0.1%
MnCl$_2$	2 mM
pH of 6.95	

Modified Schneider's Medium

We used Seecof and Donady's (1972) modification of Schneider's *Drosophila* medium, except that the insulin, which does not appear to affect the development of neuroblasts, was omitted.

Acknowledgments

I am especially appreciative of the many colleagues who have worked with me over the years and who have helped in developing and perfecting these protocols. I am thankful to Elizabeth Travaglini and Leonard Cohen, in whose laboratory we developed the protocol for mass rearing of flies; to Deborah Holland, for developing the yeast paste mixture; to David Allis and Gail Waring, for developing methods for isolating pole cells; to Allan Furst, for the elutriation procedure for isolating neuroblasts; and to John Fredieu and Randy Huff for their detailed analysis of the function of neuroblasts in culture. I am grateful for the critical comments on this chapter by John Fredieu. My work has been supported by Grants HE-17607 and HD-17608 from the NIH.

References

Allis, C. D., Waring, G. L., and Mahowald, A. P. (1977). Mass isolation of pole cells from *Drosophila melanogaster*. *Dev. Biol.* **56,** 372–381.

Allis, C. D., Underwood, E. M., Caulton, J. H., and Mahowald, A. P. (1979). Pole cells of *Drosophila melanogaster* in culture: Normal metabolism, ultrastructure and functional capabilities. *Dev. Biol.* **69,** 451–465.

Asburner, M. (1989). "*Drosophila, A Laboratory Manual.*" Cold Spring Harbor, NY: Cold Spring Habor Laboratory Press.

Eugene, O. M., and Fristrom, J. W. (1978). The mass isolation of imaginal discs. *In* "The Genetics and Biology of *Drosophila* (M. Ashburner and T. R. F. Wright, Eds.), Vol. 2a, pp. 121–126. London: Academic Press.

Fredieu, J., and Mahowald, A. P. (1989). Glial interactions with neurons during neurogenesis in the *Drosophila* embryo. *Development* **106,** 739–748.

Furst, A., and Mahowald, A. P. (1985a). Differentiation of primary embryonic neuroblasts and purified neural cell cultures from *Drosophila*. *Dev. Biol.* **109,** 184–192.

Furst, A., and Mahowald, A. P. (1985b). Cell division cycle of cultured neural precursor cells from *Drosophila*. *Dev. Biol.* **112,** 467–476.

Furst, A., Brown, E. H., Braunstein, J. D., and Schildkraut, C. L. (1981). Alpha-globin sequences are located in a region of early-replicating DNA in murine erythroleukemia cells. *Proc. Natl. Acad. Sci. U.S.A.* **78,** 1023–1027.

Fyrberg, E., Donady, J. J., and Bernstein, S. (1977). Isolation of myoblasts from primary mass cultures of embryonic *Drosophila* cells. *Tissue Culture Assoc. Manual* **3,** 689–690.

Huff, R., Furst, A., and Mahowald, A. P. (1989). Drosophila embryonic neuroblasts in culture: Autonomous differentiation of specific neurotransmitters. *Dev. Biol.* **134,** 146–157.

Lam, D. M. K., Furrer, R., and Bruce, W. R. (1970). The separation, physical characterization and differentiation kinetics of spermatogonial cells of the mouse. *Proc. Natl. Acad. Sci. U.S.A.* **65,** 192–197.

Mahowald, A. P. (1980). Improved method for dissecting late ovarian stages. *Drosophila Information Services* **55,** 157.

Mahowald, A. P., Goralski, T. J., and Caulton, J. H. (1993). *In vitro* activation of *Drosophila* eggs. *Dev. Biol.* **98,** 437–445.

Noll, E., Perkins, L. A., Mahowald, A. P., and Perrimon, N. (1993). Approaches to identify genes involved in *Drosophila* CNS development. *J. Neurosci.* **24,** 701–722.

Perkins, L. A., Doctor, J. S., Zhang, K., Stinson, L., Perrimon, N., and Craig, E. A. (1990). Molecular and developmental characterization of the *heat shock cognate 4* gene of *Drosophila melanogaster*. *Mol. Cell. Biol.* **10,** 3232–3238.

Petri, W. H., Wyman, A. R., and Kafatos, F. C. (1978). Mass isolation of *Drosophila* follicles and eggshells. *In* "The Genetics and Biology of Drosophila (M. Ashburner and T. R. F. Wright, eds.), Vol. 2a, pp. 136–140. London: Academic Press.

Seecof, R. L., and Donady, J. J. (1972). Factors affecting Drosophila neuron and myocyte differentiation in vitro. *Mech. Aging Dev.* **1,** 165–174.

Spradling, A. C. (1993). Developmental genetics of oogenesis. *In* "The Development of *Drosophila melanogaster*" (M. Bate and A. Martinez Arias, eds.), Vol. 1, pp. 1–70. Cold Spring Harbor, New York: Cold Spring Harbor Laboratory Press.

Spradling, A. C., and Mahowald, A. P. (1979). Identification and genetic localization of messenger RNAs from the ovarian follicle cells of *Drosophila melanogaster*. *Cell* **16,** 589–598.

Travaglini, E. C., and Tartof, D. (1972). "Instant" *Drosophila:* A method for mass culturing large numbers of Drosophila. *Drosophila Information Service* **48,** 157.

Wyman, R. (1979). The temporal stability of the *Drosophila* oocyte. *J Embryol. Exp. Morphol.* **50,** 137–144.

Zweidler, A., and Cohen, L. H. (1971). Large-scale isolation and fractionation of organs of *Drosophila melanogaster* larvae. *J. Cell Biol.* **51,** 240–248.

CHAPTER 8

Preparation and Analysis of Pure Cell Populations from *Drosophila*

Susan Cumberledge[1] and Mark A. Krasnow

Department of Biochemistry
Stanford University School of Medicine
Stanford, California 94305

I. Introduction

As the genetic analysis of development and cell function in *Drosophila mela-nogaster* has burgeoned over the last 15 years, so has our ability to distinguish various cell types in developing tissues, using molecular cell markers that have become available mostly through gene cloning. As our understanding of development and cell function *in vivo* becomes more sophisticated, it is increasingly important to isolate the various cell types so that they can be more fully analyzed and manipulated in various ways. This allows one to test the emerging models of the underlying cellular and molecular processes and to characterize these processes biochemically and discover new components.

[1] Present address: Department of Biochemistry and Molecular Biology, University of Massachusetts, Amherst, MA 01003.

What has been needed is a convenient, reliable way to purify large quantities of different cell types from *Drosophila*.

A wealth of knowledge has emerged from studies of purified cells and continuous cell lines from vertebrates, with the mammalian immune system perhaps the most dramatic example (Parks *et al.*, 1989). In contrast, there have been only a few serious attempts to isolate and study pure populations of *Drosophila* cells. Mahowald and his colleagues have shown that highly enriched populations of pole cells (germ-line precursors) and neuroblasts can be obtained in reasonable quantity from embryos (Allis *et al.*, 1977; Furst and Mahowald, 1985), and other groups (Bernstein *et al.*, 1978; Storti *et al.*, 1978) have described procedures for the isolation of myoblasts (see Mahowald (Chapter 7) and Ashburner (1989a) for reviews). This pioneering work demonstrated the feasibility of cell purification from *Drosophila* embryos, and it showed that purified cells can retain the ability to differentiate appropriately into morphologically distinct cell types. The fractionation schemes relied primarily on differences in general physical characteristics of the cells, such as their size, shape, density, or adhesive properties. For example, pole cells, because they tend to have a low lipid content and are larger than most embryonic cells, can be purified by equilibrium density centrifugation followed by sedimentation velocity centrifugation (Allis *et al.*, 1977). Neuroblasts also tend to be large and can be selectively enriched by centrifugal elutriation and adherence to glass (Furst and Mahowald, 1985). However, most *Drosophila* embryonic cells, at least during early embryogenesis, are rather unexceptional in morphology and hence may not be amenable to purification by methods based solely on such physical characteristics. Methods for purifying these cells must rely on other properties of the cells, such as expression of cell type-specific molecular markers.

Surface markers have been widely used in mammalian systems to isolate specific cell types, particularly cells of the immune system (Parks *et al.*, 1989). Antibodies that recognize specific cell surface antigens are commonly employed in the purification by using the antibodies to fluorescently label the cells followed by flow cytometry/fluorescence-activated cell sorting (FACS) or by coupling the antibodies to a solid phase and selectively resorbing the cells of interest ("panning") (Wysocki and Sato, 1978). These techniques have not been applied to *Drosophila,* at least in part because few antibodies to cell type-specific surface antigens have been available until recently. However, in *Drosophila,* many intracellular markers are known, perhaps the most important of which is the *Escherichia coli lacZ* (β-galactosidase) gene, which is not normally present but is easily introduced by P-element-mediated transformation. Thousands of different strains expressing *lacZ* under control of various cell- and tissue-specific promoters and regulatory elements have been constructed, many by random insertion of a *lacZ* transposon such that *lacZ* expression comes under the control of an endogenous enhancer or regulatory element ("enhancer trap") (O'Kane and Gehring, 1987; Bier *et al.*, 1989; Bellen *et al.*, 1989). We have established a method, called whole animal cell sorting (WACS), for purifying the β-galactosidase expressing cells from such transgenic strains by FACS

(Krasnow *et al.*, 1991). The key technical innovation that opened the way to this approach was the development of a viable, fluorogenic β-galactosidase substrate (fluorescein di-β-D-galactopyranoside) that was shown to be effective in the analysis and purification of cultured mammalian cells engineered to express β-galactosidase (Nolan *et al.*, 1988; Fiering *et al.*, 1991).

The general scheme for WACS is as follows (Fig. 1). (1) Embryos carrying a *lacZ* transgene expressed in a specific cell type are grown to the desired developmental stage. (2) Cells of the developing embryos are dissociated and stained with FDG and then stained with a viable cell stain and a dead cell stain.

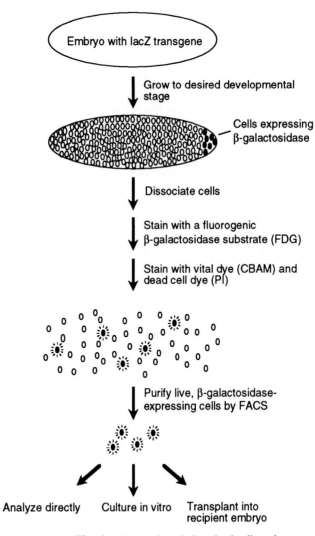

Fig. 1 Scheme for whole-animal cell sorting.

(3) Live cells expressing β-galactosidase are purified away from dead cells and nonexpressing cells by FACS. (4) The purified cells can then be analyzed directly, cultured *in vitro,* or reintroduced into a recipient embryo. The scheme requires only basic embryo handling and cell culture techniques, except for cell sorting, which is usually carried out in conjunction with a trained FACS technician. Once there is a large, healthy fly population with the appropriate *lacZ* expression pattern, the cells of interest are obtained in 2–3 hr from the time of embryo harvesting and cell dissociation.

This scheme has been used to obtain reasonable quantities ($\sim 10^5$ cells) of various types of *Drosophila* embryonic cells and to study their development *in vitro.* The same approach should be applicable to any cell type provided an appropriate *lacZ* strain exists and the cells can be efficiently dissociated and stained without disrupting their viability and development. Neuronal precursor cells, and cells expressing the segmentation genes *fushi tarazu, engrailed,* and *wingless* have all been purified by this approach (Krasnow *et al.,* 1991; Cumberledge and Krasnow, 1993). The cell populations obtained are completely viable and highly pure (typically 90% or greater), and they can continue their development *in vitro.* For example, the purified neuronal precursor cells were shown to divide and differentiate into neurons at high efficiency in culture (Krasnow *et al.,* 1991). We have also used the purified cells to study the dynamics of *engrailed* expression in isolated cells (Cumberledge and Krasnow, 1993), to reconstruct and analyze developmental signaling events between *engrailed*-expressing cells and the neighboring *wingless*-expressing cells (Cumberledge and Krasnow, 1993), and to investigate the homophilic and heterophilic adhesion properties of *engrailed*-expressing cells and *wingless*-expressing cells (S. Cumberledge and M. Krasnow, unpublished results). There are many potential applications of the purified cells that have not yet been explored, such as transplantation of the cells back into living animals, biochemical studies of cellular processes and purification of cell constituents, and the generation of cell type-specific cDNA libraries and monoclonal antibodies. While sufficient numbers of cells can be obtained by the current procedures for most cell-based analyses, a major challenge in the future will be to find ways to increase purification speed and yields so that preparative quantities of purified cells can be easily obtained at reasonable cost (see Section IV).

II. Purifying Embryonic Cells by Fluorescence-Activated Cell Sorting

A. Equipment and Reagents

1. Flow Cytometer/FACS Instrument

We have used a modified Becton Dickinson FACStar Plus flow cytometer, equipped with two argon–ion lasers. Dual laser flow cytometry, data collection, and multiparameter analysis are performed essentially as described by Parks *et al.* (1986,1989). One argon–ion laser (488 nm, 400 mW output) is used to generate four signals: forward light scatter, large angle light scatter, fluorescein (detected through a 530/30-nm bandpass filter), and propidium iodide (detected through a 575/26-nm bandpass filter). A second argon–ion laser was used as an ultraviolet light source (351–363 nm, 50 mW) to excite calcein blue, whose emission was detected through a 405/20-nm filter. Data collection and multi-parameter analysis are carried out on a Digital VAX computer system using the FACS/DESK software (Moore and Kautz, 1986). For applications in which the highest degree of cell purity and viability are not required, calcein blue staining can be omitted and a single laser flow cytometer (488 nm excitation) used for cell isolation.

2. Fluorescent Dyes and β-Galactosidase Substrates

The structures of the compounds described below are shown in Fig 2. These structures, as well as some of the information provided below, are from a very useful catalog and handbook provided by Molecular Probes, Inc. (Haugland, 1992). Additional information about the compounds including literature references can be found in Haugland (1992) and in materials provided by the supplier with the reagents.

a. Fluorescein di-β-ᴅ-Galactopyranoside (FDG) (Fig. 2A)

Prepare a solution of 2 mM FDG (Molecular Probes, No. F-1179) in dH$_2$O. Mix the solution vigorously and heat it to 37°C because 2 mM is close to the solubility limit. (It is also possible to prepare 2 mM FDG by diluting a 200-mM FDG stock solution prepared in dH$_2$O/dimethylsulfoxide (1:1), but the effects of the residual dimethylsulfoxide on *Drosophila* cells are unknown.) Store small aliquots of 2 mM FDG at −20°C in the dark, where it is stable for months. After thawing an aliquot, heat it to 37°C for 10 min to ensure complete solubilization of the FDG. FDG is less stable in powder form than in solution, so prepare the stock solution as soon as the powder is received. The working solution should have a faint yellow color. FDG is not fluorescent, but it is converted into free fluorescein (via fluorescein mono-β-ᴅ-galactopyranoside) by β-galactosidase. Background fluorescence present in some FDG preparations, due to contaminants, can be eliminated by irradiating the working stock with the 488-nm argon laser beam of the flow cytometer for 1 min. Use protective eye wear when working near an exposed laser beam. For fluorescein, λ_{abs} = ~490 nm and λ_{em} = ~514 nm.

b. Phenylethyl β-ᴅ-Thiogalactoside (PETG) (Fig. 2B)

Prepare a stock solution of 50 mM PETG (Molecular Probes, No. P-1692) in dH$_2$O; store at 4°C, where it is stable for months. PETG is a reversible inhibitor of β-galactosidase, used to stop the β-galactosidase reaction. It readily enters cells, even at 4°C.

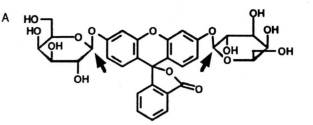

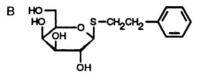

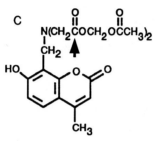

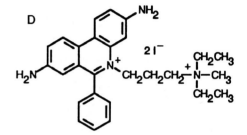

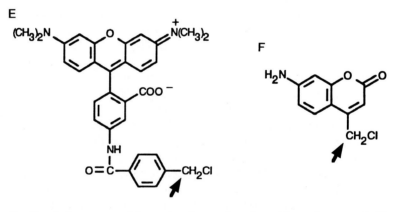

Fig. 2 Chemical structures of fluorescent dyes and β-galactosidase substrates. (A) Fluorescein di-β-D-galactopyranoside (FDG). The arrows show the positions of attack by H_2O in the hydrolysis reactions catalyzed by β-galactosidase, which liberates a free galactose group and generates a phenolic group on the dye. (B) Phenylethyl β-D-thiogalactoside (PETG). (C) Calcein blue acetoxy-methyl ester (CBAM). The arrow shows the position of attack by H_2O in the hydrolysis reaction catalyzed by cellular esterases, which generates a carboxylate group on the dye. (D) Propidium iodide (PI). (E) 5- (and- 6-)-(((4-chloromethyl)benzoyl)amino)tetramethylrhodamine (CMTMR). The arrow shows the position of attack by intracellular thiol groups, which displaces the chloride group and forms a thioester linkage to the dye. (F) 7-Amino-4-chloromethylcoumarin (CMAC). The arrow shows the position of attack by intracellular thiol groups as in E.

c. Calcein Blue Acetoxymethyl Ester (CBAM) (Fig. 2C)

Prepare a 10-mM stock solution of CBAM (Molecular Probes, No. C-1429) in anhydrous dimethylsulfoxide; store small aliquots in the dark at $-20°C$. CBAM is a fluorescent coumarin derivative used to label live cells. It readily enters cells, and once inside it is converted by cellular esterases into calcein blue, a fluorescent molecule that is retained within live cells with intact membranes but lost from dead cells. For calcein blue, the longest-wavelength absorption maximum (λ_{abs}) is ~355 nm, and the wavelength of maximal fluorescence emission (λ_{em}) is ~440 nm.

d. Propidium Iodide (PI) (Fig. 2D)

Prepare a stock solution of PI at 0.5 mg/ml (Molecular Probes, No. P-1304) in dH$_2$O; store at 4°C, where it is stable for months. PI is excluded from live cells, but it readily enters dead cells (and cells with compromised membranes), where it binds DNA and fluorescently stains the nucleus. PI is a mutagen, so use caution when handling. For PI, $\lambda_{abs} = $ ~536 nm and $\lambda_{em} = $ ~617 nm.

e. 5- (and 6-)-(((4-Chloromethyl)benzoyl)amino)tetramethylrhodamine (CMTMR) (Fig. 2E)

Prepare a stock solution of 80 μM CMTMR (Cell Tracker Orange, Molecular Probes, No. C-2927) in dH$_2$O. Store in the dark at $-20°C$ in small, single-use aliquots. CMTMR and CMAC (see below) are used for stable marking of the purified cells (Section IIIC). They are membrane-permeant fluorescent dyes with a chloromethyl reactive group. They readily enter cells and are stably retained, probably by conjugation of the chloromethyl moiety to intracellular thiols (Zhang et al., 1992). For CMTMR, $\lambda_{abs} = $ ~541 nm and $\lambda_{em} = $ ~565 nm.

f. 7-Amino-4-chloromethylcoumarin (CMAC) (Fig. 2F)

Prepare a stock solution of 80 μM CMAC (Cell Tracker Blue, Molecular Probes, No. A2110) in dH$_2$O; store at $-20°C$ in small, single-use aliquots. For CMAC, $\lambda_{abs} = $ ~354 nm and $\lambda_{em} = $ ~469 nm.

3. Media and Buffers

a. S2 Medium

Schneider's Drosophila medium (Gibco, No. 350-1720AJ; Ashburner, 1989b) supplemented with 11% fetal bovine serum (heat treated at 56°C for 30 min), penicillin at 50 units/ml, and streptomycin sulfate at 50 μg/ml.

b. SPC Medium

Schneider's Drosophila medium + 10 μg/ml PI + 10 μM CBAM, both added just before use.

c. SPP Medium

Schneider's *Drosophila* medium + 10 μg/ml PI + 1 mM PETG, both added just before use.

d. Phosphate-buffered saline (PBS)

137 mM NaCl, 2.7 mM KCl, 10.1 mM Na$_2$HPO$_4$, 1.8 mM KH$_2$PO$_4$; pH 7.4.

e. PBSB

PBS + 0.5% bovine serum albumin.

B. Methods

1. Embryo Preparation

The embryo preparation and cell dissociation protocol is a modification of the method of Furst and Mahowald (Furst and Mahowald, 1985; Ashburner, 1989b). Refer to Ashburner (1989a,b) for related protocols and for general information on embryo preparation and cell culture. Most of our studies have used germ band elongation stage embryos (3–5 hr after egg lay at 25°C), as described below. Except for pole cells, it may not be possible to obtain intact, viable cells from embryos much younger than 3 hr because cellularization is not complete. We have also used somewhat older embryos (up to 7 hr after egg lay) successfully, but we have not attempted purification from embryos older than this or from animals at later stages of development, which may require additional methods, such as treating the cells with proteases or calcium-chelating agents such as EGTA to facilitate their dissociation.

Transgenic flies of the appropriate *lacZ* strain are maintained at 25°C in a large population cage (~50 gm, or ~5 × 10^4, adult flies in an ~12-liter cage) and fed daily with trays of standard yeast/glucose/agar medium seeded with live yeast paste (see Elgin and Miller (1978) or Roberts (1986) for basic techniques for maintaining large fly populations). These conditions are usually adequate for obtaining ~10^3 or more embryos in a 2-hr collection; the size of the population should be adjusted according to the egg laying rate of the strain and the number of embryos that are needed for the purification.

On the day of the collection, prefeed the flies for 1 hr or more with a fresh yeast tray. This helps reduce the number of "held" (overly mature) eggs obtained in the collection.

Replace the prefeeding tray with a fresh yeast tray to begin egg collection. Use heat-killed yeast (e.g., treated for 60 min at 80°C) to decrease the likelihood of yeast contamination if the cells are to be cultured after purification. Remove the plate after 2 hr and allow the embryos to age for 3 hr at 25°C. The embryos obtained are 3 to 5 hr old; to obtain embryos at other stages, adjust the aging period accordingly.

Rinse the embryos from the food tray with a solution of 0.7% NaCl in a squirt bottle and a soft-hair paintbrush. Filter the embryos through a ~440-μm mesh Nitex sieve to remove any adult flies and large debris, and collect the embryos on a ~110-μm mesh Nitex sieve.

Wash embryos extensively with 0.7% NaCl to remove the remaining yeast. If cells are to be cultured after FACS, all subsequent steps are performed under sterile conditions and in a tissue culture hood whenever possible.

Transfer embryos from the sieve to a small wire mesh basket using a spatula or paintbrush or by rinsing them in with a squirt bottle. The basket can be constructed by gluing or melting a 1-inch section of plastic tube (cut from a 15-ml polypropylene tube) onto a wire mesh screen (Thomas Scientific No. 3435-B95). Immerse the embryos in a solution of 2.5% sodium hypochlorite for 2 min, swirling gently, to dechorionate the embryos.

Wash the embryos four times with 0.7% NaCl and then two times with Schneider's *Drosophila* medium, to remove all traces of sodium hypochlorite. Drain most of the liquid and then transfer the embryos to a 15-ml (17 × 120 mm) polypropylene screw-cap tube with printed graduations (Sarstedt No. 62.554.002). Bring the total volume of embryos and medium to ~10 ml.

2. Embryo Disruption and Cell Dissociation

All subsequent steps (except where indicated) are carried out at 4°C to prevent further development of the embryos.

Allow the embryos to settle in the tube, then aspirate off the supernatant. Resuspend the embryos in ~20 times their settled volume of S2 medium (e.g., 10 ml of S2 medium for 0.5 ml of settled embryos), using a Pasteur pipet. The final concentration of embryos can affect the efficiency of cell dissociation and the amount of cell lysis that occurs during douncing; for other developmental stages the optimal concentration may have to be determined empirically.

Transfer ~5-ml aliquots of the resuspended embryos to a 7-ml glass dounce tissue homogenizer. Disrupt the embryos and dissociate the cells using seven slow strokes of a type A (loose) pestle. Cell yields are usually about one-third the estimated number of embryo cells; the rest are probably destroyed during douncing. Transfer the cell suspension to a 15-ml polypropylene tube.

Wash the cells by pelleting them in a tabletop centrifuge (5 min, 500 *g*). Carefully decant the supernatant and gently resuspend the cells in the residual medium. Cells isolated from early embryos can be fragile, so cells should be resuspended gently, but thoroughly, by briskly tapping or flicking the tube. Add 20 vol of S2 medium to the resuspended cells (e.g., 10 ml of medium for 0.5 ml of settled embryos). Repeat the wash and resuspend the cells in 10 vol of Schneider's *Drosophila* medium (e.g., 5 ml of medium for 0.5 ml of settled embryos). Schneider's *Drosophila* medium is used here and in the next step because S2 medium may have higher background fluorescence, which could interfere with the FACS analysis.

Pellet the cells and resuspend them in Schneider's *Drosophila* medium. Use a volume of medium equal to the volume of embryos used (e.g., 0.5 ml of medium for 0.5 ml of settled embryos). The final cell concentration should be ~10^7 cells/ml. Pass the sample through a 50-μm mesh Nitex filter to remove large cell clumps and large debris.

3. Staining Cells with Fluorescent Dyes

The FDG staining protocol is a modification of the procedure developed by Nolan *et al.* (1988) for staining mammalian cells in culture. Because it is large and hydrophilic, FDG does not readily enter cells. To allow efficient loading of the substrate, the cells are briefly treated with high concentrations of FDG under hypotonic conditions at 25°C. β-Galactosidase converts FDG to free fluorescein, which readily passes through the plasma membrane at 25°C but not at 4°C. Thus, immediately after FDG loading, the cells are chilled to 4°C to prevent leakage of fluorescein from the cells. Refer to Nolan *et al.* (1988), Fiering *et al.* (1991), Roederer *et al.* (1991), and information provided with the reagent by Molecular Probes for additional information about FDG staining.

Staining reactions are carried out with 200-μl aliquots of cells in 6-ml (12 × 75 mm) polystyrene round-bottom tubes (Falcon No. 2058). Add 10 μl of 50 mM PETG to one of the aliquots; this inhibits β-galactosidase activity and serves as a negative control (see Step 4).

Equilibrate the cells at 25°C for 5 min. Add 200 μl of 2 mM FDG and mix gently but thoroughly. This dilutes the Schneider's medium to approximately half its normal concentration.

After 2 min at 25°C, add 4.5 ml cold (4°C) SPC medium and mix well to restore isotonicity and chill the cells. SPC contains the dyes PI and CBAM, which are used to distinguish the live cells during cell sorting (Step 4).

Incubate the cells at 4°C to allow the β-galactosidase reaction to proceed. Typically, we allow the reaction to proceed for ~30 min, but this can vary from a few minutes to an hour or more, depending on the levels of β-galactosidase activity in the cells, and the optimal time may have to be determined for each *lacZ* strain. (Also during this incubation, cellular esterases convert the CBAM to calcein blue.) Cells must be kept at 4°C at all times from this point on to prevent leakage of fluorescein from the cells.

Pellet the cells (5 min, 500 g) and resuspend the cell pellet in 1 ml cold SPP medium. SPP medium contains PETG to stop the β-galactosidase reaction, which is sometimes necessary to keep the background fluorescence from increasing in the cells that don't express *lacZ*. This step also serves to remove the free CBAM in the medium.

4. Flow Cytometry and Cell Sorting

For what follows, the reader may wish to consult a general introduction to flow cytometry (e.g., Parks *et al.*, 1989; Radbruch, 1993). Carry out the standard calibration of the FACStar instrument (Parks *et al.*, 1986,1989), set the forward light scatter amplifier to 8×, and set the forward light scatter gates to help exclude small debris and cell aggregates from the analysis. Typically, gates from 200 to 820 (scale of 0 to 1000) are used on this flow cytometer. *Drosophila* embryonic cells are heterogenous in size (and many are smaller than mammalian cells that are commonly analyzed by flow cytometry), and their large-angle light scatter values are generally well correlated with their forward light scatter values (Fig. 3, right). Analyze the fluorescence of a small number of cells in the forward light scatter range noted above. After correcting for cellular autofluorescence and for fluorescein spectral overlap into the PI channel by standard methods (Parks *et al.*, 1989), there should be at least two distinct subpopulations differing in their calcein blue fluorescence and PI fluorescence (Fig. 3, left). Set the calcein blue fluorescence and PI fluorescence gates to include only the live cell population (high calcein blue, low PI fluorescence). To set the sorting gates for the β-galactosidase-expressing cells, analyze the fluorescein fluorescence of the live cells (forward light-scatter 200–820, high calcein blue fluorescence, low PI fluorescence). (Note: To prevent fluorescein leakage from the stained cells, they must be kept cold at all times from FDG staining until after cell sorting, even in the cell sorter, where they should be surrounded by ice or chilled water in the input reservoir.) The fluorescein profile

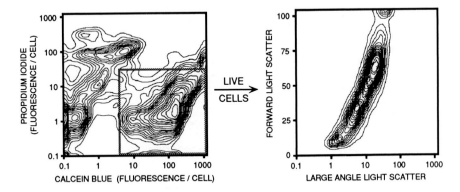

Fig. 3 Flow cytometric analysis of *D. melanogaster* embryo cells. Embryos 4 to 6 hr old were disrupted by dounce homogenization and the cells were stained with PI and CBAM and analyzed by flow cytometry. (Left) Contour plot of cellular fluorescence measurements. The live cell population (low PI fluorescence, high calcein blue fluorescence) is indicated by the stippled box. (Small cell debris and large cell clusters were excluded from the analysis by setting the forward light scatter gates as described in the text.) (Right) Light scatter measurements of the live cells. Reproduced from Krasnow *et al.* (1991).

is usually rather heterogenous, presumably due to heterogeneity in the levels of β-galactosidase in the individual cells and in the amount of FDG taken up by each, and there is overlap between the populations of β-galactosidase-expressing and nonexpressing cells (Fig. 4B; see also Krasnow *et al.,* 1991). In the best cases, there may be a distinguishable peak or "shoulder" of cells with high fluorescein fluorescence. More commonly, the overlap with the much larger population of cells that does not express β-galactosidase obscures the peak, and the β-galactosidase-expressing cells appear as a "tail" of highly fluorescent cells on the distribution of nonexpressing cells. It is therefore often useful to compare the fluorescein fluorescence profile of the experimental sam-

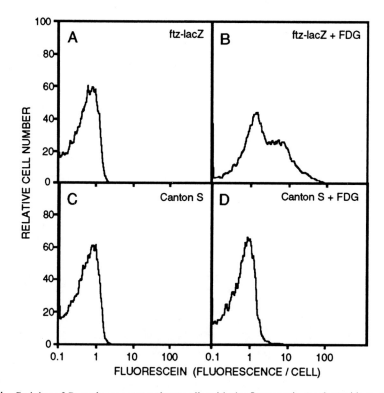

Fig. 4 Staining of *D. melanogaster* embryo cells with the fluorogenic β-galactosidase substrate FDG. Cells were prepared from 4.5- to 6.5-hr old embryos harboring a *ftz–lacZ* transgene (expressed in cells of the even-numbered parasegments) or from 4.5- to 6.5-hr old wild-type (Canton S) embryos. The washed, filtered cells were mock-treated with hypotonic Schneider's medium, or treated with 1 mM FDG in hypotonic Schneider's medium, for 1.5 min at 23°C. Isotonicity was restored, cells were stained with PI, and the β-galactosidase reactions were continued for 45 min at 4°C and stopped by the addition of PETG to 1 mM. Fluorescence values of the live cells (see Fig. 3) are shown. (A) *ftz–lacZ* cells, mock-treated. (B) *ftz–lacZ* cells, treated with FDG. (C) Canton S cells, mock-treated. (D) Canton S cells, treated with FDG. Reproduced from Krasnow *et al.* (1991).

ples with that of negative control cells (cells in which PETG was added before FDG to inhibit the β-galactosidase reaction, see Section IIB3 above), or FDG-treated cells from a strain that does not express *lacZ* (Fig. 4D), and to set the fluorescence gate above that of the highest fluorescein fluorescence detected in this sample to maximize purity.

Up to a few thousand cells can be sorted per second, or ~10^7 cells per hour. If, for example, 1% of the cells express β-galactosidase, then theoretically ~10^5 cells can be isolated per hour. However, when we have determined the number of purified cells by microscopy immediately after sorting, the number is much less, typically only 10 to 40% of the number of sorting events registered by the cell sorter. (The reasons for the low yield are unclear, especially since many mammalian cell types give yields of 60% or greater on the FACS (Parks *et al.*, 1986). To determine the number of cells by microscopy, add an equal volume of a solution containing 3 μg/ml acridine orange and 5 μg/ml ethidium bromide to the cells and view under a fluorescence microscope; acridine orange stains live and dead cells and appears green, and ethidium bromide stains the nuclei of dead cells and appears orange/red. Alternatively, cells can be stained with an equal volume of 0.25% (w/v) solution of trypan blue and viewed under a light microscope. Dead cells stain dark blue, while viable cells exclude the dye and remain unstained; by this method, though, it is sometimes difficult to distinguish live cells from debris.

III. Culturing and Analysis of Purified Cells

A. Short-Term Culturing

Cells are sorted directly into a tissue culture chamber slide (Lab-Tek, 16-well, available from Nunc Inc., No. 178599) containing 75 μl of S2 medium, and the cells are allowed to settle on the slide. The medium is carefully replaced with fresh S2 medium (to remove the sheath fluid that accumulates during sorting), and the cells are cultured at 25°C in a humidified chamber. Optimal culture conditions should be determined empirically and will depend on the type of cells and the length of the culture period. We have cultured purified cells for a few days, but we have not yet attempted to establish long-term cultures from purified cells. It may help to coat the slide or culture vessel with extracellular matrix components before culturing; we often pretreat the slides by growing *D. melanogaster* Schneider line 2 (S2) cells to near confluence in the chamber slides for 16–20 hr and then removing the cells by washing the slides with S2 medium (Cumberledge and Krasnow, 1993). Conditioned S2 medium, prepared by culturing cells from dissociated *Drosophila* embryos for 1 day and then removing the cells by centrifugation and filtration through a 0.45-μm filter, can also aid in culturing (Krasnow *et al.*, 1991).

B. Fixation and Staining with Antibodies

If cells are to be stained with antibodies, they are collected directly onto polylysine-coated (Ashburner, 1989b) chamber slides or transferred to such slides after culturing. (Culturing on polylysine-coated slides is not generally recommended as it may adversely affect the cells.) Cells are gently pelleted onto the slide by low-speed centrifugation and then fixed in a solution of 4% paraformaldehyde at room temperature for 10 min. Even after fixation, attachment of the cells to the slide is usually weak and extreme care must be taken during all subsequent incubation and washing steps to keep cells from dislodging from the slide. When changing buffers, slowly aspirate off most of the old solution using a drawn-out capillary tube or long micropipet tip and add new solutions by slowly streaming down along the wall of the chamber well. For staining intracellular antigens, cells can be permeabilized by incubating in PBS + 0.1% Triton X-100 for 5 min. Wash the cells for 1 min with PBSB five times and then incubate twice in PBSB for 15 min each. Remove the PBSB, add the primary antibody diluted in PBSB, and incubate at room temperature for 1 hr. Wash the cells again with PBSB as above. The cells are now ready for staining with a secondary antibody and detection. For antigens present in low abundance, we have used a biotinylated secondary antibody and horseradish peroxidase immunocytochemistry (Vectastain ABC, Vector Laboratories, No. PK-4001).

C. Stable Fluorescent Marking of Purified Cells

For applications in which different populations of purified cells are intermixed, such as cell adhesion or signal transduction assays, it is sometimes useful to distinguish the different cell populations during culturing or to reisolate the different cell types after culturing to determine the effects of intermixing. We have used the dyes CMTMR and CMAC to stably label purified cells for these purposes, and they also may be useful for marking purified cells before introducing them into recipient embryos. These are membrane-permeant fluorescent dyes that contain a chloromethyl reactive group. They readily enter cells, where they are stably retained, probably by reaction of the choromethyl group with intracellular glutathione (Zhang et al., 1992). After sorting, the purified cells are incubated in S2 medium + 80 nM CMTMR, or S2 medium + 160 nM CMAC, for 5 min at 25°C. After the cells are washed twice with S2 medium to remove unbound dye, two differentially marked populations can be cultured together (or cultured with unmarked cells) and viewed by fluorescence microscopy, or they can be resorted according to their CMTMR or CMAC fluorescence to reisolate the two populations. Note that the residual fluorescent products from the CBAM staining may interfere with the detection of the CMAC fluorescence because of spectral overlap. This is not a problem if the cells are cultured for several hours before analysis, as the products of the CBAM staining eventually dissipate, but if cells are to be analyzed soon after sorting CBAM staining may have to be omitted.

IV. Conclusions

WACS is a relatively new technique that has been successfully used to isolate and study a number of different early embryonic cell types *in vitro*. In its current form, it can be used to prepare analytical quantities of almost any early embryonic cell type, provided a *lacZ* strain that expresses β-galactosidase specifically in the cells of interest is available. So far, we have used WACS to purify cells from early stages of embryogenesis; applying WACS to later stages of development will probably require the use of other methods for cell dissociation besides the mechanical procedure (douncing) described here, as most cells become more adherent as the embryo matures. Although traditional methods such as treatment with trypsin or EGTA can be used, we can also envision the introduction of a transgene that can be conditionally activated to disrupt cell associations when needed. It will also be necessary to determine whether endogenous β-galactosidase activity is present at these later developmental stages, which could interfere with the procedure unless it can be specifically inhibited, for example by treatment with chloroquine (Fiering *et al.,* 1991).

Although FDG works well in this procedure, better substrates for detecting *lacZ* expression might improve both the yield and purity of cells obtained by WACS. FDG is not freely permeable across plasma membranes, requiring hypotonic conditions to introduce the substrate into cells and making it difficult to load high concentrations. Also, the fluorescent product of the β-galactosidase reaction is not efficiently retained at 25°C and the cells must be kept at 4°C until sorting is complete. New substrates with better uptake properties that give rise to products that are retained in the cells at 25°C would be useful, as would substrates that are converted into products of greater fluorescence, to allow clean separation of β-galactosidase-expressing cells from nonexpressing cells; new FDG analogs that have been described (Zhang *et al.,* 1991; Haugland, 1992) should be evaluated. Also, development of whole new ways of marking and purifying cells, such as using antibodies against cell surface antigens (see below), developing viable, fluorogenic or luminescent probes for other intracellular markers such as luciferase, or expression of *Aequorea victoria* green fluorescent protein, would also be useful adjuncts to the current method as they would allow combinations of markers to be used and thereby increase the variety of cell types that could be isolated.

Perhaps the major limitation of the current methodology, at least for some applications, is in the number of cells that can be obtained in a reasonable period of time and at reasonable cost. Adequate numbers of cells can be obtained for most cell-based analyses such as antibody staining of fixed cells, and for protein immunoblots, but for many applications it would be extremely useful to have an essentially unlimited supply of purified cells, particularly for biochemical analysis and for preparative purposes such as protein isolation from the purified cells. The limiting factor is the speed of the cell sorter, which can analyze and sort cells at rates of up to $\sim 10^7$ cells per hour. At these rates, it would be impractical to purify adequate quantities of even the most abundant cell types in a preparation for most preparative purposes. We are considering two possible solutions. One is to use another method for isolating marked cells that does not rely on flow cytometry. For example, "panning" with cell-type-specific antibodies immobilized on a solid support would increase throughput, and it would also avoid the expense of cell sorting, which becomes prohibitive for large-scale purifications. However, this approach requires an antibody against a cell-type-specific surface marker, or the production of a transgenic strain that expresses an exogenous cell surface marker in the cells of interest, and it would not take advantage of the large battery of existing β-galactosidase-expressing strains. An alternative approach is to attempt to immortalize cells purified by the standard procedure, for example by transforming them with activated oncogenes, to establish cell lines of identified cell type.

Acknowledgments

We thank Garry Nolan and David Parks for many helpful discussions, Molecular Probes for providing the structures shown in Fig. 2, and Garry Nolan, David Parks, and Evelyn Parker for comments on the manuscript. This work was supported by a postdoctoral fellowship from the National Institutes of Health to S.C. and a Lucille P. Markey Scholar Award, a National Science Foundation Presidential Young Investigator Award, and a grant from the National Institutes of Health to M.A.K.

References

Allis, C. D., Waring, G. L., and Mahowald, A. P. (1977). Mass isolation of pole cells from *Drosophila melanogaster*. *Dev. Biol.* **56**, 372–381.

Ashburner, M. (1989a). "*Drosophila:* A Laboratory Handbook." Cold Spring Harbor, NY: Cold Spring Harbor Laboratory Press.

Ashburner, M. (1989b). "*Drosophila:* A Laboratory Manual." Cold Spring Harbor, NY: Cold Spring Harbor Laboratory Press.

Bellen, H. J., O'Kane, C. J., Wilson, C., Grossniklaus, U., Pearson, R. K., and Gehring, W. J. (1989). P element-mediated enhancer detection: A versatile method to study development in *Drosophila*. *Genes Dev.* **3**, 1288–1300.

Bernstein, S. I., Fyrberg, E. A., and Donady, J. J. (1978). Isolation and partial characterization of *Drosophila* myoblasts from primary cultures of embryonic cells. *J. Cell Biol.* **78**, 856–865.

Bier, E., Vaessin, H., Shepherd, S., Lee, K., McCall, K., Barbel, S., Ackerman, L., Carretto, R., Uemura, T., Grell, E., Jan, L. Y., and Jan, Y. N. (1989). Searching for pattern and mutation in the *Drosophila* genome with a *P-lacZ* vector. *Genes Dev.* **3**, 1273–1287.

Cumberledge, S., and Krasnow, M. A. (1993). Intercellular signalling in *Drosophila* segment formation reconstructed in vitro. *Nature* **363**, 549–552.

Elgin, S. C. R., and Miller, D. W. (1978). Mass rearing of flies and mass production and harvesting of embryos. In "The Genetics and Biology of Drosophila" (M. Ashburner and T. R. F. Wright, eds.), Vol. 2a, pp. 112–121. New York: Academic Press.

Fiering, S. N., Roederer, M., Nolan, G. P., Micklem, D. R., Parks, D. R., and Herzenberg, L. A. (1991). Improved FACS-Gal: Flow cytometric analysis and sorting of viable eukaryotic cells expressing reporter gene constructs. *Cytometry* **12**(4), 291–301.

Furst, A., and Mahowald, A. P. (1985). Differentiation of primary embryonic neuroblasts in purified neural cell clusters from *Drosophila*. *Dev. Biol.* **109**, 184–192.

Haugland, R. P. (1992). "Handbook of Fluorescent Probes and Research Chemicals." Eugene, OR: Molecular Probes, Inc.

Krasnow, M. A., Cumberledge, S., Manning, G., Herzenberg, L. A., and Nolan, G. P. (1991). Whole animal cell sorting of *Drosophila* embryos. *Science* **251**(4989), 81–85.

Moore, W. A., and Kautz, R. A. (1986). Data analysis in flow cytometry. *In* "The Handbook of Experimental Immunology" (D. M. Weir, L. A. Herzenberg, C. C. Blackwell, and L. A. Herzenberg, eds.), Vol. 1, pp. 30.1–30.11. Oxford: Blackwell Scientific Publications.

Nolan, G. P., Fiering, S., Nicolas, J. F., and Herzenberg, L. A. (1988). Fluorescence-activated cell analysis and sorting of viable mammalian cells based on β-D-galactosidase activity after transduction of *Escherichia coli lacZ*. *Proc. Natl. Acad. Sci. U.S.A.* **85**, 2603–2607.

O'Kane, C. J., and Gehring, W. J. (1987). Detection *in situ* of genomic regulatory elements in *Drosophila*. *Proc. Natl. Acad. Sci. U.S.A.* **84**, 9123–9127.

Parks, D. R., Herzenberg, L. A., and Herzenberg, L. A. (1989). Flow cytometry and fluorescence-activated cell sorting. *In* "Fundamental Immunology" (W. E. Paul, ed.), pp. 781–802. New York: Raven.

Parks, D. R., Lanier, L. L., and Herzenberg, L. A. (1986). Flow cytometry and fluorescence activated cell sorting (FACS). *In* "The Handbook of Experimental Immunology" (D. M. Weir, L. A. Herzenberg, C. C. Blackwell, and L. A. Herzenberg, eds.), Vol. 1, pp. 29.1–29.21. Oxford: Blackwell Scientific Publications.

Radbruch, A. (1993). "Flow Cytometry: A Laboratory Manual." Heidelberg: Springer-Verlag.

Roberts, D. B. (1986). Basic *Drosophila* care and techniques. *In* "*Drosophila*: A Practical Approach" (D. B. Roberts, ed.), pp. 1–38. Washington, DC: IRL Press.

Roederer, M., Fiering, S., and Herzenberg, L. A. (1991). FACS-Gal: Flow cytometric analysis and sorting of cells expressing reporter gene constructs. *Methods: A Companion to Methods in Enzymology* **2**, 248–260.

Storti, R. V., Horovitch, S. J., Scott, M. P., Rich, A., and Pardue, M. L. (1978). Myogenesis in primary cell cultures from *Drosophila melanogaster*: Protein synthesis and actin heterogeneity during development. *Cell* **13**, 589–598.

Wysocki, L. J., and Sato, V. L. (1978). "Panning" for lymphocytes: A method for cell selection. *Proc. Natl. Acad. Sci. U.S.A.* **75**, 2844–2848.

Zhang, Y.-Z., Naleway, J. J., Larison, K. D., Huang, Z., and Haugland, R. P. (1991). Detecting lacZ gene expression in living cells with new lipophilic, fluorogenic β-galactosidase substrates. *FASEB J.* **5**, 3108.

Zhang, Y.-Z., Olson, N., Mao, F., Roth, B., and Haugland, R. P. (1992). New fluorescent probes for long-term tracing of living cells. *FASEB J.* **6**, A1835.

CHAPTER 9

Transformation Techniques for *Drosophila* Cell Lines

Lucy Cherbas,[*] Robert Moss,[†] and Peter Cherbas[*]

[*] Department of Biology
Indiana University
Bloomington, Indiana 47405

[†] Department of Biology
Wofford College
Spartanburg, South Carolina 29301

I. Introduction

By far, the most widespread use of *Drosophila* cell lines is as living "test tubes" in which to assess the activities of exogenous genes. For example, the functions of mammalian and *Drosophila* transcription factors have been reconstructed by cotransfections, in *Drosophila* cells, of factor genes with suitable target promoter–reporter constructs (Krasnow *et al.*, 1989; Yoshinaga and Yamamoto, 1991). Sequence elements influencing the choice of initiation codons have been analyzed by expressing the *E74A* transcript in S2 and Kc167 cells (Boyd and Thummel, 1993) and the function of a mammalian signal peptide was assessed by determining the intracellular distribution of a reporter in transfected S2 cells (Gibson *et al.*, 1993). Similar transient expression experiments have facilitated the analysis of many *Drosophila* promoters; one example from our laboratory was the identification of ecdysone response elements (EcREs) based upon their activities in Kc cells (Cherbas *et al.*, 1991). In a distinct approach, stably transformed *Drosophila* cells can be useful for producing proteins at high levels (Johansen *et al.*, 1989), and a creative use of this approach has given rise to novel assays, using transformed cells, for the interactions of the neurogenic gene products *Delta* and *Notch* (Fehon *et al.*, 1990; Rebay *et al.*, 1991).

In line with this predominant interest, this chapter focuses virtually exclusively on techniques for transforming *Drosophila* cell lines. Along with data from theses and unpublished work in our lab, we have included occasional anecdotal or incomplete observations which may, we hope, prove helpful to others. For general introductions to *Drosophila* cell culture the interested reader should consult the cited references; we note especially the thorough coverage in Ashburner's lab manual (Ashburner, 1989b).

Although dozens of *Drosophila* cell lines exist [Ashburner (1989a) lists 94 independent lines] and at least a dozen may be obtained fairly readily, most investigators have used the Schneider lines (S2 and S3) or one of several derivatives of Echalier and Ohanessian's Kc line. These, like the great majority of the extant lines, were originally obtained as immortal outgrowths from primary cultures of minced late embryos (see reviews by Schneider and Blumenthal, 1978; Sang, 1981). Other lines, for example those generated from tumorous hemocytes (Gateff *et al.*, 1980) or from imaginal discs (Ui *et al.*, 1989), exist, but their uses have so far been quite specialized and they are not mentioned further here.

A detailed examination of the expression patterns of the ecdysone-inducible genes *Eip28/29* and *Eip40* led us to hypothesize that Kc cells are derived from hematopoietic cells or their developmental precursors (Andres and Cherbas, 1992). On the basis of comparable evidence (expression patterns), Abrams *et al.* (1992) have concluded that S2 cells, but not Kc cells, are related to macrophages. Thus, on current evidence it seems plausible that most of the common lines are derived from one or another category of hemocyte or hematopoietic precursor. Similar origins would be consistent with the generally similar (i.e., lymphoid) growth habits of all these lines and with the similarities of their hormone responses (see e.g., Cherbas and Cherbas, 1981).

From a practical standpoint, the investigator choosing a line as transformation host will be interested not so much in gross similarities as in the particularities of individual lines, which can often be important. For example, individual lines differ substantially in their transfectabilities by hydroxyapatite–DNA (see Section IIIA1). They may also differ in their precise medium requirements, in their susceptibilities to the toxins used for selection, in superficial and changeable properties like surface adherence, and—most importantly—in the precise catalog of expressed genes. Finally, maintenance and record-keeping for cell lines is often a fairly casual matter—a name does not necessarily specify a set of properties.

Nothing can replace appropriate pilot experiments, but here are some general guidelines:

- Lines vary greatly in their growth rates and in their ability to tolerate abuse; Kc, S2, and S3 cells are extremely robust and fast-growing. Use them whenever possible.
- Electroporation is probably useful for all lines, but so far as we are aware, parameters for its use have, so far, been optimized only for Kc167 cells. Calcium phosphate-DNA coprecipitation is useful for many lines, but fails utterly for many Kc sublines and for the S3 cells in culture in our laboratory.
- Although it is certainly not universally true, the activities of transfected promoters often parallel the activity of the corresponding endogenous promoter (Moss, 1985). Northern evidence that the relevant promoter is active in cells can be a useful precursor to transformation work.
- Finally, individual lines differ in their expression patterns in ways that may matter for particular experimental goals. For example, the S2 line has a low concentration of ecdysone receptor (EcR) and a much weaker ecdysone response than Kc or S3 cells. To identify a line that combined their strong ecdysone response with the ready transfectability of S2 cells required systematic pilot experiments (Cherbas *et al.*, 1991).

II. Maintaining and Cloning Cells

A. Culture Media

Drosophila cell lines can be maintained in a variety of media; those most commonly used are Schneider's medium (Schneider, 1964), D-22 (Echalier and Ohanessian, 1970), and M3 (Shields and Sang, 1977), all of which are now commercially available (Sigma sells all three, and GIBCO sells Schneider's medium). We have used M3 extensively, because most cell lines adapt readily to growth in M3 supplemented with 5–10% fetal calf serum (FCS). Some lines require additional supplementation with a mixture of bactopeptone (BP) (2.5 g/liter) and extra yeast extract (YE) (2 g/liter final concentration; M3 itself contains 1 g/liter). [Note that M3 as sold by Sigma contains 2 g/liter YE, while Shields and Sang's original formulation (1977) contains 1 g/liter YE. Increasing the YE concentration has no deleterious effect on cell growth, but it can interfere with methotrexate selection (see the following).]

We have tested HyQ-CCM3, a serum-free medium marketed by HyClone for use with *Spodoptera* cell lines; it is an excellent complete medium for *Drosophila* cells. Kc167 cells grown in M3/BPYE/5% FCS can be transferred to HyQ-CCM3 without any period of adaptation. S2, S3, and GM-3 cells adapt readily from M3/BPYE/10% FCS to HyQ-CCM3, whereas D cells grow in HyQ-CCM3 only if it is supplemented with 2% FCS (R. Sadowski and L. Cherbas, unpublished). HyQ-CCM3 is less expensive than serum-supplemented media, and it is preferable to M3 for methotrexate and G418 selection (see following). From our perspective, the major disadvantage of HyQ-CCM3, in some circumstances, is that its composition is a proprietary secret.

B. Cell Maintenance

Drosophila cells are grown at 25°C, with air as the gas phase. We maintain cells in tissue culture-grade plastic petri dishes (10 ml per 100-mm plate), which are placed in air-tight plastic boxes (Tupperware or equivalent) to minimize evaporation. Growth rates vary enormously from line to line; the fastest grow-ing lines (e.g., Kc and S2) double in about 24 hr. Cells should be trans-ferred frequently enough to keep them within the exponential range; for most lines this corresponds to ca. 10^6 to 10^7 cells/ml. Most *Drosophila* cell lines adhere very loosely to the surface and can be dislodged by blowing a gentle stream of medium at the surface from a Pasteur pipet or by using a cell scraper; we have never found it necessary to trypsinize *Drosophila* cells. Cells can be maintained for at least several weeks at 18°C with ca. 10-fold reduced growth rates. This is often a convenient short-term alternative to freezing and thawing.

In our experience, most *Drosophila* cell lines can be grown in spinner flasks with little adaptation. The stirring rate should be adjusted for each line; usually the minimum rate that keeps cells suspended is optimum. Stirring at excessive speed will break cells. For best results the spinner flask should be relatively empty: the height of the culture should not exceed one-quarter flask height, i.e., the culture volume should not exceed ca. half the designated "capacity" of the spinner. We have used spinner flasks successfully to scale up cultures of Kc, S2, and S3 cells. We do not maintain lines in spinners: during prolonged growth in spinner flasks most lines become unhealthy and selective cell growth is obvious. To avoid this evolution in the properties of the population, we use spinners for short-term growth from seed cultures in plates.

C. Cloning Cells

No standard culture medium we have tried supports growth of *Drosophila* cells at very low cell densities. Nevertheless, it is possible to clone cells. Lindquist (in Ashburner, 1989b) has described one technique using conditioned medium. We continue to find it more convenient and reproducible to supplement the culture medium with irradiation-sterilized feeder cells (protocol in Ashburner, 1989b). We generally use irradiated cells at 10^6/ml. However, Kc167 cells in HyQ-CCM3 medium are extremely sensitive to low density; to achieve reasonable cloning efficiencies in this medium, we use irradiated feeder cells at a concentration of at least 2×10^6/ml. Some lines (e.g., S2) can be cloned at high efficiency (on the order of 50%) in soft agar (see protocol in Ashburner, 1989b). Others (e.g., Kc167) grow very poorly in soft agar with correspondingly poor cloning efficiencies; they must be cloned by dilution using 96-well plates.

1. Protocol for Cloning in Soft Agar

Our protocol for cloning S2 cells in soft agar using proliferatively dead feeder cells is described in detail in Ashburner (1989b); a brief summary of the procedure follows: For feeder cells, select a line that grows in the medium to be used for cloning. If a selective agent (e.g., methotrexate, ecdysone) is to be included in the cloning, the feeder cells should be resistant to that agent. An appropriate number of cells [suspended in 5 ml Robb's saline (Robb, 1969)] is irradiated with 24 kR X-rays or γ-rays. It is most convenient to use a high-intensity cesium source. The irradiated feeder cells are then resuspended at 1.25×10^6/ml in 1.25× medium, containing any desired selective agents. Effectively, they become components of the medium in which viable cells are then cloned. Viable cells are added at an appropriate concentration and 8 ml of the resulting suspension is then placed on one sector of a 100-mm bacteriological petri plate. Two milliliters of molten 1.5% Noble agar (cooled to 45–50°C) is placed in another sector of the plate and the plate is rocked gently to mix the cell suspension with the agar. The concentration of viable cells should be selected to achieve a final concentration (following selection) of ca. 200 clones/ plate. Plates should be kept in a tightly sealed plastic box at 25°C. The clones are usually visible to the naked eye within 2 weeks and should be picked when they reach about 1 mm in diameter. Clones are transferred from the agar plate into 0.1 ml of medium in a 96-well plate. We use a capillary pipet and attempt in the process to break up the agar plug, dissociating the clumped cells. Freshly picked clones include many dead and dying cells (from the interior of the spherical clump), but the population resumes normal growth within a few days.

2. Protocol for Cloning in 96-Well Plates

Cells to be cloned in 96-well plates are diluted in a feeder cell suspension (see previous discussion of concentration) to give a final concentration of about 1 viable cell/ml. If the cloning is done under selective conditions, it is best to use feeder cells that are resistant to the selective agent. If you do not know the concentration of cells to be cloned (e.g., when cloning transfected cells under selective conditions), make several cell suspensions at different dilutions of the test cells. Dispense the cells into 96-well plates at 0.1 ml/well; seal the edges of the plates with Parafilm to avoid dessication. Clones should be easily visible in 1–2 weeks. Once the clones have reached a diameter approximately one-quarter that of the well, they can be transferred to larger volumes. Use 0.3 ml/well in 48-well plates, 0.5 ml/well in 24-well plates, 1 ml/well in 12-well plates, and 2 ml/well in 6-well plates. Because it is difficult to estimate cell number in clones that are not well-dispersed, and because the cells in the interior of a compact clone are frequently dead, be conservative in transferring the cells; we generally dilute the cells only two- to threefold at each of the first few transfers.

D. Freezing Cells

It is imperative to freeze viable cells as soon as possible when a new line is created or brought into the lab. Frozen stocks in liquid N_2 appear to be infinitely stable. To freeze cells, start with a healthy culture in mid-exponential growth. Pellet the cells and resuspend them at a concentration of about $1.5–2 \times 10^6$ cells/ml in freezing medium: normal growth medium (including serum) supplemented with 10% dimethylsulfoxide. Dispense the suspension in 0.5-ml aliquots into plastic freezing ampoules. Place the ampoules into a tightly corked Dewar flask. Then place the Dewar flask into a $-80°C$ freezer to permit the temperature to drop slowly. This slow decrease in external temperature permits supercooling and the abrupt freezing of the cells that is desirable. After about 2 days, transfer the frozen ampoules to liquid nitrogen. To recover the frozen cells, gently thaw the ampoule (for example, in your hand) and transfer the contents to 5 ml of medium in a 25-cm² T-flask. Freshly thawed cells are usually ready to be transferred within about 3 days, but the proportion of viable cells (and, consequently, the immediate proliferation rate) varies somewhat from line to line.

III. Transforming Cells

A. Transfection Procedures

A number of transfection procedures have been applied to *Drosophila* cell lines; Table I lists many of the relevant references. Here, we elaborate on two procedures with which we have had extensive experience.

1. Calcium Phosphate–DNA Coprecipitation

This technique, adapted with minor modification from the original mammalian cell transformation procedure (Wigler *et al.*, 1979), has been used for a wide variety of *Drosophila* cell lines. It is simple and inexpensive, requires no special

Table I
Transformation Systems for *Drosophila* Cell Lines

Transfection agent	Cell line	Selected references
No agent	1182-4	Saunders *et al.*, 1989
	D1	Saunders *et al.*, 1989
Calcium phosophate–DNA coprecipitate	1006-2	Ayer and Benyajati, 1990; Ayer *et al.*, 1993
	1182-4	Saunders *et al.*, 1989
	1182-6	Saunders *et al.*, 1989
	67j52D	Braude-Zolotarjova and Schuppe, 1987
	789f7Dv3g	Braude-Zolotarjova and Schuppe, 1987
	D	L. Cherbas, unpublished
	D1	Sinclair *et al.*, 1983; Burke *et al.*, 1984; Allday *et al.*, 1985; Moss, 1985; Sinclair and Bryant, 1987; Saunders *et al.*, 1989
	D2	Imhof *et al.*, 1993
	DH14	Burke *et al.*, 1984
	DH15	Burke *et al.*, 1984
	DH33	Burke *et al.*, 1984; Sinclair *et al.*, 1986; Saunders *et al.*, 1989
	D. immigrans	DiNocera and Dawid, 1983
	DM1	L. Cherbas, unpublished
	E-Adh[N1]	L. Cherbas, unpublished
	EH34A3	Cherbas *et al.*, 1991
	G2	L. Cherbas, unpublished
	Gm1-S1c11	Kurata and Marunouchi, 1988
	GM3	L. Cherbas, unpublished
	Kc	Burke *et al.*, 1984; Davies *et al.*, 1986
	Kc167	Bourouis and Jarry, 1983; Cherbas *et al.*, 1991; Boyd and Thummel, 1993; Tourmente *et al.*, 1993
	S2	DiNocera and Dawid, 1983; Benyajati and Dray, 1984; Burke *et al.*, 1984; Moss, 1985; Rio and Rubin, 1985; Bunch *et al.*, 1988; and many others
	S3	Martin and Jarry, 1988; Saunders *et al.*, 1989; Walker *et al.*, 1991; Berger *et al.*, 1992
DEAE-dextran	S3	Lawson *et al.*, 1984
Electroporation	Kc167	This chapter
Liposome	mbm-2	Kappler *et al.*, 1993
Polybrene	DM1	Walker *et al.*, 1991
	Kc-H	Walker *et al.*, 1991
	S2	Walker *et al.*, 1991
	S3	Walker *et al.*, 1991
Polyornithine	Kc	Morganelli and Berger, 1984
	S3	Morganelli and Berger, 1984

equipment, and works for most cell lines. Its major drawbacks are (1) The optimal precipitate contains 20 μg DNA/ml and maintaining this concentration can place incovenient constraints on experimental design. (2) The procedure is cytotoxic. For example, induction (by ecdysone) of the endogenous *Eip28/29* gene is only 3-fold in mock-transformed Kc cells, compared to ca. 10-fold in untreated cells (Cherbas *et al.*, 1991). (3) Cell lines (and perhaps clones or sublines) differ enormously in their susceptibility to transfection. For example, we have never succeeded in detecting transfection of S3 cells by calcium phosphate–DNA coprecipitates; in our hands, Kc167 cells are easily transfected, but other Kc sublines (Kc-H, Kc_0) are not. That published reports to the contrary exist (Table I) probably reflects the diversity of the cells maintained in different laboratories and should reinforce our general warnings to that effect.

Protocol

Note: In this protocol we use BBS (Chen and Okayama, 1987) in place of the original HBS at Wigler *et al.* 1979); this change of buffer gives roughly 10-fold higher expression in transient expression, as measured by the number of expressing cells or by reporter activity.

Solutions

2x BES-buffered saline (2x BBS): 280 mM NaCl/1.5 mM Na_2HPO_4/50 mM BES, pH 6.95, filter-sterilized, and stored at $-20°$C.

2 M $CaCl_2$, filter-sterilized and stored at $-20°$C.

plasmid DNA, dissolved in TE and stored at $-20°$C. We use plasmid DNA purified on a CsCl gradient; comparably pure DNA made by other procedures will probably work equally well. We do not in general take any special precautions to sterilize the DNA. If necessary, sterilize the DNA by ethanol precipitation overnight.

Procedure

We use 10-ml cultures (in 100-mm tissue-culture grade petri plates) for transient expression and 3 ml cultures (in 60-mm plates) for stable transformation, using 1 ml of precipitate per plate in either case. The transformation efficiency is higher when 1 ml of precipitate is added to 3 ml of cells, but the cells take several days to recover from the trauma, and we therefore compromise on lower transfection efficiencies but healthier cells for transient expression. The cell concentration should be ca. 3×10^6/ml at the time of the transfection; we have found no substantial effect of varying the time since transfer to fresh medium, as long as the cells are healthy and in mid-exponential growth.

For each dish to be transformed, prepare 1 ml of precipitate as follows: In a 15-ml sterile plastic tube, mix 62 μl 2 M $CaCl_2$, 20 μg DNA, and water to 0.5 ml. Agitate the solution by bubbling air through a sterile cotton-plugged

Pasteur pipet whose opening is at the bottom of the tube. Adjust the rate of bubbling so that at least two or three bubbles are always in transit; we have not found higher rates of bubbling to be deleterious. While the aeration continues, add 0.5 ml 2x BBS dropwise (about one or two drops/sec). Remove the pipet, cap the tube, and let it sit 30 min. The precipitate should be visible by its faint, blue (Tyndall effect) coloration as soon as the aeration is stopped. If there is any visible setting of the precipitate during the 30 min of maturation, the precipitate is too coarse and transfection will be poor. If you are transfecting multiple plates with the same DNA preparation, scale up the precipitation procedure and use an aliquot of the precipitate for each plate.

Add 1 ml of precipitate dropwise to a plate of cells and then mix gently by tipping the plate back and forth. After it has had a few hours to settle, the precipitate should be easily visible under a microscope, looking not unlike bacterial contamination. We generally wash the cells 12–16 hr after addition of the precipitate, but we have obtained very similar results washing as early as 2 hr after addition of DNA. Try to disturb the cells as little as possible while washing; simply replace the medium twice with fresh medium, without dislodging cells stuck to the plate. If there are a significant number of floating cells, collect them by centrifugation and add them back to the plate.

For transient expression experiments, we normally incubate the cells for an additional 12–24 hr (i.e., until 24–40 hr after the addition of the DNA precipitate) before harvesting. For stable transformation, incubate the cells 48 hr after washing before adding the selective agent.

2. Electroporation

Electroporation has been used for transfection of a wide variety of cells, requiring only that the parameters be optimized for each cell type. We have optimized electroporation conditions for transient expression in Kc167 cells (S. Bodayla, M. Vaskova, and P. Cherbas, unpublished) and used the same protocol for stably transforming Kc167 cells; the protocol is given below. With minor modifications, this protocol can probably be applied to any other *Drosophila* cell line. Following electroporation of Kc167 cells with 20 μg DNA/plate, the levels of expression in transient assays and the frequencies of stable transformants are comparable to those obtained by calcium phosphate–DNA coprecipitation in the same line. The level of expression and the frequency of stable transformants are approximately linear over at least a 10-fold range of input DNA concentrations, up to at least 20 μg/plate. Judging by their appearance and rate of growth, the cells recover very rapidly from any ill effects of the procedure. We have used this procedure successfully for the introduction of supercoiled and linear DNAs into Kc167 cells for both transient expression and stable transformation, but we have not yet characterized the structure of the plasmid DNA in stably transformed cells.

Protocol

Cells: Kc167 cells growing in M3 medium supplemented with BPYE and 5% FBS.

Apparatus: Hoefer: PG200 Progenitor power supply with PG250 electroporation chamber and PG220C cuvette electrode (3.5 mm gap)

Setup

Cells at the dense end of exponential growth should be diluted to ca. 10^6 cells/ml (in complete medium) 48 hr before electroporation. Prepare one 10-ml plate for each plate of transfected cells that you wish to prepare. Incubate at 25°C. The cells will be about 4×10^6/ml at the time of electroporation. (The time between transfer and electroporation is important; a 24-hr change in either direction leads to a three- to fivefold decrease in transfection efficiency.)

Electroporation

All steps are performed at room temperature. Remove the cells from the plates, spin, and wash twice in half the original volume medium without serum. Resuspend at approx. 5×10^7 cells/ml medium without serum. Use 0.8 ml (4×10^7 cells)/cuvette. Add DNA and let set 5 min. (Note: Varying the concentration of cells in the electroporation curvette does not seem to matter. The concentration specified here yields 1 plate of cells per transfection; if you wish to transfect more cells, use a higher concentration of cells in the cuvette and after electroporation dispense them among several plates, at 4×10^7 cells/ plate. It is not clear that the 5-min wait matters. Removal of serum prior to electroporation increases reporter activity about twofold; electroporation at room temperature gives about the same activity as electroporation on ice.)

Shock cells at 440 V/cm; 1200 μF; 1 sec (i.e., complete decay). Allow cells to recover for 10 minutes, and then add the contents of the cuvette to 9.2 ml complete medium in a 100 mm plate. Return the cells to 25°. (Note: The optimum field strength is 400-500 V/cm, with a rapid fall-off of reporter activity outside this range. The capacitance has a broad optimum between 800 and 1800 μF.)

B. Transient Expression

Transient expression is normally used either to assess the properties of a promoter of interest, generally measured by means of a convenient reporter, or to assess the properties of a protein expressed from a constitutive promoter, or a combination of the two. Although some workers have used inducible promoters for transient expression, this is usually for the convenience of the experimenter (i.e., the plasmid was already constructed for other purposes) rather than to avoid the toxic effects of the expressed protein during the short time course of these experiments.

1. Promoters

Table II lists three constitutive promoters that have been used for the high-level expression of proteins in *Drosophila* cell lines. All three are reasonably strong; in our experience, the *actin5C* promoter gives the highest levels of expression. The *Eip28/29* promoter listed in the table is derived from the ecdysone-inducible gene, but is lacking functional ecdysone response elements and hence is not ecdysone-inducible.

A variety of weak promoters have been used for testing the function of heterologous promoter elements; a few examples are listed in the table.

2. Reporters

In general, any of the reporters used for mammalian and yeast expression studies can be employed with *Drosophila* cell lines: chloramphenicol acetyltransferase (CAT); β-galactosidase (*lacZ*), and firefly luciferase have been used extensively. *lacZ* should be used with caution; the endogenous *Drosophila* β-galactosidase gene, encoding a lysomal enzyme, is expressed at low levels in

Table II
Promoters Used for Expression in *Drosophila* Cell Lines

Promoter type	Promoter	References
Constitutive	Dm*act5C*	Krasnow *et al.*, 1989; Ayer and Benyajati, 1990; Koelle *et al.*, 1991; Yoshinaga and Yamamoto, 1991; Imhof *et al.*, 1993
	Dm*copia*	Bourouis and Jarry, 1983
	Dm*Eip28/29*	Bunch and Goldstein, 1989; Feder *et al.*, 1992
Strongly inducible	Dm*hsp70*	Ananthan *et al.*, 1993; Cumberledge *et al.*, 1993
	Dm*Mt*	Bunch *et al.*, 1988; Bunch and Goldstein, 1989; Johansen *et al.*, 1989; Fehon *et al.*, 1990; Ivey-Hoyle and Rosenberg, 1990; Koelle *et al.*, 1991; Rebay *et al.*, 1991; Feder *et al.*, 1992; Ayer *et al.*, 1993
	Various ecdysone-inducible promoters	Lee, 1990
Minimal	Dm*Adh*(distal)	Krasnow *et al.*, 1989
	Dm*en*	Yoshinaga and Yamamoto, 1991
	dm*Mt*[a]	Ananthan *et al.*, 1993
	HSV*tk*	Berger *et al.*, 1992

[a] This is a shorter version of the *Mt* promoter than the inducible promoter used above; it is lacking the elements required for the metal response.

some (perhaps most) of the cultured lines. Its activity can interfere with the analysis of weak promoters. Moreover, its activity is elevated following ecdysone treatment (Best-Belpomme *et al.,* 1978). This obviously limits its utility for studies of ecdysone regulation, but it also suggests a crucial control in other work using the *lacZ* reporter.

C. Stable Transformations

Stable transformation is usually used to express an exogenous gene at high levels. The level of expression from a given plasmid is often two to three orders of magnitude higher in a line derived from stably transformed S2 cell clone than in transiently expressing cells. Stable transformants can be characterized to a level impossible in transient experiments. However, in addition to the obvious drawbacks of time and effort, there are some problems with stable transformants that limit their utility. First, exogenous proteins may be toxic to the expressing cells and prevent recovery of transformed lines; this can be alleviated to some extent by the use of inducible promoters, but we know of no *Drosophila* promoter whose uninduced level is negligible (see following). Toxicity is often very difficult to predict; in parallel attempts to transform S2 cells with the closely related EGF-like cell surface proteins Notch and Delta, both expressed on a *Mt* promoter, we found Delta-expressing clones to be numerous and healthy, while Notch-expressing clones were rare and slow-growing (L. Cherbas, R. Fehon, and C. R. Regan, unpublished). Even CAT is toxic at high levels of expression (L. Cherbas, unpublished). Second, while most of the exogenous DNA is probably extrachromosomal in transient experiments, DNA that is stably integrated is subject to structural and/or position effects which may prove inconvenient (see below).

We have had extensive experience with stably transformed S2 cells and more limited experience with stably transformed Kc167 cells. In what follows, we will emphasize the properties of S2 cells stably tranformed with supercoiled plasmid DNA by calcium phosphate–DNA coprecipitation (using HBS), with some comments on the properties of Kc167 cells where these are known.

1. Characteristics of Stable Transformants

a. Copy Number: Array Formation

Stably transformed S2 cells generally contain several hundred to several thousand copies of the input plasmid DNA, most of it in the form of very long head-to-tail arrays. The large amount of exogenous DNA is not a consequence of selection for high copy number of the selectable marker; cells containing only a few copies of the selectable marker (pHGCO, a copia-DHFR construct conferring resistance to methotrexate) are fully resistant to the selection conditions. When two plasmids are cotransformed, nearest-neighbor analysis leads to the conclusion that the array is formed primarily by homologous recombination rather than by rolling circle replication (Moss, 1985). Stably transformed Kc167 cells are also reported to carry very long head-to-tail arrays (Bourouis and Jarry, 1983); in our experience, the number of copies of the exogenous DNA is lower than in S2 cells, but we do not have extensive data on this point.

b. Cotransformation

A useful consequence of the tendency of S2 and Kc167 cells to form long arrays is the fact that it is not necessary to link covalently the selectable marker and the gene to be expressed prior to transformation. After cotransfection with equal amounts of pHGCO (carrying methotrexate resistance) and a second plasmid, S2 cells selected for resistance to methotrexate always contain the second plasmid as well (Moss, 1985). Cotransformation has also been used successfully with α-amanitin (A. Bieber, personal communication) and hygromycin B (Johansen et al., 1989) selection systems. By varying the ratio of plasmids in the input mixture, it is possible to control the approximate copy number of a given plasmid in the transformed cells. Thus, not only does cotransfection save the labor of constructing plasmids containing both the selectable marker and the gene to be expressed, but it permits the experimenter to exercise some control over the final copy number.

c. Uniformity of Clones

When individual transformed S2 clones (see Section 3) are examined, the copy number for a given plasmid is found to be roughly proportional to the concentration of that plasmid in the input calcium phosphate–DNA coprecipitate. However, the copy number may vary by 10-fold or more in a series of clones derived from a single transfected plate. Furthermore, although the expression of the reporter is roughly proportional to the number of copies of the plasmid in a stably transformed clone, the activity per copy may also vary by 10-fold or more in the products of a single transfection (Moss, 1985).

The variability in expression levels no doubt reflects position effects; even though the plasmids form long mixed arrays, there are statistical variations in the array structure, and the chromosomal insertion sites may have some influence on expression. When Kc167 cells were stably transformed with ecdysone-inducible constructs, we found a systematic variation in the extent of the ecdysone response; cells transfected with a 1:1 mixture of p8HCO (a plasmid conferring methotrexate resistance) and an ecdysone-inducible construct showed a much stronger ecdysone induction of the reporter than cells transfected with a 10:1 mixture of the same plasmids (K. Lee and L. Cherbas, unpublished). This phenomenon is most easily interpreted as an effect of the input ratio of plasmids on the nearest-neighbor frequencies in the array, suggesting that ecdysone response elements in closely spaced plasmid copies may act cooperatively. Position effects of this sort severely reduce the value of stable transformation for the analysis of promoter function; transient expression is far preferable for this purpose.

Although the cloned stable transformants are generally quite stable in their properties even in the absence of continuous selection (Moss, 1985), the cells within a clone are frequently not uniform in their level of expression of the exogenous DNA. There is substantial variation in antibody staining of clones of S2 cells transformed with Notch or Delta expressed on a *Mt* promoter (L. Cherbas, R. Fehon, and C. R. Regan, unpublished) and in β-galactosidase staining of Kc167 clones transformed with an ecdysone-inducible *lacZ* construct (Lee, 1990).

d. Frequency of Transformants

Estimates of transformation frequency can be derived from cloning transfected cells under selective conditions. Because several days intervene between transfection and cloning, these estimates are only approximate. Moss (1985) estimated that when S2 cells were transfected with supercoiled pHGCO (a plasmid conferring methotrexate resistance; Bourouis and Jarry, 1983) by calcium phosphate–DNA coprecipitation (using HBS in place of BBS), approximately 1 cell in 1500 became methotrexate-resistant. Kc167 cells electroporated with 1–10 μg/plate of linearized act-DHFR (another plasmid conferring methotrexate resistance) yielded approximately 1 transformant per 10^4 cells per microgram of DNA (L. Cherbas, unpublished).

e. Promoters

Inducible promoters may permit the selection of a stable line that is capable of expressing a protein potentially toxic to the cells. Ideally, such a promoter should have very low basal expression coupled with high induced expression, and the conditions used for its induction should cause minimal effects on the cell other than the induction of the desired gene. We know of no promoter that meets these criteria perfectly, but the three types of inducible promoters listed in Table 2 have been found useful.

The heat-inducible *hsp70* promoter has proved rather disappointing. Using S2 cells stably transformed with the *hsp70–Adh* fusion plasmid pHAP (Bonner *et al.*, 1984), Bunch (1988) found that the *hsp70* promoter is induced about 100-fold at the transcriptional level, but the reporter protein was induced only 3- to 4-fold, presumably because heat shock inhibits splicing and the translation of non-heat-shock genes. This small translational induction, combined with pleiotropic effects of the heat-shock used to induce the promoter, limits the utility of the promoter.

The *Mt* promoter has been widely used; induction with Cu^{2+} rather than Cd^{2+} minimizes the toxicity of the inducing conditions (Bunch *et al.*, 1988). The endogenous *Mt* promoter is 30- to 100-fold inducible by Cu^{2+} (Bunch *et al.*, 1988); in our experience, individual clones carrying *Mt*–reporter fusions vary somewhat in their inducibility, but rarely exceed 30-fold.

A variety of ecdysone-inducible promoters have been made in our laboratory, and some have been used for stable transformation (Lee, 1990). Although the induction is often higher than for *hsp70* and *Mt*, the utility of these promoters is limited by the fact that the inducing agent, ecdysone, has profound effects on the physiology of most *Drosophila* cell lines.

2. Selection Systems

Table III lists five systems that have been used for selection of stably transformed *Drosophila* lines. Of these, the *gpt*/HAT system is relatively unwieldy and has not come into general use. The other four are relatively simple systems, in which the selective drug is simply added to normal medium, and cells not carrying the rescue plasmid die more or less promptly. Selection protocols for methotrexate, hygromycin, and G418 have been published by Ashburner (1989b); α-amanitin should be added 2 days after transfection, at a concentration of 5 μg/ml for S2 cells, or 10 μg/ml for Kc cells (A. J. Bieber, personal communication). These drugs vary widely in cost and in convenience of use. At current American prices, the cost of the selective drug per 10-ml plate of selective medium is \$3.90 for α-amanitin (5 μg/ml), \$.70 for G418 (1 mg/ml), \$.20 for hygromycin B (200 μg/ml), and \$.05 for methotrexate (90 ng/ml). The comments below reflect the extensive use in our laboratory of methotrexate and our very limited experience with G418 and α-amanitin.

Methotrexate works well in a variety of media, but as one might expect, its action depends on the concentration of nucleotides in the medium, normally contributed by yeast extract. We have generally used M3 medium, containing either 1 or 2 g/liter YE; the cells take several days to stop dividing, and about a week to start lysing, but eventually all of the untransformed cells die. If the

Table III
Selection Systems for Stable Transformation of *Drosophila* Cell Lines

Selection	Rescue plasmid	Selected references
gpt/HAT medium or equivalent	pCV2gpt	Sinclair *et al.*, 1983; Allday *et al.*, 1985; Sinclair *et al.*, 1986
	DSV-ARS	Kurata and Marunouchi, 1988
Methotrexate	pHGCO	Bourouis and Jarry, 1983; Moss, 1985; Martin and Jarry, 1988; Bunch and Goldstein, 1989; Johansen *et al.*, 1989; Feder *et al.*, 1992
	p8HCO	Rebay *et al.*, 1991
	act-DHFR	L. Cherbas, unpublished
Hygromycin B	pcophyg	Rio *et al.*, 1986; Koelle *et al.*, 1991; Cumberledge and Krasnow, 1993
	pcodhygro	Johansen *et al.*, 1989
α-amanitin	pPC4	Thomas and Elgin, 1988; A. J. Bieber, personal communication
G418	pcopneo	Rio and Rubin, 1985; Rio *et al.*, 1986
	hsp82-neo	Sass, 1990

YE content is raised to 3 g/liter, large numbers of untransformed cells continue to grow. We have found methotrexate selection to work in Schneider's medium about as well as in M3; in the serum-free medium HyQ-CCM3 methotrexate effects appear more rapidly than in M3. Note that when a suitable medium is used, we find no methotrexate-resistant cells without transformation; hence, methotrexate resistance is conferred exclusively by the addition of a methotrexate-resistant DHFR gene rather than by amplification of the endogenous DHFR gene.

G418 in our hands is also somewhat medium-dependent; it works faster and at lower concentrations in HyQ-CCM3 than in M3. It is reputed to vary in its efficacy from lot to lot (Ashburner, 1989b).

α-Amanitin works very quickly, and equally well, in M3 and HyQ-CCM3. So far as we can tell, its only defect is its very high price, and—for some applications—the large size of the resistance gene.

3. Cloning

Because there is substantial clone-to-clone variation in the expression of a transformed plasmid, and because bulk selection of a transfected population generally leads to the retention of only the fastest-growing transformed clone within the starting population, it is frequently useful to clone cells as soon as possible after transfection. We allow the cells 2 days to express the selectable marker before adding the selective agent and then allow 4 days for the selective agent to act before cloning in its presence. If the last step is omitted, cloning in methotrexate gives a large number of small abortive clones derived from methotrexate-sensitive cells; these appear to inhibit the growth of methotrexate-resistant clones by overcrowding.

References

Abrams, J. M., Lux, A., Steller, H., and Krieger, M. (1992). Macrophages in *Drosophila* embryos and L2 cells exhibit scavenger receptor-mediated endocytosis. *Proc. Natl. Acad. Sci. U.S.A.* **89,** 10375–10379.

Allday, M. J., Sinclair, J. H., MacGillivray, A. J., and Sang, J. H. (1985). Efficient expression of an Epstein–Barr nuclear antigen in *Drosophila* cells transfected with Epstein–Barr virus DNA. *EMBO J.* **4,** 2955–2959.

Ananthan, J., Baler, R., Morrissey, D., Zuo, J., Lan, Y., Weir, M., and Voellmy, R. (1993). Synergistic activation of transcription is mediated by the N-terminal domain of *Drosophila* fushi tarazu homeoprotein and can occur without DNA binding by the protein. *Mol. Cell. Biol.* **13,** 1599–1609.

Andres, A. J., and Cherbas, P. (1992). Tissue-specific ecdysone responses: Regulation of the *Drosophila* genes *Eip28/29* and *Eip40* during larval development. *Development* **116,** 865–876.

Ashburner, M. (1989a). "*Drosophila:* A Laboratory Handbook." Cold Spring Harbor, NY: Cold Spring Harbor Laboratory Press.

Ashburner, M. (1989b). "*Drosophila:* A Laboratory Manual." Cold Spring Harbor, NY: Cold Spring Harbor Laboratory Press.

Ayer, S., and Benyajati, C. (1990). Conserved enhancer and silencer elements responsible for differential *Adh* transcription in *Drosophila* cell lines. *Mol. Cell. Biol.* **10**, 3512–3523.

Ayer, S., Walker, N., Mosammaparast, M., Nelson, J. P., Shilo, B.-Z., and Benyajati, C. (1993). Activation and repression of *Drosophila* alcohol dehydrogenase distal transcription by steroid hormone receptor superfamily members binding to a common response element. *Nucleic Acids Res.* **21**, 1619–1627.

Benyajati, C., and Dray, D. F. (1984). Cloned *Drosophila* alcohol dehydrogenase genes are correctly expressed after transfection into *Drosophila* cells in culture. *Proc. Natl. Acad. Sci. U.S.A.* **81**, 1701–1705.

Berger, E. M., Goudie, K., Klieger, L., Berger, M., and DeCato, R. (1992). The juvenile hormone analogue, methoprene, inhibits ecdysterone induction of small heat-shock protein gene expression. *Dev. Biol.* **151**, 410–418.

Best-Belpomme, M., Courgeon, A., and Rambach, A. (1978). β-Galactosidase is induced by hormone in *Drosophila melanogaster* cell cultures. *Proc. Natl. Acad. Sci. U.S.A.* **75**, 6102–6106.

Bonner, J. J., Parks, C., Parker-Thornburg, J., Mortin, M. A., and Pelham, R. B. (1984). The use of promoter fusions in *Drosophila* genetics: Isolation of mutations affecting the heat shock response. *Cell* **37**, 979–991.

Bourouis, M., and Jarry, B. (1983). Vectors containing a prokaryotic dihydrofolate reductase gene transform *Drosophila* cells to methotrexate resistance. *EMBO J.* **2**, 1099–1104.

Boyd, L., and Thummel, C. S. (1993). Selection of CUG and AUG initiation codons for *Drosophila* E74A translation depends on downstream sequences. *Proc. Natl. Acad. Sci. U.S.A.* **90**, 9164–9167.

Braude-Zolotarjova, T. Ya., and Schuppe, N. G. (1987). Transient expression of hsp-CAT1 and copia-CAT1 hybrid genes. *Drosophila Information Service* **66**, 33.

Bunch, T. A. (1988). Characterization and use of inducible promoters in *Drosophila* cells. Ph.D. thesis, Harvard University.

Bunch, T. A., and Goldstein, L. S. B. (1989). The conditional inhibition of gene expression in cultured *Drosophila* cells by antisense RNA. *Nucleic Acids Res.* **17**, 9781–9782.

Bunch, T. A., Grinblat, Y., and Goldstein, L. S. B. (1988). Characterization and use of the *Drosophila* metallothionein promoter in cultured *Drosophila melanogaster* cells. *Nucleic Acids Res.* **16**, 1043–1061.

Burke, J. F., Sinclair, J. H., Sang, J. H., and Ish-Horowicz, D. (1984). An assay for transient gene expression in transfected *Drosophila* cells, using [³H]guanine incorporation. *EMBO J.* **3**, 2549–2554.

Chen, C., and Okayama, H. (1987). High-efficiency transformation of mammalian cells by plasmid DNA. *Mol. Cell. Biol.* **7**, 2745–2752.

Cherbas, L., and Cherbas, P. (1981). The effects of ecdysteroid hormones on *Drosophila melanogaster* cell lines. *Adv. Cell Culture* **1**, 91–124.

Cherbas, L., Lee, K., and Cherbas, P. (1991). Identification of ecdysone response elements by analysis of the *Drosophila Eip28/29* gene. *Genes Dev.* **5**, 120–131.

Cumberledge, S., and Krasnow, M. A. (1993). Intercellular signalling in *Drosophila* segment formation reconstructed *in vitro*. *Nature* **363**, 549–552.

Davies, J. A., Addison, C. F., Delaney, S. J., Sunkel, C., and Glover, D. M. (1986). Expression of the prokaryotic gene for chloramphenicolacetyl transferase in *Drosophila* under the control of larval serum protein 1 gene promoters. *J. Mol. Biol.* **189**, 13–24.

DiNocera, P. P., and Dawid, I. G. (1983). Transient expression of genes introduced into cultured cells of *Drosophila*. *Proc. Natl. Acad. Sci. U.S.A.* **80**, 7095–7098.

Echalier, G., and Ohanessian, A. (1970). *In vitro* culture of *Drosophila melanogaster* embryonic cells. *In Vitro* **6**, 162–172.

Feder, J. H., Rossi, J. M., Solomon, J., Soloman, N., and Lindquist, S. (1992). The consequences of expressing hsp70 in *Drosophila* cells at normal temperatures. *Genes Dev.* **6**, 1402–1413.

Fehon, R. G., Kooh, P. J., Rebay, I., Regan, C. L., Xu, T., Muskavitch, M. A. T., and Artavanis-

Tsakonas, S. (1990). Molecular interactions between the protein products of the neurogenic loci *Notch* and *Delta,* two EGF-homologous genes in Drosophila. *Cell* **61,** 523–534.

Gateff, E., Gissmann, L., Shrestha, R., Plus, N., Pfister, H., Schröder, J., and zur Hausen, H. (1980). Characterization of two tumorous blood cell lines of *Drosophila melanogaster* and the viruses they contain. *In* "Invertebrate Systems *In Vitro*" (E. Kurstak, K., Maramorosch, and A. Dübendorfer, eds.), pp. 517–533. Amsterdam: Elsevier/North Holland Biomedical Press.

Gibson, K. R., Vanek, P. G., Kaloss, W. D., Collier, G. B., Connaughton, I. F., Angelichio, M., Livi, B. P., and Fleming, P. J. (1993). Expression of dopamine β-hydroxylase in *Drosophila* Schneider 2 cells: Evidence for a mechanism of membrane binding other than uncleaved signal peptide. *J. Biol. Chem.* **268,** 9490–9495.

Imhof, M. O., Rusconi, S., Lezzi, M. (1993). Cloning of a *Chironomus tentans* cDNA encoding a protein (cEcRH) homologous to the *Drosophila melanogaster* ecdysteroid receptor (dEcR). *Insect Biochem. Mol. Biol.* **23,** 115–124.

Ivey-Hoyle, M., and Rosenberg, M. (1990). Rev-dependent expression of human immunodeficiency virus type 1 gp160 in *Drosophila melanogaster* cells. *Mol. Cell. Biol.* **10,** 6152–6159.

Johansen, H., van der Straten, A., Sweet, R., Otto, E., Maroni, G., and Rosenberg, M. (1989). Regulated expression at high copy number allows production of a growth-inhibitory oncogene product in *Drosophila* Schneider cells. *Genes Dev.* **3,** 882–889.

Kappler, C., Meister, M., Lagueux, M., Gateff, E., Hoffmann, J. A., and Reichhart, J.-M. (1993). Insect immunity: Two 17 bp repeats nesting a [κ]B-related sequence confer inducibility to the diptericin gene and bind a polypeptide in bacteria-challenged *Drosophila*. *EMBO J.* **12,** 1561–1568.

Koelle, M. R., Talbot, W. S., Segraves, W. A., Bender, M. T., Cherbas, P., and Hogness, D. S. (1991). The Drosophila *EcR* gene encodes an ecdysone receptor, a new member of the steroid receptor superfamily. *Cell* **67,** 59–77.

Krasnow, M. A., Saffman, E. E., Kornfeld, K., and Hogness, D. S. (1989). Transcriptional activation and repression by *Ultrabithorax* proteins in cultured Drosophila cells. *Cell* **57,** 1031–1043.

Kurata, N., and Marunouchi, T. (1988). Retention of autonomous replicating plasmids in cultured *Drosophila* cells. *Mol. Gen. Genet.* **213,** 359–363.

Lawson, R., Mestril, R., Schiller, P., and Voellmy, R. (1984). Expression of heat shock β-galactosidase hybrid genes in cultured *Drosophila* cells. *Mol. Gen. Gen.* **198,** 116–124.

Lee, K. (1990). The identification and characterization of ecdysone response elements. Ph.D. thesis, Indiana University.

Martin, M., and Jarry, B. (1988). Expression of transfected genes in a *Drosophila* cell line in transient assay and stable transformation. *J. Insect Physiol.* **34,** 691–699.

Morganelli, C., and Berger, E. M. (1984). Transient expression of homologous genes in *Drosophila* cells. *Science* **224,** 1004–1006.

Moss, R. E. (1985). Analysis of a transformation system for *Drosophila* tissue culture cells. Ph.D. thesis, Harvard University.

Rebay, I., Fleming, R. J., Fehon, R. G., Cherbas, L., Cherbas, P., and Artavanis-Tsakonas, S. (1991). Specific EGF repeats of Notch mediate interactions with Delta and Serrate: Implications for Notch as a multifunctional receptor. *Cell* **67,** 687–699.

Rio, D. C., and Rubin, G. M. (1985). Transformation of cultured *Drosophila melanogaster* cells with a dominant selectable marker. *Mol. Cell. Biol.* **5,** 1833–1838.

Rio, D. C., Laski, F. A., and Rubin, G. R. (1986). Identification and immunological analysis of biologically active *Drosophila* P element transposase. *Cell* **44,** 21–32.

Robb, J. A. (1969). Maintenance of imaginal discs of *Drosophila melanogaster* in chemically defined media. *J. Cell Biol.* **41,** 876–885.

Sang, J. H. (1981). *Drosophila* cells and cell lines. *Adv. Cell Culture* **1,** 125–182.

Sass, H. (1990). P-transposable vectors expressing a constitutive and thermoinducible *hsp82-neo* fusion gene for *Drosophila* germline transformation and tissue-culture transfection. *Gene* **89,** 179–186.

Saunders, S. E., Rawls, J. M., Wardle, C. J., and Burke, J. F. (1989). High efficiency expression of transfected genes in a *Drosophila melanogaster* haploid (1182) cell line. *Nuclei Acids Res.* **17,** 6205–6216.

Schneider, I. (1964). Differentiation of larval *Drosophila* eye-antennal discs in vitro. *J. Exp. Zool.* **156,** 91–103.

Schneider, I., and Blumenthal, A. B. (1978). *Drosophila* cell and tissue culture. *In* "The Genetics and Biology of *Drosophila*" (M. Ashburner and T. R. F. Wright, eds.), Vol. 2a, pp. 265–315. London: Academic Press.

Shields, G., and Sang, J. H. (1977). Improved medium for culture of *Drosophila* embryonic cells. *Drosophila Information Service* **52,** 161.

Sinclair, J. H., and Bryant, L. A. (1987). 20-Hydroxyecdysone increases levels of transient gene expression in transfected *Drosophila* cells. *Nucleic Acids Res.* **15,** 9255–9261.

Sinclair, J. H., Sang, J. H., Burke, J. F., and Ish-Horowicz, D. (1983). Extrachromosomal replication of *copia*-based vectors in cultured *Drosophila* cells. *Nature* **306,** 198–200.

Sinclair, J. H., Burke, J. F., Ish-Horowicz, D., and Sang, J. H. (1986). Functional analysis of the transcriptional control regions of the *copia* transposable element. *EMBO J.* **5,** 2349–2354.

Thomas, G. H., and Elgin, S. C. R. (1988). The use of the gene encoding the α-amanitin-resistant subunit of RNA polymerase II as a selectable marker in cell transformation. *Drosophila Information Service* **67,** 85.

Tourmente, S., Chapel, S., Dreau, D., Drake, M. E., Bruhal, A., Couderc, J. L., and Dastugue, B. (1993). Enhancer and silencer elements within the first intron mediate the transcriptional regulation of the β_3 tubulin gene by 20-hydroxyecdysone in *Drosophila* Kc cells. *Insect Biochem. Mol. Biol.* **23,** 137–143.

Ui, K., Ueda, R., and Miyake, T. (1989). *In vitro* culture of cells from dissociated imaginal discs of *Drosophila melanogaster*. *In* "Invertebrate Cell System Applications" (J. Mitsuhashi, ed.), Vol. II, pp. 213–221. Boca Raton, FL: CRC Press.

Walker, V. K., Schreiber, M., Purvis, C., George, J., Wyatt, G. R., and Bendena, W. G. (1991). Yolk polypeptide gene expression in cultured *Drosophila* cells. *In Vitro Cell Dev. Biol.* **27A,** 121–127.

Wigler, M., Pellicer, A., Silverstein, S., Axel, R., Urlaub, G., and Chasin, L. (1979). DNA mediated transfer of the adenine-phosphoribosyltransferase locus into mammalian cells. *Proc. Natl. Acad. Sci. U.S.A.* **76,** 1373–1376.

Yoshinaga, S. K., and Yamamoto, K. R. (1991). Signaling and regulation by a mammalian glucocorticoid receptor in *Drosophila* cells. *Mol. Endocrinol.* **5,** 844–853.

PART III

Biochemical Preparation Methods

For many years, *Drosophila* workers suffered comments to the effect that flies are less than stellar subjects for biochemical studies. While the small size of *Drosophila* can be an impediment, objective consideration of this issue reveals little supporting evidence. In fact, what is most remarkable is how much excellent biochemistry is represented in recent fly research. These developments can be traced to a number of changes in methodological and philosophical approaches to *Drosophila* work. Prominent among them are methods for rearing large numbers of flies, invention of a variety of microassays, development of *in situ* methods for nucleic acid hybridization and antibody decoration, and the ability to express proteins encoded by cloned genes in bacteria, yeast, and cultured cells. As demonstrated in the following chapters, these improvements, collectively and in combination with the genetic and cytological advantages offered by flies, make *Drosophila* an excellent subject for biochemical investigations.

Shaffer *et al.* begin this section with a protocol for preparing *Drosophila* nuclei (Chapter 10). Adaptations for either large- or small-scale isolations are included, and the method will serve as a starting point for the isolation of proteins, DNA, or RNA.

Matunis *et al.* follow with a chapter describing the isolation of *Drosophila* RNA-binding proteins (Chapter 11). The authors have successfully employed chromatography on single-stranded DNA or RNA homopolymers for this purpose. Descriptions of several methods for characterizing this interesting class of proteins as well as their two-dimensional gel electrophoretic profile are included. Genetic studies, which are just beginning, have already correlated the *squid* gene with one such protein, hrp40.

Becker *et al.* next describe the preparation of chromatin assembly extracts from fly embryos (Chapter 12). Included are protocols for several assays, including DNA supercoiling, nuclease digestion, and incorporation of exogenously added histone H1. The authors finally provide instructions for monitoring chromatin organization on specific genes.

Kamakaka and Kadonaga (Chapter 13) make a different kind of extract, a soluble nuclear fraction, that is highly active in gene transcription *in vitro*. This extract can be made in only 2.5 hr, using 0.5–150 g of embryos and appears to be particularly useful when used in conjunction with chromatin assembly extracts of the sort described by Becker *et al.* Included in this chapter are instructions for preparing extract, setting up *in vitro* transcription, and monitoring RNA quality using primer extension.

The chapter by Fyrberg *et al.* (Chapter 14) is not strictly biochemical, but describes methods for working with muscle, a tissue whose biochemistry is known in considerable detail. Included are instructions for monitoring *Drosophila* muscle quality using a battery of methods and a list of several reporter gene constructs and antibodies available to track aspects of both fiber development and myofibril assembly.

Miller and Kellogg have undertaken the large task of isolating and characterizing a broad sample of *Drosophila* cytoskeletal proteins and generously provide here a wealth of information for using affinity columns and characterizing the proteins thus isolated using immunologic methods (Chapter 15). Their contribution is all the more useful in view of recent advances in the study of the *Drosophila* cytoskeleton, which is at last being linked to a variety of cellular phenomena, including cellularization during blastoderm formation, cell shape changes during gastrulation, aquisition of cellular polarity, and directed cellular transport both in oocytes and somatic cells such as neurons.

Saxton follows with a chapter focused on isolation and analysis of *Drosophila* motor proteins (Chapter 16). *Drosophila* kinesin and its relatives have played a central role in the microtubule motor field, particularly in relating new proteins to mitotic and meiotic defects. More recent research has demonstrated that motor proteins provide at least partial explanations for both the *shibire* and *Glued* mutants, as well. Saxton's chapter includes techniques for preparing microtubules and kinesin-related proteins, as well as for setting up motility assays. The protocols described here will serve as a solid starting point for discovering and analyzing novel classes of microtubule-based motors.

Considering the prominent roles played by membrane proteins in cell–cell interactions, it is surprising that more cell and developmental biologists have not elected to characterize *Drosophila* membrane proteins. Hortsch addresses this deficiency in Chapter 17, which describes how to isolate *Drosophila* membranes and affinity purify the component proteins. The techniques described here will allow readers to explore the intricacies of protein modifications and their relationships with various developmental events. Recent successes with *Drosophila* fasciclins and the PS antigens (integrins) bode well for this research.

In the final chapter (Chapter 18), a natural companion to that preceding it, Fessler *et al.* discuss the preparation of *Drosophila* extracellular matrix. This comprehensive treatise runs from isolation to characterization and then to assays for evaluating the interactions of matrix with cells in culture. The information included here is of enormous importance for further delineating a variety of cellular interactions that figure prominently in many aspects of *Drosophila* development. Particularly informative are descriptions of cell lines that secrete ECM proteins, thus facilitating their isolation.

CHAPTER 10

Preparation of *Drosophila* Nuclei

Christopher D. Shaffer, Joann M. Wuller, and Sarah C. R. Elgin

Department of Biology
Washington University
St. Louis, Missouri 63130

I. Introduction

With the techniques described in previous chapters to rear large amounts of *Drosophila,* it is easy to obtain sufficient material for biochemical analysis. The protocols we discussed in our earlier chapter (Chapter 5) have been designed to limit the amount of time required from highly trained individuals and thus reduce the total cost of the material produced.

We have found embryos 6–18 hr old to be a good source of material for the preparation of nuclei, chromatin, DNA, RNA, proteins, etc. Embryos have several advantages over other life stages for work involving nuclei, DNA, etc. The relative number of nuclei per gram of material is high, and virtually all nuclei are diploid (Rudkin, 1972). If the embryos are dechorionated before use, contamination with yeast, etc. is nil. Levels of endogenous nucleases and proteases are not particularly high. There are no feeding costs for the embryos, so the overall cost per gram of material is relatively low (ca. $1.50/gram).

The protocols below describe large-scale procedures for the isolation of nuclei in three main steps. The isolation begins with the removal of the chorion. This is followed by disruption of the embryos and recovery of nuclei (Elgin and Miller, 1978). Gloves should be worn at all times to avoid contaminating the nuclei with degradative enzymes found on the hands. All buffers are prepared fresh and kept on ice.

II. Dechorionation of Embryos

Twenty to forty grams of embryos obtained as described and stored at $-80°C$ are used. Any large clumps should be broke up while the material is still frozen. (Fresh embryos may also be used.) The chorion of the embryos is removed by treatment with 200 ml of half-strength Clorox for 2 min at room temperature with stirring (Hill, 1945). The embryos are collected on a nylon filter (Nitex 116, see Fig. 1 in Chapter 5), washed thoroughly with cold water, and washed with 70% alcohol followed by more water. The embryos are immediately resuspended in buffer (see following) and placed on ice. All subsequent operations are carried out at 0–4°C.

III. Disruption of Embryos

The initial disruption of the embryos and subsequent filtration are carried out in a cold room. Large quantities of embryos are most conveniently disrupted by rapid decompression in a nitrogen bomb as follows. The dechorionated embryos are resuspended in 300 ml of *Drosophila* buffer [0.05M Tris–HCl (pH 7.6), 25 mM KCl, 5 mM MgOAc, 0.35 M sucrose] using a magnetic stirrer. The bomb is taken up to 300 lb/square inch with N_2 and the embryos allowed to equilibrate for 15 min with stirring. At the end of this period, the material is collected through the outlet valve while maintaining a 300 lb/square inch pressure in the bomb. This procedure works well for standard preparations of proteins such as histones. However, some proteins apparently leak out of the nuclei in this procedure, as the recovered nuclei give poor results in a transcription assay. The embryos can also be disrupted by homogenization. Embryos are resuspended in a 10-fold volume of buffer A-1M [60 mM KCl, 15 mM NaCl, 1 mM EDTA, 0.1 mM EGTA, 15 mM Tris–HCl (pH 7.4), 0.15 mM spermine, 0.5 mM spermidine, 0.5 mM dithiothreitol, 0.1 mM phenylmethylsulfonyl fluoride (PMSF), 1 M sucrose] and homogenized in a stainless-steel homogenizer with teflon pestle. (Glass homogenizers are not recommended for this procedure due to the high frequency of breakage and potential danger.)

IV. Isolation of Nuclei

The disrupted embryos are filtered through two layers of Miracloth (Chicopee Mills, Inc.). The filtrate is centrifuged at low speed (480g) for 9 min and the pellet of debris is discarded. The supernatant is collected and spun at 4300g for 10 min to pellet the crude nuclei. The supernatant is removed by aspiration making sure to remove the top fatty layer, and the nuclear pellet is resuspended in 20 ml of buffer A*-1M [Buffer A*-1M is buffer A-1M without EGTA or EDTA]. Ten milliliters of the nuclear suspension is carefully placed on top of 20 ml of buffer A*-1.7M [60 mM KCl, 15 mM NaCl, 15 mM Tris–HCl (pH 7.4), 0.15 mM spermine, 0.5 mM spermidine, 0.5 mM dithiothreitol, 0.1 mM PMSF, 1.7 M sucrose] in a 30-ml centrifuge tube. The interface is mixed a bit to ensure proper collection of the nuclei. The nuclei are then pelleted by spinning in a swinging bucket rotor at 19,600g for 30 min. The supernatant is then removed by aspiration and the pellet resuspended in buffer appropriate for subsequent analysis.

Nuclei isolated as described have been used successfully in many protocols. However, some chromosomal proteins (including HMG proteins) are known to be extracted in buffers that contain spermine and spermidine (Pfeifer and Riggs, 1991; Wagner et al., 1992). The described protocols have been successfully modified for nuclear isolation without polyamines for studies where this was necessary (Thomas and Elgin, 1988). In these cases, "nuclear buffer" [60 mM KCl, 15 mM NaCl, 15 mM Tris–HCl (pH 7.4), 5 mM MgCl$_2$, 0.1 mM EGTA, 0.5 mM dithiothreitol, 0.1 mM PMSF, 1.0 M sucrose] is used. Other variations on the described protocol have also been reported (Gilmour et al., 1988; Biggin and Tjian, 1988).

V. Nuclei from Other Developmental Stages

For a variety of problems it may be necessary to isolate large amounts of nuclei from stages other than embryos. Following are small-scale procedures that could be used as a starting point for developing large-scale protocols for larvae and adults. For experiments involving a large number of strains (e.g., P-element transformants), we have found that the small-scale protocols described here give sufficient nuclei for a variety of protocols. The procedures have been described previously (Lu et al., 1993) in greater detail.

A. Isolation of Nuclei from Small Amounts of Embryos and Larvae

1. One gram of larvae or washed embryos is pulverized in liquid nitrogen, using a mortar and pestle.

2. After the liquid nitrogen has evaporated, the homogenate is transferred

to a beaker containing 3 ml of buffer A + NP [60 mM KCl, 15 mM NaCl, 13 mM EDTA, 0.1 mM EGTA, 15 mM Tris–HCl (pH 7.4), 0.15 mM spermine, 0.5 mM spermidine, 0.5 mM dithiothreitol (DTT), 0.5% Nonidet P-40 (NP-40)] and mixed.

3. The suspension is transferred to a 7-ml ground glass homogenizer (Pyrex) and disrupted with five strokes of the pestle.

4. The homogenate is transferred to a 15-ml dounce homogenizer (Wheaton, Millville, NJ), disrupted with 10 strokes of the pestle (type B), and filtered through two layers of 89-nm nylon mesh cloth (Nitex).

5. Buffer A + NP (2 ml) is used to rinse the ground glass homogenizer and the dounce homogenizer, and the suspension is filtered through the same two layers of nylon mesh cloth; it is necessary to squeeze the cloth to complete filtration.

6. The filtrate is loaded over 1 ml of buffer AS [60 mM KCl, 15 mM NaCl, 1 mM EDTA, 0.1 mM EGTA, 15 mM Tris–HCl (pH 7.4), 0.15 mM spermine, 0.5 mM spermidine, 0.5 mM dithiothreitol, 0.3 M sucrose] and centrifuged in a Sorvall (Norwalk, CT) HB-4 swinging bucket rotor at 4°C, 3000 rpm (1500 g) for 5 min.

7. The recovered pellet is resuspended in 3 ml of buffer A + NP, transferred to a 7-ml dounce homogenizer (Wheaton), and dispersed with five strokes of the pestle (type A). The mixture is loaded over 1 ml of buffer AS and centrifuged as before.

8. The recovered pellet is resuspended in 3 ml of buffer A [60 mM KCl, 15 mM NaCl, 1 mM EDTA, 0.1 mM EGTA, 15 mM Tris–HCl (pH 7.4), 0.15 mM spermine, 0.5 mM spermidine, 0.5 mM dithiothreitol], transferred to the 7-ml dounce homogenizer, and dispersed with five strokes of the pestle (type B).

9. Nuclei are collected by centrifugation in the same rotor at 2000 rpm (700g) for 5 min and are resuspended in the appropriate buffer.

B. Isolation of Nuclei from Adult Flies

The following protocol for the isolation of nuclei from adult flies is a modification of the isolation procedure for larval nuclei described above, using the same buffers.

1. One gram of adult flies is pulverized in liquid nitrogen, using a mortar and pestle.

2. After the liquid nitrogen has evaporated, the homogenate is transferred to a beaker containing 5 ml of buffer A + NP and mixed.

3. The suspension is transferred to a 7-ml ground glass homogenizer (Pyrex) and disrupted with 10 strokes of the pestle.

4. The homogenate is then transferred to a 15-ml dounce homogenizer (Wheaton) and disrupted with 20 strokes of the pestle (type B). The homogenate is filtered through one layer of Miracloth (Habco Products, NJ). It is necessary to squeeze the Miracloth to complete filtration.

5. Buffer A + NP (3 ml) is used to rinse the ground glass homogenizer and the dounce homogenizer, and the suspension is filtered through the same Miracloth.

6. The filtrate is loaded over 2 ml of buffer AS and centrifuged in a Sorvall HB-4 swinging bucket rotor at 4°C, 3000 rpm (1500g) for 5 min.

7. The recovered pellet is resuspended in 5 ml of buffer A + NP, transferred to a 7-ml dounce homogenizer (Wheaton), and dispersed with 10 strokes of the pestle (type A). The mixture is loaded over 2 ml of buffer AS and centrifuged as before.

8. The recovered pellet is resuspended in 5 ml of buffer A, transferred to the 7-ml dounce homogenizer, and dispersed with 10 strokes of the pestle (type B).

9. The nuclei are collected by centrifugation in the same rotor at 3000 rpm (1484g) for 5 min, as previously. The pellet is washed with another 5 ml of buffer A and centrifuged at 4000 rpm (2600g) for 5 min. The pellet is resuspended in the appropriate buffer for use.

References

Biggin, M. D., and Tijan, R. (1988). Transcription factors that activate the *Ultrabithorax* promoter in developmentally staged extracts. *Cell* **53,** 699–711.

Elgin, S. C. R., and Miller, D. W. (1978). Preparation of nuclei, chromatin, and chromosomal proteins of *Drosophila* embryos. *In* "The Genetics and Biology of *Drosophila*" (M. Ashburner and T. R. F. Wright, eds.), Vol. 2a, pp. 145–150. New York: Academic Press.

Gilmour, D. S., Dietz, T. J., and Elgin, S. C. R. (1988). TATA box-dependent protein-DNA interactions are detected on heat shock and histone gene promoters in nuclear extracts derived from *Drosophila melanogaster* embryos. *Mol. Cell. Biol.* **8,** 3204–3214.

Hill, D. L. (1945). Chemical Removal of the Chorion from *Drosophila* eggs. *Dros. Inf. Serv.* **19,** 62.

Lu, Q., Wallrath, L. L., and Elgin, S. C. R. (1993). Using *Drosophila* P-Element-Mediated Germ Line Transformation to Examine Chromatin Structure and Expression of *in Vitro*-Modified Genes. *In* "Methods in Molecular Genetics" (K. W. Adolph, ed.), Vol. 1, pp. 333–357. San Diego, Academic Press.

Pfeifer, G. P., and Riggs, A. D. (1991). Chromatin differences between active and inactive X chromosomes revealed by genomic footprinting of permeabilized cells using DNase I and ligation-mediated PCR. *Genes Dev.* **5,** 1102–1113.

Rudkin, G. T. (1972). Replication in Polytene Chromosomes. *In* "Developmental Studies on Giant Chromosomes" (W. Beermann, ed.), pp. 58–85. New York: Springer-Verlag.

Thomas, G. H., and Elgin, S. C. R. (1988). Protein/DNA architecture of the DNase I hypersensitive region of the *Drosophila hsp26* promoter. *EMBO J.* **7,** 2191–2201.

Wagner, C. R., Hamana, K., and Elgin, S. C. R. (1992). A high-mobility-group protein and its cDNAs from *Drosophila melanogaster*. *Mol. Cell Biol.* **12,** 1915–1923.

CHAPTER 11

Isolation and Characterization of RNA-Binding Proteins from *Drosophila melanogaster*

Michael J. Matunis,[‡,*] **Erika L. Matunis,*** and **Gideon Dreyfuss**[†]

[‡]Laboratory of Cell Biology
[*]Rockefeller University
New York, New York 10021

[†]Howard Hughes Medical Institute and
Department of Biochemistry and Biophysics
University of Pennsylvania School of Medicine
Philadelphia, Pennsylvania 19104

I. Introduction

In the eukaryotic cell, a significant portion of the post-transcriptional regulation of gene expression is mediated by RNA-binding proteins. Members of this growing family of proteins are required for multiple steps during mRNA metabolism, including pre-mRNA processing (e.g., the splicing, capping, poly-adenylation, and transport of pre-mRNA) and mRNA localization, translation, and stability (Dreyfuss *et al.*, 1993). An increasing number of developmental events in *Drosophilia melanogaster* are known or postulated to require RNA-binding proteins (Rio, 1992; Ding and Lipshitz, 1993; MacDonald, 1992; Kelley, 1993; Matunis, 1992; Matunis *et al.*, 1994), providing unique opportunities for studying these factors and the mechanisms by which processes such as mRNA processing and/or localization are governed.

We have devised several biochemical approaches that can be used to purify and characterize RNA-binding proteins in *D. melanogaster*. These methodologies were originally developed to characterize vertebrate heterogeneous nuclear ribonucleoproteins (hnRNPs), the abundant nuclear proteins that associate with RNA polymerase II transcripts and are involved in their processing and transport (Dreyfuss *et al.*, 1993; Piñol-Roma, 1990a,b). As an initial step, affinity chromatography on ssDNA cellulose provides an extremely clean separation of hnRNP proteins from other cellular factors and an enriched source for further purification of specific proteins (Piñol-Roma, 1990a). Using ssDNA affinity chromatography, we have purified more than 20 proteins from *D. melanogaster* and have characterized several of the major proteins (Matunis *et al.*, 1992a,b). Two-dimensional gel analysis of the proteins purified from embryos by ssDNA chromatography is presented in Fig. 1. Using the procedures outlined in this review, including immunolocalization, UV-cross-linking to poly(A)+ RNA *in vivo,* and immunopurification of ribonucleoprotein complexes, we have identified the major *D. melanogaster* hnRNP proteins and have designated them according to their relative size by SDS–polyacrylamide gel electrophoresis preceded by hrp (Fig. 1). All of the hrp proteins are associated with hnRNA in the nucleus and can be immunopurified as an RNAase-sensitive complex with monoclonal antibodies to hrp40 (Matunis *et al.*, 1992a). Several additional proteins purified by ssDNA chromatography do not have these properties and may belong to other classes of single-stranded nucleic acid-binding proteins (Matunis *et al.*, 1992a). In conjunction with data arising from genetic studies, the biochemical definition of the *D. melanogaster* hnRNP proteins has proven useful in determining how they may function. In particular, the *squid* gene, which was identified in a screen of female-sterile mutants and is required for dorsoventral patterning during oogenesis, encodes hrp40 (Kelley, 1993; Matunis, 1992; Matunis *et al.*, 1992, 1994).

In addition to aiding in the characterization of the most abundant hnRNP proteins, the techniques outlined here can also be applied to characterize and define novel RNA-binding proteins. For example, the Protein on Ecdysone Puffs (PEP) binds preferentially to a subset of chromosomal sites that are

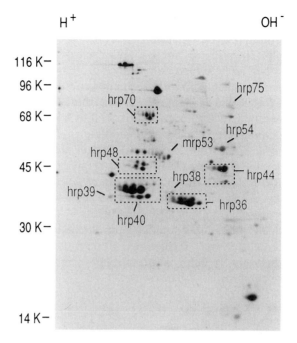

Fig. 1 Two-dimensional gel electrophoresis of proteins purified by ssDNA chromatography. S2 cell lysate was bound to ssDNA-cellulose at 0.1 M NaCl. The column was washed with 1 mg/ml heparin and proteins were eluted with 2 M NaCl. Proteins were resolved by two-dimensional gel electrophoresis (NEPHGE in the first dimension, SDS–PAGE in the second dimension), transferred to immobilon-P membrane, and stained with Coomassie blue. Nomenclature for the indicated proteins is described in Matunis *et al.* (1992). Protein standards are indicated on the left. Reproduced from the *Journal of Cell Biology* (1992) **116**, 244–255, by copyright permission of the Rockefeller University Press.

activated by ecdysone and shares sequence homology to both transcription factors and RNA-binding proteins. Analysis of the behavior of PEP in the assays detailed here demonstrates that this protein is found in a subset of the hnRNP complexes and binds RNA *in vivo* (Amero *et al.*, 1993). The procedures outlined here should, therefore, aid in identifying and characterizing newly discovered RNA-binding proteins.

II. Isolation of RNA-Binding Proteins by Affinity Chromatography

The most common affinity chromatography procedure used in our laboratory for the isolation of RNA-binding proteins from *D. melanogaster* is ssDNA-cellulose chromatography. Overall, purification can be accomplished rapidly and on a large scale. It does not depend on the integrity of RNA and does not require protein denaturing conditions. Furthermore, individual RNA-binding proteins can be separated from one another on the same column, owing to their different affinities for ssDNA. This procedure has been used extensively to isolate and characterize hnRNP proteins from human HeLa cells, and a more detailed discussion of its principles and caveats has been presented previously (Piñol-Roma *et al.*, 1990a). As a note of caution, proteins not related to RNA metabolism may also bind to ssDNA-cellulose (several dehydrogenases for example are avid ssDNA-binding proteins, as are several proteins involved in DNA replication, see Piñol-Roma *et al.*, 1990a) and conversely, not all RNA-binding proteins will necessarily bind to ssDNA. Therefore, binding to ssDNA is not by itself an absolute diagnostic criterion for RNA-binding proteins, and the identification of genuine RNA-binding proteins must be supported by complimentary data as described below.

The procedures outlined below detail the large-scale isolation of RNA-binding proteins from *D. melanogaster* embryos. The same procedures are also applicable to the isolation of proteins from other tissues, other organisms, and from cultured cells.

A. Preparation of Embryo Lysates

All of the following steps should be performed at 4°C unless otherwise noted.

1. Embryos should be collected from population cages, dechorionated, and stored at −80°C as previously described (Ashburner, 1989).
2. In a 50-ml beaker, thaw 10 g of embryos directly in 20 ml of cold buffer A:

    100 mM NaCl
    10 mM Tris–HCl (pH 7.4)
    2.5 mM MgCl$_2$
    0.5% Triton X-100 (New England Nuclear, Boston, MA)
    0.5% Aprotinin (Sigma, St. Louis, MO)
    2 μg/ml leupeptin and 2 μg/ml pepstatin A (Sigma)

3. While on ice, lyse the embryos by sonication using a macro-tipped sonicator at setting 6 (Heat System/Ultrasonics, Plainview, NY). Sonicate four times for 10 sec each, with 1 min of cooling between sonications.
4. Centrifuge the resulting lysate at 1000 *g* for 10 min to remove debris.
5. Collect the supernatant and layer over a 10 ml 30% sucrose cushion [30% sucrose (w/v) in buffer A]. Centrifuge for 15 min at 5000 *g*. This step pellets insoluble cellular structures without pelleting large ribonucleoprotein complexes.
6. Collect the material overlaying the sucrose cushion. This material is operationally defined as unfractionated embryo lysate.
7. Add micrococcal nuclease (Pharmacia, Piscataway, NJ) to 100 units/ml and CaCl$_2$ to 1 mM. Incubate at 30°C for 10 min to digest endogenous RNAs.

8. Add EGTA to 5 mM and chill on ice to inhibit the nuclease activity. This step is important to carry out prior to the addition of lysate to any nucleic acid resin, as micrococcal nuclease will otherwise digest the column.

9. Clarify the digested lysate of any precipitated proteins by centrifugation at 10,000 g for 15 min.

B. ssDNA Affinity Chromatography

1. Prior to preparation of the lysate, prepare a 15-ml ssDNA-cellulose (United States Biochemical Corp., Cleveland, OH) column. All column buffers contain 50 mM sodium phosphate (pH 7.4) and the desired NaCl concentration. For the initial loading of lysate, the column should be equilibrated with buffer containing 100 mM NaCl. A flow rate of approximately 100 ml/hr is desirable. To minimize protein degradation, columns should be run at 4°C.

2. Load the embryo lysate and wash the column extensively with buffer containing 100 mM NaCl. The column should be washed until the OD$_{280}$ returns to near zero.

3. Wash the column with 50 ml of buffer containing 100 mM NaCl and 1.0 mg/ml heparin (Sigma). This step significantly contributes to the purity of the isolated fractions, presumably by removing proteins nonspecifically bound to the column via ionic interactions with the DNA phosphate backbone.

4. Elute the proteins bound to the column by increasing the NaCl concentration. This can be done either by running a gradient from 100 mM to 2 M NaCl or by batch elution. All of the major proteins can be obtained in one fraction by eluting with 2 M NaCl after the heparin wash.

5. Wash the column with 50 ml of buffer containing 4 M guanidine HCl and reequilibrate with buffer containing 100 mM NaCl. This step removes proteins that have presumably aggregated on the column during the loading step. Although this eluted fraction is not considered useful, this step is essential for regenerating the ssDNA-cellulose column for the next use. We have found that columns can be reused numerous times if stored at 4°C in buffer containing 100 mM NaCl and sodium azide.

C. Ribohomopolymer Chromatography

Because most if not all RNA-binding proteins have different specificities and affinities for different RNA sequences, chromatography on more defined single-stranded nucleic acid matrices can be used in conjunction with ssDNA chromatography. Chromatography on ribohomopolymers has proven extremely useful for both defining (Swanson and, Dreyfuss, 1993a,b) and purifying vertebrate hnRNP proteins (Matunis et al., 1992c,1994), and several proteins purified from D. melanogaster have unique ribohomopolymer binding preferences (Matunis

et al., 1992a; M. J. Matunis and G. Dreyfuss, unpublished results). Because
RNA-matrices are more labile than ssDNA, extra caution must be used to guard
against RNases. Using partially purified proteins, such as protein fractions from
ssDNA chromatography, reduces the chances of nucleases being a problem.
The procedure outlined here can be used to determine the ribohomopolymer
binding preferences of a protein and can be scaled up for protein purification.

1. Prepare four tubes of poly(rG), four tubes of poly(rA), four tubes of
poly(rU), and four tubes of poly(rC)-agarose (Sigma). Each tube should contain
50 μl of slurry (25 μl of packed beads). For each ribohomopolymer, label one
tube 0.1 *M,* one 0.5 m*M,* one 1 *M,* and one 2 *M.* Rinse the ribohomopolymer-
agarose beads two times with buffer containing the appropriate salt concentration:

> X *M* NaCl
> 10 m*M* Tris–HCl (pH 7.4)
> 2.5 m*M* MgCl$_2$.

The beads can be pelleted by briefly spinning in a microcentrifuge.

2. Add 750 μl of protein sample to each tube. Starting material can be unfractionated lysate as described above, or preferably, partially purified protein.
Each binding reaction should contain at least 2 μg of the protein of interest, if
the protein is to be detected by SDS–PAGE followed by Coomassie blue staining. Before adding the sample to the ribohomopolymer beads, NaCl should be
added to the sample to the appropriate concentration and heparin should be
added to 1 mg/ml. Heparin reduces nonspecific binding to the ribohomopolymers and serves as a nuclease inhibitor.

3. Incubate for 10 min at 4°C with continuous rocking.

4. Pellet the beads in a microcentrifuge and wash four times with buffer
containing the appropriate salt concentration. The first wash should also contain
1 mg/ml heparin.

5. Elute the bound proteins with sample buffer or 4.0 *M* guanidine HCl. Such
extreme elution conditions are essential, as some proteins interact strongly with
specific ribohomopolymers. If proteins are to be analyzed by SDS–PAGE,
samples should be removed from the beads prior to boiling and boiled in a
clean Eppendorf tube, as the agarose beads tend to melt upon boiling. Having
established the binding preferences and the binding conditions of a specific
protein, it is a simple matter to scale-up the above procedure and purify larger
quantities of protein by column chromatography.

III. Characterization of Putative RNA-Binding Proteins

Depending on the type and possible functions of the RNA-binding proteins
of interest, several criteria can be applied to further define candidate RNA-
binding proteins. The procedures described below rely on antibody probes and

have been designed with the identification of hnRNP proteins in mind. However, depending on the results of each of the methods discussed, other classes of RNA-binding proteins can also be identified.

A. Preparation of Antisera and Monoclonal Antibodies

One of the most important considerations for the production of antibodies is the source of antigen. We have successfully produced monoclonal antibodies to seven of the most abundant proteins isolated from *D. melanogaster* embryos by ssDNA affinity chromatography, a particularly useful source because of the amounts of antigen that can be isolated (Matunis *et al.*, 1992a). The majority of these monoclonal antibodies were obtained from BALB/c mice immunized with the protein fraction illustrated in Fig. 1. Antibodies are not available for some of the less abundant proteins that are also present in this fraction; however, they should be readily obtainable by immunization of animals with more highly purified protein fractions. Although antibodies specific for vertebrate RNA-binding proteins have in general been difficult to generate because of their conservation, we have found that the *D. melanogaster* proteins are excellent antigens, as would be predicted from the degree of divergence seen so far among the major hnRNP proteins (Matunis *et al.*, 1992b). Monoclonal antibody ascites fluid, hybridoma cultured media, or polyclonal antisera are all suitable for the procedures described below.

B. Isolation of *in Vivo* Cross-Linked RNP Complexes

Proteins associated with poly(A)$^+$ RNA *in vivo* can be identified based on their ability to covalently cross-link to RNA after exposure to UV-light (Van Eekelen *et al.*, 1981; Mayrand *et al.*, 1981; Dreyfuss *et al.*, 1984). After UV-irradiation of intact cells, covalent poly(A)$^+$ RNA-protein complexes can be isolated by oligo(dT)-chromatography, and cross-linked proteins can be identified by immunoblotting. When identified by immunoblotting, cross-linked proteins generally migrate slightly slower and as a broader band compared to the uncross-linked protein as a result of residual RNase-resistant nucleotides that remain covalently bound to the protein. The major limitation of this method is the dependence on the photoreactivity of the particular protein and the RNA sequences to which it is bound and the generally low yield. Because of this limitation, not all proteins are cross-linked with equal efficiency and negative results are difficult to interpret. However, because cross-linking of the proteins to RNA is carried out in intact, living cells, and isolation is done under protein-denaturing conditions, this is a powerful approach to identify bona fide poly(A)$^+$ RNA-binding proteins. An important control that must be included is the "minus UV" experiment in which all steps, including RNase digestion, were carried out except exposure of the cell to UV-light.

1. Start with four plates (40 ml) of cultured *D. melanogaster* cells (we rou-

tinely use Schneider's line 2) in early to mid log phase (approximately 5×10^6 cells/ml). Cells are grown at 25°C in Schneider's *Drosophila* media (Gibco BRL, Gaithersburg, MD) supplemented with penicillin, streptomycin, and 10% heat-inactivated fetal calf serum. We have noted that heat inactivating the serum greatly facilitates the growth of these cells. Spin the cells down at 600 *g* and wash 1 time with cold phosphate-buffered saline (PBS). Resuspend the cells in cold PBS at approximately 1×10^7 cells/ml. Keep one-half of the cells in a tube on ice—these will serve as a negative control.

2. Pipet the other half of the cells into three 10-cm plastic cell culture dishes. Do not put more than 3 ml of cells into each dish.

3. Place the dishes (without their lids) under a 15-W germicidal lamp (Sylvania, G15T8). The bottom of the plates should be 4 cm from the bulb. To increase the efficiency of the irradiation, the entire assembly should be covered with an aluminum foil tent. Irradiate the cells for 3 min.

4. Spin down the cells (both irradiated and nonirradiated) at 600 *g* and resuspend in 5 ml of lysis buffer:

> 20 m*M* Tris–HCl (pH 7.4)
>
> 50 m*M* LiCl
>
> 1 mg/ml heparin (Sigma)
>
> 10 m*M* vanadyl-ribonucleoside complex (Gibco BRL)
>
> 2 µg/ml leupeptin and 2 µg/ml pepstatin A (Sigma)
>
> 1% aprotinin (Sigma)
>
> 1% SDS
>
> 1% β-mercaptoethanol

5. Sonicate the lysate three times for 5 sec each using a micro-tipped sonicator at setting 3 (Heat System/Ultrasonics, Plainview, NY). Cool the lysate on ice for 1 min between each sonication.

6. Clarify the lysate by centrifugation at 5000 *g* for 10 min at 4°C.

7. Heat the samples to 65°C for 5 min to remove secondary structures in the RNAs that may inhibit binding to the column.

8. Cool the samples on ice for 5 min.

9. Add LiCl to the samples to a final concentration of 500 m*M*.

10. Batch bind the samples to 0.5 ml of oligo(dT) cellulose (Type 3, Collaborative Research, Inc.) previously equilibrated with oligo(dT)-binding buffer:

> 10 m*M* Tris–HCl (pH 7.4)
>
> 1 m*M* EDTA
>
> 0.5% SDS
>
> 500 m*M* LiCl

Incubate at room temperature for 15 min with continuous rocking. This step can be done directly in a small chromatography column if both ends are sufficiently sealed. Use two columns, one for the cross-linked sample, and one for the control sample.

11. In column mode, wash the oligo(dT)-cellulose with 10 ml of binding buffer. Columns can be run at room temperature.

12. Elute bound RNP complexes from the column with 10 ml of oligo(dT)-elution buffer:

>10 mM Tris (pH 7.4)
>
>1 mM EDTA
>
>0.05% SDS

Collect the first 3 ml of the eluate in one tube.

13. Add SDS to the eluted samples to a final concentration of 0.5% and repeat Steps 7 through 11.

14. After repeating Step 11, the columns should again be eluted with elution buffer and 0.5-ml fractions should be collected. Peak fractions can be identified by monitoring the absorbance of the fractions at 260 nm or, if the samples are radioactive, by scintillation counting. In most cases, fractions two and three contain the bulk of the eluted ribonucleoprotein complexes.

15. Concentrate the samples to a volume of 300 μl by extracting several times with sec-butanol.

16. Adjust the LiCl concentration of the reduced samples to 200 mM.

17. Add 2.5 vol of cold absolute ethanol to the samples and incubate at $-80°C$ for at least 1 hr.

18. Spin down the samples in a microcentrifuge for 15 min at 4°C.

19. Discard the supernatant and dry the precipitated pellet.

20. Resuspend the pellet in 75 μl of RNase digestion cocktail:

>10 mM Tris–HCl (pH 7.4)
>
>1% aprotinin
>
>2 μg/ml leupeptin and 2 μg/ml pepstatin A
>
>1 mM $CaCl_2$
>
>400 units/ml micrococcal nuclease
>
>0.1 mg/ml RNase A

Incubate the samples for 1 hr at 37°C.

21. Repeat the ethanol precipitation described above. After precipitation, samples can be resuspended directly in SDS–PAGE sample buffer. Antibodies against suspected poly(A)$^+$ RNA-binding proteins can be used to probe immunoblots of the isolated cross-linked proteins. Cross-linked and uncross-linked samples should be run simultaneously along with a fraction containing the protein of interest in a non-cross-linked form. In many cases (but not all), the cross-linked protein will run slightly slower than the non-cross-linked form.

C. Immunolocalization in Cultured Cells

Because different classes of RNAs are localized to different compartments within the cell, an obvious approach toward further defining RNA-binding proteins is to determine their cellular localization. We have successfully used cultured *D. melanogaster* Schneider's cells (line 2) and monoclonal antibodies to distinguish between nucleolar, nucleoplasmic, and cytoplasmic RNA-binding proteins (Matunis *et al.*, 1992a). Because of the nature of most cultured *D. melanogaster* cells, which grow in suspension and are hence rounded, we have found it helpful to spin the cells onto glass slides to attach and flatten them, making the distinction between the nucleus and cytoplasm clearer. Although the localization of a protein in cultured cells can be very helpful in thinking about its possible functions, one caveat is that this localization may not reflect the localization in tissues of the organism. Of particular interest is the finding that many of the major hnRNP proteins of *D. melanogaster* are not restricted to either the nucleus or the cytoplasm during oogenesis (in contrast to their nuclear localization in cultured cells) and that the distribution of these proteins between the nucleus and cytoplasm can vary (Matunis, 1992; Matunis *et al.*, 1994). Localization in the organism should, therefore, also be considered if possible.

1. We routinely use cultured Schneider's cells (line 2) in early to mid log phase (approximately 5×10^6 cells/ml). Cells are cultured as described previously. Other cell lines such as Kc have also been used successfully. Cells should be pelleted at $600g$, washed one time in cold PBS and resuspended in PBS at a concentration approximately 1×10^6 cells/ml).

2. Cells are attached to glass slides by centrifuging at 200 rpm for 30 sec in a cytospin 3 centrifuge (Shandon Inc., Pittsburgh, PA). The amount of cells loaded per slide should be adjusted to give a semiconfluent layer of cells.

3. After centrifugation, cells are immediately fixed in PBS containing 2% formaldehyde for 30 min. It is important not to let the cells dry out during this, and subsequent steps.

4. Rinse the slides in a beaker of PBS and then permeabilize the cells by incubation in $-20°C$ acetone for 3 min.

5. Rinse the slides again in a beaker of PBS. Slides can be used immediately or stored for several days in PBS at 4°C.

6. Incubate the cells with primary antibody for 1 hr at room temperature in a moist, covered chamber. Excess PBS should be removed from around the cells using a kimwipe. Antibody can be in the form of ascites fluid or polyclonal antisera (both diluted in PBS containing 3% bovine serum albumin) or undiluted cultured cell supernatant. Generally, 50 μl of antibody is sufficient to cover the cells.

7. Rinse the slides in a beaker of PBS and apply the secondary antibody. We generally use FITC-conjugated goat anti-mouse F(ab')2 (Cappel Laboratories, Malvern, PA) diluted 1 : 50. Incubate as before for 30 min.

8. Rinse the slides with PBS and mount with a coverslip. To prevent fading of the fluorescent signal, the cells are mounted in the following:

75% glycerol

25 mM Tris–HCl (pH 8.0)

0.1% p-phenylenediamine (Sigma)

9. Observe the slides using a microscope equipped for fluorescence microscopy. Slides should be observed on the same day as they are prepared, as the signal tends to diffuse over extended periods.

D. Immunopurification of RNP Complexes

Many RNA-binding proteins such as RNP complexes exist in the cell in association with RNA and other proteins. Different types of RNP complexes include hnRNP, snRNP, and rRNP complexes and each can be distinguished from the other by their uniquely associated proteins and RNAs (Dreyfuss *et al.*, 1993; Lührmann *et al.*, 1990). By immunopurification with antibodies to a specific RNA-binding protein it is, therefore, possible to isolate unique RNP complexes and, by examining the associated proteins, further define the class of RNA-binding protein in question. This procedure has been used extensively to characterize vertebrate hnRNP complexes and a thorough discussion of its advantages and shortcomings has been printed elsewhere (Piñol-Roma *et al.*, 1990b). Two main considerations are that the procedure requires that the antigen recognized by the antibody be accessible in the RNP complex and that the RNA remains intact throughout the procedure. The former problem can best be circumvented by using polyclonal antisera or a number of different monoclonal antibodies, and the latter can be aided by working quickly and on ice. The procedures described here are used for the immunopurification of RNP complexes from cultured Schneider's cells but can be easily adapted to other types of lysates or extracts. In our hands, isolated nuclei from *D. melanogaster* cells in culture have not been a good source for isolating hnRNP complexes, presumably because of leakage to the cytoplasm, which may be aided by nuclease digestion. We therefore routinely immunopurify hnRNP complexes from whole cell lysates, a quicker procedure that provides much higher yields and does not appear to compromise purity.

1. Prepare protein-A Sepharose (Pharmacia, Piscataway, NJ) antibody beads: To a 50 μl slurry of protein-A Sepharose beads (25 μl packed volume), add 750 μl of PBS and 1 μl of antibody (either ascites fluid or polyclonal antisera). Alternatively, one can add 750 μl of undiluted cultured cell supernatant. Incubate at 4°C for at least 1 hr.

2. Four immunopurifications can be performed from one plate of cultured Schneider's cells (approximately 5×10^7 cells/plate). Cells can be labeled with [35S]methionine if desired; however, we have found that hnRNP complexes

can easily be detected by silver staining. Cells should be pelleted at 600 g and washed one time with cold PBS. All of the following steps should be performed on ice; we describe conditions for one plate of cells.

3. Resuspend the cells in 3 ml of buffer A:

　　100 mM NaCl

　　10 mM Tris–HCl (pH 7.4)

　　2.5 mM MgCl$_2$

　　0.5% Triton X-100 (New England Nuclear)

　　0.5% Aprotinin (Sigma)

　　2 μg/ml leupeptin and 2 μg/ml pepstatin A (Sigma)

4. Lyse the cells by sonicating three times for 5 sec using a micro-tipped sonicator at setting 3 (Heat System/Ultrasonics, Plainview, NY). Cool the lysate on ice for 1 min between each sonication.

5. Layer the lysate over a 30% sucrose cushion [30% sucrose (w/v) in buffer A]. This can be done in four microcentrifuge tubes, each containing a 500-μl cushion.

6. Centrifuge at 5000 g for 15 min to pellet insoluble cellular structures.

7. Remove the overlaying phase, which is operationally defined as the whole cell lysate.

8. Spin down the protein-A Sepharose/antibody beads briefly in a microcentrifuge tube and rinse two times with buffer A. Add 750 μl of whole cell lysate to each immunopurification.

9. Incubate for 15 min at 4°C (the time may be adjusted depending on how labile the complex of interest is).

10. Pellet and rinse the beads five times with buffer A.

11. After thoroughly draining the beads, add SDS–PAGE sample buffer directly to the beads. Samples can be analyzed by SDS–PAGE followed by silver staining or by immunoblotting.

12. To control for antibody specificity, the above procedure can be repeated with 1% Empigen BB (Calbiochem, San Diego, CA) in place of Triton X-100. This ionic detergent dissociates RNA–protein and protein–proteins interactions without interfering with antibody binding (Choi and Dreyfuss, 1984). Under this condition, only proteins directly recognized by the antibody are isolated. If Empigen BB is used, the last wash before sample buffer is added should be done with buffer A containing no detergent, as ionic detergents will interfere with SDS–PAGE.

E. Immunolocalization of RNA-Binding Proteins on Nascent Transcripts

The *D. melanogaster* salivary gland polytene chromosomes provide an excellent system in which to observe proteins associated with specific chromosomal loci *in situ* (for more information on this system, please refer to Chapters

19 and 20). Standard techniques for immunolocalizing proteins on polytene chromosomes such as those described by S.C.R. Elgin (in Ashburner, 1989) are useful for visualizing hnRNP proteins and snRNPs on nascent transcripts (Matunis *et al.*, 1992a, 1993). Typically, both categories of proteins are bound to most, if not all, nascent transcripts. However, the relative amounts of different hnRNP proteins vary on different transcripts (Matunis, 1993). A striking example is the PEP protein, which is present only on the ecdysone-inducible subset of transcripts (Amero *et al.*, 1993). In addition, all of these proteins can be dissociated from the nascent RNA by treating the salivary glands with RNase A prior to fixation and squashing. In this case, the only modification to the standard protocol is the addition of preboiled RNase A (Sigma) at 100 μg/ml in the buffer in which the glands are dissected and allowing incubation in this buffer for 15 min at room temperature prior to fixation (Matunis *et al.*, 1992a; Amero *et al.*, 1993). These conditions may need to be modified, depending on the protein of interest.

IV. Discussion

The methods described here provide powerful approaches for the isolation and characterization of RNA-binding proteins from *D. melanogaster*. These methods have been particularly useful for characterizing hnRNP proteins and should be equally valuable for identifying and characterizing other types of RNA-binding proteins. Although the major proteins purified by ssDNA chromatography appear to be hnRNP and mRNP proteins, other RNA-binding proteins can also be identified using this same procedure (Matunis *et al.*, 1992a). Furthermore, numerous less abundant proteins purified by ssDNA chromatography have not yet been characterized. The characterization of these additional proteins would be greatly facilitated by the use of assays designed for isolating specific factors. For example, assays could be devised to identify factors that bind specifically to defined elements essential for RNA localization (Ding and Lipshitz, 1993; MacDonald, 1992). Factors influencing pre-mRNA processing can also be assayed using *in vitro* splicing extracts (Rio, 1988,1992). The use of assays to further identify and purify specific factors isolated by ssDNA should help eliminate those proteins which are not authentic RNA-binding proteins (Piñol-Roma *et al.*, 1990a). Because the ssDNA purification does not involve the denaturation of proteins or high salt concentrations (the majority of the proteins are eluted by 500 mM NaCl), it should be a simple matter to further fractionate these proteins by additional affinity or conventional chromatography procedures. Affinity chromatography on ribohomopolymers is particularly useful for further purifying hnRNP proteins and also reveals identifying characteristics for many of these proteins (Piñol-Roma *et al.*, 1990b).

In addition to purifying RNA-binding proteins, we have also described a number of procedures that can be used to further characterize these proteins. It is important to keep in mind that none of these procedures is conclusive on its own, as each has its limitations. Because each protein is likely to subscribe to only a limited set of these criteria, it is important to consider all of the data as a whole. As an example, while all hnRNP proteins characterized to date are localized in the nucleus (but not necessarily exclusively confined to the nucleus because of shuttling), and are present in immunopurified hnRNP complexes, not all hnRNP proteins necessarily cross-link to poly(A)$^+$ RNA or bind to ssDNA (Dreyfuss *et al.*, 1993). Because of the large number of processes described in *D. melanogaster* that involve presumed, but in most cases not identified, RNA-binding proteins, the biochemical approaches described here will be helpful in purifying and characterizing these factors. The merging of genetic, biochemical, and morphologic approaches to characterize RNA-binding proteins should continue to provide exciting information on the functions of this dynamic group of proteins.

References

Amero, S. A., Matunis, M. J., Matunis, E. L., Hockensmith, J. W., Raychaudhuri, G., and Beyer, A. L. (1993). A unique ribonucleoprotein complex assembles preferentially on ecdysone-responsive sites in *Drosophila melanogaster*. *Mol. Cell. Biol.* **13**, 5323–5330.

Ashburner, M. (1989). In "Drosophila, a Laboratory Manual." New York: Cold Spring Harbor Laboratory Press.

Choi, Y. D., and Dreyfuss, G. (1984). Isolation of the heterogeneous nuclear RNA-ribonucleoprotein complex (hnRNP): A unique supramolecular assembly. *Proc. Natl. Acad. Sci. U.S.A.* **81**, 7471–7475.

Ding, D., and Lipshitz, H. D. (1993). Localized RNAs and their functions. *BioEssays* **15**, 651–658.

Dreyfuss, G., Matunis, M. J., Piñol-Roma, S., and Burd, C. G. (1993). hnRNP proteins and the biogenesis of RNA. *Annu. Rev. Biochem.* **62**, 289–321.

Dreyfuss, G., Choi, Y. D., and Adam, S. A. (1984). Characterization of heterogeneous nuclear RNA-protein complexes in vivo with monoclonal antibodies. *Mol. Cell. Biol.* **4**, 1104–1114.

Kelley, R. L. (1993). Initial organization of the *Drosophila* dorsoventral axis depends on an RNA-binding protein encoded by the squid gene. *Genes Dev.* **7**, 948–960.

Lührmann, R., Kastner, B., and Bach, M. (1990). Structure of spliceosomal snRNPs and their role in pre-mRNA splicing. *Biochim. Biophys. Acta* **1087**, 265–292.

MacDonald, P. M. (1992). The means to the ends: Localization of maternal messenger RNAs. *Semin. Dev. Biol.* **3**, 413–424.

Matunis, E. L. (1992). Characterization of the major heterogeneous nuclear ribonucleoproteins of *Drosophila melanogaster*. Ph.D. dissertation, Northwestern University, Evanston, IL.

Matunis, E. L., Kelley, R., and Dreyfuss, G. (1994). Essential role for an hnRNP protein in oogenesis: hrp40 is absent from the germline in the dorsoventral mutant *squid*. *Proc. Natl. Acad. Sci. U.S.A.* **91**, 2781–2784.

Matunis, M. J., Matunis, E. L., and Dreyfuss, G. (1992a). Isolation of hnRNP complexes from Drosophila melanogaster. *J. Cell Biol.* **116**, 245–255.

Matunis, E. L., Matunis, M. J., and Dreyfuss, G. (1992b). Characterization of the major hnRNP proteins from *Drosophila melanogaster*. *J. Cell Biol.* **116**, 257–269.

Matunis, M. J., Michael, W. M., and Dreyfuss, G. (1992c). Characterization and primary structure of the poly(C)-binding heterogeneous nuclear ribonucleoprotein complex K protein. *Mol. Cell. Biol.* **12**, 164–171.

Matunis, E. L., Matunis, M. J., and Dreyfuss, G. (1993). Association of individual hnRNP proteins and snRNPs with nascent transcripts. *J. Cell. Biol.* **119**, 219–228.

Matunis, M. J., Xing, J., and Dreyfuss, G. (1994). The hnRNP F protein: Unique primary structure, nucleic acid-binding properties, and subcellular localization. *Nucleic Acids Res.* **22,** 1059–1067.

Mayrand, S., Setyono, B., Greenberg, J. R., and Pederson, T. (1981). Structure of nuclear ribonucleoprotein: Identification of proteins in contact with poly(A)$^+$ heterogeneous nuclear RNA in living HeLa cells. *J. Cell Biol.* **90,** 380–384.

Neuman-Silberberg, F. S., and Schupbach, T. (1993). The *Drosophila* dorsoventral patterning gene *gurken* produces a dorsally localized RNA and encodes a TGFα-like protein. *Cell* **75,** 165–174.

Piñol-Roma, S., Swanson, M. S., Matunis, M. J., and Dreyfuss, G. (1990a). Purification and characterization of proteins of heterogeneous nuclear ribonucleoprotein complexes by affinity chromatography. *Methods Enzymol.* **181,** 326–331.

Piñol-Roma, S., Choi, Y. D., and Dreyfuss, G. (1990b). Immunological methods for purification and characterization of heterogeneous nuclear ribonucleoprotein particles. *Methods Enzymol.* **181,** 317–325.

Rio, D. C. (1992). RNA binding proteins, splice site selection, and alternative pre-mRNA splicing. *Gene Expression.* **2,** 1–5.

Rio, D. C. (1988). Accurate and efficient pre-mRNA splicing in *Drosophila* cell-free extracts. *Proc. Natl. Acad. Sci. U.S.A.* **85,** 2904–2908.

Swanson, M. S., and Dreyfuss, G. (1988a). Classification and purification of proteins of heterogeneous nuclear ribonucleoprotein particles by RNA-binding specificities. *Mol. Cell. Biol.* **8,** 2237–2241.

Swanson, M. S., and Dreyfuss, G. (1988b). RNA binding specificity of hnRNP proteins: A subset bind to the 3′ end of introns. *EMBO J.* **11,** 3519–3529.

van Eekelen, C. A., Riemen, T., and van Venrooij, W. (1981). Specificity in the interaction of hnRNA and mRNA with proteins as revealed by in vivo crosslinking. *FEBS Lett.* **130,** 223–226.

CHAPTER 12

Chromatin Assembly Extracts from *Drosophila* Embryos

Peter B. Becker,[*] Toshio Tsukiyama,[†] and Carl Wu[†]

[*] Gene Expression Programme
European Molecular Biology Laboratory
69117 Heidelberg, Germany

[†] Laboratory of Biochemistry
National Cancer Institute
National Institutes of Health
Bethesda, Maryland 20892

I. General Introduction

A deeper understanding of the major processes that constitute nucleic acid metabolism, replication, transcription, recombination and DNA repair, requires the reconstitution of these phenomena *in vitro* in the context of chromatin, their natural substrate. The first level in the hierarchy of chromatin folding, the association of DNA with the core histones to form nucleosomes and the binding of the linker histone H1, is currently amenable to *in vitro* reconstitution. One salient feature of natural chromatin, the retention of a defined distance between nucleosome core particles in an extended array of nucleosomes (nucleosome spacing), can be reconstituted only under physiological conditions using crude, chromatin assembly extracts derived from tissue culture cells (Banerjee and Cantor, 1990), *Xenopus* eggs and oocytes (Almouzni and Méchali, 1988; Shimamura *et al.*, 1988), and *Drosophila* embryos (Becker and Wu, 1992). Unlike extracts prepared from *Xenopus* eggs or oocytes, which are sometimes subject to seasonal variation (Rodríguez-Campos *et al.*, 1989), extracts prepared from *Drosophila* embryos harvested from population cages reproducibly yield extracts of high activity. As *Drosophila* can be raised cheaply in mass culture requiring minimal attention, large quantities of staged embryos can be harvested routinely in amounts sufficient for biochemical manipulations (see Chapters 5, 7, and 10). In addition, extracts of fly embryos are a rich source of various biochemical activities, including factors necessary for *in vitro* transcription (see Chapter 13).

Nelson, Hsieh, and Brutlag (1979) initially exploited *Drosophila* embryos as a source of extracts for chromatin assembly. While extracts prepared according to their procedure were capable of assembling nucleosomes on plasmid DNA, the assembled nucleosomes, in our experience, lacked defined spacing in extended nucleosome arrays and these extracts lost activity upon storage. In exploring alternative procedures for the preparation of chromatin assembly extracts, we found that ionic conditions developed for extract preparation from the *Xenopus* oocyte system (Shimamura *et al.*, 1988, Rodriguez-Campos *et al.*, 1989) yielded a sturdy reconstitution extract with consistently high activity for the assembly of long arrays of spaced nucleosomes (Becker and Wu, 1992).

The *Drosophila* chromatin assembly extracts are essentially cytoplasmic supernatants generated by centrifugation of embryo homogenates in a low ionic strength buffer at 150,000 g (S-150). The assembly extracts are prepared from preblastoderm embryos [collected from a window of 0–90 min, or 0–120 min after egg laying (AEL)]. These early embryos, which are largely inactive for transcription, undergo DNA replication and chromatin assembly at maximal rates, relying on stores of maternal precursors. Chromatin reconstitution *in vitro* with the S-150 extracts utilizes this endogenous pool of maternal histones and should therefore resemble preblastoderm chromatin. The chromatin of preblastoderm embryos differs from that of later developmental stages in that the linker histone H1 is absent or substituted with an as yet undefined, alternate linker histone. Kamakaka *et al.* (1993) have employed similar ionic conditions to prepare chromatin assembly extracts from slightly older embryos (0–6 hr AEL). At this stage, the maternal pool of stored histones is exhausted, and these extracts require supplementation with exogenous, purified core histones. Both preblastoderm and postblastoderm extracts contain little histone H1. However, exogenous H1, purified from late embryo chromatin can be introduced in the assembly reaction, where it becomes incorporated into chromatin and increases the nucleosome repeat length from ~180 to ~200 bp.

Cytoplasmic extracts of the kind described here contain high levels of enzymes and cofactors required for DNA synthesis and are thus capable of synthesizing the DNA strand complementary to a single-stranded (ss) circular DNA template; this reaction is also accompanied by chromatin assembly (Becker and Wu, 1992; Kamakaka *et al.*, 1993). The extracts can therefore be used to perform coupled replication/assembly reactions similar to those shown with extracts of Xenopus eggs (Almouzni *et al.*, 1990) and tissue culture cells (Krude and Knippers, 1993). The coupled replication/assembly reactions may more closely approximate the physiological deposition of histones on DNA during replication *in vivo*.

In this chapter, we describe protocols for the preparation of chromatin assembly extracts from *Drosophila* embryos and procedures for the assembly of regularly spaced nucleosomes on plasmid DNA and for the analysis of the resulting chromatin.

II. Extract Preparation

The preparation of active chromatin assembly extracts according to Becker and Wu (1992) is schematically described in Fig. 1. Briefly, dechorionated embryos are homogenized with minimal dilution in a low salt buffer. The crude homogenate is supplemented with additional $MgCl_2$ before the embryonic nuclei are pelleted by low-speed centrifugation. The cytoplasmic extract is then cleared by centrifugation at high speed (150,000g), which effectively removes (floating) lipids and pellets the yolk granules, organelles, and other cellular debris. This clarified supernatant (S-150) is the chromatin assembly extract. We have prepared over one hundred S-150 extracts in our laboratories; the extracts have consistently high activity and are stable upon storage for many months and after several cycles of freeze and thaw.

A. Preparation of Extracts from Preblastoderm Embryos

Embryos (0–90 min or 0–120 min (preblastoderm) are harvested from three population cages each containing 50,000 flies maintained on a 12-hr day/night cycle. To purge older embryos retained by females during the overnight period, embryos deposited in the first hour of the daily collection are discarded.

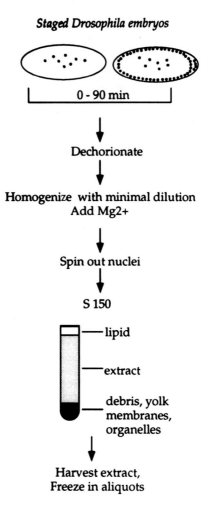

Staged *Drosophila* embryos

0 - 90 min

Dechorionate

Homogenize with minimal dilution
Add Mg2+

Spin out nuclei

S 150

— lipid

— extract

— debris, yolk
membranes,
organelles

Harvest extract,
Freeze in aliquots

Fig. 1   Schematic representation of the steps required for preparation of chromatin assembly extracts from *Drosophila* embryos.

Throughout the day, embryos are harvested in successive 90- or 120-min intervals and stored in embryo wash buffer (0.7% NaCl, 0.05% Triton X-100) on ice, where further development is arrested. Pooled embryos are allowed to settle once in 1.0-liter embryo wash buffer at room temperature, and the volume of the suspension is adjusted to 200 ml. Dechorionation occurs by either adding 200 ml of Clorox bleach and vigorous stirring for 90 sec (USA) or adding 60 ml 13% hypochloric acid and stirring for 2.5–3 min (Europe). Embryos are collected on a fine sieve and rinsed vigorously with a sharp stream of tap water. They are then allowed to settle once or twice in at least 500 ml of embryo wash buffer, and the supernatant containing broken chorions is removed by aspiration. The substantial loss of material due to floating embryos in this and the following washes is usually a sign of insufficient dechorionation or inadequate rinsing.

The embryos are resuspended and allowed to settle in 500 ml 0.7% NaCl, followed by resuspension and settling in 350 ml cold EX buffer (10 mM Hepes, pH 7.6/10 mM KCl/1.5 mM MgCl$_2$/0.5 mM EGTA/10% glycerol/10 mM β-glycerophosphate) to which 1 mM dithiothreitol and 0.2 mM phenylmethylsulfonyl fluoride has been added freshly. Finally, they are transferred to a 60-ml glass homogenizer containing EX buffer and allowed to settle for 15 min on ice. We routinely obtain ~20 ml of packed embryos from the pool of four successive embryo collections. The supernatant is aspirated, leaving behind about 2 ml of buffer on the surface of the packed embryos. All further manipulations are performed at 4°C.

Embryos are homogenized by 6–10 complete strokes at 1500 rpm using a teflon pestle connected to a motor-driven drill press (or 1000 rpm in a B. Braun homogenizer). The volume of the homogenate is determined and MgCl$_2$ is quickly mixed in from a 1 M stock solution to increase the MgCl$_2$ concentration by 5 mM to a final concentration of 6.5 mM. This concentration of MgCl$_2$ has not been optimized, and a final concentration of 5 mM MgCl$_2$ has also been found adequate (T.T.). Nuclei are then pelleted by centrifugation for 5 min at 5000 rpm in a chilled rotor HB4 (Sorvall) or JA14 (Beckmann). The supernatant is clarified by centrifugation for 2 hr at 150,000g (40,000 rpm in a SW 50.1 rotor; Beckman or equivalent).

The homogenate splits into three fractions after centrifugation, comprising a solid pellet, a mostly clear supernatant, and a floating layer of lipid (see Fig. 1). The clear extract is collected with a syringe by puncturing the tube with the syringe needle just above the tight pellet, thus avoiding the floating lipid layer. Occasionally, a white flocculant material is present in the otherwise clear extract. Much of this can be removed by centrifugation for 5 min in an Eppendorf microcentrifuge. The presence of some turbid material in an extract appears to be correlated with insufficient dechorionation. It usually does not affect the chromatin assembly reaction but may result in an increased background of copurifying proteins when reconstituted chromatin is partially purified on sucrose gradients. The S-150 extracts are flash-frozen in suitable aliquots and stored at −80°C. They can be thawed and refrozen two or three times and stored at −80°C for a year without significant loss of activity. The protein concentration of these extracts is usually around 20 mg/ml.

B. Preparation of Extracts from Postblastoderm Embryos

Kamakaka *et al.* (1993) have reported a procedure similar to the protocol described above to prepare S-190 extracts for chromatin assembly from slightly older embryos (0–6 hr AEL; mostly postblastoderm embryos). We have prepared S-190 extracts from 0- to 6-hr embryos using the above protocol and concur with their results. Kamakaka *et al.* (1993) have noted that additional precipitation of material is observed upon freezing and thawing of the S-190 extract; this material is pelleted by recentrifugation at 45,000 rpm for 2 hr. Although this additional step appears not to be necessary for the proper assembly of nucleosomes per se, it may have important consequences depending on the desired functional assay for the reconstituted chromatin and may therefore be incorporated in the overall protocol by the user.

While the properties of nucleosome structure reconstituted with extracts of preblastoderm and postblastoderm embryos should be similar at a gross level, it is likely that the chromatin assembled using these differently staged embryo extracts may possess differences in histone modification peculiar to each embryonic stage. These and other potential differences should be taken into consideration when evaluating the structural and functional properties of the reconstituted chromatins.

III. Chromatin Assembly Reaction

Chromatin assembly in the S-150 (or S-190) embryo extract is usually performed at 26–27°C, near the optimal temperature for *Drosophila* development. The reaction also occurs efficiently at slightly elevated temperatures (30°C). The assembly of regularly spaced nucleosomes requires magnesium, ATP, an energy regenerating system, and a defined concentration of monovalent cations. The conductivity of a standard assembly reaction is equivalent to 65 mM KCl. The optimal amount of extract needed for the assembly of spaced chromatin is determined empirically by a micrococcal nuclease (MNase) digestion assay and is roughly similar when different extract preparations are compared.

A. Chromatin Assembly Using Preblastoderm Embryo Extracts

We present conditions for the assembly of 900 ng plasmid DNA; the reaction can be scaled up or down 10-fold.

Prepare 10× MCNAP buffer for energy regeneration by adding:

46 μl H$_2$O
30 μl 1 M creatine phosphate (in water)
10 μl 300 mM ATP, pH 7.0
10 μl 100 ng/μl creatine phosphokinase (Sigma)
3 μl 1 M MgCl$_2$
1 μl 1 M DTT

(Creatine phosphate, creatine phosphokinase, and ATP should be stored in aliquots at −80°C and thawed only once before use).

Combine in a 1.5 ml reaction tube:

12 μl 10× MCNAP buffer

108-x-y μl EX buffer/50 mM KCl (EX 50)

y μl extract (~75 μl; titrate for each extract preparation)
x μl DNA (900 ng)
Incubate at 26°C for up to 6 hr.

B. Incorporation of Exogenous Histone H1

Extracts from preblastoderm embryos contain very little of the major linker histone H1 and no H1 is detected in reconstituted chromatin. When purified H1 is added to the assembly reaction, it is incorporated into chromatin. The binding to linker sequences results in an increased repeat length in a micrococcal nuclease digestion analysis (see following, Fig. 3B). Histone H1 is easily purified from late embryo chromatin using the protocol of Croston et al. (1991). We dilute purified histone H1 with EX buffer containing 0.01% NP-40 to prevent aggregation and mix it with the assembly extract prior to addition of the DNA. Given the difficulties in determining the precise concentrations of histone H1 with standard dye-binding assays, we empirically titrate the amounts of H1 required to increase the nucleosome repeat length from (~180 to ~200 bp (Fig. 3B). If excess H1 is added, it competes with the core histones in binding to DNA, leading to improper nucleosome spacing.

C. Chromatin Assembly Using Postblastoderm Embryo Extracts and Exogenous Histones

Exogenous histones can be used for chromatin reconstitution in conjunction with extracts from postblastoderm embryos that have depleted the maternal pools of histones (Kamakaka et al., 1993). Core histones are purified according to the method of Simon and Felsenfeld (1979) but commercially available calf thymus histones (Boehringer Mannheim, Catalog No. 223 656) can also be used. The appropriate amount of histones is determined empirically, using as a guide a stoichiometry of histones to DNA of ~0.8:1 (w/w) (Albright et al., 1979). The following protocol assembles 900 ng of DNA in chromatin.

Combine and incubate for 20–30 min at 26–27°C (to ensure that histones associate with carrier molecules in the crude extract):

55 μl extract from 0- to 6-hr embryos

1 μl of core histones (amount determined by titration)

Add to this mixture:

7 μl of 10× MCNAP buffer

900 ng of plasmid DNA

Adjust the volume to 70 μl with EX 50 buffer and incubate at 26°C for 1 to 6 hr. Properly spaced chromatin should be assembled by 3 hr of incubation, as visualized by the ladder of DNA fragments produced by partial MNase digestion.

D. Coupled Replication/Chromatin Assembly

If the double-stranded (ds) DNA in the assembly reaction is replaced by ss DNA, the complementary strand will be synthesized by the DNA replication enzymes that are abundant in the S-150 or S-190 extracts. To exploit this reaction for coupled replication/assembly, the DNA template is cloned into a phagemid such as pBluescript (Stratagene, La Jolla, CA), and ssDNA is obtained according to standard procedures (Sambrook *et al.*, 1989).

The protocol for the coupled replication/assembly reaction is similar to the standard reaction for dsDNA except that the 900 ng of plasmid is replaced by 450 ng of ss DNA. If random labeling of the resulting dsDNA plasmid is desired, 1 μl of [α^{32}P]dCTP (NEN, 2000- 3000 Ci/mmole) is included. Priming of DNA synthesis is random, presumably from RNA primers synthesized in the extract, but for the purpose of site-specific labeling, terminally labeled oligonucleotides can be incorporated into the resulting plasmid if annealed to the ss DNA prior to addition to the assembly reaction (Becker and Wu, 1992; Kamakaka *et al.*, 1993).

IV. Analysis of Reconstituted Chromatin

We describe two standard procedures for the initial characterization of reconstituted chromatin that are also useful to monitor the efficiency of an assembly reaction. The supercoiling assay is based on topological changes that accompany the wrapping of DNA around a particle. The winding of the DNA around a nucleosome core introduces one positive superhelical turn in the plasmid DNA, which is relaxed by topoisomerase I activity present in the embryo extracts. When nucleosomes are removed by proteinase K digestion and DNA purification, one negative superhelical turn corresponding to each assembled nucleosome appears in the closed circular DNA (Germond *et al.*, 1975). The superhelical density of a plasmid, i.e., the absolute number of superhelical turns, can be directly counted by visualization of the plasmid topoisomers on two-dimensional agarose gels (Peck and Wang, 1983) or by resolving duplicate samples on multiple agarose gels containing different chloroquine concentrations (Keller, 1975). As a rapid, but crude indicator of nucleosome reconstitution, the introduction of supercoils into a plasmid can simply be visualized by agarose gel electrophoresis (Fig. 2A).

DNA supercoiling measures the wrapping of DNA around a particle but does not necessarily imply the reconstitution of a full octamer of core histones.

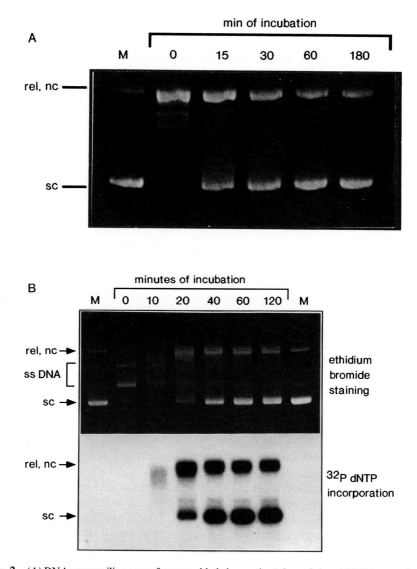

Fig. 2 (A) DNA supercoiling assay for assembled chromatin. 1.5 μg of plasmid DNA was relaxed with topoisomerase I and then incubated with 120 μl preblastoderm embryo extract under assembly conditions. At the times indicated, 40-μl aliquots were removed and the DNA was purified for analysis on a 1.2% agarose gel. Sc, supercoiled DNA (form I); rel, relaxed, closed circles (form II); nc, nicked circles. (B) Supercoiling assay showing chromatin assembly coupled with DNA synthesis in embryo extracts. Duplicate reactions containing 150 ng of ss phagemid DNA (6.2 kb) were incubated under assembly conditions in the presence of 0.5 μCi of [^{32}P]dCTP with 15 μl preblastoderm extract in a volume of 30 μl. Reactions were terminated at the indicated times and analyzed for supercoiling on an 0.8% agarose gel. DNA was visualized by staining with ethidium bromide (upper panel). The gel was then blotted and dried onto DE81 paper (Whatman). The lower panel shows an autoradiography of the dried gel.

Indeed, the winding of DNA around a complete histone octamer or a tetramer of histones H3 and H4 cannot be distinguished by this method. The MNase digestion assay, although more time consuming, is much more informative because it provides information on the nature of the nucleosome core particle as well as on the average distance between particles. This assay relies on the ability of MNase to preferentially cleave the linker DNA between nucleosome core particles. After the initial endonucleolytic attack of linker DNA, the trimming activity associated with enzyme progressively removes the linker DNA. Extensive digestion of chromatin with MNase will bring the size of the mononucleosome from 160–220 bp to the 146-bp DNA fragment protected by the nucleosome core particle (Bavykin *et al.*, 1990) whereas a partial digest results in a ladder of fragments representing oligonucleosomal DNAs (Fig. 3).

When the extent of chromatin reconstitution in the course of an assembly reaction is analyzed by measuring DNA supercoils, the rate at which supercoils are introduced is rapid: about 80% of the maximal number of supercoils are introduced within the first 30–60 min of incubation (Fig. 2A). By contrast, a regular pattern of digestion with MNase is obtained only after the reaction is allowed to proceed for an extended period of time. An appreciable improvement in quality of the MNase digestion ladder is observed when a 6-hr incubation is compared with a 3-hr incubation, even though few additional supercoils are introduced during this time interval. It is possible that the rapid introduction of DNA supercoils reflects the early assembly of subnucleosomal particles (H3–H4 tetramers) and that the complete assembly of the histone octamer requires an extended period of incubation *in vitro*.

A. Supercoiling Assay

We allocate 150–300 ng of plasmid DNA for each time point to be analyzed. A total of 1.5 μg plasmid DNA are incubated under assembly conditions (see Section IIA) with 120 μl of chromatin assembly extract in a total volume of 200 μl. After 15, 30, 60, and 180 min of incubation at 26°C, 40-μl aliquots of the reaction are added to 10 μl Stop Mix (2.5% *N*-lauroylsarcosine (Sigma), 100 m*M* EDTA).

The purification of the DNA essentially follows the procedure of Shimamura *et al.* (1988). Each sample is incubated with 1 μl of 10 mg/ml RNase A (DNase-free; Sambrook *et al.*, 1989) for 15 min at 37°C. Then, 6.5 μl of each of 2% SDS and 10 mg/ml proteinase K are added, and incubation at 37°C is continued for 30 min. The reaction is adjusted to 3*M* ammonium acetate by addition from a 7.5 *M* stock solution, 10 μg glycogen (Boehringer Mannheim) is added, and the sample is mixed. After addition of 2 vol of ethanol, DNA is precipitated for 15 min on ice. The addition of glycogen helps visualization of the pellet but is not generally required for nucleic acid precipitation. After centrifugation for 10 min in a microcentrifuge, the pellet is washed with 1 ml of 80% ethanol. The ethanol is removed completely and the DNA is dried for 3 min in a Speed

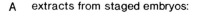

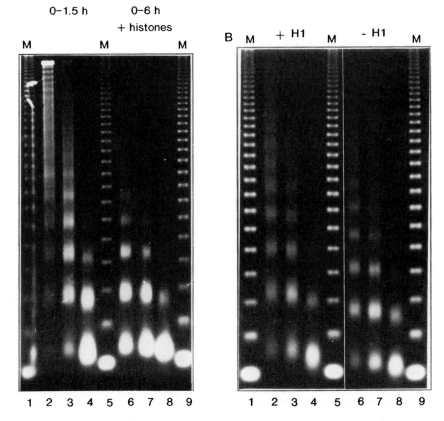

Fig. 3 (A) Micrococcal nuclease digestion assay of assembled chromatin. Nine hundred nanograms of plasmid DNA, assembled into chromatin in extracts from 0- to 2-hr embryos (lanes 2–4) or 0- to 6-hr embryos with supplemented histones (lanes 6–8) were analyzed by digestion with micrococcal nuclease as described. Purified DNA fragments were resolved in a 1.3% agarose gel stained with ethidium bromide. M, 123-bp ladder (BRL). (B) Incorporation of exogenous histone H1 in assembled chromatin. Chromatin was reconstituted in an extract from 0- to 90-min embryos in the presence (lanes 2–4) or absence (lanes 6–8) of H1. The MNase assay reveals an increase in repeat length upon incorporation of H1.

Vac concentrator. The DNA pellet is dissolved in 8 μl TE (10 mM Tris/Cl, pH 8.0; 1 mM EDTA), 2 μl of (5×) blue loading buffer (30% glycerol, 0.25% each of bromophenol blue and xylene cyanol) is added, and the sample is analyzed on an agarose gel in 1× Tris-glycine buffer (5× buffer: 144 g glycine, 30 g Tris/liter). The entire electrophoresis apparatus should be kept clean of ethidium bromide because intercalation of the dye during electrophoresis will cause additional DNA supercoiling (Keller, 1975). The agarose concentration of the gel is determined by the plasmid size (we use 1.2% agarose gels for 3- to 4-kb plasmids, and 0.8% agarose gels for 6- to 7-kb plasmids). The best resolution of topoisomers is achieved during overnight runs at 1 V/cm; however, satisfactory results are obtained at up to 5 V/cm. After electrophoresis, the gel is stained for 15–30 min in 1 gel volume of water containing 2 μg/ml ethidium bromide and destained for 15 min in deionized water.

Figure 2A shows a typical supercoiling assay. For better illustration of supercoiling, 1.5 μg of plasmid DNA was previously relaxed with topisomerase I (purified according to Javaherian *et al.*, 1982) in a total volume of 60 μl of EX-50/0.05% NP40 (Fig. 2A, 0 min). This prerelaxation is generally not required in practice, since topoisomerase I activity in embryo extracts almost immediately relaxes supercoiled plasmid DNAs upon incubation in the assembly reaction. After 30 min of incubation in the assembly extract, the purified plasmid DNAs are observed to be highly supercoiled again. The resolution of topoisomers of higher superhelical densities requires electrophoresis in the presence of chloroquine (Peck and Wang, 1983).

Figure 2B shows a supercoiling assay of a reaction starting from ss DNA in a coupled DNA synthesis/assembly reaction (Section IIID). The ss DNA (0 min incubation) is converted into supercoiled ds DNA in a reaction that is essentially complete by 30–40 min (top panel). If the histones are removed from such a reaction prior to addition of the ss DNA, relaxed plasmids are obtained, indicating that the DNA synthesis reaction can be uncoupled from the nucleosome assembly reaction (P.B.B., unpublished observations). The incorporation of [^{32}P]dCTP during the synthesis of the second strand can be followed by exposure of the dried gel to X-ray film (Fig. 2B, lower panel).

B. Micrococcal Nuclease Digestion

Nine hundred nanograms of plasmid DNA is assembled into chromatin as described (Section IIA). After incubation for 5–6 hr at 26°C, 180 μl of a premix containing 168 μl EX buffer, 9 μl 0.1 M CaCl$_2$, 3 μl MNase (50 u/μl) is added and the samples are again incubated at 26°C. (The concentration of CaCl$_2$ can also be decreased by half; T.T.) After 0.5, 1, and 5 min of incubation, a 100-μl aliquot is added to a fresh tube containing 25 μl Stop Mix (2.5% *N*-laurylsarcosine (Sigma), 100 mM EDTA). One microliter of 10 mg/ml RNase A is added and the reaction is incubated for 30 min at 37°C. The reaction is adjusted to 0.2% SDS and 300 μg/ml proteinase K and incubated overnight at 37°C. DNA is precipitated, pelleted, and washed as described above for the supercoiling assay (IVA). All traces of ethanol are removed and the pellets are air dried for 15–20 min on the bench. The pellets are dissolved in 4 μl of TE/50 mM NaCl. One microliter of loading buffer (50% glycerol/5 mM EDTA/0.3% orange G (Sigma) is added to each sample, and the samples are electrophoresed on a 1.3% agarose gel in Tris/glycine buffer. Superior resolution of long oligonucleosomal fragments can be achieved with narrow gel slots (Shimamura *et al.*, 1988). Samples are electrophoresed at 3 V/cm until the orange dye reaches the bottom of the gel, and the gel is stained with ethidium bromide as above (Section IVA). Alternatively, DNA is electrophoresed on a 1.3% agarose gel in 0.5× TBE at 7 V/cm until the Orange G dye has migrated 10 cm.

Figure 3A shows the result of MNase digestions of chromatin reconstituted with a preblastoderm embryo extract (lanes 2–4) or with an extract of postblastoderm embryos and supplemental core histones (lanes 6–8). The nucleosome repeat lengths are determined by comparison of the largest visible oligonucleosome-sized fragment with the marker DNA fragments (123-bp ladder, BRL; see Rodriguez-Campos *et al.* (1989) for a discussion of how the repeat lengths are determined). The introduction of an appropriately titrated amount of histone H1 to the assembly extract prior to the addition of the DNA results in an increased repeat length in the MNase digestion assay (Fig. 3B; compare lanes 2–4 to lanes 6–8).

V. Nucleosome Organization at Specific Sites

The presence of nucleosome organization at specific locations on the recombination DNA clone is analyzed by Southern blotting of the DNA fragments produced by digestion with MNase followed by hybridization with unique oligonucleotide probes. Since MNase initially cleaves within the stretch of linker DNA between nucleosome core particles, followed by progressive trimming to the core from both ends of the nucleosome, the presence of a nucleosome core particle can be gauged by the accumulation of the canonical, 146-bp resistant fragment towards the limit of MNase digestion.

Figure 4 shows the assembly of an intact nucleosome at sequences corresponding to −115 to −132 of the *Drosophila* hsp70 promoter, to the 3′ end of the hsp70 gene, and to the ampicillin resistance gene of the plasmid vector. In addition to the 146-bp fragment derived from the nucleosome core particle observed after extensive MNase digestion, a ladder of discrete fragments corresponding to nucleosome oligomers can be observed at intermediate stages of MNase digestion. This characteristic pattern of cleavage indicates that those DNA sequences are organized in a regularly spaced (but not necessarily positioned) array of nucleosomes. The determination of nucleosome positioning and nuclease hypersensitive sites in chromatin can be revealed by the technique of indirect end-labeling (for technical protocols, see Wu, 1989).

A. Southern Blotting

The DNA products of MNase digestion are electrophoresed along with radiolabeled DNA markers on agarose gels as described above (Section IVB). The gel is treated for 45 min in 0.5 M NaOH, 1.5 M NaCl at room temperature with constant agitation to denature DNA. After briefly rinsing with water twice, neutralize the gel for 45 min in 1 M Tris–HCl, pH 7.5, 1.5 M NaCl at room

hsp70 promoter   3' end of hsp70   vector

gene

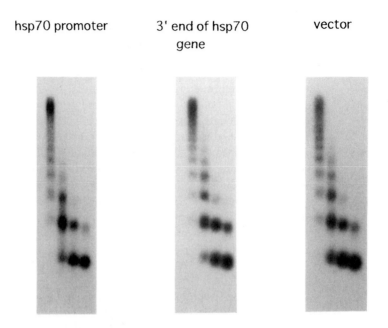

Fig. 4 Nucleosome organization at specific sites on a 6-kb *hsp70* plasmid as revealed by sequential oligonucleotide hybridization. The same DNA blot of a MNase digestion series was sequentially hybridized with oligonucleotides corresponding to −115 to −132 of *hsp70* promoter, +1803 to +1832 of the *hsp70* gene, and 2499 to 2528 of the pBluescript SK- vector.

temperature with constant agitation. DNA is transferred onto the hybridization membrane by capillary blotting overnight (Sambrook *et al.*, 1989). For the purpose of sequential hybridization with oligonucleotide probes (see following), we use nylon membrane without a surface charge (Gene Screen, DuPont/NEN) because of the ease of handling and low-background signals. DNA fragments are fixed onto the membrane by cross-linking with UV light, according to the manufacturer's instructions.

B. Oligonucleotide Hybridization

We routinely use oligonucleotide probes 17 to 40 bases in length. Usually, purification of oligonucleotides by sequencing gel electrophoresis or HPLC is not required. Oligonucleotides are 5′ end-labeled as follows:

5 pmole of oligonucleotide

1.5 μl of 10× T$_4$ polynucleotide kinase buffer (Sambrook *et al.*, 1989)

1 μl of [γ-^{32}P]ATP (7000 Ci/mmole) 166.7 μCi/μl, ICN Catalog 35020)

1 μl (10 u) of T$_4$ polynucleotide kinase

Make up to 15 μl with water

Incubate the reaction mixture at 37°C for 45 to 60 min. Terminate the reaction by adding SDS to 1% and purify the oligonucleotide through a spin column (Bio-spin 6, Bio-Rad) according to the manufacturer's protocol.

Wet the DNA blot with water and prehybridize for >30 min in 6× SSC (1× SSC: 0.15 M NaCl, 0.015 M Na citrate), 2% SDS, 100 μg/μl denatured salmon sperm DNA at the hybridization temperature. After introducing the labeled oligonucleotide directly into the prehybridization mixture, allow the probe to hybridize for 2–10 hr at 40–55°C, depending on the T_m of the probe (we usually set the temperature at $T_m - 10$°C). Wash the membrane at hybridization temperature two to three times in 6× SSC, 0.5% SDS for 15–30 min each. The hybridization solution can be stored at 4°C and reused several times within a week. After wrapping with SaranWrap, expose the membrane to film for several hours to overnight. Be careful not to let the membrane dry out during exposure.

To strip off the probe for rehybridization, incubate the membrane in 0.5 M KOH at 40–50°C for 1 hr. If the background signal from the previous hybridization is high, we wash the membrane in the same solution overnight at 65°C. After rinsing with water, the membrane can be stored dry or it can be hybridized with another probe. We have successfully rehybridized the same membrane more than eight times.

VI. Conclusions and Perspectives

The *in vitro* chromatin assembly system from *Drosophila* embryos enables the reconstitution of cloned genes in chromatin with regularly spaced nucleosomes. As the reconstituted chromatin is transcriptionally repressed (Becker and Wu, 1992; Kamakaka *et al.,* 1993), this template approaches the *in vivo* structure of inert chromatin that is the substrate for interaction with transcription factors, RNA polymerase, and other sequence-specific-binding proteins. Thus, the ability to reconstitute transcriptionally repressed chromatin provides a starting point for investigations on the mechanism of action of these proteins in a near-physiological context. The assembly system should also be useful for the analysis of the pathway of histone deposition, for an analysis of the higher orders of chromatin folding, and more generally for mechanisms of DNA replication and recombination in a chromatin context. The evolutionary conservation of core histone structures, and the feasibility of incorporating the species-specific linker histone H1 exogenously suggest that the *Drosophila* system may additionally serve to mimic the chromatin structure of mammalian genes.

While the present *in vitro* system is efficient and reliable, there is the disadvantage of working with a crude, unfractionated extract. Hence, a considerable challenge for the future will be the purification and characterization of the individual components required for the complex process of nucleosome assembly and spacing. Notwithstanding the lack of a purified system, the crude extract has proved useful in addressing the question of nucleosome positioning on a mammalian α-fetoprotein gene (McPherson *et al.*, 1993) and in the mechanism of nucleosome disruption by a constitutively active GAGA transcription factor (Tsukiyama *et al.*, 1994). A further elaboration of the crude system toward the assembly of chromatin on magnetic beads promises to greatly extend its utility for solid phase analyses (Sandaltzopoulos *et al.*, 1994). The *in vitro* assembly system may also aid biochemical studies of *Drosophila* mutants with phenotypes suggesting an involvement of chromatin structure, i.e., the suppressors and enhancers of position effect variegation (Shaffer *et al.*, 1993) and the *polycomb* and *trithorax* group genes (Paro, 1990; Tamkun *et al.*, 1992).

Acknowledgments

We acknowledge postdoctoral fellowship support from Deutsche Forschungsgemeinschaft (P.B.B.) and the NIH Fogarty Center (P.B.B. and T.T). This work was supported by the Intramural Research Program of the National Cancer Institute, National Institutes of Health.

References

Albright, S. C., Nelson, P. P., and Garrard, W. T. (1979). Histone molar ratios among different electrophoretic forms of mono- and dinucleosomes. *J. Biol. Chem.* **254,** 1065–1073.

Almouzni, G., and Méchali, M. (1988). Assembly of spaced chromatin promoted by DNA synthesis in extracts from Xenopus eggs. *EMBO J.* **7,** 665–672.

Almouzni, G., Méchali, M., and Wolffe, A. (1990). Competition between transcription complex assembly and chromatin assembly on replicating DNA. *EMBO J.* **9,** 573–582.

Banerjee, S., and Cantor, C. R. (1990). Nucleosome assembly of simian virus 40 DNA in a mammalian cell extract. *Mol. Cell. Biol.* **10,** 2863–2873.

Bavykin, S. G., Usachenko, S. I., Zalensky, A. O., and Mirzabekov, A. D. (1990). Structure of nucleosomes and organization of internucleosomal DNA in chromatin. *J. Mol. Biol.* **212,** 495–511.

Becker, P. B., and Wu, C. (1992). Cell-free system for assembly of transcriptionally repressed chromatin from Drosophila embryos. *Mol. Cell. Biol.* **12,** 2241–2249.

Croston, G. E., Lira, L. M., and Kadonaga, J. T. (1991). A general method for purification of H1 histones that are active for repression of basal RNA polymerase II transcription. *Protein Expression Purif.* **2,** 162–169.

Germond, J. E., Hirt, B., Oudet, P., Gross-Bellard, M., and Chambon, P. (1975). Folding the DNA double helix into chromatin-like structures from simian virus 40. *Proc. Natl. Acad. Sci. U.S.A.* **72,** 1843–1847.

Javaherian, K., Tse, Y., and Vega, J. (1982). Drosophila topoisomerase I: Isolation, purification and characterization. *Nucleic Acids Res.* **10,** 6945–6955.

Kamakaka, R. T., Bulger, M., and Kadonaga, J. T. (1993). Potentiation of RNA polymerase II transcription by Gal4-VP14 during but not after DNA replication and chromatin assembly. *Genes Dev.* **7,** 1779–1795.

Keller, W. (1975). Determination of the number of superhelical turns in simian virus 40 DNA by gel electrophoresis. *Proc. Natl. Acad. Sci. U.S.A.* **72,** 4876–4880.

Krude, T., and Knippers, R. (1993). Nucleosome assembly during complementary DNA strand synthesis in extracts from mammalian cells. *J. Biol. Chem.* **268,** 14432–14442.

McPherson, C. E., Shim, E.-Y., Friedman, D. S., and Zaret, K. (1993). A active tissue-specific enhancer and bound transcription factors existing in a precisely positioned nucleosome array. *Cell* **75,** 387–398.

Nelson, T., Hsieh, T., and Brutlag, D. (1979). Extracts of Drosophila embryos mediate chromatin assembly in vitro. *Proc. Natl. Acad. Sci. U.S.A.* **76,** 5510–5514.

Paro, R. (1990). Imprinting a determined state into the chromatin of Drosophila. *Trends Genet.* **6**, 416–421.

Peck, L. J., and Wang, J. C. (1983). Energetics of B- to Z- transition in DNA. *Proc. Natl. Acad. Sci. U.S.A.* **80**, 6206–6210.

Rodríguez-Campos, A., Shimamura, A., and Worcel, A. (1989). Assembly and properties of chromatin containing H1. *J. Mol. Biol.* **209**, 135–150.

Sambrook, J., Fritsch, E. F., and Maniatis, T. (1989). "Molecular Cloning, A Laboratory Manual," 2nd ed., Cold Spring Harbor, NY: Cold Spring Laboratory Press.

Sandaltzopoulos, R., Blank, T., and Becker, P. B. (1994). Transcriptional repression by nucleosomes but not H1 in reconstituted preblastoderm Drosophila chromatin. *EMBO J.* **13**, 373–379.

Shaffer, C. D., Wallrath, L. L., and Elgin, S. C. R. (1993). Regulating genes by packaging domains: Bits of heterochromatin in euchromatin? *Trends Genet.* **9**, 35–37.

Shimamura, A., Tremethick, D., and Worcel, A. (1988). Characterization of the repressed 5S DNA minichromosomes assembled in vitro with a high-speed supernatant of *Xenopus laevis* oocytes. *Mol. Cell. Biol.* **8**, 4257–4269.

Simon, R. H., and Felsenfeld, G. (1979). A new procedure for purifying histone pairs H2A + H2B and H3 + H4 from chromatin using hydroxylapatite. *Nucleic Acids Res.* **6**, 689–696.

Tamkun, J. W., Deuring, R., Scott, M. P., Kissinger, M., Pattatucci, A. M., Kaufmann, T. C., and Kennison, J. A. (1992). brahma: A regulator of homeotic genes structurally related to the yeast transcriptional activator SNF2/SWI2. *Cell* **68**, 561–572.

Tsukiyama, T., Becker, P. B., and Wu, C. (1994). ATP-dependent nucleosome disruption at a heat-shock promoter mediated by binding of GAGA transcription factor. *Nature* **367**, 525–532.

Wu, C. (1989). Analysis of hypersensitive sites in chromatin. *Methods Enzymol.* **170**, 269–289.

CHAPTER 13

The Soluble Nuclear Fraction, a Highly Efficient Transcription Extract from *Drosophila* Embryos

Rohinton T. Kamakaka and James T. Kadonaga[1]

Department of Biology and
Center for Molecular Genetics
University of California, San Diego
La Jolla, California 92093

I. Summary

We describe the preparation and use of a nuclear extract derived from *Drosophila* embryos that is highly active for transcription *in vitro* by RNA polymerase II. This extract, which is termed the soluble nuclear fraction (SNF), can support multiple rounds of transcription and generate about 0.45 transcripts per template per 30 min. Furthermore, the SNF is deficient in nonspecific DNA

[1] To whom correspondence should be addressed.

binding factors that inhibit transcription, such as histone H1, and can be used for the analysis of transcriptional regulation by promoter- and enhancer-binding factors with either naked DNA or chromatin templates.

II. Introduction

To understand the molecular mechanisms by which RNA synthesis is controlled, it is necessary to perform transcription reactions *in vitro* to elucidate the identity and function of the participating factors. Thus, the development of cell-free extracts that are capable of accurate initiation of transcription by RNA polymerase II (Manley *et al.*, 1980; Dignam *et al.*, 1983) was a major advance in the analysis of eukaryotic gene regulation. At present, a variety of *in vitro* transcription extracts have been prepared from diverse sources ranging from cultured mammalian cells (Manley *et al.*, 1980; Dignam *et al.*, 1983) to *Drosophila* embryos (Heiermann and Pongs, 1985; Soeller *et al.*, 1988) and yeast (Lue and Kornberg, 1987; Woontner and Jaehning, 1990) [for recent reviews of basal transcription by RNA polymerase II, see Conaway and Conaway, 1993; Drapkin *et al.*, 1993].

We have been studying basal and regulated transcription by RNA polymerase II with transcription factors derived from *Drosophila* embryos. The use of *Drosophila* as the experimental organism was based on the high activity of the *Drosophila* extracts (Parker and Topol, 1984; Heiermann and Pongs, 1985; Soeller *et al.*, 1988; Kadonaga, 1990), the ease of obtaining kilogram quantities of embryos at a reasonable cost, and the functional interchangeability of the *Drosophila* basal transcription factors with their human counterparts (Wampler *et al.*, 1990). In addition, the *Drosophila* transcription extracts are ideally suited for the biochemical analysis of regulatory events that occur in the development of the embryo.

In general, transcription extracts are prepared by salt extraction (such as 0.42 *M* NaCl or 0.36 *M* ammonium sulfate) of either whole cells or nuclei, which leads to extraction of histones (especially histone H1) and other nonspecific DNA binding proteins (such as ABF2 in yeast) that inhibit transcription (see, for example, Croston *et al.*, 1991). Typically, with such extracts, the efficiency of transcription is low (less than 1% template usage in a single round of transcription). Hence, we felt that it would be useful to increase the efficiency of *in vitro* transcription reactions with a highly active preparation of basal transcription factors that was deficient in inhibitors such as histone H1. This has led to the development of a low-salt nuclear extract that has been officially termed the soluble nuclear fraction (SNF; Kamakaka *et al.*, 1991), but is more often referred to by its sobriquet, "nuke juice" (as it is literally the juice of nuclei).

In this chapter, we describe the preparation and use of the SNF from *Drosoph-*

ila embryos. The SNF has been employed to study the mechanisms of transcription initiation and reinitiation (Kamakaka *et al.,* 1991) and the activation of transcription by sequence-specific factors with either naked DNA (Kamakaka *et al.,* 1991; Croston *et al.,* 1991) or reconstituted chromatin templates (Laybourn and Kadonaga, 1991, 1992; Croston *et al.,* 1992; Kamakaka *et al.,* 1993). The SNF is prepared by rapid isolation and washing of nuclei followed by ultracentrifugation of the resulting nuclei at 100,000 *g* in a low-salt medium (0.1*M* KCl) to yield the soluble fraction of the nuclei (Fig. 1). The use of a low salt concentration during the high-speed centrifugation appears to be critical because it minimizes the extraction of nonspecific DNA binding proteins. Transcription with the SNF is five- to sixfold more efficient than that with standard *Drosophila* extracts, as determined by the template usage in a single round of transcription, which has been shown to be as high as 20% with the SNF (Kamakaka *et al.,* 1991). The SNF is capable of transcribing a broad range of class II genes (both non-*Drosophila* and *Drosophila* genes) and is able to reinitiate transcription for multiple rounds. In addition, transcriptional activation by promoter- and enhancer-binding factors can be attained *in vitro* with the SNF.

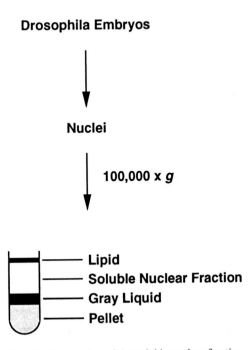

Fig. 1 Preparation of the soluble nuclear fraction.

III. Preparation of the Soluble Nuclear Fraction

The critical features of this method are the rapid isolation of nuclei to avoid the loss of soluble transcription factors, the low-salt extraction of the nuclei to minimize the extraction of nonspecific DNA binding factors that inhibit transcription, and the centrifugation of the nuclei in a minimal volume to maximize the concentration of basal transcription factors.

The SNF can be prepared from embryos of wild-type Canton S flies. For general background information regarding the growth and maintenance of *Drosophila,* see other chapters in this volume as well as Roberts (1986) and Ashburner (1989). Canton S flies are grown at 25°C at 70 to 80% humidity in population cages, and embryos are deposited onto molasses–agar trays that are spread with yeast paste. Embryos are collected from 0 to 12 hr after fertilization and then stored at 4°C before use. In the preparation of the SNF, embryos from up to 3 days (72 hr) of successive collections are harvested. The time of storage at 4°C, which is from 0 to 72 hr, does not appear to affect the properties of the SNF. Typically, 100 to 150 g of embryos will yield roughly 30 to 50 ml of SNF. The protein concentration of the SNF prepared with KCl is usually about 7 to 10 mg/ml, whereas the protein concentration of the SNF that is prepared with potassium glutamate is about 30 mg/ml, as determined with a Coomassie blue protein assay (Pierce Catalog No. 23200) by using bovine serum albumin as the reference.

A. Materials and Reagents

1. 50% (v/v) bleach: commercially available bleach, which contains 5.25% (w/v) sodium hypochlorite, is diluted 1 : 1 (v/v) with water.

2. Embryo wash solution: 0.7% (w/v) NaCl, 0.04% (v/v) Triton X-100 (Sigma Catalog No. T-6878).

3. Buffer I: 15 mM Hepes, K$^+$, pH 7.6; 10 mM KCl; 5 mM MgCl$_2$; 0.1 mM EDTA; 0.5 mM EGTA; 350 mM sucrose. Just before use, add the following (final concentrations are indicated): 1 mM dithiothreitol, 1 mM sodium metabisulfite, 0.2 mM phenylmethylsulfonyl fluoride, 1 mM benzamidine–HCl.

4. Buffer AB: 15 mM Hepes, K$^+$, pH 7.6; 110 mM KCl; 5 mM MgCl$_2$; 0.1 mM EDTA. Just before use, add the following (final concentrations are indicated): 2 mM dithiothreitol, 1 mM sodium metabisulfite, 0.2 mM phenylmethylsulfonyl fluoride, 1 mM benzamidine–HCl.

5. HEMG + 0.1 M potassium chloride: 25 mM Hepes, K$^+$, pH 7.6; 0.1 M KCl; 12.5 mM MgCl$_2$; 0.1 mM EDTA; 20% (v/v) glycerol. Just before use, add the following (final concentrations are indicated): 1.5 mM dithiothreitol, 1 mM sodium metabisulfite, 0.1 mM phenylmethylsulfonyl fluoride, 1 mM benzamidine–HCl.

6. HEMG + 0.4 M potassium glutamate: 25 mM Hepes, K$^+$, pH 7.6; 0.4 M potassium glutamate (L-glutamic acid, monopotassium salt; Sigma Catalog No.

G-1501); 12.5 mM MgCl$_2$; 0.1 mM EDTA; 20% (v/v) glycerol. Just before use, add the following (final concentrations are indicated): 1.5 mM dithiothreitol, 1 mM sodium metabisulfite, 0.1 mM phenylmethylsulfonyl fluoride, 1 mM benzamidine–HCl.

7. Stock solutions to be added to the buffers:

a. 0.5 M dithiothreitol (USB Catalog No. 15397); store at $-20°C$.

b. 0.5 M sodium metabisulfite (Sigma Catalog No. S-9000); prepare a fresh solution, store at 4°C, and use within 10 hr.

c. 0.2 M phenylmethylsulfonyl fluoride (Sigma Catalog No. P-7626), 0.2 M solution in absoute ethanol; store at $-20°C$.

d. 0.5 M benzamidine–HCl (Sigma Catalog NO. B-6506); store at $-20°C$.

B. Procedure for the Preparation of the *Drosophila* Soluble Nuclear Fraction

1. Grow *Drosophila melanogaster* (we use wild-type Canton S flies) in population cages as described above. Harvest the embryos by washing the trays over a collection apparatus that consists of two Nitex nylon screens. The top screen (500 μm, Tetko Catalog No. 3-500/49) retains the adult flies while the lower finer screen (70 μm Tetko Catalog No. 3-70/43) retains the embryos and allows the yeast to flow through. Wash the embryos extensively with water to remove traces of yeast.

2. Dechorionate the embryos by immersion for 90 sec in 50% (v/v) bleach solution at room temperature. Quickly rinse the embryos with embryo wash solution (1 liter) and then rinse thoroughly with running water at room temperature. Dry the embryos with paper towels—place the paper towels under the fine nylon mesh of the embryo collection apparatus to absorb the excess water.

NOTE: ALL SUBSEQUENT STEPS SHOULD BE PERFORMED AS QUICKLY AS POSSIBLE AT 4°C.

3. Transfer the embryos into a chilled, preweighed 800-ml beaker and determine the mass of the embryos. Combine 3 ml of chilled (4°C) Buffer I per gram of embryos, and disperse the embryos in the buffer with a glass rod.

4. Disrupt the embryos with a single passage through the Yamato LH-21 homogenizer at 1000 rpm. (If a Yamato LH-21 homogenizer is not available, homogenize the embryos with six to eight strokes in a motorized Potter-Elvehjem homogenizer with a serrated teflon pestle and a glass vessel. This apparatus often does not yield complete disruption of the embryos, however, and it may be necessary to carry out further homogenization by using a 40-ml Wheaton Dounce homogenizer with a B pestle.)

5. Filter the homogenate through a single layer of Miracloth (Calbiochem Catalog No. 475855) into a Sorvall GSA centrifuge bottle. Wash the Miracloth with additional Buffer I (2 ml per gram embryos). The final volume of Buffer

I should be about 5 ml/g embryos. *Once the embryos are homogenized, work quickly—proteins can leak out of the nuclei into the buffer.*

6. Pellet nuclei in a Sorvall GSA rotor at 8000 rpm for 15 min. Carefully decant the supernatant (the pellet is very loose). Wipe off lipids from the walls of the bottle with Kimwipes. The nuclei appear as a loose, tan-colored pellet. At the very bottom of the bottle, there is a small, yellow yolk pellet. Suspend the nuclei gently by swirling in Buffer I (3 ml/g embryos). Do not suspend the yellow yolk pellet.

7. Use a 40-ml Wheaton Dounce homogenizer with a loose B pestle to disperse the nuclei with a single stroke of the pestle. Transfer to clean GSA centrifuge bottles, and repellet the nuclei in a Sorvall GSA rotor at 8000 rpm for 15 min. Once again, carefully decant the supernatant and wipe off lipids from the walls of the bottle with Kimwipes.

8. Resuspend the nuclei gently by swirling in Buffer AB (1 ml/g embryos). Use a 40-ml Wheaton Dounce homogenizer with a loose B pestle to disperse the nuclei with two to three strokes of the pestle. Transfer the supernatant to one preweighed GSA centrifuge bottle and repellet the nuclei in a Sorvall GSA rotor at 8000 rpm for 10 min.

9. Decant the supernatant and weigh the bottle with the pellet. Determine the mass of the nuclei in the bottle.

10. Add Buffer HEMG + 0.1 M KCl to the nuclei (0.5 ml buffer/g nuclei). [Alternatively, HEMG + 0.4 M potassium glutamate buffer may be used, as described originally in Kamakaka *et al.* (1991). At present, however, we prefer to use 0.1 M KCl instead of 0.4 M potassium glutamate. The use of 0.1 M KCl instead of 0.4 M potassium glutamate results in less extraction of nonspecific DNA binding proteins, such as histone H1.] Suspend the nuclei by gentle swirling and shaking. Do not use the Dounce homogenizer. Place the nuclear suspension on ice for at least 15 min (15 to 60 min) with occasional swirling and shaking.

11. Subject the mixture to centrifugation in a Beckman SW28 rotor at 24,000 rpm (100,000 g) for 1 hr at 4°C. (Note: A fixed-angle rotor is not recommended for this step.)

12. After centrifugation is complete, there will be four distinct layers in the centrifuge tubes (Fig. 1). From top to bottom, they are as follows.

   a. A thin gray–white lipid layer. Remove it with a spatula and discard.

   b. A yellow liquid layer, which comprises about 50 to 60% of the total volume of the tube—this is the soluble nuclear fraction. Remove it with a pipet. Quick-freeze the SNF in liquid nitrogen in 0.1- to 1.0-ml aliquots and store at −80°C.

   c. A gray layer. Avoid this material, as it will inhibit transcription.

   d. At the bottom, there is a solid off-white layer composed of DNA and nuclear debris.

IV. *In Vitro* Transcription/Primer Extension Analysis

In vitro transcription with the SNF and primer extension analysis of the resulting transcripts are described in this section.

A. Materials and Reagents

1. Transcription buffer: 12.5 mM Hepes, K$^+$, pH 7.6; 6.25 mM MgCl$_2$; 0.05 mM EDTA; 5% (v/v) glycerol; 0.5 mM dithiothreitol.

2. Transcription stop: 20 mM EDTA, Na$^+$, pH 8.0; 0.2 M NaCl; 1% (w/v) SDS; 0.25 mg/ml glycogen (Sigma Catalog No. G-0885). Store at room temperature.

3. Annealing buffer: 10 mM Tris–HCl, pH 7.8; 1 mM EDTA; 0.25 M KCl. [Note: We normally prepare 5× annealing buffer (10 mM Tris–HCl, pH 7.8; 1 mM EDTA; 1.25 M KCl), store at −20°C, and dilute to 1× annealing buffer with TE (10 mM Tris–HCl, pH 7.8; 1 mM EDTA) just before use.]

4. Extension mix: Add the following components in the indicated order. The recipe given is for 1 ml of extension mix.

   a. One hundred twenty-five micrograms of actinomycin D (U.S. Biochemicals, Catalog No. 10415; 62.5 μl of a 2-mg/ml stock in ethanol is stored at −20°C and evaporated to dryness in a Speedvac rotary concentrator).

   b. Add glass-distilled water (864 μl).

   c. Add 1 M Tris–HCl, pH 8.3, buffer (62.5 μl). Vortex thoroughly to dissolve the actinomycin D.

   d. Add 100 mM MnCl$_2$ (12.5 μl).

   e. Add a stock solution containing 10 mM of each of the four dNTPs (35.6 μl; stock prepared from Pharmacia Catalog No. 27-2035-01). Vortex briefly to mix components.

   f. Add 0.5 M dithiothreitol (25 μl) to a final volume of 1 ml. Mix by vortexing. Store in the dark at −20°C.

5. Reverse transcriptase mix: Just before use, combine reverse transcriptase from Moloney Murine Leukemia Virus (Strategene Catalog No. 600084) with extension mix on ice (~4°C). Gently mix by flicking and inverting the tube. For each reaction, use 10 units (typically 0.5 μl) of reverse transcriptase with 40 μl of extension mix.

B. *In Vitro* Transcription with the *Drosophila* Soluble Nuclear Extract and Primer Extension Analysis of the RNA

Transcription reactions are usually performed with either 5 or 20 μl SNF in a total volume of 25 μl.

1a. For reactions with 5 μl SNF, combine the following at 4°C (over ice):

  15 μl   transcription buffer containing 50 mM KCl (or 0.1 potassium glutamate, if the glutamate version of the SNF is used).

  5 μl   SNF

  2 μl   template DNA (usually a supercoiled plasmid); 50 ng/μl in 10 mM Tris–HCl, pH 7.8, 1 mM EDTA

  3 μl   rNTP solution (5 mM of each of the four rNTPs; prepared from Pharmacia Catalog Nos. 27-2056-01, 27-2066-01, 27-2076-01, and 27-2086-01)

1b. Alternatively, for reactions with 20 μl SNF, combine the following at 4°C (over ice):

  20 μl   SNF

  2 μl   template DNA (usually a supercoiled plasmid); 50 ng/μl in 10 mM Tris–HCl, pH 7.8, 1 mM EDTA

  3 μl   rNTP solution (5 mM of each of the four rNTPs)

2. Incubate the reactions at 25°C in a water bath for 30 min. (Note: In an alternative procedure, the reactions can be incubated in the absence of ribonucleoside 5-triphosphates for 30 min at 25°C to allow assembly of the preinitiation complexes. The rNTPs can then be added and the reaction allowed to proceed at 25°C for another 30 min.)

3. Terminate the reactions by the addition of 100 μl transcription stop containing 0.125 μg/μl proteinase K (U.S. Biochemicals Catalog No. 20818; used as a 2.5-mg/ml stock in 10 mM Tris–HCl, pH 7.8, 1 mM EDTA buffer that is stored at −20°C).

4. Let the proteinase K digestion proceed at room temperature for 3 to 5 min. Add 300 μl of 0.3 M sodium acetate and then extract the mixture with 400 μl phenol : chloroform : isoamyl alcohol (25 : 24 : 1, v/v). Transfer the aqueous phase to a new tube. Be careful to avoid any material at the aqueous-organic interfere.

5. Extract the aqueous phase with 400 μl chloroform : isoamyl alcohol (24 : 1, v/v). Transfer the aqueous phase to a new tube. Once again, avoid any material at the interface.

6. Add 0.01 to 0.03 pmole of 5′ ^{32}P-labeled oligonucleotide (typically the oligonucleotide is 5′-phosphorylated with T4 polynucleotide kinase and labeled ATP). Mix by vortexing and then add 1 ml ethanol. Mix by inversion and pellet the nucleic acids for 15 min in a microcentrifuge at 13,000 rpm at room temperature. Remove as much of the supernatant as possible.

7. Suspend the radioactive pellet in 100 μl of 0.3 M sodium acetate. Dissolve the pellet by vortexing and add 250 μl ethanol. Mix by inversion and pellet the nucleic acids for 15 min in a microcentrifuge at 13,000 rpm. Remove the supernatant.

8. Add 0.3 ml of 75% (v/v) ethanol. Mix by vortexing. Pellet the nucleic acids for 5 min in a microcentrifuge. Remove the supernatant and dry the pellet in a Speedvac rotary concentrator.

9. Suspend the dried pellet in 10 μl of annealing buffer. Mix by vortexing. Briefly (1 sec) spin the sample in a microcentrifuge to pellet the liquid.

10. Incubate the mixture at 75°C for 90 sec and then at 58°C for 45 min (Note: The optimal temperature for the annealing of the primer to the RNA will vary depending upon the oligonucleotide used as the primer).

11. Let the mixture cool to room temperature. Briefly (1 sec) spin the sample in a microcentrifuge to pellet the liquid. To each reaction, then add 40 μl of reverse transcriptase mix.

12. Gently mix the liquid by flicking the tube (and/or slow vortexing and then incubate the samples at 37°C for 60 min.

13. Terminate the reaction by the addition of 300 μl ethanol directly to the reaction mixture. Mix by inversion and pellet the nucleic acids for 15 min in a microcentrifuge at room temperature. Remove the supernatant.

14. Add 400 μl of 75% (v/v) ethanol. Mix by vortexing. Pellet the nucleic acids for 5 min in a microcentrifuge. Remove the supernatant and dry the pellet in a Speedvac rotary concentrator.

15. Suspend the pellet in 9 μl of formamide loading buffer (such as a loading buffer that is used for DNA sequencing). Boil 3 min, quick-chill on ice, and load 3 to 4 μl per lane onto a standard 6 or 8% polyacrylamide–urea DNA sequencing-style gel.

V. Notes on the Use of the Soluble Nuclear Fraction

The SNF is a nuclear extract that can be used as a rich source of basal RNA polymerase II transcription factors. It can be used for *in vitro* transcription studies of both *Drosophila* and mammalian genes. A few comments on its properties are as follows.

1. Preparation of the SNF is simple and rapid—it can be easily made in 2.5 hr. In addition, we have found that this procedure works well for amounts of embryos ranging from less than 0.5 g to greater than 150 g. In all instances, however, we recommend using a swinging bucket rotor rather than a fixed angle rotor for the high-speed centrifugation.

2. Because the SNF is prepared with low-salt buffers, it can be used directly in transcription reactions without prior dialysis or desalting.

3. It appears that many sequence-specific transcription factors are extracted from nuclei in the preparation of the SNF. Hence, it should not be presumed that the SNF is deficient in promoter- and enhancer-binding factors. To deplete the SNF of any particular sequence-specific DNA binding factor, we recom-

mend incubation of the SNF with a sequence-specific DNA affinity resin, as described previously by Kerrigan *et al.* (1991).

4. It is not known whether the SNF contains transcription factors associated with RNA polymerases I or III. Preliminary results suggest, however, that the SNF does not contain significant RNA polymerase III transcriptional activity.

5. In our studies with the SNF, we have monitored RNA synthesis by using a primer extension assay. Thus, it is not known if the SNF can be used in conjunction with a runoff assay, which has different constraints than the primer extension assay.

6. A soluble nuclear fraction can also be prepared from HeLa cells. The HeLa SNF, however, appears to be of comparable activity to that of conventional HeLa extracts prepared by salt treatment of nuclei (Dignam *et al.*, 1983).

7. Because the SNF is deficient in core histones and H1, it is particularly useful in the transcriptional analysis of DNA templates that have been reconstituted into chromatin (for recent review, see Paranjape *et al.*, 1994). It should be noted that conventional transcription extracts that have been prepared by salt treatment of nuclei (for example, Dignam *et al.*, 1983) contain high levels of histone H1 (or ABF2 with yeast extracts), which inhibit basal transcription in vitro (see, for example, Croston *et al.*, 1991). In addition, commonly used *Drosophila* extracts, such as that described by Soeller *et al.* (1988), contain moderate amounts of the core histones, including H2B (Kerrigan *et al.*, 1993). Thus, the use of the SNF would minimize the unintended addition of histones along with the basal transcription factors in the analysis of chromatin templates.

Acknowledgments

We thank Deborah Frank, Elizabeth Blackwood, Tom Burke, and Alan Kutach for reading the manuscript. J.T.K. is a Presidential Faculty Fellow. This work was supported by grants from the National Institutes of Health, the National Science Foundation, the Council for Tobacco Research, and the Lucille P. Markey Charitable Trust.

References

Ashburner, M. (1989). "*Drosophila:* A Laboratory Manual." Cold Spring Harbor, NY: Cold Spring Harbor Laboratory Press.

Conaway, R. C., and Conaway, J. W. (1993). General initiation factors for RNA polymerase II. *Annu. Rev. Biochem.* **62**, 161–190.

Croston, G. E., Kerrigan, L. A., Lira, L., Marshak, D. R., and Kadonaga, J. T. (1991). Sequence-specific antirepression of histone H1-mediated inhibition of basal RNA polymerase II transcription. *Science* **251**, 643–649.

Croston, G. E., Laybourn, P. J., Paranjape, S. M., and Kadonaga, J. T. (1992). Mechanism of transcriptional antirepression by GAL4-VP16. *Genes Dev.* **6**, 2270–2281.

Dignam, J. D., Lebovitz, R. M., and Roeder, R. G. (1983). Accurate transcription initiation by RNA polymerase II in a soluble extract from isolated mammalian nuclei. *Nucleic Acids Res.* **11**, 1475–1489.

Drapkin, R., Merino, A., and Reinberg, D. (1993). Regulation of RNA polymerase II transcription. *Curr. Opin. Cell Biol.* **5,** 469–476.

Heiermann, R., and Pongs, O. (1985). In vitro transcription with extracts of nuclei of *Drosophila* embryos. *Nucleic Acids Res.* **13,** 2709–2730.

Kadonaga, J. T. (1990). Assembly and disassembly of the *Drosophila* RNA polymerase II complex during transcription. *J. Biol. Chem.* **265,** 2624–2631.

Kamakaka, R. T., Tyree, C. M., and Kadonaga, J. T. (1991). Accurate and efficient RNA polymerase II transcription with a soluble nuclear fraction derived from *Drosophila* embryos. *Proc. Natl. Acad. Sci. U.S.A.* **88,** 1024–1028.

Kamakaka, R. T., Bulger, M., and Kadonaga, J. T. (1993). Potentiation of RNA polymerase II transcription by Gal4-VP16 during but not after DNA replication and chromatin assembly. *Genes Dev.* **7,** 1779–1795.

Kerrigan, L. A., Croston, G. E., Lira, L. M., and Kadonaga, J. T. (1991). Sequence-specific transcriptional antirepression of the *Drosophila Krüppel* gene by the GAGA factor. *J. Biol. Chem.* **266,** 574–582.

Kerrigan, L. A., and Kadonaga, J. T. (1993). Periodic binding of individual core histones to DNA: Inadvertent purification of the core histone H2B as a putative enhancer-binding factor. *Nucleic Acids Res.* **20,** 6673–6680.

Laybourn, P. J., and Kadonaga, J. T. (1991). Role of nucleosomal cores and histone H1 in regulation of transcription by RNA polymerase II. *Science* **254,** 238–245.

Laybourn, P. J., and Kadonaga, J. T. (1992). Threshold phenomena and long-distance activation of transcription by RNA polymerase II. *Science* **257,** 1682–1685.

Lue, N. F., and Kornberg, R. D. (1987). Accurate initiation at RNA polymerase II promoters in extracts from *Saccharomyces cerevisiae*. *Proc. Natl. Acad. Sci. U.S.A.* **84,** 8839–8843.

Manley, J. L., Fire, A., Cano, A., Sharp, P. A., and Gefter, M. L. (1980). DNA-dependent transcription of adenovirus genes in a soluble whole-cell extract. *Proc. Natl. Acad. Sci. U.S.A.* **77,** 3855–3859.

Paranjape, S. M., Kamakaka, R. T., and Kadonaga, J. T. (1994). Role of chromatin structure in the regulation of transcription by RNA polymerase II. *Annu. Rev. Biochem.* **63,** 265–297.

Parker, C. S., and Topol, J. (1984). A *Drosophila* RNA polymerase II transcription factor contains a promoter-region-specific DNA-binding activity. *Cell* **36,** 357–369.

Roberts, D. B. (1986). "*Drosophila,* A Practical Approach." Oxford: IRL Press Limited.

Soeller, W. C., Poole, S. J., and Kornberg, T. (1988). In vitro transcription of the *Drosophila engrailed* gene. *Genes Dev.* **2,** 68–81.

Wampler, S. L., Tyree, C. M., and Kadonaga, J. T. (1990). Fractionation of the general RNA polymerase II transcription factors from *Drosophila* embryos. *J. Biol. Chem.* **265,** 21,223–21,231.

Woontner, M., and Jaehning, J. A. (1990). Accurate initiation by RNA polymerase II in a whole cell extract from *Saccharomyces cerevisiae*. *J. Biol. Chem.* **265,** 8979–8982.

CHAPTER 14

Basic Methods for *Drosophila* Muscle Biology

Eric A. Fyrberg,* Sanford I. Bernstein,[†] and K. VijayRaghavan[‡]

* Department of Biology
Johns Hopkins University
Baltimore, Maryland 21218

[†] Department of Biology and Molecular Biology Institute
San Diego State University
San Diego, California 92182

[‡] National Centre for Biological Sciences
Tata Institute of Fundamental Research
Indian Institute of Science Campus
Bangalore 560012
India

I. Introduction

Muscle is one of the most abundant *Drosophila* tissues, and one of the most amenable to cell biological investigations. These advantages, together with our detailed understanding of *Drosophila* genetics and embryology, afford unique opportunities for muscle biologists. Accordingly, much progress in delineating the tortuous paths of cell signaling, gene expression, and protein assembly that leads from initial inductive events to functional muscles can be attributed to studies of *Drosophila* and another organism whose genetics and embryology are well understood, the nematode *Caenorhabditis elegans* (see, for example, Epstein and Fischman, 1991; Epstein and Bernstein, 1992).

Our goal is to avail fly muscle research to the cell biological community by briefly recapitulating its history and then offering a collection of protocols that will enable readers to do their own exploring. We begin by pointing out several excellent reviews of the basic biology and genetics of *Drosophila* muscles. Miller's anatomical descriptions in *Biology of Drosophila* (Miller, 1950) remain a very useful roadmap of the adult muscles. Crossley (1978) provides an outstanding discussion of insect muscle anatomy and development in his chapter in *Genetics and Biology of Drosophila*. Particularly valuable are his outlines of larval muscle histology and development. Bernstein *et al.* (1993) considerably extend the former two treatises by integrating two decades of molecular genetic analyses into discussions of *Drosophila* muscle development, morphology, and function. Finally, Bate (1993) has recently reviewed the development of *Drosophila* mesoderm and its derivatives. His discussions of the embryological origins of muscle are of special interest to those studying muscle determination and patterning. These four chapters are essential reading for anyone investigating insect muscles.

Historically, the main foci of *Drosophila* muscle molecular genetics were to deduce how contractile protein isoforms are specified and to correlate muscle abnormalities with defects in particular sarcomeric components. The former endeavor basically has been an exercise in gene cloning, but one that has yielded many interesting findings, not least of which were the discoveries of the unexpected diversity of contractile protein isoforms and the fact that these can be specified by gene families (Fyrberg *et al.*, 1983), differential splicing (Bernstein *et al.*, 1983; Rozek and Davidson, 1983), or a combination thereof (Karlik and Fyrberg, 1985). The latter research took advantage of several unique features of *Drosophila* flight muscles, including their large size, unusually regular myofibrillar organization, and the ease with which flightless mutants are isolated (Hotta and Benzer, 1972; Deak, 1976). Most of the major contractile protein genes, specifically those encoding α-actinin, actin, myosin heavy and light chains, paramyosin, projectin (mini-titin), tropomyosin, and troponins-T, -I, and -C, have been cloned in the course of this research, and in all but two cases, troponin-C and paramyosin, these genes have been correlated with muscle mutants. Electron microscopy of the abnormal muscles has taught us elements of myofibril assembly and pathology (Fyrberg and Beall, 1990; Bernstein *et al.*, 1993). In the future, it will be possible to use directed mutagenesis to perturb any myofibrillar protein at will and to evaluate the structural and functional consequences. In addition, it will be possible to investigate the effects of eliminating any of a number of nonmyofibrillar proteins, such as integrins and vinculins, on muscle development (refer to Volk *et al.*, 1990; Volk and VijayRaghavan, 1994; for example).

A newer and rapidly blossoming activity of muscle biologists is to identify myogenic cell populations and delineate how these are determined and patterned. In addition, the innervation of muscles and their attachments to the epidermis can be followed. Previously, these were arduous undertakings due to a lack of cell-specific markers, although Bate (1990) and his predecessors made remarkable progress using conventional histology and vital dye marking methods. The use of specific markers has allowed several aspects of larval and adult muscle development to be followed in detail (Bate *et al.*, 1991; Michelson *et al.*, 1990; Paterson *et al.*, 1991; Fernandes *et al.*, 1991). It is now known that only larval and adult myoblasts continue to express the *twist* gene after gastrulation (Bate *et al.*, 1991, Currie and Bate, 1991). The *nautilus/Dmyd* locus encodes the *Drosophila* homologue of vertebrate MyoD (Michelson *et al.*, 1990; Paterson *et al.*, 1991) and marks cells that are the precursors of specific larval muscles. The gene *S59* encodes a homeodomain-containing protein and marks the precursors of some larval muscles (Dohrmann *et al.*, 1990). Antibodies that label the nervous system have been used to follow the development of innervation in the embryo (see Keshishian and Chiba, 1993, for a review) and in the adult (Fernandes and VijayRaghavan, 1993), thus providing a basis for investigating this aspect of patterning. Antibodies and reporter gene constructs have also been used to study the development of embryonic muscle attachment in *Drosophila* (Volk and VijayRaghavan, 1994). The ability to visualize all these aspects of muscle development facilitates the isolation and analysis of mutants with abnormal larval and adult muscle pattern, innervation, and attachment. Since there are a variety of novel methods for cloning *Drosophila* genes defined by mutation, we anticipate that this collective work will rapidly improve our understanding of the biochemical bases of muscle patterning. In addition, effects of extant neurological, intracellular transport, and cell signaling mutations on muscle now can be evaluated conveniently.

In sum, intrinsic advantages, together with our sophisticated knowledge of genetics and embryology, make *Drosophila* a choice experimental system for extending our comprehension of muscle development and function. In the past, the preoccupation has been to describe the synthesis and assembly of the major contractile proteins. In the future, attention will likely focus on the cues that mold muscle pattern. Continued refinement of models for myofibril assembly and contraction may be expected as well (see, for example, Reedy *et al.*, 1989; Reedy and Beall, 1993a,b).

II. Specific Methods

A. Examining Muscle Pattern and Integrity

1. Using Polarized Light Microscopy

Accurate assessment of muscle integrity can be made using polarized light microscopy, as described by Drysdale *et al.* (1993) for the embryo or Fleming *et al.* (1983) for the adult.

a. Embryos

1. Collect eggs on fruit juice–agar plates.
2. Age until the muscle system is formed; overnight at 25°C is convenient.
3. Dechorionate individual embryos and mount on a microscope slide in two to three drops of water. Overlay with a coverslip.
4. Examine in a polarizing microscope. By rotating the stage through 90° all muscles can be seen, and contractility, presence of striations, and pattern can be assessed. View muscle attachments by removing water, thus lowering the coverslip and flattening the sample.

b. Adults

Adults can be stored in 70% ethanol or 100% ethanol for an indefinite period. When necessary to examine them, they are dehydrated in 100% ethanol and then in isopropanol overnight. If the thorax is sliced open, the dehydration time can be shortened to about 15 min. Snipping off the abdomen and head also allows the separated parts to dehydrate rapidly. Dissected animals can be cleared by placing them in methyl salicylate (oil of wintergreen) for 15 min. Whole animals or whole heads, thoraces, and abdomen require longer periods of clearing (overnight). Cleared preparations are mounted in Gary's mounting medium (GMM; a 1 : 1 solution of powdered Canada Balsam in methylsalicylate; Lawrence *et al.*, 1986). The preparations are examined in transmitted polarized light using a dissecting microscope. They can also be examined using polarized light or Nomarski optics in a compound microscope. Polarizing filters to be attached to a dissecting microscope usually are available from the manufacturer of the microscope. Alternatively, sheets of polarizing filters can be bought from a local photo shop and attached to the dissecting microscope with tape.

2. Using Light or Fluorescence Microscopy

Muscles may be viewed in the light microscope using a number of protocols. View developing larval muscles after making "flat" preparations of embryos as described by Bate (1990).

1. Dissect embryos from vitelline membranes using fine needles in Ringer's solution.

2. Transfer to a poly-L-lysine coated coverslip.

3. Slit embryo along the dorsal midline and remove all yolk and gut.

4. View unstained preparations using Nomarski optics at this point, or briefly stain with 0.5–1% toluidine blue. After incubating in the stain for 1 hr, fix and destain embryos in Bodian's fixative (5 ml formalin, 6 ml glacial acetic acid, 90 ml 80% ethanol), dehydrate through an ethanol series, clear in xylene, and mount in Canada balsam.

View adult muscles after sectioning pupae or recently eclosed adults (see for example, Barbas *et al.,* 1993).

1. Fix adults in alcoholic Bouin's and then embed in paraffin.

2. Section at 10 μm.

3. Stain with toluidine blue or by the Milligan trichrome method (Humason, 1972).

Sections from pupae or adults may also be cut using a cryotome.

1. Embed flies in OCT compound on a suitable chuck as described by Deak (1977a,b).

2. After cutting, pick up sections on subbed slides.

3. Fix sections in 10% formalin for 5 min. If antibody decoration is desired, fix using 2% formalin in 75 mM phosphate buffer (pH 7.0) for 30 min and then rinse for 5 min in TBS (10 mM Tris–HCl, pH 7.5, 130 mM NaCl, 5 mM KCl, 5 mM NaN$_3$, and 1 mM EGTA).

4. Stain muscles using any of a number of histological techniques or by reacting with a suitable muscle-specific antibody (refer to following sections for more details). Cover sections with antibody and continue to process as described by Fujita *et al.,* (1982).

In some instances it may be advantageous to view only myofibrils. Flight or leg muscle myofibrils may be inspected conveniently using light or fluorescence microscopy.

1. Excise head, abdomen, legs, and wings using a razor blade or surgical scissors.

2. Dissect thorax in Ringer's on a depression slide.

3. Remove blocks of flight or leg muscles (the largest leg muscle, the TDT, will be attached to the mesothoracic leg after pulling it from the thorax). Transfer fibers to a glass slide.

4. Macerate muscle blocks using fine needles; add coverslip.

5. Inspect using a 40–100× phase objective. The sarcomeric pattern will be readily apparent in wild-type muscles.

6. React myofibrils with antibodies and view using fluorescence microscopy (see Lakey *et al.,* 1990, for example).

7. If larger quantities of myofibrils are required, refer to the mass isolation and storage protocol of Saide *et al.* (1989).

3. Using Electron Microscopy

State of the art electron microscopy of *Drosophila* tissues is described in Chapter 23. Below are techniques that have yielded very good to excellent preservation of *Drosophila* muscle in a number of laboratories. All reagents should be EM grade.

For developing embryo muscles, we recommend the protocol of Volk *et al.* (1990).

1. Collect embryos on fruit juice–agar plates and age until the proper stage of development is reached. Dechorionate in 50% bleach.
2. Fix in a 5% glutaraldehyde, 0.1 *M* cacodylate buffer (pH 7.2) and heptane (1 : 1) mixture for 20 min.
3. Transfer to fresh 2% glutaraldehyde, 0.1 *M* cacodylate buffer (pH 7.2) and remove vitelline membranes by hand peeling. Continue fixation for 4–6 hr.
4. Wash embryos in 0.1 *M* cacodylate, and postfix in 1% osmium tetroxide for 1 hr.
5. Wash again in cacodylate and stain in 2% uranyl acetate for 2 hr.
6. Dehydrate embryos through an ethanol series and infiltrate with propylene oxide 15 min and then a 1 : 1 mixture of propylene oxide and Epon–Araldite.
7. Embed embryos in Epon–Araldite, section, and stain.

For larvae, see O'Donnell and Bernstein (1988). Two key points should be noted:

1. Because the cuticle is impermeable, larvae must be punctured, or alternatively, the two "ends" cut off, to ensure fixative penetration.
2. During fixation, slices should be made in head and abdomen, or the animal should be bisected, to ensure fixative penetration.

Work with adults is focused on leg and flight muscles. Our recommended protocol for preparing those tissues is from Michael and Mary Reedy and is based upon their extensive experience with *Lethocerus* and *Drosophila* flight muscles (Reedy and Reedy, 1985).

1. Prepare ca. six thoraces of recently eclosed flies by removing heads, abdomens, legs, and wings using a razor blade or scissors.
2. Fix overnight in 3% glutaraldehyde, 0.2% tannic acid, 110 m*M* NaCl, 2 m*M* KCl, 3 m*M* MgCl$_2$, and 20 m*M* KMops (pH 6.8) in a 1.5-ml microcentrifuge tube (4°C).

3. The next day, split thoraces with fine needles and continue primary fixation for several hours or overnight at 4°C.
4. Rinse three times with cold Ringer's (110 mM NaCl, 2 mM KCl, 3 mM MgCl$_2$, and 20 mM KMops, pH 6.8).
5. Rinse three times in 100 mM KPO$_4$, 10 mM MgCl$_2$, pH 6.1.
6. Postfix 1 hr in 1% osmium tetroxide, 100 mM KPO$_4$, and 10 mM MgCl$_2$ (pH 6.1).
7. Rinse three times in deionized water.
8. Block stain overnight in 2% uranyl acetate in water at 4°C.
9. Dehydrate in 50, 95, and 100% ethanol.
10. Infiltrate with propylene oxide for 15 min.
11. Infiltrate with propylene oxide : embedding mix. Open tubes and allow propylene oxide to evaporate in hood.
12. Dissect muscles from thoraces in embedding mix lacking catalyst and then place in drops of embedding mix plus catalyst on a plastic coffee can lid. Cure and then invert a capsule full of resin over the drop and cure again. Trim this block and use for cutting longitudinal sections. Trim other drops, reembed ''longitudinally'' in coffin molds, and use to cut cross sections.

B. Identifying Muscle Precursor, or "Founder Cells"

For some time it has been known that *Drosophila* myoblasts are derived from the midventral blastoderm. It is clear that this same population of cells expresses the *twist* gene, which encodes a basic helix-loop-helix protein (Thisse *et al.*, 1987, 1988). As development proceeds, *twist* expression declines, but persists longest in a small group of cells that Bate *et al.* (1991) have discovered to be the precursors of adult muscles. Hence, ''transient'' *twist* perdurance is a convenient marker for cells determined to become the adult muscles, although it will not mark them after myoblast fusion. A *twist–lacZ* reporter construct will, however, mark muscles during later stages of development due to the much greater perdurance of β-galactosidase (Thisse and Thisse, 1992).

A second gene that prefigures muscle pattern is *nautilus/Dmyd* (Michelson *et al.*, 1990; Paterson *et al.*, 1991). This gene, like *twist,* encodes a basic helix-loop-helix protein that is expressed in muscle founder cells just prior to their fusion. Later in development, *nautilus/Dmyd* is expressed in some, but not all, growing and mature muscle fibers.

Three additional genes, *NK1/S59, NK-4/msh-2,* and *NK3* (Kim and Nirenberg, 1989; Bodmer *et al.*, 1990, Dohrmann *et al.*, 1990), are expressed in developing mesoderm in patterns that are clearly related to muscle development. All three contain homeobox sequences; hence, they may serve as master regulators of aspects of muscle determination and differentiation.

C. Following Muscle Differentiation *in Situ*

Because the expression of basic helix-loop-helix and homeodomain proteins is for the most part transient, other markers must be used to follow postfusion stages of *Drosophila* muscle differentiation. Most methods that are in current use involve staining for β-galactosidase activity. Some staining protocols are described at the end of this section. Many variations of these protocols exist and are described in Ashburner (1989).

1. Histochemical Markers

There are several enzymes whose activity can be examined histochemically in muscle fibers (Deak, 1977a,b). The most useful, activity staining for myosin Ca^{2+} ATPase activity, can be done on cryostat sections. Unfixed cryostat sections, 8–10 μm thick, of late embryos, larvae, or adult flies are collected on subbed slides. The slides are stained for myosin Ca^{2+} ATPase activity by dipping them in staining solution [10 mM $CaCl_2$, 4 mM ATP in 0.2 M sodium barbiturate buffer at pH 9.4 (Padykula and Herman, 1975)]. After 10 min in the staining solution, the slides are washed in 10% $CaCl_2$ (three times, 2 min each time) and then in 2% $CaCl_2$ (2 min), rinsed thoroughly in water, and then dipped in a 2% ammonium sulfide solution for 2 min. This results in a blackish coloration at the sites of Ca^{2+} ATPase activity. The slides are rinsed in water, dehydrated in an alcohol series, and mounted in Permount.

2. Reporter Genes

Flies carrying reporter gene constructs that express β-galactosidase in all or some muscles are useful for marking fibers. Muscle-specific reporter genes have been used successfully by Fernandes *et al.* (1991) to follow larval–adult muscle transitions. They used a fusion of the β-galactosidase gene to the myosin heavy-chain gene promoter (Hess *et al.*, 1989) to follow the degeneration of larval muscles, while simultaneously monitoring developing flight muscles with β-galactosidase driven by the promoter of the flight-muscle-specific actin gene, *Act88F* (Hiromi *et al.*, 1986). The myosin heavy-chain (*MHC36B*)-*lacZ* strain marks all the muscles of the larvae and adult (Hess *et al.*, 1989) and can also be used to study the formation of adult muscles (Fernandes *et al.*, 1991). The flight muscle-specific actin (*Actin88F*)–lacZ fusion construct labels only the developing indirect flight muscles (Hiromi *et al.*, 1986) and can be used to follow the development of these muscles.

Since many contractile protein genes of *Drosophila* are cloned, it is possible to construct reporter genes that will label any number of muscle cell patterns. To date, only a few candidate genes have been used to drive β-galactosidase reporters (Hiromi *et al.*, 1986; Courchesne-Smith and Tobin, 1989; Tobin *et al.*, 1990; Hess *et al.*, 1989; Gremke *et al.*, 1993; Meredith and Storti, 1993).

In addition to those discussed above, Courchesne-Smith and Tobin (1989) employ the *Act79B* promoter to label tubular adult muscles, and tropomyosin reporter constructs of Gremke *et al.* (1993) can be used to label all muscles. As mentioned earlier, the *twist–lacZ* strain is particularly useful, since the reporter gene is expressed in myoblasts but the perdurance of the β-galactosidase activity allows the differentiated muscles to be visualized as well (Thisse and Thisse, 1992).

3. Enhancer Trap Lines

Several enhancer trap strains that express the reporter β-galactosidase in the mesoderm or muscles are available. A catalog of the available enhancer trap lines and their expression patterns is available from the *Drosophila* stock center at Bloomington, Indiana (see also Bernstein *et al.,* 1993). The best documented to date is strain 1A122 (Perrimon *et al.,* 1991).

4. GAL4–Upstream Activating Sequence (UAS) Strains

Transformant strains that express the yeast transcription factor GAL4 under the control of regulatory sequences from the *twist* gene are available (Greig and Akam, 1993). These transformants allow the expression of any gene that has the GAL4 target UAS cloned in front of it. For example, when a *twist*–GAL4 is crossed to a UAS–*lacZ* strain, β-galactosidase activity can be visualized in the mesoderm of progeny containing *twist*–GAL4 and UAS–*lacZ*.

D. Dissecting Pupae and Staining Developing Adult Muscle Preparations

Muscle development can be studied readily in the embryo by examining whole mounts stained with antibodies or by histochemistry. Embryonic fillets are also easily made and several protocols are available (see, for example, Ashburner, 1989). However, methods for examining the development of the adult muscles are not so widely used. We therefore describe the dissection, staining, and examination of pupae.

Developing muscles can be visualized using *Actin88F–lacZ, MHC36B–lacZ,* and *twist–lacZ* strains. The developing innervation (Fernandes and VijayRaghavan, 1993) can be revealed using antibodies to horseradish peroxidase (HRP) or the monoclonal antibody MAb22C10 (Fujita *et al.,* 1982). The developing muscle attachment in the adult can be seen using antibodies to the PS antigens or an enhancer trap insert in the *stripe* locus (J. Fernandes, S. Celinker, and K. VijayRaghavan, unpublished; Volk and VijayRaghavan, 1994). In the cases in which antibodies are used, appropriate secondary antibodies must be used for detection. The protocol using *twist* antibody is described under Section 3 below.

1. Tissue Preparations

Pupae of desired ages are dissected in *Drosophila* Ringer's (in g/liter: 6.5 NaCl, 0.14 KCl, 0.2 $NaHCO_3$, 0.12 $CaCl_2$, 0.1 NaH_2PO_4). The white prepupal stage lasts about 1 hr at 25°C and is used as the 0-hr timepoint. The animals are cut open along the ventral midline, using a fine pair of Iris scissors (Fine Science Tools, U.S.A.) and pinned on Sylgard (Dow Corning Corp., U.S.A.). The insides are cleaned by gently blowing Ringer's on the preparation with a fine-glass pipet attached to a rubber-mouth tube. Dissected animals are fixed in a drop of 4% paraformaldehyde for 5–10 min. The tissue is subsequently rinsed in Ringer's for β-galactosidase histochemistry or phosphate-buffered saline (PBS; in g/liter: 20.0 NaCl, 5.0 KCl, 5.0 KH_2PO_4, 27.8 $Na_2HPO_4 \cdot 2H_2O$) for immmunocytochemistry with the desired antibody.

2. Histochemistry

Developing muscles can be stained in the *MHC36B–lacZ* and *Actin88F–lacZ* transformants using the chromogenic substrate. X-Gal (Sigma, St. Louis, MO). Dissected tissue is treated as described above and incubated in staining solution {10 mM $NaH_2PO_4 \cdot H_2O$/Na_2 $HPO_4 \cdot 2H_2O$ (pH 7.2), 150 mM NaCl, 1.0 mM $MgCl_2 \cdot 6H_2O$, 3.5 mM $K_4[FeII(CN)_6]$, 3.5 mM $K_3[FeIII(CN)_6]$, 0.3% Triton X-100 at 37°C (Simon *et al.*, 1985) for 15–30 min. The preparations are then washed in Ringer's, dehydrated in an alcohol series, cleared in xylene, and mounted in DPX (Sisco Research Labs, Bombay, India).

3. Immunocytochemistry

Antibody raised in rabbit against the *twist* protein (Thisse *et al.*, 1988) can be used to mark myoblasts. This antibody can be used effectively at a dilution of 1 : 500. Pupal tissue fixed for 0.5–1 hr is washed thoroughly with PBS, blocked with 0.5% bovine serum albumin (BSA), and incubated in the diluted *twist* antibody for 36 hr at 4°C. The tissue is then washed in PBT (0.3% Triton X-100 in PBS), for 1 hr and incubated for 1 hr in 0.2% goat serum. The preparations are then bathed in biotinylated anti-rabbit second antibody for 1 hr, washed in PBT, and amplified by reacting HRP-coupled avidin from the ABC Vectastain kit (Vector Laboratories, U.S.A.) as described by the manufacturer. The bound peroxidase is revealed using 0.5% diaminobenzidene (DAB; Sigma) 0.09% H_2O_2 in PBS until the color reaction develops. After the reaction is stopped (by washing in PBS), the tissue is dehydrated in an alcohol series, cleared in xylene, and mounted in DPX.

E. Screening for Mutants Having Abnormal Muscle Development

Traditional screens for muscle mutants involved inspection of dechorionated embryos for lack of muscle movements using an inverted phase or dissecting microscope. This method, although tedious, served to identify a number of interesting mutants, including *lethal(1) myospheroid* (Wright, 1960), which Mackrell *et al.* (1988) subsequently showed to lack β-integrin.

A somewhat different approach was taken by Drysdale *et al.* (1993). They analyzed embryos hemizygous for 67 deletions that collectively remove roughly 85% of the X chromosome, concluding that 31 of them had defects in muscle contractility or pattern. The phenotypic assay of living organisms was examination of muscles using polarized light. A higher resolution assessment of muscle pattern involved crossing female heterozygotes with males having the MHC : β-galactosidase fusion gene (described in the previous section) integrated into the second chromosome and then staining progeny with an anti-β-galactosidase antibody.

The screening method of Drysdale *et al.* (1993) offers considerably more resolution than the inspection assay used by Wright, yet does not require undue time or expense. As the nature of the muscle defect can be identified in some detail using this approach, Drysdale *et al.* (1993) could quickly sort the mutants into four categories: muscle absences, incomplete muscle fusions, failure of muscle attachments, and incorrect positioning of attachments. The report of these authors almost certainly will stimulate efforts to identify the genes conferring these phenotypes.

F. Screening for Flightless Mutants

Another traditional means for identifying muscle mutants is to screen for those unable to fly (Deak, 1977a,b; Koana and Hotta, 1978; Mogami and Hotta, 1981). In some instances, flightless individuals can be identified in vials. However, it is more efficient, and oftentimes reliable, to use a flight tester. Operationally, this involves pouring mutagenized fly offspring, or those having a muscle-specific pattern of enhancer trap staining, into a tall cylinder (a 1-liter graduated cylinder or comparably sized plexiglass tube works well). Walls are coated with mineral oil or lined with adhesive-coated acetate sheets (Tangle-trap insect trapping adhesive. Tanglefoot Co., Grand Rapids, MI). Flightless individuals fall to the bottom of the tester and into a collection funnel and vial. Muscles of individual strains subsequently can be examined for defects.

Alternative flight testers described by Sheppard (1974) and Green *et al.* (1986) permit the recovery of fly populations that can fly. Using either apparatus it is possible to mutate flightless strains and select revertants for subsequent analyses. A simple method for flight testing was also described by Drummond *et al.* (1991). Flies are released from a vial into the center of a plexiglass box that is divided into four sections: those flying up, horizontal, down, or not at all. A strong light is shown through the top of the box to elicit a positive phototrophic response. Wild-type flies fly up or, occasionally, horizontally, while flies with severe flight muscle defects usually fly down or not at all. Assaying 100 or so flies in this manner gives a reasonable estimate of flight ability for a particular strain.

Whereas screens for flightless adults have yielded many useful mutants, it should be kept in mind that flight muscles have exceptionally high rates of respiration and metabolism. Hence, a variety of hypomorphic mutations affecting these processes can result in flight loss without perturbing muscles specifically. Confirmation that the phenotype disrupts muscle, for example, by simple examination of muscle histology, is thus a wise investment before attempting to clone the candidate gene, map the mutation, etc. Finally, it is worth noting that any nonconditional *Drosophila* mutation that disrupts functioning of all muscles will result in lethal arrest, typically during late embryo or early larval stages. Hence, such mutants would be missed using the flight muscle screen exclusively.

G. Making Mosaic Muscles

1. Mitotic Clones

Mitotic recombination has been used extensively to genetically mark clones of epidermal cells (see Lawrence *et al.,* 1986, for a review). The use of this method to mark muscle cells has been limited by the paucity of markers for internal tissues. The availability of reporter gene constructs and constructs that express nuclear and membrane markers (Xu and Rubin, 1993) allows the generation of mitotic clones either by X-ray-induced mitotic recombination or by the use of the FLP recombination system. Before the availability of such transformant strains, muscle mitotic clones were studied using mutations in the gene succinate dehydrogenase (*sdh*) (Lawrence *et al.,* 1986). *sdh* mutants that make a heat labile enzyme are available. Animals that are *sdh−/sdh+* are irradiated during development and clones of marked cells visualized in the adult after heat incubating the tissue to inactivate the mutant enzyme. Mutant tissue is unstained, whereas the heterozygous tissue stains blue.

2. Nuclear Transplantation

Clones of marked muscle cells have been generated by transplanting nuclei from the preblastoderm *sdh+* embryos into *sdh*[8] embryos. *sdh*[8] is a temperature-sensitive mutant. By choosing host and donor of appropriate genotypes, both the epidermis and internal organs can be marked. Here, too, as in the case of generating clones by mitotic recombination, recently developed internal markers greatly expand the applicability of this method.

3. Transplanting Myoblasts and Surface Transplants of Imaginal Discs

The *sdh* internal marker has also been used to monitor myoblasts from wing imaginal discs transplanted into *sdh*[8] third instar larvae hosts (Lawrence and Brower, 1982). Such transplant experiments allow the examination of the cellular phenotype of muscle mutants. We describe here another transplantation protocol that results in ectopic appendages, whose donor-derived muscles are marked with reporter gene constructs.

For surface transplantation of imaginal discs, a modified protocol (N. Gendre and R. Stocker, University of Freibourg, Switzerland, in preparation) of the methods by Schubiger (1982) and Schmid *et al.* (1986) can be used. Wing discs or second leg discs from the white prepupal donor strains *MHC36B–lacZ, Actin88F–lacZ* or *twist–lacZ* are dissected and transplanted with a "constriction capillary" (Ashburner, 1989) onto an incision made in the cuticle of light brown wild-type prepupal hosts. Use as a transplantation site the lateral wall of abdominal segments, the ventral midline at the meso/metathoracic segment border, or the dorsal midline just anterior to the host wing discs (avoiding damage to the heart). If the discs are placed correctly under the host epidermis, they will evaginate together with those of the host and form supernumerary appendages.

4. Staining and Mounting Transplants

Newly emerged adults are decapitated, the tip of the abdomen as well as the legs and wings of the host is snipped off, and the specimen is washed in *Drosophila* Ringer's solution. In case of eclosion problems, flies can be dissected from the pupal case. The preparation should be cut along the dorsal or ventral midline, i.e., far from the site of the ectopic appendage, and then pinned on a Sylgard dish immersed in *Drosophila* Ringer's. Using a Pasteur pipet, a gentle flow of Ringer's is sprayed on the preparation to cleanse off gut and fat. The tissue is then fixed in 4% paraformaldehyde in PBS for 10 min, on ice, and rinsed in Ringer's for 1–2 hr. Preparations are then transferred to an X-Gal-containing staining solution: 1–2 mg X-Gal dissolved in 200 μl DMSO and 25 μl of this added to 1 ml of staining solution 150 mM NaCl, 1 mM MgCl$_2$, 3 mM K$_4$[FeII(CN)$_6$], 3 mM K$_3$[FeIII(CN)$_6$], 0.3% Triton X-100 and kept at 37°C just prior to incubation. After staining, the preparations are washed twice in Ringer's for 10 min, partially dehydrated in an ethanol series, and stored in 70% ethanol until mounted. For mounting, the tissue is further dissected in 70% ethanol, dehydrated to a final concentration of 100% ethanol, washed in isopropanol, cleared in methyl salicylate, and mounted in Gary's Magic Mountant (GMM; Canada Balsam powder dissolved in methyl salicylate, Lawrence *et al.*, 1986). Preparations can be examined in a compound microscope using Nomarski optics.

H. Primary Cultures of *Drosophila* Muscle

Primary cultures of *Drosophila* can be generated to examine embryonic muscle and its attachment to the epidermis (Volk *et al.*, 1990; Volk and VijayRaghavan, 1994). The following procedure is from Talila Volk (Weizmann Institute, Israel).

Six- to nine-hour-old embryos are dechorionated chemically by washing in bleach. The embryos are then washed extensively and dissociated into single-cell suspensions using a 7-ml Pyrex homogenizer (Corning No. 21). Following several washes with Schneider medium (Ashburner, 1989), cells are plated on coverslips coated with *Drosophila* laminin (in this instance, provided by J. and L. Fessler; see Chapter 18). Cultures are maintained for 1 or 2 days. The cells are then washed with PBS, fixed, and permeabilized using a mixture of 3% paraformaldehyde and 0.5% Triton X-100 for 20 min and then immunofluorescently labeled.

I. Hallmarks of Muscle Dysfunction

Here, we discuss general phenotypes of muscle mutants in the hope that our comments will aid in their recognition. Obviously, in many instances additional information can be gained by inspection using polarized light microscopy, as described earlier.

1. Hatching Failure

A number of mutations that disrupt embryonic muscles result in failure to hatch. The important point is that such failures do not typically result from an inability to move. Rather, the problem in most cases seems to be that the animals move actively, yet cannot muster sufficient contractile force to puncture the egg membranes and escape. This is the case, for example, in *l(1)102CDa*, in which the gene that specifies projectin (mini-titin), a myosin-associated filamentous protein, is disrupted (Fyrberg *et al.*, 1992), as well as in one myosin heavy-chain mutant, *Mhc⁴*, described by Mogami *et al.* (1986).

2. Crawling Speed

Assuming that a particular muscle mutant hatches as a first instar larva, crawling speed may be reduced. This is the case for α-actinin null mutants, for example, where newly hatched larvae move well initially, but stop moving and die by the second day of larval life (Roulier *et al.*, 1992). One interpretation is that the rapid growth that typifies the larval phase of the *Drosophila* life cycle places more severe mechanical demands on the muscle system, exacerbating any initial weakness. A second possibility is that the initial maternal complement of muscle protein allows the larval system to develop, but failure to replace it with zygotic protein leads to muscle failure.

A convenient way of measuring larval movement is to use olfactory cues such as acetone, acetic acid, or isobutanol to induce rapid migration toward the odorant source. Movement over a grid can be recorded conveniently using a dissecting microscope equipped with a video camera.

3. Jumping Assay for TDT Function

The tergal depressor of trochanter (TDT) fibers of the mesothoracic legs are the principal jumping muscles of the fly. When affected by mutations either specifically or nonspecifically, the ability of the fly to leap several centimeters may be compromised or eliminated. This can be quantified by removing the wings of anesthetized flies and evaluating their ability to jump after gently touching their abdominal regions upon revival. Such measurements are most convenient if one induces flies to jump onto an oil-coated surface calibrated in millimeters (O'Donnell *et al.*, 1989). Normal flies jump 3–7 cm, while many mutants fail to jump more than 2 cm (Deak, 1977a,b).

4. Measurement of Wingbeat Frequency

The indirect flight muscles are composed of 26 large fibers that occupy about half of the thoracic volume. If their structure and/or function is impaired, animals generally are unable to fly. In fact, a relatively slight impairment can render *Drosophila* flightless because the kinetics of shortening are "tuned" to the mechanical properties of the thorax and wings (Pringle, 1978).

There are two simple means for assessing flight muscle impairment. The first is to use a flight tester as described above. The second is to measure wingbeat frequency using a strobe. Anesthetize flies slightly and glue the end of a fine wire to their anterior thorax using nail polish. Wire should be fine gauge but sufficiently stiff so as not to bend when the fly beats its wings. After the cement sets, suspend the fly a few inches off the benchtop, being very careful that neither the wire nor the small drop of nail polish impedes wing movement or movement of the thoracic cuticle. Once the fly has recovered from anesthesia, it will generally begin to beat its wings spontaneously, air directed from the front generally will help to initiate wingbeats. Once wings are beating, the frequency quickly can be evaluated using a stroboscope. Wild-type flies generally beat their wings at 200 cycles per second, and a 10% or greater reduction, relative to the wild-type control, is a good indication of dysfunction. An alternative testing apparatus, the optical tachometer, is described by Unwin and Ellington (1979).

J. Evaluating Muscle Protein Accumulation and Filament Assembly

In many instances, perturbations of muscle protein accumulation and assembly are valuable hallmarks for abnormal muscle development. Protein and mRNA accumulation are easily monitored, while isolation and characterization of abnormally assembled filaments remains an underemployed but potentially very informative approach.

1. Radiolabeling of Flight Muscle Proteins

Accumulation of flight muscle proteins is most easily monitored by referring to prior work of Mogami *et al.* (1982). These authors have prepared numbered two-dimensional gel electrophoretic protein separations that serve as useful guides. In practice, the highest resolution separations are achieved using radiolabeled proteins, because less sample is loaded and because several different exposure times can be used. Proteins are conveniently labeled by injection of small volumes of high-specific activity [^{35}S]methionine into thoraces of recently eclosed adults. Incorporation of isotope is linear for several hours. Inject label using a fine glass micropipet by pushing through the soft thoracic cuticle just behind the head.

2. Preparation of Flight Muscle Proteins for Gel Electrophoresis

In the cases of flight and leg muscles, excellent results can be achieved by using the freeze–dry method of Mogami *et al.* (1982). Anesthetized flies are immersed in ice cold acetone and further dehydrated by changing the acetone solution daily for 3–5 days. Ten to twenty flies in a 1.5-ml Eppendorf tube are adequate for several standard one- or two-dimensional PAGE analyses and/or immunoblotting. Decant the acetone, transfer flies to a card or piece of blotting paper, and allow a few minutes for the last traces of the solvent to evaporate. Muscles are dessicated, hardened, and well defined by water loss. Individual fibers can be gently dissected without contamination from other fiber types or nonmuscle tissue. We typically homogenize fibers directly in O'Farrell equilibration buffer (O'Farrell, 1975), boil for 5 minutes, spin out insoluble matter in a microcentrifuge, add dye, and load onto standard or "mini" polyacrylamide gels. Add an equal volume of O'Farrell lysis buffer to run samples on isoelectric focusing gels.

For other muscle types, we can offer little helpful advice. One can easily make an enriched larval body wall muscle preparation by squashing larvae with a small rolling pin and homogenizing the cuticle with attached muscles in O'Farrell buffer as above. We know of no efficient means to prepare small visceral and skeletal muscles of the fly.

3. Western Blots

There are no special steps required for successful immunoblots of *Drosophila* muscle proteins. Table I lists the available antibodies known to work for immunoblotting or immunohistochemical protocols. Note that many were raised by G. Butcher and B. Bullard by immunizing rats with waterbug (*Lethocerus*) flight muscle proteins. Most of these react well with *Drosophila* homologs.

4. Immunohistochemistry

Immunohistochemical protocols for *Drosophila* are described elsewhere in this chapter (and volume). To date, we have not determined whether all of the antibodies described in Table I will react with proteins *in situ,* either in frozen sections or whole mounts.

5. Protocols for Filament Isolation

Isolation and characterization of thick and thin filaments, or their component proteins, from various *Drosophila* mutants should be a worthwhile pursuit, at least in theory. There are considerable differences in filament structure and

Table I
Antibodies for *Drosophila* Muscle Work

Antigen recognized	Description	Reference
I. Thick Filaments		
Projectin (mini-titin)	Rat monoclonal	Lakey *et al.*, 1990
Projectin (mini-titin)	Mouse monoclonal	Saide *et al.*, 1989
Myosin heavy chain	Rat monoclonal	Fyrberg *et al.*, 1990
Myosin heavy chain	Mouse monoclonal	Volk *et al.*, 1990
Myosin heavy chain	Rabbit polyclonal	Kiehart and Feghali, 1986
Alkali myosin light chain	Rabbit polyclonal	Falkenthal *et al.*, 1987
Myosin light chain-2	Rat monoclonal	Warmke *et al.*, 1992
Paramyosin	Rabbit polyclonal	Becker *et al.*, 1992
		Vinos *et al.*, 1991
II. Thin Filaments		
Actin	Mouse monoclonal	Ball *et al.*, 1987
Tropomyosin	Rat monoclonal	Bullard *et al.*, 1988
Troponin-H	Rat monoclonal	Bullard *et al.*, 1988
Troponin-I	Rabbit polyclonal	Barbas *et al.*, 1993
Troponin-T	Rat monoclonal	Bullard *et al.*, 1988
III. Z-discs, others		
α-actinin	Rat monoclonal	Roulier *et al.*, 1992
Integrins	Rat monoclonal	Bogaert *et al.*, 1987
	Rabbit monoclonal	Leptin *et al.*, 1989
Kettin	Rat monoclonal	Lakey *et al.*, 1993
Flightin	Mouse monoclonal	Vigoreaux *et al.*, 1993
Various Z-disc proteins	Mouse monoclonal	Saide *et al.*, 1989
IV. Mesoderm and muscle founder cells		
twist	Rabbit polyclonal	Thisse *et al.*, 1988
Nautilus/Dmyd	Rabbit polyclonal	Paterson *et al.*, 1991
S59	Rabbit polyclonal	Dohrmann *et al.*, 1990
V. Muscle Attachment		
MSP300	Rabbit polyclonal	Volk, 1992
Groovin	Mouse monoclonal	Volk and VijayRaghavan, 1994

function among myofibrils of various phyla, and these "natural" variants are presently being utilized to elaborate contractile processes further. Mutant filaments and proteins further extend this resource, as they are, arguably, the most reliable source of poorly functioning or nonfunctioning variants. They, like the naturally occurring variants described above, can be analyzed using structural approaches or used for *in vitro* motility assays and in this fashion further contribute to our understanding of actomyosin based motility.

If one chooses to pursue the mutant filament approach in *Drosophila,* one logically would isolate filaments from indirect flight muscles, because they are the largest fibers in this organism and because many examples of mutant flight muscles are already documented. Although *Lethocerus* flight muscle thick and thin filaments have been isolated and analyzed in some detail (Reedy *et al.,* 1981; Bullard *et al.,* 1988; Newman *et al.,* 1992), as yet mutant *Drosophila* filaments have not been so employed. The principal technical problem is that of isolating sufficient quantities of pure mutant muscle fibers or myofibrils. Means for isolating pure thoraces exist (Mogami *et al.,* 1982), but indirect flight muscles compose less than half of thoracic volume, the remainder being a mixture of smaller flight muscles and various organs. In the protocol for isolating indirect flight muscle proteins described earlier, flies were initially freeze dried in acetone to facilitate dissection of pure fiber populations. Generally, dessication denatures proteins and hence precludes meaningful structural and functional microassays. Likewise, purification of flight muscle myofibrils on the basis of their resistance to mechanical shear (Mogami *et al.,* 1982) is not possible in most mutants due to their structural abnormalities.

These technical problems notwithstanding, we feel that it may be possible to isolate pure filament populations from flight muscles of actin and myosin heavy-chain null mutants (Beall *et al.,* 1989). The advantage associated with such mutants is that only one filament type or the other accumulates in flight muscles of these strains, eliminating the need to dissociate and purify them from native myofibrils. Hence, it should be possible to gently homogenize purified thoraces and isolate the intact filaments using an appropriate centrifugation protocol.

Acknowledgments

We are grateful to colleagues for critically reading the manuscript and offering suggestions. Research of E.A.F. and S.I.B. has been supported by the NIH and Muscular Dystrophy Association. Research of K.V.R. is supported by the Indian Department of Science and Technology and by the Department of Biotechnology of the Government of India.

References

Ashburner, M. (1989). "*Drosophila:* A Laboratory Manual." Cold Spring Harbor, NY: Cold Spring Harbor Laboratory Press.
Ball, E., Karlik, C. C., Beall, C. J., Saville, D. L., Sparrow, J. C., Bullard, B., and Fyrberg,

E. A. (1987). Arthrin, a myofibrillar protein of insect flight muscle, is an actin–ubiquitin conjugate. *Cell* **51**, 221–228.

Barbas, J. A., Galceran, J., Torroja, L., Prado, A., and Ferrus, A. (1993). Abnormal muscle development in the *heldup*[3] mutant of *Drosophila melanogaster* is caused by a splicing defect affecting selected troponin-I isoforms. *Mol. Cell. Biol.* **13**, 1433–1439.

Bate, M. (1990). The embryonic development of larval muscles in *Drosophila*. *Development* **110**, 791–804.

Bate, M., Rushton, E., and Currie, D. A. (1991). Cells with persistent *twist* expression are the embryonic precursors of adult muscles in *Drosophila*. *Development* **114**, 79–89.

Bate, M. (1993). The mesoderm and its derivatives. *In* "The Development of *Drosophila*" (A. Martinez-Arias and M. Bate, eds.), Cold Spring Harbor, NY: Cold Spring Harbor Laboratory Press.

Beall, C. J., Sepanski, M. A., and Fyrberg, E. A. (1989). Genetic dissection of *Drosophila* myofibril formation: Effects of actin and myosin heavy chain null alleles. *Genes Dev.* **3**, 131–140.

Becker, K. D., O'Donnell, P. T. Heitz, J. M., Vito, M., and Bernstein, S. I. (1992). Analysis of *Drosophila* paramyosin-identification of a novel isoform which is restricted to a subset of adult muscles. *J. Cell Biol.* **116**, 669–681.

Bernstein, S. I., Mogami, K., Donady, J. J., and Emerson, C. P., Jr. (1983). *Drosophila* muscle myosin heavy chain encoded by a single gene in a cluster of muscle mutations. *Nature* **302**, 393–397.

Bernstein, S. I., O'Donnell, P. T., and Cripps, R. M. (1993). Molecular genetic analysis of muscle development, structure, and function in *Drosophila*. *Int. Rev. Cytology* **143**, 63–152.

Bodmer, R., Jan, L. Y., and Jan, Y. N. (1990). A new homeobox-containing gene, *msh*-2, is transiently expressed early during mesoderm formation of *Drosophila*. *Development* **110**, 661–669.

Bogaert, T., Brown, N., and Wilcox, M. (1987). The *Drosophila* PS2 antigen is an invertebrate integrin that, like the fibronectin receptor, becomes localized to muscle attachments. *Cell* **51**, 929–940.

Bullard, B., Leonard, K., Larkins, A., Butcher, G., Karlik, C., and Fyrberg, E. (1988). Troponin of asynchronous flight muscle. *J. Mol. Biol.* **204**, 621–637.

Courchesne-Smith, C. L., and Tobin, S. L. (1989). Tissue-specific expression of the 79B actin gene during *Drosophila* development. *Dev. Biol.* **133**, 313–321.

Crossley, A. C. (1978). The morphology and development of the *Drosophila* muscular system. *In* "The Genetics and Biology of Drosophila" (M. Ashburner and T.R.F. Wright, eds.), Vol. 2b, pp. 499–560. New York: Academic Press.

Currie, D., and Bate, M. (1991). Development of adult abdominal muscles in *Drosophila:* Adult myoblasts express *twist* and are associated with nerves. *Development* **113**, 91–102.

Deak, I. I. (1976). Use of *Drosophila* mutants to investigate the effect of disuse on the maintenance of muscle. *J. Insect Physiol.* **22**, 1159–1165.

Deak, I. I. (1977a). Mutations of *Drosophila melanogaster* that affect muscle. *J. Embryol. Exp. Morphol.* **40**, 35–63.

Deak, I. I. (1977b). A histochemical study of the muscles of *Drosophila melanogaster*. *J. Morphol.* **153**, 307–316.

Dohrmann, C., Azpiazu, N., and Frasch, M. (1990). A new *Drosophila* homeobox gene is expressed in mesodermal precursor cells of distinct muscles during embryogenesis. *Genes Dev.* **4**, 2098–2111.

Drummond, D. R., Hennessey, E. S., and Sparrow, J. C. (1991). Characterisation of missense mutations in the *Act88F* gene of *Drosophila melanogaster*. *Mol. Gen. Genet.* **226**, 70–80.

Drysdale, R., Rushton, E., and Bate, M. (1993). Genes required for embryonic muscle development in *Drosophila melanogaster*. *Rouxs. Arch. Dev. Biol.* **202**, 276–295.

Epstein, H. F., and Bernstein, S. I. (1992). Genetic approaches to understanding muscle development. *Dev. Biol.* **154**, 231–244.

Epstein, H. F., and Fischman, D. A. (1991). Molecular analysis of protein assembly in muscle development. *Science* **251**, 1039–1044.

Falkenthal, S., Graham, M., and Wilkinson, J. (1987). The indirect flight muscle of *Drosophila* accumulates a unique myosin alkali light chain isoform. *Dev. Biol.* **121**, 263–272.

Fernandes, J., Bate, M., and VijayRaghavan, K. (1991). Development of the indirect flight muscles of *Drosophila*. *Development* **113**, 67–77.

Fernandes, J., and VijayRaghavan, K. (1993). The development of indirect flight muscle innervation in *Drosophila melanogaster*. *Development* **118**, 215–227.

Fleming, R. J., Zusman, S. B., and White, K. (1983). Developmental genetic analysis of lethal alleles at the *ewg* locus and their effects on muscle development in *Drosophila melanogaster*. *Dev. Genet.* **3**, 347–363.

Fujita, S. C., Zipursky, S. L., Benzer, S., Ferrus, A., and Shotwell, S. (1982). *Proc. Natl. Acad. Sci. U.S.A.* **79**, 7929–7933.

Fyrberg, E. A., Mahaffey, J. W., Bond, B. J., and Davidson, N. (1983). Transcripts of the six *Drosophila* actin genes accumulate in a stage- and tissue-specific manner. *Cell* **33**, 115–123.

Fyrberg, E., and Beall, C. (1990). Genetic approaches to myofibril form and function in *Drosophila*. *Trends Genet.* **6**, 126–131.

Fyrberg, E., Fyrberg, C., Beall, C., and Saville, D. (1990). *Drosophila* melanogaster troponin-T mutations engender three distinct syndromes of myofibrillar abnormalities. *J. Mol. Biol.* **216**, 657–675.

Fyrberg, C. C., Labeit, S., Bullard, B., Leonard, K., and Fyrberg, E. (1992). *Drosophila* projectin: Relatedness to titin and twitchin and correlation with *lethal(4)102CDa* and *bent-Dominant* mutants. *Proc. R. Soc. London B.* **249**, 33–40.

Green, C. C., Sparrow, J. C., and Ball, E. (1986). Flight testing columns. *Drosophila Information Service* **63**, 141.

Greig, S., and Akam, M. (1993). Homeotic genes autonomously specify one aspect of pattern in the *Drosophila* mesoderm. *Nature* (*London*) **362**, 630–632.

Gremke, L., Lord, P. C. W., Sabacan, L., Lin, S.-C., Wohlwill, A., and Storti, R. V. (1993). Coordinate regulation of *Drosophila* tropomyosin gene expression is controlled by multiple muscle-type-specific positive and negative enhancer elements. *Dev. Biol.* **159**, 513–527.

Hess, N., Kronert, W. A., and Bernstein, S. I. (1989). Transcriptional and post-transcriptional regulation of *Drosophila* myosin heavy chain expression. *In* "Cellular and Molecular Biology of Muscle Development" (L. H. Kedes and F. E. Stockdale, eds.), pp. 621–631. New York: A. R. Liss.

Hiromi, Y., Okamoto, Y., Gehring, W., and Hotta, Y. (1986). Germline transformation with *Drosophila* mutant actin genes induces consitutive expression of heat shock genes. *Cell* **4**, 293–301.

Hotta, Y., and Benzer, S. (1972). Mapping of behaviour in *Drosophila* mosaics. *Nature* **240**, 527–535.

Humason, G. L. (1972). "Animal Tissue Techniques," 3rd ed., pp. 56–63 and 197–200. San Francisco: Freeman.

Karlik, C. C., and Fyrberg, E. A. (1985). An insertion within a variably spliced *Drosophila* tropomyosin gene blocks accumulation of only one encoded isoform. *Cell* **41**, 57–66.

Keshishian, H., and Chiba, A. (1993). Neuromuscular development in *Drosophila:* Insights from single neurons and single genes. *Trends Neurosci.* **16**, 278–282.

Kiehart, D. P., and Feghali, R. (1986). Cytoplasmic myosin from *Drosophila melanogaster*. *J. Cell Biol.* **103**, 1517–1525.

Kim, Y., and Nirenberg, M. (1989). *Drosophila* NK homeobox genes. *Proc. Natl. Acad. Sci. U.S.A.* **86**, 7716–7720.

Koana, T., and Hotta, Y. (1978). Isolation and characterization of flightless mutants in *Drosophila melanogaster*. *J. Embryol. Exp. Morphol.* **45**, 123–143.

Lakey, A., Ferguson, C., Labeit, S., Reedy, M., Larkins, A., Butcher, G., Leonard, K., and Bullard, B. (1990). Identification and localization of high molecular weight proteins in insect flight and leg muscle. *EMBO J.* **9**, 3459–3467.

Lakey, A., Labeit, S., Gautel, M., Ferguson, C., Barlow, D., Leonard, K., and Bullard, B. (1993). Kettin, a large modular protein in the Z-disc of insect muscles. *EMBO J.* **12**, 2863–2871.

Lawrence, P. A., and Brower, D. L. (1982). Myoblasts from *Drosophila* wing discs can contribute to developing muscles throughout the fly. *Nature* **295**, 55–57.

Lawrence, P. A., Johnston, P., and Morata, G. (1986). Methods of marking cells. *In* "*Drosophila*: A practical approach" (D. B. Roberts, ed.), pp. 229–242. Oxford: IRL Press.

Leptin, M., Bogaert, T., Lehmann, R., and Wilcox, M. (1989). The function of PS integrins during *Drosophila* embryogenesis. *Cell* **56**, 401–408.

Mackrell, A. J., Blumberg, B. Haynes, S. R., and Fessler, J. H. (1988). The *lethal myospheroid* gene of *Drosophila* encodes a membrane protein homologous to vertebrate integrin beta subunits. *Proc. Natl. Acad. Sci. U.S.A.* **85**, 2633–2638.

Meredith, J., and Storti, R. V. (1993). Developmental regulation of the *Drosophila* tropomyosin II gene in different muscles is controlled by muscle-type-specific enhancer elements and distal and proximal promoter control elements. *Dev. Biol.* **159**, 500–512.

Michelson, A. M., Abmayr, S. M., Bate, M., Martinez-Arias, A., and Maniatis, T. (1990). Expression of a *MyoD* family member prefigures muscle pattern in *Drosophila* embryos. *Genes Dev.* **4**, 2086–2097.

Miller, A. (1950). The internal anatomy and histology of the imago of *Drosophila melanogaster*. *In* "Biology of Drosophila" (M. Demerec, ed.), pp. 420–534. New York: Wiley.

Mogami, K., and Hotta, Y. (1981). Isolation of *Drosophila* flightless mutants which affect myofibrillar proteins of indirect flight muscle. *Mol. Gen. Genet.* **183**, 409–417.

Mogami, K., Fujita, S. C., and Hotta, Y. (1982). Identification of *Drosophila* indirect flight muscle myofibrillar proteins by means of two-dimensional electrophoresis. *J. Biochem.* **91**, 643–650.

Mogami, K., O'Donnell, P. T., Bernstein, S. I., Wright, T. R. F., and Emerson, C. P., Jr. (1986). Mutations of the *Drosophila* myosin heavy chain gene: Effects on transcription, myosin accumulation, and muscle function. *Proc. Natl. Acad. Sci. U.S.A.* **83**, 1393–1397.

Newman, R., Butcher, G. W., Bullard, B., and Leonard, K. R. (1992). A method for determining the periodicity of a troponin component in isolated insect flight muscle thin filaments by gold/ Fab labelling. *J. Cell Sci.* **101**, 503–508.

O'Donnell, P. T., and Bernstein, S. I. (1988). Molecular and ultrastructural defects in a *Drosophila* myosin heavy chain mutant: Differential effects on muscle function produced by similar thick filament abnormalities. *J. Cell Biol.* **107**, 2601–2612.

O'Donnell, P. T., Collier, V. L., Mogami, K., and Bernstein, S. I. (1989). Ultrastructural and molecular analyses of homozygous-viable *Drosophila melanogaster* muscle mutants indicates that there is a complex pattern of myosin heavy-chain isoform distribution. *Genes Dev.* **3**, 1233–1246.

O'Farrell, P. H. (1975). High resolution two-dimensional electrophoresis of proteins. *J. Biol. Chem.* **250**, 4007–4021.

Padykula, H. A., and Herman, E. (1975). Specificity of the histochemical method for adenosine triphosphate. *J. Histochem. Cytochem.* **3**, 170–190.

Paterson, B. M., Walldorf, U., Eldridge, J., Dubendorfer, A., Frasch, M., and Gehring, W. J. (1991). The *Drosophila* homologue of vertebrate myogenic determination genes encodes a transiently expressed nuclear protein marking primary myogenic cells. *Proc. Natl. Acad. Sci. U.S.A.* **88**, 3782–3786.

Perrimon, N., Noll, E., McCall, K., and Brand, A. (1991). Generating lineage-specific markers to study *Drosophila* development. *Dev. Genet.* **12**, 238–252.

Pringle, J. W. S. (1978). Stretch activation of muscle: Function and mechanism. *Proc. R. Soc. London B Biol. Sci.* **201**, 107–130.

Reedy, M. K., Leonard, K. R., Freeman, R., and Arad, T. (1981). Thick myofilament mass determination by electron scattering measurements with the scanning transmission electron microscope. *J. Musc. Res. Cell Motil.* **2**, 45–64.

Reedy, M. K., and Reedy, M. C. (1985). Rigor crossbridge structure in tilted single filament layers and flared-X formations from insect flight muscle. *J. Mol. Biol.* **185**, 145–176.

Reedy, M. C., Beall, C., and Fyrberg, E. (1989). Formation of reverse rigor chevrons by myosin heads. *Nature* **339**, 481–483.

Reedy, M. C., and Beall, C. J. (1993a). Ultrastructure of developing flight muscle in *Drosophila*. I. Assembly of myofibrils. *Dev. Biol.* **160**, 443–465.

Reedy, M. C., and Beall, C. J. (1993b). Ultrastructure of developing flight muscle in *Drosophila*. II. Formation of the myotendon junction. *Dev. Biol.* **160**, 466–479.

Roulier, E. M., Fyrberg, C., and Fyrberg, E. (1992). Perturbations of *Drosophila* alpha-actinin cause muscle paralysis, weakness, and atrophy but do not confer obvious nonmuscle phenotypes. *J. Cell Biol.* **116**, 911–922.

Rozek, C. E., and Davidson, N. (1983). *Drosophila* has one myosin heavy chain gene with three developmentally regulated transcripts. *Cell* **32**, 23–34.

Saide, J. D., Chin-Bow, S., Hogan-Sheldon, J., Busquets-Turner, L., Vigoreaux, J. O., Valgeirsdottir, K., and Pardue, M. L. (1989). Characterization of components of Z-bands in the fibrillar flight muscles of *Drosophila melanogaster*. *J. Cell Biol.* **109**, 2157–2167.

Schmid, H., Gendre, N., and Stocker, R. F. (1986). Surgical generation of supernumerary appendages for studying of neuronal specificity in *Drosophila melanogaster*. *Dev. Biol.* **113**, 160–173.

Schubiger, M. (1982). Technique for everted imaginal disc transplants. *Drosophila Information Service* **58**, 159–160.

Sheppard, D. E. (1974). A selective procedure for the separation of flightless adults from normal flies. *Drosophila Information Service* **51**, 150.

Simon, J. A., Sutton, C. A., Lobell, R. B., Galser, R. L., and Lis, J. T. (1985). Determinants of heat shock induced chromosome puffing. *Cell* **40**, 805–817.

Thisse, B., Stoetzel, C., El Messal, M., and Perrin-Schmitt, F. (1987). Genes of the *Drosophila* maternal dorsal group control the specific expression of the zygotic gene *twist* in presumptive mesodermal cells. *Genes Dev.* **1**, 709–715.

Thisse, B., Stoetzel, C., Gorostiza, T. C., and Perrin-Schmitt, F. (1988). Sequence of the *twist* gene and nuclear localization of its protein in endomesodermal cells of early *Drosophila* embryos. *EMBO J.* **7**, 2175–2183.

Thisse, C., and Thisse, B. (1992). Dorsoventral development of the *Drosophila* embryo is controlled by a cascade of transcriptional regulators. *Development* (Suppl.) 173–181.

Tobin, S. L., Cook, P. J., and Burn, T. C. (1990). Transcripts of individual actin genes are differentially distributed during embryogenesis. *Dev. Genet.* **11**, 15–26.

Unwin, D. M., and Ellington, C. P. (1979). An optical tachometer for measurement of the wingbeat frequency of free flying insects. *J. Exp. Biol.* **82**, 377–378.

VanVactor, D., Sink, H., Fambrough, D., Tsoo, R., Goodman, C. S. (1993). Genes that control neuromuscular specificity in *Drosophila*. *Cell* **73**, 1137–1153.

Vigoreaux, J. O., Saide, J. D., Valgeirs, K., and Pardue, M. L. (1993). Flightin, a novel myofibrillar protein of *Drosophila* stretch-activated muscles. *J. Cell Biol.* **121**, 587–598.

Vinos, J., Domingo, A., Marco, R., and Cervera, M. (1991). Identification and characterization of *Drosophila melanogaster* paramyosin. *J. Mol. Biol.* **220**, 687–700.

Volk, T., Fessler, L. I., and Fessler, J. H. (1990). A role for integrin in the formation of sarcomeric architecture. *Cell* **63**, 525–536.

Volk, T. (1992). A new member of the spectrin superfamily may participate in the formation of embryonic muscle attachments in *Drosophila*. *Development* **116**, 721–730.

Volk, T., and VijayRaghavan, K. (1994). A central role for epidermal segment border cells in the induction of muscle patterning in the *Drosophila* embryo. *Development* **120**, 59–70.

Warmke, J. W., Yamakawa, M., Molloy, J., Falkenthal, S., and Maughan, D. (1992). Myosin light chain-2 mutation affects flight, wing beat frequency, and indirect flight muscle contraction kinetics in *Drosophila*. *J. Cell Biol.* **119**, 1523–1539.

Wright, T. R. F. (1960). The phenogenetics of the embryonic mutant, *lethal myospheroid*, in *Drosophila melanogaster*. *J. Exp. Zool.* **143**, 77–99.

Xu, T., and Rubin, G. M. (1993). Analysis of genetic mosaics in developing and adult *Drosophila* tissues. *Development* **117**, 1223–1237.

CHAPTER 15

Isolation of Cytoskeletal Proteins from *Drosophila*

Kathryn G. Miller* and Douglas R. Kellogg[†]

* Department of Biology
Washington University
St. Louis, Missouri 63130

† Department of Physiology
University of California, San Francisco
San Francisco, California 94143

I. Introduction

The proteins that bind to actin filaments and microtubules play important roles in determining the structure and function of these filaments in eukaryotic cells. A number of approaches have been used to identify and characterize these proteins. For example, nucleic acid homology-based screens are now beginning to pay off in the identification of *Drosophila* homologs of those cytoskeletal proteins that have been cloned in other systems. Such gene identifications have the potential to add tremendous information through the use of mutational analysis of protein function with the powerful genetic techniques available in *Drosophila*. Biochemical strategies offer another important approach to the identification and characterization of cytoskeletal proteins in *Drosophila*. Biochemical approaches are not limited by the availability of homologous genes from other systems and thus have the potential to identify new and as yet undescribed cytoskeletal proteins. In addition, these methods can be applied to proteins being identified by genetic and homology-based molecular approaches to define the biochemical properties and activities of proteins whose biochemical function has not yet been demonstrated.

Biochemically, microtubule-binding proteins have been identified primarily by virtue of their ability to cosediment with microtubules in solution or by virtue of their ability to cause microtubule motility *in vitro*. Actin-binding proteins (ABPs) have been identified primarily by virtue of their ability to affect actin polymerization or actin filament structure (crosslinking, for example) *in vitro*. Such proteins have generally been purified using conventional purification schemes, using the activity of the protein to follow its fractionation through several (or many) steps. These protocols are determined empirically, and each scheme is peculiar to the particular protein being isolated. Such an approach has been successfully applied in *Drosophila* to a limited extent (for example, Kiehart and Feghali, 1986; Dubreuil *et al.*, 1987). Identification of such proteins in *Drosophila* can provide new approaches to study function of the cytoskeleton and particular proteins that compose it *in vivo*. Genetic techniques and manipulation of function through *in vitro* mutation of genes and transformation of those genes into the organism can provide significant insight into function in the context of the organism (Young *et al.*, 1993). For biochemical isolation of proteins from *Drosophila* that are homologs of previously identified proteins from other systems, purification protocols that yield the protein of interest can be adapted to suit the particular *Drosophila* tissue to be used as starting material, the particular characteristics of the homologous protein as they are defined through the use of an assay similar to those used in another system (such as those described previously), and the purification behavior of the protein. We could not possibly provide a useful guide to this type of approach in the space allowed. The reader is referred to excellent books on the subject of protein purification (for example, Scopes, 1993; Cooper, 1977).

As a complement to these approaches, we have developed affinity chromatography methods for the isolation and biochemical characterization of proteins that bind to actin filaments and microtubules and assemblies of such proteins that function together to regulate cytoskeletal structure. The use of affinity chromatography offers a number of advantages over the use of other procedures for the purification of cytoskeletal proteins. Large extents of purification are obtained in a small number of steps and without requiring an activity assay. Moreover, even proteins that bind to actin filaments or microtubules with relatively low affinity can be detected, since these proteins will be retarded as they flow through the column (see Scopes, 1993, Chapter 5 for discussion of the advantages of chromatographic techniques). Additionally, proteins present in very small amounts, which might be undetected in a biochemical assay in the presence of another more abundant protein with a similar activity, are identifiable. We focus here on these affinity techniques and their application to *Drosophila*. Through use of these techniques combined with biochemical and immunolocalization studies, new proteins that are likely to be of importance in controlling actin and microtubule organization and function in cells are likely to be discovered and the associations of known proteins more thoroughly understood.

II. Affinity Methods for Isolation of Interacting Proteins

A. Preparation of Actin and Microtubule Affinity Columns

In this section, we summarize protocols for the preparation and use of actin and microtubule affinity columns. For specific examples and more extensive discussion of these methods, the reader is referred elsewhere (Miller and Alberts, 1989; Kellogg *et al.*, 1988; Miller *et al.*, 1991; Barnes *et al.*, 1992). Protocols included are for preparation of affinity matrixes for all columns required to perform an experiment. Columns of monomeric actin and BSA provide controls for nonspecific binding.

1. Preparation of the Affinity Resin

Unless specified, all procedures are carried out at 4°C.

1. Columns are constructed in sterile plastic syringes (Beckman–Dickinson) fitted with polypropylene filter discs (Ace Glass, Vineland, NJ) as bed supports (6-ml syringes for columns with a bed volume of 3 ml and 60-ml syringes for a bed volume of 25 ml). To preserve flow properties, we have kept the length of the column bed nearly constant (3 to 5 cm) and increased the cross-sectional area when increasing column size.

2. The outlet of the syringe is fitted with an 18-gauge needle pushed through a rubber stopper that is mounted on a filter flask. Equal settled volumes of Affigel 10 (Bio-Rad) and Sepharose CL 6B (Pharmacia) are poured into the syringe and washed three times with glass-distilled H_2O and once with F-buffer (1 or more column volumes each). For each wash, the resin is gently mixed with a spatula and excess buffer is removed by applying suction. Care should be taken not to draw air into the bed during these steps. Because the active groups on the Affigel begin to decay as soon as it is transferred to aqueous solution, these washes should be completed within 10 min.

3. Remove the syringe from the filter apparatus and close the outlet at the

bottom with a needle plugged with a silicone stopper. At this point, no excess buffer remains on the resin.

4. Mix the packed resin with the appropriate protein solution as described following.

2. Preparation of Actin Filament Columns

Unless otherwise stated, all steps are carried out at 4°C.

1. Add 0.5 column vol of 2 mg/ml F-actin in F-buffer plus 10 μg/ml phalloidin or suberimidate-cross-linked actin in G-buffer (see following section listing buffers) to the syringe containing the packed washed resin described above and gently mix with a spatula. The volume and concentration of actin are important. Do not exceed 2 mg/ml. Allow the reaction to proceed for 1 to 15 hr in the syringe (no mixing). We have found no variation in final column properties over this time range.

Actin purified by standard procedures (Pardee and Spudich, 1982) from acetone powder from rabbit or chicken muscle has been extensively used in constructing columns. Nonmuscle actin from yeast (Drubin *et al.,* 1988) or *Drosophila* (unpublished observations of K. Miller), purified by the method of Zechel (1980), is also suitable for these columns. Actin is stable in the polymerized form at 4°C for several weeks or can be stored for longer periods of time as a lyophilized powder at −20 or −70°C by adding 2 mg of sucrose per milligram of actin in G-buffer prior to lyophilization. To use actin that has been stored as filaments for column construction, the actin should be spun to collect filaments (150,000g for 90 min), resuspended in G-buffer, depolymerized by extensive dialysis against this buffer, spun (150,000g for 90 min) to remove aggregates and unpolymerized filaments, and then repolymerized. To use actin stored lyophilized as monomer, the powder should be resuspended in G-buffer, dialyzed vs G-buffer overnight, spun to remove aggregates, and polymerized.

2. After 4 hr, the coupling reaction is essentially complete and the resin is inactive. If use of the column before this time period is desired, the reaction can be terminated by the addition of 3 M ethanolamine (redistilled and neutralized to pH8) to the slurry of resin to a final concentration of 50 mM. During the termination step, recirculate the free liquid in the syringe through the column bed using a peristaltic pump at a flow rate of 1–3 column vol/hr (up to a maximum of 25 ml/hr) to pack the bed. This is a critical variable. Too fast a flow rate shears the actin filaments and causes significant actin loss. Too slow a flow rate produces columns in which flow through the bed is difficult, and channeling occurs around the outside of the bed (as detected by dye flow, see following). Initially, all columns were packed in this manner.

We have also found the following method successful. After sufficient time has passed for inactivation of the reactive groups on the resin, allow the solution (containing protein that has not coupled) to flow out by gravity. (Retain this for a protein assay.) Layer a small amount of F-buffer on top of the column and go on to Step 3.

3. Wash the column with F-buffer to remove unbound actin and ethanolamine. Generally, this requires 1 hr at a flow rate of 3 column vol/hr.

4. The flow properties of the column should be tested by layering a small aliquot of phenol red in F-buffer containing 5% glycerol onto the column. When the dye band is washed through the column with F-buffer, it should move through the bed evenly. If the dye channels around the bed instead, the column should be gently mixed with a spatula in a minimum volume of F-buffer and allowed to stand undisturbed for several hours; the column bed can then be repacked by washing with F-buffer at 1–3 column vols/hr as above.

5. Once a satisfactory column has been prepared, wash with 1 M KCl, 50 mM K-Hepes, pH 7.5, 2 mM MgCl$_2$ (3–5 column vols) and subsequently with the buffer to be used for chromatography (see following). All washes (including those before and during the dye test) are saved for protein determination by the method of Bradford (1976), and the protein on the column is quantitated by subtracting the total protein eluted from the protein input. If coupled and packed properly, 90% of the initially added F-actin remains on the bed. The actin that is not directly covalently linked to the agarose can be quantitated by treating an aliquot of the matrix with SDS-containing gel sample buffer. Generally, we find that the columns contain about 0.75 mg of actin filaments per milliliter of resin, as judged by either measurement.

6. Columns are stored in F-buffer containing 10 μg/ml phalloidin and 0.02% NaN$_3$ at 4°C and are reusable for a period of at least 3 weeks. Just before an experiment, the column is washed with the buffer used to prepare the extract to be chromatographed (see following).

3. Preparation of Monomeric Actin Columns

1. Monomeric actin (G-actin) reacts strongly with the resin. To prevent an overcoupling that might denature the actin monomer, the washed mixture of Affigel-10 and Sepharose CL6B is partially inactivated by incubation for 90 min in G-buffer. The resin is then rinsed again with 2 column vol of G-buffer.

2. Monomeric actin is diluted to 3–4 mg/ml. One-half column vol of this G-actin solution is mixed with the deactivated resin. After 20 min at 4°C, drain the actin solution by gravity and add 0.5 column vol of G-buffer containing 50 mM ethanolamine to stop the coupling of actin. Mix the column bed gently with a spatula. Incubate for 2 hr or more before washing extensively with G-buffer by gravity flow.

3. Subsequent washes with high salt are performed as described previously for F-actin columns. We usually obtain approximately 1 mg/ml G-actin on the bed, about 80% of which is available for binding bovine pancreatic DNase I (Miller and Alberts, 1989). Because the actin monomer is relatively unstable, the G-actin columns should be prepared on the day of use.

4. Preparation of Albumin Control Column

Albumin-containing control columns are prepared in a manner similar to that described for the F-actin columns (no preincubation of resin), using bovine serum albumin (Sigma) at a concentration of 4 mg/ml in F-buffer (no phalloidin). Approximately 60% of the albumin is coupled to the resin under these conditions, leaving a final concentration of about 1 mg/ml on the column bed. These columns can be stored for months at 4°C in F-buffer containing 0.02% NaN$_3$ and used repeatedly.

5. Construction of Microtubule Affinity Columns

As starting material for the construction of microtubule affinity columns, we use tubulin at 2–3 mg/ml in BRB80 buffer. As for actin, the tubulin must be largely free of primary amines and sulfhydryl groups, which will interfere with coupling to the agarose matrix. Bovine brain tubulin purified according to Mitchison and Kirschner (1984), or *Drosophila* tubulin purified according to Kellogg *et al.* (1988), is suitable for use in this procedure.

1. To polymerize the tubulin into microtubules, taxol is added to the tubulin solution in BRB80 to a concentration to 0.15 μM, followed by a 10-min incubation at 25°C (*Drosophila* tubulin) or 37°C (bovine tubulin). Three additional aliquots of taxol are then added over a 25-min period to bring the final taxol concentration to 1, 5, and 20 μM, respectively. The taxol-induced polymerization of tubulin is carried out gradually to prevent the formation of aberrant structures (Schiff *et al.*, 1979). At the end of the assembly reaction, the microtubule solution is chilled on ice.

2. Because coupling of microtubules to the activated agarose matrix (see following) takes place inefficiently at pH 6.8, the pH of the microtubule solution is adjusted to approximately 7.5 by addition of small aliquots of 2 *M* KOH just prior to coupling. The pH of the tubulin solution is monitored by spotting small aliquots onto pH indicator strips.

3. As for the actin filament affinity columns (see earlier) the microtubule affinity columns are constructed in sterile plastic syringes (Becton–Dickinson) fitted with polypropylene discs (Ace Glass, Vineland N.J.) as bed supports. The cross-sectional area of the column is increased according to the column volume, using a 6-ml syringe for a 3-ml column, a 12-ml syringe for a 6-ml column, and so on.

For column construction, see the Section IIA, "Preparation of the Affinity Resin." As with actin, equal settled volumes of CL6B and Affigel 10 are poured into a syringe on a suction apparatus and washed several times with water at 4°C, with periodic stirring. However, after the final wash, the column is left at 4°C for 1 hr to inactivate the Affigel resin partially to prevent overcoupling of the tubulin. Due to variability in different batches of Affigel-10, the period of inactivation may need to be adjusted empirically (see Step 6).

4. The column bed is washed three times with C bufffer under suction at

4°C, the buffer is then drawn down to a level just above the surface of the column bed, and the column is removed from the suction apparatus and sealed at the bottom. About 0.5 column vol of a 2–3 mg/ml solution of taxol-stabilized microtubules at pH 7.5 (see earlier) is added, followed by thorough stirring with a teflon rod. The column is left undisturbed for 4 to 15 hr at 4°C to allow coupling to occur.

5. The column matrix is effectively deactivated after standing at 4°C for 10 hr. However, as an optional step, the column can be washed after 4 hr with several column volumes of C buffer containing 10 mM ethanolamine (added from a 3-M stock, redistilled and adjusted to pH 8) to block all unreacted groups.

6. To remove unbound tubulin, the columns are washed with C buffer containing 0.5 M KCl and 1.0 mM DTT at a flow rate of 1–2 column vol/hr. A peristaltic pump is used to maintain a steady rate of flow. The washes are saved and assayed for protein to determine the amount of tubulin that remains bound to the column. The procedure generally causes about 60–75% of the input microtubule protein to become bound to the column matrix. If greater than this fraction of tubulin is retained on the column, overcoupling is indicated, and testing to determine if any noncovalently bound monomers are present, by incubating resin with SDS (as described for F-actin affinity columns above), is necessary. If little monomer is released in the presence of SDS, this indicates that most subunits are covalently linked to the matrix. Such columns give low yields of microtubule-associated proteins, suggesting that overcoupling results in denaturation of the tubulin. In this case, the resin inactivation (see Step 3) should be adjusted to produce a column with more noncovalently bound tubulin.

7. The columns are stored at 4°C in BRB80 containing 10% glycerol, 1 mM DTT, 5 μM taxol, and 0.02% sodium azide, and can be used for at least five experiments over a 1-month period without a detectable change in their properties.

B. Use of Cytoskeletal Affinity Columns

1. General Considerations

Our experiments with several different types of cells and tissues have demonstrated that proteins from many of the previously defined classes of ABPs bind to the F-actin columns when crude extracts are chromatographed. These include the contractile protein myosin, the bundling proteins villin and fimbrin, the cross-linking proteins spectrin, TW260/240, and filamin, and the *Acanthamoeba* capping proteins (Miller and Alberts, 1989). Many other protein species, not previously identified as ABPs, are also specifically bound. Through studies using early *Drosophila* embryos, we have identified more than 40 potential actin-binding proteins. Localization studies using antibodies suggest that 90% of these proteins are part of the actin filament network inside the cell (Miller *et al.*, 1989). The methods described here have permitted the biochemical identification of new ABPs in yeast, where the three major ABPs that elute from a yeast F-actin column (200,000, 67,000, 85,000 da) have been shown to be actin-associated *in vivo,* as judged by both immunological and genetic criteria (Drubin *et al.,* 1988).

Similarly, microtubule affinity columns have been used successfully to identify microtubule-associated proteins in *Drosophila* and budding yeast. In *Drosophila,* over 100 different proteins are found to bind to microtubule affinity columns, and a library of antibodies that recognize 24 of the proteins has shown that the majority of the proteins are associated with microtubules within cells (Kellogg *et al.,* 1989). In yeast, approximately 25 different proteins bind to microtubule affinity columns, and a number of these have also been shown to be localized to microtubules within cells (Barnes *et al.,* 1992).

Extract conditions can be varied depending on the intent of the experiment. For example, to identify a large number of potential actin-binding proteins from crude extracts, we have used a low-salt cytoplasmic extract to promote depolymerization of the endogenous actin and the concomitant release of filament-specific proteins. However, isolation of proteins from different tissues or that are components of particular structures would be best accomplished through the design of a solubilization scheme that maximizes the recovery of the protein of interest and minimizes any competing or interfering proteins. The appropriate conditions, of course, must be determined empirically.

2. Preparation of the Extract for F-Actin Affinity Chromatography

1. Suspend the cells or tissue to be analyzed in 10 vol (w/v) of E-buffer. Note that the choice of buffers is empirical. We have used Tris buffers with identical results. One should modify the buffer to suit specific applications.

2. Add phenylmethylsulfonyl fluoride (PMSF) to 1 mM and 1/100 protease inhibitor stock.

3. Homogenize at 4°C. We have used 5 strokes of a motor-driven, loose-fitting teflon-glass homogenizer (Wheaton Glass) for *Drosophila* embryos. However, other methods, such as Dounce homogenization or sonication, can be used, if appropriate.

4. Centrifuge the homogenate at 10,000g for 20 min and save the supernatant.

5. Adjust the supernatant to 2 mM DTT and 50 mM K-Hepes (pH 7.5) by the addition of appropriate amounts of 1 M DTT and 1M K-Hepes, pH 7.5.

6. Centrifuge at 100,000g for 1 hr. The supernatant is now ready to be loaded onto the appropriate columns.

3. Affinity Chromatography Extracts

1. Equilibrate F-actin, G-actin, and/or control (albumin) columns of equal bed size and protein content with A-buffer containing 10% glycerol (or another appropriate loading buffer).

2. Apply equal volumes of the same extract to all of the columns in each experiment, using a flow rate of 1 column vol/hr or less.

3. After loading, rinse all of the columns at 1–2 column vol/hr with A-buffer containing 10% glycerol until the protein in the eluate reaches background levels for the F-actin columns (less than 10 μg/ml protein). F-actin columns generally require longer rinsing to reach this level than do the G-actin or control columns.

4. Elute stepwise with A-buffer containing 10% glycerol plus added salt and/ or 1 mM ATP plus 3 mM MgCl$_2$ (to distinguish ATP-eluting from Mg^{2+}-eluting proteins, preelute with 3 mM MgCl$_2$ in A-Buffer). Steps of 0.1, 0.5, and 1 M KCl are recommended as starting points for salt elutions.

5. Determine protein in each fraction by the method of Bradford (1976).

6. Pool the fractions containing protein in each elution step from F-actin columns and equivalent fractions from control or G-actin columns (which often have no detectable protein peak).

7. To analyze the protein present in each elution step, precipitate an aliquot of each pool by adding trichloracetic acid to 10%, incubating on ice for 10–30 min, and centrifuging in a microfuge at top speed for 10 min. Carefully remove as much of the supernatant as possible, respinning the tube briefly, if necessary. Resuspend each precipitate in SDS–polyacrylamide gel sample buffer, neutralize with 2 M Tris base, and load a volume representing an equal proportion of the eluate from each column onto a 5–15% polyacrylamide gradient gel or 8.5% polyacrylamide gel. Electrophoresis in SDS is carried out by standard techniques (Laemmli, 1970). Proteins are visualized by Coomassie blue or silver staining of the gels.

4. Preparation of *Drosophila* Embryo Extracts for Microtubule Affinity Chromatography

1. The extracts for our microtubule affinity chromatography experiments have been prepared from 2- to 3-hr collections of *Drosophila* embryos (i.e., embryos are between 0 and 3 hr postfertilization). The embryos are collected, dechorionated, and washed extensively with distilled water as previously described (Miller and Alberts, 1989). They are then suspended in 10 vol of C buffer containing 0.05% NP-40 and protease inhibitor stock (1/100). PMSF is added to 1 mM and the embryos are homogenized by several passes of a motor-driven teflon Dounce homogenizer. (Selection of a loose-fitting pestle prevents disruption of yolk granules.)

2. The embryo homogenate is centrifuged for 10 min at 12,000g, followed by 60 min at 100,000g. A thin floating layer on the surface of the final supernatant is removed by aspiration, and DTT is added to 0.5 mM before loading the extract onto affinity columns (see below). All steps are carried out at 4°C.

5. Microtubule Affinity Chromatography

1. The clarified extracts are loaded onto affinity columns at 0.5 to 1 column vol/hr. We generally load 10 column vol or more of extract. After loading, the columns are washed with 5–7 column vol of CX buffer and then eluted in succession with this buffer plus either 1 mM MgATP, 0.1 M KCl, or 0.5 M KCl. The wash and elution steps are carried out at 2–5 column vol/hr, and all chromatography steps are at 4°C.

2. The location of the protein peak is determined (Bradford, 1976), and the peak fractions are pooled. The amount of protein in each pool is determined and a portion is precipitated with 10% trichloroacetic acid as described by Miller and Alberts (1989), resuspended in gel sample buffer at a concentration of approximately 1 μg/μl, and neutralized with the vapor from a Q-tip soaked in ammonium hydroxide. The pellets are solubilized by incubation at 100°C for 3 min; approximately 15 μg of total protein from each pool is then analyzed by SDS–polyacrylamide gel electrophoresis (Laemmli, 1970), using Coomassie blue staining to visualize protein bands.

C. Use of Antibody Affinity Columns to Enrich for Proteins That Associate Inside the Cell

1. General Considerations

Many, if not most, cellular functions are carried out by multiprotein complexes. Some well-studied examples are DNA replication, transcription, protein synthesis, and translocation of proteins across the membrane of the endoplasmic reticulum. The study of each complexes can be considerably more informative than the study of their individual components to gain insight into how the process works *in vivo*. Affinity chromatography is a useful means of identifying proteins that interact to form such a biologically relevant complex. A limitation of this approach, however, is that one needs substantial amounts of purified protein to construct an affinity column, and it is generally very difficult to purify sufficient quantities of low abundance proteins. In many cases, large amounts of protein from cloned genes can be obtained from expression systems (e.g., bacterial or baculovirus), but many proteins are insoluble or degraded when expressed in such systems. This is especially true for large proteins expressed in bacteria.

Immunoaffinity chromatography offers a means of isolating interacting proteins without the need for large amounts of purified, native proteins. An immunoaffinity column is constructed by coupling antibodies to a column matrix. When a crude extract is passed over such a column, the protein that is recognized by the antibody will be retained, as well as any additional proteins that interact with the protein to form a complex. These interacting proteins can then be eluted and characterized. This approach has been used successfully to identify a complex of proteins that interact with the DMAP190/Bx63 centrosomal protein (Kellogg and Alberts, 1992). If one uses low-affinity antibodies to construct the immunoaffinity column, the protein that is recognized directly by the antibody can be eluted under relatively mild conditions, thereby allowing one to use the purified protein for biochemical studies.

Immunoaffinity chromatography has a number of advantages over immunoprecipitation techniques. By using a column, one can load large quantities of extract and thereby identify very rare proteins. Relatively large amounts of protein can be substantially purified allowing one to immediately raise antibodies or obtain peptide sequences. Since this approach uses the endogenous protein to identify interacting proteins, one avoids the concerns that a protein expressed in bacteria is incorrectly folded or does not have post-translational modifications that are necessary for binding of other proteins.

To obtain enough polyclonal antibody for immunoaffinity experiments, it is necessary to obtain milligram quantities of purified immunogen. This protein need not be native or functional, and thus, it is generally possible to obtain such quantities of protein using bacterial expression systems. In many experiments, we have used the pGEX vectors (Smith and Johnson, 1988), which express proteins fused to the enzyme glutathione-S-transferase (GST). These fusion proteins can be rapidly purified by virtue of the affinity of GST for glutathione. A disadvantage of this system is that many GST fusion proteins are insoluble and cannot be purified in large quantities, and serum from animals immunized with GST fusion proteins contains large amounts of antibodies that recognize GST. These anti-GST antibodies must be removed from the serum before specific antibodies can be purified on a GST fusion protein column. Vectors that express proteins fused to six histidine residues avoid these problems. These proteins can be purified by virtue of their affinity for nickel and other metals, and fusion proteins that are insoluble can be solubilized and purified in the presence of 8 M urea (Arnold, 1992). In addition, fusions of the protein to other bacterial sequences that allow for simple purification of the fusion protein can be made for use in affinity purifications.

Fusion proteins made from a full-length protein are sometimes degraded or insoluble. In these cases, it can be helpful to make fusion proteins that include smaller portions of the protein. For example, a GST fusion that includes full-length Xenopus cyclin B is insoluble and substantially degraded, whereas a GST fusion containing only the N-terminal third of the protein can be easily purified in large quantities (personal communication, D. Kellogg).

Polyclonal antibodies are generated by immunizing rabbits with 1–2 mg of purified fusion protein. Immunization of two rabbits according to standard protocols will usually generate antibodies that can be used for immunoaffinity chromatography after purification. Commercial services for antibody production (e.g., Berkeley Antibody Company) have been successfully used. In this section, we present protocols for obtaining antibodies suitable for application of affinity methods and for using the antibodies as affinity matrixes to isolate complexes that contain the protein of interest.

2. Affinity Purification of Antibodies

1. An antigen affinity column is constructed by coupling 10–30 mg of purified fusion protein to 5–10 ml of an activated matrix such as Affigel 10 (Bio-Rad). The purified antigen is dialyzed extensively into a buffer that is suitable for coupling to Affigel 10 (e.g., free of primary amines and sulfhydryl groups). We generally use a buffer containing 50 mM Hepes–KOH (pH 7.6), 0.25 M KCl, and 20% glycerol. The dialyzed protein (at a concentration of 2–10 mg/ml) is mixed with an equal volume of Affigel 10 that has been washed into coupling buffer. The Affigel 10 is washed batchwise by resuspending it in 10–20 vol of buffer, followed by pelleting of the beads by a brief spin in a clinical centrifuge. The wash is repeated several times. Once the protein is added to the Affigel 10, the slurry is mixed gently on a rotator at 4°C. The coupling reaction is monitored by measuring the amount of protein remaining in the supernatant after the beads have been pelleted and is allowed to proceed until approximately 75% of the protein has been coupled. The reaction takes between 10 min and 4 hr, depending on the protein. The reaction is terminated by addition of Tris buffer or ethanolamine to a concentration of 25 mM. Fusion proteins that have been purified in the presence of high concentrations of urea will usually precipitate when the urea is dialyzed away. In these cases, the protein precipitate can usually be resolubilized in 0.5% SDS and coupled to Affigel 15.

2. Before using the antigen affinity column, it is washed with several column volumes of 100 mM triethylamine (pH 11.5), followed by enough PBS to reequilibrate the pH of the column. Serum is centrifuged at 30,000g for 20 min and is then loaded onto the column at a flow rate of approximately 10–25 column vols/hr. Antibody purification is carried out at room temperature.

3. The column is washed with 50–100 column vols TTBS (50 mM Tris–HCl, pH 7.5, 0.5 M NaCl, 0.1% Tween 20) and is then washed with 0.3× PBS. The diluted PBS is a poor buffer and allows the pH to change more quickly when the column is eluted with base. This gives a more concentrated antibody peak.

4. If low-affinity antibodies are desired, the column is eluted with 50 mM Hepes (pH 7.6), 1.5 M MgCl2, and 10% glycerol and fractions that are one-fifth column volume are collected. After this elution, the column is eluted with 100 mM triethylamine (pH 11.5) to obtain high-affinity antibodies. Each fraction is immediately neutralized as it comes off the column by the addition of an appropriate (empirically determined) volume of 1 M Tris–HCl (pH 5.5, Sigma). The amount of protein in each fraction is assayed by measuring the absorbance at 280 nm; fractions having an absorbance of greater than 0.2 are pooled. Alternatively, acid (0.15 M NaCl, 0.2 M glycine, pH 3.5) can be used to elute high-affinity antibodies,

again neutralizing immediately with an appropriate (empirically determined) volume of 2 M Tris base.

After quantifying the total amount of antibody, a fivefold excess of BSA is added as a carrier and the antibody is dialyzed into PBS containing 10% glycerol and 0.05% sodium azide. In one example, we obtained 25 mg of low-affinity antibody and 42 mg of high-affinity antibody from 300 ml of serum using a column with 25 mg of antigen coupled at a concentration of 6 mg/ml. The purified antibodies are stored at 4°C until use.

3. Immunoaffinity Chromatography

Immunoaffinity columns are constructed by cross-linking antibodies to protein A Sepharose. We generally couple 3 mg of antibody to 1 ml of protein A-Affigel 10 (Bio-Rad) using dimethyl pimelimidate according to the protocol described by Harlow and Lane (1988). Much of the antibody appears to be inactivated by the coupling procedure, so it is important to use at least this much antibody.

1. A *Drosophila* embryo extract is made according to the protocol used for making extracts for microtubule affinity columns, with the exception that the extract buffer has 60 mM KCl rather than 25 mM. The extract is first loaded onto a precolumn constructed by coupling approximately 50 mg of BSA to 5 ml of Affigel 10. This column serves to filter out any aggregates that form during the loading procedure, as well as any proteins that bind nonspecifically to other proteins. Since the yield of protein from immunoaffinity columns can be relatively small, special steps need to be taken to eliminate background. After passing through the BSA precolumn, the extract is loaded onto the antibody affinity column. A control column is made with 3 mg of nonimmune IgG and is treated identically to the experimental column. The extract is loaded onto the columns at 5–10 column vols/hr.

2. After the extract has been loaded, the antibody column and the control column are disconnected from the BSA precolumns and are washed with 50 column vols of 50 mM Hepes (pH 7.6), 60 mM KCl, 1 mM EGTA, 1 mM MgCl$_2$, 10% glycerol, and 1/300 dilution of protease inhibitor stock. This is a fairly stringent wash and should select for proteins that interact with the antigen protein with relatively high affinity. A less stringent wash can be used (e.g., using lower salt in the wash buffer or a shorter wash) but may result in a some background binding.

3. The columns are first eluted with wash buffer containing 1 M KCl. This should elute most proteins that are bound to the antigen, while leaving the antigen itself bound to the antibody on the column. The elution is carried out by pipetting 0.25-ml aliquots of the elution buffer directly onto the top of the column bed. The buffer is allowed to flow through the column by gravity and is collected in 1.5-ml plastic tubes.

4. The column is then eluted using either 1.5 M MgCl$_2$ (for a column constructed using low-affinity antibodies) or 100 mM triethylamine (high-affinity antibodies). The fractions from both elutions are assayed for protein by the Bradford assay, and peak fractions are pooled, TCA precipitated, and analyzed by gel electrophoresis.

5. The columns are stored at $-20°C$ in PBS containing 50% glycerol and can be reused several times.

4. Buffers

G-buffer: 5 mM Hepes–KOH, pH 7.5, 0.2 mM CaCl$_2$, 0.2 mM ATP.

F-buffer: 50 mM Hepes–KOH, pH 7.5, 0.1 M KCl, 5 mM MgCl$_2$, 0.2 mM CaCl$_2$, 0.2 mM ATP.

A-buffer: 50 mM Hepes–KOH, pH 7.5, 2 mM DTT (dithiothreitol), 0.5 mM Na$_3$EDTA, 0.5 mM Na$_3$ EGTA, 0.5% Nonidet P-40, and protease inhibitor stock (1/1000).

C-buffer: 50 mM Hepes–KOH, pH 7.6, 1 mM MgCl$_2$, 1 mM Na$_3$ EGTA.

CX-buffer: C-buffer supplemented with 10% glycerol, 25 mM KCl, 0.5 mM DTT, and protease inhibitor stock (1/1000).

E-buffer: 5 mM Hepes–KOH, pH 7.5, 0.5% Nonidet P-40, 0.5 mM Na$_3$ EDTA, 0.5 mM Na$_3$ EGTA, and protease inhibitor stock (1/100).

BRB80 (microtubule assembly buffer): 80 mM Pipes–KOH, pH 6.8, 1 mM MgCl$_2$, 1 mM Na$_3$ EGTA.

Protease inhibitor stock: 1 mM benzamidine–HCl, 0.1 mg/ml phenanthroline, 1 mg/ml each of aprotinin, leupeptin, and pepstatin A.

Polyacrylamide gel sample buffer: 63 mM Tris–HCl, pH 6.8, 3% SDS, 5% β-mercaptoethanol, 10% glycerol.

III. Characterization of Proteins Isolated by Affinity Methods

A. Criteria for Determining Whether *in Vitro* Associations Reflect *in Vivo* Associations

Once proteins that bind to the affinity columns are identified, it is necessary to obtain some independent confirmation that the proteins actually associate with the molecules used for their isolation. First, it is important to compare fractions from affinity purifications to the starting extract to ensure that the eluate is not contaminated with the most abundant protein species in the extract. Such contamination would suggest that the columns have not been washed sufficiently to remove nonspecifically bound proteins or that the columns are not flowing evenly. Second, even with appropriate controls (i.e., comparison to extract, parallel BSA, and/or monomer columns), artifactual associations *in vitro* can occur. In addition, often many proteins are identified by such techniques, and some selective criteria are needed to determine which proteins might be most relevant for further study. The primary criterion we have used to select proteins for further study is immunofluorescence localization to expected sites within a cell. In the case of actin or microtubule affinity chromatography, we assume that proteins that associate with these filaments inside a cell are likely to be bonafide binding proteins. Proteins selected for further study might have patterns of association that suggest particular functions or suggest that they participate in a process that is of particular interest (ATP-dependent interaction for motor proteins, for example). An additional criterion for selection of proteins for further study that might be useful is the expected properties of a protein of interest. For those proteins isolated by antibody affinity methods, one expects that they should localize to the structure (or a subset of the structures) to which the molecule used to identify them localizes. These experiments require the generation of an antibody to the proteins isolated. While producing antibodies is time consuming, once generated, the antibodies can be used for protein localization, as well as to clone the genes that encode the proteins. Protocols that have successfully yielded antibodies are cutting of the bands of interest from SDS gels and immunization with the gel slice (Miller *et al.*, 1989; Kellogg *et al.*, 1988), blotting of SDS-fractionated proteins and injection of nitrocellulose strips that contain the protein of interest (unpublished observations, K. Miller), and scale up of the affinity method so that proteins can be eluted from the column in sufficient quantities that eluate fractions can be used directly for immunization.

An alternate approach to making antibodies to proteins identified in affinity screens is to obtain peptide sequence from the proteins of interest. With two reasonable-length peptide sequences, degenerate primers can be designed and PCR used to identify the gene that encodes the protein. Once cloned material is available, all subsequent studies are vastly simplified. This approach has the additional advantage of allowing one to obtain sequence data, which may be informative in itself to confirm the *in vivo* relevance of an *in vitro*-associaed protein. Large quantities of protein expressed in bacteria can be easily produced for immunization to generate antibodies and for serum affinity purification. This approach does not solve the problem of needing confirming information about association inside cells and an antibody is still required for immunolocalization. It simply makes generating the necessary reagents for further analysis easier.

B. Methods of Analysis Suitable for Observing Associations in Living Cells

Once proteins that are of interest are identified, antibodies to the proteins have been produced, and clones that encode them are isolated, a number of methods of analysis are available that can provide novel and significant functional information relatively quickly. Some techniques we have found useful in obtaining additional information about the proteins of interest are described in this section.

Fluorescently labeled proteins have found increasing use for analysis of protein distribution and function in living cells (Taylor and Wang, 1980; Kellogg *et al.*, 1989; Sullivan *et al.*, 1990; Warn *et al.*, 1987). For this approach, one can fluorescently label proteins purified from overexpression systems or antibodies that recognize a specific protein. This technique provides reagents that, when introduced into the cell, have been shown to localize correctly and thus provide a means to study distribution in living tissue. The availability of many prokaryotic and eukaryotic systems for protein expression should provide a fairly straightforward route, in many cases, to obtain enough protein for fluorescent labeling studies.

In vivo visualization of proteins offers a powerful means of studying their behavior and distribution during the cell cycle. In addition, by purifying and labeling fragments of proteins, it is possible to map protein domains that are responsible for specific subcellular localizations, and introduction of such protein domains into cells/organisms is likely to be useful for inducing dominant mutant effects. Similarly, antibodies that recognize specific domains are useful for disrupting the function of proteins *in vivo*. By labeling the injected antibody, one can follow its behavior *in vivo* to determine whether it recognizes its antigen and whether binding to the antigen causes a rearrangement that leads to a phenotypic effect. A dynamic picture of the rearrangements of cellular components in response to perturbations (by combining antibody or protein fragment injection with labeling of other cellular components that might be affected by the reagent) as well as normal changes in cellular physiology that affect proteins of interest can be obtained by these methods and such information provides a unique insight into the processes that are being studied. Such information is not obtained from static images of fixed material.

1. Labeling of Proteins for *in Vivo* Visualization in Live Animals

The procedure described here can be used for fluorescently labeling either antibodies or purified proteins.

1. The protein to be labeled should be in a buffer free of primary amines and sulfhydryls and have a pH of 7.4 or greater. We generally dialyze proteins to be labeled into 50 mM Hepes (pH 7.6), 100 mM KCl, and 20% glycerol. The protein should be at a concentration of 1–5 mg/ml.
2. Make a stock solution of N-hydroxysuccinimidyl carboxytetramethyl rhodamine (NHSR; Catalog No. 1171, Molecular Probes, Eugene, OR) that is 5 mg/ml in DMSO. Aliquots of this stock solution can be stored at −80°C for at least a year, as long as it is not contaminated with water.

3. Since each protein has different labeling characteristics, one should perform a pilot experiment to determine conditions that give a good labeling stoichiometry. For trial reactions, one can use as little as 0.1 to 0.2 mg of protein. Start three identical parallel reactions by adding NHSR to the protein solution to a final concentration of 75 μg/ml, allowing the duplicate reactions to proceed on ice for different time intervals (5, 10, and 20 min are reasonable starting point), and stop each reaction by adding potassium glutamate to 25 mM and DTT to 5 mM. The DTT should be omitted when labeling antibodies. It is likely that adjusting the reaction times to get the desired stoichiometry (see Step 4) will be necessary.

4. Load the reactions onto desalting columns to separate the protein from the unbound NHSR. We use Biogel P-10 (Bio-Rad) columns that have been equilibrated with 50 mM Hepes (pH 7.6) or some other appropriate injection buffer. The column should be at least 5 times the size of the sample that is to be desalted. The labeled protein should be visible as a light pink peak that flows ahead of the unincorporated NHSR. This peak is pooled, and the stoichiometry of labeling is determined by measuring the NHSR concentration and dividing this by the concentration of the protein. The concentration of the NHSR is determined by reading the absorbance of the protein solution at 550 nm and using an extinction coefficient of 100,000 cm-1 $M-1$. One should aim for a stoichiometry of between 0.5 and 1 NHSR/protein molecule. Coupling more dye molecules per protein can lead to denaturation or changes in the ability of the protein to form normal associations. If an *in vitro* assay that can test whether the labeling procedure has damaged the protein is available, it is highly recommended that the labeled protein be tested prior to use *in vivo*. Other labeling reagents, with different coupling chemistries (for example, rhodamine tetramethyl isothiocyanate) can be used in a similar manner, if labeling with NHSR proves unsuitable.

5. Once one has determined conditions that give a good stoichiometry, the reaction can be scaled up. For a large-scale reaction, one should use at least 0.5 mg of protein. After the desalting step, the protein is concentrated using a Centricon device (Amicon) to a final concentration of 2–15 mg/ml for injection. The appropriate concentration of protein to inject for good *in vivo* visualization must be determined empirically.

IV. Conclusions and Summary

In this summary of methods, we have attempted to present a series of protocols relevant to the study of proteins that associate with each other inside cells. These techniques have found their main application, in our hands, to cytoskeletal structures, but the utility of these techniques is limited to this application. As we learn more about the physiology and organization of eukaryotic cells, information from a variety of studies and systems indicates that many processes occur in a highly spatially organized manner in cells, using multiprotein complexes. The techniques described here will find application to the study of how organization and associations between proteins contribute to many of cellular processes.

References

Arnold, F., ed. (1992). Metal Affinity Separations. *Methods,* Vol. 4.

Barnes, G., Louie, K. A., *et al.* (1992). Yeast proteins associated with microtubules in vitro and in vivo. *Mol. Biol. Cell* **3,** 29–47.

Bradford, M. (1976). A rapid and sensitive method for the quantitation of microgram quantities of protein utilizing the principle of protein-dye binding. *Anal. Biochem.* **72,** 248–254.

Cooper, T. G. (1977). "Tools of Biochemistry." New York: Wiley.

Drubin, D. G., Miller, K. G., and Botstein, D. (1988). Yeast actin-binding proteins: Evidence for a role in morphogenesis. *J. Cell Biol.* **107,** 2551–2562.

Dubreuil, R., Byers, T. J., Branton, D., Goldstein, L. S. B., and Kiehart, D. P. (1987). *Drosophila* spectrin I. Characterization of the purified protein. *J. Cell Biol.* **105,** 2095–2102.

Harlow, E., and Lane, D. (1988). "Antibodies: A Laboratory Manual." Cold Spring Harbor, NY: Cold Spring Harbor Laboratory Press.

Kellogg, D. R., and Alberts, B. M. (1992). Purification of a multiprotein complex containing centrosomal proteins from the *Drosophila* embryo by chromatography with low-affinity polyclonal antibodies. *Mol. Biol. Cell* **3,** 1–11.

Kellogg, D. R., Mitchison, T. L., and Alberts, B. M. (1988). Behavior of microtubules and actin filaments in living *Drosophila* embryos. *Development* **103,** 675–686.

Kellogg, D. R., Field, G. M., and Alberts, B. M. (1989). Identification of microtubule-associated proteins in the centrosome, spindle, and kinetochore of the early *Drosophila* embryo. *J. Cell Biol.* **109,** 2977–2991.

Kiehart, D. P., and Feghali, R. (1986). Cytoplasmic myosin from *Drosophila melanogaster. J. Cell Biol.* **103,** 1517–1525.

Laemmli, U. K. (1970). Cleavage of the structural proteins during the assembly of the head of bacteriophage T4. *Nature* **227,** 680–685.

Miller, K. G., and Alberts, B. M. (1989). F-actin affinity chromatography: A technique for the identification of novel types of actin-binding proteins. *Proc. Natl. Acad. Sci. U.S.A.* **86,** 4808–4812.

Miller, K. G., Field, C. M., and Alberts, B. M. (1989). Actin-binding proteins from Drosophila embryos: A complex network of interacting proteins detected by F-actin affinity chromatography. *J. Cell Biol.* **109,** 2963–2975.

Miller, K. G., Field, C. M., Alberts, B. M., and Kellogg, D. R. (1991). The use of actin filament and microtubule affinity chromatography to identify proteins that bind to the cytoskeleton. *Methods Enzymol.* **196,** 303–318.

Mitchison, T. J., and Kirschner, M. W. (1984). Microtubule assembly nucleated by isolated centrosomes. *Nature* **312,** 232–237.

Pardee, J. D., and Spudich, J. (1982). Purification of muscle actin. *Methods Enzymol.* **85,** 164–181.

Schiff, P. B., Fant, J., and Horwitz, S. Z. B. (1979). Promotion of microtubule assembly in vitro by taxol. *Nature* **277,** 665–666.

Scopes, R. K. (1993). "Protein Purification: Principles and Practice," 3rd ed., New York: Springer-Verlag.

Smith, D. B., and Johnson, K. S. (1988). Single-step purification of polypeptides expressed in E. coli as fusions with glutathione S-transferase. *Gene* **67,** 31–40.

Sullivan, W., Minden, J. S., and Alberts, B. M. (1990), *daughterless-abo-like,* a *Drosophila* maternal-effect mutation that exhibits abnormal centrosome separation during the late blastoderm divisions. *Development* **110,** 311–323.

Taylor, D. L., and Wang, Y.-L. (1980). Fluorescently labeled molecules as probes of the structure and function of living cells. *Nature* **284,** 405–410.

Warn, R. M., Flegg, L., and Warn, A. (1987). An investigation of microtubule organization and functions in living Drosophila embryos by injection of a fluorescently labeled antibody against tyrosinated α-tubulin. *J. Cell Biol.* **105,** 1721–1730.

Young, P. E., Richman, A. M., Ketchum, A. S., and Kiehart, D. P. (1993). Morphogenesis in Drosophila requires nonmuscle myosin heavy chain function. *Genes Dev.* **7,** 29–41.

Zechel, K. (1980). Isolation if polymerization competent cytoplasmic actin by affinity chromatography on immobilized DNase I using formamide as eluant. *Eur. J. Biochem.* **110,** 343–348.

CHAPTER 16

Isolation and Analysis of Microtubule Motor Proteins

William M. Saxton

Department of Biology
Indiana University
Bloomington, Indiana 47405

I. Introduction

Eukaryotes employ force producing mechanochemical enzymes (motors) to move cytoplasmic components along microtubules. A few years ago, only two cytoplasmic microtubule-based motors were known: kinesin, which moves toward the plus ends of microtubules, and cytoplasmic dynein, which moves toward the minus ends of microtubules. Now, through gene sequence comparisons and biochemical analyses, it has become apparent that there are many microtubule-based motors in a wide variety of organisms. Those for which the gene sequences are known fall into two groups: the kinesin and dynein superfamilies. In *Drosophila* alone, genes for 11 members of the kinesin superfamily (Stewart *et al.*, 1991) and 9 members of the dynein superfamily have been characterized (Rasmussen *et al.*, 1994).

To understand microtubule-based motility, the functions of each motor need to be identified. Some motors, including kinesin, appear to be expressed in all cell types and at all stages of development (Saxton *et al.*, 1988; Hollenbeck, 1989; Stewart *et al.*, 1991). Others are either stage or tissue specific (Stewart *et al.*, 1991). While it is likely that some motors have unique functions, it has been shown that some have overlapping or redundant functions (Roof *et al.*, 1992; Saunders and Hoyt, 1992). Thus, determining which motor or motors are responsible for a given motility process will not always be straightforward.

Sorting out the biological functions of individual and sets of microtubule motors will require a variety of approaches including genetics. *Drosophila* provides an excellent system for such an endeavor, especially if one wishes to learn how microtubule motors are used during development and in differentiated tissues. Biochemistry and molecular techniques can be used to identify, isolate, and characterize motors, and then genetics and physiological studies can be used to analyze motor functions *in vivo*.

The isolation of *Drosophila* kinesin is described here to provide a guide for those who wish to isolate kinesin, to point out possible avenues for the isolation of other microtubule motor proteins, and to illustrate some of the gains that can result from an initial biochemical approach to studying motor proteins. The methods have evolved from pioneering work on *in vitro* microtubule polymerization (Weisenberg *et al.*, 1968; Weisenberg, 1972), the identification of nucleotide analogs that can induce rigor binding of motors to microtubules (Lasek and Brady, 1985), and the development of *in vitro* assays for microtubule-based motility (Brady *et al.*, 1982; Vale *et al.*, 1985a). The specific methods I use for isolating kinesin from *Drosophila* (Saxton *et al.*, 1988) were derived from those developed by Vale *et al.* (1985b) and Scholey *et al.* (1985) for the isolation of kinesin from squid neural tissue and sea urchin eggs, respectively.

The identification and characterization of *Drosophila* kinesin helped to establish the widespread distribution and similarity of this motor in metazoans and provided the background necessary for entry into molecular and genetic analyses. The *Drosophila* kinesin heavy-chain gene (*khc*) was the first microtubule motor gene to be cloned and characterized (Yang *et al.*, 1988). Its sequence founded the kinesin superfamily of related genes and also allowed secondary structure predictions for kinesin heavy chain (KHC) (Yang *et al.*, 1989). Another advance was the development of an *Escherichia coli* expression system that uses *khc* to produce functional KHC protein. This expression system has been used to identify fragments of KHC that are necessary and sufficient for the production of microtubule-based movement *in vitro* (Yang *et al.*, 1990), to study the structure of the "stalk domain" (deCuevas *et al.*, 1992), and to identify proteins that can bind KHC (Gauger and Goldstein, 1993). KHC produced in *E. coli* promises future contributions to our understanding of how kinesin generates force and movement by allowing biochemical studies of mutant KHCs and by providing material for high-resolution structural analyses.

Genetic tests of *Drosophila* KHC are being used to study the contributions of kinesin to the development and function of differentiated cells. Mutations in *Drosophila khc* were isolated (Saxton *et al.*, 1991) using reverse genetic methods such as those described by Wolfner and Goldberg and by Hamilton and Zinn in Chapters 3 and 4, respectively. The mutants have been used to demonstrate that KHC is essential and that it is likely to function as a motor for axonal transport in the nervous system (Saxton *et al.*, 1991; Gho *et al.*, 1992). Mutations in *khc* result in impaired action potential propagation and impaired neurotransmitter release. Based on these studies, it has been hypothesized that kinesin is involved in the delivery of ion channels to the axonal membrane (Gho *et al.*, 1992). Additional studies of the neuronal physiology of *khc* mutants, identification of genetic loci that interact with *khc*, and clonal analysis (Perrimon; Chapter 34) of *khc* function in a variety of nonneuronal tissues should provide further insight into the *in vivo* functions of kinesin.

II. Isolation of Motor Proteins

A. Preparing Embryos

If one wishes to prepare *Drosophila* microtubule motors from large amounts of material, the simplest and least expensive source is a population cage. Descriptions of how to develop and maintain large populations of flies are provided by Schaffer *et al.* in Chapter 5 and by Mahowald in Chapter 7. I have prepared active kinesin from as little as 1 ml of packed embryos but normally use between 25 and 100 ml. Egg collection pads covered with a thick layer of yeast paste are placed in population cages for approximately 24 hr and then rinsed into a set of graded copper screens (40, 60, and 120 mesh) with a gentle stream of room-temperature water. Embryos trapped in the bottom screen are then rinsed three times with E-wash (0.4% NaCl/0.03% Triton X-100), immersed in a 50% solution of commercial bleach to remove the chorions (about 3 min), rinsed with E-wash until the odor of bleach is completely gone, rinsed twice with distilled water, and rinsed once with extraction buffer consisting of 0.1 M Pipes, pH 6.9, 0.9 M glycerol, 5 mM EGTA, 2.5 mM MgSO$_4$, and protease inhibitors: 1 μg/ml leupeptin, 2μg/ml aprotinin, 1 μg/ml pepstatin, 1 μM phenylmethylsulfonyl fluoride (PMSF), 1 mg/ml *p*-tosyl-L-arginine methyl ester (TAME), and 0.1 mg/ml soybean trypsin inhibitor (STI). The quality of the protease inhibitors is important. The activities of some decline rapidly in aqueous solution, so it is best to add the inhibitors to cold extraction buffer just prior to use. We use stock solutions of leupeptin and aprotinin at 1 mg/ml in H$_2$O (stored at 4°C), pepstatin at 1 mg/ml in ethanol (stored at 4°C), and PMSF at 0.1 M in DMSO (stored at −20°C). TAME and STI are added to the extraction buffer dry.

During the preceding steps, it is important to remember that the embryos are alive and sensitive to stress. To optimize results, abnormal temperatures and poor oxygenation should be avoided. Embryos should be kept in a thin layer or actively suspended in a large volume of room-temperature liquid as much as possible. In addition, one should move through embryo preparation to the point of homogenization as quickly as possible.

B. Homogenization

After embryos have been rinsed with extraction buffer, they are placed in a chilled Wheaton glass homogenizer with an equal volume of ice-cold extraction buffer. Keeping the homogenizer in ice, five strokes with the loose pestle ("A") followed by seven strokes with the tight pestle ("B") should completely disrupt the embryos. The Wheaton homogenizer is cheap and simple but not ideal. Substantial force is required and it must be applied in a controlled manner to avoid spewing homogenate out of the tube and onto one's face. We have tried Brinkman and Vertishear blade-driven tissue homogenizers as well as sonication with little success. Tim Karr's lab uses a teflon pestle driven by a drill motor in a Dounce homogenizer tube with good results (Karr *et al.*, 1982).

C. Clarification

After disruption, the homogenate is clarified by low-speed and high-speed centrifugation steps. The low-speed spin is done in open plastic tubes in a swinging bucket rotor at 15,000g for 40 min at 4°C. A hardened layer of lipid will be found overlying the sample after the spin. The lipid layer should be picked up with a cotton-tipped applicator and discarded. The supernatant should then be removed with a large volume pipet, avoiding the cloudy material that lies on top of the pellet. The supernatant is transferred to clean plastic tubes and centrifuged in a fixed angle rotor at 50,000g for 30 min at 4°C. The resulting high-speed supernatant (HSS) can be used imediately or can be placed at -70°C for long-term storage. I have had good results with HSS that has been frozen for 2 years.

D. Microtubule Polymerization and Kinesin Binding

Microtubule polymerization is induced in HSS by the addition of GTP (0.3 mM) and taxol (20 μM), followed by gentle agitation at room temperature for 20 min. At this point, if one is interested in identifying novel motors whose binding to and release from microtubules is nucleotide sensitive, two samples of HSS should be treated in parallel. In sample "A", rigor binding of kinesin and other proteins to microtubules is induced by addition of the nonhydrolyzable ATP analog 5'-adenylyl imidodiphosphate (AMPPNP) to 2.5 mM, followed by an additional 10 min of incubation at room temperature. To sample "B", ATP and $MgSO_4$ are added to 2.5 mM each in place of AMPPNP. Proteins like kinesin, which bind to microtubules in A but not in B, are candidate motor proteins. If one is simply interested in quantitative preparation of kinesin, then all of the HSS should be treated with AMPPNP as described for sample A.

E. Differential Sedimentation of Microtubules and Kinesin

After the binding step, microtubules are sedimented through a sucrose cushion (20% sucrose and 10 μM taxol in extraction buffer) by centrifugation in a fixed angle rotor at 23,000g for 30 min at 4°C. The microtubules are then washed by resuspension in extraction buffer containing 10 μM taxol and 75 mM NaCl, followed by sedimentation through another sucrose cushion. The pellet from this salt wash is then resuspended in extraction buffer with 20 μM taxol, 75 mM NaCl, 10 mM $MgSO_4$, and 10 mM ATP. The volume of buffer used in this resuspension will determine the final concentration of kinesin. My lab aims for 2.5% of the initial HSS volume. Suspended microtubules are then sedimented in a fixed angle rotor at 120,000g for 20 min at 4°C. The supernatant ("ATP extract") contains highly enriched kinesin as well as some less abundant proteins. The activity of the kinesin is greatest if used fresh, but it can be stored at −70°C for long periods if a slight decrease in activity is acceptable.

F. Additional Fractionation

Column chromatography can provide fractionation of the ATP extract if required. Good results can be obtained with a Bio-Gel A5M gel filtration column (2.5 × 35 cm, 15 ml/hr) equilibrated with extraction buffer at 4°C. Bio-Gel A1.5M resin has also been used and produces good results (Cole *et al.*, 1993). We have not explored other chromatography methods, but ion-exchange columns and FPLC can undoubtedly be used for kinesin (Hackney, 1991; Urrutia and Kachar, 1992) and may be essential for purifying less abundant motors.

Fractionation of the ATP extract can also be achieved by repetitive binding to and release from microtubules. Taxol-stabilized microtubules are added to the ATP extract (0.5–1 mg/ml tubulin), rigor binding of motors is induced by addition of AMPPNP (10 mM), the microtubules are pelleted, and the bound proteins are released with ATP as explained earlier. The cycle can then be repeated. If one wishes to reduce cost, the ATP in the ATP extract can be depleted with hexokinase/glucose (Cohn *et al.*, 1987), which allows the use of less AMPPNP (1.5 mM).

This microtubule binding/release method can provide an enrichment of proteins like kinesin that bind microtubules with relatively low k_ds in the presence of AMPPNP and relatively high k_ds in the presence of ATP. If one were attempting to purify novel motors, the use of different nucleotides, nucleotide analogs (Shimizu *et al.*, 1991, 1993), and ionic strengths might be used to differentially enrich for them.

═══════ ## III. Characterization of Microtubule Motors

A. Microscopic Assays for Motor Proteins

The development of microscope-based *in vitro* motility assays has had a huge impact on our understanding of intracellular motility (Sale and Satir, 1977; Brady *et al.*, 1982; Sheetz and Spudich, 1983; Vale *et al.*, 1985a). Any attempt to identify new microtubule-based motors will eventually go to the microscope. The most common approach is to coat a glass coverslip with the cytoplasmic fraction to be tested, add microtubules and ATP, and watch for gliding of the microtubules over the coated surface (Vale *et al.*, 1985a; reviewed by Paschal and Vallee, 1993; Cohn *et al.*, 1993). Another approach is to anchor microtubules to a coverslip, add ATP, add microscopic beads coated with a cytoplasmic fraction, and look for translocation of the beads along the microtubules (Vale *et al.*, 1985a; Gelles *et al.*, 1988).

When a cytoplasmic fraction that can induce microtubule-based movements has been identified, microscopic assays can be used to test a variety of characteristics. These include the velocity of movement (Cohn *et al.*, 1993), the kinetics of nucleotide use (Cohn *et al.*, 1993), the polarity of movement (Vale *et al.*, 1985c; Porter *et al.*, 1987; Yang *et al.*, 1990; Paschal and Vallee, 1993; Howard and Hyman, 1993), the effects of ATP analogs on movement (Porter *et al.*, 1987; Shimizu *et al.*, 1993), force production (Hall *et al.*, 1993; Kuo and Sheetz, 1993), and step size (Gelles *et al.*, 1988; Svoboda *et al.*, 1993). Some of these tests require highly specialized and expensive equipment. All of them require a video microscope. The most common system utilizes a video camera attached to a microscope that is equipped for differential interference contrast with a high-intensity source of light (Cohn *et al.*, 1993). However, with suitable video cameras, dark-field and fluorescence microscopy can also be used (Vale and Hotani, 1988; Sale *et al.*, 1993; Howard and Hyman, 1993).

B. A Simple Motility Assay

If one wishes to assay microtubule motors *in vitro* with minimal fuss and little or no video equipment, sea urchin sperm provide a useful substrate (Lye *et al.*, 1989; Yang *et al.*, 1990). Sperm can be demembranated and treated with salt to expose dynein-free axonemal bundles of microtubules. Many axonemes will remain attached to their sperm heads via the minus ends of the microtubules. When applied to a coverslip that is coated with a cytoplasmic fraction, the axonemes can indicate both the presence of microtubule motors and the polarity of movement. An axoneme that pushes the sperm head in front is being moved by plus-end-directed motors and one that pulls the sperm head along behind is being moved by minus-end-directed motors. Although images of the demembranated sperm will be improved with video-enhanced contrast and image processing, they can be seen with regular light microscopic techniques.

This assay was developed by C. M. Pfarr and J. R. McIntosh (unpublished), building on methods described by Gibbons (1982). A male urchin injected with 0.5 M KCl is placed upside down on a small beaker that is resting on ice. After sperm are shed into the beaker, a 25-μl aliquot is mixed gently with 75 μl of extraction buffer containing 0.5 M KCl and 0.1% Triton X-100. After 10–20 min at room temperature, the suspension is diluted 50- to 100-fold with extraction buffer and placed on ice. During the preparation, vigorous mixing and pipetting should be avoided to prevent fragmentation of the axonemes. This stock solution of demembranated sperm can be used for several hours in place of microtubules in a standard gliding assay (Cohn *et al.*, 1993).

IV. Conclusions and Summary

Isolation of microtubule motor proteins is needed both for the discovery of new motors and for characterization of the products of motor-related genes. The sequences of motor-related genes cannot yet be used to predict the mechanochemical properties of the gene products. This was illustrated by the first kinesin-related gene product to be characterized. Protein expressed from the *ncd* gene moved toward the minus ends of microtubules (Walker *et al.*, 1990; McDonald *et al.*, 1990), while kinesin itself moves toward the plus ends. Until the relationship between mechanochemical function and amino acid sequence is more thoroughly understood, biochemical isolation and characterization of microtubule motor proteins will remain essential.

Two approaches for getting useful quantities of microtubule motor proteins have been used: isolation from cytosol as described under Section II above and isolation from bacteria carrying cloned motor protein genes in expression vectors. Bacterial expression of functional microtubule motors has been successful to date in only a few cases (Yang *et al.*, 1990; Walker *et al.*, 1990, McDonald *et al.*, 1990). Additional progress is expected with the expression of cloned genes from viral vectors in cultured eukaryotic cells, but broad success has not yet been reported.

Biochemical isolation of motors from their natural cytosol has some distinct advantages. One can have confidence that a given motor will be folded properly and have normal post-translational modifications. In addition, if it exists *in vivo* as a heteromultimer, a microtubule motor isolated from its native cytosol will carry with it a normal complement of associated proteins. Studies of such associated proteins will be important in learning how motors accomplish their tasks *in vivo*.

Drosophila cytosol should be a rich source of microtubule motors. *Drosophila* carry at least 11 and perhaps as many as 30 genes that are related to kinesin (Stewart *et al.*, 1991; Endow and Hatsumi, 1991). The work of Tom Hays' lab indicates that *Drosophila* carry more than nine dynein related genes (Rasmussen *et al.*, 1994). Relatively little effort to isolate the products of these genes from cytosol has been made. The only work that I am aware of has produced a kinesin-like microtubule motor (D. G. Cole, K. B. Sheehan, W. M. Saxton, and J. M. Scholey, in progress) that may be the *Drosophila* homolog of *Xenopus* eg5 (Sawin *et al.*, 1992). This isolation was straightforward, and efforts to identify additional motors are almost assured of success.

Acknowledgments

This work was supported in part by a grant from the NIH (GM-46295) to W. M. Saxton. The author is an Established Investigator of the American Heart Association.

References

Brady, S. T., Lasek, R. J., and Allen, R. D. (1982). Fast axonal transport in extruded axoplasm from squid giant axon. *Science* **218,** 1129–1131.

Cohn, S. A., Ingold, A. L., and Scholey, J. M. (1987). Correlation between the ATPase and microtubule translocating activities of sea urchin egg kinesin. *Nature* **328,** 160–163.

Cohn, S. A., Saxton, W. M., Lye, R. J., and Scholey, J. M. (1993). Analyzing microtubule motors in real time. *In* "Methods in Cell Biology" (J. M. Scholey, ed.), Vol. 39, pp. 75–88. New York: Academic Press.

Cole, D. G., Chinn, S. W., Wedaman, K. P., Hall, K., Vuong, T., and Scholey, J. M. (1993). Novel heterotrimeric kinesin-related protein purified from sea urchin eggs. *Nature* **366,** 268–270.

deCuevas, M., Tao, T., and Goldstein, L. S. B. (1992). Evidence that the stalk of *Drosophila* kinesin heavy chain is an alpha-helical coiled coil. *J. Cell Biol.* **116,** 957–965.

Endow, S. A., and Hatsumi, H. (1991). A multimember kinesin gene family in *Drosophila*. *Proc. Natl. Acad. Sci. U.S.A.* **88,** 4424–4427.

Gauger, A. K., and Goldstein, L. S. B. (1993). The *Drosophila* kinesin light chain: Primary structure and interaction with kinesin heavy chain. *J. Biol. Chem.* **268,** 13,657–13,666.

Gelles, J., Schnapp, B. J., and Sheetz, M. P. (1988). Tracking kinesin-driven movements with nanometre-scale precision. *Nature* **331,** 450–453.

Gho, M., McDonald, K., Ganetzky, B., and Saxton, W. M. (1992). Effects of kinesin mutations on neuronal functions. *Science* **258,** 313–316.

Gibbons, B. H. (1982). Reactivation of sperm flagella: Properties of microtubule-mediated motility. *In* "Methods in Cell Biology" (L. Wilson, ed.), Vol. 25, pp. 253–271. New York: Academic Press.

Hackney, D. D. (1991). Isolation of kinesin using initial batch ion exchange. *In* "Methods in Enzymology" (R. B. Vallee, ed.), Vol. 196, pp. 175–181. New York: Academic Press.

Hall, K., Cole, D. G., Yeh, Y., Scholey, J. M., and Baskin, R. J. (1993). Force-velocity relationships in kinesin-driven motility. *Nature* **364,** 457–459.

Hollenbeck, P. J. (1989). The distribution, abundance and subcellular localization of kinesin. *J. Cell Biol.* **108,** 2335–2342.

Howard, J., and Hyman, A. A. (1993). Preparation of marked microtubules for the assay of the polarity of microtubule-based motors by fluorescence microscopy. *In* "Methods in Cell Biology" (J. M. Scholey, ed.), Vol. 39, pp. 105–113. New York: Academic Press.

Karr, T. L., White, H. D., Coughlin, B. A., and Purich, D. L. (1982). A brain microtubule protein preparation depleted of mitochondrial and synaptosomal components. *In* "Methods in Cell Biology" (L. Wilson, ed.), Vol. 24, pp. 51–60. New York: Academic Press.

Kuo, S. C., and Sheetz, M. P. (1993). Force of single kinesin molecules measured with optical tweezers. *Science* **260,** 232–234.

Lasek, R. J., and Brady, S. T. (1985). Attachment of transported vesicles to microtubules in axoplasm is facilitated by AMP-PNP. *Nature* **316,** 645–647.

Lye, R. J., Pfarr, C. M., and Porter, M. E. (1989). Cytoplasmic dynein and microtubule transloca-tors. *In* "Cell Movement: Kinesin, Dynein, and Microtubule Dynamics" (F. D. Warner and J. R. McIntosh, eds.), Vol. 2, pp. 141–154. New York: Alan R. Liss.

McDonald, H. B., Stewart, R. J., and Goldstein, L. S. B. (1990). The kinesin-like *ncd* protein of Drosophila is a minus end-directed microtubule motor. *Cell* **63**, 1159–1165.

Paschal, B. M., and Vallee, R. B. (1993). Microtubule and axoneme gliding assays for force production by microtubule motor proteins. *In* "Methods in Cell Biology" (J. M. Scholey, ed.), Vol. 39, pp. 65–74. New York: Academic Press.

Porter, M. E., Scholey, J. M., Stemple, D. L., Vigers, G. P., Vale, R. D., Sheetz, M. P., and McIntosh, J. R. (1987). Characterization of the microtubule movement produced by sea urchin egg kinesin. *J. Biol. Chem.* **262**, 2794–2802.

Rasmussen, K., Serr, M., Gepner, J., Gibbons, I., and Hays, T. S. (1994). A family of dynein genes in *Drosophila melanogaster*. *Mol. Biol. Cell* **5**, 45–55.

Roof, D. M., Meluh, P. B., and Rose, M. D. (1992). Kinesin-related proteins required for assembly of the mitotic spindle. *J. Cell Biol.* **118**, 95–108.

Sale, W. S., and Satir, P. (1977). Direction of active sliding of microtubules in *Tetrahymena* cilia. *Proc. Natl. Acad. Sci. U.S.A.* **74**, 2045–2049.

Sale, W. S., Fox, L. A., and Smith, E. F. (1993). Assays of axonemal dynein-driven motility. *In* "Methods in Cell Biology" (J. M. Scholey, ed.), Vol. 39, pp. 89–104. New York: Academic Press.

Saunders, W. S., and Hoyt, M. A. (1992). Kinesin-related proteins required for structural integrity of the mitotic spindle. *Cell* **70**, 451–458.

Sawin, K. E., LeGuellec, K., Philippe, M., and Mitchison, T. J. (1992). Mitotic spindle organization by a plus-end-directed microtubule motor. *Nature* **359**, 540–543.

Saxton, W. M., Porter, M. E., Cohn, S. A., Scholey, J. M., Raff, E. C., and McIntosh, J. R. (1988). *Drosophila* kinesin: Characterization of microtubule motility and ATPase. *Proc. Natl. Acad. Sci. U.S.A.* **85**, 1109–1113.

Saxton, W. M., Hicks, J., Goldstein, L. S., and Raff, E. C. (1991). Kinesin heavy chain is essential for viability and neuromuscular functions in Drosophila, but mutants show no defects in mitosis. *Cell* **64**, 1093–1102.

Scholey, J. M., Porter, M. E., Grissom, P. M., and McIntosh, J. R. (1985). Identification of kinesin in sea urchin eggs, and evidence for its localization in the mitotic spindle. *Nature* **318**, 483–486.

Sheetz, M. P., and Spudich, J. A. (1983). Movement of myosin-coated fluorescent beads on actin cables *in vitro*. *Nature* **303**, 31–35.

Shimizu, T., Furusawa, K., Ohashi, S., Toyoshima, Y. Y., Okuno, M., Malik, F., and Vale, R. D. (1991). Nucleotide specificity of the enzymatic and motile activities of dynein, kinesin, and heavy meromyosin. *J. Cell Biol.* **112**, 1189–1197.

Shimizu, T., Toyoshima, Y. Y., and Vale, R. D. (1993). Use of ATP analogs in motor assays. *In* "Methods in Cell Biology" (J. M. Scholey, ed.), Vol. 39, pp. 167–177. New York: Academic Press.

Stewart, R. J., Pesavento, P. A., Woerpel, D. N., and Goldstein, L. S. B. (1991). Identification and partial characterization of six members of the kinesin superfamily in *Drosophila*. *Proc. Natl. Acad. Sci. U.S.A.* **88**, 8470–8474.

Svoboda, K., Schmidt, C. F., Schnapp, B. J., and Block, S. M. (1993). Direct observation of kinesin stepping by optical trapping interferometry. *Nature* **365**, 721–727.

Urrutia, R., and Kachar, B. (1992). An improved method for the purification of kinesin from bovine adrenal medulla. *J. Biochem. Biophys. Methods* **24**, 63–70.

Vale, R. D., Schnapp, B. J., Reese, T. S., and Sheetz, M. P. (1985a). Organelle, bead, and microtubule translocations promoted by soluble factors from the squid giant axon. *Cell* **40**, 559–569.

Vale, R. D., Reese, T. S., and Sheetz, M. P. (1985b). Identification of a novel force-generating protein, kinesin, involved in microtubule-based motility. *Cell* **42**, 39–50.

Vale, R. D., Schnapp, B. J., Mitchison, T., Steuer, E., Reese, T. S., and Sheetz, M. P. (1985c). Different axoplasmic proteins generate movement in opposite directions along microtubules in vitro. *Cell* **43,** 623–632.

Vale, R. D., and Hotani, H. (1988). Formation of membrane networks in vitro by kinesin-driven microtubule movement. *J. Cell Biol.* **107,** 2233–2241.

Walker, R. A., Salmon, E. D., and Endow, S. A. (1990). The *Drosophila claret* segregation protein is a minus-end directed motor molecule. *Nature* **347,** 780–782.

Weisenberg, R. C., Borisy, G. G., and Taylor, E. W. (1968). The colchicine-binding protein of mammalian brain and its relation to microtubules. *Biochemistry* **7,** 4466–4479.

Weisenberg, R. C. (1972). Microtubule formation *in vitro* in solutions containing low calcium concentrations. *Science* **177,** 1104–1105.

Yang, J. T., Saxton, W. M., and Goldstein, L. S. B. (1988). Isolation and characterization of the gene encoding the heavy chain of *Drosophila* kinesin. *Proc. Natl. Acad. Sci. U.S.A.* **85,** 1864–1868.

Yang, J. T., Laymon, R. A., and Goldstein, L. S. B. (1989). A three-domain structure of kinesin heavy chain revealed by DNA sequence and microtubule binding analyses. *Cell* **56,** 879–889.

Yang, J. T., Saxton, W. M., Stewart, R. J., Raff, E. C., and Goldstein, L. S. B. (1990). Evidence that the head of kinesin is sufficient for force generation and motility *in vitro*. *Science* **249,** 42–47.

CHAPTER 17

Preparation and Analysis of Membranes and Membrane Proteins from *Drosophila*

Michael Hortsch

Department of Anatomy and Cell Biology
University of Michigan
Ann Arbor, Michigan 48109

I. Introduction

Although the primary strength of *Drosophila* as a research organism lies in its powerful genetics, there are limits to the usefulness of purely genetic approaches. This is especially true for the investigation of cell–cell interactions. More and more membrane and membrane-associated proteins that play important roles during cellular differentiation and during development are being identified and cloned in *Drosophila*. Initially, only an amino acid sequence and a mutant phenotype may be known. To better understand the biological functions and molecular interactions of these proteins, a more detailed biochemical characterization is often necessary. Functionally important modifications such as protein phosphorylation, glycosylation, proteolytic processing, and the characterization of protein–protein interactions with ligands, cytoskeletal elements, and/or other membrane proteins are difficult or impossible to approach using solely genetic means.

Furthermore, an increasing number of *Drosophila* membrane proteins starting with a monoclonal antibody rather than a mutation are being cloned. Examples include the fasciclins, neurotactin, neuroglian, chaoptin, and the PS antigens. For some membrane proteins, which are recognized by a specific monoclonal antibody, expression cloning procedures have been successful (for example, Patel *et al.*, 1987; Bieber *et al.*, 1989; de la Escalera *et al.*, 1990; Hortsch *et al.*, 1990a,b; Bedian *et al.*, 1991). For others the antigen had to be isolated from insect protein extracts by immunoaffinity procedures and their cDNAs cloned using oligonucleotides designed from protein sequence data. Among these membrane proteins are chaoptin (Zipursky *et al.*, 1985), the PS2 α-integrin subunit (Bogaert *et al.*, 1987), fasciclin I and II (Bastiani *et al.*, 1987; Snow *et al.*, 1988), and fasciclin IV (Kolodkin *et al.*, 1992).

This chapter describes a tested method for the immunoaffinity purification of *Drosophila* proteins, especially membrane proteins, from detergent extracts. It also provides protocols for the isolation of membrane preparations from *Drosophila* embryos and the initial biochemical characterization of resident membrane and membrane-associated glycoproteins.

II. Immunoaffinity Purification of *Drosophila* Membrane Proteins

A. Preparation of Detergent Protein Extracts from *Drosophila* Embryos

The maintenance of a large-scale fly house to collect abundant quantities of staged *Drosophila* embryos is described by Shaffer and Elgin in Chapter 5 of this volume. To isolate 100 μg or more of a rare membrane protein from *Drosophila* embryonic protein extracts, several hundred grams of embryos may be needed. Since more than 30 g of fly embryos per generation can be harvested from one large fly cage, such quantities are obtainable. For the preparation of protein extracts, staged *Drosophila* embryos are homogenized and extracted using detergents. Although this protocol uses *Drosophila* embryos as starting material, with few adaptations other developmental stages could be used as well. Protein detergent extracts prepared as described here can be used for immunoaffinity chromatography as well as for lectin columns and other types of affinity matrices (e.g., see Montell and Goodman, 1988; Johansen *et al.*, 1989).

Make an embryo collection of the desired time window and stage the embryos as required. Wash the embryos in a standard mesh sieve (mesh size 150) with water to remove yeast cells and dechorionate in 50% bleach for 2 min. Wash the embryos well with water to remove any residual bleach. Dry the embryos by blotting off excess water and weigh. Until further use, store the moist embryo cake in a freezer at $-20°C$, or better still at $-80°C$.

Homogenize the collected embryos with 6 to 10 strokes in a tight 40-ml Dounce glass homogenizer in cold extraction buffer containing protease inhibitors (extraction buffer: 10 mM triethanolamine (TEA) (pH 8.2), 150 mM NaCl, 1 mM EDTA, 2% (v/v) NP-40 detergent, 0.5% (w/v) deoxycholic acid detergent (DOC), 20 μg/ml phenylmethylsulfonyl fluoride (PMSF), 1$\times$ protease inhibitor mix).[1] Use about 40 ml of extraction buffer for every 10 g of embryos. Perform the homogenization step with all aliquots and collect the homogenates in a 200-ml bottle or beaker. Put a stir bar into the flask and stir the homogenate at low speed for 1 hr at 4°C. Put the homogenate into ultracentrifuge tubes and spin for 1 hr at 4°C and 100,000g in a type 45 Ti (30,000 rpm) or a comparable rotor. Because some proteins are only extractable and soluble at higher salt concentrations, you might wish to extract the pellet a second time with extraction buffer containing 500 mM NaCl (mark the change in salt concentration on the storage flask!).

Store the supernatant at -80°C in a 200-ml bottle or tissue culture flask until use. Mark the flask with collection time window and date. The same extract can be used repeatedly on different affinity supports (the extract tolerates repeated freeze/thaw). However, record the type of column or immunoaffinity matrix you have used with this extract on the storage flask. It is also important to clear the lysate of any aggregated material prior to any type of column chromatography. Therefore, repeat the ultracentrifugation step every time before applying the protein extract to a column.

B. Immunoaffinity Isolation of *Drosophila* Membrane Proteins

There are several types of affinity matrices that are suitable for the coupling of immunoglobulins (e.g., CNBr-activated Sepharose, Affi-Gel 10, Protein A or Protein G Sepharose or agarose). The method developed by Schneider *et al.* (1992) uses a Protein A or G Sepharose or agarose support that will bind the Fc portion of the IgG to be coupled. Bound IgGs are then covalently cross-linked to the Protein A or G matrix. One of the advantages of this approach is the favorable orientation of the bound IgG for the binding of macromolecules displaying the correct epitope. Many subclasses of IgGs from different species will specifically bind to Protein A and no further purification of the antibody will be necessary prior to coupling. Since Protein A will not bind mouse IgG1 at physiological pH, use Protein G for monoclonal antibodies from this mouse

---

[1] Prepare the PMSF stock solution by dissolving 20 mg PMSF crystals (Sigma No. P-7626) in 1 ml isopropanol and put on a shaker for a few minutes. This stock solution is good for at least a few days. PMSF is rapidly hydrolyzed and inactivated in aqueous solutions. For a 500$\times$ protease inhibitor mix, dissolve the following protease inhibitors (all from Sigma Co.) in H$_2$O at a concentration of 500 μl/ml each and keep frozen at -20°C: Antipain (A-6271), Aprotinin (A-1153), Chymostatin (C-7268), Leupeptin (L-2884), Pepstatin A (P-4265), TLCK (T-7254), and TPCK (T-4376). Since not all inhibitors will go into solution at this concentration, thaw out the stock solution completely and vortex well before removing an aliquot.

IgG subclass. Pharmacia Biotech is a good source for Protein A Sepharose 4 Fast Flow (Pharmacia Catalog No. 17-0974-01) or Protein G Sepharose 4 Fast Flow (Pharmacia Catalog No. 17-0618-01).

For the coupling of a monoclonal antibody to a Protein A or G support both ascites fluids as well as cell culture supernatants can be used. It is advisable to first determine the suitability of the antibody for affinity purification procedures by using it for immunoprecipitations and if possible by testing its stability to the various loading, washing, and elution conditions as outlined below. Elution with high or low pH buffers is usually necessary for most monoclonal antibodies. If you want to purify a protein from detergent protein extracts, make sure none of the extract components will interfere with the affinity purification. We found one monoclonal antibody that is unable to bind its antigen in the presence of 0.5% DOC, a component of the standard extraction buffer. If the antibody is working on Western blots, this can be tested by a brief preincubation of the antibody in elution conditions before neutralizing the pH and performing the primary antibody-binding step in extraction buffer.

Wash 1 to 2 ml of Protein A or G Sepharose or agarose beads twice with phosphate-buffered saline (PBS). Add 2 ml of PBS to the beads and about 0.5 ml of ascites fluid or antiserum or resuspend the beads directly in 20 to 30 ml of tissue culture supernatant containing the monoclonal antibody. Incubate the beads overnight with this antibody solution while slowly rotating at 4°C. Pellet the beads by centrifugation at low speed (<500 rpm) and wash them twice with 30 ml of 0.2 M triethanolamine (TEA) buffer (pH 8.2). To covalently couple the antibody to the beads, resuspend the beads in 30 ml of the same buffer containing freshly dissolved 25 mM dimethyl pimelimidate dihydrochloride (DMPI)[2] and incubate slowly rotating at room temperature for 45 min.

Pellet the beads and resuspend them in 20 mM ethanolamine (pH 8.2) to inactivate any remaining DMPI. Incubate for 5 min at room temperature. Wash the beads twice with 50 ml of PBS and store them in a small volume of PBS with 0.01% NaN$_3$ at 4°C until further use.

All the following steps are best carried out in a cold room or a cooling cabinet at 4°C. Pack the prepared immunobeads in a small column (0.5 to 2 ml) and preelute the column with 5 ml elution buffer at 4°C to remove antibodies that have not been cross-linked to the Protein A/G column matrix (high pH elution buffer: 50 mM diethanolamine buffer (pH 11.5), 500 mM NaCl, 1 mM EDTA, 0.5% (v/v) NP-40; low pH elution buffer: 20 mM glycine/HCl (pH 2.2), 500 mM NaCl, 1 mM EDTA, 0.5% (v/v) NP-40). Neutralize the column immediately by washing with cold PBS. These steps can be carried out by gravitational flow without the help of a peristaltic pump.

Before loading the protein extract, connect a precolumn filled with 1 to 2 ml

[2] DMPI can be obtained from Sigma Co. (D-8388) or from other suppliers. Since DMPI will hydrolyze over time once the original bottle is opened, order multiples of the smallest quantity available and discard open bottles of DMPI.

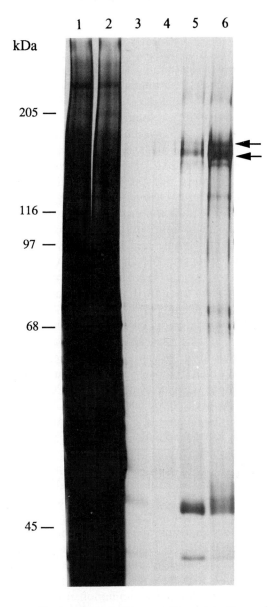

Fig. 1 Immunoaffinity purification of *Drosophila* neuroglian proteins from embryonic detergent protein extracts. An immunoaffinity matrix was prepared by cross-linking the anti-neuroglian monoclonal antibody 1B7 (Bieber *et al.*, 1989) to a Protein G Sepharose support. Two hundred milliliters of an embryonic (10–14 hr of development) detergent extract was passed over the 1B7 column. The column was washed with column washing buffer and preeluted with washing buffer containing 0.5 *M* NaCl. Bound polypeptides were recovered by elution with high pH buffer. Aliquots of the various fractions were separated on a 7.5% SDS–PAGE gel and proteins were visualized by silver staining. Lane 1, starting material; lane 2, flow through; lanes 3 and 4, washing fractions; lane 5, high-salt preelution fraction; lane 6, high pH eluate. The two arrows indicate the two different protein products of the *Drosophila* neuroglian gene (Hortsch *et al.*, 1990a).

of an inert column matrix material (e.g., Sephadex G-10) in front of the affinity column. Equilibrate the precolumn and the affinity column with cold column washing buffer (column washing buffer: PBS containing 1 mM EDTA and 0.5% (v/v) NP-40) and pass the detergent protein extract slowly over the precolumn and the affinity column using a peristaltic pump (e.g., 150 to 200 ml of extract overnight at 4°C). Before loading, centrifuge the protein extract for 30 min at approximately 100,000g and 4°C in an ultracentrifuge. Both the centrifugation of the protein extract and the precolumn will prevent protein aggregates from being trapped nonspecifically on the immunoaffinity support. This will result in significantly cleaner eluate fractions.

After loading the protein extract, disconnect the precolumn and discard it. Wash the column with 30 to 50 column vols of cold column washing buffer. The pump can be set five times faster than for the loading step. The following steps again can be carried out by gravitational flow without the help of a peristaltic pump. Preelute the column with 5 ml cold high-salt washing solution and collect the flow through fractions for analysis (high-salt washing solution: PBS with 500 ml NaCl, 1 mM EDTA, and 0.5% (v/v) NP-40). Elute the column with 5 ml of cold high or low pH elution buffer and collect the flow through in five 1-ml fractions. The protein peak will be usually eluted in fractions 2 and 3. Neutralize the column immediately with PBS or column washing buffer and store with 0.01% NaN$_3$ at 4°C for reuse. Neutralize the pH of the eluate fractions by adding 0.1 vol of 0.5 M NaH$_2$PO$_4$ for high pH elution or 0.5 M Na$_2$HPO$_4$ for low pH elution conditions. The neutralized eluate fractions can be analyzed by silver staining of SDS–PAGE gels and Western blotting and positive fractions can be pooled. Load about 30 μl of each eluate fraction onto a SDS–PAGE gel.

Yield and purity of such an immunoaffinity procedure will depend on the quality of the antibody and the abundance of the antigen. Even for a rare membrane protein several micrograms of pure protein can be purified from one 200-ml protein extract. An example for the immunoaffinity purification of a *Drosophila* membrane protein is shown in Fig. 1. The 1B7 monoclonal antibody was used for the isolation of *Drosophila* neuroglian polypeptides. Neuroglian protein purified by this procedure has been used without further purification for the generation of amino acid sequence data (Bieber *et al.*, 1989; Hortsch *et al.*, 1990a) and as immunogen for the production for new anti-neuroglian antibodies in rats and mice.

III. Isolation and Characterization of Cellular Membranes in *Drosophila* Embryos

The exact membrane topology of a protein, its mode of membrane association, and its glycosylation state is often not readily deducible from the primary amino acid sequence. Several experimental approaches are outlined for the further biochemical characterization of membrane proteins.

A. Isolation of Membrane Preparations from *Drosophila* Embryos

Drosophila embryos provide a convenient starting source for the isolation of cellular membranes. The following protocol can be scaled down to prepare membranes from tissue culture cells as well. However, the limited amount of starting material will reduce the yield.

Make an embryo collection of the desired time window, remove yeast cells by washing with water in a standard mesh sieve (mesh size 150), and dechorionate in 50% bleach for 2 min. Again, wash the embryos well with water after the incubation in bleach. Homogenize the embryos (approximately 5 g) with 6 to 10 strokes in a tight Dounce glass homogenizer in 10 to 20 ml of cold hypotonic medium containing protease inhibitors (hypotonic medium: 10 mM Tris–HCl, pH 8.0, 1 mM EDTA, 20 μg/ml PMSF, 1$\times$ protease inhibitor mix, see earlier). Centrifuge at 300 g for 10 min at 4°C. Save the supernatant on ice, homogenize the pellet again in 10 ml of cold hypotonic medium, and spin at 300 g. Pool the supernatants and spin for 10 min at 4°C and 4000 g. Save the supernatant, rehomogenize the pellet in 10 ml of cold hypotonic medium, and spin at 4000 g. Pool both supernatants and overlay onto a 5-ml cushion of cold hypotonic medium containing 0.25 M sucrose. Centrifuge for 1 hr at 4°C and 100,000 g. Aspirate the supernatant and homogenize the membrane pellet in a small Dounce glass homogenizer in a small volume of cold PBS with 20 μg/ml PMSF. Measure the OD_{280} and adjust to 5 to 50 OD_{280}/ml. Analyze 1 to 20 μl on a SDS–PAGE gel.

B. Alkaline Washes of *Drosophila* Membranes

One major question for newly identified membrane proteins is often their type of membrane association, integral versus peripheral. One easy experimental procedure for discriminating between membrane protein embedded in the lipid bilayer and peripheral membrane proteins is the exposure of membrane preparation to a high pH, which will not jeopardize bilayer integrity but will dissociate most peripheral membrane proteins from the lipid membrane.

The alkaline wash method as described by Fujiki *et al.* (1982) for vertebrate microsomal membrane preparations uses 0.1 M Na$_2$CO$_3$ at pH 11. I found that a pH of 11 can disrupt membrane integrity of *Drosophila* membranes and often leads to the partial release of genuine integral membrane proteins. I therefore recommend titration of the 0.1 mM Na$_2$CO$_3$ solution to pH 10 using dilute HCl.

Dilute an aliquot (e.g., 50 μl) of a *Drosophila* membrane preparation with at least 5 vol ($\geq$250 μl) of cold carbonate buffer (0.1 M Na$_2$CO$_3$(pH 10)) and incubate for 10 min on ice. Overlay this mixture onto 50 μl of carbonate buffer containing 0.25 M sucrose in a small ultracentrifuge tube. Centrifuge at 100,000 g for 1 hr at 4°C. This is best done in a Beckman table top ultracentrifuge or a Beckman Airfuge equipped with an A-95 rotor. If the total volume of this mixture is kept under 175 μl, the centrifugation steps can also be performed in an A-100 or an A-110 type Airfuge rotor. After pelleting the membranes,

carefully remove the supernatant (pH 10 soluble phase) with a pipet and neutralize the pH with 0.1 vol of 0.1 N HCl. This fraction will contain the pH 10-soluble,
peripheral membrane proteins. Wash the pellet by thoroughly resuspending it
in 300 μl of cold carbonate buffer and layering it onto a 50-μl alkaline sucrose
cushion. Centrifuge as described above and discard the supernatant. The pellet
containing only lipids and integral membrane proteins can be directly dissolved
in SDS–PAGE sample buffer. Load corresponding fractions of the starting
material, the first pH 10 supernatant fraction, and the pH 10 washed pellet
fraction onto a SDS–PAGE gel and analyze the distribution of your membrane
protein by Western blotting.

Alkaline washes to determine membrane attachment have been used for a
number of *Drosophila* membrane proteins including chaoptin (Reinke *et al.*,
1988), fasciclin I (Hortsch and Goodman, 1990), neuroglian, and neurotactin
(Hortsch *et al.*, 1990a,b). An example of the resistance of a *Drosophila* membrane protein, fasciclin I, to release by pH 10 from the lipid bilayer phase
is shown in Fig. 2. Fasciclin I is anchored in the plasma membrane by a
glycosyl–phosphatidylinositol–lipid moiety. As shown by alkaline washes of
Drosophila embryonic membranes this membrane attachment of fasciclin I
appears to be regulated during embryonic development (Hortsch and Goodman,
1990).

An alternative method for determining membrane association is detergent
cloud point precipitation using Triton X-114 (Bordier, 1981). This method was
used for the biochemical characterization of *Drosophila* PS-antigen complexes
(Leptin *et al.*, 1987) and for the *Drosophila* cell adhesion molecules chaoptin
and fasciclin I (Reinke *et al.*, 1988; Hortsch and Goodman, 1990).

However, high pH-resistant membrane attachment by itself is not a final
proof for an integral membrane protein. Although the exception, peripheral
membrane proteins that cannot be dissociated from their membrane receptor
by pH 11, have been found. To determine whether a protein is truly embedded in the lipid bilayer, a hydrophobic, photoactivatable probe such as 3-
(Trifluoromethyl)-3-(m[^{125}]iodophenyl) diazirine (^{125}I-TID) can be used to specifically label all integral membrane proteins with a radioactive tag (Brunner
and Semenza, 1981). Specific proteins can be immunoprecipitated after the
solubilization of the lipid bilayer with detergent and analyzed on SDS–PAGE
for the incorporation of this label. ^{125}I-TID is available from Amersham Corp.
(Catalog No. IM.148). However, since most of the radioactive label is cross-
linked to lipids and not to membrane proteins, this labeling method is not very
efficient.

C. Deglycosylation of *Drosophila* Membrane Proteins

Many membrane proteins are also glycosylated and their carbohydrate moiety
may be functionally important. A range of endoglycosydases for the further
characterization of the carbohydrate portion of glycoproteins is being offered
by a number of different suppliers. In the presence of detergents, *Drosophila*

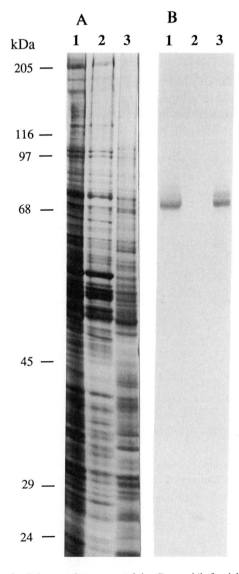

Fig. 2 pH 10 wash of cellular membranes containing *Drosophila* fasciclin I protein. A fasciclin I cDNA under the control of a *Drosophila* heat-shock promoter was transfected into and expressed in *Drosophila* Schneider 2 cells (see Chapter 35). A membrane preparation was isolated from cells expressing fasciclin I protein, subjected to a pH 10 wash, and analyzed on a 10% SDS–PAGE gel. (A) Protein pattern of the starting material (lane 1), the pH 10 soluble supernatant (lane 2), and the pH 10-resistant pellet (lane 3) as visualized by the silver staining method of Ansorge (1985). (B) Western blot of the same fraction that was incubated with an anti-fasciclin I monoclonal antibody.

membrane preparations can be directly incubated with these deglycosylating enzymes (see Fig. 3). I suggest following the recommendations as supplied by the manufacturer. However, depending on the structure of the carbohydrate moiety and the type of carbohydrate-protein linkage, these enzymes might not completely cleave off all carbohydrates groups from the polypeptide. A chemical method developed by Edge *et al.* (1981) that has worked well for the complete deglycosylation of various *Drosophila* glycoproteins is described here (Snow *et al.*, 1988; Bieber *et al.*, 1989; Hortsch *et al.*, 1990a,b).

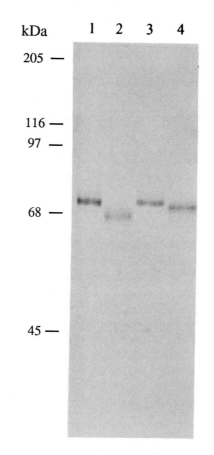

Fig. 3 Deglycosylation of *Drosophila* fasciclin I glycoprotein. A membrane preparation from *Drosophila* Schneider 1 tissue culture cells was treated with TFMS as described in the text, separated on a 10% SDS–PAGE gel, and analyzed by Western blotting using an anti-fasciclin I monoclonal antibody. Schneider 1 cells endogenously express fasciclin I protein (Elkins *et al.*, 1990). Lane 1, starting material; lane 2, sample treated with TFMS. Lane 4 contains the same material treated with 15 mU of endoglycosydase H (Boehringer and Mannheim) according to the suppliers recommendations. Proteins in lane 3 were incubated in parallel without the enzyme.

Lyophilize your glycoprotein or protein mixture (up to 5 to 10 mg protein) to complete dryness. Mix trifluoromethanesulfonic acid (TFMS) and anisol in a ratio of 2:1 and keep on ice.[3] Add 200 μl of the precooled TFMS:anisol mixture to the dry protein pellet and dissolve the pellet completely. Keep for 2 hr on ice and subsequently for 2 hr at $-20°C$. Add 10 μl of 2 M Tris base and 700 μl TEA-buffer (10 mM triethanolamine (pH 7.8), 150 mM NaCl, 1% (v/v) NP-40, 0.2% (w/v) deoxycholic acid). Add 100 μl of cold 100% trichloroacetic acid (TCA) and leave for 20 min on ice. Centrifuge for 5 min at full speed in a cold microcentrifuge. Extract the pellet with 0.5 ml of ice-cold acetone. Dry the pellet in a Speedvac and resuspend in an appropriate buffer (e.g., SDS–PAGE sample buffer). The deglycosylated protein sample can be directly loaded onto a SDS–PAGE gel and analyzed by silver staining (in the case that a purified glycoprotein was used as starting material) or by Western blotting. An example for the deglycosylation of a *Drosophila* membrane glycoprotein, fasciclin I, by TFMS and endoglycosydase H is displayed in Fig. 3. Because TFMS is a strong acid, the epitope recognized by a particular monoclonal antibody might be destroyed. Antibody binding will also be lost if the epitope recognized is part of a carbohydrate moiety.

IV. Conclusion

Using large-scale colonies of adult flies, sufficient quantities of developmentally staged material to analyze the biochemical characteristics of many *Drosophila* membrane proteins can be isolated. Preparative and analytical biochemical methods as described in this chapter will facilitate the investigation of many protein–ligand interactions and secondary protein modifications (e.g., such as glycosylation, protein phosphorylation, proteolysis and modifications in the mechanism of membrane association). Any of these modifications might be a potential target for the developmental or constitutive regulation of protein function.

Acknowledgments

I thank Drs. M. Burmeister and S. Brown for many useful suggestions on the manuscript. The work in the author's laboratory was supported by grants from the University of Michigan Cancer Center's Institutional Grant from the American Cancer Society and the National Institutes of Health Grant HD29388 to M.H.

[3] Because TFMS is a very strong acid, be extremely careful and perform the procedure in a chemical hood. TFMS (Sigma Co., T-1394) is prone to hydrolysis and you might want to use a fresh unopened vial for each new experiment. Anisol (Sigma Co., A-4405) can be stored at room temperature.

References

Ansorge, W. (1985). Fast and sensitive detection of protein and DNA bands by treatment with potassium permanganate. *J. Biochem. Biophys. Methods* **11**, 13–20.

Bastiani, M. J., Harrelson, A. L., Snow P. M., and Goodman, C. S. (1987). Expression of fasciclin I and II glycoproteins on subsets of axon pathways during neuronal development in the grasshopper. *Cell* **48**, 745–755.

Bedian, V., Jungklaus, C. E., Cardoza, L., and von Kalm, L. (1991). Kinase activity and genetic characterizations of a growth related antigen of *Drosophila*. *Dev. Genet.* **12**, 188–195.

Bieber, A. J., Snow, P. M., Hortsch, M., Patel, N. H., Jacobs, J. R., Traquina, Z. R., Schilling, J., and Goodman, C. S. (1989). *Drosophila* neuroglian: A member of the immunoglobulin superfamily with extensive homology to the vertebrate neural adhesion molecule L1. *Cell* **59**, 447–460.

Bogaert, T., Brown, N., and Wilcox, M. (1987). The *Drosophila* PS2 antigen is an invertebrate integrin that, like the fibronectin receptor, becomes localized to muscle attachments. *Cell* **51**, 929–940.

Bordier, C. (1981). Phase separation of integral membrane proteins in Triton X-114 solution. *J. Biol. Chem.* **256**, 1604–1607.

Brunner, J., and Semenza, G. (1981). Selective labeling of the hydrophobic core of membranes with 3-(trifluoromethyl)-3-(m[^{125}I]iodophenyl)diazirine, a carbene-generating reagent. *Biochemistry.* **20**, 7174–7182.

de la Escalera, S., Bockamp, E.-O., Moya, F., Piovant, M., and Jimenez, F. (1990). Characterization and gene cloning of neurotactin, a *Drosophila* transmembrane protein related to cholinesterases. *EMBO J.* **9**, 3593–3601.

Edge, A. S. B., Faltynek, C. R., Hof, L., Reichart, L. E., Jr., and Weber, P. (1981). Deglycosylation of glycoproteins by trifluoromethanesulfonic acid. *Anal. Biochem.* **118**, 131–137.

Elkins, T., Hortsch, M., Bieber, A. J., Snow, P. M., and Goodman, C. S. (1990). *Drosophila* fasciclin I is a novel homophilic adhesion molecule that along with fasciclin III can mediate cell sorting. *J. Cell Biol.* **110**, 1825–1832.

Fujiki, Y., Hubbard, A. L., Fowler, S., and Lazarow, P. B. (1982). Isolation of intracellular membranes by means of sodium carbonate treatment: Application to endoplasmic reticulum. *J. Cell Biol.* **93**, 97–102.

Hortsch, M., and Goodman, C. S. (1990). *Drosophila* fasciclin I, a neural cell adhesion molecule, has a phosphatidylinositol lipid membrane anchor that is developmentally regulated. *J. Biol. Chem.* **265**, 15,104–15,109.

Hortsch, M., Bieber, A. J., Patel, N. H., and Goodman, C. S. (1990a). Differential splicing generates a nervous system-specific form of *Drosophila* neuroglian. *Neuron* **4**, 697–709.

Hortsch, M., Patel, N. H., Bieber, A. J., Traquina, Z. R., and Goodman, C. S. (1990b). *Drosophila* neurotactin, a surface glycoprotein with homology to serine esteraes, is dynamically expressed during embryogenesis. *Development* **110**, 1327–1340.

Johansen, K. M., Fehon, R. G., and Artavanis-Tsakonas, S. (1989). The Notch gene product is a glycoprotein expressed on the cell surface of both epidermal and neuronal precursor cells during *Drosophila* development. *J. Cell Biol.* **109**, 2427–2440.

Kolodkin, A. L., Matthes, D. J., O'Connor, T. P., Patel, N. H., Admon, A., Bentley, D., and Goodman, C. S. (1992). Fasciclin IV: Sequence, expression, and function during growth cone guidance in the grasshopper embryo. *Neuron* **9**, 831–845.

Leptin, M., Aebersold, R., and Wilcox, M. (1987). *Drosophila* positon-specific antigens resemble the vertebrate fibronectin-receptor family. *EMBO J.* **6**, 1037–1043.

Montell, D. J., and Goodman, C. S. (1988). *Drosophila* substrate adhesion molecule: Sequence of laminin B1 chain reveals domains of homology with mouse. *Cell* **53**, 463–473.

Patel, N. H., Snow, P. M., and Goodman, S. C. (1987). Characterization and cloning of fasciclin III: A glycoprotein expressed on a subset of neurons and axon pathways in *Drosophila*. *Cell* **48**, 975–988.

Reinke, R., Krantz, D. E., Yen, D., and Zipursky, S. L. (1988). Chaoptin, a cell surface glycoprotein required for *Drosophila* photoreceptor cell morphogenesis, contains a repeat motif found in yeast and human. *Cell* **52**, 291–301.

Schneider, C., Newman, R. A., Sutherland, D. R., Asser, U., and Greaves, M. F. (1982). A one-step purification of membrane proteins using a high efficiency immunomatrix. *J. Biol. Chem.* **257**, 10766–10769.

Snow, P. M., Zinn, K., Harrelson, A. L., McAllister, L., Schilling, J., Bastiani, M. J., Makk, G., and Goodman, C. S. (1988). Characterization and cloning of fasciclin I and fasciclin II glycoproteins in the grasshopper. *Proc. Natl. Acad. Sci. U.S.A.* **85**, 5291–5295.

Zipursky, S. L., Venkatesh, T. R., and Benzer, S. (1985). From monoclonal antibody to gene for a neuron-specific glycoprotein in *Drosophila*. *Proc. Natl. Acad. Sci. U.S.A.* **82**, 1855–1859.

CHAPTER 18

Preparation of Extracellular Matrix[*]

J. H. Fessler, R. E. Nelson, and L. I. Fessler

Molecular Biology Institute and
Department of Biology
University of California, Los Angeles
Los Angeles, California 90024

I. Introduction

Research on *Drosophila* extracellular matrix (ECM) has mostly concerned the pericellular matrix and its relation to cells during development (reviewed in Fessler and Fessler, 1989; Hortsch and Goodman, 1991; Brown, 1993; Bunch and Brower, 1993; Fessler *et al.,* 1994). Current evidence suggests that *Drosophila* and vertebrates share fundamental problems and molecular mechanisms of cell–matrix interactions. The hemolymph of *Drosophila's* open circulatory

[*]This chapter has been modified and reprinted by permission from "*Drosophila* Extracellular Matrix" by L. I. Fessler, R. E. Nelson, and J. H. Fessler in *Methods in Enzymology,* Volume 245: Extracellular Matrix Components, edited by Erkki Ruoslahti and Eva Engvall. Copyright © 1994 by Academic Press, Inc.

system serves the joint functions of vertebrate interstitial fluid and blood, and ECM components that are primarily concentrated next to cells are also detectable in hemolymph. Wandering *Drosophila* blood cells, called hemocytes, and cells of the fat body are major producers of the ECM components considered here.

This review primarily concerns the preparation and properties of individual *Drosophila* ECM proteins. Homologies with the corresponding vertebrate proteins are notable for at least two major *Drosophila* ECM components, basement membrane collagen IV (Monson *et al.*, 1982; Natzle *et al.*, 1982; Blumberg *et al.*, 1987; Cecchini *et al.*, 1987; Blumberg *et al.*, 1988; Lunstrum *et al.*, 1988) and laminin (Fessler *et al.*, 1987; Chi and Hui, 1988,1989; Montell and Goodman, 1988,1989; Chi *et al.*, 1991; Garrison *et al.*, 1991; Kusche-Gullberg *et al.*, 1992; Henchcliffe *et al.*, 1993; MacKrell *et al.*, 1993). The most highly conserved portions are the sites of molecular interactions with other ECM components. The slit protein (Rothberg *et al.*, 1988,1990; Rothberg and Artavanis-Tsakonas, 1992) and two *Drosophila* genes belonging to the vertebrate tenascin family, ten[a] (Baumgartner and Chiquet-Ehrismann, 1993) and ten[m] (Baumgartner *et al.*, 1994) have been identified by molecular cloning based on homologies. We speculate that some ECM proteins that currently seem to be unique to *Drosophila* may eventually be found to have human homologs, specifically glutactin (Olson *et al.*, 1990), the slit protein (Rothberg *et al.*, 1988) tiggrin (Fogerty *et al.*, 1994), and peroxidasin (Nelson *et al.*, 1994), whose sequence, domain structure and occurrence during development have been established.

Partial characterizations of the following materials have been published. The large proteoglycan-like protein, papilin (Campbell *et al.*, 1987) is synthesized faily early in embryogenesis and then becomes a part of basement membranes. A proteoglycan has been detected in developing imaginal discs by immunolocalization with a monoclonal antibody (Brower *et al.*, 1987). A sulfated proteoglycan isolated from larvae has been identified as a heparin sulfate proteoglycan (Cambiazo and Inestrosa, 1990). Very interesting immunolocalization with antibodies recognizing a 240-kDa protein is seen during the formation of cell membranes following the initial syncytial blastoderm stage and during subsequent cell sheet organization, and this protein may be an ECM component (Garzino *et al.*, 1989). A 350-kDa protein, recognized by a monoclonal antibody, has a restricted distribution in the embryonic and pupal neuropil of the central nervous system (CNS), and it is suggested that this protein is a special component of the CNS ECM (Go and Hotta, 1992). It is not clear whether the secreted amalgam protein (Seeger *et al.*, 1988), found in the nervous system, is to be considered as an ECM protein. Other secreted proteins, with specialized functions, might also be included as ECM proteins, but these are not discussed here. Early studies of *Drosophila* ECM suggested that there is a homolog of vertebrate fibronectin (Gratecos *et al.*, 1988). However, extensive investigations failed to extend these observations.

Drosophila integrins, potential receptors of ECM molecules, were first identified in a monoclonal antibody screen (Wilcox *et al.*, 1981; Brower *et al.*, 1984).

The ubiquitous $\alpha_{PS2}\beta_{PS}$ and $\alpha_{PS1}\beta_{PS}$ integrins and their splice variants have extensive homologies to vertebrate integrin α- and β-chains (Bogaert *et al.*, 1987; Leptin *et al.*, 1987; MacKrell *et al.*, 1988; Brown *et al.*, 1989; Wehrli *et al.*, 1993; Zusman *et al.*, 1993). Additional integrins are present at more restricted sites (Yee and Hynes, 1993). Cell adhesion, receptors, and integrins of *Drosophila* were recently reviewed (Hortsch and Goodman, 1991; Brown, 1993; Bunch and Brower, 1993). As more ECM components are being identified, so further receptor-ligand interactions are being defined.

Three types of cell culture are particularly useful for *Drosophila* ECM research. Several immortal cell lines have been established from *Drosophila* embryos and can be grown in mass culture (Echalier and Ohanessian, 1969; Kakpakov *et al.*, 1969; Schneider, 1972; Cherbas and Cherbas, 1981) (for details see Sang, 1981; Cherbas *et al.*, 1994). They secrete ECM components into their culture media, and this review describes how biochemical amounts of several ECM proteins are isolated from the conditioned media. Some of the progenitor cells from which these cell lines were derived may have been related to hemocytes or cells of the fat body.

Short-term primary cultures are readily set up from dissociated *Drosophila* embryos (Sang, 1981; Mahowald, 1994). Differentiation of several cell types proceeds *in vitro* in these mixed cultures, over approximately the same time period as this differentiation occurs in whole embryos. These primary cultures can be initiated on substrate coatings of specific, individual *Drosophila* or vertebrate ECM glycoproteins, in the absence of fetal calf serum (Volk *et al.*, 1990; Gullberg *et al.*, 1994). This allows study of the effects of a given ECM component on cell adhesion, spreading, and differentiation. Additional ECM components are subsequently synthesized by differentiated cells that arise in the culture.

A third type of cell culture utilizes *Drosophila* S2 cell lines that have been permanently transformed with one or more DNA constructs that either code for an ECM receptor, such as an integrin (Bunch and Brower, 1992; Zavortink *et al.*, 1993), or for a part of an ECM molecule, such as a portion of a laminin chain (this laboratory, unpublished results). Cells that specifically express high concentrations of an ECM receptor are then tested for spreading on coatings of isolated ECM components.

II. Isolation of ECM Proteins

A. Isolation from Cell Cultures

Two Kc cell lines[1] (Echalier and Ohanessian, 1969; Kakpakov *et al.*, 1969) have been especially advantageous, because appreciable quantities of ECM

[1] Many established *Drosophila* cell lines are kept by Dr. Peter Cherbas at the Department of Biology, University of Indiana, Bloomington, IN 47405, and at the PHLS Center for Applied Microbiology and Research, European Collection of Animal Cell Cultures, Division of Biologics, Porton Down, Salisbury, Wiltshire SP4 0JG, United Kingdom.

proteins are secreted into the medium. Different sublines of Kc cells secrete somewhat different mixtures of ECM proteins. Several Kc cell lines, e.g., the Kc 7E10 cell line (Landon *et al.*, 1988), produce the well-characterized proteins (Fessler and Fessler, 1989) collagen IV, laminin, papilin, glutactin, and peroxidasin.[2] These cells are nonadherent and do not express PS integrins. The Kc 167 cell line is more adherent; it synthesizes integrins and a new ECM protein, tiggrin, which has been identified as a ligand of $\alpha_{PS2}\beta_{PS}$ integrin (Fogerty *et al.*, 1994). The Kc 167 cells also secrete papilin, collagen IV, laminin, and small amounts of peroxidasin and glutactin. In addition, many unidentified and higher or mostly lower molecular weight proteins are secreted. The Schneider S2 and S3 cell lines (Schneider, 1972) produce lower levels of ECM proteins. The slit protein is secreted into the medium and deposited into the matrix in cultures of S2 cells (Rothberg *et al.*, 1990).

Stock Kc cells are maintained in tissue culture flasks, but some cells, such as the Kc 7E10 cells, have been adapted for growth in suspension in spinner and roller cultures at room temperature in D22 medium (Ashburner, 1989) (Sigma) plus penicillin and streptomycin and without fetal calf serum (FCS). The Kc 167 cells attach to tissue culture flasks and can be maintained in D22 medium plus antibiotics with or without 2% FCS. These cells also grow in M3 medium (Sigma), which has a defined amino acid composition (Ashburner, 1989). For isolation of radiolabeled proteins, the amino acid to be added in labeled form is omitted from M3 medium and the amounts of yeastolate and bactopeptone are decreased to one-tenth of the normal additions. To label sulfated glycoproteins, $H_2{}^{35}SO_4$ is added to medium containing 1/20 the normal concentration of $SO_4^=$. Omission of unlabeled $SO_4^=$ yields undersulfated proteins (Campbell *et al.*, 1987). Bovine ECM components are potential contaminants of *Drosophila* proteins purified from cell cultures grown in the presence of FCS. The *Drosophila* origin of a newly isolated protein should be proven by obtaining it in labeled form from cell cultures supplied with radioactive amino acids.

1. Salt Precipitation of Media Proteins

A flow diagram outlines the purification procedure (Fig. 1). Each step, which gives only a partial separation of any of the proteins, is discussed in turn. The methods can be combined according to the reader's needs. We strongly recommend the initial velocity sedimentation step, as it improves the resolution achieved during subsequent chromatography. Higher molecular weight complexes also become apparent during velocity sedimentation.

Kc 7E10 cells grown at room temperature or at 25°C in suspension are maintained in a spinner bottle for several weeks and diluted three- to four-fold

[2] Peroxidasin was previously referred to as protein X. Peroxidasin is described in Nelson *et al.*, 1994.

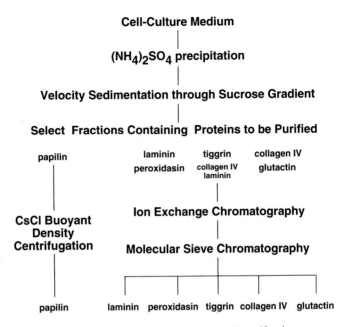

Fig. 1 Flow diagram for ECM protein purification.

with new medium weekly. These cultures can reach a density of ca. $1-5 \times 10^6$ cells/ml. Aliquots are then placed into roller bottles, diluted 1:1 with new medium, and grown to a cell density of 10^7 cells/ml. They are then diluted 1:2 and grown to high density two or three more times.[3] For isolation of proteins from the medium, the cells are removed by centrifugation at $1000g$ for 10 min and the medium is clarified by centrifugation at $6000g$ for 60 min. For isolation of proteins from adherent cell cultures, e.g., Kc 167 cell cultures, the medium is decanted and treated as above. Protease inhibitors are added 30 mM EDTA, pH 7.5, 2 mM N-ethylmaleimide (NEM), and 1 mM phenylmethylsulfonyl fluoride (PMSF) and proteins are precipitated by slow addition of solid $(NH_4)_2SO_4$ to 45% saturation at 4°C. Most of the ECM proteins are precipitated with this concentration of $(NH_4)_2SO_4$. Differential precipitation of proteins with lower $(NH_4)_2SO_4$ concentrations can be useful for partial separation of some ECM molecules. The precipitate is dissolved in 1/100 ml original culture medium

---

[3] In these cell cultures, collagen IV is folded and secreted normally even though the hydroxylation of proline and lysine residues is incomplete. To obtain good hydroxylation of collagen IV, the cultures need to be supplemented daily with 25 μg/ml ascorbic acid. This yields collagen IV with the same degree of hydroxylation as made *in vivo*: pepsin treated, fully hydroxylated collagen IV has 35% of the total pro and 60% of the total lys hydroxylated. Upon reduction, the hydroxylated collagen molecule separates into a major $\alpha 1$ IV and a minor $\alpha 1'$ IV component (Lunstrum *et al.*, 1988).

in 0.05 M Tris–HCl, pH 7.5, 0.15 M NaCl (TBS) plus inhibitors, (20 mM EDTA, 10 mM NEM, 1 mM PMSF, or 2 mM aminoethylbenzene sulfonyl fluoride (AEBSF). Following extensive dialysis against TBS, any undissolved residue is removed by centrifugation at 8000g for 30 min.

2. Partial Separation by Velocity Sedimentation

The protein mixture (1 ml/centrifuge tube) is partially separated by velocity sedimentation at 4°C on a 5–20% sucrose gradient in TBS in a Beckman SW 41 rotor at 39,000 rpm for 15 to 22 hr, depending on the proteins to be isolated. A cushion of 0.75 ml 60% sucrose in TBS is placed at the bottom of the tube to assure recovery of more rapidly sedimenting material and helps to maintain the stabilizing sucrose gradient. The sedimentation coefficients, given in Table I, help in the choice of centrifugation conditions for the isolation of any given protein. Fractions are collected from the bottom of the centrifuge tube and aliquots of individual fractions are analyzed by SDS–PAGE to identify the components in each fraction. An example of the separation of the proteins secreted by Kc 7E10 cells is shown in Fig. 2A. Individual fractions, peak fractions, or larger pools of material can be chosen for further purification: for example proteoglycans are in fractions 1–3; laminin and peroxidasin peak in fractions 5–8; collagen IV is present as monomeric molecules together with glutactin in fractions 13–15, and higher aggregates are found in faster sedimenting fractions; glutactin peaks in fractions 14–16. To isolate tiggrin, the proteins in the medium of Kc 167 cells are partially resolved by velocity sedimen-

Table I
Sedimentation Coefficients of ECM Proteins

	S_{20w}	Reference
Collagen IV	4.1 S	Lunstrum *et al.*, 1988
Laminin	11.0 S	Fessler *et al.*, 1987
Peroxidasin	11.6 S	Nelson *et al.*, 1994
Tiggrin	6.45 S	Fogerty *et al.*, 1994

Note. Sedimentation coefficients are determined with purified proteins dissolved in TBS plus 0.1% Triton X-100. (This detergent is added to avoid losses due to protein adherence to the tube.) The gradient is a 5–20% sucrose gradient in TBS plus 0.1% Triton, and 0.1-ml sample is layered on the top of the gradient. Sedimentation is in a Beckman SW 60 rotor at 4°. The position of the protein is determined by SDS–PAGE and densitometric analysis. Sedimentation coefficients are calculated as described (Fessler and Fessler, 1974).

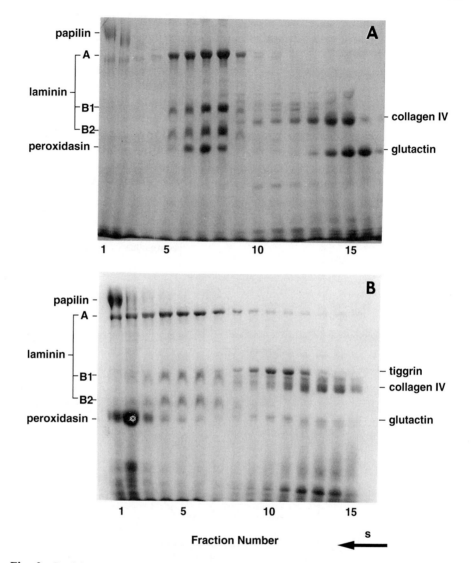

Fig. 2 Partial separation of proteins secreted by Kc cells by velocity sedimentation. (A, upper panel) The media proteins secreted by Kc 7E10 cells were concentrated by $(NH_4)_2SO_4$ precipitation and a 1-ml sample was sedimented in a Beckman SW 41 rotor at 4°C, 39,000 rpm for 19 hr and then separated into 21 fractions. The Coomassie blue-stained SDS–PAGE electrophoretogram of the reduced proteins is shown for the successive fractions of a 5–20% sucrose gradient. The electrophoretic band positions of several key polypeptides are indicated. Note that all separations are only partial, and further chromatographic purifications are required. (B, lower panel) shows the partial separation of the proteins secreted into the medium by Kc 167 cells. In this experiment, the proteins were sedimented at 39,000 rpm for 20 hr. An abundant serum protein from the fetal calf serum that was added to the medium for this culture of Kc 167 cells is shown by an asterisk.

tation, as shown in Fig. 2B. Tiggrin sediments between laminin and collagen, fractions 9–12, and is purified from these intermediate fractions.

3. Ion Exchange and Molecular Sieve Chromatography

The peak fractions of the proteins to be purified further are pooled and dialyzed against Mono Q buffer: 0.3 *M* sucrose, 0.05 *M* Tris–HCl, pH 8.0, 5 m*M* EDTA, 1 m*M* PMSF, and 0.05% Triton X-100 containing NaCl at a concentration that permits binding of the proteins at 4°C to the Mono Q ion-exchange column in an FPLC system (Pharmacia LKB Biotechnology). A low concentration of NaCl (between 0.05 and 0.15 *M* NaCl) prevents the partial,

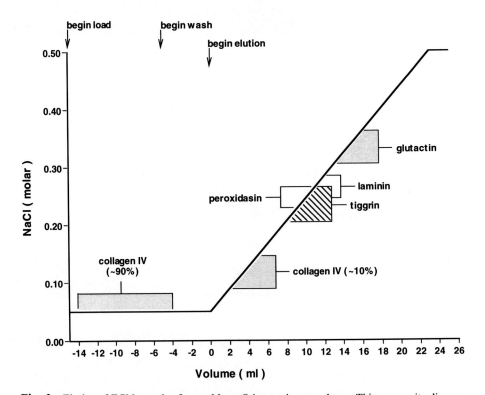

Fig. 3 Elution of ECM proteins from a Mono Q ion-exchange column. This composite diagram shows the ranges of NaCl concentrations required for release of several ECM proteins from a Mono Q column eluted with a linearly increasing gradient of 0.02 *M* NaCl/1 ml in Mono Q buffer. Starting samples for this chromatography were separate sets of selected fractions from more than one velocity sedimentation experiment. Each set of pooled sedimentation fractions is indicated by a different shading as follows: Kc 167 cell culture media proteins, tiggrin sedimentation pool ▨; Kc 7E10 cell culture media proteins, glutactin and collagen IV sedimentation pool ▢, and laminin plus peroxidasin sedimentation pool ▢. Note that additional proteins, mostly of smaller molecular size, contaminate both the velocity sedimentation and the ion-exchange fractions. They are removed by subsequent molecular sieve chromatography.

slow precipitation of collagen IV, if it is present in the mixture. Alternatively, DEAE columns DE 52 (Whatman) give satisfactory separations. The proteins are eluted with a linear gradient of NaCl, up to 0.5 M. Figure 3 shows the NaCl concentrations required for elution of the identified proteins.

Each protein is purified further by gel filtration on a Superose 6 or 12 column (Pharmacia LKB Biotechnology) or on an agarose A-50 sizing column. The appropriate fractions containing the proteins to be separated are dialyzed against Mono Q buffer containing 0.15 M NaCl. The samples are concentrated either by centrifugation through Centricon 30 microconcentrators (Amicon) or by vacuum dialysis. A 2-ml vol is applied to a 1.6 × 50-cm Superose 6 column and 1.0-ml fractions are eluted with this buffer at an elution rate of 0.13 ml/ min. Fractions are stored frozen in this buffer. Figure 4 shows separations of groups of proteins on a preparative Superose 6 column. Protein separation on a

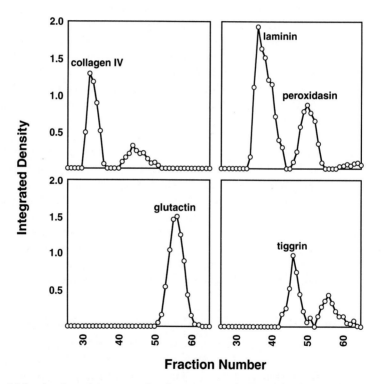

Fig. 4 Molecular sieve chromatography. Separate sets of selected fractions obtained by the ion-exchange chromatography in Fig. 3 were applied to a preparative Superose 6 column and eluted at 0.13 ml/min with 100 ml of buffer. Aliquots of consecutive 1-ml fractions were reduced and analyzed by SDS–PAGE and stained with Coomassie blue. Each electrophoretogram contained the electrophoresis buffer front and was scanned densitometrically along each complete lane. The integrated, normalized optical densities are plotted. At each named peak, only bands corresponding to that protein were seen. Unnamed peaks and material eluted after fraction 63 represent other proteins and contaminants. Totally excluded marker eluted at fraction 29.

Superose 6 analytical column (1.0 × 30 cm), calibrated with globular molecular weight standards, can provide an estimate of the molecular weights of globular proteins. However, many ECM proteins, such as collagen, and laminin, are elongated, asymmetric, and variably flexible molecules and elute from the sizing column in positions that do not correspond to their molecular weights.

4. Equilibrium CsCl Buoyant Density Centrifugation

Papilin (Campbell *et al.*, 1987), a proteoglycan-like ECM protein, is synthesized by both Kc cell lines. It is a sulfated protein and is readily detected in the medium of cells labeled with $H_2{}^{35}SO_4$. By velocity sedimentation of total medium proteins it is found together with other ECM aggregates in the fastest sedimenting fractions at the bottom of the tube. To obtain native papilin, the fast sedimenting fractions are dialyzed into TBS and resedimented for 10 hr at 39,000 rpm at 4°C. This places the monomeric and dimeric papilin molecules within the gradient around fractions 5–10. By ion-exchange chromatography on a Mono Q column, it is eluted with a NaCl gradient of 0.25 to 0.45M. Papilin elutes as a fairly broad peak, following laminin, reflecting heterogeneity in glycosylation and sulfation.

Papilin in denatured form is further purified from the mixture of proteins in the fast sedimenting fractions near the bottom of the gradient (Fig. 2, fractions 1–3) by equilibrium, CsCl buoyant density sedimentation in guanidine HCl (GuCl) solution. GuCl is needed to avoid salting out of proteins when CsCl is added. Samples are dialyzed into TBS followed by dialysis against 4 M GuCl in TBS. CsCl, 1.40 g, is dissolved in 3.56 ml of sample, and this sample is centrifuged in an SW 60 Beckman rotor at 10°C for 48 hr at 40,000 rpm. The rotor is not spun at full speed to avoid overstraining with this CsCl load. The gradient is divided into fractions by drop collection from the bottom of the tube. Fractions are dialyzed against TBS and analyzed by SDS–PAGE or on an SDS agarose gel. The peak of the proteoglycans is buoyant at a density of 1.40–1.42 g/ml with some spread of the molecules to lower density fractions. The other proteins are buoyant at or near the top of the gradient.

B. Isolation of ECM Proteins from Embryos, Larvae, Pupae, and Flies

Some ECM proteins can be obtained from embryos by extraction with 1 M NaCl, 0.05 M Tris–HCl, 0.02 M EDTA, 0.5% Triton X-100 buffer, pH 7.5, or from embryo lysates. The method is illustrated by Montell's extraction of laminin (Montell and Goodman, 1988). *Drosophila* embryos are dechorionated with 30% Clorox, washed well with 0.7% NaCl solution, and either extracted directly or frozen in liquid N_2 and powdered in a precooled mortar and pestle. For extraction at 0°C, all solutions contain protease inhibitors 1 mM PMSF, 20 mM EDTA, and 1 µg/ml of the following: aprotinin, pepstatin, leupeptin, and possibly other inhibitors, antipain, chymostatin, N-tosyl-L-phenylalanine-chloromethylketone (TPCK), and Na-*p*-tosyl-L-lysine chloromethylketone (TLCK). The embryos are solubilized in IPB: 10 mM triethanolamine, 1% NP-40, 0.5% deoxycholate, and 0.15 M NaCl, plus protease inhibitors (pH 8.2) by homogenization (5 strokes) in a glass Dounce homogenizer with a loose teflon pestle (1 g embryos per 2–5 ml buffer). The lysate is stirred for 1 hr at 0°C. The lysate is sedimented at 13,000g for 30 min and then clarified at 100,000g for 2 hr at 4°C. Some of the laminin is then bound on a peanut lectin–agarose column (Vector Lab) equilibrated with IPB buffer. The lysate is passed over the column at 5–10 ml/hr. Unbound proteins, including some of the laminin and a number of ECM proteins, are removed by successive washes with IPB, IPB with 0.5 M NaCl, and IPB. Laminin and some other proteins are eluted with IPB containing 0.5 M D-galactose. These proteins can be visualized by SDS–PAGE and silver staining of the gel or on Western blots with antibodies. Laminin also binds quantitatively to wheat germ agglutinin Sepharose (Sigma) and heparin Sepharose (Pharmacia) (Fessler *et al.*, 1987). Other ECM proteins glutactin, collagen IV, and tiggrin do not bind to heparin Sepharose. Further separation of ECM proteins extracted from embryos is achieved by chromatography on Mono Q and Superose columns. Affinity chromatography on lectin columns is a useful tool that should be explored further when unidentified proteins are to be purified.

Quantitative extraction of matrix proteins, followed by identification of specific proteins on Western immunoblots, is useful for evaluation of the temporal expression during development (Fessler and Fessler, 1989). Staged dechorionated embryos, larvae, pupae, and adult flies are frozen and powdered in liquid N$_2$. Equal weights of tissue are homogenized in 5–10 vols of 2% SDS, 0.15 M NaCl, 0.05 M Tris–HCl, pH 7.5, 0.01 M EDTA, 1 mM PMSF at 100°C, heated 5 min at 100°C and sonicated to fragment nucleic acid. Complete extraction in most cases requires reduction with 10 mM dithiothreitol at 100°C for 5 min. The insoluble residue is sedimented and the concentration of soluble protein in each sample is determined with the Bio-Rad protein assay. Known quantities of protein extracted at each stage are then electrophoresed on SDS–PAGE and a Western blot with antibodies to specific ECM proteins is developed. The antigen–antibody complex can then be recognized by binding of ^{125}I-Protein A and autoradiography, followed by counting of the cut out radioactive bands. To obtain accurate quantitation, increasing volumes of extract from each time period are electrophoresed, and the Western blot is developed with an excess of antibody and [125]I Protein A. It is important to determine that the quantity of antigen electrophoresed, and the antigen–antibody–Protein A complex formed are within a linear response range of signal. An example of developmental Western analyses of collagen IV is shown in Fig. 5. Eggs and early embryos already contain a very small quantity of ECM proteins, but after zygotic transcription is initiated much higher levels of proteins are synthesized at the indicated times. Other systems for recognition of the antigen–antibody complex, such as the use of alkaline phosphatase-conjugated secondary IgG

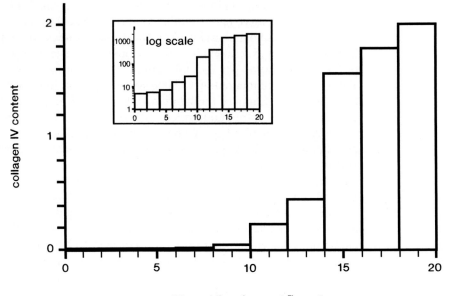

Fig. 5 Developmental Western analysis of collagen IV accumulation during embryogenesis. SDS extracts were made of embryos that had developed at 25°C. After electrophoretic separation, Western blots were developed with antibodies to collagen IV (Lunstrum *et al.*, 1988) on several dilutions of each sample to assure a linear response range of signal. The relative content of collagen IV at 2-hr intervals is plotted on a linear scale and a logarithmic scale (inset). Nominally, the linear scale indicates micrograms and the logarithmic scale nanograms of collagen IV per milligram of embryo, but the absolute calibration has some uncertainty. The histograms show that although trace amounts of the protein are present in early embryos, the major accumulation of collagen IV occurs during later embryogenesis.

or enhanced chemiluminescence systems are more difficult to quantitate, but are also adequate for general analyses.

III. Characterization of ECM Molecules

A. Electrophoretic Mobilities of ECM Proteins

The relative electrophoretic mobilities of the nonreduced and reduced proteins that we have characterized are shown in Fig. 6. The mobilities of ECM components may vary significantly from those of standard, commercial "marker proteins" of the same M_r for various reasons, e.g., the average mass residue of collagens is significantly less than that of marker proteins (Noelken *et al.*, 1981) and highly charged proteins, such as glutactin, bind SDS abnormally. Collagen IV (Lunstrum *et al.*, 1988) molecules are disulfide-linked trimers, which on reduction separate into a major α1 IV and a minor α1′ IV chain.[3]

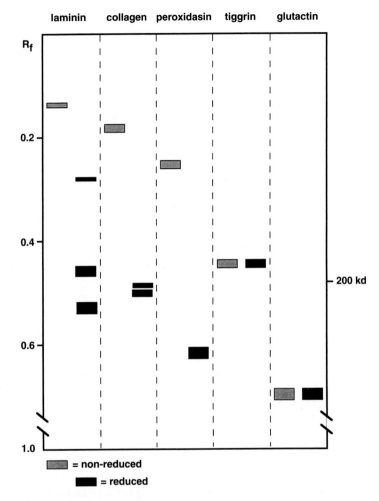

Fig. 6 Relative electrophoretic mobilities of ECM proteins. Diagrammatic representation of the mobilities of purified disulfide-linked and reduced proteins electrophoresed on a 24-cm SDS 5% polyacrylamide gel is shown. The molecular weights of the subunits of each protein before post-translational modification are given in Table II. The 200-kDa molecular weight marker is myosin.

Collagen IV can associate into higher aggregates, as is shown by SDS–agarose gel electrophoresis, Fig. 7. Laminin (Fessler *et al.*, 1987; Montell and Goodman, 1988), a disulfide-linked heterotrimer, consists of one A, B1, and B2 chain. Each of these chains also contains multiple intramolecular S–S bridges. Peroxidasin (Nelson *et al.*, 1994) is a disulfide-linked homotrimer of 170-kDa subunits. Papilin (Campbell *et al.*, 1987) exists as a disulfide-linked monomer or as higher oligomers. The size of the reduced, monomeric core protein is in the range of 400 kDa. The electrophoretic mobility of tiggrin is not altered by reduction

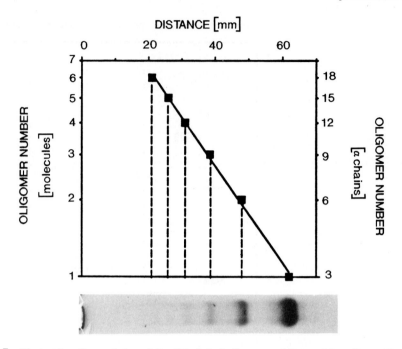

Fig. 7 Electrophoretic resolution of disulfide-linked oligomers of *Drosophila* collagen IV on an SDS–agarose gel. The electrophoretogram stained with Coomassie blue is shown together with a plot of the distance migrated by each band vs the logarithm of the presumed number of collagen IV molecules (left ordinate) or α chains (right ordinate) that are disulfide linked in the complexes. The statistical regression line is shown; its correlation coefficient is 0.85. This figure is reproduced from Lunstrum *et al.* (1988).

(Fogerty *et al.*, 1994). Since the tiggrin subunit of ca. 260 kDa has only one cysteine, intrachain S–S bridges cannot form and most of the molecules isolated from conditioned medium have no intermolecular bonds. Glutactin (Olson *et al.*, 1990) has intramolecular S–S bridges and exists as a monomeric chain. Slit proteins of different electrophoretic mobilities, in the range of 200 kDa, are formed from alternatively spliced mRNAs (Rothberg *et al.*, 1990).

B. Higher Complexes and Characterization by Agarose Gel Electrophoresis

Oligomers of ECM proteins such as collagen IV or papilin are too large for separation by SDS–PAGE, but can be separated partly by velocity sedimentation in native form or, better, by SDS–agarose slab gel electrophoresis (Duncan *et al.*, 1983): 1.5% agarose (Type I, low EEO, Sigma) is dissolved in 0.19 *M* Tris–HCl, pH 8.8, 0.25% SDS, and a horizontal 200-mm-long, 130-mm-wide, × 4-mm thick gel is cast. The gel is placed in a horizontal electrophoresis apparatus (just like a gel for nucleic acid separation), and reservoir buffer, 0.2% SDS, 0.19 *M* Tris–glycine, pH 8.3, is added so that the gel is just submerged in the buffer. The samples in TBS, 0.2% SDS and 4% glycerol and bromophenol blue are heated at 100°C for 2 min. Electrophoresis is carried out for about 16 hr at 20 mA, constant current. The gels are fixed in 50% methanol, 7% acetic acid and stained with 0.2% Coomassie blue. For radiolabeled proteins, fluorography is carried out after the gels are soaked in 50% methanol, 7% acetic acid for 1 hr, washed with running water, soaked in 0.7 M sodium salicylate (adjusted to about pH 6.5) for 30 min, and dried. The electrophoretic resolution of disulfide-linked oligomers of *Drosophila* collagen IV is shown in Fig. 7. The semilogarithmic relationship of the mobilities of these bands to a simple integer sequence is consistent with a series of oligomers of internally disulfide-linked, triple helical molecules. Oligomers of myosin can be used as a size marker ladder for SDS–agarose gel electrophoresis.

C. Electron Microscopy and Velocity Sedimentation

Electron microscopy of large ECM molecules can be very informative (Engel and Furthmayr, 1987). Materials are usually sprayed from a glycerol solution containing a volatile buffer onto a mica surface, adsorbed onto pentylamine-treated films (Campbell *et al.*, 1987) and rotary shadowed, or visualized by the technique of negative contrast with uranyl acetate (for examples of micrographs of *Drosophila* ECM molecules, see Fessler *et al.*, 1994). Although the thicknesses of thread-like portions of the images are not meaningful, their lengths and flexing and branching patterns are instructive. Oligomers can also be visualized, but it becomes increasingly difficult to trace overlapping threads of a complex. Initial, partial separation of successively higher complexes by velocity sedimentation in H_2O/D_2O density gradients and volatile buffer salts helps and allows direct spraying of the fractions after addition of glycerol (Lunstrum *et al.*, 1988).

Informative velocity sucrose gradient sedimentation analyses can be made with purified or crude ECM proteins, even with mere radioactive amounts. (Table I). Fractions are analyzed by SDS–PAGE and densitometry of the stained or fluorographed bands. An integrated plot of the data allows determination of the sedimentation peak position to better resolution than individual fractions, and sedimentation coefficients $s_{20,w}$ are either calculated with a program that corrects for the varying sucrose solution (Fessler and Fessler, 1974) or obtained by comparison with marker proteins added to the sample. The hydrodynamic frictional ratio, f/f_o, can be indicative of highly nonglobular, threadlike molecules, and is obtained from $s_{20,w}$ and M_r, which may be known from cDNA sequence analysis (Fogerty *et al.*, 1994).

D. Modifications of ECM Proteins

Table II gives the molecular weights of ECM proteins, based on DNA sequence analyses. Post-translational modifications of these proteins occur. Collagen IV contains hydroxyproline and hydroxylysine. Glutactin and papilin, and to a much lesser extent the A-chain of laminin, are sulfated. Sulfation of glutactin

Table II
Molecular Weights of ECM Proteins Derived from DNA Sequences

	kDa	Reference
Collagen IV	172	Blumberg *et al.*, 1988
Laminin		
A-chain	409	Kusche-Gullberg *et al.*, 1992, Henchcliffe *et al.*, 1993
B1-chain	196	Montell and Goodman, 1988
B2-chain	179	Montell and Goodman, 1989; Chi and Hui, 1988
Tiggrin	255	Fogerty *et al.*, 1994
Glutactin	117	Olson *et al.*, 1990
Peroxidasin	168	Nelson *et al.*, 1994
Slit protein	ca.200	Rothberg *et al.*, 1990
Ten[a]	85	Baumgartner and Chiquet-Ehrismann, 1993
Ten[m]	281	Baumgartner *et al.*, 1994

is on tyrosines and in papilin-sulfated glycosaminoglycan (GAG) side chains are present. Most of the proteins are N- or O-glycosylated. However, carbohydrate analyses are incomplete.

Potential asparagine N-glycosylation sites are predicted from the derived amino acid sequences. Comparative SDS–PAGE before and after removal of N-linked carbohydrates with peptide:N-glycosidase F (PNG-ase F) (New England Biolabs) provides a measure of the extent of N-glycosylation of proteins such as peroxidasin, glutactin, and laminin (Fessler *et al.*, 1994). Removal of O-linked carbohydrates with O-glyconase (Genzyme) has not been studied extensively with these proteins. Some characterization of the carbohydrate side chains of these proteins on Western blots is available by affinity binding of a large variety of biotinylated lectins and recognition of the bound lectins with the biotin–avidin–peroxidase system (Vector Labs).

The GAG side chains of the proteoglycans can be removed by a number of methods. Enzymatic treatment with chondroitinase ABC, heparinase, heparatinase, keratinase, neuraminidase, and PNG-ase F and H aid in the identification of the GAG side chains. To identify the O-linked carbohydrate chains, β elimination in 0.1 N NaOH with 1.0 M NaBH$_4$ at 47°C for 48 hr releases the GAG chains, which can then be characterized further. Treatment of the protein with HNO$_2$ destroys heparan sulfate. Deglycosylation with trifluoromethanesulfonic acid (Edge *et al.*, 1981) yields the core protein.

E. DNA Cloning and Analysis of Domain Structure

Antibodies prepared against purified *Drosophila* ECM proteins have been used to isolate cDNA clones from expression libraries (e.g., laminin A, B1 and B2 chains (Montell and Goodman, 1988; Garrison *et al.*, 1991), glutactin (Olson *et al.*, 1990), and tiggrin (Fogerty *et al.*, 1994). Heterologous antibodies have not been used successfully thus far. With the availability of sufficient quantities of proteins, N-terminal amino acid sequence analyses of intact or fragmented proteins has been possible. This is essential for confirmation of the identity of the cDNA isolated, as above. Oligonucleotide mixtures were used successfully for the isolation of cDNA clones coding for peroxidasin (Nelson, *et al.*, 1994).

By low-stringency screens with a chicken procollagen I cDNA clone, a number of *Drosophila* genes coding for collagen IV (Dcg1) (Monson *et al.*, 1982; Natzle *et al.*, 1982; Blumberg *et al.*, 1987; Cecchini *et al.*, 1987) and Dcg2 (Natzle *et al.*, 1982; Le Parco *et al.*, 1986a) and additional collagen-like genes mapping to different locations on the chromosomes were isolated (Le Parco *et al.*, 1986a). Some vertebrate proteins are classified as collagens, even though only a minor portion of their amino acid sequence has the canonical $(Gly-X-Y)_n$ sequence required for folding into a triple helix. Correspondingly, there may be *Drosophila* proteins with relatively short $(Gly-X-Y)_n$ repeats. A family of loci homologous to the EGF-like portion of *Notch* has been identified and one of these, the cDNA coding for the secreted slit protein, has been sequenced (Rothberg *et al.*, 1988,1990).

Vertebrate ECM proteins have conserved domains and PCR approaches may lead to successful isolations of *Drosophila* cDNAs with homologous domains. Comparisons of *Drosophila* and vertebrate amino acid sequences derived from cDNA analyses have led to the overall conclusion that functional domains of proteins, such as putative cell attachment sequences and portions involved in connections with other ECM components, are more conserved than structural regions (laminin, collagen IV, slit protein, ten[a] and ten[m]). Molecular cloning of ten[a] (Baumgartner and Chiquet-Ehrismann, 1993) and ten[m] (Baumgartner *et al.*, 1994) genes, homologous to tenascin, was achieved using the polymerase chain reaction (PCR) with degenerate primers derived from the EGF portion of chicken tenascin.

IV. Interactions of ECM Proteins with Cells *in Vitro*

A. Interactions of ECM Substrate Coatings with Integrin-Expressing Cells

The interaction of cells with ECM proteins involves multiple cell surface receptors, especially integrins. To identify which ECM proteins are ligands of *Drosophila* integrins, Bunch and Brower used transfected S2 cells (Bunch and Brower, 1992; Zavortink *et al.*, 1993). For example, the integrin β_{PS} subunit and the two splice variants, $\alpha_{PS2(c)}$ or $\alpha_{PS2(m8)}$, of the α_{PS2} subunit are expressed in these cells under the control of a heat-shock promoter. Cells stimulated to express integrins at their cell surfaces attach and spread on immobilized ligands, such as vertebrate fibronectin or vitronectin (Bunch and Brower, 1992; Zavortink *et al.*, 1993) and *Drosophila* tiggrin (Fogerty *et al.*, 1994). These ligand–integrin interactions utilize an arginine–glycine aspartic acid (RGD)-containing sequence, and RGD peptides inhibit this interaction. Antibodies recognizing the functional domain of the integrin β-chain (Hirano *et al.*, 1991) block this ligand–receptor interaction. Thus far, only one *Drosophila* ECM protein, tiggrin, has been demonstrated to be a ligand for the $\alpha_{PS2}\beta_{PS}$ integrins, but other investigations are in progress.

B. Primary Embryo Cell Culture on ECM Substrate Coatings

Two major ECM proteins of vertebrate sera that facilitate cell attachment and spreading of *Drosophila* cells are fibronectin (Gratecos *et al.*, 1988) and vitronectin (Hirano *et al.*, 1991; Gullberg *et al.*, 1994). In the absence of serum, *Drosophila* laminin supports attachment, spreading, and partial differentiation of a broad spectrum of primary *Drosophila* embryo cells (Volk *et al.*, 1990; Gullberg *et al.*, 1994), including epithelial cells, neurons, hemocytes, and myocytes. Mouse laminin cannot substitute for the *Drosophila* protein. Coatings of vertebrate vitronectin best support *Drosophila* embryo cells that express $\alpha_{PS2}\beta_{PS}$ integrins, in particular myocytes (Gullberg *et al.*, 1994). Tiggrin, which is a ligand of this integrin, also supports the differentiation of primary *Drosophila* embryo cells and myotube formation (Fogerty *et al.*, 1994).

Primary *Drosophila* embryo cell cultures grown on an ECM substrate and provided with radioactive amino acids synthesize the labeled *Drosophila* ECM proteins papilin, laminin, collagen IV, glutactin, tiggrin, and peroxidasin (Gullberg *et al.*, 1994). Immunohistology indicates that these ECM components are largely made by the hemocytes that differentiate in the cultures. The ECM products elaborated by differentiating primary embryo cell cultures influence the culture.

V. Temporal and Spatial Expression of ECM Genes and Location of ECM Products

In situ hybridization (Langer-Safer *et al.*, 1982) on polytene chromosomes gave the locations of the ECM proteins listed in Table III. Transcription from these genes during development is followed by Northern blots (e.g., Fig. 8) of RNA extracted (Chirgwin *et al.*, 1979) from embryos, larvae, pupae, and adults. For quantitative comparison of a pair of mRNAs, which differ in electrophoretic mobility, a single antisense RNA probe containing segments complementary to both mRNAs is helpful (Garrison *et al.*, 1991). The spatial locations of expression are found by *in situ* hybridization to whole embryos (Tautz and Pfeifle, 1989; Tautz and Lehman, 1994). Formation and deposition of the corresponding ECM proteins is followed by immunostaining of whole embryos or sections and quantitated by Western blots of detergent extracts (Fig. 5). Immu-

Table III
Chromosome Locations of Genes Coding for ECM Proteins

	Chromosome band	Reference
Collagen IV	25C	Monson *et al.*, 1982; Natzle *et al.*, 1982; Le Parco *et al.*, 1986a
Laminin		
A-chain	65A 10-11	Montell and Goodman, 1988
B1-chain	28D	Montell and Goodman, 1988
B2-chain	67C	Montell and Goodman, 1988
Glutactin	29D	Olson *et al.*, 1990
Tiggrin	26D 1-2	Fogerty *et al.*, 1994
Peroxidasin	62E8-F2	Nelson *et al.*, 1994
Slit Protein	52D	Rothberg *et al.*, 1988
Ten[a]	11A 6-9	Baumgartner and Chiquet-Ehrisman, 1993
Ten[m]	79E1-2	Baumgartner *et al.*, 1994

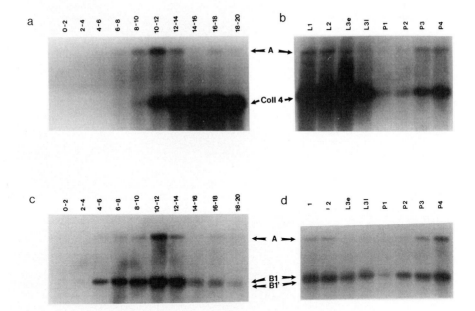

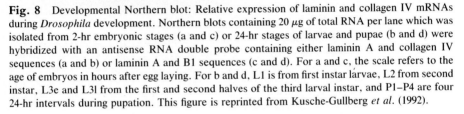

Fig. 8 Developmental Northern blot: Relative expression of laminin and collagen IV mRNAs during *Drosophila* development. Northern blots containing 20 μg of total RNA per lane which was isolated from 2-hr embryonic stages (a and c) or 24-hr stages of larvae and pupae (b and d) were hybridized with an antisense RNA double probe containing either laminin A and collagen IV sequences (a and b) or laminin A and B1 sequences (c and d). For a and c, the scale refers to the age of embryos in hours after egg laying. For b and d, L1 is from first instar larvae, L2 from second instar, L3e and L3l from the first and second halves of the third larval instar, and P1–P4 are four 24-hr intervals during pupation. This figure is reprinted from Kusche-Gullberg *et al.* (1992).

nolocalization of collagen IV in the widely distributed hemocytes and in the fat bodies is illustrated in Fig. 9. Polyclonal antibodies are preferably raised in rats and mice, as many rabbit preimmune sera give background staining.

Monoclonal antibodies have been prepared, using purified antigens [e.g., antibodies to laminin (Montell and Goodman, 1989)], or by immunizations with whole tissues [e.g., *Drosophila* heads (Fujita *et al.*, 1982), discs (Wilcox *et al.*, 1981; Brower *et al.*, 1987), etc.] and identification of antibodies localized to specific sites. Brower and associates (Wilcox *et al.*, 1981; Brower *et al.*, 1987) generated monoclonal antibodies against imaginal discs and identified an extra-cellular matrix proteoglycan on the basal surface of imaginal disc epithelia and in a filamentous network between epithelial cells. Fujita *et al.* (1982) injected

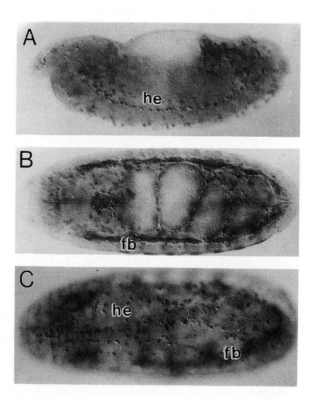

Fig. 9 Whole-mount staining of *Drosophila* embryos. Embryos were stained with antibodies to a fragment of collagen IV (Fessler *et al.*, 1993) and biotinylated secondary antibodies and the biotin–avidin–peroxidase amplification system (Vector Lab). A shows a side view of an embryo at stage 14, and B and C show 2 levels of focus of the same embryo at stage 16. In all cases, the focus was adjusted to highlight the staining of the hemocytes (he) and fat bodies (fb) rather than basement membrane staining.

Drosophila heads and isolated many monoclonal antibodies, including a number of antibodies that recognize basement membranes.

The principal sites of synthesis of ECM components are hemocytes, the cells of the fat body, and some glial cells, as detected both by *in situ* hybridization for RNA and by immunostaining (Fogerty *et al.*, 1994; Nelson *et al.*, 1994; Le Parco *et al.*, 1986b; Lunstrum *et al.*, 1988; Mirre *et al.*, 1988; Le Parco *et al.*, 1989; Montell and Goodman, 1989; Olson *et al.*, 1990; Kusche-Gullberg *et al.*, 1992). Other tissues, such as epithelia or muscle, may be able to synthesize these materials at a lower level, but this is difficult to document definitively. While immunostaining of a hemocyte might equally show a newly synthesized or a phagocytosed protein, the correspondence with *in situ* hybridization is consistent with *de novo* production. A polyclonal antibody made to denatured laminin chains primarily stains intracellular laminin, suggesting a significant intracellular presence of incompletely assembled molecules (Kusche-Gullberg *et al.*, 1992). Conversely, relatively poor staining of some cells for glutactin (Olson *et al.*, 1990) may indicate a short intracellular residence time for completed molecules of this ECM component. Double immunostaining for pairs of ECM molecules (with antibodies marked by two different fluorescent labels) indicates that individual hemocytes synthesize several ECM components, specifically: collagen IV, laminin, papilin, tiggrin, and peroxidasin. Quantitative confocal microscopy can provide a measure of the relative production of a pair of ECM proteins in hemocytes at different locations within one embryo. Different ECM proteins that are all found in a common structure, such as basement membranes, are not necessarily made at the same time. Indeed, all evidence points to a successive deposition of basement membrane ECM components (Fessler and Fessler, 1989).

ECM proteins have been detected during oogenesis (Gutzeit *et al.*, 1991) and low maternal contributions of some ECM proteins are found in extracts of very young embryos (see Fig. 5) and by immunostaining (Knibiehler *et al.*, 1990). Major subsequent accumulations of ECM are in basement membranes that surround the CNS and muscles and underlie epithelia (Fogerty *et al.*, 1994; Lunstrum *et al.*, 1988; Montell and Goodman, 1989; Olson *et al.*, 1990; Kusche-Gullberg *et al.*, 1992). More restricted synthesis of the slit protein by glial cells and its deposition along axon tracts, as well as its presence in the developing dorsal vessel, is to be noted (Rothberg *et al.*, 1990). ECM proteins are concentrated at muscle insertions (Fogerty *et al.*, 1994; Olson *et al.*, 1990; Rothberg *et al.*, 1990) and at locations of mechanical stress. In larvae, ECM proteins are synthesized by the fat body and hemocytes, in the lymph glands, and by adepithelial cells (Mirre *et al.*, 1988). Special problems of ECM modification (Fessler *et al.*, 1993) and deposition (Brower *et al.*, 1987; Fristrom *et al.*, 1993) arise in imaginal discs. Immunoelectron microscopy with antibodies to collagen IV and laminin shows localization to excessive layers of basement membranes in the tumor-like imaginal discs of lethal (1) discs-large-1 mutants (Abbott and Natzle, 1992).

VI. Miscellaneous Aspects and Conclusions

Soluble ECM molecules are easier to study than the insoluble matrices into which they associate. The earliest soluble forms are intracellular, unassembled chains and vesicle-stored proteins, but only cell culture allows separation of intra- and extracellular soluble forms of an ECM component. Caution must be exercised when only RNA-based evidence is used to deduce the time, place, and relative rate of appearance of an ECM component. The rate of increase of extractable protein, taken as a measure of protein synthesis, is often not directly related to the relative level of the cognate RNA. For example, in later embryos, a steady increase of extractable tiggrin occurs in spite of decreasing levels of tiggrin RNA (Fogerty *et al.*, 1994). Newly secreted proteins may be transported in the hemolymph. To determine whether glutactin is primarily a hemolymph or a matrix protein, Olson (this laboratory, unpublished) determined its distribution between the hemolymph and washed carcasses of third instar larvae and found that at least 90% is matrix-associated. Little is known about the covalent links that stabilize mature matrix. While reduction, presumably of disulfide links, facilitates extraction, Western blots of the extracts indicate higher associations of known ECM chains held by unknown covalent linkages. To avoid the potential formation of lysyl-derived cross-links, which arise between vertebrate collagen molecules, the lysyl oxidase inhibitor 2-aminopropionitrile was fed to larvae before extracting collagen (Murray and Leipzig, 1986). A group of tyrosine-based cross-links occurs in the insect elastomer resilin (Andersen, 1966) and, potentially, this type of cross-link could be formed in *Drosophila* by the action of peroxidasin, although this has not yet been investigated.

Modifications of ECM can lead to apparently new proteins. For example, the targeted, hormone-controlled cleavage of the basement membrane collagen IV of imaginal discs produces specific fragments (Fessler *et al.*, 1993).

Our current knowledge of *Drosophila* ECM components is unduly influenced by the few cell lines that have been used as practical sources of ECM components. For several research applications, it may eventually be easier to produce a key fragment of a given *Drosophila* ECM protein in a bacterial expression system, e.g., a fusion protein of that portion of tiggrin that contains the integrin-binding site retains the ability to interact with cells (Fogerty *et al.*, 1994). *Drosophila* cells can also be used to express transfected DNA coding for an ECM component, when correct post-translational modifications are important. Finally, while the methodologies described in this review were developed to characterize large ECM proteins, there could be important small components for which no search has been made.

Acknowledgments

Some of the research described in this review was supported by grants from the Muscular Dystrophy Association and USPHS, Grant AG02128, which also supported the cost of writing this review. We thank Kathryn Brill for typing the manuscript.

References

Abbott, L. A., and Natzle, J. E. (1992). Epithelial polarity and cell separation in the neoplastic l(1)dlg-1 mutant of Drosophila. *Mech. Dev.* **37,** 43–56.

Andersen, S. O. (1966). Covalent cross-links in a structural protein, resilin. *Acta Physiol. Scand.* **66** (Suppl. 263), 5–81.

Ashburner, M. (1989). "Drosophila: A Laboratory Manual." Cold Spring Harbor, NY: Cold Spring Harbor Laboratory.

Baumgartner, S., and Chiquet-Ehrismann, R. (1993). Tenª, a Drosophila gene related to tenascin, shows selective transcript localization. *Mech. Dev.* **40,** 165–176.

Baumgartner, S., Martin, D., Hagios, C., and Chiquet-Ehrismann, R. (1994). Submitted for publication.

Blumberg, B., MacKrell, A. J., and Fessler, J. H. (1988). Drosophila basement membrane procollagen alpha 1(IV). II. Complete cDNA sequence, genomic structure, and general implications for supramolecular assemblies. *J. Biol. Chem.* **263,** 18328–18337.

Blumberg, B., MacKrell, A. J., Olson, P. F., Kurkinen, M., Monson, J. M., Natzle, J. E., and Fessler, J. H. (1987). Basement membrane procollagen IV and its specialized carboxyl domain are conserved in Drosophila, mouse, and human. *J. Biol. Chem.* **262,** 5947–5950.

Bogaert, T., Brown, N., and Wilcox, M. (1987). The Drosophila PS2 antigen is an invertebrate integrin that, like the fibronectin receptor, becomes localized to muscle attachments. *Cell* **51,** 929–940.

Brower, D. L., Piovant, M., Salatino, R., Brailey, J., and Hendrix, M. J. (1987). Identification of a specialized extracellular matrix component in Drosophila imaginal discs. *Dev. Biol.* **119,** 373–381.

Brower, D. L., Wilcox, M., Piovant, M., Smith, R. J., and Reger, L. A. (1984). Related cell-surface antigens expressed with positional specificity in Drosophila imaginal discs. *Proc. Natl. Acad. Sci. U.S.A.* **81,** 7485–7489.

Brown, N. H. (1993). Integrins hold Drosophila together. *Bioessays* **15,** 383–390.

Brown, N. H., King, D. L., Wilcox, M., and Kafatos, F. C. (1989). Developmentally regulated alternative splicing of Drosophila integrin PS2 alpha transcripts. *Cell* **59,** 185–195.

Bunch, T. A., and Brower, D. L. (1992). Drosophila PS2 integrin mediates RGD-dependent cell-matrix interactions. *Development* **116,** 239–247.

Bunch, T. A., and Brower, D. L. (1993). Drosophila cell adhesion molecules. *Curr. Top. Dev. Biol.* **28,** 81–123.

Cambiazo, V., and Inestrosa, N. C. (1990). Proteoglycan production in Drosophila egg development: Effect of β-D-xyloside on proteoglycan synthesis and larvae motility. *Comp. Biochem. Physiol.* **97,** 307–314.

Campbell, A. G., Fessler, L. I., Salo, T., and Fessler, J. H. (1987). Papilin: A Drosophila proteoglycan-like sulfated glycoprotein from basement membranes. *J. Biol. Chem.* **262,** 17605–17612.

Cecchini, J. P., Knibiehler, B., Mirre, C., and Le Parco, Y. (1987). Evidence for a type-IV-related collagen in Drosophila melanogaster: Evolutionary constancy of the carboxyl-terminal noncollagenous domain. *Eur. J. Biochem.* **165,** 587–593.

Cherbas, L., and Cherbas, P. (1981). The effects of ecdysteroid hormones on Drosophila melanogaster cell lines. *In* "Advances in Cell Culture" (K. Maramorosch, ed.), Vol. 1, pp. 91–124. New York: Academic Press.

Cherbas, L., Moss, R., and Cherbas, P. (1994). "*Drosophila melanogaster:* Practical Uses in Cell and Molecular Biology" *In* (L. S. B. Goldstein and E. A. Fyrberg, eds.). New York: Academic Press.

Chi, H. C., and Hui, C. F. (1988). cDNA and amino acid sequences of Drosophila laminin B2 chain. *Nucleic Acids Res.* **16,** 7205–7206.

Chi, H. C., and Hui, C. F. (1989). Primary structure of the Drosophila laminin B2 chain and comparison with human, mouse, and Drosophila laminin B1 and B2 chains. *J. Biol. Chem.* **264,** 1543–1550.

Chi, H. C., Juminaga, D., Wang, S. Y., and Hui, C. F. (1991). Structure of the Drosophila gene for the laminin B2 chain. *Dna Cell Biol.* **10,** 451–466.

Chirgwin, J. M., Przybyla, A. E., MacDonald, R. J., and Rutter, W. J. (1979). Isolation of biologically active ribonucleic acid from sources enriched in ribonuclease. *Biochemistry* **18,** 5294–5299.

Duncan, K. G., Fessler, L. I., Bachinger, H. P., and Fessler, J. H. (1983). Procollagen IV. Association to tetramers. *J. Biol. Chem.* **258,** 5869–5877.

Echalier, G., and Ohanessian, A. (1969). Isolation, in tissue culture, of Drosophila melangaster cell lines. *C. R. Acad. Sci. Hebd Seances Acad. Sci. D* **268,** 1771–1773.

Edge, A. S., Faltynek, C. R., Hof, L., Reichert, L., Jr., and Weber, P. (1981). Deglycosylation of glycoproteins by trifluoromethanesulfonic acid. *Anal. Biochem.* **118,** 131–137.

Engel, J., and Furthmayr, H. (1987). Electron microscopy and other physical methods for the characterization of extracellular matrix components: Laminin, fibronectin, collagen IV, collagen VI, and proteoglycans. *Methods Enzymol.* **145,** 3–78.

Fessler, J. H., and Fessler, L. I. (1989). Drosophila extracellular matrix. *Annu. Rev. Cell Biol.* **5,** 309–339.

Fessler, L. I., Campbell, A. G., Duncan, K. G., and Fessler, J. H. (1987). Drosophila laminin: Characterization and localization. *J. Cell Biol.* **105,** 2383–2391.

Fessler, L. I., Condic, M. L., Nelson, R. E., Fessler, J. H., and Fristrom, J. W. (1993). Site-specific cleavage of basement membrane collagen IV during Drosophila metamorphosis. *Development* **117,** 1061–1069.

Fessler, L. I., and Fessler, J. H. (1974). Protein assembly of procollagen and effects of hydroxylation. *J. Biol. Chem.* **249,** 7637–7646.

Fessler, L. I., Nelson, R. E., and Fessler, J. H. (1994). Drosophila extracellular matrix. *Methods Enzymol.* **245,** in press.

Fogerty, F. J., Fessler, L. I., Bunch, T. A., Yaron, Y., Parker, C. G., Nelson, R. E., Brower, D. L., Gullberg, D., and Fessler, J. H. (1994). Tiggrin, a novel Drosophila extracellular matrix protein that functions as a ligand for Drosophila $\alpha_{PS2}\beta_{PS}$ integrins. *Development* **120,** 1747–1758.

Fristrom, D., Wilcox, M., and Fristrom, J. (1993). The distribution of PS integrins, laminin A and F-actin during key stages in Drosophila wing development. *Development* **117,** 509–523.

Fujita, S. C., Zipursky, S. L., Benzer, S., Ferrus, A., and Shotwell, S. L. (1982). Monoclonal antibodies against the Drosophila nervous system. *Proc. Natl. Acad. Sci. U.S.A.* **79,** 7929–7933.

Garrison, K., MacKrell, A. J., and Fessler, J. H. (1991). Drosophila laminin A chain sequence, interspecies comparison, and domain structure of a major carboxyl portion. *J. Biol. Chem.* **266,** 22899–22904.

Garzino, V., Berenger, H., and Pradel, J. (1989). Expression of laminin and of a laminin-related antigen during early development of Drosophila melanogaster. *Development* **106,** 17–27.

Go, M. J., and Hotta, Y. (1992). An antigen present in the Drosophila central nervous system only during embryonic and metamorphic stages. *J. Neurobiol.* **23,** 890–904.

Gratecos, D., Naidet, C., Astier, M., Thiery, J. P., and Semeriva, M. (1988). Drosophila fibronectin: A protein that shares properties similar to those of its mammalian homologue. *EMBO J.* **7,** 215–223.

Gullberg, D., Fessler, L. I., and Fessler, J. H. (1994). Differentiation, extracellular matrix synthesis, and integrin assembly by Drosophila embryo cells cultured on vitronectin and laminin substrates. *Dev. Dyn.* **199,** 116–128.

Gutzeit, H. O., Eberhardt, W., and Gratwohl, E. (1991). Laminin and basement membrane-associated microfilaments in wild-type and mutant Drosophila ovarian follicles. *J. Cell Sci.* **100,** 781–788.

Henchcliffe, C., Garcia Alonso, L., Tang, J., and Goodman, C. S. (1993). Genetic analysis of Laminin-A reveals diverse functions during morphogenesis. *Development* **118,** 325–337.

Hirano, S., Ui, K., Miyake, T., Uemura, T., and Takeichi, M. (1991). Drosophila PS integrins recognize vertebrate vitronectin and function as cell-substratum adhesion receptors in vitro. *Development* **113,** 1007–1016.

Hortsch, M., and Goodman, C. S. (1991). Cell and substrate adhesion molecules in Drosophila. *Annu. Rev. Cell Biol.* **7,** 505–557.

Kakpakov, V. T., Gvozdev, V. A., Platova, T. P., and Polukarova, L. G. (1969). Embryonic cell lines of Drosophila melanogaster in vitro. *Genetika* **5,** 67–75.

Knibiehler, B., Mirre, C., and Le Parco, Y. (1990). Collagen type IV of Drosophila is stockpiled in the growing oocyte and differentially located during early stages of embryogenesis. *Cell. Differ. Dev.* **30**, 147–157.

Kusche-Gullberg, M., Garrison, K., MacKrell, A. J., Fessler, L. I., and Fessler, J. H. (1992). Laminin A chain: Expression during Drosophila development and genomic sequence. *EMBO J.* **11**, 4519–4527.

Landon, T. M., Sage, B. A., Seeler, B. J., and O'Connor, J. D. (1988). Characterization and partial purification of the Drosophila Kc cell ecdysteroid receptor. *J. Biol. Chem.* **263**, 4693–4697.

Langer-Safer, P. R., Levine, M., and Ward, D. C. (1982). Immunological method for mapping genes on Drosophila polytene chromosomes. *Proc. Natl. Acad. Sci. U.S.A.* **79**, 4381–4385.

Le Parco, Y., Cecchini, J. P., Knibiehler, B., and Mirre, C. (1986a). Characterization and expression of collagen-like genes in Drosophila melanogaster. *J. Biol. Cell* **56**, 217–226.

Le Parco, Y., Knibiehler, B., Cecchini, J. P., and Mirre, C. (1986b). Stage and tissue-specific expression of a collagen gene during Drosophila melanogaster development. *Exp. Cell Res.* **163**, 405–412.

Le Parco, Y., Le Bivic, A., Knibiehler, B., Mirre, C., and Cecchini, J. P. (1989). Dcg1 Alpha-IV collagen chain of Drosophila melanogaster is synthesized during embryonic organogenesis by mesenchymal cells and is deposited in muscle basement membranes. *Insect Biochem.* **19**, 789–802.

Leptin, M., Aebersold, R., and Wilcox, M. (1987). Drosophila position-specific antigens resemble the vertebrate fibronectin-receptor family. *EMBO J.* **6**, 1037–1043.

Lunstrum, G. P., Bachinger, H. P., Fessler, L. I., Duncan, K. G., Nelson, R. E., and Fessler, J. H. (1988). Drosophila basement membrane procollagen IV. I. Protein characterization and distribution. *J. Biol. Chem.* **263**, 18318–18327.

MacKrell, A. J., Blumberg, B., Haynes, S. R., and Fessler, J. H. (1988). The lethal myospheroid gene of Drosophila encodes a membrane protein homologous to vertebrate integrin beta subunits. *Proc. Natl. Acad. Sci. U.S.A.* **85**, 2633–2637.

MacKrell, A. J., Kusche-Gullberg, M., Garrison, K., and Fessler, J. H. (1993). Novel Drosophila laminin A chain reveals structural relationships between laminin subunits. *FASEB J.* **7**, 375–381.

Mahowald, A. (1994). *In "Drosophila melanogaster:* Practical Uses in Cell and Molecular Biology" (L. S. B. Goldstein and E. A. Fyrberg, eds.), New York: Academic Press.

Mirre, C., Cecchini, J. P., Le Parco, Y., and Knibiehler, B. (1988). De novo expression of a type IV collagen gene in Drosophila embryos is restricted to mesodermal derivatives and occurs at germ band shortening. *Development* **102**, 369–376.

Monson, J. M., Natzle, J., Friedman, J., and McCarthy, B. J. (1982). Expression and novel structure of a collagen gene in Drosophila. *Proc. Natl. Acad. Sci. U.S.A.* **79**, 1761–1765.

Montell, D. J., and Goodman, C. S. (1988). Drosophila substrate adhesion molecule: Sequence of laminin B1 chain reveals domains of homology with mouse. *Cell* **53**, 463–473.

Montell, D. J., and Goodman, C. S. (1989). Drosophila laminin: Sequence of B2 subunit and expression of all three subunits during embryogenesis. *J. Cell Biol.* **109**, 2441–2453.

Murray, L. W., and Leipzig, G. V. (1986). Collagen from Drosophila melanogaster larvae and adults is similar to vertebrate type IV collagen. *J. Cell Biol.* **103**, 389a.

Natzle, J. E., Monson, J. M., and McCarthy, B. J. (1982). Cytogenetic location and expression of collagen-like genes in Drosophila. *Nature* **296**, 368–371.

Nelson, R. E., Fessler, L. I., Takagi, Y., Blumberg, B., Olson, P. F., Parker, C. G., Keene, D., and Fessler, J. H. (1994). Peroxidasin: a novel enzyme-matrix protein of Drosophila development. *EMBO J.* **13**, in press.

Noelken, M. E., Wisdom, B., Jr., and Hudson, B. G. (1981). Estimation of the size of collagenous polypeptides by sodium dodecyl sulfate-polyacrylamide gel electrophoresis. *Anal. Biochem.* **110**, 131–136.

Olson, P. F., Fessler, L. I., Nelson, R. E., Sterne, R. E., Campbell, A. G., and Fessler, J. H. (1990). Glutactin, a novel Drosophila basement membrane-related glycoprotein with sequence similarity to serine esterases. *EMBO J.* **9**, 1219–1227.

Rothberg, J. M., Hartley, D. A., Walther, Z., and Artavanis-Tsakonas, S. (1988). Slit: An EGF-homologous locus of D. melanogaster involved in the development of the embryonic central nervous system. *Cell* **55,** 1047–1059.

Rothberg, J. M., Jacobs, J. R., Goodman, C. S., and Artavanis-Tsakonas, S. (1990). Slit: An extracellular protein necessary for development of midline glia and commissural axon pathways contains both EGF and LRR domains. *Genes Dev.* **4,** 2169–2187.

Rothberg, J. M., and Artavanis-Tsakonas, S. (1992). Modularity of the slit protein: Characterization of a conserved carboxy-terminal sequence in secreted proteins and a motif implicated in extracellular protein interactions. *J. Mol. Biol.* **227,** 367–370.

Sang, J. H. (1981). Drosophila cells and cell lines. *In* "Advances in Cell Culture" (K. Maramorosch, ed.), Vol. 1, pp. 125–181. New York: Academic Press.

Schneider, I. (1972). Cell lines derived from late embryonic stages of Drosophila melanogaster. *J. Embryol. Exp. Morphol.* **27,** 353–365.

Seeger, M. A., Haffley, L., and Kaufman, T. C. (1988). Characterization of amalgam: A member of the immunoglobulin superfamily from Drosophila. *Cell* **55,** 589–600.

Tautz, D., and Lehman, R. (1994). *In* "*Drosophila melanogaster:* Practical Uses in Cell and Molecular Biology" (L. S. B. Goldstein and E. A. Fyrberg, eds.). New York: Academic Press.

Tautz, D., and Pfeifle, C. (1989). A non-radioactive in situ hybridization method for the localization of specific RNAs in Drosophila embryos reveals translational control of the segmentation gene hunchback. *Chromosoma* **98,** 81–85.

Volk, T., Fessler, L. I., and Fessler, J. H. (1990). A role for integrin in the formation of sarcomeric cytoarchitecture. *Cell* **63,** 525–536.

Wehrli, M., Di Antonio, A., Fearnley, I. M., Smith, R. J., and Wilcox, M. (1993). Cloning and characterization of α_{PS1}, a novel Drosophila melanogaster integrin. *Mech. Dev.,* **43,** 21–36.

Wilcox, M., Brower, D. L., and Smith, R. J. (1981). A position-specific cell surface antigen in the Drosophila wing imaginal disc. *Cell* **25,** 159–164.

Yee, G. H., and Hynes, R. O. (1993). A novel, tissue-specific integrin subunit, Beta-Nu, expressed in the midgut of Drosophila melanogaster. *Development* **118,** 845–858.

Zavortink, M., Bunch, T. A., and Brower, D. L. (1993). Functional properties of alternatively spliced forms of the Drosophila PS2 integrin alpha subunit. *Cell Adh. Comm.* **1,** 251–264.

Zusman, S., Grinblat, Y., Yee, G., Kafatos, F. C., and Hynes, R. O. (1993). Analyses of PS integrin functions during Drosophila development. *Development* **118,** 737–750.

PART IV

Cytological Methods

The importance of cytological methods is perhaps best measured by inspecting the covers of contemporary leading scientific journals. Of late, a disproportionate number seem to show specifically labeled embryos, cells, or subcellular compartments. The ability to visualize such patterns directly, without the necessity of reconstructing serial sections, is revolutionizing biology. One gleans much information from such images, and they in turn prompt biochemical, physiological, or developmental extensions of the work that ultimately permit the role of the molecules in question to be evaluated precisely. Noteworthy also is the rapidity with which microscopy, photographic acquisition and display, and image enhancement is proceeding. With each passing month, it seems, comes an advance that facilitates extracting more information from micrographs. Finally, panels of monoclonal antibodies recognizing particular proteins or protein classes are proliferating rapidly, as are reporter constructs for particular gene activities. In the face of this sort of exposure, can *Drosophila* retain any of its remaining secrets much longer?

Pardue initiates the section with an update on the condition of a familiar friend, the polytene chromosome (Chapter 19). This chapter is comprehensive, describing the multitude of applications and special tricks, and also very forward thinking, underscoring the importance of polytene chromosomes in mapping and sequencing the fly genome. It appears that these special chromosomes will continue to play a prominent role in future research.

Andrew and Scott follow in Chapter 20 with a related article detailing how one uses polytene chromosomes to map chromatin domains and chromosomal protein distributions. Originally, this analysis could be carried out only for abundant chromosomal proteins (histones, RNA polymerases, topoisomerases, etc.), but recently it has been extended to regulatory proteins such as *Polycomb* and *twist*. Andrew and Scott provide instructions for identifying binding sites and for minimizing problems arising from cross reactivity.

Gatti *et al.* stay with the chromosome theme, describing the preparation and analysis of mitotic chromosomes (Chapter 21). These methods are indispensable, both for examining mutants having putative mitotic and cytokinesis defects and for analysis of heterochromatin. Gatti *et al.* also discuss larval neuroblast chromosome preparation, various banding techniques, and *in situ* hybridization.

Gunawardena and Rykowski extend interphase chromosome analysis through visualization of sequences within the nuclear context after *in situ* hybridization of fluorescently tagged probes (Chapter 22). This approach allows the positions of genes to be related to nuclear compartments and landmarks; hence, it will be extremely useful for further revealing how a functioning genome is organized.

McDonald follows with a description of electron microscopy methods designed for *Drosophila* (Chapter 23). He points out the importance of good tissue fixation and offers a novel alternative to standard chemical fixation, namely high pressure freezing. What is particularly valuable here is that the author has carried out careful comparisons of alternate methods and discusses the pros and cons of each. He extends his treatise to immuno-EM methods, comparing Lowicryl, LR White, and LR Gold resins.

Patel follows with a comprehensive guide to immunocytochemical methods for imaging neuronal and other cell types in embryo and larvae

(Chapter 24). Basic whole-mount antibody labeling methods, and a catalog of antibodies, are presented. The technique of gently cracking late embryos and larvae using sonication is particularly valuable, as are methods for enhancing HRP labeling, adding various colors, and double labeling. Patel concludes with tips for photography and image handling and a trouble-shooting guide.

Theurkauf outlines methods for immunofluorescent analysis of the cytoskeleton during oogenesis and early development (Chapter 25). The detailed protocols cover egg chamber preparation and fixation, preparation of devitellinized early embryos and their fixation, whole-mount labeling of particular cytoskeletal components, and how to prevent artifacts due to anoxia.

Kiehart *et al.* describe high-resolution microscopy for monitoring cell movements (Chapter 26). This chapter comprehensively covers how one improves resolving power and contrast, how one handles embryos so as to avoid artifacts, and how one sets up a state of the art imaging facility. Excellent examples of how to process images so as to minimize degradation are presented here.

Girdham and O'Farrell describe using photoactivable, or "caged" compounds, for studies of cell lineage (Chapter 27). For examination of marked cells in real time, the methods outlined here are very versatile, and avoid some of the artifacts inherent to transplantation methods. Also worth noting is that the nuclear-localized marker (a dextran backbone conjugated to a nuclear localization signal) facilitates the identification of particular types of cells.

Verheyen and Cooley discuss a repertoire of methods for looking at oogenesis (Chapter 28). The approaches described here have begun to break down oogenic transport into its molecular components and are expected to reveal many facets of cellular localization and transport. Relevant female sterile mutants, as well as a variety of methods for localizing particular components, are described.

CHAPTER 19

Looking at Polytene Chromosomes

Mary-Lou Pardue

Department of Biology
Massachusetts Institute of Technology
Cambridge, Massachusetts 02139

I. The Beautiful Contig Maps of *Drosophila*

One goal of genome projects is construction of contig maps that will allow any DNA or RNA sequence from an organism to be mapped directly to a specific site on its genome. For most organisms, contig maps are being made by arraying libraries of cloned DNA fragments to produce a set of overlapping clones covering each of the chromosomes. For *Drosophila*, contig maps come ready-made in the form of polytene chromosomes.

The *Drosophila* contig maps are aesthetically more pleasing than DNA clone arrays. They also have a number of practical advantages. Each larva has its own contig map and any chromosomal rearrangements in that individual can be detected directly. The DNA of these *Drosophila* contigs is *in situ* and therefore still associated with protein and RNA, making it possible to map chromosomal proteins and RNAs directly on the genome. Because polytene nuclei are interphase nuclei and active in both transcription and DNA replication, these nuclei offer a unique opportunity to visualize the genomic distribution of many interesting chromosomal proteins and RNAs (such as snRNAs involved in transcript processing).

The resolution of mapping on *Drosophila* polytene chromosomes is comparable to the resolution of mapping on an arrayed DNA clone library. On the polytene chromosome, the resolution depends on the morphology of the regions of interest. If a sequence is associated with a faint band or interband, its position can be estimated within perhaps 5–10 kb. Very thick bands may contain up to 200 kb, so sequences associated with them will be less precisely positioned. When sequences are mapped on a library of arrayed DNA clones, the resolution will depend on the sizes of the clone inserts, which generally fall within the ranges mentioned above. Whether the initial mapping is done with polytene chromosomes or cloned contigs, precise mapping requires study of a cloned DNA fragment containing the sequence. Here, the clone arrays have the advantage that the needed clone is in hand. However, the community of *Drosophila* workers is rapidly eliminating this advantage. A large and growing fraction of the genome is now cloned and available. Once a site has been identified by polytene analysis, finding a clone that is at least nearby should not be difficult.

Because polytene chromosomes make such beautiful and useful research material, they are the subject of a rich body of literature. This chapter will make no attempt to review this work. Instead, it will concentrate on methods and problems with which the author has experience. The largest part of the literature on *Drosophila*, as well as most of the experience in my laboratory, concerns *Drosophila melanogaster* and this is the species implied throughout the text unless otherwise specified. One of the advantages of *Drosophila* as an experimental system is the number of related species that are being studied. The comparisons that can be made between species give important insights into many questions and reference to other species will be made when possible.

II. Sources of Polytene Chromosomes

The most useful polytene chromosomes are found in the salivary gland of the late third instar larvae. These chromosomes undergo more rounds of replication than chromosomes in other tissues and will contain 1024–2048 chromatids in some nuclei. (The largest nuclei are at the distal end of the gland.) Several other tissues can yield analyzable chromosomes. These include the prothoracic gland and hindgut, each with 256–512 chromatids, the midgut with 512–1024 chromatids (Hochstrasser, 1987), and the larval fat body with 256 chromatids (Richards, 1980).

The number of chromatids in a polytene nucleus is affected by the growth conditions of the larvae. It is also affected by the genotype of the fly and genetics can sometimes be used to advantage for studying the chromosomes. Bridges used a stock carrying the *giant* (*g*t) mutation for his polytene maps, picking the giant larvae and analyzing the nuclei that had undergone the extra round of replication (Bridges, 1935). Several years ago, when probes for *in situ* hybridization were of relatively low specific activity, we also used this trick. Because *giant* alleles have low penetrance, we used heterozygotes of the genotype gt^{xII}/gt^{I}. Some of the larvae do have truly giant chromosomes but probes are now so efficient that the giant chromosomes are unnecessary. In addition, the largest chromosomes have slightly different morphology and some experience is required to match them to the photographic maps.

Mutants have also been useful for studies of polytene chromosomes in other tissues. In the nurse cells of the ovary, chromatids are associated during early rounds of replication but lose this organization later. Certain female sterile mutations lead to polytenization of chromosomes in the nurse cells, presenting an opportunity to analyze chromosomes of germline cells (Storto and King, 1987; Heino, 1989). Larvae carrying the *DTS-3* mutation have been used to study prothoracic gland chromosomes (Holden and Ashburner, 1978).

III. Structure of Polytene Chromosomes

A. Chromatin Structure

Polytene chromosomes are larger than metaphase chromosomes both because of the multiple chromatids and because of lower compaction. In mitotic metaphase chromosomes, packing reduces the length of the DNA more than 10,000-

fold. In salivary gland chromosomes, the packing of a chromatid averages 57-fold but there are significant regional variations. In region 87E1-2, the packing ratio is 180, whereas in region 87D5-10 the ratio is 23 (Spierer *et al.*, 1983). The variations in packing reflect the chromatid structure of a given region and, because of the precise alignment of chromatids, the packing differences form bands and interbands that extend across the width of the chromosomes. Bands vary in width and compaction. Most are continuous across the width of the chromosomes but some appear to be formed of a series of dots or dashes. The character of each band is reproduced consistently from nucleus to nucleus and the pattern of bands along each chromosome identifies sites on the chromosomes easily and precisely. The characteristics of bands tend to be conserved between species also. Although rearrangements have altered the order of genes on the chromosomes, the banding pattern within the rearranged regions shows significant conservation between different species.

The structure of polytene bands has been analyzed extensively with both light microscopy and electron microscopy, but the molecular basis of this structure is still unclear. The development of techniques for inserting defined pieces of DNA into *Drosophila* chromosomes (Spradling, 1986) has opened possibilities for direct tests of the sequences involved in forming these structures. Such experiments have shown that as little as 5 kb of inserted DNA can form a new band (Semeshin *et al.*, 1989).

B. Synapsis and Asynapsis

In most polytene chromosomes, a tight association is formed, not only between the chromatids of one chromosome, but also between homologous chromosomes. What appears to be one chromosome is two tightly synapsed homologs in most cases. Major exceptions to this rule occur when an animal is heterozygous for chromosomal rearrangements that make pairing difficult; however, asynapsis can also be seen at times in regions that have no apparent structural heterozygosity. There is no evidence that such asynapsis is produced by squashing. Instead, it seems to indicate regions where pairing never took place. The two homologs separate cleanly in asynapsed regions without evidence that there is any intermingling of their chromatids. This observation leads to a picture of the synapsed homologs as tightly apposed and perhaps wound around each other, as in Bridges' 1935 map.

In situ hybridization analyses of animals known to be heterozygous for mobile elements at specific sites suggest that the organization of chromatids in synapsed chromosomes may be more intimate than might be supposed from looking at the asynapsed regions (Pardue and Dawid, 1981). When the heterozygous elements are in asynapsed regions, hybridization is to only one of the two chromosomes, as expected from other studies. Unexpectedly, in most of the synapsed regions, hybridization is detected across the width of the chromosome, although the element is present in only half of the chromatids. Only rarely is the probe found to be limited to part of the synapsed chromosomes and this is usually near a region of asynapsis, as though the homologs were beginning to separate. These observations suggest that the organization of chromatids may be complex. They also suggest an important caution for the interpretation of *in situ* hybridization experiments. A hybrid detected across a region of synapsed chromosomes is not necessarily evidence that the target sequence is present in both homologs.

C. Ectopic Pairing

Chromosome synapsis occurs between homologous regions. A second type of association is frequently seen between polytene regions that appear, at least superficially, to be nonhomologous. This association, ectopic pairing, usually occurs between single chromosome bands. In spite of the small regions involved, the associations are strong enough to survive the pressure used in making a chromosome squash and can make it difficult to achieve optimal spreading for analysis. Regions with a strong tendency for ectopic pairing, such as the base of 2L, are often difficult to study. The frequency of ectopic pairing is a site-specific characteristic (Zhimulev et al., 1982). Sites may have more than one possible pairing partner and sometimes several are associated in a cluster. Whether these ectopic associations are based on sequence homologies between the regions is still an open question. Although some experiments have detected association of regions with shared sequence, none have shown that the homology is either necessary or sufficient for ectopic pairing.

D. The Chromocenter and Other Heterochromatin

The chromocenter is a diffuse region composed of the fused pericentric regions of all of the chromosomes. Heitz (1934) concluded that this region contained the heterochromatin that surrounds the centromeres in mitotic chromosomes and, on the basis of differential staining characteristics, divided the chromocenter into α- and β-heterochromatin. The α-heterochromatin makes up a very small part of the chromocenter; most of the chromocenter is β-heterochromatin. More recent work has shown that α-heterochromatin contains the highly repeated simple sequences of classical satellite DNA and that these sequences do not take part in the multiple rounds of replication in polytene nuclei (Gall et al., 1971). β-Heterochromatin is a more complex mixture of sequences. The sequence structure of β-heterochromatin is far from understood, but this region has been shown to contain sequences cross-hybridizing with many mobile elements, as well as sequences found only in heterochromatin. The sequences of β-heterochromatin undergo at least some polytenization but the exact amount of replication has not been measured. The level of replication may be under genetic control. For instance, we see less hybridization of certain probes to β-heterochromatin in gt^{x11}/gt^1 larvae than in larvae of either parental stock. The diffuse structure of the chromocenter is in striking contrast to the elaborate structure of the banded regions that make up the euchromatic portions of the chromosome arms and allows much less precise spatial analysis (Young *et al.,* 1983).

In his studies of *Drosophila* chromosomes, Muller (1938) concluded that telomere regions had a "chromocentral" nature, based on shared characteristics. These characteristics may result from shared sequences; at least one family of sequences is found at both telomeres and the chromocenter but never in euchromatic regions (Valgeirsdottir *et al.,* 1990). The tapered shapes of many telomeres in polytene chromosomes have suggested that telomeres might, like the chromocenter, be underreplicated in polytene DNA. This suggestion is supported by experiments showing that a sequence inserted in a telomere region is underrepresented in polytene DNA (Karpen and Spradling, 1992). On the other hand, we have been able to detect the terminal 3 kb of DNA on polytene chromosomes by *in situ* hybridization (Biessmann *et al.,* 1990), so the underrepresentation does not completely remove the terminal sequences.

E. Constrictions

Local underreplication is not entirely limited to chromocentral and telomeric heterochromatin. Constrictions, or "weak spots," are also seen at several specific places in euchromatin. Their appearance suggests that they are sites of local underreplication and this appears to be so, at least for the constrictions at 89DE (Spierer and Spierer, 1984) and 39DE (Lifschytz, 1983). Outside of the constrictions, bands and interbands generally have the same level of polyteny and differ in the way that the DNA is packed, rather than in the number of DNA strands (Spierer and Spierer, 1984). Although not an invariant feature, these weak spots can occur often enough to be useful landmarks for reading the polytene chromosomes (e.g., the weak spots at 11A and 89DE of *D. melanogaster*).

F. Puffs, Transcription, and Replication

In contrast to most nuclei with condensed chromosomes, polytene nuclei are interphase nuclei. DNA replication and RNA transcription are active in these cells. Most of these chromosomal activities make no detectable change in the banded structure of the chromosomes; however, radioactive precursors for both DNA and RNA are incorporated at many sites along the chromosomes and RNA isolated from salivary glands contains transcripts from these sites (Bonner and Pardue, 1977a). Some transcriptional activity does produce dramatic changes in polytene chromosome morphology; many sites of heavy transcription are associated with a localized loosening of chromatid packing that causes the region to form a puff.

Puffs are induced and regress in a complex developmental sequence. Ashburner (1972) has published an analysis of puffing patterns in *D. melanogaster* that makes it possible to stage third instar larvae quite precisely. Puffs can also be induced by agents such as heat shock or the solutions used for salivary gland dissection (Ashburner, 1972). All of the puffs that have been studied are sites at which transcription has either been induced or much increased over the non-puffed state (Bonner and Pardue, 1977b). A high level of inducible activity may be what distinguishes puffs from other sites of transcription in polytene nuclei. *In situ* hybridization experiments show that only a portion of the puffed region is transcribed. The transcribed segment may be in the center or on one side of the puffed region but the position seems to be always the same for a puff at a specific site (Pardue *et al.*, 1977). These observations suggest that the boundaries of each puff are determined by flanking chromatin rather than by the transcription itself.

One aspect of puffing should be considered in analyzing *in situ* hybrids: puffed sequences hybridize more efficiently than do the same sequences in the nonpuffed conformation. There are several possible explanations for the efficient hybridization and all may contribute to the effect. Less compact chromatids may adhere more firmly to the slide, minimizing DNA loss during the various treatments. Chromatids may also denature more fully in the extended state. They may be more accessible to the hybridization probe. If the region of interest falls within a puff, it can be detected more easily on chromosomes that have the puff but, because chromatids are less tightly aligned in puffed regions, the sequence will be less precisely defined. For quantitative studies, hybridization efficiencies must be taken into account when comparing puffed and nonpuffed regions.

G. Nucleoli

Polytene nuclei actively transcribe ribosomal RNA (rRNA) and each cell has a large nucleolus. The nucleolus is attached to the chromocenter by a thin strand of chromatin but this strand is frequently broken when a preparation is made. The genes encoding rRNA (the rDNA) are clustered in pericentric heterochromatin on the X and Y chromosomes in *D. melanogaster* and offer a variation on the theme of underreplication in polytene chromosomes. Only one of the two clusters in each nucleus is polytenized and there is a dominance effect in the choice of the cluster (Endow, 1983). *In situ* hybridization detects rDNA within the nucleolus and on the strand leading toward the chromocenter. The nucleolus may be considered an exceptionally large and active puff with transcription and processing complexes clustered around the DNA. In Dipterans, in which the rDNA clusters are located in euchromatic regions, the nucleolus has a more typical puff shape. It seems likely that the isolation of the *D. melanogaster* nucleolus is the result of the underrepresentations of the heterochromatic sequences flanking the rDNA, leaving the rDNA attached to a thin strand.

H. Y Chromosomes

The Y chromosome does not form a banded chromosome in polytene nuclei. Any Y chromosomal sequences present in polytene nuclei must be mingled in the chromocenter. Since chromocenters in Y-containing nuclei show little, if any, size differences from those in Y-negative nuclei, Y sequences appear to be very underrepresented in polytene nuclei. The genes for rRNA can be replicated in salivary gland cells (Endow and Glover, 1979) but they may well be the only Y chromosomal sequences which polytenize.

I. X Chromosomes in Male Nuclei

In general, when homologs are asynapsed, the two partners are equal in size and half the width of synapsed regions. The unpaired X in cells of males is an exception. Although it contains only half as many chomatids as the paired X chromosomes of equivalent female chromosomes, the male X is nearly as wide as the synapsed autosomes in the same nucleus (or the synapsed X chromosomes in female nuclei). Giemsa stain, which binds rather nonspecifically to nucleoprotein, stains the male X less intensely than the autosomes, indicating that the chromatin of the male X is less compact.

The less compact chromatin of the male X chromosome may reflect the hyperactivity of the X-linked genes, which compensates for the absence of the second X chromosome (Belote and Lucchesi, 1980). This looser chromatin structure appears to increase the efficiency of hybrid detection on the male X. We have compared the relative amounts of hybridization of a P-element insert at several different chromosomal sites (Pardue *et al.*, 1987). Hybridization efficiencies over autosomal sites in both sexes and over X chromosome sites in females were not significantly different. On the single X chromosome in the male, sequences were detected 1.5 times as efficiently as the same sequences on synapsed chromosomes.

IV. Maps of Polytene Chromosomes

The most complete polytene chromosome maps are the *D. melanogaster* maps made by Bridges and Bridges (Bridges, 1935, 1938, 1941a,b, 1942; Bridges and Bridges, 1939). These maps are so detailed that electron microscopic analyses have found only minor differences. In view of evidence that inserted DNA segments as small as 5 kb can produce a new band (Semeshin *et al.*, 1989), it seems likely that many of the differences may have been due to genetic differences in the chromosomes studied. Lefevre (1976) has constructed a photographic version of the Bridges 1935 map with a summary of landmark sites and difficult regions. The photographic version is a good place to begin in identifying a site. After the site has been placed on this map, the appropriate revised map can be used to obtain finer mapping. The Ashburner (1972) maps of developmental puffing patterns also help in identifying regions on chromosomes from different stages.

Polytene maps have been published for a very large number of other *Drosophila* species (Ashburner, 1989). These maps vary in the degree of detail and sometimes in the accuracy. For some species, several maps have been made. When attempting to map a sequence in an unfamiliar species, it is important to know which map is considered the general standard.

V. Equipment for Salivary Gland Preparations

A. Subbed Slides

The surface of these slides gives better retention of cytological preparations. Slides are washed thoroughly, dipped into subbing solution, and allowed to air dry. Subbing solution is an aqueous solution of 0.1% gelatin and 0.01% chrome alum [chromium potassium sulfate $CrK(SO_4)_2 \cdot 12H_2O$]. To make the solution, dissolve gelatin in water ($\sim65°C$), allow the solution to cool, and add the chrome alum. Subbing solution can be stored for long periods at 4°C. Subbed slides can be kept indefinitely in a dry, dust-free container.

B. Siliconized Coverslips

Siliconization prevents adherence of cytological material to the coverslip. The treatment also causes solution to form a rounded droplet when placed on the coverslip so that cytological material can be incubated in a small amount of solution before squashing. Coverslips can be treated with any of the procedures used to siliconize laboratory glassware, dried, and stored. Immediately before use, coverslips should be freed of dust with microscope lens tissue.

C. Moist Chambers

When cytological preparations are incubated with a small amount of liquid, moist chambers are used to prevent evaporation. Small plastic sandwich boxes with tight lids work well. The bottom of the box is covered with a thin layer of incubation medium and slides are supported above this liquid. The liquid in the bottom of the moist chamber must have the same salt concentration as the solution over the cytological preparation to prevent distillation and subsequent changes in concentration of the solution on the chromosomes.

======   # VI. Protocols for Squash Preparations of Salivary Glands

There are many good protocols for making salivary gland preparations. The one to be used should be chosen with care because the preparation is the most important part of the experiment and will determine the quality of the final result. The goal of the experiment also makes a difference in the protocol chosen. This is especially important if the chromosomes are to be used for *in situ* hybridization. Many of the steps taken to maintain morphology are exactly the opposite of those that increase hybridization efficiency and the tradeoffs must be considered. For polytene chromosomes, large probes (such as lambda phage clones) will give a strong signal even with very modest hybridization efficiency, so, for these probes, it is possible to choose the simplest technique that maintains good morphology. On the other hand, it is often important to map very small probes or to look for small segments of cross-hybridizing sequence elsewhere in the genome. In those cases, care at each step of the protocol can significantly increase the sensitivity of the experiment. The protocols described here are those that give us the maximum sensitivity combined with reasonable morphology.

A. Larvae for Salivary Gland Chromosomes

The quality of chromosome spreads depends greatly on healthy larvae. Some strains produce the best chromosomes when grown at 18°C. Others do better at 22–25°C. Culture bottles should not be overcrowded and can be fed with extra yeast the day before making squashes. Large third instar larvae should be selected. Unless there is a special reason for wanting the male chromosome set, females should be chosen because the chromosomes are usually better. To avoid the many puffs that develop late in the larval period, choose larvae that have not everted their spiracles.

B. Isolation of Salivary Glands

Rinse the larvae in PBS (140 mM NaCl, 50 mM phosphate buffer, pH 7.4). Dissect in PBS or 45% acetic acid. Dissection in PBS should be done quickly because incubation in saline can induce some chromosome puffs. The ducts of the paired salivary glands join just behind the head segment of the larva. With two needles or a pair of watchmaker forceps, pinch off the anterior end of the larva just above the junction of the ducts. Push the glands out through the opening. The glands are clear or slightly milky and somewhat cucumber-shaped. They frequently have pieces of cream-colored fat body attached. Tease off as much fat body as possible without abusing the glands. Glands are very sensitive and it is better to err on the side of leaving fat body than to poke the gland cells. The next steps in the chromosome preparation depend on the desired use. Protocols for making preparations for *in situ* hybridization and also for making orcein-stained preparations are given below. The orcein-stained preparations are used for analyzing chromosome aberrations, although we frequently find that the *in situ* protocol, followed by Giemsa staining, can be equally useful.

C. Chromosomes for *in Situ* Hybridization

Transfer one or two salivary glands from the dissection medium (above) to ~20 μl of 45% acetic acid on a siliconized 18-mm^2 coverslip, No. 1 thickness. (Use a small coverslip so that the chromosomes will stay in a restricted area during the squash.) It is important that no PBS be carried along with the glands since dilution of the fixative leads to poor morphology. After the glands have fixed for 2–5 min, lower a subbed slide face down onto the coverslip. It is a good idea to be consistent in the place on the slide at which the glands are squashed. In later steps, small drops of liquid will be placed over the chromosomes. In the proper light, it is usually possible to see the squash but it helps to have a good idea where it will be. To save autoradiographic emulsion, it is good to have the squashes ~ $\frac{1}{2}$ inch from the end of the slide so that the slide can be dipped in a small volume.

For species other than *D. melanogaster,* chromosomes are sometimes improved by either longer fixation in 45% acetic acid or by fixing in freshly made ethanol:55% acetic acid (3:1) for a few minutes before moving to the coverslip to fix in 45% acetic acid.

After the slide is lowered onto the fixed glands, turn it over and tap lightly on the coverslip with a pointed object, such as the tip of a dried-out ball point pent. Tap directly over the glands, moving the coverslip back and forth slightly. There should be just enough liquid to allow this slippage without shearing the chromosomes; this step is intended to force the chromosomes out of the cells and to spread the chromosome arms. The preparation should be checked periodically with a phase microscope. Preparations usually end in two classes: (1) a few beautifully spread nuclei and the rest still very tangled or (2) a few beautifully spread nuclei and the rest extremely stretched. The happy medium seldom appears but actually the second class is more useful than the happy medium. The few beautifully spread nuclei can be used to decide where the probe maps and the overly stretched nuclei can be used to do a high-resolution mapping. When the chromosomes are well spread, place the slide on a paper towel with the coverslip side down and press hard with the thumb directly over the coverslip. At this step the coverslip must not slip sideways because the pressure will cause shearing. This step does not improve the spreading of the chromosomes; instead, the intent is to flatten the chromosomes tightly against the slide and thus help with the retention of good morphology.

Immediately after the final thumb press, lower the slide into liquid nitrogen to freeze the preparation. Freezing takes ~30 sec but slides can be left here for several minutes while other squashes are made. The coverslip is then flipped off by sliding a razor blade under one corner and the slide is plunged into 95%

ethanol quickly so that it does not thaw during the process. After three 10-min washes in 95% ethanol, slides are air-dried. They can now be held for long periods, although hybridization efficiency decreases slowly. The best preparations, when dry, will show a clear banding pattern and will be flat enough to refract very little light.

D. Chromosomes for Orcein Staining

Transfer glands from the dissection medium into freshly made ethanol:acetic acid (3:1) for 10 min and then move to a drop (~20 μl) of lactic–acetic orcein on a siliconized coverslip. (Lactic–acetic orcein is made by mixing 2 g of powdered orcein into 33 ml each of 85% lactic acid, glacial acetic acid, and distilled water. The mixture is heated and filtered while still hot.) Care should be taken not to carry over any of the ethanol:acetic acid. Chromosomes should stain for 5 min before being squashed. The procedure for squashing the glands is like that described for the *in situ* hybridization preparations but slides used for orcein preparations should not be subbed. Preparations are checked for proper spreading with phase microscopy because staining is faint at first. When the chromosome spread is satisfactory, seal the coverslip edges with nail polish. Staining will continue and the chromosomes will be easier to analyze the next day. Preparations may be stored frozen.

VII. Nucleic Acid Probes for *in Situ* Hybridization

A. Choice of Label

Either radioactive or nonradioactive probes can be used. Tritium is by far the best radioactive isotope because the very low-energy beta particle (0.0181 MeV) travels less than 1 μm in autoradiographic emulsion. Two other radioisotopes, ^{35}S and ^{125}I, give higher specific activities but also decreased resolution and are not advised for polytene chromosomes. The nonradioactive probes that are most used are biotin and digoxygenin.

Radioactive and nonradioactive labels have different advantages and disadvantages, so use depends on the goal of the experiment. The two major advantages of nonradioactive probes are the speed with which the hybrid is detected and the fact that these probes can be used where facilities for autoradiography are not available. The second advantage is probably the most important since little time is saved in hybrid detection. Probes large enough to be detected with nonradioactive probes can be detected in 1–3 days when labeled with ^{3}H.

Nonradioactive probes are most useful for mapping large cloned fragments when time is important because the next step in the experiment must wait for the mapping results. Probes labeled with ^{3}H are much more sensitive than nonradioactive probes and can be used to detect DNA sequences at least an order of magnitude smaller than can be detected with nonradioactive probes.

We have been able to detect sequences as short as 40 bp on polytene chromosomes with exposures of 2–4 weeks. A second major advantage of ^{3}H-labeled probes is that the amount of hybrid detected can be quantitated by grain counts. If amounts of probe and also autoradiographic exposures are carefully chosen, the measurements are approximately linear over a reasonably wide range. Thus, radioactive probes yield richer information than nonradioactive ones, not only because radioactive probes can detect small amounts of sequence, but also because they allow quantitative conclusions.

B. Type of Probe

For hybridization to DNA in polytene chromosomes, both RNA and DNA probes, double stranded and single stranded, can be used. There has been little attempt to systematically compare the different types of probes. We find that double-stranded DNA fragments, labeled by random-primed reaction or nick-translation (Pardue, 1986; Feinberg and Vogelstein, 1984) are efficient and easy to prepare. We usually cut the probe fragment out of the cloned DNA and gel-purify it before labeling. Under the high-stringency conditions that we use for ^{3}H-labeled probes, we have never detected hybridization of pBR322, lambda, or M13 vectors to *D. melanogaster* chromosomes, so removal of the vector is not necessary but does concentrate the label in sequences of interest. We have not tested for vector hybridization under the less stringent conditions that are used for nonradioactive probes.

C. Probe Concentration and Specific Radioactivity

The specific radioactivity of probes transcribed by RNA polymerases is limited only by the specific activity of the nucleotide precursors. To a first approximation, this is also true of random-primed DNA synthesis. The specific activity of DNA labeled by nick-translation depends on both the nucleotide precursors and the extent of replacement of unlabeled DNA. For targets on polytene chromosomes, we find no need to use the maximum activity. DNA probes labeled with [^{3}H]TTP (100 Ci/mmole) and/or [^{3}H]dATP (67 Ci/mmole) and RNA probes labeled only with [^{3}H]UTP (50 Ci/mmole) give very effective probes, unless the base composition of the probe is very uneven.

Typically, nonradioactive probes are biotinylated DNA made by nick-translation reactions with limiting TTP and trace amounts of Bio-11-dUTP.

The amount of hybrid detected depends both on the specific radioactivity of the probe and on how many of the target sequences have bound the probe. Therefore, it is most effective to use the probe at a concentration that will nearly saturate the target DNA. Higher concentrations will only contribute to the background without improving the signal. We have done only a little study of saturation levels for various probes on polytene chromosomes. For a 1.7-kb RNA probe from single copy DNA, 67 ng/20 μl was clearly above saturation. The 1.7-kb probe, at 1.5×10^7 cpm/μg, gave an extremely heavy labeling after a 3-day autoradiographic exposure. Small plasmids (~5 kb) carrying 1 to 2 kb inserts of single-copy DNA were at saturating concentrations between 15 and 45 ng/20 μl, probably near 20 ng/20 μl. These probes, at 3 $\times$ 10^6 cpm/μg, gave strong signals in 3 days. Bacteriophage clones with 10- to 20-kb inserts are usually used at 20–50 ng/20μl. When nick-translated to 3 $\times$ 10^6 cpm/μg, these larger probes are easily detected in 1–3 days. Nonradioactive probes have not been studied quantitatively. They are used at 15–20 ng/20 μl.

We frequently use less-than-saturating amounts of probe, both to be economical and to make it easy to obtain a series of increasing autoradiographic exposures. For quantitative comparisons, it is necessary to have silver grains over each site at a density that allows each grain to be counted. On the other hand, very short exposures with only a few grains give the best picture of the chromosome structure in the region of the hybrid. Very long exposures allow detection of secondary sites that share some sequence with the primary site. Different exposures give a more complete analysis.

VIII. Protocols for *in Situ* Hybridization to Polytene Chromosomes

A. Pretreatments

1. Increasing Adherence to the Slide

We have found that a heat treatment followed by dehydration significantly reduces the loss of morphology that results when part of a chromosome does not tightly adhere to the slide. This treatment is accomplished by incubating slides in 2$\times$ SSC (1$\times$SSC: 0.15 M NaCl, 0.015 M sodium citrate, pH 7.0) at 70°C for 30 min. Preparations are then dehydrated by incubation in ethanols: 2$\times$ 10 min in 70% and 5 min in 95%. Slides are air dried. This dehydration is repeated several times during the remaining steps to maintain adherence to the slide.

2. RNase Treatment

Endogenous RNA is digested by RNase (100 g/ml in 2$\times$ SSC) at room temperature for 2 hr. A drop (~50 μl) of solution is placed over the preparation and covered with a coverslip. Preparations are placed in a moist chamber to prevent evaporation. After the incubation, coverslips are rinsed off and slides are washed 3$\times$ 5 min in 2$\times$ SSC. They are then dehydrated through ethanol (2$\times$ 10 min in 70% and 5 min in 95%) and air dried. This digestion of endogenous RNA is optional but tests showed that it significantly improved the hybridization of ribosomal RNA, presumably because the nucleolar environment contains endogenous RNA competing with the probe (Pardue, 1970). Although the effect of RNase treatment on other sequences has not been tested, it seems worthwhile to include it if there is any reason to suspect that there might be a concentration of transcripts near the target, as in polytene chromosomes.

3. Acetylation

Slides are suspended in a dish over a magnetic stirring bar and covered with 0.1 M triethanolamine–HCl, pH 8.0. The solution is stirred vigorously and acetic anhydride is added to 5 ml/liter. Stirring is stopped when the acetic anhydride has dispersed and the slides are left for 10 min. They are then dehydrated through ethanols and air dried. This treatment is a most effective way to reduce nonspecific binding of negatively charged nucleic acid probes.

4. Denaturation

Slides are placed in 0.07 M NaOH (pH 12.5) for exactly 3 min, dehydrated through ethanols (3× 10 min in 70%, 2× 5 min in 95%), and air dried. All of the treatments that denature purified DNA also denature DNA in cytological preparations but alkali denaturation gives the best combination of morphological preservation and hybridization efficiency.

B. Hybridization

Hybridization can be performed either in aqueous salt solution at high temperature or in formamide solutions at lower temperatures. Most studies of parameters affecting *in situ* hybridization have shown that conditions of ionic strength, temperature, probe concentration, and hybridization can be chosen as they would for hybridization to DNA on nitrocellulose filters. The one exception that we have found is that we are not able to achieve the very highest stringency with *in situ* preparations; some very similar sequences that can be distinguished in filter hybridization cannot be distinguished *in situ*.

1. Probe Preparation

To reduce nonspecific binding of the probe, sheared *Escherichia coli* DNA is added to DNA probes and *E. coli* rRNA is added to RNA probes. In both cases, the concentration is ~4 μg/20 μl. If the probe is double stranded, it is dissolved in a small volume of water, denatured at 85°C for 5 min, and chilled. When the probe has cooled, concentrated buffer (and formamide, if desired) is added to the appropriate concentration and volume. Dextran sulfate can be added to the hybridization mix but we have seen no advantage to its use in our experiments.

2. Hybridization in Aqueous Salt Solution

This is carried out typically in 2× TNS (1× TNS is 0.15 M NaCl, 10 mM Tris, pH 6.8) at 67°C (relatively stringent conditions for DNA of average base composition). The probe (15–20 μl) is placed over the chromosomes and covered with a coverslip (18 mm^2). Preparations are placed over a reservoir of hybridization buffer in a moist chamber. Typical hybridization times are 12–14 hr at 67°C.

3. Hybridization in Formamide Solution

This is carried out typically in 40% formamide in 4× SSC at 40°C. Probe (5 μl) is placed over the chromosomes and covered with a coverslip. To save the expense of filling a moist chamber with formamide buffer, coverslips are sealed. This can be done with rubber cement spread from a syringe but a simple way is to wrap a strip of parafilm around the slide just over the coverslip. The parafilm should be pressed tightly to the slide around the edges of the coverslip without pressing on the coverslip itself. Typical incubations are 12–14 hr at 40°C.

C. Posthybridization Washes

Following hybridization, coverslips and remaining probe are rinsed off by dipping the slide into a beaker of 2× SSC. Nonspecifically bound DNA probes are removed by 3× 10-min washes in 2× SSC at a temperature 5°C below the hybridization temperature. These washes can be made more stringent by raising the temperature or lowering the salt concentration of the wash. Nonspecifically bound RNA probes are removed by treatment with RNase (20 μg/ml in 2× SSC) at 37°C for 1 hr, followed by 2× 10 min in 2× SSC.

D. Detection of Radioactive Probes

After the removal of nonspecifically bound probe, preparations are dehydrated by ethanol (2× 10 min 70% and 5 min 95%) and air dried. Slides are dipped in autoradiographic emulsion (Kodak NTB-2 diluted 1:1 with H$_2$O) and stored in light-tight boxes at 4°C, away from all other radioactivity, volatile chemicals, and moisture that can destroy the latent image in the emulsion. (Note: We handle autoradiographic emulsion under a safelight; Kodak Wratten series II or OA filters work well. Do not direct the safelight at the emulsion or slides.) Test slides are developed at intervals to determine the proper exposure time. The schedule for developing autoradiograms is Kodak D-19, 2.5 min; water rinse, 30 sec; Fixer, 5 min. Slides are then rinsed in several changes of distilled water. Although the exact temperature is not critical, if developing is done at 15–20°C grain size will be smaller. It is more important that the slides and all solutions be at the same temperature because temperature changes may produce wrinkles in the emulsion. After the final rinse, slides may be air dried or stained immediately.

There are several stains that will stain through autoradiographic emulsion. A useful one if Giemsa stain, which can also be used after enzymatic detection of biotinylated probes. A stock solution of Giemsa blood stain (Harelco) is diluted 1:20 with 10 mM phosphate buffer, pH 6.8, immediately before use. Staining times depend both on the cytological preparation and on the batch of stain. Polytene chromosomes generally require 2–10 min. The progress of staining can be checked by rinsing the slide in distilled water and viewing the wet slide with a light microscope. Preparations should be stained lightly so that the autoradiographic grains or enzymatic product that identify hybrids are very prominent. After staining is complete, slides should be rinsed and air dried for at least 1 hr. They can then be mounted under a coverslip with a small drop of Permount or kept unmounted. Unmounted slides can be viewed by placing a coverslip over a drop of water.

E. Detection of Biotinylated Probes

After the removal of nonspecifically bound probe, slides should not be dried but instead moved immediately into PBS for at least $2 \times$ 10-min washes. Slides can be left in PBS for several hours. Bound probe can be effectively detected by attaching alkaline phosphatase to the biotin and then using the enzyme in a reaction that yields a colored precipitate (Engels *et al.*, 1986). Slides are blocked in buffer 1 (0.1 M Tris, 0.1 M NaCl, 0.002 M MgCl$_2$, pH 7.5, 0.05% Trition X-100) plus 2% (w/v) BSA at 42°C for 20 min. They are cooled to room temperature for 10 min and drained, and each is covered with 50μl of strepavidin solution (2 μl of strepavidin from BRL DNA Detection System/ml of buffer 1]. After 5 min, a second 50 μl of strepavidin is added and left for 5 min more. Slides are rinsed in buffer 1 ($3 \times$ 3 min) and drained. Each preparation is covered with 50μl of biotinylated alkaline phosphatase (BRL DNA Detection System) mixed with buffer 1 (1 μl enzyme solution/ml of buffer freshly mixed in a plastic tube). After a 5-min incubation, a second 50 μl of biotinylated alkaline phosphatase is added and incubation continued for 5 additional min. Slides are rinsed in buffer 1 ($2 \times$ 3 min) and then $2 \times$ 3 min in buffer 2 (0.1 M Tris, pH 9.5, 0.1 M NaCl, 0.05 M MgCl$_2$).

Staining solution contains nitro-blue tetrazolium (NBT) and 5-bromo-4-chloro-3-indolyl-phosphate (BCIP). (Both are from the BRL DNA Detection System). Just before use, 4.4 μl NBT is added to 1 ml of buffer 2 and mixed. Then, 3.3 μl BCIP is added to the solution. Preparations are covered with 30 μl of stain and a coverslip. They are placed in the dark in a closed box to prevent drying and checked with a phase microscope. The staining of the hybrid may need 30 min–2 hr. When staining is satisfactory, the reaction is stopped by rinsing in distilled water. Chromosomes may be stained lightly with lacto-acetic orcein diluted 1:3 with water. Slides can be kept uncovered and viewed by placing a drop of water and a coverslip over the preparation. Slides should be stored dry in a dark box.

References

Ashburner, M. (1972). Puffing patterns in Drosophila melanogaster and related species. In "Developmental Studies on Giant Chromosomes: Results and Problems in Cell Differentiation" (W. Beermann, ed.), Vol. 4, pp. 101–151. New York: Springer-Verlag.

Ashburner, M. (1989). "Drosophila, A Laboratory Handbook," pp. 44–49. Cold Spring Harbor, NY: Cold Spring Harbor Laboratory Press.

Belote, J. M., and Lucchasi, J. C. (1980). Control of X-transcription by the maleless gene in Drosophila. *Nature* **285,** 573–575.

Biessmann, H., Mason, J. M., Ferry, K., d'Hulst, M., Valgeirsdottir, K., Traverse, K. L., and Pardue, M. L. (1990). Addition of telomere-associated HeT DNA sequences "heals" broken chromosome ends in Drosophila. *Cell* **61,** 663–673.

Bonner, J. J., and Pardue, M. L. (1977a). Polytene chromosome puffing and in situ hybridization measure different aspects of RNA metabolism. *Cell* **12,** 227–234.

Bonner, J. J., and Pardue, M. L. (1977b). Ecdysone-stimulated RNA synthesis in salivary glands of Drosophila melanogaster: Assay by in situ hybridization. *Cell* **12,** 219–225.

Bridges, C. B. (1935). Salivary chromosome maps with a key to the banding of the chromosomes of *Drosophila melanogaster. J. Hered.* **26,** 60–64.

Bridges, C. B. (1938). A revised map of the salivary gland X-chromosome of *Drosophila melanogaster. J. Hered.* **29,** 11–13.

Bridges, C. B., and Bridges, P. N. (1939). A new map of the second chromosome: a revised map of the right limb of the second chromosome of Drosophila melanogaster. *J. Hered.* **30,** 475–476.

Bridges, P. N. (1941a). A revised map of the left limb of the third chromosome in *Drosophila melanogaster. J. Hered.* **32,** 64–65.

Bridges, P. N. (1941b). A revision of the salivary gland 3R-chromosome map of *Drosophila melanogaster. J. Hered.* **32,** 299–300.

Bridges, P. N. (1942). A new map of the salivary gland 2L-chromosome map of *Drosophila melanogaster. J. Hered.* **33,** 403–408.

Endow, S. A. (1983). Nucleolar dominance in polytene cells of Drosophila. *Proc. Natl. Acad. Sci. U.S.A.* **80,** 4427–4431.

Endow, S. A., and Glover, D. M. (1979). Differential replication of ribosomal gene repeats in polytene nuclei of Drosophila. *Cell* **17,** 597–605.

Engles, W. R., Preston, C. R., Thompson, P., and Eggleston, W. B. (1986). *Focus* **8,** 6–8.

Feinberg, A. P., and Vogelstein, B. (1984). A technique for radiolabeling DNA restriction endonuclease fragments to high specific activity [Addendum]. *Anal. Biochem.* **137,** 266–267.

Gall, J. G., Cohen, E. H., and Polan, M. L. (1971). Repetitive DNA sequences in Drosophila. *Chromosoma* **33,** 319–344.

Heino, T. I. (1989). Polytene chromosomes from ovarian pseudonurse cells of the Drosophila melanogaster otu mutant. *Chromosoma* **97,** 363–373.

Heitz, E. (1934). On alpha- and beta-heterochromatin and the constancy and structure of the chromosome of Drosophila. *Biol. Zbl.* **54,** 588–609.

Hochstrasser, M. (1987). Chromosome structure in four wild-type polytene tissues of Drosophila melanogaster. *Chromosoma* **95,** 197–208.

Holden, J. J., and Ashburner, M. (1978). Patterns of puffing activity in the salivary gland chromosomes of Drosophila IX: The salivary and prothoracic gland chromosomes of a dominant temperature sensitive lethal of D. melanogaster. *Chromosoma* **68,** 205–227.

Karpen, G. H., and Spradling, A. S. (1992). Analysis of subtelomeric heterochromatin in the Drosophila minichromosome Dp1187 by single P element insertional mutagenesis. *Genetics* **132,** 737–753.

Lefevre, G., Jr. (1976). A photographic representation and interpretation of the polytene chromosomes of Drosophila melanogaster. In "The Genetics and Biology of Drosophila" (M. Ashburner and E. Novitski, eds.), Vol. 1a, pp. 31–66. New York: Academic Press.

Lifschytz, E. (1983). Sequence replication and banding organization in the polytene chromosomes of Drosophila melanogaster. *J. Mol. Biol.* **164,** 17–34.

Muller, H. J. (1938). The remaking of chromosomes. *Collect. Net.* **13,** 181–195.

Pardue, M. L. (1970). Molecular hybridization of nucleic acids in cytological preparations. Ph.D thesis. Yale University, New Haven, CT.

Pardue, M.L. (1986). In situ hybridization to DNA chromosomes and nuclei. *In* "Drosophila, a Practical Approach" (D. B. Roberts, ed.), pp. 111–137. Oxford: IRL Press.

Pardue, M. L., and Dawid, I. B. (1981). Chromosomal locations of two DNA segments that flank ribosomal insertion-like sequences in Drosophila:flanking sequences are mobile elements. *Chromosoma* **83,** 29–43.

Pardue, M. L., Bonner, J. J., Lengyel, J. A., and Spradling, A. C. (1977). Drosophila salivary gland polytene chromosomes studied by in situ hybridization. *In* "International Cell Biology" (Brinkley, B. R. and Porter, K. R. eds.), pp. 509–519. New York: Rockfeller Univ. Press.

Pardue, M. L., Lowenhaupt, K., Rich, A., and Nordheim, A. (1987). (dC–dA)·(dG–dT)$_n$ sequences have evolutionary conserved chromosomal locations in Drosophila with implications for roles in chromosome structure and function. *EMBO J.* **6,** 1781–1789.

Richards, G. (1980). The polytene chromosomes in the fat body nuclei of Drosophila melanogaster. *Chromosoma* **79,** 241–350.

Semeshin, V. F., Demakov, S. A., Alonso, M. P., Belyaeva, E. S., Bonner, J. J., and Zhimulev, I. F. (1989). Electron microscopal analysis of Drosophila polytene chromosomes. *Chromosoma* **97,** 396–412.

Spierer, A., and Spierer, P. (1984). Similar level of polyteny in bands and interbands of Drosophila giant chromosomes. *Nature* **307,** 176–178.

Spierer, P., Spierer, A., Bender, W., and Hogness, D. S. (1983). Molecular mapping of genetic and chrommeric units in Drosophila melanogaster. *J. Mol. Biol.* **168,** 35–50.

Spradling, A. C. (1986). P element-mediated transformation. *In* "Drosophila, A Practical Approach" (D. B. Roberts, ed.), pp. 175–197. Oxford: IRL Press.

Storto, P. D., and King, R. C. (1987). Fertile heteroallelic combinations of mutant alleles of the otu locus of Drosophila melanogaster. *Roux's Arch. Dev. Biol.* **196,** 210–221.

Valgeirsdottir, K., Traverse, K. L., and Pardue, M. L. (1990). HeT DNA: A family of mosaic repeated sequences specific for heterochromatin in Drosophila melanogaster. *Proc. Natl. Acad. Sci. U.S.A.* **87,** 7998–8002.

Young, B. S., Pession, A., Traverse, K. L., French, C., and Pardue, M. L. (1983). Telomere regions in Drosophila share complex DNA sequences with pericentric heterochromatin. *Cell* **34,** 85–94.

Zhimulev, I. F., Semeshin, V. F., Kulichkov, V. A., and Belyaeva, E. S. (1982). Intercalary heterochromatin in Drosophila. *Chromosoma* **87,** 197–228.

CHAPTER 20

Immunological Methods for Mapping Protein Distributions on Polytene Chromosomes

Deborah J. Andrew[*] and Matthew P. Scott[†]

[*] Department of Cell Biology and Anatomy
Johns Hopkins University School of Medicine
Baltimore, Maryland 21205

[†] Department of Developmental Biology and Genetics
Howard Hughes Medical Institute
Stanford University School of Medicine
Stanford, California 94305

I. General Introduction

The giant polytene chromosomes of the larval salivary gland have for years provided a physical map of the euchromatic genome of *Drosophila*. Here, we review uses of polytene chromosomes in mapping chromatin structures and protein accumulation sites. We also describe different procedures for chromosome staining.

The reproducible banding pattern of the polytene chromosomes has been used to map genes relative to chromosomal rearrangements, to map cloned DNAs to specific bands, to probe chromatin structure, and to map sites of protein accumulation. Polytene chromosomes are created through a process known as endoreduplication, in which multiple rounds of DNA replication occur without the concomitant strand separation and subsequent nuclear and cell division characteristic of normal mitoses. Although chromosomes of most larval tissues are polytenized, those of the salivary gland are replicated to the greatest extent, a result of the relatively early onset of endoreduplication in this tissue (Smith and Orr-Weaver, 1991). About 10 doublings lead to a thousandfold amplification of the chromosomes. The replicated strands are aligned, and the interspersion of condensed and less dense chromatin gives rise to the dark bands and light interbands. Polytene chromosomes are functionally similar to diploid interphase chromosomes as assayed by gene activity and chromatin ultrastructure (Tissièrres *et al.*, 1974; Elgin and Boyd, 1975; Bonner and Pardue, 1976; Woodcock *et al.*, 1976). Thus, polytene chromosomes are a visible representation of the longest and most transcriptionally active stage in the cell cycle.

The highly reproducible banding pattern of polytene chromosomes allowed Bridges in the 1930s to devise a reference system designating every band in the euchromatic genome (Bridges, 1935, 1938; Bridges and Bridges, 1939). This reference system is still used today. Bridges divided each of the major chromosome arms into 20 number divisions, which were subdivided into six letter divisions, A–F. Each band within the number–letter subdivision has a designated number. The telocentric X chromosome contains major number divisions 1 through 20. The metacentric second chromosome contains divisions 21 through 60. The third chromosome, which is also approximately metacentric, contains number divisions 61–100. The tiny telocentric fourth chromosome contains number divisions 101 and 102. There are approximately 5000 chromosome bands in the genome, an average of 8 bands within each number–letter division. Based on total DNA content, it has been estimated that each band contains an average of 21.6 kb of DNA, but the range in DNA content varies from a few kilobases to over 200 kb. Heterochromatin is not replicated during formation of the polytene chromosomes, so the banding pattern provides a representation of euchromatin only.

The characteristic polytene-binding pattern has been useful for mapping DNA and chromatin structure. Antibodies have been used to map regions of heterochromatization and atypical DNA structure such as non-B (Kabakov and Poverenny, 1993) and Z-DNA (Nordheim *et al.*, 1981; Arndt-Jovin *et al.*, 1983; Lancillotti *et al.*, 1985). Antibodies have been generated against various nuclear fractions that recognize the α- and/or β-heterochromatin of the chromocenter (Will and Bautz, 1980; James and Elgin, 1986; Fleischmann *et al.*, 1987; James *et al.*, 1989), junctions between α- and β-heterochromatin (Miklos and Cotsell, 1990), the highly condensed bands (Silver and Elgin, 1977; Saumweber *et al.*, 1980; Fleischman *et al.*, 1987; Frasch, 1991), the interband regions (Saumweber *et al.*, 1980), and the chromosome puffs (Silver and Elgin, 1977; Mayfield *et al.*, 1978; Saumweber *et al.*, 1980, 1990; Fleischman *et al.*, 1989; Amero *et al.*, 1991; Wieland *et al.*, 1992).

Antibodies to histones, topoisomerases, and RNA polymerases have been used to probe polytene chromosomes for physical changes in chromatin structure resulting from induced gene activity (Plagens *et al.*, 1976; Jamrich *et al.*, 1977; Greenleaf *et al.*, 1978; Fleischmann *et al.*, 1984; Heller *et al.*, 1986). The induction and distribution of chromosomal transcripts have been followed with antibodies to DNA/RNA hybrids (Alcover *et al.*, 1982). Antibodies have also been used to map accumulation sites of RNA-binding proteins known or proposed to be involved in post-transcriptional RNA processing (Christensen *et al.*, 1981; Hovemann *et al.*, 1991; Risau *et al.*, 1983; Kim *et al.*, 1992).

Antibody staining of polytene chromosomes has also provided evidence for the *in vivo* association of transcription factors with the genes they regulate (Urness and Thummel, 1990; Payre and Vincent, 1991; Thisse and Thisse, 1992; Hill *et al.*, 1993; Muralidhar *et al.*, 1993) and the *in vivo* association of proteins with other proteins of related function (DeCamillis *et al.*, 1992; Franke *et al.*, 1992; Martin and Adler, 1993; Rastelli *et al.*, 1993).

II. Mapping Known Chromosomal Components

To correlate cytological observations with the molecular components of chromatin, abundant well-understood proteins have been labeled on polytene chromosomes. Among the first proteins to be immunolocalized on the polytene chromosomes were the histones (Desai *et al.*, 1972; Alfageme *et al.*, 1976; Jamrich *et al.*, 1977). Histone antibodies label all of the DNA-rich banded regions of the chromosomes consistent with the role of histones in DNA packaging (Alfageme *et al.*, 1976; Jamrich *et al.*, 1977). Antibodies to RNA polymerase II, on the other hand, accumulate in the interbands and chromosomal puffs (Plagens *et al.*, 1976; Jamrich *et al.*, 1977; Greenleaf *et al.*, 1978) in agreement with observations that actively transcribed regions of the genome are relatively loosely packed (Bjorkroth *et al.*, 1988). Heat-shock induction leads to a rapid high-level accumulation of RNA polymerase II in all known heat-shock inducible puffs (Greenleaf *et al.*, 1978) corresponding to the increased rate of transcription at these loci.

More recently, sites of RNA polymerase III accumulation have been localized using antisera to the second largest subunit of Pol III (Kontermann *et al.*, 1989). As expected, the RNA Pol III binding pattern is distinct from that generated with antibodies to RNA Pol II. Sites of RNA Pol III accumulation include the tRNA loci (4 sites) as well as a number of sites (~10–15 sites) not yet identified as containing Pol III transcribed genes.

Antibodies to topoisomerase I and II yield complementary patterns of accumulation on chromosomes. Topoisomerase I is preferentially associated with presumed active regions of the genome as shown by prominent staining of chromosomal puffs and its rapid accumulation on heat-shock puffs following induction at 37°C (Fleischmann *et al.*, 1984). Topoisomerase II is distributed along the chromosomes paralleling the distribution of DNA with the heavily banded regions of the chromosomes staining most strongly with the topo II antibodies (Heller *et al.*, 1986).

III. Identification and Characterization of Unknown Chromosomal Components

One productive approach to characterizing unknown chromosomal proteins has been to generate antisera against nonhistone proteins isolated from the nuclei of *Drosophila* embryos or mammalian tissue culture cells. The antiserum is then screened by immunostaining *Drosophila* polytene chromosomes to identify proteins associated with specific types of chromatin (Silver and Elgin, 1977; Mayfield *et al.*, 1978; Saumweber *et al.*, 1980; Will and Bautz, 1980; Fleischmann *et al.*, 1987; Fleischmann *et al.*, 1989; James *et al.*, 1989). Once an antibody shows a desirable pattern on polytene chromosomes, the same antibody is used to screen cDNA expression libraries and isolate cDNAs corresponding to the recognized protein. The sequence of the corresponding cDNA is determined, and once the DNA is mapped to the polytene chromosomes using *in situ* hybridization techniques, a genetic analysis to determine protein function can be initiated. Genes corresponding to a variety of chromosomal proteins identified by this method have been cloned, and mapped to specific sites in the genome (James and Elgin, 1986; Saumweber *et al.*, 1990; Amero *et al.*, 1991; Champlin *et al.*, 1991; Frasch, 1991; Wieland *et al.*, 1992). This approach identified the heterochromatin-specific protein, HP-1, as the gene affected in a dominant suppressor of variegation mutation, *Su(var)205* (Eissenberg *et al.*, 1990).

IV. Mapping Sites of Accumulation of Regulatory Proteins

Although the technique of immunostaining polytene chromosomes has been extensively utilized to detect relatively abundant chromosomal proteins, only recently has the technique been applied to mapping binding sites of known regulatory proteins. The first such application involved the mapping of the *Polycomb (Pc)* protein (Zink and Paro, 1989). *Pc* is a member of a group of genes, known as the *Pc*-group genes, first identified as regulators of homeotic gene expression. As expected from *Pc* regulatory functions, antibodies to *Pc* protein bind at the sites of both *Drosophila* homeotic gene complexes as well as approximately 60–80 other positions in the genome. A *Pc* protein binding site has been further localized to a 4-kb genomic fragment of the homeotic gene *Antennapedia (Antp)*, which is known to be repressed by *Pc*. Novel *Pc* binding was detected at the insertion sites of a transgene containing 4 kb of *Antp* sequence fused to the *Escherichia coli lacZ* gene (Zink *et al.*, 1991). Consistent with *Pc*'s role as a repressor of *Antp* expression, *Pc* antibody binding to the transgenes on the salivary gland chromosomes correlated with the absence of *lacZ* expression from the transgenes in salivary glands. No *Pc* binding to the transgene was detected in the few transgenic lines in which ectopic salivary gland expression did occur.

Since the mapping of *Pc*-binding sites, a number of regulatory proteins have been localized to specific sites in the genome. Among the mapped proteins are other members of the *Pc* group, including *polyhomeotic (ph)* (DeCamillis *et al.*, 1992; Franke *et al.*, 1992) and *Posterior sex combs (Psc)* (Martin and Adler, 1993), as well as other proteins that genetically interact with *Pc*-group genes (Pirrotta *et al.*, 1988; Rastelli *et al.*, 1993). Interestingly, these proteins of related function colocalize with *Pc* and/or each other on chromosomes, providing the first suggestion that these proteins may function as multimeric regulatory complexes (Franke *et al.*, 1992; Martin and Adler, 1993; Rastelli *et al.*, 1993).

Other regulatory proteins have been mapped to the polytene chromosomes including the products of the early ecdysone genes, *E74* and *E75A* (Urness and Thummel, 1990; Hill *et al.*, 1993), the dosage compensation gene, *maleless (mle)* (Kuroda *et al.*, 1991; Gorman *et al.*, 1993), two genes involved in dorsal–ventral patterning, *twist (twi)* (Thisse and Thisse, 1992) and *single-minded (sim)* (Muralidhar *et al.*, 1993), and two related zinc-finger genes of unknown function, *sry β* and *sry δ* (Payre *et al.*, 1990; Noselli *et al.*, 1992). As predicted by the Ashburner model for ecdysone regulation of transcription during metamorphosis, *E74* and *E75A* antibodies both bind to the early and late ecdysone-induced puffs (Urness and Thummel, 1990; Hill *et al.*, 1993). Similarly, as expected from models for dosage compensation, *mle* protein accumulates on only the male single X chromosome (Kuroda *et al.*, 1991; Gorman *et al.*, 1993). Antibodies to *twi* and *sim* each bind to many sites in the genome (about 60 sites for *twi* and 23 sites for *sim*) including sites at or near suspected downstream target genes. *sry β* and *sry δ* bind to 54 and 196 sites, respectively. Despite the high sequence conservation between *sry β* and *sry δ* in the DNA-binding region of the two proteins, only 123 of the *sry β* and *sry δ* sites overlap (Noselli *et al.*, 1992; Payre *et al.*, 1990; Payre and Vincent, 1991).

V. Mapping Protein Domains

The binding of antibodies to chromosomes can be and has been used to map functional domains of proteins. This approach was elegantly used to map the region required for chromosome binding to the conserved "chromodomain" of the *Pc* protein (Paro and Hogness, 1991; Messmer *et al.*, 1992). A *Pc–β-galactosidase* (*Pc-β-gal*) fusion protein gene was introduced into flies and with antibodies to *β-galactosidase* (*α-β-gal*) was demonstrated to have the same pattern of binding sites as the endogenous *Pc* protein. A set of deletions and mutations was made in the *Pc-β-gal* fusion gene construct, introduced into flies, and tested for binding activity on polytene chromosomes using the *α-β-gal* antibody. The results indicate that the carboxy-terminal portion of the *Pc* protein is not required for binding. However, both deletions and point mutations in the amino-terminal chromodomain abolish binding of the *Pc-β-gal* fusion protein to the chromosomes.

Experiments similar to those carried out with the *Pc* protein were done to map the regions in the *sry β* and *sry δ* proteins required for chromosome binding (Noselli *et al.*, 1992). The carboxy-terminal zinc finger domain of each protein, which is required for DNA binding *in vitro,* is also absolutely required for chromosome binding. However, mutations in the N-terminal domain drastically reduce the number of sites bound by each protein. To map the regions of the *sry β* and *sry δ* proteins that confer specificity of binding on the polytene chromosomes, chimeric *sry β/δ* protein constructs were built and introduced into flies. The novel set of binding sites observed with the chimeric proteins suggested that *sry* binding specificity is a result of a combination of specificity of protein contacts with the target DNA as well as contacts with other proteins. The approaches used with both *Pc* and the *sry* proteins will be useful for mapping functional domains of other chromosome-binding proteins.

VI. Tests of Regulatory Interactions among Proteins of Related Function

Immunostaining of polytene chromosomes can also be used to study regulatory hierarchies. One example for which chromosome binding has been so applied is in studies of dosage compensation in *Drosophila* (Gorman *et al.*, 1993). Dosage compensation is achieved by transcription of the male X chromosome at approximately twice the level of each X chromosome in females. Four autosomal genes have been shown to be absolutely required for hypertranscription of the X chromosome in males. In females, hypertranscription of the X chromosome is blocked by the activity of the gene *Sex lethal* (*Sxl*). In wild-type animals, antibodies to the autosomal dosage compensation gene *maleless* (*mle*) bind specifically to the male and not the female X chromosome, although the protein is expressed in both sexes (Kuroda *et al.*, 1991). The ectopic expression of functional *Sxl* protein, which is normally made only in females, results in an absence of *mle* protein binding to the male single X chromosome. With loss-of-function mutations in the remaining three autosomal dosage compensation genes, *mle* protein binding to the male X chromosome is not observed. Thus, *Sxl* prevents *mle* function in females by preventing its interaction with chromatin, and each of the remaining three autosomal dosage compensation genes is required for *mle* chromosome binding.

In similar studies with the *Pc*-group proteins, it was demonstrated that binding of *Suppressor of zeste 2 (Su(z)2)* and *Psc* proteins to polytene chromosomes depends on function of the *Enhancer of zeste (E(z))* gene, whereas binding of the *zeste* protein is unaffected by mutations in *E(z)* (Rastelli *et al.*, 1993). Other regulatory interactions involving proteins that bind to polytene chromosomes can be studied using this approach.

VII. History of Antibody-Staining Methods for Polytene Chromosomes

The first immunostaining of *Drosophila* polytene chromosomes was with antisera directed against whole histone and histone fractions isolated from human leukemic lymphoblasts (Desai *et al.*, 1972). Chromosomal binding of histone antibodies, made in rabbit, was detected with a goat anti-rabbit secondary antibody conjugated to fluorescein. The chromosomes used in this early study were from second rather than third instar larvae, and the staining pattern with the histone antibodies was somewhat murky. By the mid to late 1970s several laboratories were using fluorescently tagged secondary antibodies to map accumulation sites of a variety of known and unknown proteins on third instar larval chromosomes (Elgin and Boyd, 1975; Alfageme *et al.*, 1976; Plagens *et al.*, 1976; Silver and Elgin, 1976, 1977; Jamrich *et al.*, 1977; Greenleaf *et al.*, 1978; Mayfield *et al.*, 1978). The quality of preparations was sufficient to map binding sites to the resolution of individual bands or interbands. The major improvements in the basic technique were in the fixation of the chromosomes. It was demonstrated that acid fixation alone often resulted in a loss of chromosomal proteins and that an early aldehyde fix prior to the acid fixation prevented such protein loss (Silver and Elgin, 1976). Both the duration of fixation and concentration of aldehydes used today are a compromise between obtaining chromosome preparations of good morphology and maintaining bound proteins in an immunologically reactive state. The fixation protocol outlined in the next section is currently the most commonly used (Zink and Paro, 1989; Urness and Thummel, 1990; Kuroda *et al.*, 1991; Zink *et al.*, 1991; Franke *et al.*, 1992; Gorman *et al.*, 1993; Martin and Adler, 1993; Rastelli *et al.*, 1993), although other fixation protocols yield similarly good results (James and Elgin, 1986;

Pirrotta *et al.*, 1988; Payre *et al.*, 1990; Saumweber *et al.*, 1990; Clark *et al.*, 1991; Messmer *et al.*, 1992; Hill *et al.*, 1993; Muralidhar *et al.*, 1993). Slight variations in fixation time may affect binding of some proteins. A typical preparation of fluorescently stained chromosomes is shown in Figure 1.

A major technical advance in mapping proteins to sites on chromosomes was to use enzyme-linked rather than fluorescently labeled secondary antibodies (Zink and Paro, 1989). This advance made it possible to map binding sites directly onto chromosomes that have been counterstained with the DNA stain Giemsa. The red/brown signal detected with the enzyme-linked antibody is obvious against the blue staining of the chromosomal DNA. Thus, sites can be mapped directly on the microscope using bright-field optics. Moreover, unlike fluorescently labeled chromosome preparations, those made using enzyme-linked detection systems are permanent. One drawback with the enzyme-linked detection method, however, is the difficulty of obtaining stained chromosomes with good morphology. The reaction products of the staining procedure are harsh to chromosomes. Often, a compromise must be made between the intensity of signal at each site and the final morphology of the chromosomes. Thus, for many purposes, especially photography, the fluorescently labeled chromosomes are more easily prepared.

VIII. Procedure for Enzyme-Linked Detection of Proteins on Polytene Chromosomes

This procedure is a modification of the original one of Zink and Paro (1989). The addition of spermine and spermidine to the staining and washing solutions prevents some degradation of chromosome morphology resulting from the enzymatic staining reactions. The use of azide to stop the enzyme reaction at the desired time also helps maintain good chromosome morphology.

A. Solutions to Prepare

1. 10× PBS (*Phosphate-Buffered Saline*)

   10.68 g K_2HPO_4 (dibasic)

   5.28 g KH_2PO_4 (monobasic)

   81.8 g NaCl

   Add to 800 ml of ddH_2O. pH to 7.4 with HCl. Bring the volume to 1 liter.

2. 37% formaldehyde

   3.7 g paraformaldehyde

   Volume to 9.95 ml with ddH_2O. Add 50 μl 3 *M* NaOH. Dissolve by boiling. Place 100 μl aliquots in 1.5-ml tubes. Store at $-20°C$.

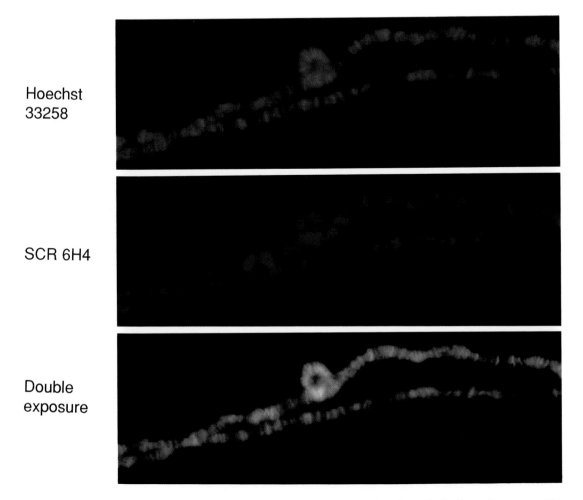

Fig. 1 Antibodies to the SCR protein accumulate at specific reproducible sites on the salivary gland polytene chromosomes. The top panel shows Hoechst 33258 staining (DNA) of salivary gland polytene chromosomes isolated from HS-SCR larvae that were incubated at 37°C for 60 min followed by a 30-min recovery at 25°C. The middle panel shows the same chromosomes stained with the monoclonal Scr antibody, αSCR-6H4 (Glicksman and Brower, 1988) detected with a secondary goat anti-mouse antibody conjugated to Texas Red. The lower panel is a double exposure of the DNA (Hoechst) staining and the SCR antibody staining. Note that SCR antibody accumulation occurs both in banded and unbanded regions of the chromosomes. We do not detect staining with αSCR-6H4 on salivary gland polytene chromosomes from wild-type control larvae not containing the HS-SCR construct (Glicksman and Brower, 1988).

3. PBSTBSS (*PBS* + *T*ween + *BSA* + *S*permine + *S*permidine)

    200 ml 10× PBS

    4 g BSA (Fraction V from Sigma)

    4 ml Tween 20 (polyoxyethylene-sorbitan monolaurate)

    20 μl 2.5-*M* spermine

    50 μl 2.0-*M* spermidine

    1800 ml H_2O

4. PBSTBSS-AZIDE (*PBS* + *T*ween + *BSA* + *S*permine + *S*permidine + *A*zide)

    200 ml 10× PBS

    4 g BSA

    4 ml Tween 20

    20 μl 2.5-*M* spermine

    50 μl 2.0-*M* spermidine

    2 g NaAzide (very poisonous)

    1800 ml H_2O

5. Solution I (dissecting solution)

    1 ml 10× PBS

    1 ml 10% Tween 20

    8 ml H_2O

6. Solution II (first fixation solution). Make this solution fresh each time.

    100 μl 37% formaldehyde

    696 μl H_2O

    100 μl 10× PBS

    100 μl 10% Tween 20

    2 μl 0.25-*M* spermine

    2 μl 0.5-*M* spermidine

7. Solution III (second fixation solution). Make this solution fresh each time.

    100 μl 37% formaldehyde

    296 μl H_2O

    500 μl glacial acetic acid

    100 μl 10% Tween 20

    2 μl 0.25-*M* spermine

    2 μl 0.5-*M* spermidine

8. Siliconized cover slips

    Dip 22- or 18-mm^2 coverslips in Sigmacote (Sigma, Catalog No. SL-2) twice. Air dry.

9. Gelatinized slides

Melt 0.1 g of gelatin (Difco, Catalog No. 0143-01) in 50 ml H$_2$0. When the solution is cooled to about 60°C, add 0.1 g of chromalum (chromium potassium sulfate dodecahydrate) (Aldrich Catalog No. 24,336-1). Dip the slides (frosted on one side at one end) twice (a 50-ml disposable plastic tube is a useful container for dipping slides), wipe off the back of the slide, and air dry. Store at room temperature in a slide box that prevents contact between the slides.

B. Dissection of Salivary Glands and Antibody Staining

The single most critical element in the preparation of good polytene chromosome squashes is the health and well being of the larvae. Set up bottles of food lightly sprinkled with fresh baker's yeast with about 25 adult flies of each sex. Larvae grow extremely well on "instant" fly food (Formula 4-24; Carolina Biological Supplies). Transfer the adults every 2 days. Once the adults are removed, add more yeast and some H$_2$O to bottles to make a paste-like substance on the food. Larvae will be ready for dissection 3–5 days after the removal of the parents.

1. Dissect salivary glands in Solution I.
2. Transfer the two glands from one larva to 20 μl solution II in a droplet on a siliconized cover slip. Leave for ~30 sec.
3. Transfer the glands to 23 μl of Solution III on a nonsiliconized coverslip. Leave the glands for 2–4 min. Place a gelatinized slide on coverslip and pick up the glands. Invert the slide and gently tap around the glands to create a wave-like motion of solution over the glands (watch, using a dissecting microscope, while you do this). Finally, tap directly over glands lightly for several seconds. If the coverslip begins to stick to the slide, quit tapping because continued tapping will only shatter the chromosomes. Blot away excess liquid and squash the slide with all your weight between the leaves of a folded piece of 3MM paper.
4. Examine the chromosomes under a compound microscope using phase optics. Continue only with preparations that are stellar. Chromosome arms should be well spread and completely flat (i.e., no light refraction). Unless chromosomes are perfect at this stage, they will be impossible to work with later. If several preparations from a single bottle are not good, either use a different bottle of larvae or quit for the day. Usually if a bottle yields larvae with good chromosomes from the first few preparations, it will continue to do so for the rest of that day.
5. Freeze the slide in liquid N$_2$ and remove the coverslip with a razor blade. Mark the position of the coverslip on the slide with a diamond pen.
6. Place the slide in a rack in a Wheaton jar filled with 200 ml of PBSTBSS.

Slides can be accumulated at this stage until you have enough for your experiment.

7. Rinse the slides twice in PBSTBSS (10 min each wash).

8. Set up an incubation chamber. A large flat plastic container, with a lid that seals, containing very damp paper towels, works well. Set up parallel plastic 1-ml pipets on which the slides will be placed.

9. Dilute the primary antibody into PBSTBSS. (The appropriate dilution of the antibody has to be empirically determined but will be similar to concentrations required for staining embryos). Place ~50 μl of diluted antibody on each slide and set a 22 × 40-mm coverslip over chromosomes making sure to avoid air bubbles. Incubate 1 hr at RT in the incubation chamber assembled in Step 8.

10. Rinse the slides twice in PBSTBSS (10 min each wash).

11A. Incubate the slides for 30 min in 5% normal serum (obtained from the same species as the secondary antibody to be used). Dilute the secondary antibody (1/100 to 1/1000) into PBSTBSS containing 5% normal serum. If the protein is relatively abundant, a secondary antibody directly conjugated to horseradish peroxidase (HRP) can be used. If the protein is less abundant, a biotin–avidin amplification step may be appropriate and the secondary antibody should be one conjugated to biotin. Add ~50 μl/slide of secondary antibody. Cover sample with a 22 × 40-mm coverslip. Leave the slides for 40 min at RT in an incubation chamber. If the secondary antibody is directly conjugated to HRP, proceed to Step 15. If not, continue with Step 12.

12A. Preincubate the biotin with avidin–HRP to allow biotin–avidin–HRP complexes to form prior to applying the solution to the chromosomes. Forty microliters Solution A and 40 μl of Solution B from the Vectastain Elite ABC kit (Vector Labs, Catalog No. PK6100) diluted into 1 ml PBSTBSS.

13A. Rinse the slides twice in PBSTBSS (10 min each wash).

14A. Place ~50 μl of preincubated biotin–avidin–HRP (A + B) under a coverslip on each slide in the incubation chamber. Incubate the slides for 40 min at RT.

15A. Rinse the slides twice in PBSTBSS (10 min each wash).

16A. The HRP-labeled sites on the chromosomes are detected by applying a 1/4000 dilution of 30% H_2O_2 + 0.5 mg/ml 3,3'-diaminobenzidine tetrahydrochloride (DAB) (Sigma Catalog No. D-5905) diluted in PBSTBSS (25 μl of 1/100 dilution of H_2O_2 and 50 μl of 10 mg/ml DAB in 1 ml PBSTBSS). Apply ~50 μl to the sample on a slide under a coverslip. Watch the reaction products develop using a compound microscope. As soon as there is sufficient signal, place the slide in a Wheaton jar containing PBSTBSS–azide. The azide solution will stop the reaction immediately, which is important because reactions that go too far destroy chromosome morphology.

CAUTION: DAB is a potent carcinogen; wear gloves and dispose of properly.

17A. Rinse slides several times in PBSTBSS–azide followed by PBSTBSS.

18A. Counterstain in a 1/100 dilution of Giemsa (Baker, Catalog No. M708-01) in H_2O for 30 sec. Check Giemsa staining on the microscope using bright-field optics. Staining should be sufficient to show banding pattern of polytenes without obscuring the HRP signal. Stain longer if necessary.

19A. Rinse the slides several times in distilled water.

20A. Air dry the slides and mount in Entellan (EM Science, Catalog No. 7961-2) or another preferred permanent mounting medium.

IX. Procedure for Fluorescent Labeling of Proteins on Polytene Chromosomes

A. Solutions to Prepare

1. $10\times$ PBS (see previous recipe)
2. 37% formaldehyde (see previous recipe above)
3. PBSTB (*PBS + Tween + BSA*)

   200 ml $10\times$ PBS

   4 g BSA

   4 ml Tween 20 (polyoxyethylene-sorbitan monolaurate)

   1800 ml H_2O
4. Solution I (dissecting solution)

   1 ml $10\times$ PBS

   1 ml 10% Tween 20

   8 ml H_2O
5. Solution II (first fixation solution). Make this solution fresh each time.

   100 μl 37% formaldehyde

   700 μl H_2O

   100 μl $10\times$ PBS

   100 μl 10% Tween 20
6. Solution III (second fixation solution). Make this solution fresh each time.

   100 μl 37% formaldehyde

   300 μl H_2O

   500 μl glacial acetic acid

   100 μl 10% Tween 20

7. Poly-lysine-coated slides
Make a 10% solution of poly-L-lysine (Sigma, Catalog No. P-8920) (dilute commercial solution 1/10). Apply 100 μl to the slide that is frosted on the end. Place another slide face down on the slide with the poly-lysine. Separate slides and air dry. Store the slides at room temperature in a box that prevents contact between slides.

B. Dissection of Salivary Glands and Antibody Staining

Follow the same protocol as outlined for enzymatic detection of proteins until Step 11A, with the exception of using poly-lysine-coated slides rather than gelatinized slides in Step 3. Gelatin will quite effectively quench fluorescent signals. Continue with Step 1 here.

11B. Incubate slides for 30 min in 5% normal serum (obtained from the same species as the secondary antibody to be used). Dilute the secondary antibody (1/100 to 1/1000) into PBSTB containing 5% normal serum. The secondary antibody should be conjugated to the fluorescent dye of choice; e.g., fluorescein, rhodamine, Texas Red. Add ~50 μl/slide of secondary antibody. Cover the sample with a 22 × 40-mm coverslip. Leave the slides for 40 min at RT in an incubation chamber.

12B. Wash the slides twice in PBSTB.

13B. If desired, counterstain DNA with Hoechst 33258 (5 min in 10 μg/ml Hoechst in 1× PBS). Do not counterstain with Giemsa. Giemsa, like gelatin, quite effectively quenches fluorescence.

14B. Wash the slides twice in distilled H_2O.

15B. Dry the slides and mount the preparation in 90% glycerol with 2 mg/ml N-propyl gallate (N-propyl gallate retards photobleaching).

X. Double-Labeling of Proteins on Polytene Chromosomes Using Fluorescently Tagged Antibodies

Follow exactly the steps outlined for fluorescent labeling of proteins on chromosomes, except use two primary antibodies, made in different organisms, and the appropriate secondary antibodies conjugated to fluorescent tags that do not overlap in their absorption and emission spectra.

XI. Interpretations and Limitations

One major caution in immunostaining chromosomes for studies of protein function is antibody cross-reactivity. Given the ease of generating antibodies against different chromosomal components, it is not surprising that antisera from a variety of organisms contain antibodies that react well with specific chromosomal epitopes even without prior immunization. Several commercially available secondary antibodies give beautiful distinct patterns on polytene chromosomes. Controls against such artifactual binding include testing preimmune serum and testing each batch of secondary antibody. We strongly recommend that for each protein being localized to chromosomes antibodies be made to nonoverlapping regions of the protein to detect cross-reactivity to chromosomal proteins with epitopes similar to those in the protein being studied. Alternatively, it may be possible to genetically remove the protein of interest. A third approach to avoiding epitope cross-reactivity is to tag the protein with a peptide epitope from a different source such as the human *myc* tag or the influenza hemagglutinin tag (HA1) (Munro and Pelham, 1984; Field *et al.*, 1988). Monoclonal antibodies specific to both peptide epitopes are available commercially.

A second concern in using polytene chromosome localization for functional studies is the overinterpretation of results. A site of accumulation near a suspected regulatory target gene is simply that, a site of nearby antibody (presumably protein) accumulation. The accumulation may be without functional importance or may not be exactly at the location of a gene of interest. In identifying genes regulated by a transcription factor, for example, it must be demonstrated that the binding site corresponds to a particular gene using the same methodology as is used for demonstrating that a cloned piece of DNA corresponds to a particular gene. A transposed copy of a gene should create a new site of protein accumulation on the chromosomes. To date, this methodology has been applied in only a few published instances (Saumweber *et al.*, 1990; Zink *et al.*, 1991; Muralidhar *et al.*, 1993). It is also important to further demonstrate that protein binding at a particular locus is functional. A correlation between function and binding has been shown in only two cases (Zink *et al.*, 1991; Muralidhar *et al.*, 1993).

Another potential problem with chromosome immunostaining is a cell-type problem. The protein of interest may not be expressed in salivary glands. Experimentally, this problem can and has been gotten around by using inducible ubiquitously active promoters, such as hsp70 (Muralidhar *et al.*, 1993), or salivary gland-specific promoters, such as sgs 3 (Payre *et al.*, 1990; Noselli *et al.*, 1992), to ectopically express proteins of interest. The second cell-type problem is not as easily circumvented. The binding sites in third instar larval salivary gland cells may not reflect binding sites in other tissues nor at other developmental stages. Thus, the lack of binding to a particular site does not argue against direct physical interaction between a protein and its partners, be they protein, RNA, or DNA. There is simply too little known at this time to decide whether a binding protein can interact with its full repertoire of partners in a single tissue type at a single developmental stage. Thus, binding of proteins in a variety of cell types and stages needs to be assayed. Since many, if not most, larval tissues are polytenized to some degree, this is not an impossible task. It will be difficult, however, because, aside from some rare exceptions (Richards, 1980), chromosome maps do not exist for most cell types.

Using antibodies to map proteins to polytene chromosomes has proven to be of great value in studying the function and potential interactive partners of different proteins. The approach takes advantage of the unique cytology of insect chromosomes. Because proteins of heterologous species have been shown to function in *Drosophila,* this approach can also be used to study chromosome associated proteins from a wide variety of organisms.

Acknowledgments

We thank the many people with whom we have consulted in our own work with polytene chromosomes, especially Renato Paro, Will Talbot, and Toshi Watanabe. Our research was supported by a grant from NIH (RO1 No. 18163). MPS is an Investigator with the Howard Hughes Medical Institute.

References

Alcover, A., Izquierdo, M., Stollar, D., Kitagawa, Y., Miranda, M., and Alonso, C. (1982). In situ immunofluorescent visualization of chromosomal transcripts in polytene chromosomes. *Chromosoma* **87,** 263–277.

Alfageme, C. R., Rudkin, G. R., and Cohen, L. H. (1976). Locations of chromosomal proteins in polytene chromosomes. *Proc. Natl. Acad. Sci. U.S.A.* **73,** 2038–2042.

Amero, S. A., Elgin, S. C. R., and Beyer, A. L. (1991). A unique finger protein is associated preferentially with active ecdysone-responsive loci in *Drosophila. Genes Dev.* **5,** 188–200.

Arndt-Jovin, D. J., Robert-Nicoud, M., Zarling, D. A., Greider, C., Weimer, E., and Jovin, T. M. (1983). Left-handed Z-DNA in bands of acid-fixed polytene chromosomes. *Proc. Natl. Acad. Sci. U.S.A.* **80,** 4344–4348.

Bjorkroth, F., Ericsson, D., Lamb, M. M., and Daneholt, F. (1988). Structure of the chromatin axis during transcription. *Chromosoma* **96,** 333–340.

Bonner, J. J., and Pardue, M. L. (1976). The effect of heat shock on RNA synthesis in Drosophila tissues. *Cell* **8,** 43–50.

Bridges, C. B. (1935). Salivary chromosome maps. *J. Hered.* **26,** 60–64.

Bridges, C. B. (1938). A revised map of the salivary gland X chromosome of *Drosophila melanogaster. J. Hered.* **29,** 11–13.

Bridges, C. B., and Bridges, P. N. (1939). A new map of the second chromosome: A revised map of the right limb of the second chromosome of *Drosophila melanogaster. J. Hered.* **30,** 475–479.

Champlin, D. T., Frasch, M., Saumweber, H., and Lis, J. T. (1991). Characterization of a *Drosophila* protein associated with boundaries of transcriptionally active chromatin. *Genes Dev.* **5,** 1611–1621.

Christensen, M. E., LeStourgeon, W. M., Jamrich, M., Howard, G. C., Serunian, L. A., Silver, L. M., and Elgin, S. C. R. (1981). Distribution studies on polytene chromosomes using antibodies directed against hnRNP. *J. Cell Biol.* **90,** 18–24.

Clark, R. F., Wagner, C. R., Craig, D. A., and Elgin, S. C. R. (1991). Distribution of chromosomal proteins in polytene chromosomes of Drosophila. *Methods Cell. Biol.* **35,** 203–227.

DeCamillis, M., Cheng, N., Pierre, D., and Brock, H. W. (1992). The *polyhomeotic* gene of *Drosophila* encodes a chromatin protein that shares polytene chromosome-binding sites with *Polycomb. Genes Dev.* **6,** 223–232.

Desai, L. S., Pothier, L., Foley, G. E., and Adams, R. A. (1972). Immunofluorescent labeling of chromosomes with antisera to histones and histone fractions. *Exp. Cell Res.* **70,** 468–471.

Eissenberg, J. C., James, T. C., Foster-Hartnett, D. M., Hartnett, T., Ngan, V., and Elgin, S. C. R. (1990). Mutation in a heterochromatin-specific chromosomal protein is associated with suppression of position-effect variegation in *Drosophila melanogaster*. *Proc. Natl. Acad. Sci. U.S.A.* **87**, 9923–9927.

Elgin, S. C. R., and Boyd, J. B. (1975). The proteins of polytene chromosomes of *Drosophila hydei*. *Chromosoma* **51**, 135–145.

Field, J., Nikawa, J.-I., Broek, D., Macdonald, B., Rodgers, L., Wilson, I. A., Lerner, R. A., and Wigler, M. (1988). Purification of a *RAS*-responsive adenylyl cyclase complex from *Saccharomyces cervisiae* by use of an epitope addition method. *Mol. Cell. Biol.* **8**, 2159–2165.

Fleischmann, B., Filipski, R., and Fleischmann, G. (1989). Isolation and distribution of a *Drosophila* protein preferentially associated with active regions of the genome. *Chromosoma* **97**, 381–389.

Fleischmann, G. Filipski, R., and Elgin, S. C. R. (1987). Isolation and distribution of a *Drosophila* protein preferentially associated with inactive regions of the genome. *Chromosoma* **96**, 83–90.

Fleischmann, G., Pflugfelder, G., Steiner, E. K., Javaherian, K., Howard, G. C., Wang, G. C., and Elgin, S. C. R. (1984). *Drosophila* DNA topoisomerase I is associated with transcriptionally active regions of the genome. *Proc. Natl. Acad. Sci. U.S.A.* **81**, 6958–6962.

Franke, A., DeCamillis, M., Zink, D., Cheng, N., Brock, H. W., and Paro, R. (1992). *Polycomb* and *polyhomeotic* are constituents of a multimeric protein complex in chromatin of *Drosophila melanogaster*. *EMBO J.* **11**, 2941–2950.

Frasch, M. (1991). The maternally expressed *Drosophila* gene encoding the chromatin-binding protein FJ1 is a homolog of the vertebrate gene *Regulator of Chromatin Condensation, RCC1*. *EMBO J.* **10**, 1225–1236.

Glicksman, M. A., and Brower, D. L. (1988). Expression of the *Sex combs reduced* protein in *Drosophila* larvae. *Dev. Biol.* **127**, 113–118.

Gorman, M., Kuroda, M. I., and Baker, B. S. (1993). Regulation of the sex-specific binding of the maleless dosage compensation protein to the male X chromosome. *Cell* **72**, 39–49.

Greenleaf, A. L., Plagens, U., Jamrich, M., and Bautz, E. K. F. RNA polymerase B (or II) in heat induced puffs of *Drosophila* polytene chromosomes. *Chromosoma* **65**, 127–136.

Heller, R. A., Shelton, E. R., Dietrich, V., Elgin, S. C. R., and Brutlag, D. L. (1986). Multiple forms and cellular localization of *Drosophila* DNA topoisomerase II. *J. Biol. Chem.* **261**, 8063–8069.

Hill, R. J., Segraves, W. A., Choi, D., Underwood, P. A., and Macavoy, E. (1993). The reaction with polytene chromosomes of antibodies raised against the E75A protein. *Insect Biochem. Mol. Biol.* **23**, 99–104.

Hovemann, B. T., Dessen, E., Mechler, H., and Mack, E. (1991). *Drosophila* snRNP associated protein P11 which specifically binds to heat shock puff 93D reveals strong homology with hnRNAP core protein A1. *Nucleic Acids Res.* **19**, 4909–4914.

James, T. C., Eissenberg, J. C., Craig, C., Dietrich, V., Hobson, A., and Elgin, S. C. R. (1989). Distribution patterns of HP1, a hetrochromatin-associated nonhistone chromosomal protein of *Drosophila*. *Eur. J. Cell Biol.* **50**, 170–180.

James, T. C., and Elgin, S. C. R. (1986). Identification of a nonhistone chromosomal protein associated with heterochromatin in *Drosophila melanogaster* and its gene. *Mol. Cell. Biol.* **6**, 3862–3872.

Jamrich, M., Greenleaf, A. L., and Bautz, E. K. F. (1977). Localization of RNA polymerase in polytene chromosomes of *Drosophila melanogaster*. *Proc. Natl. Acad. Sci. U.S.A.* **74**, 2079–2083.

Kabakov, A. E., and Poverenny, A. M. (1993). Immunochemical probing of DNA structure with monoclonal antibody to OsO4/2,2'-bipyridine adduct. *Anal. Biochem.* **211**, 224–232.

Kim, Y.-J., Zuo, P., Manley, J. L., and Baker, B. S. (1992). *Drosophila* RNA-binding protein RBP1 is localized to transcriptionally active sites of chromosomes and shows a functional similarity to human splicing factor ASF-SF2. *Genes Dev.* **6**, 2569–2579.

Kontermann, R., Sitzler, S., Seifarth, W., Petersen, G., and Bautz, E. K. F. (1989). Primary structure and functional aspects of the gene coding for the second-largest subunit RNA polymerase III of Drosophila. *Mol. Gen. Genet.* **219**, 373–380.

Kuroda, M. I., Kernan, M. J., Kreber, R., Ganetzky, B., and Baker, B. S. (1991). The *maleless* protein associates with the X chromosome to regulate dosage compensation in *Drosophila*. *Cell* **66**, 935–947.

Lancillotti, F., Lopez, M. C., Alonso, C., and Stollar, B. D. (1985). Locations of Z-DNA in polytene chromosomes. *J. Cell. Biol.* **100**, 1759–1766.

Martin, E. C., and Adler, P. N. (1993). The *Polycomb* group gene *Posterior sex combs* encodes a chromosomal protein. *Development* **117**, 641–655.

Mayfield, J. E., Serunian, L. A., Silver, L. M., and Elgin, S. C. R. (1978). A protein released by DNAaseI digestion of Drosophila nuclei is preferentially associated with puffs. *Cell* **14**, 539–544.

Messmer, S., Franke, A., and Paro, R. (1992). Analysis of the functional role of the *Polycomb* chromo domain in *Drosophila melanogaster*. *Genes Dev.* **6**, 1241–1254.

Miklos, G. L., and Cotsell, J. N. (1990). Chromosome structure at interfaces between major chromatin types: α- and β-heterochromatin. *Bioessays* **12**, 1–6.

Munro, S., and Pelham, H. R. B. (1984). Use of peptide tagging to detect proteins expressed from cloned genes: Deletion mapping functional domains of *Drosophila* hsp70. *EMBO J.* **3**, 3087–3093.

Muralidhar, M G., Callahan, C. A., and Thomas, J. B. (1993). *Single-minded* regulation of genes in the embryonic midline of the *Drosophila* central nervous system. *Mech. Dev.* **41**, 129–138.

Nordheim, A., Pardue, M. L., Lafer, E. M., Möller, A., Stollar, B. D., and Rich, A. (1981). Antibodies to left-handed Z-DNA bind to interband regions of *Drosophila* polytene chromosomes. *Nature* **294**, 417–422.

Noselli, S., Payre, F., and Vincent, A. (1992). Zinc fingers and other domains cooperate in binding of *Drosophila sry* β and δ proteins at specific chromosomal sites. *Mol. Cell. Biol.* **12**, 724–733.

Paro, R., and Hogness, D. S. (1991). The Polycomb protein shares a homologous domain with a heterochromatin-associated protein of *Drosophila*. *Proc. Natl. Acad. Sci. U.S.A.* **88**, 263–267.

Payre, F., Noselli, S., Lefrere, V., and Vincent, A. (1990). The closely related *Drosophila sry* β and *sry* δ zinc finger proteins show differential embryonic expression and distinct patterns of binding sites on polytene chromosomes. *Development* **110**, 141–149.

Payre, F., and Vincent, A. (1991). Genomic targets of the *serependipity* β and δ zinc finger proteins and their respective DNA recognition sites. *EMBO J.* **10**, 2533–2541.

Pirrotta, V., Bickel, S., and Mariani, C. (1988). Developmental expression of the *Drosophila zeste* gene and localization of *zeste* protein on polytene chromosomes. *Genes Dev.* **2**, 1839–1850.

Plagens, U., Greenleaf, A. L., and Bautz, E. K. F. (1976). Distribution of RNA polymerase on *Drosophila* polytene chromosomes as studied by indirect immunofluorescence. *Chromosoma* **59**, 157–165.

Rastelli, L., Chan, C. S., and Pirrotta, V. (1993). Related chromosome binding sites for *zeste*, suppressors of *zeste* and *Polycomb* group proteins in *Drosophila* and their dependence on *Enhancer of zeste* function. *EMBO J.* **12**, 1513–1522.

Richards, G. (1980). The polytene chromosomes in the fat body nuclei of Drosophila melanogaster. *Chromosoma* **79**, 241–253.

Risau, W., Symmons, P., Saumweber, H., and Frasch, M. (1983). Nonpackaging and packaging proteins of hnRNA in Drosophila melanogaster. *Cell* **33**, 529–541.

Saumweber, H., Frasch, M., and Korge, G. (1990). Two puff-specific proteins bind within the 2.5 kb upstream region of the *Drosophila melanogaster* Sgs-4 gene. *Chromosoma* **99**, 52–60.

Saumweber, H., Symmons, P., Kabisch, R., Will, H., and Bonhoefer, F. (1980). Monoclonal antibodies against chromosomal proteins of *Drosophila melanogaster*. *Chromosoma* **80**, 255–288.

Silver, L. M., and Elgin, S. C. R. (1976). A method for determination of the *in situ* distribution of chromosomal proteins. *Proc. Natl. Acad. Sci. U.S.A.* **73**, 423–427.

Silver, L. M., and Elgin, S. C. R. (1977). Distribution patterns of three subfractions of Drosophila nonhistone chromosomal proteins: Possible correlations with gene activity. *Cell* **11**, 971–983.

Smith, A. V., and Orr-Weaver, T. L. (1991). The regulation of the cell cycle during *Drosophila* embryogenesis: The transition to polyteny. *Development* **112**, 997–1008.

Thisse, C., and Thisse, B. (1992). Dorsoventral development of the *Drosophila* embryo is controlled by a cascade of transcriptional regulators. *Development (Suppl.)* 173–181.

Tissièrres, A., Mitchell, H. K., and Tracy, U. M. (1974). Protein synthesis in salivary glands of *Drosophila melanogaster:* Relation to chromosome puffs. *J. Mol. Biol.* **84,** 389–398.

Urness, L. D., and Thummel, C. S. (1990). Molecular interactions within the ecdysone regulatory hierarchy: DNA binding proteins of the Drosophila ecdysone-inducible *E74A* protein. *Cell* **63,** 47–61.

Wieland, C., Mann, S., vonBesser, H., and Saumweber, H. (1992). The *Drosophila* nuclear protein B × 42, which is found in many puffs on polytene chromosomes, is highly charged. *Chromosoma* **101,** 517–525.

Will, H., and Bautz, E. K. F. (1980). Immunological identification of a chromocenter-associated protein in polytene chromosomes of *Drosophila. Exp. Cell Res.* **125,** 401–410.

Woodcock, C. L. F., Safer, J. P., and Stanchfield, J. E. (1976). Structural repeating units in chromatin. *Exp. Cell Res.* **97,** 101–110.

Zink, B., Engström, Y., Gehring, W. J., and Paro, R. (1991). Direct interaction of the *Polycomb* protein with *Antennapedia* regulatory sequences in polytene chromosomes in *Drosophila* melanogaster. *EMBO J.* **10,** 153–162.

Zink, B., and Paro, R. (1989). *In vivo* binding of a *trans*-regulator of homoeotic genes in *Drosophila melanogaster. Nature* **337,** 468–471.

CHAPTER 21

Looking at *Drosophila* Mitotic Chromosomes

Maurizio Gatti, Silvia Bonaccorsi, and Sergio Pimpinelli

Istituto Pasteur
Fondazione Cenci-Bolognetti
and Centro di Genetica Evoluzionistica del CNR
Dipartimento di Genetica e Biologia Molecolare
Universitá di Roma "La Sapienza"
00185 Rome, Italy

METHODS IN CELL BIOLOGY, VOL. 44

I. Introduction

Mitotic chromosome analysis is an essential tool for several aspects of *Drosophila* research. Mitotic cytology is routinely needed for definition and correct interpretation of chromosome rearrangements that involve heterochromatin. Heterochromatic breakpoints that are difficult or impossible to define by polytene chromosome analysis can be readily determined in mitotic chromosome preparations. In addition, mitotic cytology plays a fundamental role in at least two specific research lines: the mutational dissection of mitosis (for recent reviews, see Glover, 1990; Gatti and Goldberg, 1991) and the cytogenetic analysis of heterochromatin (reviewed by Gatti and Pimpinelli, 1992).

In the past 15 years, more than 60 mutants that disrupt different aspects of chromosome behavior have been isolated. The cytological analysis of these mutants has shown that they exhibit a wide range of abnormalities, which in many cases are sufficiently distinctive to suggest possible roles for their wild-type gene products. On the basis of their mitotic phenotypes, the mitotic mutants thus far isolated can be divided into six main classes: (1) those that produce high frequencies of spontaneous chromosome aberrations (Gatti, 1979; Baker *et al.*, 1982; reviewed by Gatti and Goldberg, 1991); (2) those that affect chromosome condensation (Gatti *et al.*, 1983; Gatti and Baker, 1989; Axton *et al.*, 1990); (3) those that exhibit metaphase arrest, no anaphases, and frequent polyploid cells and are therefore defective in one of the components of the mitotic apparatus (Ripoll *et al.*, 1985; Gonzalez *et al.*, 1988; Sunkel and Glover, 1988; Gatti and Baker, 1989; Rosengard *et al.*, 1989; Heck *et al.*, 1993); (4) those that exhibit frequent failures in the segregation of individual chromosomes and produce aneuploid cells (Smith *et al.*, 1985; Karess and Glover, 1989; Williams *et al.*, 1992); (5) those that affect chromatid separation during anaphase (Gatti and Goldberg, 1991; Girdham and Glover, 1991; Mayer-Jaekel *et al.*, 1993); (6) those that exhibit high frequencies of large polyploid cells but a normal frequency of anaphases and are thus defective in the completion of cell division or cytokinesis (Gatti and Baker, 1989; Karess *et al.*, 1991). Several genes specified by these diverse mutants have been cloned, providing information about the biochemical basis of the observed cytological phenotypes. For example, the *KLP61F* gene, specified by mutations that arrest cell division at metaphase and produce polyploid cells, encodes a kinesin-like protein that mediates centrosome migration to the spindle poles (Heck *et al.*, 1993), while *spaghetti squash,* identified by mutants that produce giant polyploid cells without arresting the cell cycle, encodes a myosin regulatory light chain that appears to regulate the assembly of the contractile ring during cytokinesis (Karess *et al.*, 1991).

Another research line for which mitotic chromosome cytology proved to be essential is the cytogenetic analysis of heterochromatin. Heterochromatin cannot be cytogenetically dissected by polytene chromosome analysis because the bulk of this material is included in the chromocenter. However, a series of high-resolution chromosome-banding techniques such as quinacrine, Hoechst, and N-banding provides a very fine differentiation of mitotic heterochromatin (Holmquist, 1975; Gatti *et al.*, 1976; Pimpinelli *et al.*, 1976b; Gatti and Pimpinelli, 1983). These techniques differentiate the heterochromatic segments of prometaphase chromosomes of larval neuroblasts into an array of discrete regions with different cytological features. This allows precise determination of heterochromatic breakpoints, rendering heterochromatin amenable to cytogenetic analysis (Pimpinelli *et al.*, 1986; Gatti and Pimpinelli, 1992; Pimpinelli *et al.*, 1994). The application of such banding techniques in conjunction with genetic analysis has provided substantial insights into the organization of heterochromatic functions such as the Y chromosome fertility factors (Kennison, 1981; Hazelrigg *et al.*, 1982; Gatti and Pimpinelli, 1983; Bonaccorsi *et al.*, 1988), the *crystal* (or *Suppressor of Stellate*) locus (Hardy *et al.*, 1984), the *Responder* element of the *Segregation Distortion* system (Pimpinelli and Dimitri, 1989), and the second chromosome heterochromatic loci (Dimitri, 1991).

A good mitotic cytology is also essential for fine mapping of repetitive DNA sequences along heterochromatin by *in situ* hybridization (Peacock *et al.*, 1976,1977; Steffensen *et al.*, 1981; Hilliker and Appels, 1982; Lohe and Roberts, 1988). The resolution of these experiments has been improved by employing chromosome rearrangements previously characterized by Hoechst and N-banding. Following this approach, several cloned satellite repeats (Lohe and Brutlag, 1986) were precisely mapped to specific heterochromatic blocks defined by these banding techniques (Bonaccorsi and Lohe, 1991; Lohe *et al.*, 1993). A further improvement in the molecular mapping of heterochromatic sequences can be obtained by fluorescence *in situ* hybridization (FISH) followed by DAPI staining. Digital images of either FISH or Hoechst fluorescence, recorded by a CCD camera, can be pseudocolored and merged using suitable computer programs, so that the hybridization signal can be readily localized on banded chromosomes (Palumbo *et al.*, 1994).

In the present paper, we describe in detail the protocols routinely used in our laboratories to obtain mitotic chromosome preparations. In addition, we describe the main staining protocols for heterochromatin differentiation and a FISH technique. These experimental protocols will be consistently related with the needs of the research that is currently being done on *Drosophila* chromosomes.

II. Preparation of Larval Neuroblast Chromosomes

Although good mitotic chromosome preparations can be obtained from male and female gonial cells (Bridges, 1916), the tissue that provides the best mitotic figures is the larval brain (Kaufmann, 1934). Many cytological techniques have been developed for neuroblast chromosome preparations since the early days of *Drosophila* genetics, some of which are described in detail in the recently published Ashburner's manual (Ashburner, 1989). Here, we present a series of squashing protocols that are minor modifications of a basic technique developed

20 years ago (Gatti *et al.*, 1974). These procedures have a wide applicability and can be successfully used for preparing neuroblast, gonial, and meiotic chromosomes of various *Drosophila* and mosquito species (Gatti *et al.*, 1976, 1977; Pimpinelli *et al.*, 1976a,b; Bonaccorsi *et al.*, 1980,1981; Ripoll *et al.*, 1985).

Mitotic chromosomes squashes from larval brains can be prepared by three different experimental regimes: (1) with colchicine and hypotonic pretreatment; (2) with hypotonic pretreatment only; (3) without any pretreatment. As described below, each of these regimes permits certain aspects of mitotic chromosome behavior to be precisely defined. Together, these three experimental procedures allow full characterization of the cytological phenotype of mitotic mutants (Gatti and Baker, 1989; reviewed by Gatti and Goldberg, 1991).

A. Protocol 1: Hypotonic and Colchicine Pretreatment

To obtain large numbers of well-spread metaphase figures, dissected larval brains are incubated *in vitro* with colchicine, treated with a hypotonic solution, and then fixed and squashed in aceto-orcein according to the following protocol:

1. Dissect larval brains from actively crawling third instar larvae in saline (0.7% NaCl in distilled water); there is no need to remove the imaginal discs associated with the brain.

2. Wash the dissected brains for 1 to 5 min in a drop of saline placed on a siliconized slide.

3. Place the brains in a small, covered petri dish (35 × 10 mm) containing 2 ml saline supplemented with 100 μl of 10^{-3} colchicine. Incubate for 1.5 hr at 25°C [the cell cycle of larval neuroblasts takes 8 hr, with the G2 phase lasting about 1.5 hr (Gatti *et al.*, 1974; Pimpinelli *et al.*, 1976a)].

4. Transfer the brains to a hypotonic solution of 0.5% sodium citrate (0.5 g of sodium citrate · $2H_2O$ added to 100 ml of distilled water) and incubate for 10 min at room temperature. Hypotonic treatments that exceed 10 min may induce sister chromatid separation in metaphase chromosomes.

5. Fix brains for 10–20 sec at room temperature in a mixture of acetic acid, methanol, and distilled water in the ratio 11:11:2. This fixative should be freshly prepared because it deteriorates within a few hours.

6. Transfer individual fixed brains to small drops (2 μl) of 2–3% aceto-orcein placed on very clean 20 × 20- or 22 × 22-mm nonsiliconized coverslips. One to four aceto-orcein drops can be placed on each coverslip.

7. Leave the brains in aceto-orcein for 1–2 min and then lower a very clean slide onto the coverslip; surface tension will cause the coverslip to adhere to the slide. Invert the slide and place it under three or four sheets of blotting paper; remove the excess of aceto-orcein by exerting a gentle pressure and then squash very hard. The gentle presquashing usually prevents coverslip sliding during squashing and the consequent damage to the preparation.

8. Seal the edges of the coverslip with a suitable material. We have found that the best sealing material is depilatory wax, which can be found in most cosmetic shops (at least in Europe). This wax, which at room temperature is solid, can be easily melted with a heated scalpel that can be then used to seal the edges of the coverslip. Well-sealed slides can be stored for 1–2 months at 4°C without substantial deterioration.

The quality of the squash preparations obtained with the above protocol depends on the type of the aceto-orcein employed. Good results are obtained with synthetic Gurr orcein. The orcein powder is dissolved in 45% acetic acid and then boiled for 45 min in a reflux condenser. We usually prepare 5% orcein, which is subsequently diluted with 45% acetic acid to obtain the right staining intensity. Before use, properly diluted aceto-orcein is filtered through blotting paper to remove particulate sediments that may prevent uniform squashing of the preparations.

Good squashes of colchicine-treated brains usually contain 200–400 well-spread metaphases that can be scored for the presence of chromosome aberrations and defective chromosome condensation (Fig. 1). In addition, this colchicine-treated material can be examined for the presence of aneuploid and polyploid cells (Fig. 1). However, since colchicine induces metaphase arrest followed by chromosome overcontraction, colchicine treatment must be omitted if the degree of chromosome condensation has to be evaluated. This can be done by omitting Step 3 of Protocol 1 and leaving the other steps unchanged.

Omission of both colchicine treatment (Step 3) and incubation in hypotonic solution (Step 4) allows the observation of all phases of mitosis and permits evaluation of the mitotic index and the frequency of anaphases (Gatti and Baker, 1989). Anaphases are almost absent in hypotonically treated brains because hypotonic shock disrupts these mitotic figures (Brinkley et al., 1980). However, in squash preparations without colchicine and hypotonic pretreatments chromosome morphology is poorly defined.

B. Protocol 2: Permanent Preparations

Although well-sealed aceto-orcein squashes remain in good condition for 1–2 months, there are cases in which permanent chromosome preparations are needed. This can be done according to the following protocol:

1. Follow Steps 1–5 of Protocol 1 [colchicine treatment (Step 3) and hypotonic swelling (Step 4) may be omitted depending on the desired type of preparation].
2. Transfer fixed brains to small drops (2 μl) of 45% acetic acid placed on very clean 20 × 20-mm siliconized coverslips. Very good siliconized coverslips can be obtained using the SurfaSil siliconizing fluid (Pierce). Clean 20 × 20-mm coverslips are immersed in this fluid, dipped in liquid soap, and thoroughly washed under tap water. After air drying, the coverslips are rubbed with a lint-free cloth to remove excess of silicon.

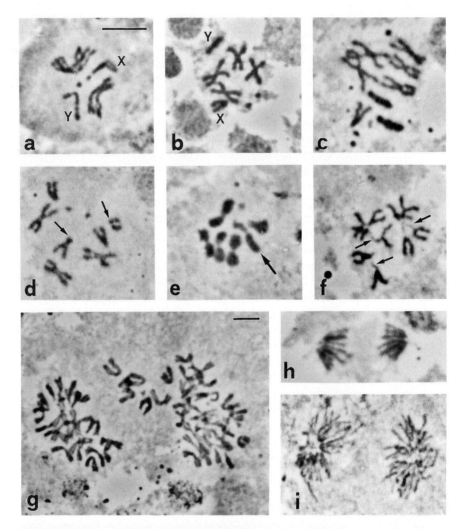

Fig. 1 Examples of orcein-stained mitotic chromosome preparations of *D. melanogaster* larval neuroblasts. (a) Wild-type male metaphase from a brain without colchicine and hypotonic pretreatments. (b) Colchicine-treated, hypotonically swollen normal male metaphase. (c) Colchicine- and hypotonic-treated aneuploid metaphase from *l(1)zw10* (Smith *et al.*, 1985; Williams *et al.*, 1992). (d) Colchicine- and hypotonic-treated metaphase from *mei-41* (Gatti, 1979) showing an isochromatid deletion of a major autosome (arrows). (e) Abnormally condensed metaphase chromosomes from a *mit(1)4* brain (Smith *et al.*, 1985), treated with hypotonic solution but not with colchicine; note the elongated Y chromosome (arrow). (f) Colchicine- and hypotonic-treated metaphase from *mus-101^{ts}* (Gatti *et al.*, 1983), showing undercondensation of the heterochromatic regions (arrows). (g) Noncolchicine, hypotonic-treated polyploid cells from *l(3)7m-62* (Gatti and Baker, 1989). (h) A wild-type anaphase from a noncolchicine, nonhypotonic-treated brain. (i) Polyploid anaphase from *l(3)7m-62*. Magnification is the same in all panels except g; the polyploid metaphases in g have a lower magnification. Bars, 5 μm.

3. Squash brains as in Protocol 1.

4. Freeze slides on dry ice or in liquid nitrogen.

5. Flip off the coverslip with a razor blade and air dry.

6. Stain in 4% Giemsa in a phosphate buffer at pH 7 and differentiate by washing with tap water. The timing of Giemsa staining varies with the Giemsa brand and should be adjusted to obtain the desired staining. Remember that Giemsa stain is additive, so that insufficiently stained chromosomes can be stained again.

7. Mount in Euparal or similar medium.

III. Chromosome-Banding Techniques

About 35% of the *Drosophila melanogaster* male genome is composed of constitutive heterochromatin. The entire Y chromosome, the proximal 40% of the X chromosome, the centric 25% of the major autosomes, and most of the fourth chromosome exhibit typical heterochromatic properties (Heitz, 1933; Kaufmann, 1934; Cooper, 1959; Pimpinelli *et al.*, 1978). These regions are more condensed than euchromatin at prophase, exhibit sister chromatid apposition at prometaphase and metaphase (see Fig. 1), and are positively stained by the C-banding technique that specifically identifies constitutive heterochromatin (Hsu, 1971; Pimpinelli *et al.*, 1976b). Moreover, in favorable aceto-orcein-stained prometaphase chromosomes, prepared without colchicine and hypotonic pretreatments, heterochromatic regions are segmented into a number of blocks separated by secondary constrictions (Cooper, 1959). Additional properties of *Drosophila* heterochromatin are late replication with respect to euchromatin (Barigozzi *et al.*, 1966) and a particular richness in highly repetitive simple sequence satellite DNAs, which are virtually absent from euchromatin (reviewed by Lohe and Roberts, 1988).

The chromosome-banding techniques developed for mammalian chromosomes do not differentiate *Drosophila* euchromatin into G or R bands. However, most of these techniques produce a sharp banding of the heterochromatin portions of *Drosophila* chromosomes (Holmquist, 1975; Gatti *et al.*, 1976). Here, we describe three banding techniques, the Hoechst-, quinacrine-, and N-banding procedures, that produce a very fine longitudinal differentiation of *D. melanogaster* heterochromatin and allow precise localization of heterochromatic breakpoints (see Fig. 2). These techniques can also be successfully employed to differentiate heterochromatin of various *Drosophila* and mosquito species (Gatti *et al.*, 1976,1977; Bonaccorsi *et al.*, 1980,1981).

Hoechst-, quinacrine-, and N-banding produce different staining patterns of *D. melanogaster* heterochromatin. Some regions that are poorly differentiated by one banding technique are sharply defined by another banding procedure. Thus, some heterochromatic areas are clearly differentiated only by sequential

378 Maurizio Gatti et al.

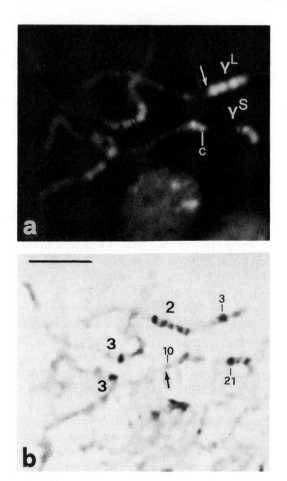

Fig. 2 A partial prometaphase from a male carrying the Y-autosome translocation T(Y;2) P57. The chromosome preparation from a noncolchicine, hypotonic-treated brain is sequentially stained with Hoechst 33258 (a) and N-banding (b). The large numbers indicate the autosomes; the small numbers identify Y-chromosome regions stained by N-banding. Note the Y-chromosome breakpoint at the junction between regions h9 and h10, the N-band on chromosome 3, and the five N-bands on chromosome 2. Bar, 5 μm. Reproduced from Gatti and Pimpinelli (1983) with permission. © Springer-Verlag.

application of two banding procedures. This situation is illustrated in Fig. 3, which shows a complete cytological map of D. melanogaster heterochromatin, suggesting which banding procedure(s) should be applied for the best differentiation of each heterochromatic region.

The degree of resolution of the banding techniques depends on the degree of elongation of the heterochromatic regions. We have found that the most suitable chromosome preparations for banding analysis are those prepared without colchicine pretreatment but with hypotonic shock. Thus, for all the banding

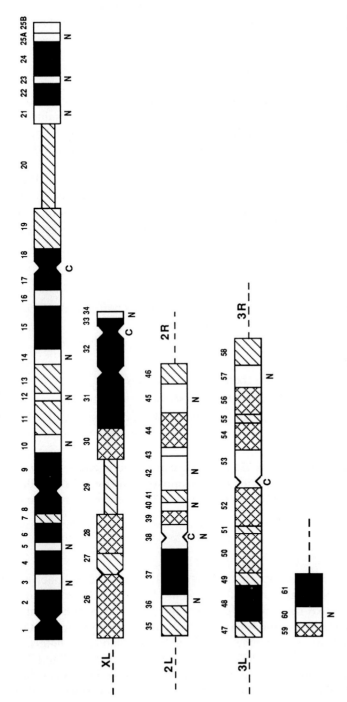

Fig. 3 Cytological map of *D. melanogaster* heterochromatin. The entirely heterochromatic Y chromosome, the X chromosome, and the second, the third, and the fourth chromosome heterochromatin are schematically represented from top to bottom. Only the heterochromatic portions of chromosomes are shown; euchromatin is depicted as a broken line. C indicates the position of the centromere; the location of the fourth chromosome centromere has not been precisely determined. The diagrams are representative of prometaphase neuroblast chromosomes stained with Hoechst 33258. Filled segments indicate bright fluorescence, cross-hatched segments moderate fluorescence, hatched segments dull fluorescence, and open segments no fluorescence. The N letters below each diagram indicate the N-banded regions.

procedures chromosomes are prepared following Protocol 2 to Step 9, with the omission of Step 3. Air-dried chromosome preparations can be kept at 4°C for a few days before processing.

A. Hoechst Banding

Hoechst staining consistently produces a highly reproducible banding pattern that provides a heterochromatin differentiation finer than that obtained by quinacrine or N-banding. Thus, it should be considered the basic banding technique to define heterochromatic breakpoints. Depending on the localization of these breakpoints on Hoechst-stained chromosomes, a choice can be made on whether another banding technique should be sequentially applied. The Hoechst staining protocol described below has been developed by Latt (1973) and applied to *Drosophila* chromosomes with slight modifications (Gatti *et al.*, 1976; Gatti and Pimpinelli, 1983). Other Hoechst staining protocols have been described by Holmquist (1975) and Hazelrigg and co-workers (1982).

1. Hoechst Staining Protocol

1. Rehydrate air-dried slides for 5 min in Hoechst buffer (HB: 150 mM NaCl, 30 mM KCl, 10 mM Na$_2$HPO$_4$) at pH 7.
2. Stain for 10 min in 0.5 μg/ml Hoechst 33258 dissolved in HB.
3. Wash quickly (~5 sec) in HB.
4. Air dry slides, keeping them in a vertical position.
5. Mount in 160 mM Na$_2$HPO$_4$, 40 mM Na citrate at pH 7.
6. Seal the coverslip with rubber cement and store slides in the dark at 4°C for 1–2 days before observation. This treatment reduces fluorescence fading.
7. Analyze slides with a Zeiss fluorescence microscope equipped with a mercury light source for incident illumination and the 0.1 Zeiss filter combination (BP 365/11 excitation filter, FT 395 dichroic mirror, and LP 397 barrier filters).

B. Quinacrine Banding

The array of fluorescent and dark bands observed after quinacrine staining is similar but not identical to the Hoechst banding. The major differences between these two banding patterns are in region h18 of the Y chromosome and in region h31–h32 of the X chromosome, which are both Hoechst bright and quinacrine dull. In addition, the quinacrine banding along the heterochromatic regions of the autosomes is less sharp than the Hoechst banding (Gatti *et al.*, 1976; Gatti and Pimpinelli, 1983; Pimpinelli *et al.*, 1994). The quinacrine staining protocol that follows has been developed by Gatti and co-workers (1976); other quinacrine-banding techniques for *Drosophila* chromosomes have been described by Vosa (1970), Ellison and Barr (1971), and Faccio Dolfini (1974).

1. Quinacrine Staining Protocol

1. Immerse slides for 5 min in absolute ethanol.
2. Stain slides for 10 min in 0.5% quinacrine dihydrocloride (Gurr) in absolute ethanol.
3. Wash twice (5 sec each) in absolute ethanol; air dry, keeping the slides in vertical position.
4. Mount in distilled water.
5. Seal the coverslip with rubber cement and store slides in the dark at 4°C for 1–2 days before observation. This treatment improves the degree of differentiation and reduces fluorescence fading.
6. Examine slides with a Zeiss fluorescence microscope equipped with a mercury light source for incident illumination and the 0.9 Zeiss filter combination (BP 450–490 excitation filter, FT 510 dichroic mirror, and LP 420 barrier filter).

C. N-banding

The N-banding procedure developed by Matsui and Sasaki (1973) and modified by Funaki and co-workers (1975) specifically stains the nucleolus organizing regions (NORs) of several animal and plant species. In *D. melanogaster*, there are 15 heterochromatic regions that are positively stained by the N-banding technique, 8 on the Y chromosome, 1 on the X chromosome, 5 on the second chromosome, 1 on the third chromosome, and 1 on the fourth chromosome (see Fig. 3; Pimpinelli *et al.*, 1976b; Gatti and Pimpinelli, 1983; Dimitri, 1991; Pimpinelli *et al.*, 1994). These regions are not fluorescent after Hoechst or quinacrine staining and do not correspond to the NORs; the nucleolar constrictions of the X and Y chromosomes are not N-banded (Pimpinelli *et al.*, 1976b). The N-banding procedure presented below is essentially that of Funaki and co-workers (1975), with minor modifications (Pimpinelli *et al.*, 1976b).

1. N-banding Protocol

1. Keep air-dried slides for 2–5 days at 4°C before processing; 2- to 5-day-old slides usually give better results than either fresh or older slides.
2. Incubate slides in 1 M NaH_2PO_4 for 15 min at 85°C.
3. Rinse in distilled water.
4. Stain for 20 min in 4% Giemsa dissolved in phosphate buffer, pH 7.

5. Rinse in tap water and air dry.
6. Mount in Euparal.
7. Examine slides using a phase contrast microscope.

D. Sequential Banding

The Hoechst-, quinacrine-, and N-banding procedures can be sequentially applied in two combinations: Hoechst, quinacrine, and N-banding or quinacrine, Hoechst, and N-banding.

Sequential staining with Hoechst and quinacrine is performed according to the following protocol:

1. Peel off the rubber cement from the coverslip edges, trying to completely remove it.
2. Place the slides in a Coplin jar containing Hoechst buffer and keep them there until the coverslip slides off. To facilitate coverslip sliding, the jar may be shaked or tapped against the bench.
3. Air dry.
4. Destain by washing for 30 sec in a 3 : 1 solution of methanol/acetic acid.
5. Air dry and stain with quinacrine using the standard protocol.

A similar procedure is used to sequentially stain with quinacrine and Hoechst:

1. Peel off the rubber cement from the coverslip edges as described above.
2. Dismount slides as described above by immersion in distilled water.
3. Air dry.
4. Wash in 70% ethanol for 5 min.
5. Destain by washing three times (30 min each) in a 3 : 1 solution of ethanol/acetic acid.
6. Air dry and stain with Hoechst using the standard protocol.

The N-banding technique can be successfully applied to slides previously stained with one or more fluorochromes. The slides, dismounted in distilled water as indicated above, do not need to be destained; after air drying, they are rapidly washed with 95% ethanol and air dried again, stored for 2–5 days at 4°C, and then processed with the N-banding technique, using the standard protocol.

E. The Cytochemical Basis for Banding

Quinacrine binds to DNA by intercalation without an appreciable specificity for DNA base composition. However, quinacrine fluorescence is enhanced by AT-rich DNA, whereas GC-rich DNA tends to quench quinacrine fluorescence (for reviews see Verma and Babu, 1989; Sumner, 1990). Hoechst 33258 binds in the minor groove of the DNA molecule with preference for AT sequences.

In addition, *in vitro* studies have shown that Hoechst fluorescence is markedly enhanced by AT-rich DNA, whereas the enhancement is less with GC-rich DNA. Thus, both quinacrine and Hoechst are thought to be general indicators of AT-richness along the chromosomes (Verma and Babu, 1989; Sumner, 1990).

The N-banding procedure involves the extraction of DNA, RNA, and histones from the fixed chromosomes. It has been therefore suggested that Giemsa staining reveals certain nonhistone proteins that remain associated with the chromosomes after extraction of both nucleic acids and histones (Matsui and Sasaki, 1973; Funaki *et al.,* 1975).

Hoechst, quinacrine, and N-banding identify four main types of heterochromatic blocks in *D. melanogaster* chromosomes: (1) those that are Hoechst and quinacrine brightly fluorescent and N-banding negative; (2) those that are Hoechst bright, quinacrine dull, and N-banding negative; (3) those that exhibit intermediate Hoechst and quinacrine fluorescence and are not stained by N-banding; and (4) those that are Hoechst and quinacrine negative but are positively stained by N-banding (see Fig. 3). Based on the fluorescence properties of quinacrine and Hoechst, and on studies on satellite DNA distribution along *D. melanogaster* heterochromatin, it has been suggested that quinacrine- and Hoechst-bright blocks are enriched in AT-rich satellite DNAs (Holmquist, 1975; Gatti *et al.,* 1976; Gatti and Pimpinelli, 1983). Moreover, the correspondence between the location of the N-bands and the distribution of the 1.705 g/cm^3 satellite DNA suggested the hypothesis that the N-banding procedure stains chromosomal proteins specifically associated with this type of highly repetitive DNA (Gerlach, 1977; Gatti and Pimpinelli, 1983). These inferences have been substantiated by *in situ* hybridization studies with cloned satellite DNAs (Lohe and Brutlag, 1986), which showed that the quinacrine- and Hoechst-bright regions contain large numbers of AATAT, AATAC, AATAAAC, and AATAACATAG repeats, whereas the N-banded regions are invariably enriched in the AAGAG repeats cloned from the 1.705 g/cm^3 satellite DNA (Bonaccorsi and Lohe 1991; Lohe *et al.,* 1993). *In situ* hybridization studies have also shown that the Hoechst-bright, quinacrine-dull region h31-h32 of the X chromosome accommodates the relatively AT-rich 359-bp satellite DNA (Hilliker and Appels, 1982; Lohe and Roberts 1988; Lohe *et al.,* 1993). Most of the Hoechst and quinacrine-dull regions do not contain known simple sequence satellite DNAs. Thus, they may either accommodate satellite sequences that have not yet been identified or contain long stretches of middle repetitive DNAs (Bonaccorsi and Lohe 1991; Lohe *et al.,* 1993).

IV. *In Situ* Hybridization

In situ hybridization techniques with either tritiated probes detected by autoradiography or biotinylated probes detected by nonfluorescent staining do not allow simultaneous visualization of the signal and the chromosome banding. Thus, to assign repetitive DNA sequences to the heterochromatic blocks defined by the banding techniques, *in situ* hybridization has to be performed on cytologically determined chromosome rearrangements (Bonaccorsi and Lohe, 1991; Lohe *et al.*, 1993). The main limitation of this approach is the availability of suitable breakpoints in the heterochromatic regions that contain the DNA repeats to be mapped. An additional limitation is the considerable amount of time and experimental efforts needed to map DNA sequences with multiple chromosomal locations. Precise mapping of these sequences requires the execution and the analysis of several *in situ* hybridization experiments, performed on a large series of well-characterized chromosome rearrangements.

To overcome these limitations, we have adapted to *Drosophila* chromosomes a fluorescence *in situ* hybridization technique that can be coupled with DAPI staining, to produce a Hoechst-like banding pattern. Digital images of the FISH signal and the DAPI fluorescence are recorded separately by a CCD camera, pseudocolored, and merged using suitable computer programs, so that DAPI banding and the hybridization signal can be simultaneously visualized on the computer screen (Fig. 4). Using this technology, we have localized several middle-repetitive DNAs along *D. melanogaster* heterochromatin, assigning them to specific regions of the banding map (Palumbo *et al.*, 1994; our unpublished results). Another FISH-CCD camera method for the analysis of *Drosophila* chromosomes has been described by Hostenbach and co-workers (1993).

The fluorescence *in situ* hybridization procedures described below are essentially those of Lichter *et al.* (1990), with the modifications of D'Aiuto *et al.* (1993). We have introduced a few additional modifications to these basic procedures to adapt them to *Drosophila* preparations.

A. Probe Preparation and Denaturation

1. Label 1 μg of probe (linearized plasmids or excised fragments) by nick-translation using either biotin-11-dUTP (Enzo) or digoxigenin-11-dUTP (Boehringer, Mannheim) following standard procedures (for a detailed protocol, see Trask, 1991). The DNase/polymerase ratio is adjusted to obtain labeled fragments of 300–500 bp.

2. Precipitate the desired amount of labeled DNA (40–80 ng per slide) with the addition of sonicated salmon sperm DNA (3 μg per slide), 0.1 vol of 3 *M* sodium acetate, pH 4.5, and 2 vol of cold (−20°C) absolute ethanol. Place at −80°C for 15 min and spin precipitated DNA at 13,000 rpm for 15 min. Dry the pellet in a Savant centrifuge.

3. Prepare the hybridization mixture (10 μl per slide) by mixing 5 μl ultrapure formamide (J. Baker), 2 μl 50% dextran sulfate, 2 μl distilled water, and 1 μl 20× SSC.

4. Resuspend the DNA pellet in the hybridization mixture (10 μl per slide) by vortexing.

5. Denature probe mixture at 80°C for 8 min; place probe on ice until used.

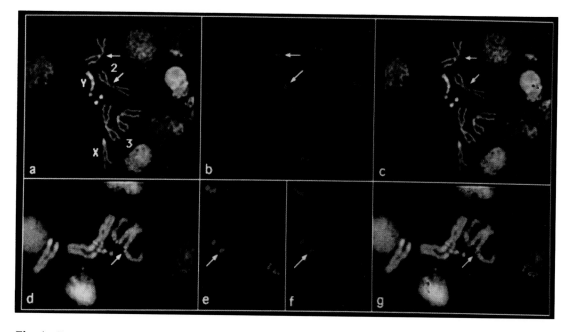

Fig. 4 Chromosomal localization of the *Bari-1* (Caizzi *et al.,* 1992) and *Responder* (*Rsp*) (Wu *et al.,* 1988) repetitive sequences by fluorescent *in situ* hybridization and CCD camera analysis. The *Bari-1* probe was labeled with digoxigenin and detected by a rhodamine-conjugated antibody, while the *Rsp* probe was labeled with biotin and detected by FITC-conjugated avidin. (a–c) Localization of the *Bari-1* sequences; (d–g) simultaneous localization of the *Bari-1* and *Rsp* sequences. (a) DAPI-stained male metaphase, pseudocolored in blue. (b) *In situ* hybridization signal from *Bari-1* sequences, pseudocolored in red. (c) Merging of (a) and (b) by the Gene Join computer program. (d) A DAPI-stained female metaphase. (e) *In situ* hybridization signal from *Rsp*, pseudocolored in yellow. (f) *In situ* hybridization signal from *Bari-1*, pseudocolored in red. (g) Merging of (d), (e), and (f). The two sequences are both localized in region h39 of the second chromosome heterochromatin, with the *Rsp* sequence proximal to the centromere, in agreement with the breakpoint/*in situ* hybridization mapping performed by Caizzi and co-workers (1993).

B. Slide Denaturation

1. Prepare 50 ml denaturation solution by mixing 35 ml 70% ultrapure formamide (J. Baker), 5 ml 20× SSC, and 10 ml distilled water. Pour this solution in a Coplin jar and place in a water bath at 70°C.

2. Dehydrate 2- to 3-day-old slides (prepared according to Protocol 2 up to Step 9, with the omission of Step 3) by immersion in 70, 90, and 100% ethanol (3 min each time). Air dry slides after dehydration.

3. Immerse slides in the denaturation solution at 70°C for exactly 2 min. To avoid a temperature drop of this solution, do not immerse more than three to four slides at any given time.

4. Transfer the slides quickly to 70% ethanol at −20°C, incubate for 3 min, and then dehydrate through ice-cooled 90 and 100% ethanol (3 min each time). Air dry slides.

C. Hybridization and Posthybridization Washes

Do not allow slides to dry during the following passages.

1. Apply 10 μl hybridization mixture to denatured slides and cover with 24 × 24-mm, clean coverslips, trying to avoid the formation of air bubbles. Seal the edges of the coverslip with rubber cement.

2. Incubate slides overnight in a moist chamber at 37°C.

3. Peel off the rubber cement, gently remove the coverslip, and wash slides three times (5 min each) in 50% formamide, 2× SSC at 42°C.

4. Wash slides three times (5 min each) in 0.1× SSC at 60°C and blot excess liquid from the slide edges.

5. Apply 100 μl of blocking solution (4× SSC, 3% BSA, 0.1% Tween 20) to each slide, cover with a 24 × 24-mm coverslip, and incubate at 37°C for 30 min.

The hybridization temperature we routinely use for middle repetitive probes is 37°C. However, for simple sequence satellite DNAs other hybridization temperatures should be used, depending on their nucleotide composition and sequence. In *D. melanogaster,* different satellite repeats show up to 80% sequence homology. Thus, to avoid cross-hybridization among different classes of satellite sequences, for each satellite probe the hybridization temperature is maintained at 8–12°C below its T_m value [for a detailed account on the hybridization conditions for satellite DNAs, see Bonaccorsi and Lohe (1991) and Lohe *et al.* (1993)]

D. Biotin-Labeled DNA Detection

1. Remove coverslips and blot excess liquid from the slide edges.

2. Apply to each slide 80 μl of 3.3 μg/ml fluorescein isothiocyanate (FITC)-

conjugated avidin (Vector Laboratories) diluted in 4× SSC, 1% BSA, 0.1% Tween 20; cover with a 24 × 24-mm coverslip and incubate for 30 min at 37°C in a dark, moist chamber.

3. Remove coverslips and wash three times (5 min each) in 4× SSC, 0.1% Tween 20, at 42°C.

4. Stain with 0.16 µg/ml 4,6-diamino-2-phenylindole-dihydrochloride (DAPI) in 2× SSC for 5 min at room temperature.

5. Rinse in 2× SSC at room temperature.

6. Mount slides in 20 mM Tris–HCl, pH 8, 90% glycerol containing 2.3% of DABCO [1,4-diazo-bicyclo-(2,2,2)octane; Merck] antifade. Seal the coverslip with rubber cement and store in the dark at 4°C. Slides can be stored for weeks without substantial deterioration.

E. Digoxigenin-Labeled DNA Detection

1. Remove coverslips and blot excess liquid from the slide edges.

2. Apply to each slide 80 µl of 2 µg/ml rhodamine-conjugated anti-digoxigenin sheep IgG, Fab fragments (Boehringer, Mannheim), diluted in 4× SSC, 1% BSA, 0.1% Tween 20; cover with a 24 × 24-mm coverslip and incubate for 30 min at 37°C in a dark, moist chamber.

3. Remove coverslips and wash three times (5 min each) in 4× SSC, 0.1% Tween 20, at 42°C.

4. Stain with DAPI as in D, Step 4.

5. Rinse as in D, Step 5.

6. Mount as in D, Step 6.

F. Double Labeling

To perform simultaneous *in situ* hybridizations with biotin- and digoxigenin-labeled probes, 40–80 ng per slide of each probe are mixed together and ethanol precipitated as described in A, Step 2. Probe and slide denaturation, hybridization, and posthybridization washes are the same as for single probes. For signal detection, slides are incubated for 30 min with 80 µl 3.3 µg/ml FITC-conjugated avidin (Vector Laboratories), 2 µg/ml rhodamine-conjugated anti-digoxigenin sheep IgG, Fab fragments (Boehringer, Mannheim), diluted in 4× SSC, 1% BSA, 0.1% Tween 20; they are then processed following Steps 3–6 in D.

G. Fluorescence Microscopy and Image Analysis

Chromosome preparations are analyzed using a computer-controlled Zeiss Axioplan epifluorescence microscope equipped with a cooled CCD camera (Photometrics). FITC, rhodamine, and DAPI fluorescence are visualized using the Pinkel No 1 filter set combination (Chroma Technology) that allows detection of these fluorochromes without any image shifting. The fluorescent signals, recorded separately as gray-scale digital images by IP Spectrum Lab Software, are pseudocolored and merged using the Adope Photoshop program.

V. Summary and Conclusions

The repertoire of cytological procedures described in the present paper permits full analysis of brain neuroblast chromosomes. Moreover, if brains are cultured for 13 hr in the presence of 5-bromo-2'-deoxy-uridine, our fixation and Hoechst staining protocols allow visualization of sister chromatid differentiation and the scoring of sister chromatid exchanges (Gatti et al., 1979). Finally, we note that our cytological procedures can be successfully employed for preparation and staining of gonial cells of both sexes and male meiotic chromosomes (Ripoll et al., 1985; our unpublished results). Good chromosome preparations of female meiosis are obtained with the procedure described by Davring and Sunner (1977,1979), Nokkala and Puro (1976), and Puro and Nokkala (1977).

In this chapter, we have focused on the organization and behavior of Drosophila mitotic chromosomes, describing a repertoire of cytological techniques for neuroblast chromosome preparations. We have not considered the numerous excellent cytological procedures for embryonic chromosome preparations (for an example, see Foe and Alberts, 1985; Foe, 1989), because these chromosomes are usually less clearly defined than those of larval neuroblasts. In addition, we have not included the whole-mount and squashing techniques that allow chromosome visualization and spindle immunostaining of neuroblast cells (Axton et al., 1990; Gonzalez et al., 1990), male meiotic cells (Casal et al., 1990; Cenci et al., 1994), and female meiotic cells (Theurkauf and Hawley, 1992), because the fixation methods used in these procedures alter chromosome morphology.

Fixation methods for antibody staining result in poorly defined chromosomes, whereas the methanol/acetic acid fixation techniques, such as those described here, preserve very well chromosome morphology but remove a substantial fraction of chromosomal proteins. Thus, one of the major technical breakthroughs in Drosophila mitotic cytology will be the development of fixation procedures that maximize chromosomal quality with minimal removal of proteins. This will be particularly useful for precise immunolocalization of heterochromatic proteins, including those associated with the centromere.

Acknowledgments

We are grateful to Mariano Rocchi and Maria Berloco for advice in the FISH/CCD camera procedures. We also thank Enzo Marchetti for help in computer image analysis. This work has been in part supported by a grant from Progetto Finalizzato Ingegneria Genetica.

References

Ashburner, M. (1989). "*Drosophila:* A Laboratory Handbook." Cold Spring Harbor, NY: Cold Spring Harbor Laboratory Press.

Axton, J. M., Dombradi, V., Cohen P. T. W. and Glover, D. M. (1990). One of the protein phosphatase 1 isoenzymes in *Drosophila* is essential for mitosis. *Cell* **63**, 33–46.

Baker, B. S., Smith, D. A., and Gatti, M. (1982). Region specific effects on chromosome integrity of mutants at essential loci in *Drosophila melanogaster*. *Proc. Natl. Acad. Sci. U.S.A.* **79**, 1205–1209.

Barigozzi, C., Dolfini, S., Fraccaro, M., Rezzonico Raimondi, G., and Tiepolo, L. (1966). In vitro study of the DNA replication patterns of somatic chromosomes of *Drosophila melanogaster*. *Exp. Cell Res.* **43**, 231–234.

Bonaccorsi, S., and Lohe, A. (1991). Fine mapping of satellite DNA sequences along the Y chromosome of *Drosophila melanogaster:* Relationships between satellite sequences and fertility factors. *Genetics* **129**, 177–189.

Bonaccorsi, S., Pimpinelli, S. and Gatti, M. (1981). Cytological dissection of sex chromosome heterochromatin of *Drosophila hydei*. *Chromosoma* **84**, 391–403.

Bonaccorsi, S., Pisano, C., Puoti, F., and Gatti, M. (1988). Y chromosome loops in *Drosophila melanogaster*. *Genetics* **120**, 1015–1034.

Bonaccorsi, S., Santini, G., Gatti, M., Pimpinelli, S., and Coluzzi, M. (1980). Intraspecific polymorphism of sex chromosome heterochromatin in two species of the *Anopheles gambiae* complex. *Chromosoma* **76**, 57–64.

Bridges, C. B. (1916). Non-disjunction as a proof of the chromosome theory of heredity. *Genetics* **1**, 1–52 and 107–163.

Brinkley, B. R., Cox, S. M., and Pepper, D. A. (1980). Structure of the mitotic apparatus and chromosomes after hypotonic treatment of mammalian cells in vitro. *Cytogenet. Cell Genet.* **26**, 165–176.

Caizzi, R., Caggese, C., and Pimpinelli, S. (1993). *Bari-1,* a new transposon-like family in *Drosophila melanogaster* with a unique heterochromatic organization. *Genetics* **133**, 335–345.

Casal, J., Gonzalez, C., Wandosell, F., Avila, J., and Ripoll, P. (1990). Spindles and centrosomes during male meiosis in *Drosophila melanogaster*. *Eur. J. Cell Biol.* **51**, 38–44.

Cenci, G., Bonaccorsi, S., Pisano, C., Verni, F., and Gatti, M. (1994). Chromatin and microtubule organization during premeiotic, meiotic, and early postmeiotic stages of *Drosophila melanogaster* spermatogenesis. Submitted.

Cooper, K. W. (1959). Cytogenetic analysis of major heterochromatic elements (especially *Xh* and *Y*) in *Drosophila melanogaster* and the theory of "heterochromatin." *Chromosoma* **10**, 535–588.

D'Aiuto, L., Antonacci, R., Marzella, R., Archidiacono, N., and Rocchi, M. (1993). Cloning and comparative mapping of a human chromosome 4-specific alpha satellite DNA sequence. *Genomics* **18**, 230–235.

Davring, L., and Sunner, M. (1977). Late prophase and first metaphase in female meiosis of *Drosophila melanogaster*. *Hereditas* **85**, 25–32.

Davring, L., and Sunner, M. (1979). Cytological evidence for procentric synapsis of meiotic chromosomes in female *Drosophila melanogaster*. The behavior of an extra Y chromosome. *Hereditas* **91**, 53–64.

Dimitri, P. (1991). Cytogenetic analysis of the second chromosome heterochromatin of *Drosophila melanogaster*. *Genetics* **127**, 553–564.

Ellison, J. R., and Barr, H. J. (1971). Differences in the quinacrine staining of the chromosomes of a pair of sibling species: *Drosophila melanogaster* and *Drosophila simulans*. *Chromosoma* **44**, 424–435.

Faccio Dolfini, S. (1974). The distribution of repetitive DNA in the chromosomes of cultured cells of *Drosophila melanogaster*. *Chromosoma* **44**, 383–391.

Foe, V. (1989). Mitotic domains reveal early commitment of cells in *Drosophila* embryos. *Development* **107**, 1–22.

Foe, V., and Alberts, B. M. (1985). Reversible chromosome condensation induced in *Drosophila* embryos by anoxia: Visualization of interphase nuclear organization. *J. Cell Biol.* **100**, 1623–1636.

Funaki, K., Matsui, S., and Sasaki, M. (1975). Location of nucleolar organizers in animal and plant chromosomes by means of an improved N-banding technique. *Chromosoma* **49**, 357–370.

Gatti, M. (1979). Genetic control of chromosome breakage and rejoining in *Drosophila melanogaster:* Spontaneous chromosome aberrations in X-linked mutants defective in DNA metabolism. *Proc. Natl. Acad. Sci. U.S.A.* **76**, 1377–1381.

Gatti, M., and Baker, B. S. (1989). Genes controlling essential cell cycle functions in *Drosophila melanogaster*. *Genes Dev.* **3**, 438–453.

Gatti, M., and Goldberg, M. L. (1991). Mutations affecting cell division in *Drosophila*. *In* "Methods in Cell Biology" (B. A. Hamkalo and S. C. R. Elgin, eds.), Vol. 35, pp. 543–585. San Diego, Academic Press.

Gatti, M., and Pimpinelli, S. (1983). Cytological and genetic analysis of the Y chromosome of *Drosophila melanogaster*. I. Organization of the fertility factors. *Chromosoma* **88**, 349–373.

Gatti, M., and Pimpinelli, S. (1992). Functional elements in *Drosophila melanogaster* heterochromatin. *Annu. Rev. Genet.* **26**, 239–275.

Gatti, M., Pimpinelli, S., and Santini, G. (1976). Characterization of *Drosophila* heterochromatin. I. Staining and decondensation with Hoechst 33258 and quinacrine. *Chromosoma* **57**, 351–375.

Gatti, M., Santini, G., Pimpinelli, S., and Coluzzi, M. (1977). Fluorescence banding techniques in the identification of sibling species of the *Anopheles gambiae* complex. *Heredity* **38**, 105–108.

Gatti, M., Santini, G., Pimpinelli, S., and Olivieri, G. (1979). Lack of spontaneous sister chromatid exchanges in somatic cells of *Drosophila melanogaster*. *Genetics* **91**, 255–274.

Gatti, M., Smith, D. A., and Baker, B. S. (1983). A gene controlling condensation of heterochromatin in *Drosophila melanogaster*. *Science* **221**, 83–85.

Gatti, M., Tanzarella, C., and Olivieri, G. (1974). Analysis of the chromosome aberrations induced by X-rays in somatic cells of *Drosophila melanogaster*. *Genetics* **77**, 701–719.

Gerlach, W. L. (1977). N-banded karyotypes of wheat species. *Chromosoma* **62**, 49–56.

Girdham, C. H., and Glover, D. M. (1991). Chromosome tangling and breakage at anaphase result from mutations in *lodestar*, a *Drosophila* gene encoding a putative nucleotide triphosphate binding protein. *Genes Dev.* **5**, 1786–1799.

Glover, D. (1990). Abbreviated and regulated cell cycles in *Drosophila*. *Curr. Opin. Cell Biol.* **2**, 258–261.

Gonzalez, C., Casal, J., and Ripoll, P. (1988). Functional monopolar spindles caused by the mutation in *mgr*, a cell division gene of *Drosophila melanogaster*. *J. Cell Sci.* **89**, 34–47.

Gonzalez, C., Saunders, R. D. C., Casal, J., Molina, I., Carmena, M., Ripoll, P., and Glover, D. M. (1990). Mutations at the *asp* locus of *Drosophila* lead to multiple free centrosomes in syncytial embryos, but restrict centrosome duplication in larval neuroblasts. *J. Cell Sci.* **96**, 605–616.

Hardy, R. W., Lindsley, D. L., Livak, K. J., Lewis, B., Siversten, L., Joslyn, G. L., Edwards, J., and Bonaccorsi, S. (1984). Cytogenetic analysis of a segment of the Y chromosome of *Drosophila melanogaster*. *Genetics* **107**, 591–610.

Hazelrigg, T., Fornili, P., and Kaufman, T. C. (1982). A cytogenetic analysis of X-ray induced male steriles on the Y chromosome of *Drosophila melanogaster*. *Chromosoma* **87**, 535–559.

Heck, M. M. S., Pereira, A., Pesavento, P., Yannoni, Y., Spradling, A. C., and Goldstein, L. S. B. (1993). The kinesin-like protein KLP61F is essential for mitosis in *Drosophila, J. Cell Biol.,* **123**, 665–679.

Heitz, E. (1933). Dies somatische heteropyknose bei *Drosophila melanogaster* und ihre genetische bedeutung, *Z. Zellforsch.* **20**, 237–287.

Hilliker, A. J., and Appels, R. (1982). Pleiotropic effects associated with the deletion of heterochromatin surrounding rDNA on the X chromosome of *Drosophila*. *Chromosoma* **86**, 469–490.

Holmquist, G. (1975). Hoechst 33258 fluorescent staining of *Drosophila* chromosomes. *Chromosoma* **49**, 333–356.

Hostenbach, R., Wilbrink, M., Suijkerbuijk, R., and Hennig, W. (1993). Localization of the lampbrush loop pair *Nooses* on the Y chromosome of *Drosophila hydei* by fluorescence in situ hybridization. *Chromosoma* **102,** 546–552.

Hsu, T. C. (1971). Heterochromatin pattern in metaphase chromosomes of *Drosophila melanogaster. J. Hered.* **62,** 285–287.

Karess, R. E., Chang, X-J., Edwards, K. A., Kulkarni, S., Aguilera, I., and Kiehart, D. P. (1991). The regulatory light chain of non-muscle myosin is encoded by *spaghetti-squash,* a gene required for cytokinesis in *Drosophila. Cell* **65,** 1177–1189.

Karess, R. E., and Glover, D. M. (1989). *rough deal:* A gene required for proper mitotic segregation in *Drosophila. J. Cell Biol.* **109,** 2951–2961.

Kaufmann, B. P. (1934). Somatic mitoses of *Drosophila melanogaster. J. Morphol.* **56,** 125–155.

Kennison, J. A. (1981). The genetical and cytological organization of the Y chromosome of *Drosophila melanogaster. Genetics* **98,** 529–548.

Latt, S. A. (1973). Microfluorometric detection of deoxyribonucleic acid replication in human metaphase chromosomes. *Proc. Natl. Acad. Sci. U.S.A.* **70,** 3395–3399.

Lichter, P., Tang, C.-J. C., Call, K., Hermanson, G., Evans, G. A., Housman, D., and Ward, D. C. (1990). High-resolution mapping of human chromosome 11 by in situ hybridization with cosmid clones. *Science* **247,** 64–69.

Lohe, A. R., and Brutlag D. L. (1986). Multiplicity of satellite DNA sequences in *Drosophila melanogaster. Proc. Natl. Acad. Sci. U.S.A.* **83,** 696–700.

Lohe, A. R., Hilliker, A. J., and Roberts, P. A. (1993). Mapping simple repeated DNA sequences in heterochromatin of *Drosophila melanogaster. Genetics* **134,** 1149–1174.

Lohe, A. R., Roberts, P. A. (1988). Evolution of satellite DNA sequences in *Drosophila. In* "Heterochromatin: Molecular and Structural Aspects" (R. Verma, ed.), pp. 148–186. Cambridge: Cambridge Univ. Press.

Matsui, S., and Sasaki, M. (1973). Differential staining of nucleolus organizers in mammalian chromosomes. *Nature* **246,** 148–150.

Mayer-Jaekel, R. E., Okhura, H., Gomes, R., Sunkel, C., Baumgartner, S., Hemmings, B. A., and Glover, D. M. (1993). The 55kd regulatory subunit of *Drosophila* protein phosphatase 2A is required for anaphase. *Cell* **72,** 621–633.

Nokkala, S., and Puro, J. (1976). Cytological evidence for a chromocenter in *Drosophila melanogaster* oocytes. *Hereditas* **83,** 265–268.

Palumbo, G., Bonaccorsi, S., Robbins, L., and Pimpinelli, S. (1994). Genetic analysis of *Stellate* elements of *Drosophila melanogaster. Genetics,* in press.

Peacock, W. J., Appels, R., Dunsmuir, P., Lohe, A. R., and Gerlach, W. L. (1976). Highly repeated DNA sequences: Chromosomal location and evolutionary conservation. *In* "International Cell Biology 1976–1977" (B. K. Brinkley and K. R. Porter, eds.), pp. 494–506. New York: Rockfeller Univ. Press.

Peacock, W. J., Lohe, A. R., Gerlach, W. L., Dunsmuir, P., Dennis, E. S., and Appels, R. (1977). Fine structure and evolution of DNA in heterochromatin. *Cold Spring Harbor Symp. Quant. Biol.* **42,** 1121–1135.

Pimpinelli, S., Bonaccorsi, S., Dimitri, P., and Gatti, M. (1994). A cytological map of *Drosophila melanogaster* heterochromatin, in preparation.

Pimpinelli, S., Bonaccorsi, S., Gatti, M., and Sandler, L. (1986). The peculiar genetic organization of *Drosophila* heterochromatin, *Trends Genet.* **2,** 17–20.

Pimpinelli, S., and Dimitri, P. (1989). Cytogenetic organization of the *Rsp (Responder)* locus in *Drosophila melanogaster. Genetics* **121,** 765–772.

Pimpinelli, S., Pignone, D., Gatti, M., and Olivieri, G. (1976a). X-ray induction of chromatid interchanges in somatic cells of *Drosophila melanogaster:* Variation through the cell cycle of the pattern of rejoining. *Mutat. Res.* **35,** 101–110.

Pimpinelli, S., Santini, G., and Gatti, M. (1976b). Characterization of *Drosophila* heterochromatin. II. C- and N-banding. *Chromosoma* **57,** 377–386.

Pimpinelli, S., Santini, G., and Gatti, M. (1978). ^{3}H-Actinomycin-D binding to mitotic chromosomes of *Drosophila melanogaster. Chromosoma* **66,** 389–395.

Puro, J., and Nokkala, S. (1977). Meiotic segregation of chromosomes in *Drosophila melanogaster* oocytes. *Chromosoma* **63,** 273–286.

Ripoll, P., Pimpinelli, S., Valdivia, M. M., and Vaila, J. (1985). A cell division mutant of *Drosophila* with a functionally abnormal spindle. *Cell* **41,** 907–912.

Rosengard, A. M., Krutzsch, H. C., Shearn, A., Biggs, J. R., Barker, E., Margulies, I. M. K., King, C. R., Liotta, L. A., and Steeg, P. S. (1989). Reduced NM23/awd protein in tumor metastasis and aberrant *Drosophila* development. *Nature* **342,** 177–180.

Smith, D. A., Baker, B. S., and Gatti, M. (1985). Mutations in genes encoding mitotic functions in *Drosophila melanogaster*. *Genetics* **110,** 647–670.

Steffensen, D. L., Appels, R., and Peacock, W. J. (1981). The distribution of two highly repeated DNA sequences within *Drosophila melanogaster* chromosomes. *Chromosoma* **82,** 525–541.

Sumner, A. T. (1990). "Chromosome Banding." London: Unwin Hyman Ltd.

Sunkel, C., and Glover, D. M. (1988). *polo,* a mitotic mutant of *Drosophila* displaying abnormal mitotic spindles. *J. Cell Sci.* **89,** 25–38.

Theurkauf, W. E., and Hawley, R. W. (1992). Meiotic spindle assembly in *Drosophila* females: Behavior of nonexchange chromosomes and the effects of mutations in the kinesin-like protein. *J. Cell Biol.* **116,** 1167–1180.

Trask, B. J. (1991). DNA sequence localization in metaphase and interphase cells by fluorescence *in situ* hybridization. *In* "Methods in Cell Biology" (B. A. Hamkalo and S. C. R. Elgin, eds.), Vol. 35, pp. 3–35. San Diego: Academic Press.

Verma, R. S., and Babu, A. (1989). "Human Chromosomes: Manual of Basic Techniques." New York: Pergamon.

Vosa, C. G. (1970). The discriminating fluorescence patterns of the chromosomes of *Drosophila melanogaster*. *Chromosoma* **31,** 446–451.

Williams, B. C., Karr, T. L., Montgomery, J. M., and Goldberg, M. L. (1992). The *Drosophila* *l(1)zw10* gene product, required for accurate mitotic chromosome segregation, is redistributed at anaphase onset. *J. Cell Biol.* **118,** 759–773.

Wu, C.-I., Lyttle, Wu, M.-L., and Lin, G.-F. (1988). Association between a satellite DNA sequence and the *Responder (Rsp)* of *Segregation Distorter* in *Drosophila melanogaster*. *Cell* **54,** 179–189.

CHAPTER 22

Looking at Diploid Interphase Chromosomes

Shermali Gunawardena and Mary Rykowski

Departments of Anatomy and Molecular and Cellular Biology and
Graduate Program in Genetics
College of Medicine
University of Arizona
Tucson, Arizona 85724

I. Introduction

Partly as a result of improved reagents and optical equipment, there has been an increasing interest in nuclear substructure. From the localization of transcription machinery to understanding the mechanism of meiosis, it is necessary to look at the nucleus directly. New light microscopic techniques, including *in vivo* three-dimensional wide-field and confocal microscopy and sensitive *in situ* hybridization and immunofluorescence reagents are helping to map the nucleus both structurally and functionally. Once the poor relation in the cell biologist's family, the relatively small nuclei of *Drosophila* are now amenable

METHODS IN CELL BIOLOGY, VOL. 44

to the same kinds of analysis possible for larger mammalian cells using these powerful techniques.

The purpose of this article is to present a procedure for two-color *in situ* hybridization to DNA in embryonic nuclei. This procedure, with the appropriate changes, should be valuable for hybridization to imaginal discs and cultured cells, although we have not attempted this. DNA *in situ* hybridization is best used in combination with immunofluorescence (see Chapters 21, 24, and 25) and RNA *in situ* hybridization (see Chapters 24, 30, and 31) to build a complete picture of the working nucleus. Moreover, because it is still a new technique, many more observations will be required to determine the maximum resolution of two-color interphase DNA *in situ* hybridization. *In vivo* observations (see Chapter 26) wherever possible, are also required to understand the temporal relationship between different states and to corroborate observations of fixed material.

Before presenting the technique, we will present a short introduction to what is already known about diploid chromosome structure, and for this it is necessary to review what is known about the structure of polytene chromosomes.

A. Polytene Chromosomes, the First Interphase System

The nuclei of early *Drosophila* embryos are diploid and undergo synchronous divisions until the 14th nuclear cycle, at which point they become cellularized (for review, see Ashburner, 1989). Most cells undergo no more than two or three more divisions, the major exceptions being the cells that form the imaginal discs and the nervous system. The 10-fold increase in size that occurs over the 5-day course of larval life is accomplished by cell growth and genome amplification. There are two modes of genome amplification, polyploidy and polyteny; the bulk of the body mass of the emerging larva is polytene, including the gut, fat body, and salivary gland.

In polyteny, the homologously paired interphase chromosomes are replicated but never separate or condense. Instead, round after round, the individual chromatids remain associated, first as ribbon-like arrays (Ananiev and Barsky, 1985) and then as cables, eventually containing up to 1000–2000 individual fibers (Hochstrasser and Sedat, 1987). With the exception of heterochromatin at the centromeres, telomeres, and a few other loci interspersed in euchromatin, polytene chromosomes are likely to compromise bundles of continuous fibers that contain the same chromatin structures as their diploid counterparts.

Polytene chromosomes are typically used to help paint a picture of interphase chromatin, primarily through the observation of fixed, squashed material (see Chapters 19 and 20). However, the global, three-dimensional organization of polytene nuclei has also been examined in some detail using high-resolution optical microscopy and conventional physical sectioning. Using the obvious banding pattern of polytenes, it is possible to identify and localize specific regions of the chromosome. As a result of this work, we know that polytene nuclei in the salivary glands and some other tissues show the classic Rabl orientation, with the chromocenter (the specialized centromeric region of polytenes) diametrically opposed to the telomeres (Mathog *et al.*, 1984). In addition, the arms of the chromosomes coil in a right-handed sense from the chromocenter to the telomeres. Chromosome arms generally occupy separate radial segments of the nucleus, never crossing over or intertwining (but see following). The order of the chromosome arms is nonrandom, but is not strongly correlated between putative sister cells (Mathog and Sedat, 1989). Likewise, the orientation of the chromosomes is not related to the axis of the tissue nor to the position of the lumen. Some chromosomal sites appear to associate at high frequency with the nuclear envelope (Hochstrasser *et al.*, 1986); the association sites are conserved even in rearranged chromosomes (Mathog and Sedat, 1989).

How many of these global characteristics are required for polytene function? Probably very few if any. In cells of the gut, which undergo extreme changes in shape during peristalsis, the order is lost to a great extent (Hochstrasser and Sedat, 1987). Chromosome arms tear free of their moorings at the chromocenter and nuclear envelope, appearing to float free in the nucleus. Moreover, the genome can be rearranged significantly, while manifesting defects easily explained as local effects on genes lying near the chromosomal breakpoints. Changes in gene expression had no obvious effect on chromosome position (Hochstrasser and Sedat, 1986). These indications suggest that the global organization of the more ordered polytene nuclei may be a vestige of previous diploid order and not a requirement for normal gene expression (Hochstrasser and Sedat, 1987).

B. Diploid Embryonic Chromosomes

Polytene chromosomes are functional interphase chromosomes, but they are not ideal models for chromosomes of vertebrates, which are usually diploid or occasionally polyploid. To what extent do diploid nuclei reflect the global order of polytene nuclei? This question can be answered only by direct observations of diploid nuclei, the systems of choice being the nuclei of embryos and neuroblasts (see Chapter 21). Until recently, this task has been limited by the inherent lack of resolution in diploid interphase nuclei. Light microscopy has obvious limitations, but even in electron micrographs, it has been difficult to follow individual chromosome fibers for very long and impossible to identify individual genes. Nevertheless, observation of anoxic nuclei (Foe and Alberts, 1985) (in which chromosomes condense prematurely) and localization of centric heterochromatin showed that diploid embryonic chromosomes have a Rabl orientation and that this orientation is regular with respect to the embryo axis. [This means that a particular gene tends to lie in the same plane in adjacent nuclei (see Fig. 2).] Chromosome-nuclear envelope attachment sites are transiently observed along the nuclear equator as chromosomes condense at mitosis (Hiraoka *et al.,* 1989), but the chromosome morphology at this time is too diffuse to confirm that the associated sites in diploids are identical to the ones in polytenes. Homolog pairing in diploids has been inferred (Metz, 1916; Hilliker, 1986) and recently observed directly (Hiraoka *et al.,* 1993). Based on these observations, it seems safe to conclude that polytene and diploid nuclei are constructed in much the same way, their differences reflecting the consequences of a few changes in the regulation of cell cycle events in the two types of cells.

There are questions about diploid chromosomes that cannot be approached by looking at polytenes, especially with respect to early development and meiosis. Techniques are needed to identify specific DNA sequences among the tangle of fibers left after chromosomes decondense after mitosis. Such a technique is presented here. This procedure is based on published work (Hiraoka *et al.,* 1993); similar techniques have been presented for mammalian cells (Johnson *et al.,* 1991) and *Drosophila* RNA (Chapter 30).

II. Two-Color Fluorescence *in Situ* Hybridization to Single-Copy DNA Sequences in Diploid Nuclei

In assembling this procedure, we have assumed very little in terms of experience with embryo preparation or molecular techniques. Unless noted, procedures for making solutions can be found in Sambrook *et al.* (1989). At the risk of being tiresome, we have tried to be as explicit as possible about how we actually do the preparation as a reasonable starting point for others. There is no implication that we have exhaustively optimized at every step. The known pitfalls are noted under Section E, but the list may be incomplete.

A. Nick-Translation of Probe DNA

1. Equipment and Apparatus

> G50 Sephadex spin columns
> Clinical centrifuge
> Radiation monitor
> Microcentrifuge tubes

2. Procedure (enough for one hybridization in one color):

   1. The labeling reaction for 1 μg plasmid, phage, or fragment DNA (Table I) is carried out for 60 min at 25°C.
   2. Purify the labeled DNA and remove unincorporated nucleotides by spinning through a G50 Sephadex-spun column (Sambrook *et al.,* 1989). The incorporation should be around 30–50%, judged by counting radiation from the column and the eluate with a hand-held meter.

Table I
Nick–Translation Reaction

Solutions	Add volume
10× nick-translation buffer	2.5 μl
ACG mix	2.5 μl
1 : 400 dilution DNaseI[a]	1.5 μl
DNA polymeraseI	1 μl
1 : 10 dilution [^{32}P]dATP	1 μl
DNA	1 μg
Biotin-16-dUTP	0.6 μl
or	or
Dig mix	1 μl
Milli-Q water	to 25 μl

[a] 1 : 400 dilution of DNase into 1× DNase dilution buffer is made the day of experiment and stored on ice until used.

3. Precipitate the labeled DNA by adding 20 μg of herring testis carrier DNA, 1/10 vol 3M sodium acetate, and 2.5 vol absolute ethanol. Incubate 20 min in a dry ice/ethanol bath. Spin at 4°C for 10 min at 14,000 rpm. Wash in 70% ethanol and air dry. Resuspend in 10 μl hybridization buffer. Mix digoxigenin- and biotin-labeled probes for two-color hybridization.

4. To determine the single-strand length of the labeled probe, run on a denaturing polyacrylamide gel (Sambrook *et al.*, 1989), transfer electrophoretically to nylon membrane (Sambrook *et al.*, 1989), and visualize using either avidin or anti-digoxigenin linked to alkaline phosphatase (Boehringer Mannheim) using manufacturer's instructions. The single-strand length should be 75–150 bp.

B. Collection and Fixation of Embryos

1. Equipment Needed in Preparation Area (see Tables III and IV for recipes)

Clean fine paint brush (2 inch)

Detergent saline in 1-liter squeeze bottle, aerated

Glass baking dish 8 × 8 inch or equivalent

200 ml household bleach and 200 ml 1× Buffer A, aerated, in a 500-ml beaker

40-mesh (to retain dead flies) and 20-mesh (to catch eggs) stainless steel sieves ("Cellector" from Bellco Biotechnology, Vineland NJ)

500-ml beaker with detergent saline, aerated

10 ml heptane/10 ml 1× buffer A with 4% formaldehyde in a screw-capped glass test tube (KIMAX 16× 125 mm)

Stopwatch

Powdered dry ice

10 ml heptane/10 ml methanol in a screw-capped glass test tube (KIMAX
16× 125 mm), cooled to −70°C

5–7 ml of methanol in a screw-capped small glass test tube (KIMAX 13×
100 mm) for embryo storage

Platform shaker (New Brunswick, Edison, NJ)

Water bath at 37°C

Table II
Stock Solutions for Nick-Translation

Stock solution	Stock concentration	Final concentration	Volume
10× nick-translation buffer[a]			
Tris, pH 7.6	1 M	0.5 mM	7.5 ml
MgSO$_4$	1 M	0.1 M	1.5 ml
dithiothreitol	1 M	1 mM	15 μl
bovine serum albumin	10 mg/ml	1 mg/ml	7.5 μl
Milli-Q water			5.9 ml
10× DNase dilution buffer[a]			
Tris, pH 7.5	1 M	10 mM	0.1 ml
MgCl$_2$	1 M	10 mM	0.1 ml
Milli-Q water			9.8 ml
ACG mix[a]			
dATP	100 mM	0.3 mM	30 μl
dCTP	100 mM	0.3 mM	30 μl
dGTP	100 mM	0.3 mM	30 μl
Milli-Q water			910 μl
dTTP[a]			
dTTP	100 mM	1 mM	1 μl
Milli-Q water			99 μl
[^{32}P]dATP			
[^{32}P]dATP (DuPont, Boston, MA)	10 μCi/ml	1 μCi/ml	1 μl
Milli-Q water			9 μl
Dig mix			
Digoxigenin-11-dUTP	1 mM	0.49 mM	25 μl
dTTP	1 mM	0.26 mM	12.5 μl
Milli-Q water			12.5 μl
Other stocks			
Biotin-16-dUTP	1 mM		
DNase I (Worthington, DPFF grade)	1 mg/ml		
DNA polymerase I (New England Biolabs, Beverly, MA)	10 U/ml		

Note. 100 mM dNTP solutions (already neutralized) were from Pharmacia LKB Biotechnology, Piscataway NJ. Bovine serum albumin, Biotin-16-dUTP, and digoxigenin-11-dUTP were from Boehringer Mannhiem Biochemicals, Indianapolis, IN.
[a] Stock solutions were stored at −20°C in aliquots.

Table III
Preparations for Embryo Collection

	Add/final volume
Feeding plates	
Agar	88 gm
Molasses	360 ml
Milli-Q water	2.5 liter
Autoclave for 30 min, cool to 50°C and add 20 ml ethyl acetate. Pour food into shallow 6 × 8.5-inch styrofoam meat plates or equivalent and leave to set.	
Yeast paste	
Active dry yeast	~1/3
Water	~2/3
Mix until paste forms a consistency of creamy peanut butter.	

2. Procedure:

Steps 3 through 6 should not exceed 6–7 min. If these steps take longer, the embryo morphology will suffer.

1. Harvest embryos from a population case (see Chapter 10) fed with molasses medium (see Table II for recipe). Before collection, prefeed the population cage using two feeding plates and generously spread with yeast paste for 90 min to encourage flies to lay eggs that they have been holding.

2. Collect eggs on three feeding plates with a thin spread of yeast paste, for 60–80 min and age for 20–30 min (depending on temperature) to ensure embryos are in cycles 10–14.

3. Wash the eggs into the glass dish with aerated detergent saline and collect them using the two sieves, the coarse sieve inside the fine one (Fig. 1A).

4. Put embryos into 50% bleach in 1× buffer A for 90 sec to dechorionate, stirring constantly to keep embryos from sinking to the bottom. Collect in the fine sieve (Fig. 1B).

5. Rinse embryos with a stream of detergent saline until all smell of bleach is gone.

6. Rinse the embryos to the corner of the sieve (Fig. 1C) using a small amount of the heptane phase of the fixation medium. Transfer them to the two phase permeabilization and fixation medium using a Pasteur pipet avoiding the aqueous phase. Detergent from this phase will degrade morphology.

7. Shake embryos vigorously (350 rpm) in the fixation medium in a horizontal position at room temperature for 15–20 min.

8. Using a Pasteur pipet, remove embryos to the tube containing heptane/methanol cooled to −70°C with dry ice.

Table IV
Solutions for Fixation of Embryos

	Stock concentration	Final concentration	Add/final volume
Detergent saline[a]			
Triton X-100		0.05%	1 ml
NaCl		0.7%	14 g
Milli-Q water			2 liters
10× buffer A*[b]			
Pipes		150 mM	22.68 g
KCl		800 mM	29.82 g
NaCl		200 mM	5.85 g
Add 400 ml Milli-Q water, pH to 6.8,[c] using 10 N NaOH			
Milli-Q water			to 500 ml
Autoclave			
40% fresh formaldehyde[d]			
Paraformaldehyde (Polysciences, Warrington, PA)			1 g
Milli-Q water to 2.5 ml			
NaOH (freshly made)	10 N		3.5 μl
Boil to dissolve, cool, and filter through 0.2 μm Nalgene filter			
Two-phase permeabilization and fixation medium[d]			
10× buffer A*		1×	1 ml
EGTA, pH 7	250 mM	0.5 mM	20 μl
EDTA, pH 7	250 mM	2 mM	80 μl
Spermidine	0.5 M	0.5 mM	10 μl
Spermine	0.5 M	0.2 mM	4.0 μl
2-mercaptoethanol	14.5 M	0.1%	10.5 μl
Fresh formaldehyde	40%	4%	1 ml
Milli-Q water			7.88 ml
Heptane			10 ml
1× buffer A[d]			
10× buffer A*		1×	40 ml
EGTA, pH 7	250 mM	0.5 mM	40.8 ml
EDTA, pH 7	250 mM	2 mM	3.2 ml
Spermidine	0.5 M	0.5 mM	0.4 ml
Spermine	0.5 M	0.2 mM	160 μl
2-mercaptoethanol	14.5 M	0.1%	420 μl
Milli-Q water			315 ml
50% buffer A/50% bleach[d]			
Household bleach	5% NaClO		200 ml
1× buffer A			200 ml

[a] Solutions that can be stored at room temperature.

[b] Solutions that are stored at 4°C.

[c] Use 1× buffer A, pH 7.2 for mounting when using fluoroscein for visualization.

[d] Solutions freshly made on the day of prep.

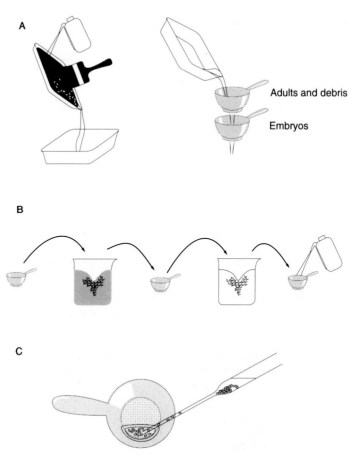

Fig. 1 Embryo fixation procedure. See text for explanation.

9. Pack the tube in powdered dry ice and shake at 350 rpm in a horizontal position for 10 min.

10. Quickly transfer test tube to the 37°C water bath and shake vigorously for 30 sec until almost all of the devitellinized embryos sink to the methanol phase.

11. Transfer the devitellinized embryos to the tube containing methanol and step through a series (5:0, 4:1, 3:2, 2:3, 1:4, 0:5) of methanol/buffer A washes (about 5 min at room temperature) into 1× buffer A.

12. Freshly prepared embryos are stained with 0.1 μg/ml DAPI in 1× buffer A to assess chromosome morphology and then used for DNA–DNA *in situ* hybridization.

C. *In Situ* Hybridization to Whole–Mount Embryos

1. ˙Equipment and Apparatus (see Table V for recipes)

> 6 × 50 mm glass culture tubes (Kimax)
> Pasteur pipets
> Parafilm
> Rotator for washes
> Heating block at 70°C
> Hot water bath
> Ice bucket
> Incubator at 37°C

2. Procedure:

1. Rinse (on rotator) freshly fixed embryos three times with 1× PBT for 10 min each.
2. Remove as much PBT as possible using a pulled Pasteur pipet. Treat the embryos with 100 μl RNase A solution for 15 min.

Table V
Solutions and Buffers for *in Situ* Hybridization

Solution	Stock concentration	Final concentration	Add/final volume
10× PBS and PBT[a]			
NaCl		0.13 N	37.9 g
Na$_7$HPO$_4$		7 mM	9.34 g
NaH$_2$PO$_4$		3 mM	2.1 g
Milli-Q water			500 ml
Autoclave			
For 10× PBT add			
Surfact-Amp 20 (Pierce, Rockford, IL)		10%	0.1%
Hybridization buffer[a]			
Formamide (distilled)	100%	50%	5 ml
Herring testis DNA	5 mg/ml	100 μg/ml	0.2 ml
Surfact-Amp 20	10%	0.1%	0.1 ml
SSC	20×	4×	2 ml
Milli-Q water			2.7 ml
RNase A solution[b]			
RNase A[b, c] (DNase free)	20 mg/ml	10 mg/ml	50 μl
PBT (diluted from 10× above)	2×	1×	50 μl
1× buffer A[b] (Table 4)			
40% formaldehyde (Table 4)			

  [a] Solutions that can be made and stored.
  [b] Solutions made on the day of experiment.
  [c] RNase A (Boehringer Mannhiem Biochemicals, Indianapolis, IN). Boil as per Sambrook *et al.* (1989).

3. Rinse embryos again in 1× PBT three times for 10 min each.

4. Remove as much PBT as possible using a pulled Pasteur pipet.

5. Add 50% hybridization buffer/50% 1× PBT; incubate 10 min.

6. Add 100% hybridization buffer for 1 hr at 37°C.

7. Denature embryo chromosomes by heating to 70°C for 15 min and chill immediately.

8. Denature DNA probes just before use by boiling in a hot water bath for 5 min and chill immediately in ice water.

9. Remove as much buffer from the embryos as possible, add 20 μl of denatured DNA probes to embryos, add heavy mineral oil to fill the tube, cover with Parafilm, and incubate on a tube rotator at 37°C for 24 to 36 hr.

10. After hybridization, remove oil layer and add 1/3 vol 1× PBT, incubate 10 min at 37°C, and remove as much as possible; repeat five times. Wash five times in 1× PBT for 10 min at room temperature.

D. Visualization of Hybridization Signals

1. Procedure:

All incubations and washes are done using a tube rotator.

1. Hybridization signals are detected by incubating embryos at room temperature for 1 hr in anti-digoxigenin FITC and Avidin–Texas Red (Boehringer Mannheim) diluted to the appropriate dilutions. Generally, we find that dilutions of 1/1000 to 1/10,000 are optimal.

2. Wash embryos again three times 10 min each and postfix with 4% formaldehyde in PBT for 10 min.

3. Rinse again in 1× PBT twice for 10 min.

4. The embryos are mounted in 1× buffer A containing 0.1 μg/ml DAPI and observed on a fluorescent microscope using a 40×, 1.3 NA lens as described (Rykowski, 1991). An image of the hybridization is shown in Fig. 2. Notice that signals are seen in a narrow band along the embryo axis. This is because the gene is localized at a constant distance below the embryo surface. The curvature of the embryo and depth of focus of the lens allow only a small portion of the signals to be seen at one time. Note that in this and subsequent figures, the images are flat-field corrected but are not deconvolved and are not confocal images. Thus, the images are presented as they would look in any fluorescent microscope.

E. Notes on Techniques

To obtain enough embryos to hybridize efficiently and with a minimum of fixation artifact, it is preferable to use mass isolation procedures. For this reason, it is easiest to obtain staged embryos from population cages. Of course,

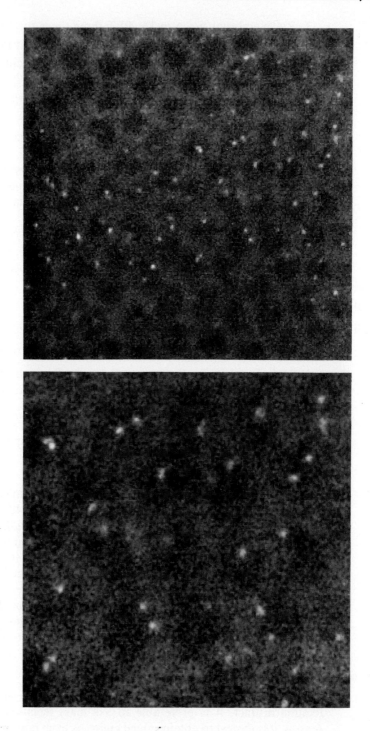

it is sometimes necessary to use smaller amounts, as when examining mutants, but reducing the scale will increase the extent or loss of embryos in rinsing steps. Less than 300–500 μl of eggs will be inconvenient using the procedures described. Hand dechorionation and devitellinization can be used (Ashburner, 1989), but may be impractical to produce sufficient numbers to be used in *in situ* hybridization as described. Single embryo procedures using specially designed vessels for rinsing and incubations should also be possible.

The most important consideration in the preparation is the elimination of fixation artifacts, which significantly alter the native structure of the chromosome. The most important source of fixation artifact is anoxia (Foe and Alberts, 1985), in which nuclei swell and DNA condenses abnormally. Examples of such preparations are shown in Fig. 3 and in Foe and Alberts (1985). To avoid this artifact, work quickly and aerate all solutions on the day of the preparation.

Another source of artifacts is incorrect concentrations of salts, detergents or solvents, and soapy glassware. These are manifest by obvious problems with morphology in mitotic figures. Normal mitotic chromosomes are shown in Fig. 4 and can also be seen in Hiraoka *et al.* (1989, 1990, 1993). They compare favorably with *in vivo* recorded mitotic figures (Hiraoka *et al.*, 1990). Abnormal chromosomes, most obvious with anaphase, look stringy, misshapen, and kinked; anaphases are often separated into several clumps along the mitotic axis. A common problem is that detergent solution is transferred with the embryos from the last rinse into the fixative. This should be avoided, and if a small amount of aqueous phase is taken up, it is better to sacrifice a few embryos than to transfer detergent.

Another potential source of artifact is hybridization to RNA rather than to DNA. Controls should be done in every preparation to demonstrate that hybridization requires denaturation of the embryonic chromosomes and that hybridization is DNase sensitive using RNase-free DNase. Also, yolk cells, pole cells, and mitotic figures should all hybridize if the hybridization is to DNA and not to RNA.

We find that denaturation at higher temperatures can slightly increase signal intensity, but only at the expense of chromosome morphology. In our hands, nuclear morphology after hybridization is indistinguishable from that immediately after fixation.

Fig. 2 Hybridization of a 7.8-kb single-copy DNA fragment to a cycle 14 interphase embryo. The probe is labeled with biotin-16-dUTP and the hybridization signal is visualized with avidin–Texas Red. A single, wide-field optical section bisects the embryo through a plane containing the hybridization signals. Notice that the curvature of the embryo allows only the topmost signals to be in focus. This and subsequent images were obtained using a 40×, 1.3 NA lens and a charge-coupled device (CCD) camera (Photometrics). Other than flat-field correction, the images are unprocessed. An enlargement of the top photo is shown below.

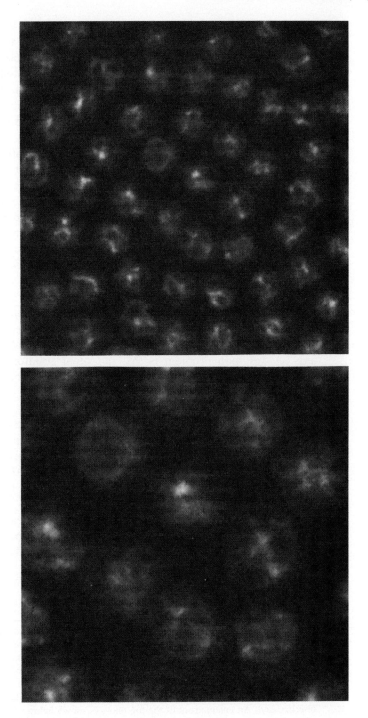

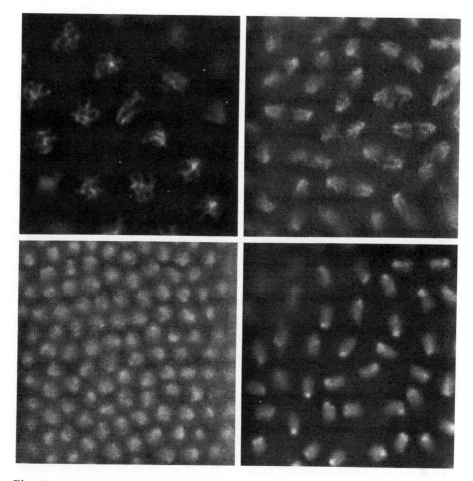

Fig. 4 A montage of normal nuclei. Clockwise from upper left: cycle 12 metaphase, cycle 13 anaphase, cycle 13 telophase, cycle 14 interphase. Notice that the mitotic figures are compact and that the density gradients along the chromosome arms are smooth; the mitoses are synchronous. The interphase nuclei are, save the spots of heterochromatin near the embryo surface, uniform in density, and at the same depth within the embryo. There are no patches of missing nuclei or other abnormalities.

Fig. 3 Anoxic nuclei. A cycle 13 embryo was fixed as described, except that the embryo was deprived of oxygen. Notice the clumpy appearance of the DNA and the enlarged nuclei. A portion of the top photo is enlarged below.

III. Summary and Future Directions

Using this two-color *in situ* hybridization protocol, we can resolve two signals using a minimum probe size of 7.8 kb, which are separated from each other by 25 kb. We would be unable to resolve these signals if they were the same color. Shorter probes have been used in mammalian cells (Xing *et al.*, 1993). Presumably, this problem will be solved with subtle improvements in technique, and perhaps, by sacrificing preservation to some extent. The question now is, what is the protocol good for? The results of the hybridization are clear enough and what we see is at least consistent with what we expect. The signal is in a position in the nucleus indicative of its position on the chromosome and there are one or two signals in each nucleus. When two probes that lie near each other on the molecular map are chosen, they appear close together in the nucleus. However, interpretation of all DNA *in situ* hybridization techniques poses several problems. The first is that conditions used to denature nucleic acids might alter cellular structures, so material must be fixed. Fixation artifacts may be undetected until a large number of data are produced and may be subtly different in different stages of the cell cycle as the level of DNA modification and the complement of associated proteins changes. It would be ideal to find a procedure that allows visualization to DNA that does not require denaturation or fixation. Simultaneous visualization of RNA and DNA has been achieved by others (Xing *et al.*, 1993), but we have yet to attempt it.

References

Ananiev, E. V., and Barsky, V. E. (1985). Elementary structures in polytene chromosomes of *Drosophila melanogaster. Chromosoma* **93,** 104–112.

Ashburner, M. (1989). "Drosophila, A Laboratory Manual." Cold Spring Harbor, NY: Cold Spring Harbor Laboratory Press.

Foe, V. E., and Alberts, B. M. (1985). Reversible chromosome condensation induced in *Drosophila* embryos by anoxia: Visualization of interphase nuclear organization. *J. Cell Biol.* **100,** 1623–1636.

Hilliker, A. J. (1986). Assaying chromosome arrangement in embryonic interphase nuclei of *Drosophila melanogaster* by radiation induced interchanges. *Genet. Res.* **47,** 13–18.

Hiraoka, Y., Agard, D. A., and Sedat, J. W. (1990). Spatial and temporal coordination of chromosome movement, spindle formation and nuclear envelope breakdown during prometaphase in *Drosophila melanogaster* embryos. *J. Cell Biol.* **111,** 2815–2828.

Hiraoka, Y., Dernberg, A. F., Parmelee, S. J., Rykowski, M. C., Agard, D. A., and Sedat, J. W. (1993). The onset of homologous chromosome pairing during *Drosophila melanogaster* embryogenesis. *J. Cell Biol.* **120,** 591–600.

Hiraoka, Y., Minden, J. S., Swedlow, J. R., Sedat, J. W., and Agard, D. A. (1989). Focal points for chromosome condensation and decondensation revealed by three-dimensional in-vivo time-lapse microscopy. *Nature* **342,** 293–296.

Hochstrasser, M., Mathog, D., Gruenbaum, Y., Saumweber, H., and Sedat, J. W. (1986). Spatial organization of chromosomes in the salivary gland nuclei of *Drosophila melanogaster. J. Cell Biol.* **102,** 112–123.

Hochstrasser, M., and Sedat, J. W. (1986). Three-dimensional organization of *Drosophila melanogaster* interphase nuclei. II. Chromosome spatial organization and gene regulation. *J. Cell Biol.* **104,** 1471–1483.

Hochstrasser, M., and Sedat, J. W. (1987). Three-dimensional organization of *Drosophila melanogaster* interphase nuclei. I. Tissue-specific aspects of polytene nuclear architecture. *J. Cell Biol.* **104,** 1455–1470.

Johnson, C. V., Singer, R. H., and Lawrence, J. B. (1991). Fluorescent detection of nuclear RNA and DNA: Implications for genome organization. *Methods Cell Biol.* **35,** 73–99.

Mathog, D., Hochstrasser, M., Gruenbaum, Y., Saumweber, H., and Sedat, J. (1984). Characteristic folding pattern of polytene chromosomes in *Drosophila* salivary gland nuclei. *Nature* **308,** 414–421.

Mathog, D., and Sedat, J. W. (1989). The three-dimensional organization of polytene nuclei in male *Drosophila melanogaster* with compound XY or ring X chromosomes. *Genetics* **121,** 293–311.

Metz, C. W. (1916). Chromosome studies on the Diptera II. The paired association of chromosomes in the Diptera and its significance. *J. Exp. Zool.* **21,** 213–279.

Rykowski, M. C. (1991). Optical sectioning and three-dimensional reconstruction of diploid and polytene nuclei. *Methods Cell Biol.* **35,** 253–286.

Sambrook, J., Fritsch, E. F., and Maniatis, T. (1989). "Molecular Cloning: A Laboratory Manual," 2nd ed. Cold Spring Harbor, NY: Cold Spring Harbor Laboratory Press.

Xing, Y., Johnson, C. V., Dobner, P. R., and Lawrence, J. B. (1993). Higher level organization of individual gene transcription and RNA splicing. *Science* **259,** 1326–1330.

CHAPTER 23

Electron Microscopy and EM Immunocytochemistry

Kent L. McDonald

Electron Microscope Laboratory
26 Giannini Hall
University of California, Berkeley
Berkeley, California 94720-3330

METHODS IN CELL BIOLOGY, VOL. 44

I. Introduction

Drosophila is among that group of organisms that are difficult to fix well by conventional electron microscope (EM) methods. This group includes any organism or stage of development for which there is a barrier between the cells and the environment that significantly hinders the rapid diffusion of fixatives and other solutions used in EM specimen preparation. The cuticle layers of some plants, fungi, and animals are examples of such barriers, as are the specialized layers surrounding many animal embryos. In *Drosophila* embryos, the best ultrastructural preservation has always been obtained when the chorion and vitelline membranes are removed prior to, or during, the primary fixation (Fullilove and Jacobsen, 1971; Zalokar and Erk, 1977; Mahowald *et al.*, 1979; Stafstrom and Staehelin, 1984; Jacobs and Goodman, 1989). In adult tissues, the results are better when organs are dissected in fixative, allowing penetration through soft tissues, rather than through cuticle. While these procedures are adequate for many different kinds of EM studies, they may not be suitable for high-resolution work or for certain problems in EM immunolocalization. Fortunately, there is a workable alternative, thanks to the development and availability of new technology for ultrarapid freezing of large (up to 0.5-mm samples). How this new technology can be applied to problems of fly ultrastructure and immunoelectron microscopy is one of the main subjects of this chapter.

The best and easiest way to circumvent the problems raised by diffusion barriers is to rapidly freeze samples and then fix and dehydrate them at low temperatures (so-called freeze-substitution). This strategy has been used to good effect by mycologists since the late 70s and early 80s (Howard and Aist, 1979; Hoch and Staples, 1983), and they find that ultrastructural preservation is dramatically improved. In comparative fixation studies, it was found that certain organelles, such as bundles of microfilaments, could be preserved only by rapid freezing–freeze-substitution (RF–FS) (Hoch and Staples, 1983) or that numbers of microtubules differed, depending on the fixation methods used (Heath *et al.*, 1984; Heath and Rethoret, 1982). The explanation for these differences probably can be traced to the speed of fixation. The slow penetration of chemical fixatives allows enough time for cytoskeletal organelles to disassemble, for membranes to rupture, and for proteolytic enzymes to be activated. When the fixative finally reaches the cell interior, it can be quite different in structure from what it was at the beginning of fixation. In contrast, rapid freezing immobilizes all molecules within milliseconds. Freeze-substitution occurs at temperatures (-78 to $-90°C$) that are too low to allow rearrangement of large molecules. When the temperature is "warm" enough for fixatives to react (-20 to $-50°C$), it is still cold enough so that molecular rearrangements are minimized.

So, if rapid freezing methods are so much better, why are they not used for all EM specimen preparation? The answer can be found in the physics of heat transfer. At some depth in the specimen, there will be a point at which the heat cannot be removed fast enough to prevent water molecules from rearranging into ice. At that point, the cell ultrastructure is destroyed to a greater extent than by slow chemical fixation. Commonly available methods of freezing such as plunging into propane, or against a cooled copper block, give good freezing only to a depth of about 10 μm, perhaps twice that in favorable samples. Double propane jet freezing increases the depth by a factor of four and to greater than 100 μm if cryoprotectants are used (Ding *et al.*, 1991). Thus, for most fly tissues, either embryos or adults, these methods are not very useful, unless the area of interest is very near the surface of the sample.

To successfully freeze samples greater than 100 μm in size, it is necessary to use the technology of high-pressure freezing (Moor, 1987; Dahl and Staehelin, 1989), which has been shown to work well on some fly tissues (McDonald and Morphew, 1993). After freeze-substitution and embedding in conventional resins, the ultrastructure is noticeably better than samples prepared by other methods.

II. Criteria for Judging Ultrastructure Quality

To the casual student of electron microscopy, it is not always clear why some EM images are judged better than others. In fact, there are some simple guidelines one can follow. The first thing to look for is the state of the cell membranes because these are delicate structures that are easily deformed by adverse conditions. In most instances, membranes should show smooth, continuous contours as opposed to discontinuous, wavy contours. This is illustrated in Figs. 1A and 1B, which compare membranes in cells fixed by high-pressure freezing (A) and freeze-substitution (HPF–FS) with similar membranes prepared by conventional methods (B). The HPF–FS sample (A) would be considered by most microscopists to be the better preservation. Careful light microscopy of nuclei in living *Drosophila* cells (Kellogg *et al.*, 1988) suggests that the nucleus should be round and not irregular.

A second criterion has to do with the density of the cytoplasm (and nucleoplasm). The ground cytoplasm should not show signs of extraction or coagulation. Areas that are white in micrographs are usually areas that remain after proteins have coagulated or have been extracted. It should be emphasized that poor freeze-substitution (Fig. 2), or poor freezing (Fig. 3) will result in problems that are even worse than those generated by conventional fixation and embedding. Sometimes chemically fixed samples (Fig. 4) appear sharper than those prepared by low temperature methods (Fig. 5). This is because the combination of extraction and coagulation results in a higher contrast image, and for some studies, these specimen preparation methods might actually be preferable. How-

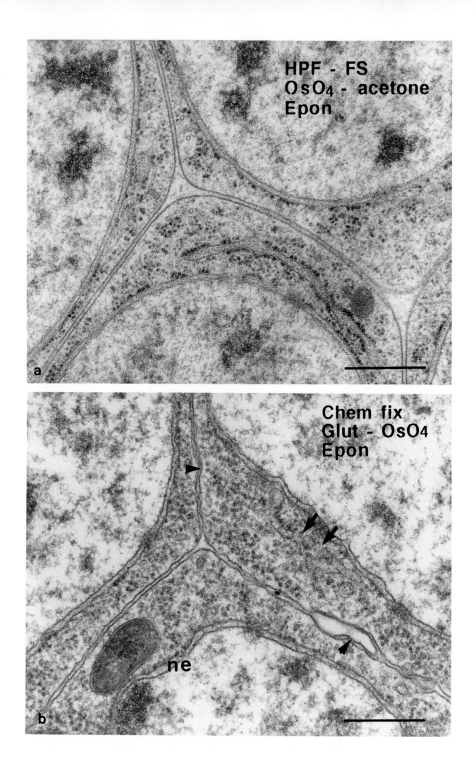

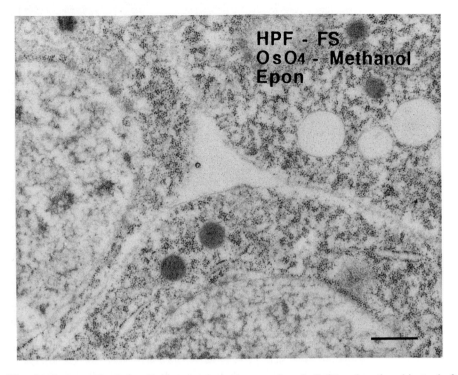

Fig. 2 Same as Fig. 1, but the freeze substitution was done in OsO4 and methanol instead of OsO4–acetone. For this particular tissue, the methanol gives poorer quality preservation than acetone. Bar, 0.5 μm.

ever, for immunolocalization studies such methods could lead to loss and/or rearrangement of antigens.

The final criterion for judging EM quality concerns the cytoskeleton. Cytoskeletal elements should be straight or show smooth arcs where they are under tension. This feature is not illustrated here, but can be seen to some extent in the article by McDonald and Morphew (1993) on high-pressure freezing of fly tissues. Among cytoskeletal components, actin is the most likely to be

Fig. 1 Comparison of *Drosophila* embryo cells prepared by high-pressure freezing-freeze substitution (A) and by conventional room-temperature methods (B). In A, all the membranes are smooth and continuous and the ground cytoplasm is reasonably uniform and dense. Bar, 0.5 μm. In B, membrane distortion is the most obvious problem, with both collapsed nuclear envelope membranes (ne) and fragmented endoplasmic reticulum (arrows) evident, as well as "wavy" plasma membranes (arrowheads). More subtle artifacts include extraction of the nucleoplasm and cytoplasm, seen as clear white spaces. Bar, 0.5 μm.

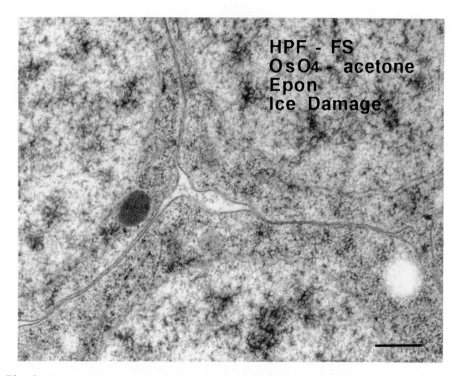

Fig. 3 Same as Fig. 1, but these cells show evidence of ice crystal damage, i.e., a reticulate pattern of white spaces (where the ice was) and darker areas of aggregated cytoplasm. This is relatively mild ice damage because one can still make out membrane profiles. Bar, 0.5 μm.

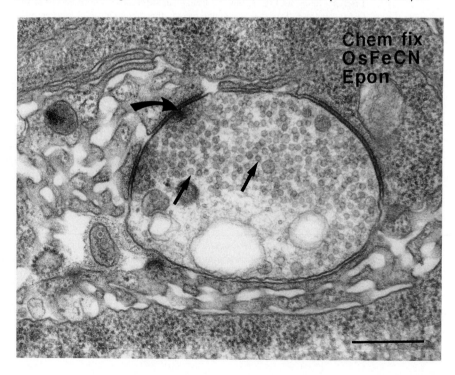

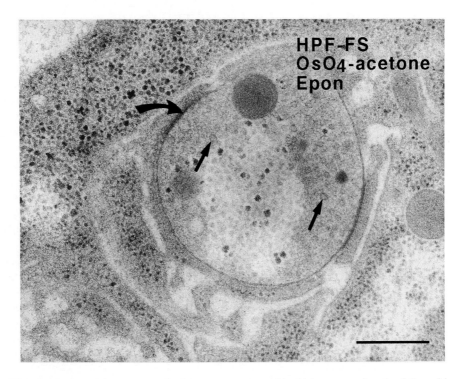

Fig. 5 Same as Fig. 4, except the larvae were prepared by high-pressure freezing followed by freeze substitution in OsO4–acetone. The neuroplasm is less extracted, so the contrast with synaptic vesicles (arrows) is reduced. It is also possible to distinguish a central area of neuroplasm that is lighter density than the outer regions. Synapses (curved arrows) are clearly visualized. Bar, 0.4 μm.

altered by poor specimen preparation, and intermediate filaments are the most resistant to problems.

Ultimately, the choice of specimen preparation method will be determined by what one needs to study. Fast freezing and freeze-substitution are best for questions requiring spatial quantification using high magnification, high-resolution imaging, e.g., measuring the spacing between different classes of microtubules in the mitotic spindle (Ding *et al.*, 1993). These methods are also good for immunocytochemistry because there is usually less extraction of the ground cytoplasm. If one is doing low to medium magnification work, and/or qualitative studies, e.g., identifying certain cell types, then conventional methods will be suitable.

Fig. 4 Nerve terminals (boutons) from the junction of muscles in third instar wild-type larvae prepared by OsFeCN fixation. The synaptic vesicles (arrows) are well contrasted with the neuroplasm, and synapses (curved arrows) are distinct. Bar, 0.4 μm.

III. Aim and Scope of This Chapter

The aim of this chapter is to provide detailed, but easy-to-follow descriptions of the procedures required to obtain the best possible ultrastructure and electron microscope immunolocalization in *Drososphila*. We have attempted to make the instructions sufficiently clear so that anyone, not just electron microscopists, can follow them and get satisfying results. The main difficulty facing the users of these methods will be the lack of access to the specialized equipment needed to obtain the best results. In particular, it will be necessary to use a machine that freezes samples while under pressures of up to 30,000 lb per square inch. These machines are relatively rare at the moment, but it is hoped that they will be more readily available. In any case, this is the way to get the best results and will therefore be the primary focus in this chapter. Other, more conventional methods will also be included, but these cannot be expected to give images of the same quality.

Our experience is mostly with embryos, and the instructions in this article will reflect that bias. However, we have some limited experience with dissected adult tissues, such as ovaries and testes, and the results with these tissues suggest that they will be well preserved by the methods described here for embryos. In principle, dissected larval tissues should also work well. Certain tissues, such as adult eyes, have not worked well by our methods, and we describe some alternative procedures for these problematic organs.

IV. Embryos

A. Fixation by High-Pressure Freezing

1. Materials and Equipment Needed

1. High-pressure freezer and accessories (liquid nitrogen, work station for unloading samples, storage dewar for samples, etc.). Consult with high-pressure freezer source for these details.
2. Dissecting scope.
3. Flies at the correct stage of development.
4. 50% bleach for dechorionating embryos.
5. Plastic mesh screen, e.g., 80–100-μm Nitex.
6. Small (6 cm) petri dish.
7. Size 00 camels hair paintbrush.
8. 10% methanol solution.
9. Dry baker's yeast.

10. 1.5-ml Eppendorf tubes and rack.

11. Fly food plate caps (small petri dishes filled with apple juice or grape juice in agar; for example see recipes in the article by Elgin *et al.* in Chapter 5).

12. Fine forceps.

13. Filter paper or lab tissues.

2. Preparing the Materials for Freezing

1. Mix up dry yeast with 10% methanol solution in an Eppendorf tube so that the yeast make a stiff paste (should form peaks when pulled apart). Place this tube plus a tube of 10% methanol in a rack near the dissecting scope.

2. Dechorionate the embryos by rinsing for 2 min in 50% bleach. We use a 50-ml disposable plastic centrifuge tube from which the conical end has been cut off and a large hole cut in the cap. A square (8 cm) of 100-μm-mesh Nitex is placed over the threads of the tube and the cap is screwed on. The embryos are rinsed from their plates into the tube so they accumulate on the Nitex. They are rinsed with 50% bleach from a squeeze bottle for 2 min and then rinsed with water to remove excess bleach. The cap is unscrewed and the Nitex plus embryos is transferred to the small petri dish half-filled with water so that the embryos remain wet.

3. Place a fly food cap on the dissecting scope stage. This will be the working surface for loading the samples into the high-pressure freezer specimen holders.

4. Onto the food cap place a top and bottom high-pressure freezer specimen holder. Also place a little of the methanol yeast paste on the cap.

5. Use the paintbrush to collect about 10–12 embryos from the Nitex screen and then put them on the food cap next to the specimen holders and yeast paste. If there is excess water carried over from the dish, use filter paper or lab tissue to wick off excess moisture.

6. Using the paintbrush, paint a little yeast paste in the top and bottom specimen holders. Then, on the food plate, use the paintbrush to mix the embryos with just enough yeast paste to hold them together and to fill in the spaces between them. Place this "ball" of embryos in the bottom specimen holder and add more yeast paste if it looks as if there is not enough to fill the specimen holder. If the yeast paste is too thick, thin it out with a little 10% methanol. Place the top holder on the bottom and squeeze gently together with forceps. If using the interlocking specimen holders designed by Craig *et al.* (1987), you should see just a little yeast paste extruded from the vent slot. (The goal is to fill the chamber inside entirely with sample so that there are no air spaces. Air collapses under the high pressure and leads to collapse and damage of the embryos.) The samples are now ready to be frozen.

3. High–Pressure Freezing

1. Place the filled specimen holder into the specimen carrier. The main source of concern here is that the specimen carrier and specimen holders are compatible. You need to make sure that the specimen holder will not move once the specimen carrier is locked in place.

2. Freeze in high-pressure freezer as per instructions and store in liquid nitrogen.

3. Separate top and bottom of specimen holder if the sample is to be processed immediately. If not, leave the top and bottom together until just before processing. This reduces the likelihood of small ice crystals being deposited on the samples while in storage.

4. Storage is usually in 1.8-ml cryotubes in a Dewar or other liquid nitrogen storage device.

B. Freeze-Substitution

Freeze-substitution is the process whereby the cell water is replaced, or "substituted," with an organic solvent. It serves the same function as dehydration at room temperature, i.e., replacing the cell water with a substance that is miscible with resins. However, FS must be done below a temperature of $-70°C$ so the frozen water in the cell doesn't "recrystallize" into ice. There are a number of different solvents that can be used for FS (Robards and Sleytr, 1985; Echlin, 1992); however, acetone and methanol are the most popular. Methanol replaces water much faster than acetone and can tolerate up to 10% water in the sample (compared to 1% for acetone), but in comparative studies on fly embryos, we found that the acetone gave a much better final image. The methanol-substituted samples (Fig. 2) appeared extracted and sometimes collapsed by comparison with the acetone-treated samples (Fig. 1A). For other tissue types, however, methanol seems to work well as a substitution fluid (Hippe, 1985; Meissner and Schwartz, 1990).

1. Fixatives

Although it is possible to freeze-substitute in the absence of fixatives and then embed in a low-temperature resin (Villiger, 1991), it is more common to have some chemical fixative in the substitution fluid. The choice of fixative will be determined by whether one is going to look at the samples for morphology alone or after immunolabeling with electron-dense probes. For standard EM morphological analysis, the most common fixative is 1% OsO_4; for immunocytochemistry 2% paraformaldehyde plus 0.05% glutaraldehyde is a good place to start (Robertson et al., 1992).

a. Materials and Equipment Needed

1. Ice bucket or styrofoam box.
2. 1–2 lb dry ice.
3. 1.8-ml cryotubes.
4. Plastic disposable pipets.
5. Crystalline osmium tetroxide in 0.1- or 0.25-g amounts.
6. Anhydrous glutaraldehyde.
7. EM grade acetone, or acetone dried over molecular sieves.
8. Liquid nitrogen refrigerator or storage dewar.
9. 50-ml disposable polypropylene centrifuge tube.
10. Disposable gloves.
11. Ampule breaker or paper towels.

b. Preparation of Osmium Fixatives in Acetone

Crystalline osmium tetroxide in sealed ampules is available from most vendors of EM supplies. Some vendors (Ted Pella, Inc.; Electron Microscopy Sciences) sell amounts in less than 0.5-g quantities, which can be useful when one does not want to make up a large volume at one time. On the other hand, it is possible to make stock solutions of 1–4% and keep them stored in liquid nitrogen.

1. Measure 25 ml of dry acetone into a 50-ml polypropylene centrifuge tube and chill on dry ice.
2. In a fume hood (OsO_4 vapors are quite toxic), break open an ampule containing 0.25 g OsO_4 and submerge in chilled acetone. When handling OsO_4, be sure to wear protective disposable gloves. When breaking the ampule, be sure to use a rubber ampule breaker (available from EM vendors) or wrap the ampule in several layers of moist paper towels. Most ampules are prescored, but if not, use a triangular file or diamond marker to score the neck prior to breaking. Take care not to let small bits of glass get into the ampule as they may show up in the embedded sample and wreak havoc with diamond knives during sectioning.
3. The OsO_4 will dissolve quite rapidly even at dry ice temperatures ($-78°C$); however, most of it will be in the ampule, so use a disposable plastic (polypropylene) pipet to flush the ampule into the rest of the tube.
4. Dispense the OsO_4–acetone into labeled cryotubes (about 1 ml per tube is usually sufficient). Cap the tubes and then immerse in liquid nitrogen to freeze the solution. Keep the tubes upright during freezing so most of the fixative stays in the bottom of the tubes. Placing the tubes in freezer canes and lowering into a large dewar flask filled with nitrogen is an easy way to freeze the fixative. Store until ready to use.

c. Preparation of Glutaraldehyde Fixatives in Acetone

Fast freezing followed by low-temperature embedment in resins such as Lowicryl is an excellent way to prepare samples for postembedment immuno-labeling. The morphology of such samples is often improved if there is even a small amount of glutaraldehyde in the freeze-substitution solvent. We routinely use 0.1–0.2% glutaraldehyde in acetone as our fixative for immunocytochemical studies. The fixative is made up in essentially the same way as the OsO_4, except that you use a glutaraldehyde stock in acetone (available from EMS) or 70% aqueous glutaraldehyde instead of crystalline osmium. The diluted fixatives can be frozen in liquid nitrogen and stored like the OsO_4 fixatives.

d. Preparation of Paraformaldehyde Fixatives in Methanol

Sometimes antibodies will not tolerate any glutaraldehyde fixatives and if that is the case, one can try using paraformaldehyde as the fixative in the freeze-substitution fluid. To make up this fixative, you will need to have a hot plate, dry methanol, paraformaldehyde powder, a pH meter, NaOH pellets, a 50- to 100-ml beaker, and cryotubes. Weigh out enough paraformaldehyde to make a 2% final concentration in methanol. Heat the solution on a hot plate until you see vapor above the liquid phase. Carefully add a few NaOH pellets to bring the pH to near neutrality. Try not to overshoot the pH beyond 7.2. This entire operation should be done in a fume hood. The final mixture should look clear. At this point, it can be aliquoted into cryotubes and frozen for later use.

2. Freeze-Substitution Equipment and Procedure

a. Equipment

Although a number of vendors sell equipment specifically designed for freeze-substitution, the occasional user is better off using equipment that can be put together from materials found in most every lab or purchased at low cost. If a lab were doing freeze-substitution runs nearly every day, then purchase of a specialized machine might be worth the investment. In the meantime, try the following: (1) Find a styrofoam shipping box (with lid) that fits inside the freezer compartment of a refrigerator or will fit in a −20°C freezer. (2) Line the bottom of the box with dry ice. (3) Place an aluminum block with 12- to 13-mm-diameter holes drilled in it to a depth of about 45 mm on the dry ice and pack more dry ice around the block. While these blocks can be simply made in any machine shop, a cheap source is the heating blocks for dry bath incubators used in molecular biology. These can be ordered for less than $50 from any general biological supplier. When the block has cooled to dry ice temperature (around −78°C), it is ready to receive the sample in freeze-substitution fixative.

b. Procedure for Freeze Substitution in Osmium–Acetone

1. Remove the frozen sample from its storage vial and transfer to the fixative vial under liquid nitrogen. Try to work fast so that the vial is not open to

air because the cooled surfaces, including the interior, will condense water vapor from the air. Excess water inside the vial will interfere with or at least slow down the substitution process, especially with acetone. The entire transfer can be done under liquid nitrogen, but before the lid is screwed on the vial, it will be necessary to remove the liquid nitrogen because when it warms up to −78°C, the liquid will expand into gas. It is important to leave the cap slightly loose at this point.

2. Transfer the sample plus frozen fixative in the cryotube to a hole in the cooled aluminum block.

3. Place the styrofoam box plus sample into the freezer compartment. When the fixative has become liquid, screw down the cap on the cryotube.

4. Leave the box/sample in the freezer until the dry ice has sublimated and the temperature has risen to −20°C. This should take 2–3 days.

5. Transfer the sample to the refrigerator compartment and let it equilibrate to that temperature.

6. Transfer the sample to room temperature. If the osmium solution turns black, this will not harm the samples. In fact, prolonged incubation in osmium at this stage can improve the contrast of certain organelles, particularly membranes.

7. Rinse the samples in 2–3 changes of fresh 100% acetone and embed in epoxy resin (see Section C).

c. Procedure for Freeze-Substitution in Aldehyde Fixatives

The operation is essentially the same as for freeze-substitution in osmium except that the sample needs to be kept cold if the final embedding is into a low-temperature resin polymerized by UV light. Typically, for embedding in Lowicryl K4M, we warm the sample up to −40°C. In the apparatus described, this is achieved by monitoring the warm-up temperature after the dry ice has sublimated. Details of processing after that will be discussed below.

C. Embedding

1. Materials and Equipment Needed

1. Epon 812 or substitute (Embed 812, Epox 812, etc.).
2. Araldite 502.
3. Dodecenylsuccinic anhydride (DDSA).
4. Benzyldimethylamine (BDMA). (All of these can be obtained from any EM vendor.)
5. Top-loading balance reading at least to 0.01 g.
6. Plastic disposable beakers (usually 50–100 ml are good).

7. Half-inch magnetic stir bar.

8. Magnetic stirrer plate.

9. Embedding oven set to 60°C.

10. Disposable vinyl or latex gloves.

11. Fume hood.

12. Glass microscope slides, frosted.

13. Parafilm.

14. Miller–Stephenson MS-122 teflon spray.

15. Razor blades.

16. Scalpel with sharp point.

17. Miller–Stephenson Epoxy 907 two-part adhesive.

18. 1-ml disposable syringe.

2. Mixing the Epoxy Resin

We find that the following Epon–Araldite formulation gives good sectioning qualities and stability under the electron beam:

Egon 812 (or epon substitute)	6.2 g
Araldite 502	4.4 g
DDSA	12.2 g
BDMA	0.8 ml

To mix up the resin, **put on gloves,** add a stir bar to a tripour beaker and tare the beaker/bar on the balance (you may want to cover the balance top with a paper towel or aluminum foil before you begin, in case you spill resin components in the subsequent steps). If your balance is easily moved and your hood air flow is not too fast, you can try doing this in the hood. Otherwise, work in a well-ventilated place. Weigh out the first three components listed above by carefully pouring from the bottle into the beaker. You may want to use disposable plastic pipets to add the final amounts. Cover the beaker with a cap or parafilm and then mix thoroughly on a stir plate taking care not to introduce too much air into the resin. If you are ready to add accelerator, proceed to the hood and add the BDMA. This component is quite volatile and potentially toxic if inhaled, so be sure to work in the hood. Use the 1-ml disposable syringe to measure out the correct amount, cover, and mix thoroughly. The resin may change color slightly (darken) but this is normal. If you are working in a region of moderate to high humidity, you may want to put the mixed resin in a dessicator under house vacuum for 15–20 min to remove excess air and water vapor. After the accelerator is added, this resin mix gets very viscous within about an hour, so if you want to incubate with accelerator for more than 2–3 hr, you may have to mix up a fresh batch at that time. Resin that is left over can be used to make "blanks" in BEEM caps or flat embed molds for remounting samples that have been flat embedded. Also, be sure to remember to remove the stir bar from the beaker before the resin hardens. You can wash it off with a special soap made for removing resin, North 214 soap available from Ted Pella, Inc.

3. Flat Embedding

We have found that for many samples, especially embryos, embedding them in a thin layer of resin that can be scanned on a good light microscope is a very useful technique. The way to do this is as follows:

1. Wipe microscope slides with a clean cloth until there is little resistance to wiping. You will need two slides for each different sample.

2. Write the experiment number on the frosted side.

3. Spray the frosted side with Miller–Stephenson MS-122 spray, holding the nozzle about 6–8 inches from the slide. Spray liberally so that when dried the slide looks opaque white.

4. Air dry the slide, or if you are in a hurry, use a hot plate set on a low setting to drive off the solvent.

5. Use a lab wipe or clean cloth to wipe the spray off. The slide should be transparent, so this is why it is important to label the coated side.

6. Put a small strip (about 2 inches long, $\frac{1}{4}$ inch wide) of fresh resin on the center of the slide (coated side).

7. Transfer your samples to this strip of resin, distributing them evenly.

8. At each end of the slide, put pieces of parafilm ($\frac{3}{4} \times \frac{1}{4}''$ in size, two layers thick for embryos) as spacers.

9. Put the other coated slide on top of the resin/samples, as if you were putting on a coverslip, except that just the clear parts of the slides are overlapping, and the frosted ends should stick out. This will make it easier to take the slides apart at the last step.

10. Put the slides/samples in a 60°C oven to polymerize. Prop them up on toothpicks or similar sticks that have been taped to a piece of stiff support (e.g., cardboard). This is in case you have used too much resin and it overflows the slides. The samples can be removed and cut after 24 hr; however, we prefer to leave them for 2 days to ensure complete polymerization of the resin.

11. Remove samples from oven and use a razor blade to trim excess resin from all edges of the slide. Put a razor blade under one corner of the overlapped slides and push gently between the slides. If you have trimmed the excess well, one slide should lift off and you will be left with your samples in a thin layer of resin on the remaining slide. Careful experimentation with Step 6 to get just the right amount of resin can make this step very easy. But if you use too little

resin, it will be hard to get the slides to split, and too much resin will flow all over and be a mess to trim. You may also lose sample if it flows off the slide.

12. Put the slide on a good light microscope with phase optics and observe your samples. If they are embryos, you can stage them pretty well and select those that you want to remount and cut. You can mark them with a diamond scribe that inserts into the objective nosepiece of your microscope. These scribes can probably be obtained from the microscope manufacturer.

13. Make up some Miller–Stephenson Epoxy 907 just before you are ready to remount the sample for sectioning.

14. You can use a sharp-pointed scalpel to cut out the selected sample. Make two parallel cuts on either side. At this point you should be able to see the resin lifting off the slide. Make the third cut perpendicular to the first two, and the last the same, except do not cut all the way through to the first mark. If you do, the sample will be completely loose and it is easy to launch the little chip of resin with sample into lab space as you try to pick it up with fine forceps. Better to leave a small corner uncut and break this off after you have a good grip on the chip.

15. Remount the chip in a known orientation for sectioning using BEEM cap or flat embed resin blanks and the 907 epoxy resin. Put in a 60–70°C oven for 45 min and it will be ready to trim and section. The 907 epoxy is based on Epon resin chemistry and can be sectioned if necessary.

D. Conventional Room-Temperature Chemical Fixation

Not all fly tissues can be prepared by fast freezing and freeze-substitution. These include any sample that is too large to fit into the high-pressure freezer specimen holder, e.g., third instar larvae and whole adults, and adult tissues that, for some reason, do not freeze well, such as eyes. Also, many researchers are probably willing to sacrifice quality of specimen preservation for convenience. In these instances, it is necessary to use traditional EM specimen preparation methods. There are a number of different published protocols that seem to work reasonably well, and these have minor differences in types of buffer used, concentration and type of fixatives, and how the sample is handled prior to fixation. For persons wanting to take this route, the best approach is to find the images in the literature that seem best and then use that particular protocol. In the text that follows, we will refer to some of the articles that show decent EM preservation and present a method that is a compilation of several different published methods plus some ideas of our own.

1. Removing the Chorion

There are two common methods for removing the chorion, and the EM images suggest that mechanical dechorionation gives consistently good results (Zalokar and Erk, 1977; Stafstrom and Staehelin, 1984; Jacobs and Goodman, 1989). Embryos are put on a piece of double sticky tape on a slide, and then another slide with double sticky tape is put on top of the embryos. To prevent undue compression of the embryos, a wire spacer of about the same diameter as the embryos is used between the slides. When the slides are separated, the chorion is torn and the embryos can be collected for further processing. See Zalokar and Erk (1977) for more details. The other method for removing the chorion is to use 50% bleach (commercial bleach diluted 1 : 1 with water). Embryos are rinsed in this for 2 min, and then rinsed with water.

2. Aldehyde Fixation

Early investigators of fly ultrastructure were able to get good results by piercing the chorion and vitelline envelope with a glass needle in the presence of fixative (Fullilove and Jacobson, 1971). After 10 min or more, the chorion and vitelline layers were removed mechanically and the embryos allowed to remain for 1–2 hr in fixative.

Most contemporary fixation schemes rely on a modification of the method developed by Zalokar and Erk (1977) for fixing the embryos in aldehydes. With this method, the fixative is dissolved in heptane and when the embryos are placed in this mixture, the fixatives penetrate the vitelline layer. The following protocol from Jacobs and Goodman (1989) is representative of the modified Zalokar and Erk (1977) method:

1. Shake equal volumes of heptane and fixative (25% glutaraldehyde plus 2% acrolein in 0.1 M cacodylate buffer). NOTE: Cacodylate buffer contains arsenic and should be treated with appropriate caution.
2. Place mechanically dechorionated embryos in the heptane phase for 10 min.
3. Remove excess heptane from embryos and immerse in the following for 1 hr: 2% paraformaldehyde and 2.5% glutaraldehyde plus 0.5% DMSO in cacodylate buffer, pH 7.3.
4. After 10–15 min in the above fixative, remove the vitelline layer with needles.

3. Postfixation and Embedding

Processing after the primary fixation is standard. The method used by Stafstrom and Staehelin (1984) is typical:

1. After aldehyde primary fixation, rinse 3 × 10 min in cacodylate buffer, pH 7.3.
2. Fix in 1% OsO_4 in cacodylate buffer for 1 hr.
3. Rinse 2 × 10 min in cacodylate buffer and 1 × 10 min in distilled water.

4. Stain in aqueous 1% uranyl acetate for 2 hr (This is a deviation from the Stafstrom–Staehelin (1984) protocol, which calls for incubation overnight at 4°C).

5. Dehydrate in an ethanol or acetone series (25, 50, 75, 95, 100, 100% for 15 min each). If ethanol is used, rinse twice in propylene oxide after the final 100% ethanol step.

6. Embed in epon–araldite (see Section C).

V. Stages Other Than Embryos

A. Fast Freezing–Freeze-Substitution

For tissues other than embryos, size will determine whether they can be reliably processed by fast freezing methods. The largest volume that a standard high-pressure freezer specimen holder will accept is a cylinder with a radius of 1 mm and a height of 0.6 mm. Dissected adult body parts or imaginal discs can be loaded into the holders and treated much like embryos. If yeast paste proves inadequate as a packing material, then there are other options such as high-molecular-weight dextrans or paraffin oil such as 1-hexadecene.

We have tried with little success to freeze adult eyes by high-pressure freezing. There is probably something about the fluid in the eye that is causing the difficulty. Other adult organs, such as wings, could probably be frozen by high-pressure methods, but Steinbrecht (1993) has achieved beautiful results with plunge freezing followed by freeze-substitution in osmium acetone.

B. Conventional Fixation

1. OsFeCN Fixation

For stages other than embryos, we would recommend using a formulation much like the one above for embryos, except the heptane fixation step might be omitted. Instead, it may be necessary to open up the organ or organism by making an incision while it is in fixative, especially if there is a cuticle covering. For fixation of third instar laravae, we had some success by pinning out the larvae at each end and slitting it open on the ventral side (Gho *et al.*, 1992). Because we were interested in neuromuscular junctions, we removed the internal organs and cannot comment on their state of fixation. However, the remaining tissues fixed well, using a fixation scheme that included reduced osmium and tannic acid (McDonald, 1984). The details of this protocol are as follows:

1. Make up 2% glutaraldehyde in buffer plus 5 mM CaCl$_2$. The buffer for this and all subsequent steps is 0.05 M cacodylate, pH 7.4. It is probably not important to use this particular buffer, or pH, or final osmolarity. Each tissue may have slightly different requirements and these may have

to be tested if this initial formulation does not work. For example, many tissues may look better if a higher osmolarity, e.g., 0.1 M, is used. In this regard, one can be guided by published protocols for the tissue of interest.

2. Rinse in buffer 3 × 5 min.

3. Incubate for 15 min in 0.5% OsO_4 + 0.8% OsFeCN in buffer. This step should be done on ice, but all other steps are at room temperature.

4. Rinse in buffer 3 × 5 min.

5. Place in 0.15% tannic acid in buffer for 1 min.

6. Rinse in buffer 2 × 5 min and in distilled water once for 5 min.

7. Incubate in 2% aqueous uranyl acetate in the dark for 2 hr (leaving it longer could lead to uranium salt precipitates in the sample).

8. Rinse in distilled water for 5 min.

9. Dehydrate in acetone (25, 50, 75, 90, 95, 100, 100, 100%) for 10 min each, longer for tissues greater than 0.5 mm on a side.

10. Embed in epon–araldite.

The key step in conventional chemical fixation is the initial cross-linking by glutaraldehyde. As a general rule, the faster this takes place, the better, i.e., there is less time for the cells to undergo autolysis and other rearrangements such as depolymerization of cytoskeletal components. Therefore, the smaller the pieces fixed the better, and if there are diffusion barriers such as cuticle, these should be removed or at least cut through in the presence of fixative.

2. PLT–Osmium Fixation

We are currently experimenting with a technique we call "Progressive Lowering of Temperature (PLT)-Osmium" that might be very useful for improving the preservation of fly ultrastructure without requiring expensive freezing equipment. The final results will not be as good as ultrarapid freezing–freeze-substitution, but they may be better than conventional room-temperature methods. PLT-osmium takes advantage of the fact that most of the damage in EM specimen preparation takes place in the dehydration step when done at room temperature. By reducing the temperature during dehydration, you can reduce the damage done to your sample. One way to do PLT-osmium is to fix the sample as if you were going into Lowicryl resin, but use higher concentrations of glutaraldehyde (1% or more) and then dehydrate by PLT. The details of how to do PLT are under Section VIC, part 3. When you get to the step at which your samples are in 100% ethanol at −35°C, put them in 1% osmium in acetone (or ethanol) and then let them warm to room temperature over about 6 hr, rinse in 100% solvent without fixative several times, and embed in the resin of your choice. Alternatively, you can fix at room temperature as you normally would (with both glutaraldehyde and osmium), and then do your dehydration by PLT.

You can warm the solution up faster (over 1–2 hr) if you use this variation. PLT-osmium has worked well for us with tissues such as mouse testis and we are just beginning to experiment with fly tissues such as adult eyes. The preliminary results are subtle, but are sufficiently encouraging that we will continue exploring the applications of this technique.

VI. Immunolabeling

One of the most exciting applications of EM technology these days is in identifying the subcellular localization of gene products by immunogold-labeled probes. The methods to achieve this goal are constantly evolving and, as yet, there is no one protocol that will satisfy all situations. An article by Kellenberger and Hayat (1991) gives a good overview of the problems and choices to consider when beginning EM immunolabeling studies. We can briefly summarize the issues they consider. The first choice to make is whether one wants to do pre- or postembedding labeling. In pre-embedding labeling, cells or tissues are made premeable with detergents or organic solvents so that the relatively large antibody molecules can freely diffuse into the cells. It is worth noting that an IgG molecule is about 10 nm in diameter, and a Fab fragment about 6 nm in its longest dimension. Thus, it may be necessary to open up rather large holes in the cytoplasm to allow free penetration of probes. This is less of a problem for cytoskeletal components than for membrane-bound organelles, and a real problem for soluble proteins. Once the labeling has occurred, the samples can be handled by standard EM preparation methods and when sectioned, the probes are in place throughout the dimension of the section. With the exception of certain cytoskeletal preparations, the big problem with preembedding labeling is the loss of ultrastructural integrity due to solubilization prior to labeling.

If one chooses postembedding labeling, then the next decision is between using cryosections and using resin sections. Excellent results can be obtained with cryosections; however, it requires using expensive cryomicrotomes and is a difficult technique to master. Should one go that route, there are excellent cryosectioning systems, e.g., the CR-21 from RMC Inc., that can be added to most modern ultramicrotomes. We believe that a detailed description of cryosectioning is beyond the scope of this chapter and refer the interested reader to the following papers (Tokuyasu, 1986; Echlin, 1992; Sitte *et al.*, 1989) for particulars on the method. It is worth mentioning that some people believe that cryosections are permeable to antibody probes and that they are therefore better than resin sections in which only surface labeling is possible. The work of Stierhof *et al.* (1991) convincingly refutes that idea, showing that the labeling of cryosections is also on the surface. However, it may still be possible that for some antibodies, there is denser labeling on cryosections. For the beginner in EM immunolabeling, we recommend postembedding labeling of resin sections as discussed in detail here.

   Many researchers will come to do EM immunocytochemistry after considerable experience with light microscope (LM) immunocytochemistry. Unfortunately, much of what one knows from LM work will not be applicable to the EM situation. Higher concentrations of antibody will need to be used, and almost certainly, the details of fixation will be different. By EM standards, LM fixation is very crude, no matter how good it looks at the LM level. This is because cytoplasm is transparent to light but not electrons. A living cell and a heavily extracted cell can look identical in the light microscope. However, at the EM level they look quite different (Figs. 6 and 7). Melan and Sluder (1992) have published a paper that should be considered by anyone doing LM immunocytochemistry. In this study, they show how the pattern of labeling of soluble proteins can be changed dramatically simply by changing the fixation conditions.

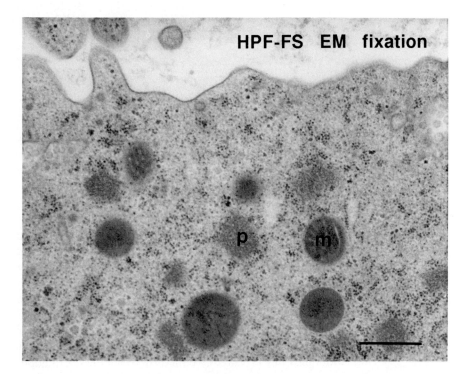

Fig. 6 Posterior pole plasm in a 1- to 2-hr wild-type embryo showing polar granules (p) and mitochondria (m) in a dense cytoplasm. The embryo was high pressure frozen and then substituted in OsO4 acetone and embedded in Epon-Araldite. Bar, 0.5 μm.

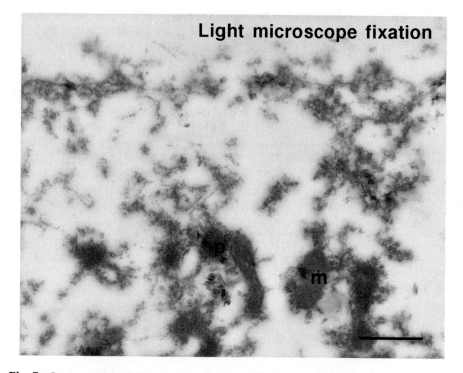

Fig. 7 Same material as in Fig. 6, except that it was first fixed for light microscopy by immersion in 1 : 1 heptane: 4% formaldehyde (in 100 mM Hepes, 2 mM MgSO$_4$, 1 mM EGTA) for 20 min at room temperature and then devitellinized by vigorous shaking in 1 : 1 heptane : methanol, followed by several rinses in absolute methanol. Embryos were then rehydrated by rinsing in 1 : 1 methanol : PBT (PBS + 0.1% Tween) and then PBT. The samples were then refixed for EM by immersion in 2% glutaraldehyde, postfixed in OsO4, dehydrated in acetone, and embedded in Epon–Araldite. Comparison with Fig. 6 shows how poorly the LM fixation scheme preserves high-resolution information compared to the EM fixation. Bar, 0.5 μm.

A. Antibodies

The real limiting factor for success or failure at EM immunolabeling is the nature of the antibody being used. Some antibodies are so potent that they can be used on samples fixed in glutaraldehyde and osmium and embedded in Epon resin. Others are so sensitive that they will not work after any type of EM processing. Most are somewhere in between. In our experience, antibodies that work at the LM level often need to be further purified and concentrated for use at the EM level. Details of antibody preparation and purification are beyond the scope of this chapter, but a recent Methods in Cell Biology volume on Antibodies in Cell Biology edited by Asai (1993) is a good place to start for this kind of information. Antigen concentration is another factor that contributes to the success or failure of EM immunolabeling. Some antigens are so locally concentrated in cells that they will work with low-affinity, purified antibodies.

B. Cryofixation and Freeze–Substitution

As with fixation for standard morphology, we recommend cryofixation as the optimum method to preserve structure and antigenicity. After high-pressure freezing, samples can be freeze-substituted in acetone containing glutaraldehyde or paraformaldehyde plus glutaraldehyde in methanol. Samples can also be substituted in organic solvent, but only if one is prepared to use a resin that can be polymerized at −70°C such as Lowicryl K11M or HM23. The preparation of fixatives in acetone is covered under Section IV,B, Part 1. After freeze-substitution at −90°C for 2 days or so, the samples are warmed to −40°C, at which point they are infiltrated with Lowicryl K4M or HM20 resin as described here.

C. Embedding

1. Lowicryl K4M

Most of what follows was previously published in the article on "Electron Microscopy Immunocytochemistry Following Cryofixation and Freeze Substitution," (Kiss and McDonald, this series, Vol. 37, 1993). It is reprinted here with minor changes in its entirety for convenience, i.e., so the reader does not have to find the secondary reference if so desired.

a. Preparation

When you order a Lowicryl kit from an EM vendor, you will receive detailed instructions on how to prepare the resin. If you borrow a kit, be sure to ask for the instruction booklet that came with it. We follow the manufacturers' recommendations when making up Lowicryl. Also, there is a very useful chapter on Lowicryls and low temperature embedding by Hobot (1989) that is worth having if you use this resin. For most applications, Lowicryls K4M and HM20 will suffice. We use the following proportions:

K4M (−35 C)		HM20 (−50 C)	
Cross-linker A	2.70 g	Cross-linker D	2.98 g
Monomer B	17.30 g	Monomer E	17.02 g
Initiator C	0.10 g	Initiator C	0.10 g

In a fume hood, weigh out the first two components of the resin into an amber bottle and mix by gently bubbling a stream of dry nitrogen gas through the resin. After about 5 min. of mixing, add initiator, and mix for an additional 10 min or until initiator is dissolved. If you don't want to use nitrogen, there are other mixing methods described in the instruction booklet. The main idea is to avoid mixing oxygen into the mixture because that will impede polymerization. Cool the tightly capped bottle to the desired temperature.

b. Infiltration

The resin infiltration is carried out in cryotubes, which are placed in holes in an aluminum block that is immersed in a methanol-dry ice bath cooled to the desired temperature. Use a Styrofoam shipping box for the bath, and place in a fume hood for all operations involving Lowicryl. The aluminum block recommended for freeze substitution will work fine for the cryotubes, but a rack or other holder needs to be placed in the bath to hold the Lowicryl and ethanol solutions. Temperature is monitored by a themocouple probe in a cryotube filled with solvent and placed in the aluminum block. Dry ice chips are added to the bath as needed to keep the temperature in the desired range. The infiltration schedule for K4M at $-35°C$ is as follows:

1. Rinse in solvent at three times over 20 min
2. Put in 1 part resin: 1 part ethanol for 2 hr
3.   "   " 2 "   "   1 "   "   " 2 "
4. 100% resin overnight*
5. 100% resin for 2 hr

* For this you will need a low temperature refrigerator set at $-35°C$, or a controlled temperature box such as the one sold by Pella, Inc.

c. Flat Embedding

Embryos and other small samples can be flat embedded in Lowicryl and this provides the same advantages as flat embedding in Epon, i.e., the samples can be viewed and evaluated with a light microscope at high magnification and remounted in precise orientation for sectioning. To flat embed in Lowicryl, the samples are sandwiched between 22 mm square Thermanox (EMS) coverslips. These plastic coverslips are transparent to UV (unlike glass) and permit UV polymerization at low temperatures. The detailed procedure is as follows:

1. Spray two coverslips with Teflow-based spray release agent (Miller-Stephenson). Let dry for 5 min then wipe off the dried spray with a cloth or tissue. Mark the sprayed side.

2. Cut the center out of a third coverslip with a scalpel, cutting in about 3–4 mm from each edge. Use a cyanoacrylate adhesive (so-called Superglue) to bond this spacer to the Teflon-coated side of one of the other coverslips. Let dry. This is the bottom coverslip.

3. Place the bottom coverslip in an aluminum weighing dish (57 × 18 mm), add a few small dry ice chips at the edges, then nest another weighing dish on top for a cover. This is to keep the coverslip dry in the next step, which is to place the dishes/coverslip in the $-35°$ chamber to cool. Repeat with the top coverslip, leaving the coated side up.

4. Transfer the samples, e.g., embryos, from the cryotubes at $-35°C$ to the well of the bottom coverslip. Try not to carry over too much resin (about 0.5 ml is good), work fast to prevent moisture condensation, then put the top

coverslip on (coated side down) so that no air bubbles are trapped and not much, if any, resin spills over. Practice this step with resin only (no samples) until you can do it quickly. If resin spillover is a problem, you can prop up the bottom coverslip on a pair of flat toothpicks so the bottom coverslip doesn't bond to the aluminum dish.

5. If flat embedding is not possible, or seems too complicated, samples can be embedded in gelatin capsules that are suspended in wire loops in the polymerization chamber (Chiovetti, 1982).

d. Polymerization

To polymerize Lowicryl resins at low temperature you will need a box that can be cooled to −35 or −50°C for 24 hr and has UV lights in it. A dedicated low temperature refrigerator will work or you can buy commercial devices ranging in price from about $800 (Pella, Inc.) to $17,000 (Leica). Or, you can make your own apparatus for about $250. Whatever one uses, the initial polymerization is at −35 or −50°C for 24 hr with indirect UV illumination, i.e., with a baffle such as a piece of aluminum foil between the lights and the coverslips and/or gelatin capsules. The samples are then placed directly under UV lights at room temperature for 2–3 days. This kind of information is covered in detail in the Lowicryl instruction booklet that comes with the resin. After the flat embedded preparations are fully polymerized, the top coverslip should peel off easily and the samples in the well of the bottom coverslip can be cut out with a scalpel and remounted for sectioning. If one wants to look at the samples with the light microscope, then it may be necessary to peel off the bottom coverslip also. This will be more critical if one is using other than brightfield optics to check the sample.

2. Other Resins

a. LR White

Many people use LR White as their resin of choice for EM immunolabeling. It is easy to use, is nontoxic, can be heat polymerized in a conventional oven, and has reasonably good sectioning properties for an immunoelectron microscopic resin. Furthermore, it can be polymerized in the presence up to 30% water and this can sometimes be important to the preservation of some antigen–antibody reactions. The articles by Newman and Hobot (1989) and Newman (1989) on LR White resins discuss in some detail the advantages and disadvantages of using LR White and Lowicryls. In our experience, LR White is satisfactory if you can use relatively high ($\geq 1\%$) concentrations of glutaraldehyde as your primary fixative. If one has to use "light" fixation, i.e., 0.1% glutaraldehyde and/or paraformaldehyde, then the LR White seems to extract the cytoplasm during processing, and the final quality of the ultrastructure is not as good as it is with the Lowicryls (Figs. 9 and 10). For this reason, we have chosen to work primarily with the Lowicryl resins. Details of how to use

LR White are available from the manufacturer or EM vendor and in the article by Newman (1989). A typical schedule might be as follows:

1. Fix in glutaraldehyde and/or formaldehyde at the concentration appropriate for your antibody.
2. Rinse in buffer.
3. Dehydrate in ethanol. You can stop anywhere between 70- and 100% but must use the final concentration, e.g., 70%, in the LR White-ethanol mixtures that follow.
4. Put in 2 parts ethanol to 1 part LR White for 1 hour on a rotating or rocking mixer. Use the mixer for all subsequent steps.
5. Put in 1 part ethanol to 2 parts LR White.
6. Two changes in pure LR White for 1 hour each.
7. LR White overnight.
8. LR White for 1 hour then embed. LR White will not flat embed by the methods described above for Lowicryl. Use of gelatin capsules is the usual embedding method.
9. Polymerize in 50–55°C oven.

b. LR Gold

This resin may be a good compromise between Lowicryl and LR White. We have not tried it with fly tissues, but we have seen results with other tissues that suggest it may be worth a try. It combines the convenience of LR White with the low temperature dehydration and embedding of Lowicryl. A schedule for fixation, dehydration and embedding is as follows:

1. Fix and rinse as for LR White.
2. Dehydrate according to the following schedule:
   a. 30% ethanol at room temperature for 15 min.
   b. 50% ethanol at 0°C for 15 min.
   c. 70% ethanol at −25°C for 45 min.
   d. 90% ethanol at −25°C for 45 min.
   e. 50% LR Gold monomer/50% pure ethanol at −25°C for 30 min.
   f. 70% LR monomer/30% pure ethanol at −25°C for 60 min.
   g. 100% LR Gold monomer at −25°C for 60 min.
   h. 100% LR Gold monomer + initiator* at −25°C for 60 min.
   i. 100% LR Gold monomer + initiator at −25°C overnight.
   j. Change to fresh LR Gold monomer + initiator at −25°C, then polymerize with 365 nm wavelength UV light at −25°C for 24 hours.**

   * Benzoin Methyl Ether at 0.5% w/v.
   ** Simply start at −25°C, and let the temperature rise to ambient over the 24 hr. As the resin polymerizes, it takes on a chartreuse color under the UV lamps.

3. Progressive Lowering of Temperature (PLT) Dehydration

Although it is best to use the low-temperature resins like Lowicryl following rapid freezing and freeze substitution, they can be used following room-temperature fixation. In fact, the original recommendation from the developers of the resin (Carlemalm *et al.,* 1982) was to fix at room temperature, then lower the temperature progressively to −35 or −50°C during the dehydration. This strategy is advantageous over cryofixation because the size of the sample is not a factor. The resulting morphology from this method is often quite good and we recommend this method as a reasonable, low-cost alternative to cryofixation/freeze substitution. The quality of the final results will depend critically on the quality of the primary fixation, and that step should be given careful consideration. It is worth experimenting with such parameters as the pH and osmolarity of the buffer used, the type of buffer used, the concentration and duration of aldehyde fixation, and so on. If these conditions are optimized, the chances are good that the final morphology in Lowicryl will be acceptable to quite good. The virtue of this method is that it reduces considerably the distortions due to room temperature dehydration and embedding (Carlemalm *et al.,* 1982). The details of the procedure are given in the Lowicryl instruction booklet. For HM20, a revised schedule is given in an article by Robertson *et al.* (1992). The following is a method slightly modified from the one recommended by the Lowicryl manufacturer:

a. 30% ethanol at 0°C for 30 min
b. 50% ethanol at −20°C for 60 min
c. 70% ethanol at −35°C for 60 min
d. 95% ethanol at −35°C for 60 min
e. 100% ethanol at −35°C for 60 min
f. 100% ethanol at −35°C for 60 min
g. Embed according to the schedule listed above in Section VI,C,1,b for Lowicryl K4M.

Notes:

1. Care needs to be taken at the changes from 30, 50 and 70% ethanol that you don't carry over too much of the lower concentration ethanol to the next lower temperature because the sample can freeze. If necessary, transfer to the next higher concentration, e.g., 50% ethanol at the temperature of the previous solution, i.e., 0°C, then rapidly cool to the lower temperature (−20°C). This will insure that the cells will not freeze.

2. Solutions should be agitated frequently at all steps, if possible.

D. Sectioning

1. Grid Preparation

Use nickel or gold grids of whatever mesh you prefer for doing immunolabeling. Copper grids will react with the labeling reagents and are not suitable for immunocytochemistry. We routinely use nickel slot grids coated with Formvar and stabilized with a thin layer of carbon. By using coated grids, you restrict the labeling to one side of the section. If you pick up sections with unfilmed grids, you can stain both sides of the section, but large areas of your sample will be under grid bars. Whether you choose filmed or unfilmed grids will probably depend on the reactivity of your antibody and/or the distribution of your antigen. If the antibody is highly reactive, then single-sided labeling is usually adequate. If the antigen is widely distributed throughout the tissue, then unfilmed grids may be preferable. The filmed grids should be lightly carbon-coated just before use. If the carbon-coated grids are more than a day old, then they should be glow discharged in an evaporator just before they are to be used.

2. Sectioning

The nonpolar Lowicryl resins (HM20 and HM23), LR White and LR Gold, are hydrophobic and section much like the conventional epoxy resins. The polar Lowicryl resins (K4M and K11M) are more hydrophilic and are more difficult to section because they tend to take up water in the sectioning boat and swell. Sectioning setup is more or less normal with the following qualifications:

1. Take care not to let the block face get wet. The water will penetrate the block and make sectioning very difficult. If this happens, change blocks.
2. To prevent wetting and to facilitate sectioning, adjust the meniscus in the boat so the water is not level with the knife edge, but slightly lower. This will help prevent wetting the block and reduce the chances of losing sections over the knife edge, and the sections will ribbon more or less normally.
3. Pick up the sections frequently (do not let a lot build up in the boat) so they do not swell too much.

E. Antibody Labeling on Sections

1. Make up blocking buffer (BB) in PBS
   1. PBS: 2.68 mM KCl, 1.50 mM KH$_2$PO$_4$, 0.49 mM MgCl$_2$ · 6H$_2$, 137 mM NaCl, 8.0 mM Na$_2$HPO$_4$ · 7H$_2$O; pH 7.4.
   2. BB: 0.1% cold water fish gelatin (v/v) (Sigma, St. Louis, MO), 0.8% bovine serum albumin (w/v) (Sigma), 0.02% tween-80 (v/v).

2. Incubate grids in BB for 30 min at room temperature. All steps will be at room temperature. Immerse the grids if they are uncoated so you wet both sides of the section.

3. Blot quickly, taking care not to let the grids dry out between steps. Apply about 20 μl of primary antibody diluted in BB. Incubate about 1–1.5 hr in a moist chamber. This time can vary according to the activity and dilution of the primary.

4. Blot and rinse once with BB and twice with PBS over a period of 5 min.

5. Incubate in a moist chamber for 1–1.5 hr in secondary gold conjugate diluted in BB.

6. Blot and rinse as in Step 3.

7. Fix in 0.5% (v/v) glutaraldehyde in PBS for 5 min.

8. Rinse once in PBS and twice in distilled H_2O over 2 min.

9. Air dry.

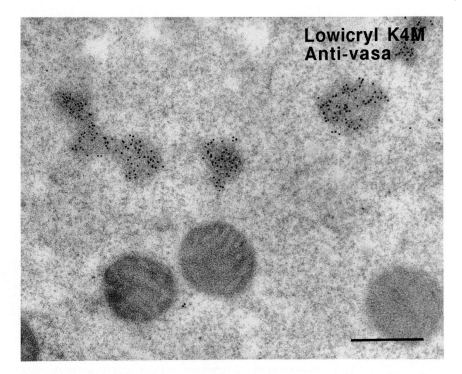

Fig. 8 Same material as in Fig. 6, except following high-pressure freezing, it was freeze substituted in 0.2% glutaraldehyde in acetone and embedded in Lowicryl K4M at −35°C. Polar granules are labeled with 10 nm gold secondary to a primary antibody (courtesy of Dr. Paul Lasko) made against vasa protein. Bar, 0.4 μm.

Poststain the grids with uranyl acetate and lead citrate. The times for poststaining will vary with the type of resin and sample. Start with 10 min in aqueous uranyl acetate and 5 min in lead citrate and adjust according to results seen in the microscope. If contrast persists in being too low, one can try methanolic uranyl acetate (1–2% in 70% methanol for 2–4 min) instead of aqueous.

F. Microscopy and Publishing

If you are using nickel grids, you will have to use so-called anti-magnetic forceps to handle them; otherwise, they will usually "stick" to stainless steel forceps. Alternatively, and probably preferably, you can demagnetize the grid before you try to pick it up and put it in the microscope. Magnetized grids are not only hard to handle, but they will interfere with the electromagnetic lenses in the electron microscope and give an astigmatic image. It is worth checking the astigmatism correction periodically when examining sections on nickel grids.

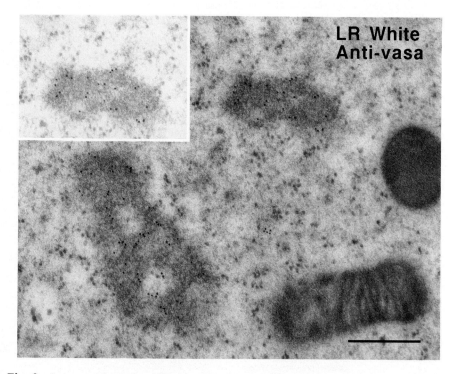

Fig. 9 Same experimental conditions as Fig. 8, except the sample was embedded in LR White resin at 55°C. Compared to the Lowicryl, the antibody staining is less dense, and there is more noticeable extraction of the cytoplasm as shown by the white areas. The ribosomes are also more densely stained with this resin. The inset shows how the gold on the polar granules is more readily seen in a print with less contrast. Bar, 0.4 μm.

At 100,000× magnification, a 10-nm gold particle is 1 mm across; at 10,000× it is only 100 micrometers in diameter. If the final magnification of your print for publication is in the 10,000× range with 10 nm gold, then it will be almost impossible to see your marker. At 50,000–100,000× it will be easy (Fig. 8). If you plan to print at this high final magnification, then you will need to take pictures on the microscope in the 15,000 to 30,000× range to permit reasonable photographic enlargement. You will probably also want to take a lower magnification view of the cell/tissue for orientation. In publishing, it may be necessary to show both views—one to illustrate the specificity of labeling and the other for reference or orientation. An alternative strategy that allows one to show both aspects of labeling in one picture is to use silver enhancement of the gold labeling. Silver will nucleate off gold particles and can be "developed" to very large sizes, such that the label is easily seen at low magnification. The drawback is that not all gold particles will serve as nucleating sites.

If your gold happens to label a region of the cell that is electron dense because of grid poststaining, it may be necessary to print that negative slightly lighter than you normally would (Fig. 9). The point is to show where the gold is located, and as long as the organelle is recognizable, it is not essential that it be as dark as it would be in a normal micrograph.

Acknowledgments

Thanks especially to Mary Morphew of the Laboratory for Three-Dimensional Fine Structure, Department of Molecular, Cellular and Developmental Biology, University of Colorado, Boulder for the many hours of technical assistance in high-pressure freezing of flies, embedding, sectioning, and immunolabeling. Thanks also to Bob Boswell of MCD Biology, Boulder, CO for providing the embryos, and to L. Andrew Staehelin of MCD Biology, Boulder, CO for unlimited use of his high-pressure freezer. Thanks to Bill Saxton of Indiana University, Bloomington, IN for processing the larvae for OsFeCN fixation shown in Fig. 5. The help of everyone in the Lab for 3-D Fine Structure is gratefully acknowledged, especially J. R. McIntosh, Director. Dr. Paul Lasko of McGill University, Toronto provided the vasa antibody. Portions of this work were supported by NIH Grant RR00592 to the Boulder Lab for 3-D Fine Structure, J. R. McIntosh, P. I.

Appendix: Vendor Information

1. General EM Supplies:

Ted Pella, Inc.
P.O. Box 492477
Redding, CA 96049-2477
800-237-3526 (U.S.A.)
800-637-3526 (California)
800-243-7765 (Canada)
916-243-3761 (FAX)

Electron Microscopy Sciences
321 Morris Road
Box 251
Fort Washington, PA 19034
800-523-5874
215-646-1566
215-646-8931 (FAX)

Ernest F. Fullam, Inc.
900 Albany Shaker Road
Latham, NY 12110-1491
800-833-4024
518-785-5533
518-785-8647 (FAX)

Polysciences, Inc.
400 Valley Road
Warrington, PA 18976
800-523-2575 (orders)
215-343-6484 (information)
800-343-3291 (FAX)

2. High-Pressure Freezers and Other Cryogenic Equipment (See Appendix in Robards and Sleytr (1985) for More Cryotechnique-Related Vendor Information):

Bal-Tec Products, Inc.
984 Southford Road
P.O. Box 1221
Middlebury, CT 06762
800-875-3713
203-598-3660
203-598-3658 (FAX)

Leica Aktiengesellschaft
Hernalser Hauptstrasse 219
A-1171 Wien, Austria
0222-4616 41-0 (Phone)
0222-460326/461579 (FAX)
114872 rja (Telex)

RMC, Inc.
4400 S. Santa Rita
Tuscon, AZ 85714
602-889-7900
602-741-2200 (FAX)

3. Teflon Spray and Epon Remount Glue:
Miller-Stephenson Chemical Co., Inc.
George Washington Highway
Danbury, CT 06810
203-743-4447
203-791-8702 (FAX)

Note added in proof. The following two references came to the author's attention after the manuscript was submitted. They should be consulted by anyone proposing to do EM immunocytochemistry.

Griffiths, G. (1993). Fine Structure Immunocytochemistry. 459 pages. Springer-Verlag, Berlin.
Newman, G. R., and Hobot, J. A. (1993). Resin Microscopy and On-Section Immunocytochemistry. 221 pages. Springer-Verlag, Berlin.

References

Asai, D. J. (1993). Antibodies in Cell Biology. *In* "Methods in Cell Biology" (D. J. Asai, ed.) Vol. 37, p. 452. San Diego: Academic Press.
Carlemalm, E., Gravito, R. M., and Villiger, W. (1982). Resin development for electron microscopy and an analysis of embedding at low temperature. *J. Microsc.* **126,** 123–143.
Chiovetti, R. (1982). "Instructions for Use, Lowicryl K4M and Lowicryl HM20." Waldkraiburg, Germany: Chemische Werke Lowi GmbH.

Craig, S., Gilkey, J. C., and Staehelin, L. A. (1987). Improved specimen support cups and auxiliary devices for the Blazers high pressure freezing appartus. *J. Microsc.* **48,** 103–106.

Dahl, R., and Staehelin, L. A. (1989). High-pressure freezing for the preservation of biological structure: Theory and practice. *J. Electron Microsc. Tech.* **13,** 165–174.

Ding, B., Turgeon, R., and Parthasarathy, M. V. (1991). Routine cryofixation of plant tissue by propane jet freezing for freeze-substitution. *J. Electron Microsc. Tech.* **19,** 107–117.

Ding, R., McDonald, K. L., and McIntosh, J. R. (1993). Three-dimensional reconstruction and analysis of mitotic spindles from the yeast *Schizosaccharomyces pombe*. *J. Cell Biol.* **120,** 141–145.

Echlin, P. (1992). "Low-Temperature Microscopy and Analysis." New York: Plenum.

Fullilove, S. L., and Jacobsen, A. G. (1971). Nuclear elongation and cytokinesis in *Drosophila montana*. *Dev. Biol.* **26,** 560–577.

Gho, M., McDonald, K., Ganetzky, B., and Saxton, W. M. (1992). Effects of kinesin mutations on neuronal functions. *Science* **258,** 313–316.

Heath, I. B., and Rethoret, K. (1982). Mitosis in the fungus *Zygorhynchus moelleri:* Evidence for stage specific enhancement of microtubule preservation by freeze substitution. *Eur. J. Cell Biol.* **28,** 180–189.

Heath, I. B., Rethoret, K., and Moens, P. B. (1984). The ultrastructure of the mitotic spindle from conventionally fixed and freeze-substituted nuclei of the fungus *Saprolegnia*. *Eur. J. Cell Biol.* **35,** 284–295.

Hippe, S. (1985). Ultrastructure of haustoria of *Erysiphe grammis* f. sp. *hordei* preserved by freeze-substitution. *Protoplasma* **129,** 52–61.

Hobot, J. A. (1989). Lowicryls and low-temperature embedding for colloidal gold methods. *In* "Colloidal Gold: Principles, Methods, and Applications" (M. A. Hayat, ed.), pp. 75–115. San Diego: Academic Press.

Hoch, H. C., and Staples, R. C. (1983). Ultrastructural organization of the non-differentiated uredospore germing of *Uromyces phaseoli* variety *typica*. *Mycologia* **75,** 795–824.

Howard, R. J., and Aist, J. R. (1979). Hyphal tip ultrastructure of the fungus *Fusarium:* Improved preservation by freeze-substitution. *J. Ultrastruct. Res.* **66,** 224–234.

Jacobs, J. R., and Goodman, C. S. (1989). Embryonic development of axon pathways in the *Drosophila* CNS. I. A glial scaffold appears before the first growth cones. *J. Neurosci.* **9,** 2402–2411.

Kellenberger, E., and Hayat, M. A. (1991). Some basic concepts for the choice of methods. *In* "Colloidal Gold: Principles, Methods, and Applications" (M. A. Hayat, ed.), Vol. 3, pp. 1–30. San Diego: Academic Press.

Kellogg, D. R., Mitchison, T. J., and Alberts, B. M. (1988). Behaviour of microtubules and actin filaments in living *Drosophila* embryos. *Development* **103,** 675–686.

Kiss, J. Z., and McDonald, K. (1993). EM immunocytochemistry following cryofixation and freeze-substitution. *In* "Methods in Cell Biology" (D. J. Asai, ed.), Vol. 37, pp. 311–341. San Diego: Academic Press.

Mahowald, A. P., Caulton, J. H., and Gehring, W. J. (1979). Ultrastructural studies of oocytes and embryos derived from female flies carrying the *grandchildless* mutation in *Drosophila subobscura*. *Dev. Biol.* **69,** 118–132.

McDonald, K. (1984). Osmium ferricyanide fixation improves microfilament preservation and membrane visualization in a variety of animal cell types. *J. Ultrastruct. Res.* **86,** 107–118.

McDonald, K., and Morphew, M. K. (1993). Improved preservation of ultrastructure in difficult-to-fix organisms by high pressure freezing and freeze substitution. I. *Drosophila melanogaster* and *Strongylocentrotus purpuratus* embryos. *Microsc. Res. Tech.* **24,** 465–473.

Meissner, D. H., and Schwarz, H. (1990). Improved cryoprotection and freeze substitution of embryonic qual retinal: A TEM study on ultrastructural preservation. *J. Electron Microsc. Tech.* **14,** 348–356.

Melan, M. A., and Sluder, G. (1992). Redistribution and differentiation extraction of soluble proteins in permeabilized cultured cells: Implications for immunofluorescence microscopy. *J. Cell Sci.* **101,** 731–743.

Moor, H. (1987). Theory and practice of high pressure freezing. *In* "Cryotechniques in Biological Electron Microscopy" (R. A. Steinbrecht and K. Zierold, eds.), pp. 175–191. Berlin: Springer-Verlag.

Newman, G. R. (1989). LR White embedding medium for colloidal gold methods. *In* "Colloidal Gold: Principles, Methods, and Applications" (M. A. Hayat, ed.), Vol. 2, pp. 47–73. San Diego: Academic Press.

Newman, G. R., and Hobot, J. A. (1989). Role of tissue processing in colloidal gold methods. *In* "Colloidal Gold: Principles, Methods, and Applications" (M. A. Hayat, ed.), Vol. 2, pp. 33–45. San Diego: Academic Press.

Robards, A. W., and Sleytr, U. B. (1985). Low temperature methods in biological electron microscopy. *In* "Practical Methods in Electron Microscopy" (A. M. Glauert, ed.), Vol. 10, p. 551. Amsterdam: Elsevier.

Robertson, D., Monaghan, P., Clarke, C., and Atherton, A. (1992). An appraisal of low-temperature embedding by progressive lowering of temperature into Lowicryl HM20 for immunocytochemical studies. *J. Microsc.* **168**, 85–100.

Sitte, P., Neumann, K., and Edelmann, L. (1989). Cryosectioning according to Tokuyasu vs. rapid freezing, freeze-substitution, and resin embedding. *In* "Immuno-Gold Labeling in Cell Biology" (A. J. Verkleij and J. L. M. Leunissen, eds.), pp. 63–93. Boca Raton, FL: CRC Press.

Stafstrom, J. P., and Staehelin, L. A. (1984). Dynamics of the nuclear envelope and of nuclear pore complexes during mitosis in the *Drosophila* embryo. *Eur. J. Cell Biol.* **34**, 179–189.

Steinbrecht, R. A. (1993). Freeze-substitution for morphological and immunocytochemical studies in insects. *Microsc. Res. Tech.* **24**, 488–504.

Stierhof, Y., Schwarz, H., Durrenberger, M., Villiger, W., and Kellenberger, E. (1991). Yield of immunolabel compared to resin sections and thawed crysections. *In* "Colloidal Gold: Principles, Methods, and Applications" (M. A. Hayat, ed.), Vol. 3, pp. 87–115. San Diego: Academic Press.

Tokuyasu, K. T. (1986). Application of cryo-ultramicrotomy to immunocytochemistry. *J. Micrcosc.* **143**, 139–149.

Villiger, W. (1991). Lowicryl resins. *In* "Colloidal Gold: Principles, Methods, and Applications" (M. A. Hayat, ed.), Vol. 3, pp. 59–71. San Diego: Academic Press.

Zalokar, M., and Erk, I. (1977). Phase-partition fixation and staining of *Drosophila* eggs. *Stain Tech.* **52**, 89–95.

CHAPTER 24

Imaging Neuronal Subsets and Other Cell Types in Whole-Mount *Drosophila* Embryos and Larvae Using Antibody Probes

Nipam H. Patel

Department of Embryology
Carnegie Institution of Washington
Baltimore, Maryland 21210

METHODS IN CELL BIOLOGY, VOL. 44

445

I. Introduction

The purpose of this chapter is to detail the immunocytochemical procedures used to view specific tissues and cells within whole-mount preparations of *Drosophila* embryos and young larvae. A special emphasis is placed on providing examples of antibodies and techniques used to examine the patterns of neurons and axons within the developing nervous system. These methods have two primary applications in the study of *Drosophila* development. First, once a gene of interest has been cloned and antibodies have been raised to the protein product, these immunocytochemical techniques can be used to determine the pattern of protein distribution during embryogenesis and larval development.

Second, antibody probes that recognize defined tissues and cells can be used as markers to analyze phenotypes in mutant embryos.

Most immunocytochemical methods for staining *Drosophila* whole-mount embryos utilize techniques first developed by Zalokar and Erk (1977) and Mitchison and Sedat (1983). For the most part, subsequent modifications reflect progress made in the production of primary and secondary antibody reagents, in the techniques developed for visualizing enzymatically and fluorescently coupled antibodies, and in the optical and dissection methods used to examine stained preparations. The procedures outlined in this chapter also work well for immunostaining a wide variety of organisms other than *Drosophila* (Patel *et al.*, 1989a, 1994).

I have chosen to detail various methods for fixing, staining, and dissecting whole-mount *Drosophila* embryos and larvae and to discuss parameters that affect the quality of the results. I have not, however, attempted to exhaustively describe the specific tissues, identified neurons, and available antibody reagents because these issues have been dealt with in great detail and in several excellent reviews that have been published recently.

The book edited by Bate and Martinez-Arias (1993) presents extensive reviews discussing embryonic pattern formation and the development of all tissues and organs. For descriptions of neural development, the following references may serve as good starting points. For details of neuroblast formation, and a few examples of markers expressed during early neurogenesis, see Campos-Ortega (1993), Goodman and Doe (1993), Gutjahr *et al.* (1993), Doe (1992), and Skeath and Carroll (1992). The position and axon projections of identified neurons and glia of the central and peripheral nervous system are described in Goodman and Doe (1993), Jan and Jan (1993), Klämbt and Goodman (1991), Sink and Whitington (1991), Jacobs and Goodman (1989a,b), Bodmer and Jan (1987), Dambly-Chaudière and Ghysen (1986), Ghysen *et al.* (1986), and Thomas *et al.* (1984). Examples of antibodies that stain specific patterns of neurons or axons within the developing central and peripheral nervous system are described in Nose *et al.* (1992), Dambly-Chaudière *et al.* (1992), Grenningloh *et al.* (1991) Kania *et al.* (1990), Blochlinger *et al.* (1990), Patel *et al.* (1989b), Patel *et al.* (1987), and White and Vallés (1985). Listed in the next section are a few specific examples of antibody reagents that are useful in imaging specific tissues and cells within developing embryos and larvae.

II. Reagents Used to Visualize Specific Tissues and Subsets of Neurons

A. Examples of Markers for Specific Tissues

Not all these antibodies are strictly specific for the tissues listed; nevertheless, they are useful for examining these tissues during certain stages of development.

1. Germline Cells

Anti-*vasa* antiserum (Lasko and Ashburner, 1988) and monoclonal antibody (MAb 46F11; Hay *et al.*, 1988). Stain pole plasm and germline cells (Fig. 1A).

2. Trachea. See Manning and Krasnow (1993) for a detailed review of tracheal development.

Monoclonal antibody 2A12. Stains lumen of tracheal tree (N. Patel and C. Goodman, unpublished; described in Manning and Krasnow, 1993; Fig. 1C).

3. Mesoderm. See Bate (1993) for a detailed review of mesoderm and muscle formation.

Anti-*snail* antiserum (Alberga *et al.*, 1991). Stains nuclei of mesodermal cells during early embryogenesis.

Anti-muscle myosin antiserum (Kiehart and Feghali, 1986). Stains cytoplasm of differentiated muscles.

Anti-fasciclin III monoclonal antibody (MAb 2D5; Patel *et al.*, 1987). Stains cell surface of visceral mesoderm cells (Fig. 1B; also see Section IIB3).

4. General Nervous System or Large Subset of Nervous System

Anti-*hunchback* antiserum (Tautz *et al.*, 1987). Stains nuclei of all neuroblasts during early neurogenesis.

Anti-*elav* antisera (Robinow *et al.*, 1988) and monoclonal antibody (MAb 44C11; Bier *et al.*, 1988). Stain nuclei of all neurons (Figs. 1D–1I).

Anti-HRP (horseradish peroxidase) antiserum (not to be confused with HRP-conjugated antibodies; Jan and Jan, 1982). Stains neural cell membranes by recognizing a carbohydrate moiety attached to a number of neural proteins (Snow *et al.*, 1987).

Fig. 1 (A) Anti-*vasa* monoclonal antibody staining of Stage 7 embryo (using protocols described in Sections IIIA, IV, VB, VIIIB, and IXA). Pole cells (arrow) are moving dorsally with the posterior midgut invagination. (B) MAb 2D5 (anti-fasciclin III) staining of Stage 12 embryo (Protocols IIIA, IV, VA, VIIIA, and IXA). Arrow points to prominent staining of the visceral mesoderm. Segmentally repeated staining is in the neuroepithelium and in the developing nervous system. (C) MAb 2A12 staining of developing trachea in a Stage 15 embryo (Protocols IIIA, IV, VA, VIIIA, and IXA). The trachea entering the nervous system (arrow) are visible in this ventral view. (D–I) Embryos stained with MAb44C11 (anti-*elav*; using Protocols IIIA, IV, VB, VIIIB, and IXA). D–F are lateral views and G–I are corresponding ventral views at Stage 10 (D and G), Stage 13 (E and H), and Stage 15 (F and I). All neurons are stained as they differentiate. Some of the isolated neurons of the peripheral nervous system are indicated by the arrows.

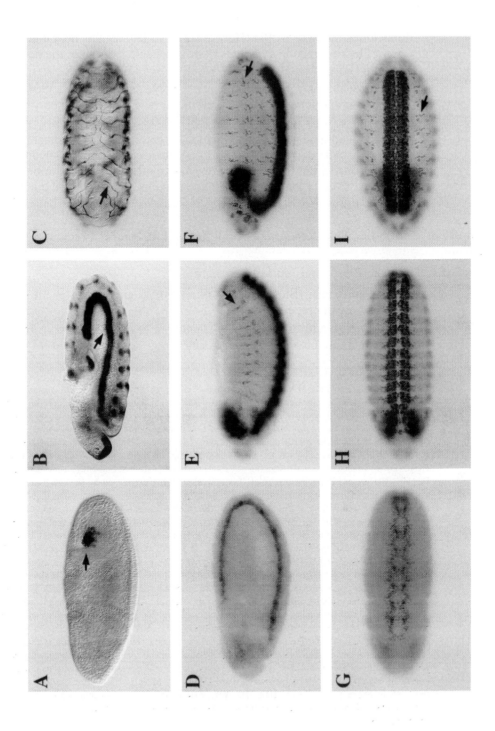

Monoclonal antibody BP104 (Hortsch *et al.*, 1990). Stains neural cell membranes by recognizing a nervous system-specific isoform of the neuroglian protein (Figs. 2B and 2C).

Monoclonal antibody 22C10 (Fujita *et al.*, 1982; Zipursky *et al.*, 1984). Stains cytoplasm and inner surface of cell membrane of all PNS neurons and a subset of CNS neurons (Figs. 4A and 4B, 6A, 6E, and 6F).

Monoclonal antibody BP102 (A. Bieber, N. Patel, and C. Goodman, unpublished; described in Seeger *et al.*, 1993). Stains axons of CNS neurons. All CNS axons within the commissures and connectives appear to be stained by Stage 13 or 14, but many longitudinal axons are not stained during their initial outgrowth in Stage 12 (Figs. 3A–3D, 5A–5F, and 6B, 6C, and 6G–6I).

B. Examples of Antibodies That Recognize Small Subsets of Neural Precursors and Neurons

1. Subsets of Cells in the Peripheral Nervous System

Anti-*cut* antiserum (Blochlinger *et al.*, 1990). Stains nuclei of all external sensory organ precursors.

Anti-*pox neuro* antiserum (Dambly-Chaudière *et al.*, 1992). Stains nuclei of poly-innervated external sensory cells.

Fig. 2 (A) Anti-achaete MAb staining of Stage 9 embryo (Protocols IIIA, IV, VB, VIIIB, and IXB). At this time, expression is in four neuroblasts per hemisegment (the column 1 and m neuroblasts of rows b and d). The arrowhead indicates the l column neuroblast of row d and the arrow indicates the m column neuroblast of row d (also called NB 7-1 and 7-4, respectively, by Doe, 1992). (B) MAb BP104 staining at Stage 11 (Protocols IIIA, IV, VA, VIIIA, and IXA). (C) MAb BP104 of the peripheral nervous system at Stage 15 (Protocols IIIA, IV, VA, VIIIB, and IXB). (D) Anti-fasciclin II MAb staining of central nervous system at mid-Stage 12 (Protocols IIIA, IV, VB, VIIIB, and IXB). At this stage, fasciclin II is highly expressed by the cell bodies of aCC (arrowhead), pCC (arrow), and the growth cone of pCC (triangle). Expression is also beginning on vMP2/MP1 (not visible in this focal plane). (E) MAb 4D9 (anti-*engrailed*) staining of Stage 11 embryo (Protocols VII, VIIIB, and IXA). The embryo is slightly flattened so that more of each stripe is in a single focal plane. Arrow points to the stripe of the mandibular segment, arrowhead points to the stripe of the first abdominal segment. (F) MAb 4D9 (anti-*engrailed*) staining of Stage 9 embryo (Protocols IIIA, IV, VB, VIIIB, and IXB). The germband has been "unfolded," the hindgut removed, and the head folded out and flattened. The three triangles indicate the three head segments that express engrailed at this stage (pre-antennal, antennal, and intercalary segments; expression in the clypeolabrum will start after this stage). Arrow and arrowhead indicate the mandibular and first abdominal stripe, respectively. (G and H) MAb 4D9 (anti-*engrailed*) staining of a subset of CNS neurons at stage 15 (Protocols IIIA, IV, VB, VIIIB, and IXB). At this stage, prominent medial and lateral clusters of *engrailed* expressing neurons are located slightly ventral to the dorsal surface of the CNS (G). At a slightly more ventral focal plane (H), *engrailed*-positive progeny of the median neuroblast (arrow) and some of the *engrailed*-positive neurons arising from *engrailed*-negative neuroblasts (arrowhead) are visible. (I) MAb 2D5 (anti-fasciclin III) staining of the CNS at Stage 14 (Protocols IIIA, IV, VA, VIIIA, and IXB). Fasciclin III is expressed on three bundles of the anterior commissure (arrows) and two pathways in the posterior commissure (arrowheads).

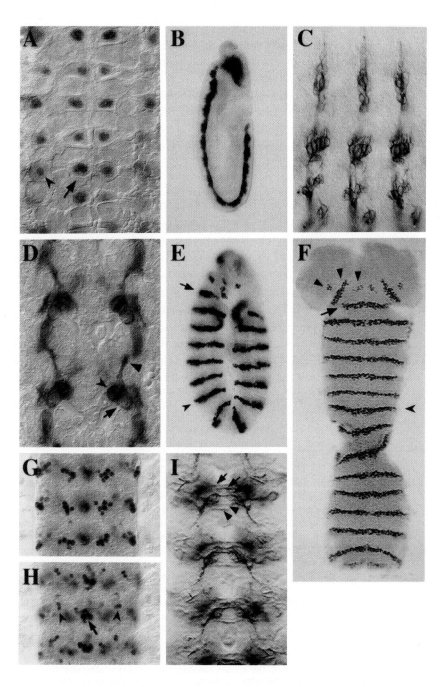

2. Subset of Central Nervous System Neuroblasts

Anti-*acheate* monoclonal antibody (Skeath and Carroll, 1992). Stains nuclei of four neuroblasts per hemisegment during the first wave of neuroblast segregation.

Anti-*engrailed* antisera (DiNardo *et al.*, 1985) and monoclonal antibody (MAb 4D9; Patel *et al.*, 1989a). Stain nuclei of all row 6 and 7 neuroblasts and a neuroblast of row 1.

Anti-*gooseberry* antiserum (Gutjahr *et al.*, 1993). Stains nuclei of all row 5 and 6 neuroblasts.

3. Subset of Central Nervous System Neurons

Anti-*even-skipped* antiserum (Frasch *et al.*, 1987) and monoclonal antibodies (MAb 3C10 and MAb 2B8; Patel *et al.*, 1992, 1994). Stain nuclei of a small subset of neurons (Figs. 3E–3G, 5H–5L, 6A, 6H, and 6I).

Anti-*engrailed* antisera (DiNardo *et al.*, 1985) and monoclonal antibody (MAb4D9; Patel *et al.*, 1989a). Stains nuclei of a small subset of neurons (Figs. 2E–2H, 4C and 4D, and 6B, 6E, and 6F).

Anti-fasciclin III monoclonal antibody (MAb 2D5; Patel *et al.*, 1987). Stains surface of a subset of neurons and axons (Figs. 1B and 2I).

Anti-fasciclin III antiserum (Grenningloh *et al.*, 1991) and monoclonal antibody (Mab1D4; G. Helt and C. Goodman, unpublished; described in Van Vactor *et al.*, 1993). Stain surface of a subset of neurons and axons (Fig. 2D).

Fig. 3 (A) MAb BP102 staining at early Stage 13 shows that the commissures have just begun to separate from one another (Protocols IIIA, IV, VA, VIIIB, and IXA) (B) Same embryo as in A, but now dissected and photographed at higher magnification. (C) MAb BP102 staining at Stage 14 (Protocols VII, VIIIB, and IXA). (D) MAb BP102 staining of a *zipper* mutant (Protocols VII, VIIIB, and IXA). The head has failed to involute, leaving the brain in an abnormal anterior position (arrow), and the commissures show some fusion. (E) MAb 3C10 (anti-even-skipped) staining of Stage 15 embryo (Protocols IIIA, IV, VB, VIIIB, and IXB). This embryo has been dissected flat so that the entire CNS and body wall are visible in a single focal plane. The brain has been unfolded so that it lies flat and the arrowhead points to one of the pairs of even-skipped expressing neurons in the brain. Arrow indicates the anal pad and the triangle points to the dorsal mesoderm expression of *even-skipped*. (F and G). Higher magnification views of the CNS of the embryo shown in E. (F) Ventral focal plane showing *even-skipped* expression in the EL neurons (arrow), the CQ neurons (triangle), and the U neurons (arrowhead). (G) Dorsal focal plane with aCC (arrow), pCC (triangle), and RP2 (arrowhead) neurons. (H) Dorsal focal plane of the CNS of an MAb 3C10 (anti-*even-skipped*) stained cyclin A mutant (compare to G). Instead of three *even-skipped* expressing neurons per hemisegment, the mutant has only two *eve*-expressing neurons per hemisegment on the dorsal surface of the CNS. This is because the ganglion mother cell that produces the siblings aCC and pCC fails to complete its division (this same failure to divide may be happening for the RP2 ganglion mother cell, but there is no data on the fate of the sibling of RP2).

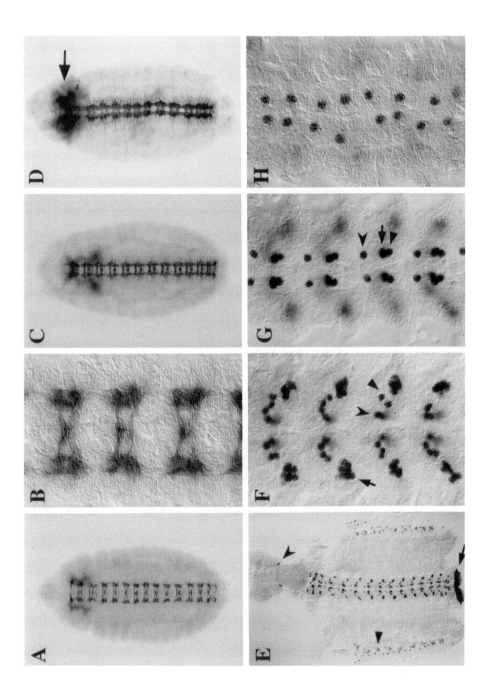

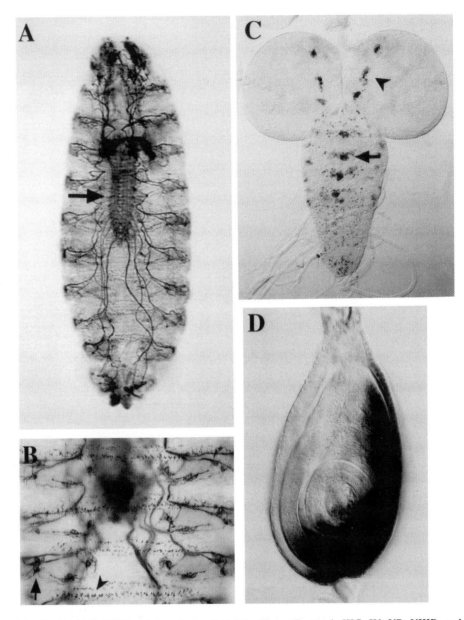

Fig. 4 (A) MAb 22C10 staining of a newly hatched larva (Protocols IIIC, IV, VB, VIIIB, and IXA) Ventral view. Arrow points to the condensed CNS. (B) MAb 22C10 staining of a 21-hr embryo (Protocols IIIB, IV, VB, VIIIB, and IXA). Arrowhead indicates the forming denticles of the ventral cuticle. Arrow points to some of the sensory neuron projections into the cuticle. (C) Third instar CNS stained with MAb 4D9 (anti-*engrailed;* Protocols IIID, IV, VA, VIIIA, and IXB). During larval development, a prominent increase in the number of engrailed expressing neurons is seen in the median neuroblast lineages of the three thoracic segments and the first abdominal segment

Anti-FMRFamide antiserum (Schneider *et al.*, 1993). Stains cytoplasm of a subset of neurons (cell body and axonal processes).

Anti-serotonin antiserum and monoclonal antibody (described in White and Vallés, 1985). Stains cytoplasm of a subset of neurons (cell body and axonal processes).

C. Segmentation Markers

Antibodies to a wide variety of segmentation gene products are available and many of these genes are also expressed in specific tissue and neurons later in development. For example, *even-skipped* and *engrailed* are both segmentation genes, but are also expressed by specific patterns of neurons (see above and Figs. 2E–H, 3E, and 5J). Further details of segmentation gene expression patterns can be found in Akam (1987), and Ingham (1988), Ingham and Martinez-Arias (1992), and Pankratz and Jäckle (1993).

D. Identification of Specific Tissues and Neurons by Enhancer Trap Lines or by Nonradioactive *in Situ* Hybridization

A large number of *lacZ* enhancer trap lines that express β-galactosidase in specific patterns of cells and tissues have been described (see, for example, Bier *et al.*, 1989; Bellen *et al.*, 1989; Ghysen and O'Kane, 1989; Hartenstein and Jan, 1992). The patterns in these lines can be visualized with a mouse monoclonal antibody against β-galactosidase (Promega, Catalog No. Z3781; use at a 1:200–1:1000 dilution) or a rabbit antiserum against β-galactosidase (Cappell, Catalog No. 55976; use at a 1:1000–1:5000 dilution; Fig. 6G). In cases in which antibodies are not available to a gene of interest, the expression pattern can be monitored in whole-mount preparations using the *in situ* procedures described in Chapter 30 (also see Patel and Goodman, 1992).

III. Fixation, Methanol Devitellinization, Storage, and Rehydration Protocols

A. Whole-Mount Preparation of 0- to 17-hr Embryos

1. Collect embryos on apple or grape juice agar plates.
2. Rinse embryos with water into a Nitex nylon mesh (Tetko Inc., Catalog No. 3-85/35XX or 3-100/47).

(arrow points to T2). There are also three prominent columns of engrailed-expressing cells in each brain lobe (arrowhead indicates the middle column of one lobe). (D) Imaginal disc stained with MAb 4D9 (anti-engrailed; Protocols IIID, IV, VA, VIIIA, and IXA). Staining is clearly seen in the posterior compartment of the disc.

3. Dechorionate with 50% bleach (2.6% sodium hypochlorite) for about 3 min.

4. Rinse embryos thoroughly with dH$_2$O. Examine embryos under a dissecting microscope to check that they are completely dechorionated; if not, extend time in 50% bleach until fully dechorionated.

5. Place embryos into a 25-ml glass scintillation vial containing 10 ml heptane plus 10 ml of PEM-FA fixative (see Section XIII). The solutions will form two layers; the lower layer is fixative, the upper layer is heptane, and the embryos will sink in the heptane to lie at the interface of the two layers. The heptane will become saturated with fixative, allowing the fixative to penetrate the hydrophobic vitelline membrane surrounding the embryos.

6. Agitate the vial gently for 10 to 60 min. The optimal fixation time varies depending on the particular antigen and antibody involved. In most cases, fixation time of 10–20 min appears to give good results. See Section XII for further comments on fixation.

7. Remove as much of the aqueous (bottom) layer as possible. Swirling the vial will cause the embryos to spin to the center but will leave the final drops of the aqueous phase at the edges, where they can then be easily removed. Make sure that there is still 8–10 ml of heptane remaining with the embryos in the vial.

8. Add 10 ml of methanol to the embryos and shake vigorously for 30 to 60 sec. The vitelline membranes should split open and the fixed embryos will fall to the bottom of the container. The empty vitelline membranes and nondevitellinized embryos will remain at the interface between the heptane and methanol. At least 80% of the embryos should be devitellinized.

9. Remove the heptane, methanol, and all material at the interface. Wash the devitellinized embryos at least three times with methanol. Shake gently between washes. These methanol washes are required to remove the last traces of heptane.

10. Embryos in methanol can be transferred to a polypropylene tube and kept at $-20°C$ for extended storage. I have embryos that have been stored for four years in absolute methanol at $-20°C$; they still retain excellent morphology and stain well with a wide variety of antibodies.

11. To proceed with antibody staining, rehydrate embryos by removing the methanol and then washing 2×5 and 1×30 min with PT or PBT (see Section XIII). At this point, the embryos will be become somewhat "sticky" and it is best to use polypropylene tubes and pipets (1000 μl "blue" tips) to prevent too many embryos from being lost. If you do use glassware, make sure to coat it with PBT before trying to transfer the embryos. Precellularized embryos (<3 hr) should be rehydrated through a

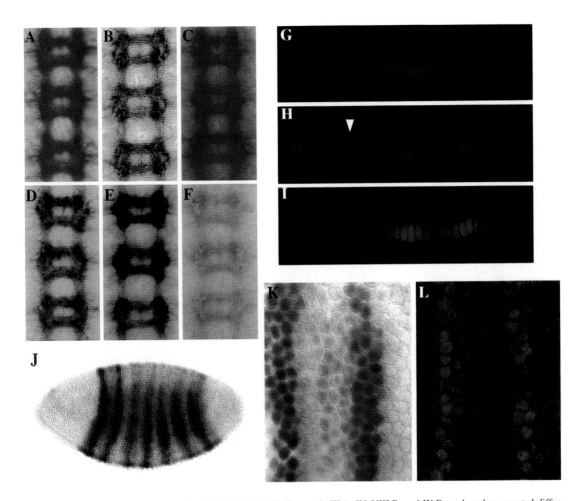

Fig. 5 (A–F) Stage 14 embryos stained with MAb BP102 (Protocols III.A, IV, VIII.B, and IX.B; each embryo reacted differently at Step V). (A) Brown HRP reaction of Section V.A. (B) Black HRP reaction of Section V.B. (C) Orange HRP reaction of Section V.C. (D) Purple alkaline phosphatase reaction of Section V.D. (E) Blue alkaline phosphatase reaction of Section V.E. (F) Red alkaline phosphatase reaction of Section V.F. (G–I) Stage 5 embryo double labeled with rabbit anti-Krüpple (RITC) and MAb 2B8 (anti-even-skipped; FITC) using the procedure described in VI.A. (G) Anti-Krüpple staining. (H) MAb8 staining. Arrowhead indicates eve stripe 2. (I) Overlap between the two patterns. (J and K) Stage 5 embryo stained with MAb 2B8 (anti-even-skipped; purple) and rabbit anti-paired (brown) as described in Section VI.B. (L) Another Stage 5 embryo stained with the same combination of primary antibodies as in J and K, but now detected fluorescently (even-skipped in green and paired in red).

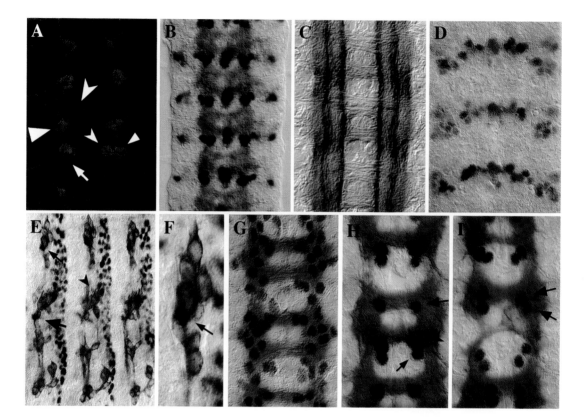

Fig. 6 (A) CNS of Stage 13 embryo stained with MAb 22C10 (green) and rabbit anti-even-skipped (red) as described in Section VI.A. U1 neuron (small arrowhead); aCC (small triangle); pCC (arrow); SP1 (large arrowhead); RP2 (large triangle). (B) CNS of a Stage 15 embryo stained with MAb 4D9 (anti-engrailed; purple alkaline phosphatase detection) and BP102 (red HRP detection). (C) Anti-fasciclin II MAb (black HRP reaction) and MAb BP102 (brown HRP reaction) staining of a Stage 16 CNS using the basic procedure of Section VI.D. At this stage, two prominent fasciclin II-expressing bundles are present near the dorsal surface of the bilaterally symmetric longitudinal connectives. Additional fasciclin II-expressing bundles are located at more ventral focal planes. (D) MAb 4D9 (anti-engrailed; brown HRP reaction) and MAb 2B8 (anti-even-skipped; black HRP reaction) staining of a Stage 15 CNS using the basic procedure of Section VI.D. (E and F) MAb 22C10 (brown HRP reaction) and MAb 4D9 (anti-engrailed; purple alkaline phosphatase reaction) staining of the peripheral nervous system of a Stage 14 embryo as described in Section VI.C. Small arrow points to the engrailed positive neuron in the dorsal PNS cluster. Arrowhead points to the chordotonal cells of the lateral cluster that express engrailed. Large arrow indicates the engrailed-positive neuron of the dorsal cluster that is located more internally within the embryo. Stripes of engrailed-positive cells are in the ectoderm. (G) Anti-β-galactosidase staining (black HRP reaction of an embryo carrying a P-*lacZ* enhancer trap in a gene expressed in longitudinal glia (repo; Xiong *et al.*, 1994) combined with MAb BP102 staining (brown HRP reaction). The numerous longitudinal glia lie on the dorsal surface of the axonal connectives. (H) Stage 14 embryo CNS stained with MAb 2B8 (anti-even-skipped) and BP102 (brown HRP reaction) as described in Section VI.D. Arrowhead points to aCC, large arrow to RP2, small arrow to pCC. (I) *gooseberry* mutant embryo stained with the same combination of antibodies as in H. Posterior commissure of each segment is reduced or eliminated. RP2 neurons (arrows) are duplicated.

methanol/PT series to maintain good morphology. Proceed to Section IV to continue antibody staining.

Keep the embryos at room temperature throughout the procedure because exposure to cold or heat just prior to fixation may create artifacts in morphology. Do not allow the embryos to dry out at any time during the procedure. The protocol can be easily scaled up to process several grams of embryos collected from population cages. Use 250 ml of heptane and 250 ml of PEM-FA in a large glass bottle with a lid impervious to heptane and methanol to fix up to 5 g of embryos. Store up to 5 ml of packed embryos in 40 ml of absolute methanol at −20°C. Remove aliquots of embryos when needed and rehydrate as described above. The fixation protocol described here preserves good cellular morphology at the level of light microscopy; see the protocols in Chapters 23 and 25 for procedures required to preserve cytoskeletal structure and to maintain good morphology at the E.M. level.

B. Whole-Mount Preparation of 17- to 22-hr Embryos

Embryos older than 17 hr develop a cuticle that impedes the penetration of antibodies. The following protocol uses the same fixation procedure as described, but adds a sonication step that allows easy penetration of antibodies through the cuticle. Precedent for sonication comes from a procedure originally developed for staining brine shrimp naupuli (Manzanares *et al.*, 1993).

1. Collect embryos aged 17–22 hr (at 25°C). This covers the period from the onset of cuticle deposition to hatching. A wider age collection will make it difficult to determine the optimal sonication period.
2. Follow Steps 1–6 of Protocol IIIA.
3. Embryos just ready to hatch (~21–22 hr) will break out of their vitelline membrane and sink to the bottom of the aqueous layer. Remove and discard the aqueous layer (PEM-FA), but retain both the embryos at the interface and those that have fallen through the interface to the bottom of the aqueous layer. Remove as much fixative as possible. Make sure that there is still 8–10 ml of heptane remaining with the embryos in the container.
4. Follow Steps 8–10 of Protocol IIIA.
5. Rehydrate by washing 3 × 5 and 1 × 30 min with PT.
6. Place about 100–150 µl of packed embryos into 500 µl of PT in an Eppendorf tube.
7. Sonicate with a probe tip sonicator. I use a Branson Sonifier 250 set at 100% duty cycle and the lowest possible output (setting 1). At these settings, 3 sec of sonication results in almost no visible damage, but antibody penetration improves considerably. After 15 sec of sonication,

about 25–50% of the embryos show obvious morphological damage. Total sonication time of 6–9 sec (two or three pulses of 3 sec duration) minimizes the percentage of damaged embryos, while maintaining excellent staining with MAb BP102, MAb 22C10, anti-serotonin, anti-fasciclin II MAb, MAb 2B8 (anti-*even-skipped*), and MAb 4D9 (anti-*engrailed*). The optimal settings for a given sonicator should be determined empirically by sonicating for a series of different times and then assaying for alterations in morphology and extent of antibody staining. Protect the embryos from excessive heating during the sonication procedure by placing the tubes on ice for 15–30 sec between sonication pulses. MAb 22C10 (1 : 10 dilution of tissue culture supernatant) is an especially useful control antibody when working out the sonication conditions. Remember to include unsonicated embryos as a control. After sonication, keep the embryos on ice until the next step. It does not appear that a postsonication fixation step is necessary, but this may be useful for some antigens.

8. Wash 2 × 5 min in PT. Proceed to Section IV to continue antibody staining.

C. Whole-Mount Preparation of Young First Instar Larvae (0–3 hr Posthatching)

Larvae have a cuticle that not only prevents antibody penetration but also requires a somewhat harsher fixation procedure.

1. Collect larvae that are 0–3 hr posthatching. When you first try this procedure, make sure to start with this narrow age collection so that you can determine the proper sonication period.

2. Rinse with 50% bleach for 1 min to remove adhering yeast.

3. Rinse well with dH$_2$O.

4. Place larvae into a scintillation vial containing 10 ml of heptane plus 10 ml of PEM-FA. Rotate container for 10–20 min. Heptane may be unnecessary for fixing larvae because there is no vitelline membrane, but I have never left it out.

5. Unlike embryos, larvae will sink to the bottom of the aqueous layer because they do not have vitelline membranes. Transfer those larvae that have sunk to the bottom of the vial to a new scintillation vial and fix them for an additional 5 min in PEM-FA containing 0.1% Tween 20. If you are staining for neurotransmitters, such as serotonin, you will need to fix for as long as 2 hr. See comments in Section XIIB,1.

6. Remove the fixative and wash the larvae several times with methanol. Washing stepwise with a dehydration series of methanol/PBS will help minimize morphological damage.

7. As with embryos, larvae can be stored in methanol at −20°C for an extended period of time. As needed, aliquots of larvae can be removed and processed as described below.

8. Remove methanol and replace with PT. Again, a methanol/PT series will help minimize tissue damage.

9. Wash for $2 \times$ 10 min in PT.

10. Place about 100–200 μl of packed larvae into 500 μl of PT in an Eppendorf tube.

11. Sonicate with a probe tip sonicator. As described earlier, I use a Branson Sonifier 250 set at 100% duty cycle and the lowest possible output (setting 1). After about 18–20 sec of sonication, about 25 to 50% of the larvae will show obvious morphological damage. Sonications times of 12–15 sec result in good CNS staining with the same antibodies described in Step 7 of Section IIIB. Also see Step 7 of section IIIB for suggestions on determining the optimal sonication time; again, MAb 22C10 serves as a good control antibody for initial attempts at this procedure. After sonication, keep the larvae on ice until the next step.

12. Wash $2 \times$ 5 min in PT. Proceed to Section IV to continue antibody staining.

I have only recently developed the protocols for sonicating and staining older embryos and larvae described in Sections IIIB and IIIC. While these protocols have worked well in my hands for staining with a half dozen or so different primary antibodies, it is possible that not all antigens can be detected with these procedures.

D. Hand Dissection of Larvae

Third instar larvae are easily dissected by hand. With practice, second instar larvae can also be hand dissected.

1. Wash larvae with water to remove adherent yeast.

2. Dissect larvae in PBS or directly into PBS-FA or PEM-FA in a glass depression well slide (precoating the wells and pipets with PBS + 1% BSA will help prevent larval tissue from getting stuck to the glass). To remove brain, central nerve cord, salivary gland, and anterior discs, use one pair of forceps to grab the mouth hooks and a second to grab the middle of the larva. Pull apart and then tease away excess tissue. Larvae can also be slit along their entire length and held flat while fixed. Alternatively, they can be cut in half and then turned inside out. If the dissection is done in PBS, transfer the tissue to PBS-FA. Fix for 15–30 min. If you are staining for neurotransmitters, you may want to fix for as long as 2 hr. See discussion of fixation time in Section XIIB, 1.

3. Wash $3 \times$ 5 min with PT. From this point on, you can treat the dissected larval tissue as you would intact larvae. Proceed to Section IV to continue antibody staining.

E. Hand Dissection of Living Embryos

In some situations, it is preferable to hand-devitellinize embryos. For example, some epitopes are denatured by methanol and the methanol devitellinization procedure would eliminate antibody recognition. For protocols for hand devitellinization of fixed embryos, see Chapter 36. It is also possible to hand dissect living embryos and then fix them to a glass slide.

1. Prepare microscope slides with silicone walls. First, wash slides thoroughly with ethanol and dry. Using silicone aquarium sealant, construct a rectangular wall about of 2×3 cm and about 2–3 mm high. Be certain that there is continuous contact between the silicone and the glass surface. Allow the silicone to dry overnight. Test the slides by adding PBT and checking that none leaks out through gaps between the silicone and the slide. Just before use, clean the slides by rubbing the glass in the well vigorously with a Kimwipe soaked in ethanol.

2. Cut a small square of double-stick tape (~0.7-cm square) and stick it down on the bottom of one side of the silicone well. Carefully place about one dozen Stage 12–15 embryos on the double stick tape.

3. Using blunt forceps, gently roll all the embryos to remove their chorions. Orient the embryos dorsal side up and make sure they are still well stuck to the double-stick tape.

4. Fill the well with $1 \times$ PBS.

5. Using a sharpened tungsten needle (see Section IXB1), carefully slit open the vitelline membrane over the dorsal side of one of the embryos. Gently lift the embryo, always keeping it well below the meniscus of the liquid, and move it to an area of clean glass.

6. Gently push the embryo against the glass. If the glass is well cleaned, the embryo will stick quite strongly to the surface (it may be helpful to coat the surface with poly-L-lysine beforehand if problems are encountered).

7. Use the needle to fold down the body wall and brain region. Lift the gut to the side to expose the underlying CNS. Repeat Steps 5 and 6 for as many embryos as possible. Try to keep all dissected embryos in the center of the well. Embryos dissected close to a wall will be difficult to look at later. Eventually, the PBS will fill with enough protein to coat the glass and prevent additional dissections from adhering to the glass.

8. Simultaneously, use two pasteur pipets, one to add PBS-FA and another to remove the $1 \times$ PBS. In this way, the dissections can be fixed without the the meniscus of the liquid touching the dissections (if the meniscus does come close to a dissected embryo, the surface tension will pull the nerve cord from the rest of the embryo or the entire embryo from the glass).

9. After 10–30 min of fixation, use two pasteur pipets to wash the embryos $3 \times$ 1 min and then $3 \times$ 10 min with $1 \times$ PBS. Stain the embryos as

described in Section IV and V. Always use two pasteur pipets to exchange the wash solutions. To add antibodies, remove enough solution so that the embryos are just below the surface of the liquid and then add the primary or secondary antibodies at the appropriate concentration.

10. When ready, the embryos can be viewed using a liquid immersion lens. Alternatively, the well can be filled with 50 and then 70% glycerol and the silicone walls can be scraped away with a razor blade and the dissections covered with a coverslip. Use a Kimwipe to blot away the excess glycerol and seal the coverslip with fingernail polish.

This dissection technique can also be used to prepare embryos for Lucifer fills of neurons (Sink and Whitington, 1991). If no detergents are used up until the time that the secondary antibody is washed off, this technique can also be used to show that an antibody epitope is on the outside of the cell membrane (i.e., accessible without detergents to permeabilize the cell membrane). MAb 2D5 (anti-fasciclin III) can be used as a positive control and anti-tubulin can be used as a negative control in this type of experiment.

IV. Primary and Secondary Antibody Incubation Procedure

1. Wash fixed, rehydrated embryos, larvae, or larval tissues for 30 min in PBT. This and all subsequent steps are performed at room temperature unless otherwise stated and can be carried out in 5-ml polypropylene, round-bottom, snap-cap tubes (Falcon No. 2063 culture tubes) or in 1.5-ml microcentrifuge tubes. Use about 25–75 μl of embryos per 5-ml tube or 5–25 μl of embryos per microcentrifuge tube. Washes are improved if the tubes are gently agitated on a rotator (Labquake; Lab Industries) or a rocking platform (Nutator; Clay Adams). Use wash volumes of 4.7 ml with 5-ml tubes and 1.2-ml with 1.5-ml tubes.

2. Incubate 30 min in 100 μl of PBT + NGS (see Section XIII). The normal goat serum (NGS) and BSA will help to block nonspecific antibody binding sites. Gently mix and, if possible, place tubes in a rack mounted on a rotary platform shaker set to approximately 100 rpm. Avoid shaking so hard that embryos splash up onto the walls of the tube.

3. Add the appropriate amount of primary antibody to achieve the desired final concentration (see Section XIIB,3). For example, use tissue culture supernatants of MAb2B8 (anti-even-skipped) at a dilution of 1 : 50 by the tube. For antisera that need to be used at very low concentrations, you can first make up a predilution in sterile 1× PBS containing 5% normal goat serum and 0.02% sodium azide. This predilution should be stable for at least several weeks at 4°C.

4. Gently mix the embryos and antibody solution and incubate overnight at 4°C. If possible, carry out this incubation on a rotary platform shaker set at about 100 rpm.

5. Wash 3 × 5 min with PBT. Before these washes are started, it is possible to recover the diluted primary antibody, and this used antibody can often be used several more times.

6. Wash 4 × 30 min with PBT.

7. Incubate 30 min in 100 μl of PBT × NGS as in Step 2 above.

8. Add appropriate secondary antibody to the desired final concentration. Listed here are the concentration ranges I routinely use for several secondary antibodies available from Jackson ImmunoResearch Labs. The antibodies will arrive in a lyophilized form. After they are reconstituted according to the enclosed instructions, centrifuge the solutions to remove any insoluble particles and then store them as 200-μl aliquots at -70°C (or in a nondefrosting -20°C freezer). Repeated freeze-thawing will rapidly lower the titer of many of these secondary antibodies. Aliquots in use can be stored at 4°C for several weeks or months.

Goat anti-mouse IgG conjugated to FITC, RITC, HRP, or alkaline phosphatase (Catalog No. 115-095-003, 115-025-003, 115-035-003, or 115-055-003). Use at 1 : 300 to 1 : 1000 final dilution. Although antibodies to IgG do cross-react with IgM, it is best to use secondary antibodies directed against both IgG and IgM if you know that the primary antibody is an IgM monoclonal.

Goat anti-rat IgG conjugated to FITC, RITC, HRP, or alkaline phosphatase (Catalog No. 112-095-003, 112-025-003, 112-035-003, or 112-055-003). Use at 1 : 300 to 1 : 1000 final dilution.

Goat anti-rabbit IgG conjugated to FITC, RITC, HRP, or alkaline phosphatase (Catalog No. 111-095-045, 111-025-045, 111-035-045, or 111-055-04). These anti-rabbit secondary antibodies are subtracted against human serum, which appears to lower the background often found with anti-rabbit secondary antibodies. Use at 1 : 500 to 1 : 2000 final dilution.

As a working stock, the secondary antibody should be first diluted to 1 : 100 in PBT + NGS before use. Then, to achieve a final dilution of 1 : 300, you would add 50 μl of 1 : 100 diluted secondary to the 100-μl PBT + NGS embryo mix. These 1 : 100 dilutions of secondary antibody should be stable for at least a week at 4°C.

9. Mix the embryos and secondary antibody solution gently. For embryo staining, incubate for 2–3 hr at room temperature. For larval staining, incubate overnight at 4°C. If possible, carry out these incubations on a rotary platform shaker set at about 100 rpm.

10. Wash 3 × 5 min with PBT.

11. Wash 4 × 30 min with PBT. If you used a fluorescently tagged secondary antibody, proceed to the clearing procedures in Section VIII. For HRP

or alkaline phosphatase-conjugated secondary antibodies, proceed to the histochemical developmental reactions in Section V.

In this protocol, the secondary antibodies are produced in goats and normal goat serum is used as a blocking agent. This ensures that the secondary antibodies will not recognize any of the immunoglobulins in the normal serum being used as a blocking agent. If you use secondary antibodies produced in other animal species, be certain that the normal serum in Steps 2, 3, 7, and 8 does not contain immunoglobulins that will be recognized by the secondary antibody being used.

V. Histochemical Development Reactions

A. Brown HRP Reaction

1. Incubate embryos in 300 μl of DAB solution (see Section XIII) for approximately 2 min. For larvae and larval tissues, extend this to 5–10 min to improve DAB penetration into the tissue.

2. Start the reaction by adding H_2O_2 to a final concentration of 0.01–0.03%. To do this, add 10 to 30 μl of 0.3% H_2O_2 to each tube. For a stock solution, 3% H_2O_2, which is available in any drug store, works well. It is stabilized with 0.001% phosphoric acid and can be stored for many years at room temperature. Dilute a small amount to 0.3% in PBS just before use.

3. Monitor the reaction by looking down into the tube with a dissecting microscope or by removing a few embryos to a depression slide. Brown signal will often be visible within 30 sec. Depending on the primary antibody being used, and the concentration at which it is used, it can take anywhere from 1–5 min for the embryos to reach an optimal signal-to-noise ratio. Generally, no improvement in signal is seen after about 10 min.

4. Stop the reaction by washing 2× 1 min with PT. Proceed to the glycerol or methyl salicylate clearing protocols in Section VIII.

B. Black HRP Reaction

1. Incubate embryos in 300 μl of DAB + Ni solution (see Section XIII) for approximately 2 min. For larvae and larval tissues, extend this incubation time to 5 to 10 min.

2. Follow Steps 2 and 3 in the protocol for the brown HRP reaction (Section VA). The reaction of the nickel ions and DAB product results in the production of a purple to black signal that will often be visible within 10 sec. This nickel enhanced procedure is severalfold more sensitive than the regular DAB procedure. While this reaction is more sensitive than the

brown HRP reaction, it produces a more granular staining that may make fine details harder to visualize.

3. Stop the reaction by washing 2×1 min with PT. Proceed to the glycerol or methyl salicylate clearing protocols in Section VIII. Silver-enhanced HRP reaction provides an even more sensitive stain. Details of this enhancement procedure can be found in Lazar and Taub (1992).

C. Orange and Red HRP Reactions

1. An orange reaction product can be obtained using a DAB plus catechol reagent available from Kirkegaard & Perry Labs (Catalog No. 54-74-00). This reaction is about as sensitive as the regular brown DAB reaction.

2. A red reaction product can be obtained using 3-amino-9-ethylcarbazole (AEC; available from Sigma, Catalog No. AEC-101). Due to the relatively poor sensitivity, however, this reaction is useful only when used with robust primary antibodies. The reaction product is soluble in organic solvents, so use only the glycerol clearing procedure of Section VIII.

D. Purple Alkaline Phosphatase Reaction

1. Wash embryos or larvae 2×5 min in A.P. buffer (see Section XIII).
2. Add 300 μl of BCIP/NBT solution (see Section XIII).
3. Monitor the reaction. An optimal signal-to-noise ratio will usually be reached anywhere from 5–15 min, but the reaction can be allowed to continue for several hours if needed. The sensitivity of this technique is equal to or slightly better than a black HRP reaction. Alkaline phosphatase reactions, however, are more prone to background problems than HRP reactions and the reaction products sometimes give diffuse staining around fine structures such as individual axons.
4. Stop the reaction by washing 2×1 min with A.P. buffer. Proceed to the glycerol clearing protocol in Section VIII.
5. Background alkaline phosphatase activity in cuticular stripes and in the trachea are sometimes observed in embryos past Stage 17. Note that this background is not effectively inhibited by Levamisole, which is often used as an inhibitor of endogenous alkaline phosphatase in vertebrate tissues.

E. Blue Alkaline Phosphatase Reaction

1. Follow Steps 1–4 in the protocol for the purple alkaline phosphatase reaction (Section VD above).
2. After stopping the reaction by washing with A.P. buffer, wash the embryos 2×5 and 1×30 min in methanol. The signal will slowly turn from purple to blue. Wash with methanol until the desired level of blue is obtained.

The color change occurs because the purple alkaline phosphatase reaction product is actually composed of two different reaction products: an alcohol-soluble purple product and an alcohol-insoluble blue product. Because this procedure lowers the level of signal, it should be used only if the starting purple reaction product is relatively strong. An alternative substrate (Fast Blue Base) system for generating a blue reaction product is available from Kirkegaard & Perry (Catalog No. 55-70-00).

3. Wash 2×5 min with A.P. buffer. Proceed to the glycerol clearing protocol in Section VIII.

F. Red Alkaline Phosphatase Reaction Product

1. A red alkaline phosphatase reaction product can be obtained using a Fuchsin dye substrate (Kirkegaard & Perry; Catalog No. 55-69-00). The reaction is somewhat weak and slow (30–60 min), but gives very good staining with antibodies that have a strong signal (MAb BP102 for example). The background will look relatively high initially, but much of the background staining will disappear during the glycerol clearing steps in Section VIII. This reaction product is soluble in organic solvents, so do not use the methyl salicylate clearing procedure of Section VIII.

VI. Labeling with Multiple Primary Antibodies

Because of the variety of possible color detection schemes, both fluorescent and histochemical, it is possible to visualize multiple antibody staining patterns simultaneously in the same embryo. Double labeling procedures have a number of useful applications. For example, by double labeling, the staining pattern of a newly isolated antibody can be compared to the patterns seen with a well-characterized antibody. Double labeling also allows the unambiguous comparison of genotype and phenotype in embryo collections (see Section XI). Fluorescent double labeling is possible only if the primary antibodies are from different animal species or if they are conjugated directly to fluorochromes. However, it is possible to use primary antibodies made in the same animal species, such as two mouse monoclonal antibodies, for histochemical double labeling (see Sections VIC and VID).

In theory, RITC and FITC signals can be collected independently in the microscope, which makes it possible to determine whether and where two staining patterns overlap. Three factors affect the ability to accurately detect overlapping fluorescent signals. First, the secondary antibodies must not recognize the inappropriate primary antibody. Appropriate "subtracted" secondary antibody reagents are now widely available from a number of commercial sources. Second, the microscope's fluorescence filter set must be able to separate the emission spectra of FITC and RITC. Secondary antibodies conjugated to Texas Red rather than RITC can be used to improve color separation. Finally, the microscope must be able to distinguish stained cells that are lying one on top of the other. Confocal microscopy would appear to be the obvious solution to this third consideration, but the small size of many of the cells inside the *Drosophila* embryo will prove a challenge to many confocal microscopes. This is especially problematic when trying to distinguish the flat glial cells from the neurons they sometimes wrap around.

Using histochemical reactions, it is possible to use primary antibodies made in the same species (two mouse monoclonal antibodies, for example) because the antibody incubations and histochemical reactions can be done sequentially (see Sections VIC and VID). When two antibodies stain the same cell, the overlap can be easily distinguished as long as the two antibodies stain different cellular compartments (nucleus vs cytoplasm vs cell membrane). The example in Section VIC shows such a case, in which one antibody recognizes a nuclear antigen (purple stain) and the other recognizes a cytoplasmic antigen (brown stain). Staining should be planned so that the darker reaction product does not obscure the product of the lighter reaction product. For example, if the colors were reversed in Section VIC below, the brown nuclear stain would not be visible through the intense purple cytoplasmic staining. With practice, and with the appropriate controls, it is possible in some cases to document histochemically detected staining overlap even if the antigens reside in the same cellular compartment.

Certain color combinations are not useful. For example, the brown HRP and orange HRP products are not sufficiently different from one another to be used together in double labeling. Usually, a combination of brown staining (DAB reaction) with black (DAB + Ni), purple (alkaline phosphatase), or blue (alkaline phosphatase plus methanol) staining provides the best results. Combinations of black or purple with red also work in some cases. A procedure for fluorescent triple labeling has been described (Paddock *et al.*, 1993), and I have had some success with a few histochemical triple-label combinations.

A large number of procedural combinations are possible for double labeling, but some specific examples will illustrate the basic approaches to double labeling. In each case, embryos should be fixed and devitellinized following the steps in Section IIIA. Antibody incubations and washes are done as described in Section IV.

A. FITC Detection of Mouse Monoclonal Antibody 22C10 Combined with RITC Detection of Rabbit Anti-*even-skipped* (Fig. 6A)

1. Incubate embryos 30 min in 100 μl PBT + NGS.
2. Incubate embryos overnight at 4°C in PBT + NGS containing a 1:5000 dilution of rabbit anti-*even-skipped* and a 1:20 dilution of MAb 22C10.
3. Wash 3× 5 min and 4× 30 min with PBT.

4. Incuate 30 min in 100 μl PBT + NGS.

5. Incubate for 2 hr at room temperature in PBT + NGS containing 1 : 300 dilution of goat anti-rabbit IgG conjugated to RITC (Jackson ImmunoResearch Catalog No. 111-025-144; this secondary antibody has been subtracted against human, mouse, and rat sera so that it will not cross-react with immunoglobulins from these species) plus a 1 : 250 dilution of goat anti-mouse IgG conjugated to FITC (Jackson ImmunoResearch Catalog No. 115-095-146; this secondary antibody has been subtracted against human, bovine, horse, rabbit, and swine sera).

6. Wash 3× 5 and 4× 30 min with PBT.

7. Clear in DABCO solution (see Section XIII) as described in Section VIIIB. Dissect and mount as described in Section IXB.

A fluorescent double-labeling procedure that is especially easy and useful when looking at neural antigens involves the use of goat anti-HRP conjugated directly to RITC. Follow Steps 1–7 in Section IV. Add the appropriate FITC-conjugated secondary antibody but also include goat anti-HRP conjugated to RITC (Jackson ImmunoResearch Catalog No. 123-025-021; see Section IIA4) at a dilution of 1 : 500. Continue with the remaining steps in Section IV. The FITC channel will reveal the pattern of the primary antibody, whereas the RITC channel will reveal all neurons and their axons.

B. Alkaline Phosphatase (Purple Reaction Product) Detection of Mouse Monoclonal Antibody 2B8 (Anti-*even-skipped*) Combined with HRP (Brown Reaction Product) Detection of Rabbit Anti-*paired* (Figs. 5J and 5K)

1. Incubate embryos 30 min in 100 μl PBT + NGS.

2. Incubate embryos overnight at 4°C in PBT + NGS containing a 1 : 1000 dilution of rabbit anti-paired and a 1 : 30 dilution of MAb 2B8.

3. Wash 3× 5 and 4× 30 min with PBT.

4. Incubate 30 min in 100 μl PBT + NGS.

5. Incubate for 2 hr at room temperature in PBT + NGS containing a 1 : 300 dilution of goat anti-mouse IgG-alkaline phosphatase conjugated (Jackson ImmunoResearch Catalog No. 115-055-146; this secondary antibody has been subtracted against human, bovine, horse, rabbit, and swine sera) plus a 1 : 300 dilution of goat anti-rabbit IgG-HRP conjugated (Jackson ImmunoResearch Catalog No. 111-035-144; this secondary antibody has been subtracted against human, mouse, and rat sera). If non-crosscreative secondary antibodies are unavailable, the sequential procedures of Sections VIC and VID may be used instead.

6. Wash 3× 5 and 4× 30 min with PBT.

7. Proceed with the brown HRP reaction as described in Section VA.

8. After the $2\times$ 1 min washes in PT, wash $2\times$ 5 min with A.P. buffer.

9. Proceed with the purple alkaline phosphatase reaction described in Section VD. Steps 7 and 9 can be interchanged if desired so that the alkaline phosphatase reaction is carried out before the HRP reaction.

10. Change the purple reaction product to blue using the protocol in Section VE. The methanol treatment will not alter the color of the brown DAB reaction product.

11. Clear embryos in glycerol as described in Section VIIIB. Mount as in Section IXA.

C. Alkaline Phosphatase (Purple Reaction Product) Detection of Mouse MAb 4D9 (Anti-*engrailed*) Combined with HRP (Brown Reaction Product) Detection of Mouse MAb 22C10 (Figs. 6E and 6F).

1. Incubate embryos 30 min in 100 μl PBT + NGS.

2. Incubate embryos overnight at 4°C in PBT + NGS containing a 1:3 dilution of MAb 4D9 (100 μl PBT + NGS plus 30 μl of MAb 4D9 tissue culture supernatant).

3. Wash $3\times$ 5 min and $4\times$ 30 min with PBT.

4. Incubate 30 min in 100 μl PBT + NGS.

5. Incubate for 2 hr at room temperature in PBT + NGS containing a 1:300 dilution of goat anti-mouse IgG conjugated to alkaline phosphatase.

6. Wash $3\times$ 5 min and $4\times$ 30 min in PBT.

7. React the embryos with NBT/BCIP (purple alkaline phosphatase reaction) as described in Section VD.

8. After the $2\times$ 1-min washes in PT, wash the embryos $2\times$ 30 min with PBT.

9. Incubate embryos in 100 μl of PBT + NGS.

10. Incubate overnight at 4°C in PBT + NGS containing a 1:10 dilution of MAb 22C10.

11. Wash $3\times$ 5 and $4\times$ 30 min in PBT.

12. Incubate 30 min in 100 μl PBT + NGS.

13. Incubate for 2 hr at room temperature in PBT + NGS containing a 1:300 dilution of goat anti-mouse IgG conjugated to HRP.

14. Wash $3\times$ 5 and $4\times$ 30 min in PBT.

15. React the embryos with DAB (brown HRP reaction) as described in Section VA.

16. Clear in glycerol solution as described in Section VIIIB. Dissect and mount as described in Section IXB.

D. HRP (Black Reaction Product) Detection of Mouse MAb 2B8 (Anti-*even-skipped*) Combined with HRP (Brown Reaction Product) Detection of Mouse MAb BP102 (Figs. 6H and 6I)

1. Incubate embryos 30 min in 100 μl PBT + NGS.
2. Incubate embryos overnight at 4°C in PBT + NGS containing a 1:50 dilution of MAb 2B8.
3. Wash 3× 5 and 4× 30 min with PBT.
4. Incubate 30 min in 100 μl PBT + NGS.
5. Incubate for 2 hr at room temperature in PBT + NGS containing a 1:300 dilution of goat anti-mouse IgG conjugated to HRP.
6. Wash 3× 5 and 4× 30 min in PBT.
7. React the embryos with DAB + Ni solution (black HRP reaction) as described in Section VB.
8. After the 2× 1 min washes in PT, wash the embryos 2× 30 min with PBT.
9. Incubate embryos in 100 μl of PBT + NGS.
10. Incubate overnight at 4°C in PBT + NGS containing a 1:10 dilution of MAb BP102.
11. Wash 3× 5 and 4× 30 min in PBT.
12. Incubate 30 min in 100 μl PBT + NGS.
13. Incubate for 2 hr at room temperature in PBT + NGS containing a 1:300 dilution of goat anti-mouse IgG conjugated to HRP.
14. Wash 3× 5 and 4× 30 min in PBT.
15. React the embryos with DAB (brown HRP reaction) as described in Section VA.
16. Clear in glycerol solution as described in Section VIIIB. Dissect and mount as described in Section IXB.

The reactions in Section VIC and VID must be done in the order described; the darker (black or purple) reaction must be done before the lighter (brown) reaction. This is because the second histochemical reaction will create some signal over the first reaction. This will not be a problem if the first reaction is much darker than the second (a little brown staining on top of a dark black signal will make no difference). This color overlap can be avoided entirely, however, if the antibodies are eluted after the first histochemical reaction. This can be accomplished by washing the embryos 2× 5 min in antibody elution buffer (10 mM glycine, pH 2.3; 500 mM NaCl; 0.1% Triton X-100; 0.1% BSA) after the 2× 1-min washes in PT in Step 8 in Sections VIC or VID. Then wash 2× 15 min in PBT and continue with Step 9 in Sections VIC or VID.

Be certain that there will be no unwanted cross-reactions between the primary and secondary antibodies. For example, if a combination of a mouse monoclonal and a rabbit polyclonal primary antibody is used, do not detect them with a combination of rabbit anti-mouse secondary antibody conjugated to FITC and a goat anti-rabbit secondary antibody conjugated to RITC. Also, be certain that the normal serum does not contain immunoglobulins that will be recognized by the secondary antibody being used.

VII. Rapid Staining Procedure

Following the embryo fixation and staining procedures described in Sections III and IV, it will take 1 to 1 1/2 days to go from embryo collection to stained embryos. However, with robust antibodies such as MAb BP102, MAb 22C10, and MAb 2B8 (anti-even-skipped), it is possibly to streamline the procedure significantly so that stained embryos are ready only 3 hr after the embryos are collected. This rapid procedure also uses somewhat simplified solutions. Embryos fixed according to the procedures in Section IIIA and then stored in methanol can enter this rapid procedure at Step 6.

1. Collect, dechorionate, and rinse embryos as in Steps 1–4 in Section III.
2. Place embryos into a scintillation vial containing 10 ml heptane plus 10 ml PBS-FA or PEM-FA. Fix for 10–20 min, agitating the vial two or three times during the fixation period.
3. Remove the aqueous layer from the bottom.
4. Add 10 ml methanol. Shake for 10–15 sec.
5. Remove devitellinized embryos from the bottom and wash them 3× 1 min with methanol.
6. Wash 2× 1 min followed by 1× 10 min with PT.
7. Incubate 10 min in 100 μl PT + NGS.
8. Add primary antibody to the appropriate final concentration (MAb BP102 at 1:10; MAb 22C10 at 1:5; MAb 2B8 at 1:30). Primary antibodies should be used about 1.5 to 2 times the "normal" concentration (i.e., that used for the staining procedure in Section IV).
9. Mix and incubate in the primary antibody at room temperature for 30 min.
10. Wash 3× 1 min with PT.
11. Wash 3× 10 min with PT.
12. Add secondary antibody. It is not necessary to preblock with PT + NGS. For HRP immunohistochemistry, use the goat anti-mouse IgG at a dilution of 1:250. Mix and incubate in the secondary antibody for 30 min at room temperature.
13. Wash 3× 1 min with PT.
14. Wash 3× 10 min with PT.

15. Add 300 μl DAB or DAB + Ni solution.
16. Add 10–30 μl of 0.3% H_2O_2.
17. When ready, stop the reaction by washing $3 \times$ 1 min with PT.
18. Wash $1 \times$ with $1 \times$ PBS.
19. Replace $1 \times$ PBS with 70% glycerol solution. After 5 min, mix the embryos gently. Embryos will be cleared sufficiently in about 10–20 min. Proceed as in Section IX.

VIII. Clearing Embryos

After washing in PT, the embryos or larvae may look good under a dissecting microscope, but they will appear relatively opaque under a compound or fluorescence microscope. By "clearing" (equalizing the refractive index of the tissue and surrounding media) the embryos or larvae with methyl salicylate or with glycerol, it is possible to view them at higher magnification. Optically, methyl salicylate clearing produces the best results. However, tissues cleared in methyl salicylate are very brittle and difficult to dissect in a controlled manner. Furthermore, methyl salicylate will solubilize some of the reaction products described in Section V. Although glycerol clearing does not produce the same level of transparency, the embryos are easy to handle, and cleared tissue can be dissected and flattened to obtain an improved view of certain internal structures. For embryos stained using DAB or DAB + Ni, it is often useful to clear half of the stained embryos in methyl salicylate and the other half in glycerol. Methyl salicylate should be used with care as it can damage plastic racks as well as cause skin irritation.

A. Methyl Salicylate Clearing

Fluorescently (RITC or FITC) and DAB (brown or black reaction) stained embryos can be cleared in methyl salicylate with no loss of signal. Methyl salicylate by itself seems to be efficient at preventing the quenching of both FITC and RITC. Methyl salicylate will entirely or partially extract the other reaction products described in Section V.

1. Wash $1 \times$ 1 min with $1 \times$ PBS. Make sure that the embryos or larvae do not get stuck to the sides of the tube.
2. Wash $1 \times$ 5 min with 50% ethanol (ethanol series is in dH_2O).
3. Wash $1 \times$ 5 min with 70% ethanol.
4. Wash $1 \times$ 5 min with 90% ethanol.
5. Wash $2 \times$ 5 min with 100% ethanol. Make certain that all traces of water are removed from the tube and cap of the tube.

6. Add 500 µl of methyl salicylate (Sigma, Catalog No. M 6752). Do not disturb the tube for the first 1–2 min or the embryos will end up stuck to the walls of the tube. After 5–10 min, the embryos should settle to the bottom. Fluorescently stained specimens will be so clear that they will be visible only as blue ghosts. Draw off the methyl salicylate and replace with a fresh 500 µl of methyl salicylate.

7. DAB-stained tissue can be stored in sealed tubes containing methyl salicylate for several months at room temperature or indefinitely at 4°C. Proceed to Section IX to mount the embryos.

DAB stained (brown or black reaction) tissue can also be prepared for sectioning. At the end of Step 5, embed the dehydrated embryos or larvae in LR White acrylic resin (medium grade) and section with a diamond knife.

B. Glycerol Clearing

All fluorescently or histochemically stained embryos and larvae can be cleared safely with glycerol.

1. Wash $1\times$ 1 min with $1\times$ PBS.

2. Add 500 µl of 50% glycerol (v:v with $1\times$ PBS) to the embryos or larvae. To counterstain nuclei so that they fluoresce blue, include DAPI (diaminophenylindole) at a concentration of 1.0–0.1 µg/ml in the 50% glycerol solution. Allow the tissue to settle. After about 30 to 60 min, remove the 50% glycerol.

3. For histochemically stained preparations, add 500 µl 70% glycerol (v:v with $1\times$ PBS). Allow the embryos to sit at room temperature for 60 min or overnight at 4°C. The embryos can be stored indefinitely in 70% glycerol in sealed tubes in the dark at either 4°C or -20°C. Proceed to Step IX when you are ready to mount the embryos. Even clearer preparations can be produced by placing the embryos in 90% glycerol, but the tissue is more difficult to dissect and mount than when in 70% glycerol.

4. For fluorescently stained preparations, add 500 µl of DABCO solution (anti-fade agent in 70% glycerol; see Section XIII). Allow the embryos to sit at room temperature for 60 min or overnight at 4°C. The embryos can be stored at least a week in the DABCO solution at -20°C in the dark. Proceed to Step IX when you are ready to mount the embryos.

If histochemically stained tissues are counterstained with DAPI in Step 2 above, the stained cells will appear to be unstained by DAPI when viewed under fluorescence. This quenching of the DAPI signal is especially prominent if the histochemical staining is nuclear. The negative staining effect is quite striking and is very useful when examining expression patterns at the blastoderm stage (Karr and Kornberg, 1989).

IX. Dissecting and Mounting Stained Embryos

Proper mounting of embryos is essential for seeing internal tissues clearly and for obtaining high-quality photographs.

A. Mounting Cleared Embryos and Larvae

1. Allow the embryos or larvae in methyl salicylate or 70% glycerol (with or without DABCO) to warm to room temperature. Place them into a glass depression slide and examine under a dissecting microscope. Select and transfer anywhere from a single embryo to two dozen embryos in a volume of about 25–30 μl to a clean glass slide. Place two 18 × 18-mm No. 1 thickness coverslips on both sides of the drop, about 1 cm apart. These act as support coverslips to keep the embryos from being crushed by the coverslip that will go on top. For larvae, No. 2 thickness coverslips should be used.

2. Place a single 18 × 18-mm No. 1 thickness coverslip over the embryos. The glycerol or methyl salicylate should just fill about 80–90% of the space bounded by the slide and three coverslips. Additional methyl salicylate or glycerol can be added from the side, or any excess can be removed by wicking it away with a Kimwipe.

3. While looking at the embryos under the compound or dissecting microscope, you can roll the embryos by gently moving the top coverslip. Excessive rapid rolling will cause the methyl salicylate-cleared embryos to fracture into pieces and the glycerol-cleared embryos to twist up. If a single embryo is mounted, it is possible to photograph all orientations of the embryo by gently rolling it between photographs.

4. It is also possible to slightly flatten an embryo so that more of it is in a single focal plane. First, orient the embryo by sliding the top coverslip and then use one hand to hold the top coverslip steady and another hand to pull away one or both of the support coverslips. The top coverslip will now press down on the embryo, flattening it somewhat. The quantity of glycerol or methyl salicylate under the coverslip (and whether one of both support coverslips have been removed) will control the extent to which the specimen is flattened. If there is insufficient glycerol or methyl salicylate, the embryo will crack or become badly distorted. If the specimen is not flat enough, a Kimwipe can be used to wick away excess methyl salicylate or glycerol.

5. To save a slide of methyl salicylate-cleared specimens, seal all the edges of the coverslips with Permount. Alternatively, the embryos can be placed in a 1 : 4 mixture of methyl salicylate:Permount (v:v) after Step VIIIA,6 and then mounted under a coverslip. Slides should be allowed to dry for 1 or 2 days. If you want to save the slides of glycerol-cleared specimens, completely fill the space under the coverslip with glycerol and seal all the edges of the coverslips with fingernail polish. Some batches of fingernail polish contain solvents that

will either cause the staining to fade or cause the embryos to become discolored after a few days, so first check your brand of fingernail polish on some less important specimens. Slides of histochemically stained preparations will last at least several years when stored in the dark at room temperature. Slides of fluorescently stained preparations will last at least several days if stored at 4°C in the dark.

B. Dissecting Glycerol Cleared Embryos and Larvae

With a bit of practice, it is possible to dissect glycerol-cleared embryos and larvae before mounting. Larvae that have been stained as whole mounts can be dissected to obtain a better view of the stained tissues. Dissecting germband-extended embryos makes it possible to visualize both the future dorsal and ventral ectoderm of the embryo at the same time. Most of the identified neurons sit on the dorsal surface of the CNS, and by dissecting open the embryo, it is possible to look directly down onto the dorsal surface of the nerve cord.

1. For the dissections, you will need a sharpened wire needle made from a 2- to 3-cm-long piece of tungsten wire (0.13 mm diameter; Ted Pella Inc., Catalog No. 27-11). Put a kink near one end of the wire and insert this end into the beveled opening of a 26-G syringe needle. The kink will create enough friction so that the base of the wire will remain securely in place within the syringe needle. The wire can then be sharpened electrically using a 1 M solution of NaOH and a direct current power supply (I use one that supplies 6.3 amps at an adjustable voltage between 2 and 6 V). Wear face and hand protection to avoid being splashed by any droplets of NaOH that might bubble up from the solution. Place one electrode in a small beaker of 1 M NaOH and clip the other onto the syringe needle. Dipping the needle in and out of the solution will cause the needle to taper to a point. Start at a setting of 4–5 V to taper the end of the wire to a blunt point and then switch to 2–3 V to put a sharp point just at the end of the needle. Avoid creating a long, thin end as this will be bent too easily when you use it to dissect embryos.

2. Transfer embryos to a depression well slide and pick out an embryo to dissect. Transfer a single embryo in 2–4 μl of 70% glycerol to a microscope slide. Use the wire needle to move the embryo out of the drop of glycerol. The surface tension of the glycerol still coating the embryo will hold the embryo down to the slide and allow you to manipulate it without it rolling around uncontrollably. The glycerol also makes it possible to dissect the embryo slowly without it drying out. For mid-stage 12 to late stage 15 embryos, use the wire tool to roll the embryo onto its ventral side. Make slits along the amnioserosa, at the base of the head, and near the posterior end of the embryo. Fold down the body walls and brain region. Lift out the gut and remove it entirely or slide it next to the embryo. For Stage 8 to early Stage 12 embryos, roll the embryo on its side, make a slit along the amnioserosa, and use the wire

tool to "unfold" the embryo. Scrape out the mesoderm if you are going to be looking at ectodermal or neural patterns. Turn the embryo so that the ectoderm faces up (away from the slide) to get the best view of ectodermal or early neuroblast patterns.

3. Place a 10–15-μl drop of 70% glycerol on a coverslip. Invert the coverslip and hold it directly over the slide so that the drop is above the dissected embryo. Drop the coverslip onto the slide. The drop should spread out and fill the area under the coverslip. If the dissected embryo flattens too much, use a larger volume of glycerol in the drop on the coverslip next time. Alternatively, place support coverslips (No. 1 or 0 thickness) to both sides of the dissected embryo, place a coverslip on top, introduce glycerol from the side, and then remove one or both supports to somewhat flatten the preparation (this support technique is advisable when looking at neuroblast patterns). With some practice, it should be possible to dissect multiple embryos on a single slide. Never let the coverslip move sideways once it has been placed on the slide—this will destroy the dissected embryos. To flatten the dissected embryos a bit more, wick away excess glycerol from the side of the coverslip with a Kimwipe. To seal the slide, first place a small drop of fingernail polish at each corner of the coverslip. Wait for these drops to dry (about 5 min) and then use the fingernail polish to seal all the edges of the coverslip.

X. Hints for Photographing Histochemically Stained Specimens

A. Optics

1. Bright-field optics will show the highest contrast between staining and background, but little detail will be visible in the unstained regions. Nomarski (DIC) optics will allow unstained tissue to be visualized and stained regions will appear sharper than in bright field, but the contrast between stained and unstained regions will be lessened. By adjusting the DIC slider, it should be possible to find a good optical balance that gives both good contrast and visualization of unstained structures.

2. Magnification in the range of 200–250$\times$ will allow the entire field to be filled by a whole embryo. I generally use 20$\times$ and 40$\times$ dry lenses and a 100$\times$ oil immersion lens for photography. Oil immersion 40$\times$ and 63$\times$ lenses are also useful in many instances.

3. Closing the aperture diaphragm of the condenser will help eliminate excessive "glow" in the center of glycerol cleared embryos, but if it is closed too much, unstained tissues will take on a grainy appearance.

4. Before taking a photograph, make sure that you have Köhler illumination. Keep the luminous field diaphragm closed as far as possible (to the edge of the photographic field) to minimize light scattering.

B. Color Balance

1. For 35-mm-slide photography, Kodak Professional Tungsten Ektachrome film ASA 64 (EPY135) works well. When taking pictures, first make sure that the light intensity is set to 3200K. Most microscopes will have either a set switch or an indicator to show when the light is set at this level.

2. Although tungsten film is designed so that 3200K illumination will provide the proper color balance without additional filters, the use of a blue color-filter may improve the quality of photographed images. Most microscopes come with a dark blue filter that is designed to convert tungsten illumination to a daylight color balance (3200K to 5500K conversion filter). Even more useful than this filter, however, are light and medium blue filters (CB3 and CB6; Zeiss Catalog Nos. 467851 and 467852). They can be placed directly on top of the luminous field diaphragm and do not necessarily need to be installed into the filter magazine.

3. Bright-field photographs will usually require the dark blue filter so that the background does not come out too yellowish. If this background is too blue, the CB6 or CB3 filters should be used instead.

4. Specimens with brown or red reaction products viewed with DIC optics look best when photographed with a medium blue background. Either the CB3 or CB6 filter should be used for this.

5. Black, purple, and blue reaction products look best when photographed on a white or even slightly gray background. Use the CB3 filter or no filter at all.

C. Video Image Capture

The majority of the figures in this chapter were captured electronically, rather than on film, using a Sony 3CCD camera (DXC-760MD) attached to a Zeiss Axioplan microscope. Images were captured on a Macintosh Quadra 800 with a NuVista Plus Video board (Source Digital Systems), using Adobe PhotoShop. Images that were originally captured on film were scanned with a Kodak RFS 2035 slide scanner. The figures were arranged and labeled with Aldus PageMaker and printed using a Kodak XL7700 dye-sublimation printer. Fluorescence images were captured on a Zeiss laser scan microscope.

XI. Hints for Looking at Mutants That Affect Central Nervous System Development

A. Identifying Genotype of Embryos

When looking at the embryos collected from a balanced lethal stock, only one-fourth of the embryos will be homozygous for the mutation being studied. To unambiguosly identify the mutant embryos, balancer chromosomes carrying *lacZ* constructs can be used. For example, mutants in *gooseberry* (*gsb*) can be balanced with a second chromosome balancer containing a *lacZ* gene under the control of the *fushi tarazu* promoter (*CyO-ftz lacZ*). When embryos from this stock are collected, they can be stained for both β-galactosidase (see Section IID) and an antibody of interest (MAb 2B8, for example) using the double-labeling protocols in Section VI. Three-fourths of the Stage 8–12 embryos will show the ftz-β-galactosidase pattern (large subset of the CNS) and the MAb 2B8 pattern, and one-fourth of the embryos, those homozygous for the *gsb* mutation, will show only the MAb 2B8 pattern. In this way, the MAb 2B8 pattern of *gsb* mutants can be studied. A wide variety of *lacZ* markers are available for the commonly used balancer chromosomes. Remember to pick a balancer whose β-galactosidase pattern will be clearly visible for the stages of embryogenesis that are under examination.

If the mutation being studied is a protein null allele, the genotype can also be determined by staining with an antibody to the protein product. For example, if the embryos for the *gsb* stock described above are double-labeled with an antibody of interest plus an antibody against gsb, the *gsb* mutants will fail to show any staining with the gsb antibody.

B. Examples from Known Mutants

1. *zipper* mutant. The *zipper* gene encodes the heavy chain of nonmuscle myosin (Young *et al.*, 1993). This product is supplied maternally, but as the maternal contribution is exhausted, several morphological defects become visible in a variety of tissues. In the nervous system, axon outgrowth is altered and commisures are "fused" together (Fig. 3D).

2. *cyclin A* mutant. Cyclin A is provided maternally, but as this maternal product runs out, cell division is disrupted (Lehner and O'Farrell, 1989). In the nervous system, this causes several changes in the normal lineage patterns. For example, the ganglion mother cell that forms aCC and pCC neurons fails to undergo its division to produce these two neurons (Fig. 3H).

3. *gooseberry* mutant. The *gooseberry* gene (*gsb-d*) encodes a transcription factor containing both a homeodomain and a *paired* domain (Baumgartner *et al.*, 1987). Mutations at this locus cause changes in neural differentiation. For example, RP2 neurons become duplicated and changes in cell differentiation also lead to alterations in the pattern of axonogenesis (Patel *et al.*, 1989b; see Fig. 6I).

4. Two notes of caution: First, when analyzing mutants with neural markers, be suspicious of axonal defects which appear only in the third to fifth abdominal segments. These defects are often related to pleiotropic defects that cause the process of germband retraction to be delayed. Second, use a variety of nonneural and segmentation markers to examine the

development of other organs to be certain that neural defects are not secondary to defects in adjacent tissues.

XII. Trouble Shooting

A. Low Percentage of Devitellinized Embryos

1. If there are too many embryos in the vial, a smaller percentage will be devitellinized. Optimally, the embryos should form no more than a monolayer at the interface between the heptane and fixative. When fixing large quantities of wild-type embryos in a large container (5 g of embryos in 250 ml heptane plus 250 ml PEM-FA), only about half the embryos will become devitellinized, but this is usually satisfactory since enough embryos for hundreds of staining experiments will still be obtained.

2. Devitellinization efficiency can be improved by a rapid temperature change. After Step 7 in Section IIIA, remove most of the remaining heptane (do not let the embryos dry out). Add 10 ml of heptane that has been chilled to −20 to −50°C. Then quickly add 10 ml of room-temperature methanol and shake the vial vigorously while holding it under a warm tapwater stream. Although this procedure is not necessary when processing *Drosophila* embryos, it is very useful for devitellinizing many other insect embryos (such as *Tribolium* embryos).

3. If the methanol has absorbed significant quantities of moisture from the air, it will be less efficient at devitellinizing embryos. Keep methanol containers tightly closed.

4. If necessary, the nondevitellinized embryos from the interface can be salvaged to some extent. At Step 8 of Section IIIA, remove the nondevitellinized embryos from the interface to a new vial containing 2–3 ml of methanol. Remove any heptane that was transferred with the embryos (it will be floating on top of the methanol). Fill the vial with methanol and shake gently. The embryos should now sink in the methanol. Wash 3× 5 min with methanol and then proceed to Step 11 of Section IIIA. Continue staining as if the embryos were devitellinized. Anywhere from 10–50% of these nondevitellinized embryos will be stained, because although the vitelline membrane has not been removed, sufficiently large holes are present to allow antibodies to enter. These nondevitellinized embryos will not, however, sink into glycerol. The remaining vitelline membranes can be dissected away while the embryos are in 50% glycerol and the embryos can then be transferred to 70% glycerol.

B. Poor Signal and/or High Background

1. Overfixation often results in low signal and underfixation often results in high background. For most of the antibodies mentioned in these protocols, fixation times of 10–20 min produce good results. Some antigens, however,

require harsher fixation. For example, the neurotransmitters serotonin and FMRFamide are small molecules that require longer periods of fixation or harsher fixatives such as Bouin's fixative (see Schneider *et al.* (1993) for a discussion of fixation parameters for FMRFamide staining).

2. Methanol also acts as a fixative. In fact, several antibodies, such as MAb 2D5, will produce acceptable results even if no formaldehyde is used in Section IIIA. Methanol, however, may extract and/or denature some antigens. If staining is methanol sensitive, embryos will have to be devitellinized by hand (see Section IIIE).

3. Adjust concentration of the primary antibody. It is important to try a range of concentrations to see what concentrations work best for any given primary antibody. On occasion, even two- to threefold changes in antibody concentration produce dramatic changes in signal to background ratios. For antisera and ascites fluid, usable dilutions may range anywhere from 1 : 200 to 1 : 50,000. Monoclonal antibodies in the form of tissue culture supernatants may work well anywhere in the range of dilutions from 1 : 1 to 1 : 200. If necessary, antibodies from tissue culture supernatants can be concentrated rapidly and without danger of denaturation using E-Z-Sep reagents (Middlesex Sciences).

4. Preabsorb the primary antibodies. Crude antisera, especially from rabbits, often contain random immunoglobulins that bind to various epitopes in *Drosophila* embryos or larvae, leading to high background. This can often be eliminated by preabsobing the antiserum with fixed *Drosophila* embryos. First, dilute the antiserum 1 : 50 with PT+NGS. In a 500-μl microfuge tube, combine fixed embryos (i.e., processed as far as the end of Step 11 in Section IIIA) with the diluted antiserum using approximately 50 μl of packed embryos per 200 μl of diluted antiserum. Mix gently at 4°C overnight. Recover the supernatant, centrifuge to remove any remaining particulate material, and store the preabsorbed antisera at 4°C. Sodium azide can be added to a final concentration of 0.02% to inhibit bacterial growth.

5. Adjust concentration of the secondary antibody. It is useful to try a range of concentrations to see what concentrations work best for any given primary antibody and detection system. If the secondary antibodies are affinity purified, preabsorption with fixed embryos makes very little difference. For example, embryos that are not incubated with a primary antibody but are incubated for 2 hr with a 1 : 300 dilution of goat-anti-mouse IgG conjugated to HRP (Jackson ImmunoResearch, Catalog No. 115-035-003) and then washed, will not show any color change after 10 min of reaction with DAB+Ni containing 0.01% H_2O_2. If preabsorption is needed, follow the procedure for preabsorbing primary antibodies but do not add sodium azide to HRP conjugated antibodies (see XII,B,11).

6. Prepare fresh substrate solutions. The various substrates used for histochemical detection will degrade with time.

7. Adjust histochemical reaction times. Staining will often appear suitable when viewed with a dissecting microscope but will seem weak when observed

with the intense transillumination of a compound microscope. If this occurs, allow histochemical reactions to proceed longer to intensify the staining reaction. If histochemical reactions proceed too rapidly, high background may result. In this case, either reduce the concentrations of the antibodies or reduce the concentration of the reaction substrates.

8. Replace directly conjugated secondary antibodies with biotin-conjugated secondary antibodies and follow with HRP, FITC, RITC, or alkaline phosphatase-conjugated streptavidin. Biotinylated secondary antibodies and streptavidin conjugates are available from a variety of sources. A system using streptavidin-biotin complexes is available from Vector Labs (VectaStain Elite Kit). Compared to directly conjugated secondary antibodies, these avidin/biotin reagents may provide increased sensitivity in situations in which the concentration of primary antibody is limiting. In my own experience, the vast majority of antibody staining of whole-mount preparations is not improved significantly by switching to biotin–streptavidin complex detection systems.

9. Increase the stringency of the washes. In Steps 6 and 11 of Section IV, replace the first of the 30-min washes in PBT with a 30-min wash in PT in which the NaCl concentration has been increased to 500 mM. This increased salt concentration (from 175 to 500 mM) will help elute potential low-affinity binding to cross-reactive epitopes. Increasing the length of the washes beyond 4× 30 min will generally not make much difference because these wash times are probably already excessive.

10. Endogenous HRP activity can be destroyed by incubating tissue for 30 min in 70% methanol containing 0.5% H_2O_2. This potential background source is rarely a problem in embryos, but it does occur in some larval tissues.

11. Make certain that HRP-conjugated reagants are not mixed with solutions containing sodium azide. Sodium azide will reduce or completely destroy the activity of the HRP enzyme. Sodium azide in primary antibody solutions will not be a problem because it will be washed away from the tissue before the addition of HRP-coupled secondary antibodies.

12. Some antigens are removed from fixed tissue by prolonged exposure to Triton X-100. The replacement of the Triton X-100 in PBT and PT with 0.02% saponin will improve staining intensity in such cases.

13. Adjust antibody incubation times. Increasing or decreasing incubation times in primary and secondary antibodies may increase signal or reduce background.

XIII. Solutions

10X PBS

18.6 mM	NaH$_2$PO$_4$	(2.56 g NaH$_2$PO$_4$ · H$_2$O per 1000 ml dH$_2$O)
84.1 mM	Na$_2$HPO$_4$	(11.94 g Na$_2$HPO$_4$ per 1000 ml dH$_2$O)
1750.0 mM	NaCl	(102.2 g NaCl per 1000 ml dH$_2$O)

Adjust pH to 7.4 with NaOH or HCl. Prepare 1× PBS by diluting 1 : 10 with dH₂O. Both 1× and 10× PBS can be kept indefinitely at room temperature.

PT

1×   PBS
0.1%   Triton X-100

Mix 100 ml 10× PBS, 899 ml dH₂O, and 1 ml Triton X-100. Store at 4°C or at room temperature

PBT

1×   PBS
0.1%   Triton X-100
0.1%   BSA (Sigma, Catalog No. A-7906)

Mix 100 ml 10× PBS, 800 ml dH₂O, 1 ml Triton X-100, and 1 g BSA. Adjust volume to 1000 ml. Store at 4°C. Solution will usually last at least 1 month. Discard if bacterial growth is detected (solution will start to turn cloudy).

PBT+NGS

1×   PBS
0.1%   Triton X-100
0.1%   BSA (Sigma, Catalog No. A-7906)
5.0%   normal goat serum (Gibco-BRL, Catalog No. 200-6210AG)

Heat inactivate the serum at 56°C for 30 min. Filter through a 0.22-μm filter while still warm. Aliquot into sterile tubes. Store aliquots at −20°C. Once thawed, aliquots are stable for several months at 4°C. To prepare the PBT+NGS solution, mix 4.75 ml PBT with 0.25 ml normal goat serum and store at 4°C. Solution will usually last at least 2 or 3 weeks. Discard if bacterial growth is detected (solution will start to turn cloudy).

PT+NGS
Same as PBT+NGS, but without the BSA.

PEM (pH 7.0)

100.0 mM Pipes (Disodium salt, Sigma Catalog No. P-3768)
2.0 mM EGTA
1.0 mM MgSO₄

Weigh out the solid Pipes, EGTA, and MgSO$_4$ into a beaker, add the appropriate volume of dH$_2$O, mix for 20 min, and then adjust the pH to 7.0 with concentrated HCl. The free acid form of Pipes is more difficult to get into solution and the pH will need to be adjusted with NaOH instead of HCl. PEM can be stored for at least 1 year at 4°C.

PEM-FA

9.0 ml   PEM
1.0 ml   37% formaldehyde (Fisher Catalog No. F79-500; this 37% stock solution can be stored at room temperature for at least 1 year)

This PEM-FA solution should be made just before use. For the immunohistochemical and immunofluorescence whole-mount procedures outlined in this chapter, it is unnecessary to start with solid paraformaldehyde. The 37% formaldehyde solution is stabilized with 10–15% methanol, however, so when using the protocols in Section IIID or IIIE to detect antigens sensitive to methanol, the fixative should be prepared from solid paraformaldehyde.

PBS-FA

Same as PEM-FA, except use 9.0 ml 1× PBS instead of 9.0 ml of PEM.

DAB Solution

1×   PBS
0.05%   Tween 20 (Sigma, Catalog No. P-1379)
0.3 mg/ml   DAB (3,3'-diaminobenzidine; Sigma Catalog No. D-5905)

The 10-mg DAB tablets sold by Sigma are very convenient and help minimize the risk of exposure. Note that DAB is a potential carcinogen and should be handled and disposed of in accordance with University regulations. Add one 10-mg DAB tablet to a 50-ml tube containing 33.0 ml PBS and 16.5 μl Tween 20. Rock gently in the dark for about 30 min. Filter through a 0.22-μm filter to remove particulate matter. Store aliquots at −70°C or in a nondefrosting −20°C freezer. Aliquots should be used immediately after thawing.

DAB+Ni Solution

Prepare an 8% solution of nickel chloride (NiCl$_2$ · 6H$_2$O; Fisher Catalog No. N54-500) in dH$_2$O. This 8% solution can be stored indefinitely at room temperature. Prepare the DAB+Ni solution by combining 1 ml of the 0.3 mg/

ml DAB solution described above with 8 μl of 8% nickel chloride. Mix well and use immediately. It is not advisable to store DAB containing nickel chloride because the nickel will precipitate out of solution (as nickel phosphate) after a few hours.

A.P. (Alkaline phosphatase) Buffer

5.0 mM	MgCl2
100.0 mM	NaCl
100.0 mM	Tris, pH 9.5
0.1%	Tween 20 (Sigma Catalog No. P-1379)

Prepare just prior to use. The solution will become cloudy after a few hours and then will not work as well for the enzymatic reaction.

BCIP/NBT Solution

1.0 ml	A.P. Buffer
4.5 μl	NBT (50 mg/ml in 70% DMF)
3.5 μl	BCIP (50 mg/ml in 70% DMF)

Mix just before use. The NBT and BCIP solutions can be purchased together from Promega (Catalog No. S3771).

Glycerol Solutions

Some batches of glycerol contain contaminants that cause nickel-enhanced DAB reactions to fade within 1 or 2 days. To avoid this, use ultrapure glycerol (Boehringer Mannheim, Catalog No. 100 647). Prepare 50, 70, and 90% glycerol solutions by mixing the appropriate volumes of glycerol with 1× PBS. Use pH paper to make certain that the pH of the glycerol solutions is around 7.4. Low pH will cause rapid fading of DAB reaction products. Glycerol solutions can be stored at room temperature. Glycerol solutions containing DAPI should be stored in the dark at 4°C.

DABCO solution

70.0% glycerol containing 2.5% DABCO (Sigma, Catalog No. D-2522)

Quenching of fluorescently stained preparations can be minimized by the addition of free radical scavengers such as DABCO (1,4-diazobicyclo-[2.2.2]-octane). DABCO is more stable and soluble than other commonly used anti-fade reagents. Prepare this solution by dissolving 1.25 g of

DABCO in 15 ml 1× PBS. Add 35 ml of ultrapure glycerol and mix gently for 1–2 hr. Store at −20°C.

Appendix: Suppliers

Middlesex Sciences
100 Foxborough Blvd. Suite 220
Foxborough, MA 02035

Jackson ImmunoResearch Laboratories
P.O. Box 9
872 W. Baltimore Pike
West Grove, PA 19390

Kirkegaard and Perry Laboratories, Inc.
2 Cessna Court
Gaithersburg, MD 20879

Gibco/BRL
P.O. Box 68
Grand Island, NY 14072

Tetko Inc.
333 South Highland Blvd.
Briarcliff Manor, NY 10510

Ted Pella, Inc.
P.O. Box 492477
Redding, CA 96049

Source Digital Systems
1420 Springhill Rd.
McLean, VA 22102

Fisher Scientific
50 Fadem Rd.
Springfield, NJ 07081

Acknowledgments

I thank Irina Orlov for technical assistance and Haifan Lin for his help with the confocal microscopy. I thank Rebecca Chasan and Susan Dymecki for comments on the manuscript, Allan Spradling, Joe Gall, and Corey Goodman for their encouragement, and the editors for their patience.

References

Akam, M. (1987). The molecular basis for metameric pattern in the *Drosophila* embryo. *Development* **101**, 1–22.

Alberga, A., Boulay, J.-L., Kempe, E., Dennefeld, C., and Haenlin, M. (1991). The *snail* gene required for mesoderm formation in *Drosophila* is expressed dynamically in derivatives of all three germ layers. *Development* **111**, 983–993.

Bate, M. (1993). The mesoderm and its derivatives. *In* "The Development of *Drosophila melanogaster*" (M. Bate and A. Martinez-Arias, eds.), pp. 1013–1090. New York: Cold Spring Harbor Laboratory Press.

Bate, M., and Martinez-Arias, A. (1993). "The Development of *Drosophila melanogaster*." New York: Cold Spring Harbor Laboratory Press.

Baumgartner, S., Bopp, D., Burri, M., and Noll, M. (1987). Structure of two genes at the *gooseberry* locus related to the *paired* gene and their spatial expression during *Drosophila* embryogenesis. *Genes Dev.* **1**, 1247–1267.

Bellen, H. J., O'Kane, C. J., Wilson, C., Grossniklaus, U., Pearson, R., and Gehring, W. J. (1989). P-element-mediated enhancer detection: A versatile method to study development in *Drosophila*. *Genes Dev.* **3**, 1288–1300.

Bier, E., Ackerman, L., Barbel, S., Jan, L. Y., and Jan, Y. N. (1988). Identification and characterization of a neuron-specific nuclear antigen in *Drosophila*. *Science* **240**, 913–916.

Bier, E., Vässin, H., Shepherd, S., Lee, K., McCall, K., Barbel, S., Ackerman, L., Carretto, R., Uemura, T., Grell, E., Jan, L. Y., and Jan, Y. N. (1989). Searching for pattern and mutation in the *Drosophila* genome with a P-lacZ vector. *Genes Dev.* **3**, 1273–1287.

Blochlinger, K., Bodmer, R., Jan, L. Y., and Jan, Y. N. (1990). Patterns of expression of Cut, a protein required for external sensory organ development, in wild-type and *cut* mutant Drosophila embryos. *Genes Dev.* **4**, 1322–1331.

Bodmer, R., and Jan, Y. N. (1987). Morphological differentiation of the embryonic peripheral neurons in *Drosophila*. *Roux's Arch. Dev. Biol.* **196**, 69–77.

Campos-Ortega, J. A. (1993). Early neurogenesis in *Drosophila melanogaster*. *In* "The Development of *Drosophila melanogaster*" (M. Bate and A. Martinez-Arias, eds.), pp. 1091–1130. New York: Cold Spring Harbor Laboratory Press.

Dambly-Chaudière, C., and Ghysen, A. (1986). The sense organs in the *Drosophila* larva and their relation to the embryonic pattern of sensory neurons. *Roux's Arch. Dev. Biol.* **195**, 222–228.

Dambly-Chaudière, C., Jamet, E., Burri, M., Bopp, D., Basler, K., Hafen, E., Dumont, N., Spielmann, P., Ghysen, A., and Noll, M. (1992). The paired box gene *pox neuro:* A determinant of polyinnervated sense organs in Drosophila. *Cell* **69**, 159–172.

DiNardo, S., Kuner, J., Theis, J., and O'Farrell, P. H. (1985). Development of embryonic pattern in *D. melanogaster* as revealed by accumulation of the nuclear *engrailed* protein. *Cell* **43**, 59–69.

Doe, C. Q. (1992). Molecular markers for identified neuroblasts and ganglion mother cells in the Drosophila nervous system. *Development* **16**, 855–864.

Frasch, M., Hoey, T., Rushlow, C., Doyle, H., and Levine, M. (1987). Characterization and localization of the *even-skipped* protein of *Drosophila*. *EMBO J.* **6**, 749–759.

Fujita, S. C., Zipursky, S. L., Benzer, S., Ferrus, A., and Shotwell, S. W. (1982). Monoclonal antibodies against the *Drosophila* nervous system. *Proc. Natl. Acad. Sci. U.S.A.* **79**, 7929–7933.

Ghysen, A., and O'Kane, C. (1989). Detection of neural enhancer-like elements in the genome of Drosophila. *Development* **105**, 35–52.

Ghysen, A., Dambly-Chaudiére, C., Aceves, E., Jan, L. Y., and Jan, Y. N. (1986). Sensory neurons and peripheral pathways in *Drosophila* embryos. *Roux's Arch. Dev. Biol.* **195**, 281–289.

Goodman, C. S., and Doe, C. Q. (1993). Embryonic development of the *Drosophila* central nervous system. *In* "The Development of *Drosophila melanogaster*" (M. Bate and A. Martinez-Arias, eds.), pp. 1131–1206. New York: Cold Spring Harbor Laboratory Press.

Grenningloh, G., Rehm, E. J., and Goodman, C. S. (1991). Genetic analysis of growth cone guidance in Drosophila: Fasciclin II functions as a neuronal recognition molecule. *Cell* **67**, 45–57.

Gutjahr, T., Patel, N. H., Li, X., Goodman, C. S., and Noll, M. (1993). Analysis of the *gooseberry* locus in *Drosophila* embryos: *gooseberry* determines the cuticular pattern and activates *gooseberry neuro. Development* **118**, 21–31.

Hartenstein, V., and Jan, Y. N. (1992). Study of *Drosophila* embryogenesis with P-lacZ enhancer trap lines. *Roux's Arch. Dev. Biol.* **201**, 194–220.

Hay, B., Ackerman, L., Barbel, S., Jan, L. Y., and Jan, Y. N. (1988). Identification of a component of Drosophila polar granules. *Development* **103**, 625–640.

Hortsch, M., Bieber, A. J., Patel, N. H., and Goodman, C. S. (1990). Alternative splicing generates a nervous system specific form of neuroglian. *Neuron* **4**, 697–709.

Ingham, P. W. (1988). The molecular genetics of embryonic pattern formation in *Drosophila*. *Nature* **335**, 25–34.

Ingham, P. W., and Martinez-Arias, A. (1992). Boundaries and fields in early embryos. *Cell* **68**, 221–236.

Jacobs, J. R., and Goodman, C. S. (1989a). Embryonic development of axon pathways in the Drosophila CNS. I. A glial scaffold appears before the first growth cones. *J. Neurosci.* **9**, 2402–2411.

Jacobs, J. R., and Goodman, C. S. (1989b). Embryonic development of axon pathways in the Drosophila CNS. II. Behavior of pioneer growth cones. *J. Neurosci.* **9**, 2412–2422.

Jan, L. Y., and Jan, Y. N. (1982). Antibodies to horseradish peroxidase as specific neuronal markers in *Drosophila* and in grasshopper embryos. *Proc. Natl. Acad. Sci. U.S.A.* **79**, 2700–2704.

Jan, Y. N., and Jan, L. Y. (1993). The peripheral nervous system. *In* "The Development of *Drosophila melanogaster*" (M. Bate and A. Martinez-Arias, eds.), pp. 1207–1244. New York: Cold Spring Harbor Laboratory Press.

Kania, M. A., Bonner, A. S., Duffy, J. B., and Gergen, J. P. (1990). The *Drosophila* segmentation gene *runt* encodes a novel nuclear regulatory protein that is also expressed in the developing nervous system. *Genes Dev.* **4**, 1701–1713.

Karr, T. L., and Kornberg, T. B. (1989). *fushi tarazu* protein expression in the cellular blastoderm of Drosophila detected using a novel imaging technique. *Development* **105**, 95–103.

Kiehart, D. P., and Feghali, R. (1986). Cytoplasmic myosin from *Drosophila melanogaster*. *J. Cell Biol.* **103**, 1517–1525.

Klämbt, C., and Goodman, C. S. (1991). The diversity and pattern of glia during axon pathway formation in the Drosophila embryo. *Glia* **4**, 205–213.

Lasko, P. F., and Ashburner, M. (1988). The product of the *Drosophila* gene *vasa* is very similar to eukaryotic initiation factor-4a. *Nature* **335**, 611–617.

Lazar, J. G., and Taub, F. E. (1992). A highly sensitive method for detecting peroxidase in in situ hybridization or immunohistochemical assays. *In* "Nonradioactive Labeling and Detection of Biomolecules" (C. Kessler, ed.), pp. 135–142. Berlin: Springer-Verlag.

Lehner, C., and O'Farrell, P. H. (1989). Expression and function of Drosophila cyclin A during embryonic cell cycle progression. *Cell* **56**, 957–968.

Manning, G., and Krasnow, M. A. (1993). Development of the *Drosophila* tracheal system. *In* "The Development of *Drosophila melanogaster*" (M. Bate and A. Martinez-Arias, eds.), pp. 609–686. New York: Cold Spring Harbor Laboratory Press.

Manzanares, M., Marco, R., and Garesse, R. (1993). Genomic organization and developmental pattern of expression of the engrailed gene from the brine shrimp Artemia. *Development* **118**, 1209–1219.

Mitchison, T. J., and Sedat, J. W. (1983). Localization of antigenic determinants in whole *Drosophila* embryos. *Dev. Biol.* **99**, 261–264.

Nose, A., Mahajan, V. B., and Goodman, C. S. (1992). Connectin: A homophilic cell adhesion molecule expressed on a subset of muscles and the motoneurons that innervate them in Drosophila. *Cell* **70**, 553–567.

Paddock, S. W., Langeland, J. A., DeVries, P. J., and Carroll, S. B. (1993). Three-color immunofluorescence imaging of Drosophila embryos by laser scanning confocal microscopy. *Biotechniques* **14**, 42–48.

Pankratz, M. J., and Jäckle, H. (1993). Blastoderm segmentation. *In* "The Development of *Drosophila melanogaster*" (M. Bate and A. Martinez-Arias, eds.), pp. 467–516. New York: Cold Spring Harbor Laboratory Press.

Patel, N. H., and Goodman, C. S. (1992). Preparation of digoxigenin-labeled single-stranded DNA probes. *In* "Nonradioactive Labeling and Detection of Biomolecules" (C. Kessler, ed.), pp. 377–381. Berlin: Springer-Verlag.

Patel, N. H., Snow, P. M., and Goodman, C. S. (1987). Characterization and cloning of fasciclin III: A glycoprotein expressed on a subset of neurons and axon pathways in Drosophila. *Cell* **48,** 975–988.

Patel, N. H., Martin-Blanco, E. Coleman, K. G., Poole, S. J., Ellis, M. C., Kornberg, T. B., and Goodman, C. S. (1989a). Expression of *engrailed* proteins in arthropods, annelids, and chordates. *Cell* **58,** 955–968.

Patel, N. H., Schafer, B., Goodman, C. S., and Holmgren, R. (1989b). The role of segment polarity genes during *Drosophila* neurogenesis. *Genes Dev.* **3,** 890–904.

Patel, N. H., Ball, E. E., and Goodman, C. S. (1992). Changing role of *even-skipped* during the evolution of insect pattern formation. *Nature* **357,** 339–342.

Patel, N. H., Condron, B. G., and Zinn, K. (1994). Pair-rule expression patterns of *even-skipped* are found in both short and long-germ beetles. *Nature* **367,** 429–434.

Robinow, S., Campos, A. R., Yao, K.-M., and White, K. (1988). The *elav* gene product of *Drosophila,* required in neurons, has three RNP consensus motifs. *Science* **242,** 1570–1572.

Schneider, L. E., Sun, E. T., Garland, D. J., and Taghert, P. H. (1993). An immunocytochemical study of the FMRFamide neuropeptide gene product in Drosophila. *J. Comp. Neurol.* **337,** 446–460.

Seeger, M. A., Tear, G., Ferres-Marco, D., and Goodman, C. S. (1993). Mutations affecting growth cone guidance in *Drosophila:* Genes necessary for guidance towards or away from the midline. *Neuron* **10,** 409–426.

Sink, H., and Whitington, P. M. (1991). Location asnd connectivity of abdominal motoneurons in the embryo and larva of *Drosophila melanogaster*. *J. Neurobiol.* **12,** 298–311.

Skeath, J. B., and Carroll, S. B. (1992). Regulation of proneural gene expression and cell fate during neuroblast segregation in the Drosophila embryo. *Development* **114,** 939–946.

Snow, P. M., Patel, N. H., Harrelson, A. L., and Goodman, C. S. (1987). Neural-specific carbohydrate moiety shared by many surface glycoproteins in *Drosophila* and grasshopper. *J. Neurosci.* **712,** 4137–4144.

Tautz, D., Lehmann, R., Schnürch, H., Schuh, R., Seifert, E., Kienlin, A., Jones, K., and Jäckle, H. (1987). Finger protein of novel structure encoded by *hunchback,* a second member of the gap class of *Drosophila* segmentation genes. *Nature* **327,** 383–389.

Thomas, J. B., Bastiani, M. J., Bate, C. M., and Goodman, C. S. (1984). From grasshopper to Drosophila: A common plan for neuronal development. *Nature* **310,** 203–207.

Van Vactor, D., Sink, H., Fambrough, D., Tsoo, R., and Goodman, C. S. (1993). Genes that control neuromuscular specificity in Drosophila. *Cell* **73,** 1137–1153.

White, K., and Vallés, A. M. (1985). Immunohistochemical and genetic studies of serotonin and neuropeptides in Drosophila. *In* "Molecular Basis of Neural Development" (G. M. Edelman, W. Einar Gall, and W. M. Cowan, eds.), pp. 547–564. New York: Wiley.

Xiong, W-C., Okano, H., Patel, N. H., Blendy, J. A., and Montell, C. (1994). *repo* encodes a glial-specific homeodomain protein required in the Drosophila nervous system. *Genes Dev.* **8,** 981–994.

Young, P. E., Richman, A. M., Ketchum, A. S., and Kiehart, D. P. (1993). Morphogenesis in Drosophila requires non-muscle myosin heavy chain function. *Genes Dev.* **7,** 29–41.

Zalokar, M., and Erk, I. (1977). Phase-partition fixation and staining of Drosophila eggs. *Stain Technol.* **52,** 89–92.

Zipursky, S. L., Venkatesh, T. R., Teplow, D. B., and Benzer, S. (1984). Neuronal development in the *Drosophila* retina: Monoclonal antibodies as molecular probes. *Cell* **36,** 15–26.

CHAPTER 25

Immunofluorescence Analysis of the Cytoskeleton during Oogenesis and Early Embryogenesis

William E. Theurkauf

Department of Biochemistry and Cell Biology
State University of New York at Stony Brook
Stony Brook, New York, 11794

I. Introduction

Drosophila oogenesis and early embryogenesis are syncytial processes that rely on establishment and maintenance of developmentally significant cytoplasmic order and asymmetry. Cytoskeletal elements play important roles in organizing the cytoplasm of eukaryotic cells, and immunocytochemical and pharmacological studies have provided insight into the functions of the cytoskeleton during oocyte formation and early embryonic development (reviewed here). However, the powerful genetic techniques that are available in *Drosophila* are only beginning to be systematically applied to the cytoskeleton during oogenesis and embryogenesis, and cytological analyses of mutant ovaries and embryos should continue to contribute to our understanding of cytoskeletal function.

Reliable cytological procedures are essential to the study of cytoskeletal biology. In this chapter, immunocytochemical techniques for the visualization of actin filaments and microtubules during oogenesis and early embryogenesis are described. Cytoskeletal and nuclear organization in living *Drosophila* embryos and oocytes have also been examined (Kellogg *et al.,* 1988; Minden *et al.,* 1989; Theurkauf, unpublished observations). In these systems, therefore, direct comparisons to *in vivo* data can be used to determine the fidelity of immunocytochemical techniques. The results obtained using the procedures described here are remarkably similar to the *in vivo* data, indicating that these techniques preserve most aspects of cytoskeletal organization (Kellogg *et al.,* 1988; Theurkauf, unpublished).

The remainder of this chapter is divided into three sections. Sections II and III describe the cytoskeletal biology of oogenesis and embryogenesis and present the tissue-specific isolation and fixation protocols used in these systems. The final section describes procedures that are common to immunolabeling both egg chambers and embryos.

II. The Cytoskeleton during Oogenesis

A. Review of Cytoskeletal Organization and Function

Drosophila ovaries are composed of a series of parallel ovarioles, each of which is divided into an anterior and posterior compartment (reviewed by Mahowald and Kambysellis, 1980). Oogenesis is initiated within the anterior compartment, or germarium, by a stem cell division that produces a cystoblast and regenerates the stem cell. The cystoblast proceeds through four incomplete mitotic divisions, producing a cyst of 16 cells that are interconnected by large cytoplasmic bridges, called ring canals. Although these 16 cells share a common cytoplasm, they go on to form a single oocyte and 15 nurse cells. The oocyte nucleus enters meiotic prophase and remains transcriptionally inactive through most of oogenesis, while the nurse cell nuclei are actively transcribed and become polyploid through multiple rounds of endomitotic replication. During the first half of oogenesis, several mRNAs that appear to be synthesized in the nurse cells accumulate specifically in the ooplasm, while the ooplasm excludes the *vasa* protein (Wharton and Struhl, 1989; Kim-Ha *et al.,* 1991; Suter and

Steward, 1991; Ephrussi *et al.*, 1991; Dalby and Glover, 1992; Hay *et al.*, 1990; Lasko and Ashburner, 1990). The oocyte is therefore a specialized cytoplasmic compartment within a syncytium.

Microtubules appear to play a key role in establishing the cytoplasmic compartment that will become the oocyte. Microtubule disassembly in the germarium leads to the production of egg chambers that contain 16 nurse cells and no oocyte (Koch and Spitzer, 1983; Theurkauf *et al.*, 1993). Microtubule assembly inhibitors also disrupt oocyte-specific accumulation of at least three mRNAs, and oocyte-specific exclusion of vasa protein (Theurkauf *et al.*, 1993). A prominent microtubule organizing center (MTOC) forms within the future oocyte of germarial cysts. Microtubules extend from this MTOC through the intracellular connections and into the surrounding nurse cells (Theurkauf *et al.*, 1993). A single microtubule cytoskeleton thus interconnects the 16-cell cysts and structurally differentiates the oocyte from the nurse cells. Mutations that lead to production of 16-nurse-cell cysts also disrupt the polarized microtubule cytoskeleton (Theurkauf *et al.*, 1993). These observations suggest that the microtubule cytoskeleton provides a polarized scaffold for vectorial transport within the syncytial cysts and that this transport system is essential to oocyte differentiation. Presumably, this polar transport system leads to accumulation of an oocyte-differentiation factor(s) in the cytoplasmic compartment that will eventually form the oocyte.

As the 16-germ-cell cysts progress through the germarium; they are surrounded by a monolayer of somatic follicle cells. The germ cell–follicle cell complex within the most posterior germarial compartment is referred to as a Stage 1 egg chamber. Most of oogenesis occurs after egg chambers bud from the germarium and enter the posterior compartment of the ovariole, or vitellarium. Vitellaria contain developmentally ordered linear arrays of egg chambers (Cummings and King, 1969). The newly formed Stage 2 egg chambers are in the region closest to the germarium, whereas fully mature Stage 14 egg oocytes are found at the posterior of the vitellarium. The polarized microtubule arrangement that is established within the germarium is maintained through Stage 6 (Theurkauf *et al.*, 1992). Microtubule disassembly during vitellarial Stages 2 through 6 inhibits oocyte growth and oocyte-specific mRNA accumulation, indicating that microtubule-dependent transport to the oocyte continues during these stages (Theurkauf *et al.*, 1993; Koch and Spitzer, 1983).

During Stages 7 and 10, morphogenetic determinants accumulate at the anterior and posterior poles of the oocyte (Ephrussi *et al.*, 1991; Kim-Ha *et al.*, 1991; Berleth *et al.*, 1988; Lasko and Ashburner, 1990; Hay *et al.*, 1990). At the transition from Stage 6 to 7, microtubules begin to associate preferentially with the anterior cortex of the oocyte, and an anterior to posterior cortical microtubule gradient has formed by Stage 9 (Theurkauf *et al.*, 1992). The oocyte microtubule cytoskeleton thus becomes polarized along the anterior–posterior axis at the time that morphogenetic molecules are first asymmetrically localized within the oocyte cortex. In addition, microtubule inhibitors disrupt anterior localization of bicoid mRNA (Pokrywka and Stephenson, 1991). These observations strongly suggest that the microtubule cytoskeleton plays a direct role in axial patterning.

Stages 10b through 12 are characterized by two dramatic forms of cytoplasmic movement. The bulk of the nurse cell cytoplasm is transferred to the oocyte and the ooplasm streams. Transfer of nurse cell cytoplasm depends on actin filament function and is described in detail in Chapter 28. Ooplasmic streaming, which mixes the nurse cell cytoplasm with the ooplasm, is inhibited by microtubule depolymerizing drugs (Gutzeit and Koppa, 1982). The onset and cessation of ooplasmic streaming is temporally coordinated with the assembly and disassembly of subcortical microtubule arrays (Theurkauf *et al.*, 1992), and analysis of microtubule behavior *in vivo* demonstrates that the direction of ooplasmic streaming parallels the subcortical microtubules (Theurkauf, unpublished observation). These observations indicate that ooplasmic streaming is microtubule-based.

Oocyte maturation occurs during Stage 13 and induces a final dramatic reorganization of the oocyte cytoskeleton. Stages 1 through 12 oocytes contain a dense layer of cortical actin filaments, but only punctate phalloidin-staining material in the cytoplasm. During Stage 13, an extensive network of cytoplasmic actin filaments assembles. The function of these cytoplasmic thin filaments is not clear. In addition, maturation causes breakdown of the subcortical microtubules that are present during cytoplasmic streaming, and short microtubules are present throughout mature Stage 14 oocytes (Theurkauf *et al.*, 1992).

Stage 14 oocytes are arrested in metaphase of the first meiotic division, and cytological analyses of maturation-induced spindle assembly have provided insight into the mechanism of chromosome segregation. These studies suggest that the *nod* kinesin-like protein provides a force directed away from the spindle pole, and that this force maintains nonexchange chromosomes on the meiotic spindle (Theurkauf and Hawley, 1992). Analysis of mutations that suppress meiotic exchange suggests that the physical linkage provided by recombination-dependent chiasmata prevents premature chromosome movement to the poles, and thus assures that the oocyte arrests in metaphase (McKim *et al.*, 1993).

B. Protocol 1: Isolating and Fixing Egg Chambers

Egg chambers are mass isolated by a modification of the procedure of Mahowald *et al.* (1983). Mass isolation is faster than hand dissection and thus minimizes the time that egg chambers are exposed to medium before fixation. This is important, because prolonged exposure to a variety of culture media activate mature oocytes and induce abnormal actin filament assembly in earlier stage egg chambers (Theurkauf, unpublished observation). Manually dissected ovaries can be fixed as described here, but isolation times should be kept to a minimum. Fixed whole ovaries should be teased apart before immunolabeling.

1. Procedure

1. Anesthetize flies (using CO_2 or ether) and transfer to a blender containing 200 ml of modified Robb's medium (55 mM potassium acetate, 40 mM sodium acetate, 100 mM sucrose, 10 mM glucose, 1.2 mM magnesium chloride, 1.0 mM calcium chloride, 100 mM Hepes, pH 7.4). Female flies do not need to be separated from males prior to transfer.

2. Pulse the blender three times, for 2 sec each, at low speed. This disrupts the cuticle and releases egg chambers and ovariole fragments.

3. Remove larger contaminants and flies that are still intact by passing the disaggregated fly parts through a 250-μm mesh filter. Rinse the material retained on the mesh into the blender with 200 ml of Robb's medium and repeat the blender treatment. Pass this material through the same screen and pool with the previously collected filtrate.

4. Let the egg chamber gravity settle while monitoring progress with a dissecting microscope. After most of the egg chambers have settled to the bottom of the beaker, which usually takes 2 to 5 min, aspirate off all but 3 to 5 ml of supernatant.

5. Swirl the beaker to resuspend the isolated egg chambers and transfer to a 10× 150-mm test tube.

6. Let the egg chambers settle and immediately aspirate off the supernatant. Resuspend in 5 ml of oocyte fixation buffer (100 mM potassium cacodylate, pH 7.2, 100 mM sucrose, 40 mM potassium acetate, 10 mM sodium acetate, 10 mM Na_3EGTA, 8% formaldehyde (electron microscopy grade, methanol free) and incubate for 5 to 10 min with gentle agitation.

7. Place the test tube upright in a rack and let the egg chambers settle. Remove the supernatant and resuspend the fixed egg chambers in phosphate-buffered saline (PBS; 1× PBS is 8 g NaCl, 0.2 g KCl, KH_2PO_4, 1.15 g $NaHPO_4$/liter; adjust to pH 7.3 with NaOH) (Karr and Alberts, 1986) containing 0.01% Triton X-100 (PBST). To remove residual formaldehyde, repeat this PBST wash two more times. After the final wash, resuspend the egg chambers in PBS containing 1% Triton X-100. Incubate for 1 to 2 hr with gentle agitation.

8. Let the egg chambers settle, remove the 1% Triton solution, and resuspend in PBST. Germaria and Stage 2 through 10b egg chambers are now ready for immunolabeling (Section IV, Protocol 2). The developing chorion and vitelline membranes must be removed from later egg chambers before the oocyte can be labeled with antibodies (Section IV, Protocol 1).

2. Comments

1. The blender treatment should be monitored to determine the length and number of pulses that will disrupt the cuticle without destroying the egg chambers. Overblending can produce high autofluorescence background.

2. This stripped down version of the mass isolation protocol is designed to get the egg chambers into fixative as quickly as possible. The fixed material is therefore contaminated with a variety of nonovarian fly parts that can generally be ignored during cytological analysis. If more highly purified preparations are required, the resuspension and gravity settling steps can be repeated several times. In addition, egg chambers can be staged by passing the material obtained at Step 3 through various pore-size nylon mesh filters. For example, Stages 13 and 14 egg chambers can be separated from earlier chambers with a 125-μm mesh filter. Prolonged incubation in Robb's medium can induce artifactual actin filament assembly and oocyte activation, however, and these additional isolation steps should be performed as quickly as possible.

3. This procedure calls for methanol-free electron microscopy grade formaldehyde. Microtubule preservation is not as consistently satisfactory when reagent grade formaldehyde is used. Paraformaldehyde should not be used in this procedure, because anti-tubulin antibodies do not penetrate beyond the follicle cell layer in paraformaldehyde-fixed egg chambers.

4. 1% Triton extraction (Step 6) improves antibody penetration to the germline cells, but is not required prior to phalloidin or 4.6-diamino-2-phenylindole (DAPI) labeling.

5. Antibodies sometimes fail to label germline cells within egg chambers. This is often due to overfixation, and shorter formaldehyde incubations should be tried if germline labeling is weak. Antigens that are very abundant in the somatic follicle cells can also be difficult to detect in the germline, because the follicle cell antigen pool serves as an antibody sink. To overcome this, primary and secondary antbody concentrations and incubation times should be increased. Alternatively, fluorescently conjugated primary antibodies can be used. Directly conjugated anti-tubulin antibodies label germline microtubules much more efficiently than the same antibodies combined with fluorescent secondary antibody conjugates (Theurkauf *et al.*, 1992).

III. The Cytoskeleton during Early Embryogenesis

A. Review of Embryonic Cytoskeletal Organization and Function

Embryogenesis in *Drosophila* begins with a series of rapid and nearly synchronous nuclear divisions that proceed without cytokinesis. During the first four divisions, the nuclei form a roughly spherical mass within the center of the embryo, toward the anterior pole. Nuclei redistribute along the anterior–posterior axis during division cycles 4 through 6, and this axial expansion forms a nuclear elipsoid that mirrors the shape of the embryonic cortex. During division cycles 8 and 9, nuclei migrate synchronously from the ellipsoid toward the surface, and an uniform monolayer of cortical nuclei is present at interphase of division 10. The syncytial blastoderm is thus formed, and the cortical nuclei of the syncytial blastoderm divide four times before membranes invaginate and a true cellular blastoderm is established. Gastrulation begins immediately after the completion of cellularization (Foe and Alberts, 1983; Zalokar and Erk, 1976).

Actin filament inhibitors disrupt nuclear spacing and chromosome segregation during the cortical divisions of the syncytial blastoderm stage and membrane invagination during blastoderm cellularization (Zalokar and Erk, 1976; Foe and Alberts, 1983). Actin filaments and associated membranes reorganize to form furrows that surround the cortical mitotic spindles during the syncytial blastoderm divisions (Karr and Alberts, 1986; Warn et al., 1985). Mutations that disrupt furrow formation lead to spindle fusions and chromosome segregation errors, suggesting that the actin-based furrows provide mechanical barriers that isolate spindles within the syncytium (Postner et al., 1992; Sullivan et al., 1993). During cellularization, actin filaments are assembled into an extended network of interlinked hexagonal units, each of which surrounds a blastoderm nucleus. Contraction of this actin network has been proposed to drive membrane invagination during cellularization (Schejter and Wieschaus, 1993).

Microtubules appear to drive the cortical nuclear movements that produce the syncytial blastoderm. Microtubule assembly inhibitors block cortical nuclear migration and the migrating nuclei are linked by an interdigitating network of microtubules (Foe and Alberts, 1983; Baker et al., 1993). This network includes regions of microtubule overlap that are morphologically similar to the overlapping microtubules of the spindle midzone. This morphological similarity suggests that force generation during cortical migration may be mechanistically similar to spindle pole separation during the anaphase B movements of mitosis (Baker et al., 1993).

B. Protocol 1: Formaldehyde/Methanol Fixation

This simple variation on the embryonic fixation protocol of Mitchison and Sedat (1983) preserves microtubule and actin organization and works well with antibodies to several nuclear and cytoskeletal proteins. This is the best general purpose fixation protocol for most cytoskeletal applications, although it is not appropriate in all circumstances (see following).

The most significant difference between this protocol and most others is that undiluted 37% formaldehyde solution is used during fixation. Lower concentrations of formaldehyde do not preserve microtubule structures during the rapid syncytial divisions, unless the microtubule-stabilizing drug taxol is used (Karr and Alberts, 1986). Taxol induces abnormal microtubule assembly, however, and should be avoided (Kellogg et al., 1988).

1. Procedure

1. Collect embryos on apple juice agar plates (Ashburner, 1989).
2. Rinse the embryos from the plates with Triton/NaCl (0.05% Triton X-100, 0.7% NaCl) and collect on a fine nylon mesh supported in filter assembly (prepared as shown in Fig. 1).

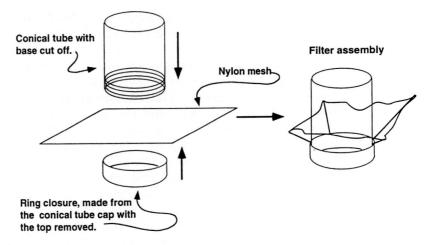

Fig. 1 Construction of an embryo filtration apparatus. Cut the bottom off of a 50-ml conical tube. To make a retaining ring, cut the top from the tube cap. Place a nylon mesh over the tube opening and screw on the modified cap.

3. Transfer the filter assembly with embryos to a beaker containing a solution of 50% commercial bleach, 50% Triton/NaCl. Incubate 2 to 3 min to dissolve the chorion.

4. Remove the filter assembly and embryos from the bleach solution. Rinse the dechorionated embryos with Triton/NaCl and then with distilled water.

5. Draw water away from the embryos by placing paper towels against the nylon mesh. Dry the inside walls of the filter assembly with a tissue.

6. Rinse the embryos from the filter assembly into a 50-ml Erlenmeyer flask with 10 ml of 100% octane. Gently swirl the embryos in the octane for 20 to 30 sec to permeabilize the vitelline membranes.

7. Add 10 ml of 37% formaldehyde (reagent grade) to the embryo/octane mixture and incubate for 3 to 5 min. The embryos float at the interface between the formaldehyde (lower phase) and the octane (upper phase).

8. Draw off the formaldehyde phase, leaving the octane and fixed embryos in the flask.

9. Add 10 ml 100% methanol, stopper the flask, and shake vigorously for 30 sec to 1 min. The vitelline membranes swell and rupture, and embryos without vitelline membranes sink to the bottom of the methanol phase. Embryos that have not lost their vitelline membranes remain trapped at the methanol/octane interface.

10. Draw off the octane, any embryos remaining at the interface, and most of the methanol.

11. Resuspend the remaining embryos in 100% methanol and incubate for 1 to 2 hr at room temperature.

12. Rehydrate the embryos by passage through a methanol series. Start with 90% methanol (10% water), and follow with 75% methanol, 50% methanol, and 100% PBS. At each solution change, allow the embryos to settle to the bottom of the flask, immediately draw off the supernatant, and resuspend in the next solution. Prolonged incubation at each methanol solution change is not required.

13. Draw off the PBS, add PBST, and incubate for 30 min.

The embryos are now ready for antibody labeling (Section IV, Protocol 2).

2. Comments

1. The most common artifacts encountered with this technique are the result of anoxia, which can develop when small containers are used during dechorionation. Under these conditions, the embryos are tightly packed and appear to rapidly deplete the oxygen supply in the bleach and wash buffers. Anoxia leads to anaphase chromosome bridging, which is easily observed in DAPI-stained preparations. Many examples of these anaphase bridges have been published. To avoid anoxia, the nylon filter apparatus should be large enough so that the embryos form a sparse monolayer during dechorionation. Similarly, the fixation vessel should be wide enough so that the embryos form a monolayer at the octane /formaldehyde interface. An Erlenmeyer flask is an ideal fixation container, because the octane/formaldehyde interface spreads out within the wide base. In general, narrow dechorionation and fixation containers should be avoided.

2. The syncytial mitoses are very rapid, and analysis of living embryos indicates that the microtubule cytoskeleton probably turns over in a matter of seconds (Theurkauf, unpublished). Rapid fixation is therefore essential to microtubule preservation. Octane prepermeabilization, combined with very high concentrations of fixative, assure rapid formaldehyde penetration and fixation. When embryos are added directly to an octane/formaldehyde mixture, microtubule preservation is poor. This is probably because the vitelline membranes are permeabilized in the presence of fixative and penetration of fixative during the initial stages of permeabilization is slow.

3. Because of the high concentration of formaldehyde that is used in this protocol, relatively short fixation times are required. If fixation is allowed to proceed too long, methanol removal of the vitelline membranes is inefficient and antibody penetration is poor. Formaldehyde fixation for as little as 2.5 min works well, and shorter times may be sufficient.

4. Incubating the fixed embryos in methanol (Step 10) clears the cytoplasm somewhat, although immediate rehydration also works well. Fixed embryos

can be stored in 100% methanol at $-20°C$ for at least 2 weeks, and more extended storage is probably fine.

C. Protocol 2: Formaldehyde Fixation

The majority of the actin in a typical cell is not filamentous, and anti-actin immunolabeling thus produces significant cytoplasmic signal that can obscure filament organization. Fluorescent conjugates of phalloidin specifically bind to filamentous actin (Cooper, 1987) and are therefore extremely useful reagents for cytological analysis of actin filament organization. These powerful reagents cannot be used to label tissues that have been treated with methanol, however. The formaldehyde fixation protocol presented here preserves microtubules and allows phalloidin labeling, although the vitelline membranes must be mechanically removed from these embryos before labeling (Section IV, Protocol 1).

1. Procedure

1. Collect embryos on apple juice agar plates (Ashburner, 1989).
2. Rinse the embryos from the plates with Triton/NaCl (0.05% Triton X-100, 0.7% NaCl) and collect on a fine nylon mesh supported in a filter assembly (prepared as shown in Fig. 1).
3. Transfer the filter assembly to a beaker containing a solution of 50% commercial bleach, 50% Triton/NaCl. Incubate 2 to 3 min.
4. Remove the filter assembly from the bleach and rinse the embryos with Triton/NaCl and then with distilled water.
5. Draw water away from the embryos by placing paper towels against the nylon filter. Dry the inside walls of the filter assembly with a tissue.
6. Rinse the embryos into a 50-ml Erlenmeyer flask with 10 ml of 100% octane. Gently swirl the embryos in the octane for 20 to 30 sec to permeabilize the vitelline membranes.
7. Add 10 ml of 37% formaldehyde (reagent grade) to the embryo/octane mixture and incubate for 3 to 5 min. The embryos float at the interface between the formaldehyde (lower phase) and the octane (upper phase).
8. Draw off the formaldehyde phase, leaving the octane and fixed embryos in the flask.
9. Add 10 ml PBS (see Section IIA, Protocol 1), swirl, and remove the PBS. Repeat this PBS wash two more times to remove residual formaldehyde.
10. Add 10 ml of fresh PBS.

The fixed embryos are present at the interface between the PBS and the octane. Vitelline membranes must now be removed as described in Section IV, Protocol 1.

2. Comment

Small numbers of embryos can be fixed using this procedure. The procedure for mass removal of vitelline membranes does not work well when very few embryos are present, however, and vitelline membranes should be manually removed as described in Chapter 36.

D. Protocol 3: Methanol Fixation

Some antibody–antigen combinations are sensitive to formaldehyde fixation. This methanol fixation protocol is a useful alternative to the formaldehyde-based procedure described above. The overall organization of the microtubules and actin filaments is preserved with this method (Kellogg *et al.,* 1988), but individual microtubules are not as well defined as in formaldehyde fixed material. In addition, phalloidin labeling is not possible.

1. Procedure

1. Collect embryos on apple juice agar plates (Ashburner, 1989).
2. Rinse the embryos from the plates with Triton/NaCl (0.05% Triton X-100, 0.7% NaCl) and collect on a fine nylon mesh supported in a filter assembly (prepared as shown in Fig. 1).
3. Transfer the filter assembly to a beaker containing solution of 50% commercial bleach, 50% Triton/NaCl. Incubate 2 to 3 min.
4. Remove the filter assembly from the bleach and rinse the embryos with Triton/NaCl and then with distilled water.
5. Draw water away from the embryos by placing paper towels against the nylon filter. Dry the inside walls of the filter assembly with a tissue.
6. Rinse the embryos into a 50-ml Erlenmeyer flask with 10 ml of 100% octane. Gently swirl the embryos in the octane for 20 to 30 sec to permeabilize the vitelline membranes.
7. Add 10 ml 100% methanol, stopper tightly, and shake vigorously for 30 sec to 1 min.
8. Remove the octane and most of the methanol, leaving the embryos that have settled to the bottom of the flask in a minimal volume of methanol.
9. Add 100% methanol and incubate for 1 to 2 hr at room temperature.
10. Rehydrate the embryos through a methanol series, starting with 90% methanol (in water), followed by 75% methanol, 50% methanol, and 100% PBS. During each solution change, allow the embryos to settle to the bottom of the flask and immediately draw off the supernatant and resuspend in the next solution. Prolonged incubation at each step is not required.

11. Draw off the PBS, add PBST, and incubate for 30 min.

Label the fixed embryos as described in Section IV, Protocol 2.

2. Comment

Vitelline membrane removal is very efficient with this simple technique. It is therefore a good choice when isolating embryos for use in preabsorbing secondary antibodies (Karr and Alberts, 1986).

IV. Labeling Cytoskeletal Elements in Fixed Oocytes and Embryos

The protocols described in this section include manual removal of vitelline membranes from late-stage egg oocytes and formaldehyde-fixed embryos, antibody labeling procedures, and tissue clearing. These protocols are used for both egg chambers and embryos.

A. Protocol 1: Manual Removal of Chorion and Vitelline Membranes from Late-Stage Egg Chambers and Early Embryos

Prior to immunolabeling, the follicle cells, chorion and vitelline membranes must be removed from Stage 13 and 14 oocytes. Similarly, the vitelline membranes must be removed from embryos that are fixed in formaldehyde as described under Section III, Protocol 2. Both of these tasks can be accomplished by rolling the fixed material between the frosted surface of a slide and a coverglass (Theurkauf et al., 1992).

1. Procedure

1. Transfer the fixed egg chambers (or embryos) to the frosted portion of a slide and remove most of the liquid. Embryos are drawn from the octane–PBS interface with a Pasteur pipet; egg chambers are in PBST.
2. Draw the edge of a 22 × 40-mm coverglass across the egg chambers and then roll the egg chambers between the slide and coverglass, using a circular motion (Fig. 2). Alternate between these two disrupting techniques while monitoring vitelline membrane removal with a dissecting microscope.
3. Rinse the "rolled" egg chambers/embryos into a test tube with PBST and allow them to gravity settle.
4. Remove the supernatant and add 1% Triton X-100 in PBS and incubate with gentle rocking for 1 to 2 hr.

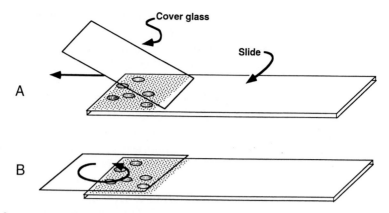

Fig. 2   Manual removal of vitelline membranes from fixed egg chambers and embryos. First, transfer egg chambers/embryos to the frosted portion of a glass slide. Then, draw off most of the buffer. Drag the edge of a 22 mm × 40mm cover glass over the egg chambers/embryos (A). Roll the egg chamber/embryos between the cover glass and slide, using a circular motion (B). Monitor removal of chorion/vitelline membranes on a dissecting microscope. Rinse egg chambers/embryos into test tube with PBST.

   5. Let the "rolled" egg chambers settle to the bottom of the tube, remove the 1% Triton solution, and resuspend in PBST.

The egg chambers are now ready to label as described here.

2. Comment

The number of egg chambers on the slide and amount of pressure applied with the coverglass influence the efficiency of chorion/vitelline membrane removal. If too few egg chambers or embryos are used, the samples tend to get crushed. If too many egg chambers or embryos are present on the slide, the vitelline membranes are not efficiently removed. Fixed egg chambers and embryos are tougher than you might expect, and efficient mechanical disruption requires fairly rough treatment. When the rolling procedure has broken open a few of the embryos or egg chambers on the slide, the vitelline membranes have probably been removed from most cells.

B. Protocol 2: Labeling Cytoskeletal Elements in Egg Chambers and Embryos

The same procedure is used to fluorescently label cytoskeletal proteins in egg chambers and embryos.

1. Procedure

   1. Transfer the fixed egg chambers/embryos to a 0.5-ml tube and resuspend in 0.2 ml PBST.

2. Add an appropriate amount of primary antibody and incubate overnight at 4°C. Constant gentle mixing is required during all incubations. A rotator set up in a cold room works well. Antibody dilution must be determined empirically. For dual label applications using antibodies from different species, both primary antibodies can be added to the sample at this time.

3. Remove the primary antibody solution and wash the embryos in four changes of PBST, incubating for 15 min at each solution change.

4. After the final PBST, wash, resuspend the embryos/egg chambers in 200 μl PBST and add fluorescently conjugated secondary antibodies that have been preabsorbed against fixed *Drosophila* embryos (Karr and Alberts, 1986). Incubate embryos/egg chambers in secondary antibody for 2 hr at room temperature or overnight at 4°C.

5. Rinse the fluorescently labeled egg chambers four times in PBST, as described in Step 3.

6. Remove the PBST and replace with PBST containing fluorescently conjugated phalloidin (1 μg/ml) and DAPI (1 μg/ml). Incubate for 10 min at room temperature.

7. Wash the embryos/egg chambers in PBST, as described in step 3. Follow these washes with two rinses in PBS.

8. Let the labeled embryos/egg chambers settle and remove the PBS. Add a few drops of mounting medium (1 mg/ml p-phenylene diamine, 9 parts glycerol, 1 part 10× PBS) to the labeled material, gently stir with the tip of a Pasteur pipet and transfer to a slide. Place a coverglass over the mixture and seal with nail polish. Alternatively, labeled egg chambers/embryos can be cleared as described here.

2. Comments

1. When monoclonal or affinity-purified primary antibodies are used, blocking with nonspecific protein is generally not required. If background is a problem, 1% bovine serum albumin or powdered milk can be added to the antibody solution. Secondary antibodies should be preabsorbed against fixed embryos (Karr and Alberts, 1986).

2. The appropriate dilution of primary antibody must be determined empirically. Strong microtubule labeling is obtained using a 1:1000 dilution of the monoclonal anti-α-tubulin antibody DM1α (Sigma Chemical, St. Louis, MO).

3. Fluorescently conjugated anti-tubulin antibodies seem to penetrate into the interior of embryos and egg chambers more efficiently than standard primary–secondary antibody combinations (Theurkauf, 1992; Theurkauf *et al.*, 1993). Direct antibody conjugates can therefore help reveal internal structures. However, direct antibody conjugates may not give adequate signal when used to detect relatively minor proteins.

C. Protocol 3: Clearing Immunolabeled Egg Chambers and Embryos

1. Procedure

Egg chamber and embryos are relatively large, and internal structures are generally difficult to image. Out of focus signal can be eliminated through the use of laser scanning confocal microscopy. However, light scattering and defraction by yolk, lipid droplets, and membrane-bound organelles lead to loss of fluorescent signal from structures deep within these samples, and this limits the depth of the optical sections that can be obtained with a confocal microscope. To overcome this limitation, fluorescently labeled embryos and egg chambers can be cleared using the following technique.

1. Remove as much of the PBS wash as possible from immunolabeled egg chambers or embryos.
2. Add 100% methanol and resuspend the embryos/egg chambers by inverting the tube several times.
3. Remove the methanol supernatant and add fresh methanol. Incubate 1 to 2 min.
4. Draw off as much of the methanol as possible and add a few drops of the clearing solution (benzyl benzoate:benzyl alcohol, 2 : 1). Stir gently with a pipet tip.
5. Transfer the cleared material, in the clearing solution, to a slide. Place a coverglass over the sample and seal with nail polish. The slides last for weeks if stored at −20°C.

2. Comments

1. Embryos and egg chambers are almost impossible to detect after clearing and do not rapidly settle to the bottom of the tube. To avoid losing your sample, therefore, a minimum volume of clearing solution should be used.
2. Both fluorescein and rhodamine bleach very rapidly in this clearing solution. It is therefore necessary to quickly photograph or electronically capture fluorescent images of cleared material.

References

Ashburner, M. (1989). "*Drosophila:* A Laboratory Manual." Cold Spring Harbor, NY: Cold Spring Harbor Press.

Baker, J., Theurkauf, W. E., and Schubiger, G. (1993). Dynamic changes in microtubule configuration correlate with nuclear migration in the preblastoderm *Drosophila* embryo. *J. Cell Biol.,* **122,** 113–121.

Berleth, T., Burri, M., Thoma, G., Bopp, D., Richtenstein, S., Fogerio, G., Noll, M., and Nüsslein-Volhard, C. (1988). The role of localization of bicoid RNA in organizing the anterior pattern of the Drosophila embryo. *EMBO J.* **7,** 1749–1856.

Cooper, J. A. (1987). Effects of cytochalasin and phalloidin on actin. *J. Cell Biol.* **105,** 1473–1478.

Cummings, M. R., and King, R. C. (1969). The cytology of the vitellogenic stages of oogenesis in *Drosophila melanogaster*. I. General staging characteristics. *J. Morphol.* **128**, 427–442.

Dalby, B., and Glover, D. M. (1992). 3′ non-translated sequences in *Drosophila* cyclin B transcripts direct posterior pole accumulation late in oogenesis and peri-nuclear association in syncytial embryos. *Development* **115**, 989–997.

Ephrussi, A., Dickinson, L. K., And Lehman, R. (1991). oskar organizes the germ plasm and directs localization of the posterior determinant nanos. *Cell* **66**, 37–50.

Foe, V. E., and Alberts, B. M. (1983). Studies of nuclear and cytoplasmic behavior during the five mitotic cycles that precede gastrulation in *Drosophila* embryogenesis. *J. Cell Sci.* **61**, 31–70.

Gutzeit, H. O., and Koppa, R. (1982). Time-lapse film analysis of cytoplasmic streaming during late oogenesis in *Drosophila*. *J. Embryol. Exp. Morphol.* **67**, 101–111.

Hay, B., Jan, L. Y., and Jan, Y. N. (1990). Localization of vasa, a component of *Drosophila* polar granules, in maternal-effect mutations and alter anteroposterior polarity. *Development* **109**, 425–433.

Karr, T. L., and Alberts, B. M. (1986). Organization of the cytoskeleton in early *Drosophila* embryos. *J. Cell Biol.* **102**, 1494–1509.

Kellogg, D. R., Mitchison, T. J., and Alberts, B. M. (1988). Behavior of microtubules and actin filaments in living *Drosophila* embryos. *Development* **103**, 675–686.

Kim-Ha, J., Smith, J. L., and MacDonald, P. M. (1991). oskar mRNA is localized to the posterior pole of the *Drosophila* oocyte. *Cell* **66**, 23–36.

Koch, E. A., and Spitzer, R. H. (1983). Multiple effects of colchicine on oogenesis in *Drosophila*: Induced sterility and switch of potential oocyte to nurse-cell developmental pathway. *Cell Tissue Res.* **228**, 21–32.

Lasko, P. F., and Ashburner, M. (1990). Posterior localization of vasa protein correlates with, but is not sufficient for, pole cell development. *Genes Dev.* **4**, 905–921.

Mahowald, A. P., and Kambysellis, M. P. (1980). Oogenesis. *In* ''The Genetics and Biology of *Drosophila*'' (M. Ashburner, ed.) pp. 141–224. New York: Academic Press.

Mahowald, A. P., Goralski, T. J., and Caulton, T. H. (1983). *In vitro* activation of *Drosophila* eggs. *Dev. Biol.* **98**, 437–445.

McKim, K. S., Jang, T. K., Theurkauf, W. E., and Hawley, R. S. (1993). Mechanical basis of meiotic metaphase arrest. *Nature* **362**, 364–366.

Minden, J. S., Agard, D. A., Sedat, J. W., and Alberts, B. M. (1989). Direct cell lineage analysis in *Drosophila melanogaster* by time lapse three dimensional optical microscopy of living embryos. *J. Cell Biol.* **109**, 505–516.

Mitchison, T. J., and Sedat, J. W. (1983). Localization of antigenic determinants in whole *Drosophila* embryos. *Dev. Biol.* **99**, 261–264.

Pokrywka, N. J., and Stephenson, E. C. (1991). Microtubules mediate the localization of bicoid RNA during *Drosophila* oogenesis. *Development* **113**, 55–66.

Postner, M. A., Miller, K. G., and Wieschaus, E. (1992). Maternal-effect mutations of the *sponge* locus affect cytoskeletal rearrangements in *Drosophila melanogaster* embryos. *J. Cell Biol.* **119**, 1205–1218.

Schejter, E. D., and Wieschaus, E. (1993). *Bottleneck* acts as a regulator of the microfilament network governing cellularization of the *Drosophila* embryo. *Cell* **75**, 373–385.

Sullivan, W., Fogarty, P., and Theurkauf, W. (1993). Mutations affecting the cytoskeletal organization of syncytial Drosophila embryos. *Development* **118**, 1245–1254.

Suter, B., and Steward, R. (1991). Requirement for phosphorylation and localization of the *Bicaudal-D* protein in *Drosophila* oocyte differentiation. *Cell* **67**, 917–926.

Theurkauf, W. (1992). Behavior of structurally divergent alpha-tubulin isotypes during *Drosophila* embryogenesis: Evidence for post-translational regulation of isotype abundance. *Dev. Biol.* **154**, 204–217.

Theurkauf, W. E., Smiley, S., Wong, M. L., and Alberts, B. M. (1992). Reorganization of the cytoskeleton during Drosophila oogenesis: Implications for axis specification and intracellular transport. *Development* **115,** 923–936.

Theurkauf, W. E., and Hawley, R. S. (1992). Meiotic spindle assembly in Drosophila females: Behavior of nonexchange chromosomes and the effects of mutations in the nod kinesin-like protein. *J. Cell Biol.* **116,** 1167–1180.

Theurkauf, W. E., Alberts, B. M., Jan, Y. N., and Jongens, T. A. (1993). A central role for microtubules in the differentiation of Drosophila oocytes. *Development* **118,** 1169–1180.

Warn, R. M., Smith, L., and Warn, A. (1985). Three distinct distributions of F-actin occur during the divisions of polar surface caps to produce pole cells in *Drosophila* embryos. *J. Cell Biol.* **100,** 1010–1015.

Wharton, R. P., and Struhl, G. (1989). Structure of the *Drosophila* BicaudalD protein and its role in localizing the posterior determinant *nanos*. *Cell* **59,** 881–892.

Zalokar, M., and Erk, I. (1976). Division and migration of nuclei during early embryogenesis of *Drosophila melanogaster*. *J. Microbiol. Cell* **25,** 97–106.

CHAPTER 26

High-Resolution Microscopic Methods for the Analysis of Cellular Movements in *Drosophila* Embryos

**Daniel P. Kiehart,* Ruth A. Montague,* Wayne L. Rickoll,†
Donald Foard,* and Graham H. Thomas‡**

* Department of Cell Biology
Duke University Medical Center
Durham, North Carolina 27710

† Department of Biology
University of Puget Sound
Tacoma, Washington 98416

‡ Department of Biology and
Department of Biochemistry and Molecular Biology
Pennsylvania State University
University Park, Pennsylvania 16802

I. Introduction

This chapter outlines methods for the observation and documentation of living fly embryos. It updates and supplements an excellent contribution from Wieschaus and Nüsslein-Volhard (1986) that focuses much more broadly on methods for handling and looking at living embryos. Here, we concentrate on high-resolution methods for observing and recording images of living embryos that take advantage of a number of advancements in video microscopy and digital image processing put into common use since their paper was published. These advances in video imaging have been detailed in a number of excellent books, the strengths of which vary somewhat. Each provides a useful resource. Slayter and Slayter's book *Light and Electron Microscopy* is a basic primer in optics at a level appropriate for most research biologists. Shinya Inoué's book *Video Microscopy* is a classic text with an invaluable chapter on microscope image formation that summarizes the basic considerations required for optimizing image quality. Practical sections on choice and care of lenses are a must for all. The book also provides an excellent description of how the eye works and a comprehensive description of how video works. Two newer books, one by Shotton (*Electronic Light Microscopy—Techniques in Modern Biomedical Microscopy*) and the other by Herman and Lemasters (*Optical Microscopy—Emerging Methods and Applications*), contain treatments of more recent software and hardware developments. In addition, papers on electronic imagers (Aikens *et al.*, 1989) and video-enhanced differential interference microscopy (Salmon *et al.*, 1989) and a book on digital image processing (Castleman, 1979) are very helpful.

II. Optimization of Resolution and Contrast

The relationship between resolution and the numerical aperture (NA) of an objective lens is detailed by Inoué (1986). We assume for the following discussion: Incoherent illumination (i.e., illumination produced by a typical microscope illuminator); proper alignment of the microscope's lenses on its optic axis; clear and dust-free lens surfaces; careful adjustment for Köhler illumination; and adjustment of the condenser diaphragm so that it is slightly smaller than the aperture of the objective. Resolution between two adjacent points can be achieved when the distance between the points is equivalent to the radius of the two-dimensional projection of the points' three-dimensional diffraction patterns (the Airy discs) produced by the objective. This is the so-called *Rayleigh criterion*. The radius of the Airy disc and the minimum resolvable distance, d_{min}, in a specimen are thus one and the same and are given by

$$r_{\text{Airy disc}} = d_{min} = 1.22\lambda/(\text{NA}_{\text{objective}} + \text{NA}_{\text{condenser}}), \quad (1)$$

where λ is the wavelength of incident light in micrometers and the NA of a lens is given by

$$\text{NA} = n \sin \phi, \quad (2)$$

where n is the refractive index of the medium between the lens and the specimen and ϕ is the half angle at which the lens can *accept* light, for the objective (or *project* light, for the condenser).

Together, Eqs. (1) and (2) establish that optimal resolution requires lenses of the highest NA. For the objective, sin ∅ would be 1 if ∅ were 90°, i.e., if the objective could collect light, diffracted by the specimen, that entered the objective perpendicular to its optic axis. For $n \sin \varnothing$ to be greater than 1, n must be greater than 1 (the refractive index of air). A typical high-performance oil immersion lens has an NA of 1.4. Immersion oils, made to match the refractive index of glass and standardized by the microscope companies, should have a refractive index of 1.515. This indicates that the maximum half-angle of light that can be accepted by the objective (or projected by a condenser of matching NA) is 67.5°. For high NA, nonimmersion (or so-called "dry") objectives (NA = 0.9, $n = 1$) the half angle of light is slightly less (64.2°) while for low NA, dry objectives or condensers the half angle of light is considerably less (e.g., for a 10×, 0.25 NA objective, it is 14.5°). Equation (1) establishes that optimal resolution requires that both condenser and objective be of high NA and that for numerical apertures above 1, the condenser, as well as the objective, should be of the oil immersion type. There is no advantage to adjusting the condenser NA to greater than the objective NA, so with the multi-immersion lenses described below (NA = 0.8 and 0.9, respectively) a high NA, dry condenser (NA = 0.9) provides sufficient condenser NA. Differences in image quality produced by objectives with various NAs are shown in Fig. 1 (especially compare Fig. 1A, 10×, 0.25 NA; Fig. 1B, 40×, 0.75 NA; Fig. 1C, 40×, 0.9 NA; and Fig. 1F, 25×, 0.8 NA. The detrimental effect of stopping down the condenser diaphragm, thereby reducing the NA, is shown in Fig. 1J).

Structures of primary interest may not be near the limit of resolution of even lower power, nonimmersion lenses. If such is the case, low NA lenses may be preferable. First, the low NA lenses have a large field size: a low resolution, larger view of the specimen may be most informative. Second, the low NA lenses have greater working distances. The working distances of the best quality, high NA lenses can be so small that they allow imaging only of the embryo surface or close, parasagittal sections and may preclude imaging deeper regions of the embryo. For example, the 100×, 1.3-NA objective can physically focus only 60 μm past the coverslip, making it impossible to image a true sagittal section.

The ability to image distinctly two adjacent objects requires not only that they be resolved but that there be sufficient contrast. That is, two white dots, separated by sufficient distance but superimposed on a white background, will not be visible because they lack contrast: the intensity of light reflected (in the microscope, transmitted) by the two dots and the background are equivalent. Formally, contrast is defined as

$$(I_B - I_S)/I_B, \tag{3}$$

where I_B and I_S are the intensities of background and specimen, respectively. Thus, as the background gets brighter, or the difference between the intensity of the background and specimen is reduced, contrast will drop. Most living biological specimens, including the *Drosophila* embryo, are not particularly

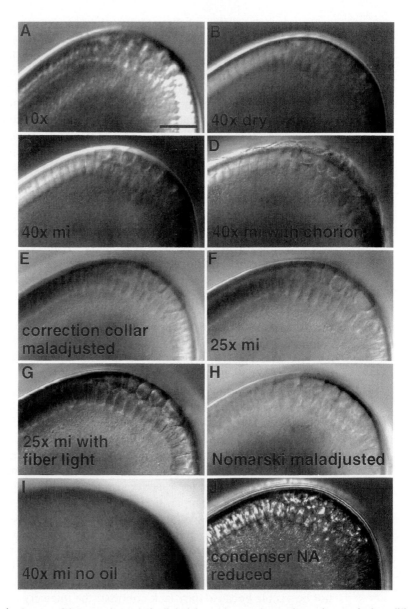

Fig. 1 Images of the posterior end of cellularizing embryos taken under various optical conditions. The optivar was used to project an image of the specimen at sufficient magnification to allow adequate digital sampling of the image (see text and Table I). Magnification has been adjusted in each panel so that the final magnification is constant. In each panel except for D and I, the embryos have been dechorionated and mounted in Halocarbon 700 oil. (A) 10×, 0.25 NA objective; scale bar is 25 μm. (B) 40×, 0.75 Plan-Neofluor dry objective. (C) 40×, 0.9 NA multi-immersion, Plan-Neofluor lens with oil immersion and the correction collar rotated to minimize spherical aberration. (D) 40×, 0.9 NA multi-immersion, Plan-Neofluor lens with oil immersion and the correction collar

light absorbing. To absorb light, the specimens have to be stained with an appropriate dye. Unfortunately, few dyes are vital and the impermeability of the vitelline envelope makes introducing dyes cumbersome at best.

Modern microscopes generate contrast through the use of the inherent *phase* (versus *intensity*) information in biological specimens, which is the consequence of inhomogeneity in the refractive index of various biological structures. The phase microscope, the differential interference microscope, the interference microscope, and the polarizing light microscope are all instruments that manipulate the light passing through the microscope in a unique way. Each microscope converts phase information (gained as wave fronts of light experience regions of different refractive index as they pass through the specimen) into variation in intensity. This produces contrast and allows a visible or recordable image to be formed. The phase microscope and the differential interference microscope (which generates intensity proportional to the first derivative, *i.e.*, the *differential* of the change in refractive index across the specimen) are both useful for nonbirefringent specimens, but the phase microscope generates phase halos that obscure high-resolution detail in all but the flattest of specimens, making

---

rotated to minimize spherical aberration (same optical parameters as in C), but the chorion was left on this embryo. (E) 40×, 0.9 NA multi-immersion, Plan-Neofluor lens with oil immersion and the correction collar maladjusted (i.e., same optical parameters as in C, but with correction collar maladjusted). (F) 25×, 0.8 NA, multi-immersion, Plan-Neofluor lens, with the correction collar rotated to minimize spherical aberration. (G) 25× 0.8 NA multi-immersion, Plan-Neofluor lens with oil immersion and the correction collar rotated to minimize spherical aberration (same optical parameters as in C), but a mercury arc lamp and a fiber light illuminator were used to illuminate the back aperture of the condenser with uniform light. (H) 40×, 0.9 NA multi-immersion, Plan-Neofluor lens with oil immersion and the correction collar rotated to minimize spherical aberration (same optical parameters as in C), but with Nomarski maladjusted. (I) 40×, 0.9 NA multi-immersion, Plan-Neofluor lens with oil immersion and the correction collar rotated to minimize spherical aberration (same optical parameters as in C), but the chamber was not filled with oil. (J) 40×, 0.9 NA multi-immersion, Plan-Neofluor lens with oil immersion and the correction collar rotated to minimize spherical aberration (same optical parameters as in C), but the condenser diaphragm was stopped way down. Images (1000 × 1018 × 12 to 14 bit) were captured with a Hamamatsu C4880 cooled CCD camera operating at about −30°C, driven by MetaMorph software operating on a Universal Imaging 486 PC clone, and saved as a TIFF file onto a Pinnacle 650 rewritable laser disk. Images were thresholded (to remove unnecessary dynamic range) and scaled to 8-bit images. The images were transferred to Photoshop where they were cropped and their gray values were adjusted (this step could be done in MetaMorph, but we have a Macintosh graphic work station that allows us to perform the digital image processing while leaving the microscope and MetaMorph free for image acquisition). We save these final, individual images as TIFF files and then export them to Canvas for montaging and labeling. While Photoshop could, in principle, be used for labeling the images, the labels in Canvas are postscript, rather than bit mapped and can therefore be sized over a continuous range without becoming pixelated. The final montaged figures were outputted onto a 300-dpi Tektronix Phaser IISDX color, dye sublimation printer for glossy hard copy or saved to a 128 Meg, 3.5-inch rewritable optical disk for submission. Low-quality work "prints" (more than sufficient for layouts or notebook copies of the panels) can be made on a laser printer.

the differential interference microscope the instrument of choice for imaging living embryos. Phase microscopy remains superior for imaging detail in many kinds of flat preparations, such as polytene chromosome squashes.

A key consideration for producing highest contrast in any specimen is illumination of the back aperture of the condenser with uniform intensity. With differential interference microscopy, which requires light that is plane polarized, conventional filament sources may prove too dim for adequate illumination and light sources that generate high luminous density may be required. Unfortunately, such sources, typically arc lamps, do not provide adequately uniform illumination of the back aperture of the condenser when conventional collectors are used. Gordon Ellis solved this problem by introducing into the optical path a bent optical fiber (detailed in Inoué, 1986). The fiber serves to scramble nonuniform input light (*i.e.*, an arc source focused on the end of the fiber) such that the output is a uniform, luminous disc of light. An appropriate collector is used to image the disc on the back aperture of the condenser. A fiber optic light scrambler that can be used in conjunction with many commercially available microscopes is available from Technical Video Ltd. (Woods Hole, MA).

The ability of an image to reproduce faithfully the contrast inherent in a specimen is described mathematically by the *contrast transfer function*, which varies with the spatial frequency of the specimen. We introduce this concept for completeness and refer to Inoué (1986) and Slayter and Slayter (1992) for more detailed treatments.

III. Thick, Living Specimens, High NA, and Spherical Aberration

The thickness of the *Drosophila* embryo and the requirement it has for a continual supply of oxygen present special problems for the microscopist: lenses are usually designed to view specimens through a coverslip and standard glass coverslips are not particularly O_2 permeable.

Most microscope lenses are designed to image specimens through a No. 1.5 thickness glass coverslip. That is, the lens is engineered so that its *front element* is really the coverslip, which is followed by the space between the coverslip and the lens (filled with air or an appropriate immersion medium) and then finally, the first lens in the objective itself. Standard coverslip glass has a refractive index of 1.515. The "0.17" stamped on most lenses specifies that the glass coverslip should be 0.17-mm thick, the thickness of a No. 1.5 coverslip. Use of a coverslip of inappropriate thickness, *and/or* a medium inappropriate for the chosen objective lens between the front lens of the objective and the focal plane of interest in the specimen causes spherical aberration, which can badly degrade image quality. Spherical aberration is the degradation in image quality observed when a perfectly spherical lens is used to image a specimen. Such lenses have different focal lengths depending on whether imaging rays pass near or far from the optic axis of the lens (see Slayter and Slayter, Fig. 6.14). Inoué (1986) discusses the importance of coverslip thickness and points out that variation by as little as 10 μm (or 6% of coverslip thickness) can seriously degrade the image. Because immersion oil is meant to match the refractive index of the coverslip, it is somewhat surprising that even oil immersion objectives can require coverslip correction collars (*i.e.*, one would think that the reduced optimal path length through glass could be compensated by "additional" oil). This suggests that the common use of thin coverslips (*e.g.*, #1 coverslips, 0.15-mm-thick) to allow imaging of deeper focal planes in thick specimens can contribute to image degradation and should be evaluated carefully. Depending on the focal plane, the aqueous medium that is the cytoplasm of the embryo itself and the medium that surrounds the embryo (aqueous medium or halocarbon oil; see following) will contribute to the degradation of the image. Thus, for standard lenses (*i.e.*, lenses without a correction collar designed to overcome spherical aberration; see following) optimal resolution is attained only with specimens pressed tightly against the coverslip, a condition impossible to achieve for all focal planes in a thick aqueous specimen, such as a *Drosophila* embryo.

Optimal image quality for the observation of embryo morphogenesis is achieved by maximizing the NA of the objective and correcting for spherical aberration through the use of an objective with a correction collar. The correction collar allows the distance between elements in the lens to be adjusted to compensate for nonideality in the optical properties of the materials between the front element of the lens and the focal plane (i.e., of the space between lens and coverslip, the coverslip, and the space between the coverslip and the focal plane in the specimen). Excellent objective lenses with such collars are the 25×, 0.8 NA and the 40×, 0.9 NA infinity-corrected, multi-immersion lenses from Zeiss (Carl Zeiss Inc., Thornwood, NY). They offer high NA with reasonable working distance (see Table I). Unfortunately, lenses with an appropriate combination of higher NA, sufficient working distance, and a correction collar are currently unavailable. Because the optical path for the specimens is fairly complex, it is necessary to determine empirically the adjustment of the correction collar for any given combination of lens, immersion medium, specimen, and focal plane. Empirical adjustment requires changing the position of the correction collar on the objective, refocusing the microscope, and evaluating the image that results. It is usually helpful to scrutinize small, immobile particles in the specimen that are close to the limit of resolution of the system to make these adjustments. In a collaboration with Inoué (see Kiehart *et al.*, 1990), we determined empirically that we could optimize the image of a sagittal or parasagittal optical section of the embryo by using an immersion oil with a refractive index of 1.515 combined with setting the correction collar to the glycerol immersion position. The effect of spherical aberration on image quality can be ascertained by comparing Fig. 1C (correction collar properly adjusted) and Fig. 1E (correction collar improperly adjusted). This correction

Table I
Objectives, Optivar Setting for Optimal Digital Sampling, and Appropriate Sliders for DIC Imaging

Objective/magnification	NA	Working distance in mm	Calculated resolution in μm	Measured field of view (on camera target) in μm	Number of resolvable units	Optivar setting	Slider for differential interference/condenser
10× Achrostigmat	0.25	6.5	1.332	1200	901	1×	444441
				655	492	1.8×	
20× Achroplan	0.45	2.07	0.74	591	799	1×	444445
				367	497	1.6×	
25× Multi-Immersion Plan Neofluor	0.8	0.13	0.416	454	1092	1×	444441
				201	483	2.25×	
40× Plan Neofluor	0.75	0.33	0.444	282	631	1×	444450
				217	489	1.3×	
40× Multi-Immersion Plan Neofluor	0.9	0.13	0.37	293	791	1×	444482
				180	486	1.6×	
40× Plan Apochromat Oil/Iris	1	0.31	0.333	302	907	1×	444450
				158	473	1.9×	
100× Plan Neofluor Pol	1.3	0.06	0.256	120	470	1×	444480

Note. In each case, a multiturret condenser with a 0.9-NA front element was used. The condenser sliders were appropriate for the NA of the lens shown except in the case of the 40× multi-immersion objective, which for proper DIC imaging requires use of the 0.3- to 0.4-NA setting on the multiturret condenser.

collar also allows excellent image quality when an immersion oil with a refractive index other than 1.515 is used. Of two oils from Cargille (Type DF, $n = 1.515$ and Type FF, $n = 1.479$; R. P. Cargille Laboratories, Cedar Grove, NJ), one has a refractive index that introduces substantial spherical aberration that cannot be corrected if the objective is not equipped with a correction collar. Such oils are useful because their viscosity is significantly less than the standard immersion oils supplied by the microscope companies. This reduces viscous coupling between the coverslip and the objective that causes the coverslip to flex and the specimen to move during focusing. In addition, these alternate oils have reduced (*i.e.*, little or no) fluorescence, which can improve greatly contrast in images of fluorescent specimens.

An additional advantage of the multi-immersion lenses is that water, instead of oil, can be used as an immersion medium. Especially with open chambers (see following), water may be preferable to immersion oil because it is not miscible with the halocarbon oils used in the chamber to surround the embryos. In addition, the water is easier to remove than oil. The majority of water can be wicked away with a piece of filter paper and any remaining droplets can be blown away with compressed gas (*e.g.*, dispensed in a MicroDuster OS, Texwipe Co., Upper Saddle River, NJ). A new objective from Nikon that combines high resolution (60×, 1.2 NA), water immersion, and long working distance (220 μm), may be optimal for imaging embryos but is expensive ($9,965 list; Brenner, 1994).

Coverslips are not O_2 permeable, so chambers designed to optimize image quality must also include special features designed to oxygenate the embryos (see Section IV, following). Foe and Alberts (1985) investigated the rapidity with which embryos could deplete oxygen from small volumes of culture medium and found that they rapidly depleted existing oxygen and become arrested in their development. One strategy designed to circumvent the oxygen impermeability of glass coverslips might be to employ a system in which a coverslip is not used and instead the front element of an immersion lens fitted with a correction collar contacts the medium that surrounds the embryo. Unfortunately, the objective is as O_2-impermeable as a coverslip and its shape makes it a barrier to the introduction of O_2 nearly equivalent to the barrier provided by a glass coverslip (we determined this empirically). Sealing the embryos between coverslip and slide in air or in humidified air might provide sufficient oxygen to allow development, but in air the vitelline envelope is highly refractile and depolarizing, which hinders observation of subcellular structure at high resolution or with high extinction (compare Fig. 1C with oil and Fig. 1I in air). Suitable image quality requires surrounding the embryo with a medium with a refractive index close to that of the vitelline envelope and the cytoplasm. Although aqueous buffers are potentially useful for observations over a brief time interval, they are not useful for long-term observation. As a consequence, halocarbon oils, like those commonly used to overlay embryos during the injection of DNA for germline transformation, are optimal from both optical and physiological considerations. Some of the relevant physical properties of commonly used halocarbon oils are shown in Table II. Optically, the oil suppresses light scattering from the refractive vitelline envelope and, for embryos that still have chorions (see following) serves to clear optically the chorions (compare Fig. 1D, chorion intact, and Fig. 1C, chorion removed). Physiologically, the halocarbons prevent desiccation and provide a source of oxygen. Such oils are now available

Table II
Physical Properties of Various Halocarbon Oils

Oil	Absolute viscosity in centipoise at 38°C	Refractive index	Density
Halocarbon 27	51	1.407	1.9
Halocarbon 700	1365	1.414	1.95
Voltalef 3s	115	1.405	1.93
Voltalef 10s	1550	1.41	1.96

through Sigma Chemical Co., (St. Louis, MO) (HC 27, and HC 700, Catalog Nos. H 8773 and H 8898, respectively).

IV. Chambers for the Observation of Embryos

We typically mount specimens in two kinds of chambers (Figs. 2 and 3). One has glass windows (standard 22 × 22-mm square, No. 1½ coverslips) on either side of the embryos; the other uses a gas permeable membrane on the condenser side of the specimen. Whereas this second configuration is fine for epi-illuminated specimens, we have not yet identified a membrane with low birefringence that permits the use of differential interference imaging methods.

Optimal resolution entails using a chamber that is open on one or more sides to allow free exchange of gas and holds two coverslips parallel to one another, separated by a distance that is not much greater than the diameter of an embryo. A stainless steel slide (Fig. 2), designed originally for micromanipulation studies by Dr. Gordon Ellis, and modified for immobilizing marine eggs for microinjection (Kiehart, 1981,1982), fits these criteria and allows us to surround the embryos with halocarbon oil. The advantage of this geometry is that both windows are made of glass that can be maintained strain-free and optical quality is unsurpassed. A disadvantage is that the embryos must be held close to the air/oil interface or a source of oxygen needs to be supplied. When oil immersion objective and condensers are in place, the embryos should be mounted away from the open edge of the coverslip to avoid contamination of the halocarbon oil with immersion oil. We have tried mounting the embryos in the middle of the chamber and filling the chamber only half full of halocarbon oil. However, we find that this arrangement is frequently physically unstable, and during video time-lapsing the oil wicks away from the embryos, leaving an unsuitable image and a high likelihood of desiccation. We have found that narrow, gas-permeable polyethylene (PE) tubing (*e.g.*, PE-50 tubing, Clay Adams No. 7411, outside diameter 0.965 mm, inside diameter 0.58 mm, also available from other suppliers) can be used to introduce pure, molecular oxygen to the chamber. A medical E-type canister of oxygen fitted with a two-stage regulator is used as a source of gas. A short-length (8–10 cm) of PE tubing passes along the backside of the embryo chamber. More flexible silastic tubing may also be used but tends to nucleate more bubbles that force the oil out of the chamber and prevent adequate imaging. When O_2 is passed through the tubing, sufficient molecular oxygen (and at times annoying bubbles) diffuses out of the PE tubing and down its concentration gradient past the embryos to the front of the chamber. A very slow flow rate (1–2 liter/hr) of O_2 is adequate to maintain viable embryos. We use fresh tubing for each chamber we assemble and reusable, larger bore Tygon tubing delivers the oxygen from the regulator to the PE tubing. The Tygon tubing is connected to the PE tubing with the appropriate hypodermic fittings (23 gauge needles for PE-50) with the beveled tips removed. A similar fitting

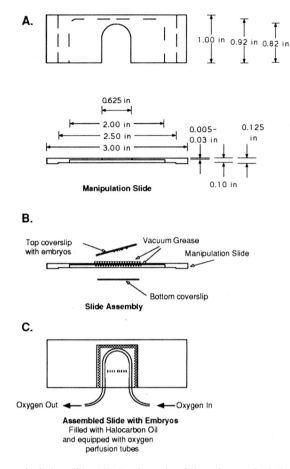

Fig. 2 Micromanipulation slide. (A) A schematic of the micromanipulation slide for use by a machinist, top and front views. It is unnecessary to reproduce the tolerances shown and the slide can be made from aluminum or stainless steel. (B) Assembly of the chamber. A small bead of vacuum grease is dispensed (from a 3-cc disposable syringe with a 16- to 18-gauge hypodermic needle) so as to surround the U-shaped opening in the slide on both the top and the bottom. A bottom coverslip is pressed into place. If O_2 is used to oxygenate the chamber, polyethylene tubing is pressed into place above the bottom coverslip, inside the U-shaped opening. A volume of halocarbon oil is added so that when the top coverslip with the embryos affixed is lowered into place the oil just fills the space. The open end of the preparation allows the introduction of manipulation/microinjection tools. (C) The assembled chamber showing the position of the polyethylene tubing and the embryos. Redesigned and drawn from Kiehart, 1982.

is attached to the other end of the PE tubing and the Tygon tubing attached to this end is placed 10 cm under a column of water to introduce back pressure, which helps force O_2 out of the PE tubing.

An alternate design for the chamber was suggested to us by Jon Minden (Carnegie Mellon, Pittsburgh, PA); it uses a glass coverslip on the objective

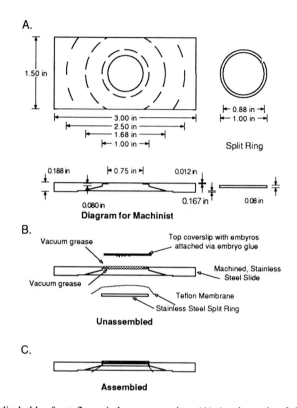

Fig. 3 Coverslip holder for teflon window preparation. (A) A schematic of the stainless steel slide use with a teflon window for use by a machinist, top and front views. It is unnecessary to reproduce the tolerances shown, except that it is necessary for the split ring to fit snugly into the recess in the bottom of the slide. (B) Sketch of the unassembled chamber. Assembly of the chamber is begun by applying a bead of vacuum grease (see legend to Fig. 2) to the inside of the recess on the bottom of the slide and onto the top of the slide, around the opening. Next, the teflon membrane is placed across the bottom of the opening and the split ring is pressed into place. The slide is returned to an upright position and halocarbon oil is placed in the well (the bottom of which is formed by the teflon membrane). A top coverslip, with 5–10 embryos glued on the bottom in a circular pattern, is lowered on to the oil (displaced to the right or left so that bubbles can escape). Finally, the top coverslip is centered over the hole and the chamber sealed. (C) Assembled chamber.

side of the embryo and a commercially available, gas-permeable window (teflon, commercially available from Fisher Scientific, Catalog No. 13-298-82) on the condenser side of the chamber (Fig. 3). We have modified his design to allow a larger viewing window, which probably should be enlarged even further if one attempts to use a high NA, short working distance condenser. The chamber allows free exchange of gas and development of embryos proceeds through hatching at a frequency comparable to controls: chorionated or dechorionated embryos on agar plates in a moist chamber. Unfortunately, the teflon window, while transparent, is very birefringent, which interferes with the differential interference imaging. Rotating the stage to align the optic axis of the birefringent membrane with the axis of plane polarized light in the microscope does not improve the image substantially. Thus, this chamber is suitable only for epi-illuminated specimens.

V. Handling the Embryos

Optimal image quality requires dechorionation of the fly embryos; although the chorions are sufficiently cleared by immersion in halocarbon oil (Wieschaus and Nüsslein-Volhard, 1986), the image of near sagittal sections can be surprisingly good (Fig. 1D). With practice, embryos can be hand peeled by placing them on double-stick tape and then using forceps or a stainless steel probe to gently rupture the chorion and remove the embryo. A variety of methods have been devised to automate mechanical dechorionation, for example, by rolling embryos between double-stick tape affixed to glass slides, held apart by short lengths of wire of correct thickness (200 μm wire, 16–17 gauge, Ashburner 1989). Alternatively, chemical dechorionation by solutions of commercial bleach (5.25% sodium hypochlorite) diluted 1 : 1 with water can be used although we regularly use shorter times (typically 1 min 10 sec to 1 min 30 sec at room temperature, ca. 22°C) than those recommended for dechorionation of embryos to be fixed shortly thereafter for antibody staining or histology. An essential part of any method devised to remove the chorions is to verify that embryonic development is not perturbed by the mechanical or chemical treatments necessary for dechorionation. We regularly evaluate the efficacy of such manipulations by lining up 20 to 40 sibling embryos, treated identically, in an orthogonal array on a grape *Drosophila* egg lay plate (Wieschaus and Nüsslein-Volhard, 1986) and overlaying them with halocarbon oil. The percentage of embryos that hatch should be scored after 24–48 hr (at 25°C, or longer at lower temperatures) and should be identical to a parallel batch of chorionated embryos that were gently manipulated by their dorsal appendages with watchmaker's forceps. We find that gentle, chemical dechorionation, performed on a 15-ml microanalysis-fritted glass and funnel filtration apparatus (e.g., Millipore Catalog No. XX10 025 00) is considerably more efficacious than any mechanical strategies and that equivalent numbers (typically 95%) of Oregon R or *klarsicht* embryos hatch whether left with chorions on or gently dechorionated chemically.

Dechorionated embryos can be placed in appropriate orientation, directly on a clean glass coverslip. When chorions are hand peeled they remain moist and will stick to a clean, dry coverslip by surface tension. Chemically dechorionated embryos can be gently dried on filter paper, which allows them to stick to clean glass. Gentle manipulation with a steel probe or the side of watchmaker's forceps allows orientation of the embryos for the observation of various morphogenic movements. Chorionated embryos do not stick well to clean glass. We have had excellent success with embryo glue prepared from the 3M double-stick tape as recommended by Whiteley and Kassis (1993). We use several drops of glue, made by extracting 7 cm of tape in 2 ml of heptane, for each 22 × 22-mm coverslip and allow the glue to dry on the coverslip for at least 10 min. Dust is a major enemy of high-resolution microscopy; consequently, coverslips are placed in covered racks and dusted prior to use.

We clean coverslips by dropping half a box, one by one, into a dilute solution of Liquinox or 7×, sonicating them in a bath sonicator, rinsing exhaustively with tap water and then double-distilled or high-quality deionized water. We constructed an apparatus for rinsing coverslips as shown in Fig. 4 and typically rinse coverslips with tap water running at a trickle for 1–2 hr and 10–20 liters of deionized water. Coverslips are stored in 70% ethanol in a tightly covered specimen jar, handled only with forceps, and dried with commercially available lab wipes (*e.g.*, Kimwipes) or on a small, centrifugal apparatus specifically designed for drying coverslips (see page 143 in Inoué, 1986).

VI. High-Fidelity Digital Electronic Imaging and Image Acquisition

We have described considerations designed to optimize the quality of the image produced by the microscope. The best images will be produced when the chorion is removed, when an objective with sufficient working distance and maximum NA is chosen, when the NA of the condenser matches the NA of the objective, when the correction collar on the objective is rotated to minimize or eliminate spherical aberration, when all optical components are strain-free and nonbirefringent, when the back aperture of the condenser is evenly illuminated, and when Köhler illumination is achieved. Under these conditions, excellent differential interference images of cellular detail in the *Drosophila* embryo can be achieved.

The next problem is how to ensure that the image is faithfully recorded, stored, and reproduced. New advances in electronic image detection greatly facilitate image acquisition and manipulation so that high-quality electronic images now compete with the best photographic film images. Given appropriate instrumentation, an image can be easily acquired, manipulated, and transferred to hard copy in a fraction of the time required to produce comparable output by photographic methods. Most microscope companies and a variety of solid state and vidicon vacuum tube camera suppliers offer a wide range of products that allow optimal attachment of electronic image devices to existing research microscopes. Inoué (1986) also details ways to jury-rig a marriage between electronic cameras and a microscope. In the next section, we outline the requirements for faithfully reproducing an image by electronic means. An excellent

A. Raw Materials

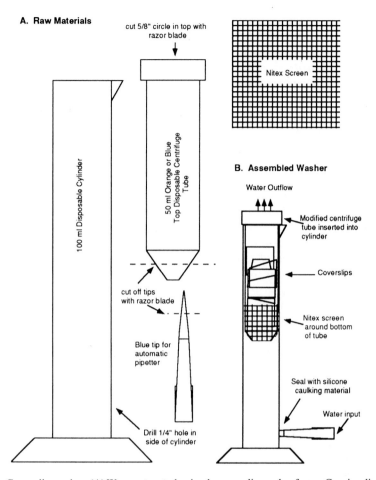

cut 5/8" circle in top with
razor blade

Nitex Screen

100 ml Disposable Cylinder

50 ml Orange or Blue
Top Disposable Centrifuge
Tube

B. Assembled Washer

Water Outflow

Modified centrifuge
tube inserted into
cylinder

Coverslips

Nitex screen
around bottom
of tube

cut off tips
with razor blade

Blue tip for
automatic
pipetter

Seal with silicone
caulking material

Water input

Drill 1/4" hole in
side of cylinder

Fig. 4   Coverslip washer. (A) We constructed a simple coverslip washer from a Corning disposable
100-ml graduate cylinder (Corning No. 25500), a 50-ml disposable, conical centrifuge tube (Corning
No. 25330-50), a blue tip from an automatic pipetter, a 2.5 × 2.5-inch sheet of Nitex screen and
a length of Tygon tubing. We cut the ends off the blue tip and the centrifuge tube and a hole in
the top of the centrifuge tube with a razor blade (shown by the dotted lines). We drilled a hole in
the cylinder (as shown) and cemented the blue tip into place with silicon caulking. We slid the
Nitex screen over the base of the conical tube and inserted it into the cylinder. The Nitex screen
is held in place between the conical tube and the walls of the cylinder. (B) The assembled washer,
with coverslips in place.

chapter on resolution in electronic imaging by Castleman (1993) in the Shotton
book gives a more detailed treatment of this subject.

   High-fidelity electronic imaging implies that resolution, ideally diffraction
limited as output from the microscope, stays optimal throughout image recording, digital image processing, and image outputting to hard copy (*i.e.,* to

a slide or print). While no lens is truly diffraction limited, modern computer design methods ensure that the finest lenses can produce images whose resolution comes close to the ideal. For a recorded image to be diffraction limited, the detector (i.e., film, video camera, or CCD camera) must record the magnified image with adequate resolution. The cost of the detectors and the expense, in terms of required memory to store and/or manipulate large images, make it essential that the magnified image is not recorded with *too much* resolution. Castleman (1993) outlines a number of important practical and theoretical considerations and introduces the sampling theory behind making an educated decision on how big a detector is necessary for each specific application. Essentially, an image will be adequately captured if it is digitally sampled at a frequency that is at least 2 times the frequency of the highest spatial frequency present in the specimen. Because the specimen is complex, the frequency of interest is simply the limit of resolution of the microscope, as given by Eq. (1). It is essential that each imaging system be empirically evaluated to establish the size of the magnified image that is required at the detector surface. The method is as follows: First, the diffraction-limited resolution is calculated for each lens using Eq. (1) (see Table I). The width of the microscope field (W) imaged by the detector is then evaluated for each lens by imaging a stage micrometer and measuring the extent (field size) of the image produced by the video or CCD system. It is important that the system is not vignetted; *i.e.*, the field displayed on the detector is smaller than the microscope field seen through the eyepieces. If some component of the system does vignette the field, the image produced by the detector will be surrounded by an out-of-focus mask (see Inoué, 1986). This mask covers part of the stage micrometer, thereby introducing error into the measurement of W such that undersampling is likely. The number of resolvable units across the field is given by

$$\text{resolvable units} = W/d_{\min}, \tag{3}$$

where W is the width of the field in micrometers and $d_{\min}$, also in micrometers, is the minimum resolvable distance calculated through the use of Eq. (1) for the objective in question.

The number of resolvable units should be approximately equal to half the resolution of the video or CCD detector used. Thus, on our system the $100\times$, 1.3 NA lens produces an image at the detector that consists of 470 resolvable units (see Table I). Because the chip in our Hamamatsu C4880 camera is a 1000×1018 channel array (25×25 mm), the resolution of the image produced will not be limited by the detector. In contrast, the $25\times$, 0.8 NA multi-immersion lens that we use will image 1092 resolvable units on the detector and the image has to be magnified at least 2.25 times to produce an image that has resolution limited by the microscope and not the detector. Because oversampling through excessive magnification causes serious drawbacks in the speed at which images can be recorded and manipulated, and because the memory they require is proportional to their area (*i.e.*, the square of an increase in linear distance), we find a continuous, zoom optivar to be invaluable. Fig. 5A shows a properly

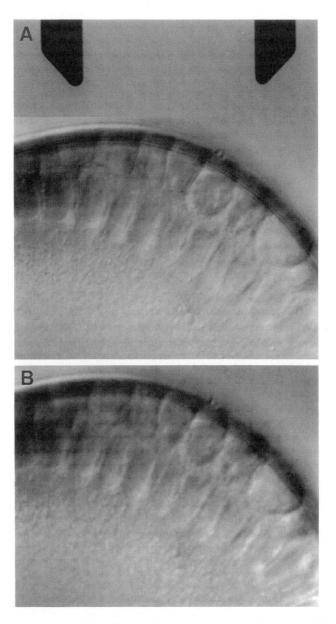

Fig. 5 Images of the posterior end of cellularizing embryos taken under various digital sampling conditions. In each case the images were generated with 25×, 0.8 NA multi-immersion, Plan-Neofluor lens with oil immersion and the correction collar rotated to minimize spherical aberration (same optical parameters as in Fig. 1F). (A) Properly sampled. The optivar (set at 2.25×) was used to magnify the real image made by the objective such that it was digitally sampled at a frequency approximately equal to twice the theoretical limit of resolution of this lens. (B) Undersampled. The image was projected on the CCD chip such that it was undersampled (the frequency was about half of what is required to prevent aliasing). Image detail around the base of the nuclei and the clarity of the granules in the cytoplasm are better preserved in the properly sampled image (A).

sampled image. Image degradation that is a consequence of inadequately sampling an image is shown in Fig. 5B. A scale calibration device (shown in the upper part of Fig. 5A) is built into the Zeiss Axioplan microscope that we use and appears superimposed on the image plane to provide a visual record of the setting of the optivar and the extent of magnification for each image we acquire.

Table I establishes that the resolution provided by a number of the lenses we use "outperforms" the resolution provided by the detector when standard magnification provided by the imaging system is used to magnify the real image produced by the objective onto the CCD chip or video detector. This table provides a model for generating a comparable table for any given set of objectives.

The conclusion is that if field size is not a consideration, a high-resolution image can be magnified onto a small detector and such a detector would be adequate for electronic sampling of the image with high fidelity. *In vitro* motility assays or small numbers of *Drosophila* cells in culture or investigations of small regions of the embryo can be imaged readily with 500×500 pixel (12×12 mm) detectors. However, if the object is to record large regions of the embryo at diffraction-limited resolution, large detectors are necessary.

Our research microscope is equipped with a Hamamatsu C4880 cooled charge coupled device (CCD) camera that dumps images into a MetaMorph Imaging System (Universal Imaging Corporation, West Chester, PA), a top-of-the line computer-controlled imaging system that allows manipulation of the camera and other hardware (shutters, filter wheels, stepper motor control of z-axis [focus] and the $x–y$ position of a mechanical stage, VCRs, and optical memory disc recorders (OMDRs)). We chose the camera because of its 1018×1000 pixel array CCD chip, essential for acquiring images of large fields, a rapid scan system that allows focusing at close to real time, 12- to 14-bit capabilities and high sensitivity in low light. We chose the MetaMorph system, a second-generation digital image processor, because of the responsiveness of Universal Imaging, the company that has developed and marketed it. In particular, Meta-Morph provides for manipulation of hardware and programming for optical and temporal sectioning. The total cost of the package as we have it configured including the cooled CCD camera, a 486 66 MHz PC computer with 64 megabytes of random access memory, a 213M Maxtor hard drive, a Pinnacle 650 megabyte magneto-optical removable drive, Uniblitz electronic shutters, a Ludl z-axis stepper assembly and a Technical Video Ltd., stepper motor attachment for our Zeiss $x–y$ mechanical stage, all with appropriate drivers and software, runs approximately $80,000. However, the educational base price for the MetaMorph software is $12,750, and applications can be acquired as necessary, so very high-quality digital acquisition can be obtained for a very modest investment.

Digital image acquisition and analysis require rapidly evolving technologies. More modest software and hardware systems are available and may be adequate for many imaging purposes. NIH-Image is a very capable public-domain program written by Wayne Rasband at the National Institutes of Health. It can be used for image acquisition and transformation and can be obtained over the Internet by anonymous FTP from ZIPPY.NIMH.NIH.GOV in the directory / PUB/NIH-IMAGE, or can be obtained on floppy disc from NTIS, 5285 Port Royal Rd., Springfield, VA 22161, part number PB93-504868. There is a very responsive electronic bulletin board centered around the use of this program for ideas and troubleshooting. To sign up for the bulletin board an internet address is required and a subscription (to ''subscribe'' or ''sub'') may be obtained by sending an e-mail, that includes only the following, single line in the body of the message ''sub NIH-image Your Name'' (where ''Your Name'' is replaced by your name), to the e-mail address ''listproc@soils.umn.edu.'' The listproc is smart enough to figure out your name and e-mail address from the message that you send. A return message will inform you that your name has been added to the bulletin board's e-mail list, and messages will start trickling in. More conveniently, a summary of all the messages on a given day can be obtained by sending a second message that reads ''set NIH-image mail digest'' to the same address. The digest will come once a day and includes an unabridged version of all the correspondence for the previous 24 hr. A good commercial alternative to NIH Image is Signal Analytics' IP Lab Spectrum; however, there are probably many other suppliers of such software. More modest electronic imaging hardware may also be adequate for a number of applications. Smaller array CCD black and white cameras are available from a number of companies including Dage, Cohu, Hamamatsu, Sony, and Photometrics. Tube-based cameras are also available and may be useful. Two framegrabbers that are supported by NIH Image are the Data Translations Quickcapture and the Scion LG3, both of which have the advantage of being compatible with the new Power Macintosh architecture. As this goes to press many companies contacted were having incompatibility problems with this new standard. Our experience is that the AV option on the Macintosh computers does not allow scientific grade image capture. A minimal system could be set up for ~$14–15,000 at today's prices that would consist of an 8-bit CCD camera, a video monitor, a midrange Macintosh computer, a framegrabber and would use NIH-Image software. Some companies also offer 'turn key' systems that are available as an integrated package.

VII. Dynamics of Image Capture

The morphogenic movements that are of interest to us are slow compared to the time it takes our camera to acquire and store an image on disc. Depending on the movements of interest, this may not always be the case. More rapid image acquisition can be achieved if short-time scale changes in the specimen are of interest. Both vidicon vacuum tubes and solid state (CCD) detectors that can image at video frame rates (30 per sec) are available. An excellent treatment by Bookman and Horrigan (1993) on solid state, video frame rate detectors appears in the Herman and Lemaster's book (1993).

VIII. Digital Image Processing

A wide variety of algorithms designed to enhance contrast are available for manipulation of a digitally recorded image. This enhanced contrast can bring out detail in images crucial to evaluating relevant elements of structure in the developing embryo. Such algorithms compare individual pixels to those that surround them. For example, a low-pass filter can be used to smooth out noise by averaging adjacent pixels, thereby suppressing high-frequency transitions in pixel intensity.

The algorithm we regularly use to enhance the image of nuclei and to observe progress of the furrow canals in the syncytial blastoderm is called *unsharp mask*. This routine brings out details at the extremes of the dynamic range of the image (shadows and highlights). It does this by mathematically producing a blurred image (the *unsharp mask*) and subtracting it from the original image. Comparison of two images before and after application of the unsharp mask routine are shown in Fig. 6. A variety of other routines for detecting edges are also available and are frequently helpful for bringing out the contrast in various regions of a complicated specimen such as a *Drosophila* embryo.

IX. Hard Copy of Electronic Images

Digital images acquired by the methods discussed here can be output on hard medium through a variety of methods. The challenge is to create a finished output without losing any of the resolution that characterizes the image created by the microscope. To do this efficiently it is necessary to understand three kinds of *resolution* in the imaging process.

First, the stored, digital image can be characterized by its *bit* and *image resolution*. The *bit resolution* indicates the dynamic range of the digitized image and is represented by the number of gray levels that represent the finished product. Typical CCD cameras will have between 8- and 16-bit imaging capabilities, indicating that they can resolve either 256 or 65,536 independent gray levels. Whereas the human eye cannot distinguish this many levels of gray, the advantage of a camera that will image with 16-bit resolution is that a useful image can be extracted from a digitized image over a wide dynamic range and that the full range of gray levels can be used for quantitative analysis of the range. That is, one can stretch the gray scale of very dark or very light regions of the image to extract detail. Unfortunately, many of the image-processing programs (*e.g.,* Photoshop (Adobe Systems, Inc., Mountain View, CA), NIH Image, and Digital Darkroom) have only 8-bit capabilities at this time.

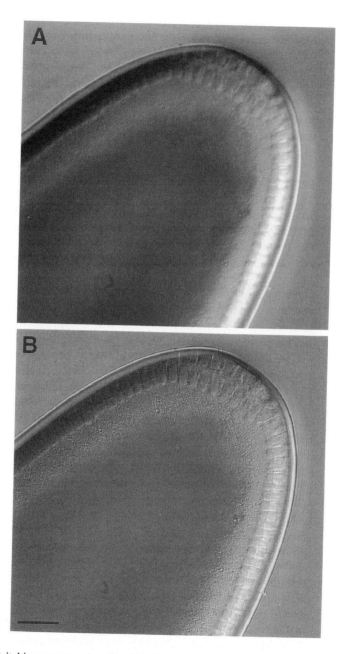

Fig. 6 Digital image processing. The digitally saved image can be processed to enhance certain characteristic features in the image. Here, the raw, digitally captured image is shown above an image that was manipulated with the unsharp mask command in MetaMorph. The obvious increase in contrast enhances our ability to pick out the position of the furrow canals.

Image resolution can best be described as the number of picture elements that are resolved in the image. Photoshop describes *image resolution* in ppi (pixels per inch), but it is often more useful to consider resolution in terms of the number of resolvable units across the entire image.

Finally, *device* or *output* resolution is the resolution provided by whatever output device is used to generate the image for viewing. This might be a computer screen (72 dots per inch (dpi) for a typical Macintosh monitor, a print generated by a dye sublimation printer (300 dpi) or the hard copy printed on a journal page.

Ultimately, the resolution of an electronically captured image is defined by the resolution of the detector. In our case, the camera has a CCD chip that is 1018×1000 (an array that is dealt with by Photoshop and the TIFF (*Tagged Image File Format*; Kay and Levine, 1992) file as 1024×1024 pixels). To adjust properly the image for output, it is necessary to understand the resolution of the output device and the size of the starting image. For example, if you start with an image that is 1024×1024 pixels and want to output to the printer that is 300 dpi, the resolution of the output will be *limited by the output device* if the image that you print is less than 3.4 inches across. As the size of the printed image is made smaller than 3.4 inches across, resolution will be lost. For example, if the size of the image was decreased to 1 square inch, resolution would decrease from the 1024×1024 pixel image recorded by the camera to 300×300 dpi outputted by the printer. That is, the 300 dpi becomes limiting and you throw away information that was originally recovered by the camera.

In contrast, if the size of the image is made larger than 3.4 inches, at 300 dpi the resolution of the printed image is *limited by the camera*. Any increase in the size of the printed image is the consequence of empty magnification. The simplest algorithms designed to deal with such empty magnification (*e.g.*, how does a digital image that is 1024×1024 pixels map to a new image that is 4 times the linear dimensions?) can result in objectionable pixelation of the image. That is, individual pixels from the original, recorded image are mapped to a number of pixels and rafts of adjacent pixels take on the same gray value (16 pixels in the example cited) in the outputted micrograph. Some of the available digital image-processing programs (*e.g.*, the one we use, Photoshop) allow a more sophisticated approach to magnification that assigns values to the additional, newly created pixels based not on a single pixel but a number of surrounding pixels. This dramatically improves the way the image looks but does not improve resolution. For example, the bicubic interpolation method (which can be accessed through the Preferences and the General Preferences Menus in Photoshop) is the most time-consuming, computationally complex method for interpolation but yields the most pleasing output.

In practical terms, we use Photoshop and then Canvas (Deneba Software, Miami, FL) to prepare all stored TIFF images for output as prints or slides or for electronic submission to journals. In Photoshop, we take images and make them 300 pixels per inch resolution by selecting the Image Size option window in the Image pull down menu and checking both the constrain proportion and constrain file size boxes and entering 300 pixels per inch resolution. This automatically shrinks the 1024×1024 pixel image (displayed on the screen at the Macintosh monitor default of 72 dpi; the resolution of the image is limited by the characteristics of the image and not low resolution of the monitor) to 3.413 square inches (*i.e.*, 1024 pixels/300 dpi = 3.413 inches) in the final output. The size of the image on the screen does not change, only the way in which Canvas or a printer handles the size of the image changes. Next, we calculate the size of the image that will actually appear on the hard copy. For this chapter, the size of the printable space is 4.5×6.5. For a collage such as Fig. 1 that includes 10 photos, each panel for the montage is made 1.372 by 2.315 inches. When calculated out, this is 675 by 390 pixels, but a glitch in one of the programs insists that TIFF files have to be expressed in even by even pixels (*i.e.*, 390 by 674 or 676, but not 675). By double clicking on the marquee tool in upper left-hand corner of the PhotoShop system, you can set the parameters that you want, i.e., 674 by 390 pixels. Next, click the marquee tool onto the image and then position it correctly by grabbing the marquee by one of its corners (a little + appears instead of the arrow) and dragging it. We found that putting an overlay of plastic wrap on the monitor and sketching the position of the outline of the subject allowed us to select the same region of many different frames, even if they are not located in the same place in the overall frame. Use the crop tool to discard that part of the image that you do not want. Remember to retain some record of the scale bar from the microscope image. Use the scale bar to make a line of known length. We found that for these images a 5-pixel-wide line works well. Put this line on one edge of the micrograph as a scale record for all your images (you can later discard this image). Finally, use Image commands and the Image adjust commands, levels, and curves to adjust the contrast appropriately. Save as a TIFF file using Macintosh labeling to make a meaningful title.

Images are manipulated and montaged further in Canvas. This is because Canvas uses postscript printing that permits continuous change of overall sizes without causing problems in the output of characters (letters, numbers, arrows, etc.). The program is highly rated but requires use of the manual for optimum results.

We use a departmentally owned dye sublimation printer (Tektronix Phaser IISD, Tektronix, Inc., Wilsonville, OR) that offers 8-bit, 300 pixel per inch addressing resolution. Such printers produce images that have "photographic quality" and indeed can generate hard copy that comes close to but does not quite match image quality from a photographic print. Specific dye-sublimation printers were reviewed by Heid in Macworld (1993,1994) and significantly outperform alternate, available technologies for producing hard copy. Confocal images, transferred to Photoshop, labeled, and montaged in Canvas and then outputted through the Phaser IISD, look equivalent to any hard copy we have produced by photographic means and show exquisite rendering of detail in complex immunofluorescent images of a *Drosophila* embryo (Fig. 7). In addition, such prints surpass the quality of an image produced in a high-quality magazine such as Audubon or National Geographic. Such magazines use a 6 by 6-pixel dithering process that reduces the resolution realized from 1200 dots per inch to approximately 200 dots per inch (depending on the complexity of the image). Two crucial technical problems currently limit a manufacturer's ability to improve the resolution on dye-sublimation prints. First, individual heating elements are required to transfer pigment from a ribbon in the printer to the paper. Three hundred dpi printers have 300 such elements per inch and

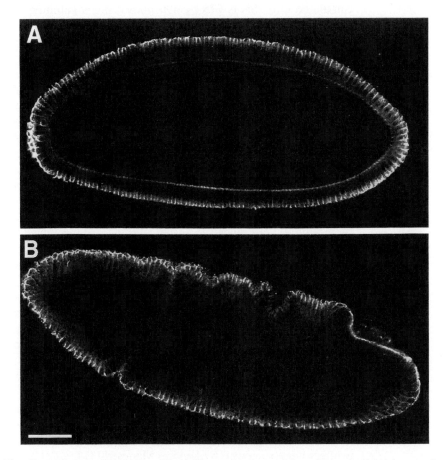

Fig. 7 Digitally manipulated confocal image of a *Drosophila* embryo. A digital image of an embryo stained with antibodies against *Drosophila* βH-spectrin was recovered with a Bio-Rad scanning confocal/Zeiss Axioplan microscope system and stored as a TIFF file. The image was opened in Photoshop, adjusted for contrast, and then exported to Canvas for montaging and labeling. (Reprinted from Thomas and Kiehart, 1994).

while it is possible to increase the spatial frequency at which such heaters are arrayed, it is not necessarily cost effective to do so. Second, the dye itself diffuses after it is applied to the page. This helps the dots blend to produce a continuous tone image, but also reduces image resolution (if every second pixel is programmed to deposit pigment on the image, the sequential spots spread sufficiently to just cover the spot in between, suggesting that the actual resolution is between 200 and 300 pixels per inch). It is not surprising that 300-dpi dye-sublimation printers can produce high-quality images. A commonly used figure for the resolution of the human eye is approximately 1 min of arc (Inoué, 1986). At comfortable reading distance, this translates into approximately 300 pixel resolution. Nevertheless, higher resolution printers do offer crisper images. This is probably because resolution of the eye under optimal conditions can approach 1 sec of arc and high resolution of edges contributes to the perception of "crispness." Finally, it is possible that aliasing can contribute to image distortion when the digital image file is interpreted by the printer. This can occur if the digital image is presented to the printer at a resolution higher than the printer's addressable resolution (*e.g.*, if a 400-dpi image is presented to the 300-dpi addressable resolution Tektronix printer that we use). The problem occurs because the software in the printer will resample the image in a fashion the user cannot control. A simple, practical solution is to set the resolution of the image file (in Photoshop, Canvas, or NIH image) to the addressable resolution of the printer. Stephen M. Pizer is currently preparing a book entitled *Medical Image Display* that will detail more completely considerations for optimizing the quality of outputted images.

X. Summary and Prospectus

Electronic, digital-image acquisition offers a facile approach for obtaining data on the morphology of living or fixed embryos. The quality of the electronically derived output approaches that available with standard photographic methods and digitally processed images can reveal aspects of morphology not visible in standard microscope images. Moreover, these strategies avoid time-consuming efforts in the darkroom and speed turn-around time between image generation by the microscope and their output as finished prints, slides, or video tapes. Rapid advances in the power of personal computers and substantial decreases in their cost ensure that these strategies will become an integral part of biological image manipulation and processing.

Acknowledgments

We thank Ted Salmon for critically reading the manuscript and making a number of valuable suggestions; members of the Kiehart lab, Ted Salmon, and Dave Miller for valuable discussions; George McNamara and Ted Inoué at Universal Imaging, Del Moore at Tektronix, and Kenneth

R. Castleman in Research and Development and Chuck Johnson, both of whom are at Perceptive
Scientific Instruments, Inc. for excellent technical advice; Dr. Claude Piantadosi, P. Owen Doar,
and Robert Schumacher at the Duke Hyperbaric Center, Duke University Medical Center, Durham,
NC, for valuable assistance in devising strategies for the oxygenation of embryos; and Gary Bishop
and Stephen M. Pizer at the University of North Carolina for discussion about the resolution of
hard copy output.

References

Aikens, R. S., Agard, D. A., and Sedat, J. W. (1989). Solid-State imagers for Microscopy. *Methods Cell Biol.* **29,** 291–313.

Ashburner, M. (1989). "*Drosophila:* A Laboratory Handbook." Cold Spring Harbor, NY: Cold Spring Harbor Laboratory Press.

Bookman, R. J., and Horrigan, F. T. (1993). Sampling characteristics of CCD video cameras. *In* "Optical Microscopy" (B. Herman and J. J. Lemasters, eds.), pp. 115–131. San Diego: Academic Press.

Brenner, M. (1994). Imaging dynamic events in living tissue using water immersion objectives. *Amer. Lab.* April, 14–19.

Castleman, K. R. (1979). "Digital Image Processing." Englewood Cliffs, NJ: Prentice Hall.

Castleman, K. R. (1993). Resolution and sampling requirements for digital image processing, analysis, and display. *In* "Electronic Light Microscopy" (D. Shotton, ed.), pp. 71–93. New York: Wiley–Liss.

Foe, V. E., and Alberts, B. M. (1985). Reversible chromosome condensation induced in *Drosophila* embryos by anoxia: Visualization of interface nuclear organization. *J. Cell Biol.* **100,** 1623–1636.

Heid, J. (1993). Dye-sublimation: Photographic-quality printing finally becomes affordable. *MacWorld,* May 1993, pp. 106–111.

Heid, J. (1994). Photo-realistic color printers. *MacWorld,* July 1994, pp. 106–113.

Herman, B., and Lemasters, J. J., eds. (1993). "Optical Microscopy." San Diego: Academic Press.

Inoué, S. (1986). Video Imaging. New York: Plenum.

Kay, D. C., and Levine, J. R. (1992). Graphics File Formats. Windcrest Books of McGraw–Hill, Inc., New York.

Kiehart, D. P. (1981). Studies on the *in vivo* sensitivity of spindle microtubules to calcium ions and evidence for a vesicular calcium-sequestering system. *J. Cell Biol.* **88,** 604–617.

Kiehart, D. P. (1982). Microinjection of echinoderm eggs: Apparatus and Procedures. *Methods Cell Biol.* **25,** 13–31.

Kiehart, D. P., Ketchum, A., Young, P., Lutz, D., Alfentito, M. R., Chang, X.-j., Awobuluyi, M., Pesacreta, T. C., Inoué, S., Stewart, C. T., and Chen, T.-L. (1990). Contractile proteins in *Drosophila* development. *In* "Cytokinesis: Mechanisms of Furrow Formation During Cell Division." *Ann. N.Y. Acad. Sci.* **582,** 233–251.

Salmon, T., Walker, R. A., and Pryer, N. K. (1989). Video-enhanced differential interference contrast light microscopy. *Biotechniques* **7,** 624–633.

Shotton, D., ed. (1993). "Electronic Light Microscopy." New York: Wiley–Liss.

Slayter, E. M., and J. S. Slayter (1992). *Light and Electron Microscopy.* New York: Cambridge University Press.

Thomas, G. H., and Kiehart, D. P. (1994). β_{Heavy}-spectrin has a restricted tissue and subcellular distribution during *Drosophila* embryogenesis. *Development* **120,** 2039–2050.

Whiteley, M., and Kassis, J. A. (1993). Double-sided sticky tape for embryo injections. *Drosophila Information Newsletter* July 1993, Vol. 11.

Wieschaus, E., and Nüsslein-Volhard, C. (1986). Looking at embryos. *In* "*Drosophila,* A Practical Approach" (D. B. Roberts, ed.). Washington, D.C.: IRL Press.

CHAPTER 27

The Use of Photoactivatable Reagents for the Study of Cell Lineage in *Drosophila* Embryogenesis

Charles H. Girdham and Patrick H. O'Farrell

Department of Biochemistry and Biophysics
University of California, San Francisco
San Francisco, California 94143

I. Introduction

This chapter describes the use of photoactivatable lineage tracers in early *Drosophila* development (as originally reported in Vincent and O'Farrell, 1992). The technique involves injection of a nuclear-targeted dextran reagent bearing caged fluorescein moieties into a precellularization embryo. All syncytial nuclei accumulate this nonfluorescent reagent. Fluorescence is induced in nuclei, during or following cellularization, by a microbeam of light that stimulates photochemical uncaging. The fluorescent cell and its daughters can be identified either in the living embryo, offering the opportunity to follow the movements of cells during embryogenesis, or in fixed material allowing the positioning of clones with respect to patterns of gene expression.

This photoactivatable tracer has several advantages over other cell marking techniques and is particularly well suited to lineage studies in the embryo. The marking method does not perturb cellular interactions so that the normal lineage can be followed. The nuclear localization of the reagent permits single cells to be identified accurately in fields of adjoining, marked cells. The marking is precise in that the stage of the embryo and the particular cell that is marked are defined visually at the time of marking. Since visualization of the marked cells does not require fixation of the embryo, dynamic processes such as gastrulation can be followed in real time. For many applications, these features compare favorably with those of another commonly used technique: transplantation of cells from a marked to an unmarked embryo. Transplantation approaches inevitably disturb cellular interactions both during the manipulation and subsequently, since the position of the cell in the donor cannot be precisely matched to its position in the recipient. Although nuclear-localized reagents that are visible in real time could be used in the transplantation approach, there are no reports of such adaptations of the technique in *Drosophila*. The use of a peptide to confer nuclear localization should be applicable to a number of different lineage tracers. We have found that this localization within the cell allows accurate determination of cell number and greatly increases resolution when positioning clones with respect to patterns of gene expression.

At present, the major limitation of the photoactivatable lineage tracer is the time frame during which the marked cells can be studied. The signal in the fluorescein channel becomes increasingly difficult to detect against background fluorescence, which is more noticeable at later stages of development. However, it is likely that this limitation can be at least partially overcome. The uncaged fluorescein can be detected with anti-fluorescein antibodies and the signal amplified with alternatively conjugated secondary antibodies, eliminating problems of background in the fluorescein channel. Calculations suggest that around 3000 fluorescein molecules are generated in a single cycle 14 nucleus in a typical photoactivation. Although direct observation of the uncaged fluorescein is mainly limited to the period of the three postblastoderm divisions, the use of antibodies to amplify the signal produced by photoactivation may also compensate for dilution of the reagent caused by growth and cell division. Alternatively, a reagent incorporating a photoactivatable molecule based on rhodamine (currently being developed by Tim Mitchison), may greatly extend the period over which the fluorescently marked cells can be observed directly, since the background from the embryo is significantly lower in this channel.

When choosing a lineage marking method, it is important to evaluate the specific experimental requirements such as the number of clones that will be needed for a definitive analysis; the importance of precision in terms of the stage of marking; the use of real time analysis; and the need for analysis of late clones. Like the photoactivatable lineage tracer, most markers are subject to dilution and are thus of limited value in analyzing clones in late imaginal tissues (see, for example, Tix *et al.*, 1989; Udolph *et al.*, 1993). In contrast, the transplantation of genetically marked cells that constitutively express β-galactosidase into nonexpressing embryos is a versatile technique in that it allows lineages to be identified at any stage of development as long as expression is under the control of a truly ubiquitous promoter (Meise and Janning, 1993). Classical X-ray-induced mitotic recombination strategies for marking cells have been adapted to create clones lacking β-galactosidase expression in an expressing background to examine, for example, compartmental boundaries in discs and wing tissue (Blair, 1992). Additionally, new approaches based on the use of the FLP site-specific recombinase are providing powerful ways to induce genetic tags (Struhl and Basler, 1993). Despite these advances, it seems likely that the photoactivatable lineage tracer will be the method of choice for applications requiring high precision, such as the tracking of the lineage of identified cells in the early embryo.

II. Preparation of the Photoactivatable Lineage Tracer

The photoactivatable lineage tracer has three components, a dextran backbone, a nuclear localization peptide, and a photoactivatable caged fluorescein. Whereas the original caged fluorescein developed by Tim Mitchison (Mitchison, 1989) can be combined with dextran to produce a cellular lineage tracer (a similar compound is available commercially from Molecular Probes Inc., OR), problems arise with insolubility when a nuclear localization peptide is introduced (Vincent and O'Farrell, unpublished). To develop a more water-soluble reagent, Tim Mitchison modified the caging groups of the original structure to enhance water solubility while retaining the activation kinetics (unpublished). The nuclear-localized photoactivatable reagent utilizes this modified version of the caged fluorescein. Unfortunately, the water-soluble caged fluorescein reagents are not available commercially and must be synthesized from simpler, readily available precursors. Full synthesis protocols are available from Tim Mitchison, Department of Pharmacology, University of California, San Francisco, CA 94143.

The caged fluorescein molecules were linked to a dextran molecule as a standard method of preventing intercellular diffusion after activation. Nuclear localization peptides were also coupled to the dextran to target the lineage tracer to the nucleus, enhancing resolution of the resultant clones and making the tracer amenable to fixation with formaldehyde through the amino groups of the amino acid side chains. All the caged fluorescein reagents and derivatives that were used for the preparation of the photoactivatable dextran were provided by Tim Mitchison.

The photoactivable lineage tracer comprises a 70-kDa dextran carrying between two and five caged fluorescein groups and about 10 nuclear localization peptides. A 70-kDa dextran was chosen, as this provided the tightest localization of the tracer to the nucleus, compared with smaller 10 and 40 kDa dextran molecules, which showed some additional cytoplasmic signal. The caged fluorescein groups were first coupled to 70-kDa amino dextran (with approximately

36 amino groups per molecule, Molecular Probes, Inc., Catalog No. D-1862) using an N-hydroxysuccinimide (abbreviated as NHS) derivative of caged fluorescein that reacts with the primary amino groups (the full structure of caged fluorescein sulfo-NHS ester is shown in Fig. 1A). The dextran was dissolved at 50 mg ml^{-1} in anhydrous dimethylsulfoxide (DMSO), together with a small amount of triethylamine as a proton acceptor, and the caged fluorescein sulfo-NHS ester was added to the desired molar ratio from a 100 mM stock in DMSO (we assume a 50% coupling efficiency; therefore, we add 4–10 mole of caged fluorescein sulfo-NHS ester/mole dextran). After 30 min at room temperature, the remaining amino groups on the dextran were reacted with iodoacetic acid-NHS-ester (Sigma) to allow the conjugation of a nuclear localization peptide. Iodoacetic- acid-NHS-ester was added from a 50 mg ml^{-1} stock solution in DMSO to give a final concentration of 50 mole per mole dextran (in excess over the amino groups). After a further 30 min, the reaction was dialyzed against 100 mM Hepes, pH 7, to remove excess iodoacetic acid-NHS ester.

The nuclear localization peptide was derived from the polyoma large T antigen [Chelsky et al., 1989; residues 189–196 underlined in the complete peptide sequence: CGYGVSRKRPRPG (kind gift of Dan Chelsky)] and had been shown to efficiently direct nuclear targeting in Drosophila. The peptide is coupled to the dextran via the thiol group on the N-terminal cysteine using a standard chemical cross-linking procedure for linking amino and thiol groups (see, for example, Rector et al., 1978). The peptide, in water, was added to the iodoacetyl-caged fluorescein dextran prepared previously to the correct molar ratio (approximately 10 peptides per dextran molecule) and allowed to react overnight at room temperature. Remaining unreacted iodacetate groups were reduced with 5% β-mercaptoethanol. All synthesis steps and subsequent manipulations of the reagent were performed in the dark or with a standard photographic safe light. The reagent was stored in small aliquots at −70°C. A schematic diagram of the final reagent is shown in Figure 1B.

III. Injection and Activation of the Photoactivatable Lineage Tracer

Syncytial embryos, 30–90 min AED, are dechorionated in 50% bleach and aligned for injection on fine threads of rubber cement or glue made from sticky tape shaken with heptane on a coverslip lightly affixed to a microscope slide. After a suitable desiccation period, the embryos are covered with a 1 : 1 mix of Series 27 and Series 700 halocarbon oils (Halocarbon Products Corporation, NJ). The tracer is injected at a concentration of approximately 1 mg ml^{-1} and injected embryos are left to develop to cellularization at 18°C, allowing the tracer time to diffuse throughout the embryo. For photoactivation, the embryo is viewed with the aid of a video camera on an epifluorescence microscope with a 63X Zeiss PlanNeofluor lens, immersed directly into the halocarbon oil

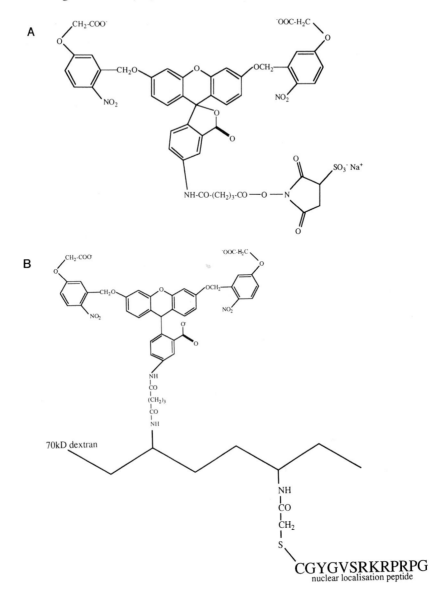

Fig. 1 (A) Structure of the reactive caged fluorescein sulfo *N*-hydroxysuccinimide ester, which allows coupling of the caged fluorescein moieties to the amino dextran. (B) Schematic diagram of the final photoactivatable lineage tracer reagent.

without a coverslip (the above mix of halocarbon oils was found to give a clear image on the photoactivation microscope using this lens and its viscosity was also found to be important for real time examination of embryos mounted on oxygen-permeable membranes under a coverslip (see following, Vincent, unpublished). Uncaging of the tracer in individual cells is carried out using ultraviolet light from a mercury vapor bulb passing through a 365-nm barrier filter and a 100-μm pinhole positioned before the diaphragm. The embryos can be viewed using red-filtered transillumination and the activating beam can be aligned on the surface of an embryo. The location of the beam can be marked with a pen on the video monitor, so that other embryos can be precisely positioned for the activation of single cells. The duration of illumination is controlled with an electronic shutter (Vincent and Assoc., New York) and optimal uncaging times are between 5 and 10 sec. Longer times may cause bleaching and significant uncaging in neighboring cells due to flare from the beam and increase the chances of cell damage. Figure 2 shows a diagram of the microscope set up for photoactivation. A number of cells can be photoactivated in the same embryo, either several cell diameters apart so that separated clones will be identifiable in the later embryo or in a particular pattern to observe cell movements, for example (see following and Fig. 3). The accuracy and efficiency of activation can be checked on a separate fluorescence microscope equipped with a sensitive CCD camera to minimize exposure to the excitation

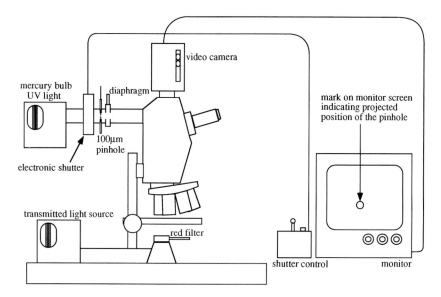

Fig. 2   Microscope setup for photoactivation.

beam (typical exposure times are 0.02 sec per image). Viewing the activated embryos down a regular epifluorescence microscope results in rapid bleaching of the fluorescein signal.

The caged reagent shows low toxicity as judged by the apparently normal development of the injected embryos. When higher concentrations of the reagent were injected, no lethality was observed, but recovery of clones was reduced (Vincent and O'Farrell, unpublished). We suspect that this is due to damage associated with uncaging of the reagent, since the photocleavage reaction produces reactive aldehydes. This effect was not observed using the reagent concentration described here (1 mg ml^{-1}); however, investigators ought to follow the recovery of marked cells as a diagnostic for this problem. Most of the cells of the epidermis undergo three rounds of postblastoderm division (Edgar and O'Farrell, 1990), such that a cell marked at cycle 14 will produce eight marked progeny.

IV. Fixation of Embryos Carrying Marked Cells

The coverslip carrying the photoactivated embryos can be removed from the supporting microscope slide used for injection (see preceding) and submerged in a petri dish containing a small amount of halocarbon oil to prevent further dessication. The dish is then placed in a humid chamber, and the embryos are allowed to develop for a suitable period of time. Aged embryos are washed off the glue into a small petri dish using a stream of heptane, which also serves to dissolve the halocarbon oil. The embryos are then transferred to a glass scintillation vial and fixed in a mix of heptane and PBS containing 10% formaldehyde (in the aqueous phase) for 30 min on a shaker. The fixed, injected embryos are refractory to simple methanol devitellinization and must be devitellinized by hand. They are pipetted onto nylon mesh, blotted free of heptane by apposing tissue paper to the opposite side of the mesh, and picked up with a piece of double-sided sticky tape. The tape is attached to the bottom of a petri dish, and the embryos are covered with PBS. They are then teased out of their vitelline membranes using a 21-gauge syringe needle.

The embryos can now be stained with antibodies using standard procedures. All procedures with the injected embryos are carried out in the dark or with a photographic safelight, and all dissecting microscope light sources are red or yellow filtered, to cut off shorter wavelengths. Taking these precautions, we have not encountered significant problems with nonspecific uncaging of the reagent. As mentioned above, it is possible to amplify the signal from the photoactivated reagent using antibodies against fluorescein that are specific for the uncaged molecule. These antibodies can then be detected with secondary antibodies conjugated to alternative fluorescent tags or enzymes.

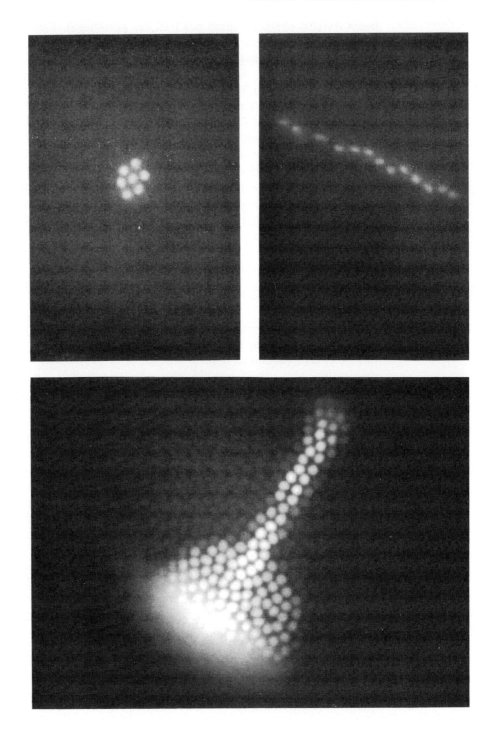

V. Analysis of Living Embryos Carrying Marked Cells

A slight modification of the injection procedure is required to view cell movements of patches of marked cells in living embryos. We have found that the clearest fluorescence images are achieved using a coverslip, but addition of a coverslip ordinarily restricts the oxygen supply to the embryo. To allow extended observation without anoxia, the embryos are mounted on oxygen-permeable membranes (Yellow Springs Instrument Co., Ohio, Catalog No. 5793) clamped tightly in a metal slide, diagrammed in Fig. 4. Dechorionated embryos are mounted on a thin thread of rubber cement drawn across the membrane. The width of the rubber cement strip should usually be similar to the length of the embryo. After injection of the lineage tracer and photoactivation, a coverslip can be placed over the marked embryos. The halocarbon oil mix, specified earlier, allows the coverslip to rest gently on the embryo, but any problems with embryo flattening can be alleviated with the use of strips of coverslips as spacers between the oxygen-permeable membrane and the coverslip. The flexibility of the membrane generates some problems with fluctuations of the plane of focus when the embryos are under continuous observation. Time-lapse movies of gastrulating embryos can be made using images compiled from the CCD camera. We have found that a software network (ISee image acquisition and analysis software, Inovision Corp., Durham, NC, running on a SUN Sparc workstation) that takes images from five focal planes separated by 2 μm (using a motorized microscope stage) and then makes a final image based on the maximal pixel value obtained at each position, produces a clear and informative movie. Data acquisition and storage for each composite image take approximately 45 sec.

VI. Summary

Photoactivatable lineage tracers represent a major advance for clonal analysis in the early embryo and the study of cell movements. Any cell in the blastoderm can be marked, and the nuclear localization of the signal allows excellent resolution in identifying the daughters of individual cells. Although the technique is limited by the availability of the water-soluble caged fluorescein and its derivatives for synthesis of the complete tracer, these may become commer-

Fig. 3   Three examples of patterns of marked cells created for real time analysis of cell movements during gastrulation. These patterns are easily produced by careful photoactivation during cellularization. In the bottom panel, the pinhole (see text) was removed and the diaphragm was partially closed down. The stage was then moved during exposure to the photoactivating light, creating a strip of fluorescent cells. All pictures were taken with a CCD camera and show the fluorescein channel. All embryos are at cellular blastoderm prior to the start of gastrulation.

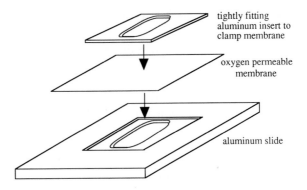

tightly fitting
aluminum insert to
clamp membrane

oxygen permeable
membrane

aluminum slide

Fig. 4 Custom microscope slide for mounting embryos for real time analysis. Basic design for tightly clamping an oxygen-permeable membrane in a machined aluminum slide. Embryos are mounted through the opening on the bottom side of the slide shown. Dimensions are not given since they should be customized to the particular microscope stage and objective lenses used for activation and examination of the embryos.

cially available in the future. The use of caged rhodamine derivatives or antibody amplification of the signal may greatly extend the developmental period over which marked clones can be identified.

Acknowledgments

We thank Jean-Paul Vincent for discussing unpublished results; Tim Mitchison for his generous provision of reagents and invaluable help with the chemistry; and Shelagh Campbell, Ilan Davis, and Tony Shermoen for their comments on the manuscript.

References

Blair, S. S. (1992). Engrailed expression in the anterior lineage compartment of the developing wing blade of Drosophila. *Development* **115**, 21–33.
Chelsky, D., Ralph, R., and Jonak, G. (1989). Sequence requirements for synthetic peptide-mediated translocation to the nucleus. *Mol. Cell. Biol.* **9**, 2487–2492.
Edgar, B. A., and O'Farrell, P. H. (1990). The three postblastoderm cell cycles of Drosophila embryogenesis are regulated in G2 by string. *Cell* **62**, 469–480.
Meise, M., and Janning, W. (1993). Cell lineage of larval and imaginal thoracic anlagen cells of Drosophila melanogaster as revealed by single cell transplants. *Development* **118**, 1107–1121.
Mitchison, T. J. (1989). Polewards microtubule flux in the mitotic spindle: Evidence from photoactivation of fluorescence. *J. Cell. Biol.* **109**, 637–652.
Rector, E. S., Schwenk, R. J., Tse, K. S., and Sehon, A. H. (1978). A method for the preparation of protein–protein conjugates of predetermined composition. *J. Immunol. Methods* **24**, 321–336.
Struhl, G., and Basler, K. (1993). Organizing activity of wingless protein in Drosophila. *Cell* **72**, 527–540.

Tix, S., Minden, J. S., and Technau, G. M. (1989). Pre-existing neuronal pathways in the developing optic lobes of Drosophila. *Development* **105,** 739–746.

Udolph, G., Prokop, A., Bossing, T., and Technau, G. M. (1993). A common precursor for glia and neurons in the embryonic CNS of Drosophila gives rise to segment-specific lineage variants. *Development* **118,** 765–775.

Vincent, J. P., and O'Farrell, P. H. (1992). The state of engrailed expression is not clonally transmitted during early Drosophila development. *Cell* **68,** 923–931.

CHAPTER 28

Looking at Oogenesis

Esther Verheyen[1] and Lynn Cooley

Department of Genetics
Yale University School of Medicine
New Haven, Connecticut 06510

I. General Background

The analysis of oogenesis in *Drosophila* has proven to be invaluable in dissecting diverse cell biological processes. Oogenesis requires carefully regulated interactions between the germline-derived nurse cells and oocyte and the somatically derived follicle cell populations. In this chapter, we will describe techniques commonly used to examine ovarian tissue. These techniques are particu-

---

[1] Present address: Howard Hughes Medical Institute, Department of Cell Biology, Yale University, New Haven, CT 06536.

METHODS IN CELL BIOLOGY, VOL. 44

larly useful for determining the phenotype of mutations, as a first step in characterizing the function of a gene required during oogenesis.

A. Oogenesis

Each of the two ovaries of *Drosophila* contains 15–17 tubular ovarioles in which egg chambers develop (Fig. 1) (for oogenesis reviews, see King, 1970; Mahowald and Kambysellis, 1980; Spradling, 1994). The egg chambers are arranged in a linear array displaying a continuum of growth and development. Individual egg chambers travel down the ovariole until mature eggs pass into the oviduct where they are fertilized and then laid. At the anterior of the ovariole is a structure called the germarium in which the egg chambers are initially formed (Fig. 2A). One to two germline stem cells reside at the tip of each germarium. Oogenesis is initiated when a stem cell daughter divides mitotically exactly four times to give rise to the 16 germline-derived cells of the egg chamber. Incomplete cytokinesis during each division results in cytoplasmic bridges interconnecting the 16 cells. After all four mitoses are complete, each arrested cleavage furrow is stabilized by the addition of several cytoskeletal proteins forming a ring canal. There are 2 cells with four ring canals, 2 with three, 4 with two, and 8 with one. The oocyte develops at the posterior of the egg chamber from 1 of the 2 cells having four ring canals. The 15 remaining cells differentiate into nurse cells that are synthetically active and contribute cytoplasm to the oocyte. The 16-cell cluster is surrounded by a layer of somatically derived follicle cells in the central region of the germarium. The cyst exits the germarium as a Stage 2 egg chamber (Fig. 2A).

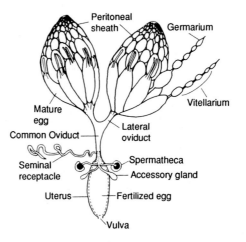

Fig. 1 Female reproductive system. A dorsal view of the internal female reproductive system (taken from Mahowald and Kambysellis, 1980). There are two ovaries that are joined by a common oviduct. The ovary is made up of 15–20 ovarioles, 2 of which are separated from the right ovary. More mature egg chambers move progressively posteriorly in the ovary.

The development of an egg chamber has been arbitrarily divided into 14 stages. Previtellogenic egg chambers in Stages 2–6 are spherical and covered by a uniform layer of follicle cells (Fig. 2A). The stages are distinguished by size and nurse cell nuclear morphology. By Stage 7, egg chambers become ovoid. During these stages, the nurse cells and oocyte are growing in a slow and uniform manner, so that the oocyte cannot be easily distinguished from the nurse cells. Early in oogenesis the nurse cells endoreplicate and reach a state of 1000–2000 polyploidy. Simultaneously, the follicle cells divide mitotically four to five times to generate approximately 1000 cells and then also become polyploid. The oocyte nucleus is transcriptionally quiescent and arrested in meiosis 1.

At Stage 8, the endocytic uptake of yolk by the oocyte begins allowing it to be visually differentiated from the nurse cells by the granular appearance of the cytoplasm (Fig. 2A). At Stage 9, the oocyte volume is roughly one-third that of the egg chamber and by Stage 10 it composes half the egg chamber volume. In addition to oocyte size, the egg chambers can be differentiated by the status of follicular migrations. During Stage 9, a specialized set of 6–10 follicle cells, the border cells, migrates from the anterior edge of the egg chamber to the border between the nurse cell and the oocyte. In addition, the majority of the follicle cells covering the egg chamber migrate along the outer surface of the germline cells to form a columnar layer covering the oocyte, leaving a few follicle cells in a thin squamous layer covering the nurse cells. At Stage 10, the follicle cells begin a centripetal migration between the oocyte and the nurse cells (Fig. 2B). At Stage 10B, dramatic morphological changes occur in the nurse cells in preparation for final cytoplasm transport. In the cytoplasm of each nurse cell, actin filament networks polymerize and form bundles that extend from the cell membrane to the large nucleus. Subsequently, in the 30 min of Stage 11, the nurse cell nuclear membranes permeabilize and the entire cytoplasmic contents of the nurse cells are rapidly transported into the oocyte (Fig. 2C) (Cooley et al., 1992). Microtubules mediate ooplasm streaming during Stage 11. Cytoplasm transport results in a doubling of oocyte volume and regression and degeneration of the nurse cell cluster. After cytoplasm transport is complete, the follicle cells secrete the egg shell, or chorion, including the specialized structures such as dorsal respiratory appendages, the micropyle, and the operculum (Fig. 2E).

B. Oogenesis Cell Biology

The ovary is ideal for the study of cell biology since relatively few cell types are involved and their morphology and behavior are fairly well characterized. The nurse cells are synthetically very active and they have a complex and dynamic cytoskeleton. The somatic follicle cells are polarized secretory epithelial cells that undergo cellular rearrangements as they carry out their various migrations around the developing oocyte.

Using genetic manipulations, many genes involved with cellular functioning during oogenesis have been described. For example, the *slbo* gene, encoding

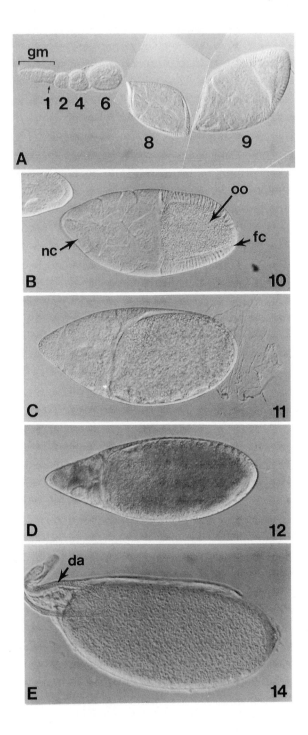

Drosophila C-EBP transcription factor, is required for border cell migration (Montell *et al.,* 1992). Mutations of the epidermal growth factor receptor disrupt communication between germline and somatic cells in the egg chamber, leading to a ventralized egg shell and embryo (Schüpbach, 1987; Price *et al.,* 1989). Mutations in several genes affect the determination of the oocyte. These include genes encoding RNA binding proteins (*orb*) (Lantz *et al.,* 1992) and cytoskeletal proteins (*hts*) (Yue and Spradling, 1992). The transport of cytoplasm from the nurse cells to the oocyte requires genes for several actin-binding proteins (Cooley *et al.,* 1992; Cant *et al.,* 1994).

II. Gross Morphological Analysis

A. Female-Sterile versus Maternal Effect Lethal

The first step in defining a given mutation as female-sterile is to examine whether any eggs are laid by mutant females. Mutations affecting the production of viable eggs fall into two categories: female-sterile and maternal effect lethal mutations. Females carrying maternal effect lethal mutations produce morphologically normal mature oocytes that are laid. Such eggs fail to proceed through embryogenesis, however, because the mother was mutant for a locus whose gene product is needed during the early stages of development before zygotic gene activity begins. On the other hand, female-sterile flies fail in the production of normal eggs. In female-sterile mutations, the defect can arise at any stage of oogenesis, and most often mutant eggs are not laid. However, occasionally mutant eggs are laid but these generally are morphologically different from wild-type or maternal effect lethal egg chambers.

B. Looking at Ovaries

1. Care and Feeding of Females

To examine the egg chambers, ovaries must be dissected out of adult flies. Newly eclosed and very young females will have mainly previtellogenic egg chambers; therefore, it is best to dissect 3- to 5-day-old flies. To stimulate egg

---

Fig. 2 Stages of oogenesis. Fixed egg chambers were photographed using DIC optics and a 20× objective. (A) A germarium (gm) in which a Stage 1 egg chamber has formed. Egg chambers from Stages 2, 4, 6, 8, and 9 are shown. The earlier stages can be distinguished by size, shape, and follicle cell morphology. (B) A Stage 10 egg chamber in which the oocyte occupies half the volume of the egg chamber. The three cell types present in an egg chamber: nurse cells (nc), oocyte (oo), and follicle cells (fc) are indicated. During Stage 11 (C), the remaining contents of the nurse cells are transported into the oocyte. During Stage 12 (D), the nurse cell remnants regress and the follicle cells continue to deposit a chorion around the oocyte. By Stage 14 (E), the nurse cell cluster is completely degenerated and the chorion with dorsal respiratory appendages (da) is complete.

production, females are fed a wet yeast paste before dissecting them. This is done by mixing baker's yeast with water until the consistency is that of toothpaste. A dab of yeast about the size of a quarter is then smeared on the inside of a food vial and females of interest and several males are added. The optimal amount of time for yeast feeding can be 24–48 hr but should be determined empirically for each mutant line. The goal is to obtain ovaries as healthy as possible to avoid assigning nonspecific consequences of undernourishment to the mutant phenotype description.

2. Removing Ovaries

Wild-type ovaries are relatively easy to dissect with the aid of a dissecting microscope since they compose roughly half the volume of the abdomen. Females anesthetized with either carbon dioxide or ether are dissected in *Drosophila* saline solution (EBR) using fine-point tweezers (we recommend Dumont No. 5). To do this, small glass dishes or spot plates (VWR, Pyrex spot plates) filled with EBR are used. Place the dish on a black background for optimal contrast. Light should be provided by a fiber optic source (goosenecks are best) and come from the sides and not the top. The fiber optic light source will not generate enough heat to rapidly evaporate the dissection solution (and the ovaries), and illumination from the sides gives much better contrast. The fly is placed on its dorsal surface, either floating on the surface or held at the bottom of the dish with forceps (Fig. 3A). One pair of forceps holds the thorax while the other teases gently at the cuticle of the ventral abdomen to produce a hole. Another option is to tear off the genitalia entirely to produce an opening (Fig. 3B). It is best not to separate the whole abdomen from the fly since that releases gut contents into the dish making it more difficult to see. Once the cuticle is opened, the forceps that were used to produce the opening can be used to gently squeeze out the ovaries. In addition to ovaries, portions of the gut and malpighian tubules will emerge. The ovaries can be easily separated from the other organs and removed from the female (Fig. 3C). Remove the carcass of the dissected female and proceed to the next. A number of ovaries can be accumulated in the bottom of the dish. You will notice peristaltic contractions of the ovarioles. Several changes of dissecting solution with a pasteur pipet may be needed to remove debris from the dish.

3. Dissecting Egg Chambers

To isolate individual egg chambers, the ovaries first must be gently teased apart into ovarioles. This can be done by holding the base of an ovary (the end with the most mature egg chambers) with one pair of forceps and "combing" the ovarioles apart with a second pair. Unfixed egg chambers are very fragile and can be easily punctured by forceps. If you plan to stain filamentous actin in egg chambers, you need to remove the muscle-containing sheath that sur-

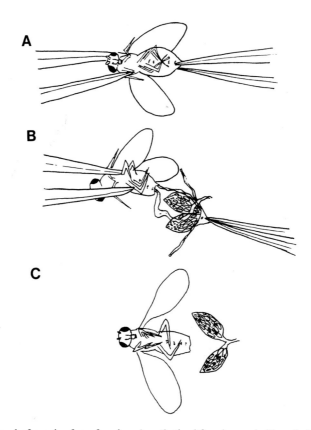

Fig. 3 Removal of ovaries from females. Anesthetized females are held on their dorsal surface with one pair of forceps while a second pair of forceps teases at the posterior abdomen (A). The ovaries and guts are often attached to the portion of abdomen torn off (B). The internal organs can be easily separated away from the ovaries (C). Dissect under a dissecting scope at 12–20×.

rounds each ovariole. This will stain brilliantly with rhodamine–phalloidin and obscure actin structures within egg chambers. To remove egg chambers from the sheath, it is easiest to grab hold of the germarium end of an ovariole and push more mature egg chambers out. This should be done on fragile unfixed tissue to take advantage of the contractile nature of the sheath; egg chambers will be squeezed out by a live sheath with a little encouragement. It is nearly impossible to remove the muscle sheath from fixed tissue.

4. To Fix or Not to Fix

For a quick look at egg chambers, no fixation is required. Separated ovarioles or individual egg chambers can be mounted on a slide in plenty of dissection solution and covered with a coverslip. Such a preparation will last long enough

(30–60 min) for examination and photography using a microscope. Differential interference optics (DIC) are much better than phase contrast for seeing egg chamber structures.

For more detailed analysis involving any one of a number of staining procedure, egg chambers should be fixed. For β-galactosidase, DAPI, and antibody staining, whole ovaries can be fixed. For rhodamine–phalloidin staining, the ovaries must first be dissected into individual egg chambers and then fixed.

C. Phenotypic Classes

Oogenesis defects associated with mutations occur at any stage of development. The phenotype can involve production of mature but abnormal egg chambers or complete arrest at a specific stage. Some commonly found phenotypic classes are illustrated in Figs. 4 and 5. Agametic (ag) or tumorous (tu) mutations cause oogenesis to arrest very early, and the ovaries are usually very small compared to wild type (Fig. 4A). Their small size makes it difficult to locate the ovaries within females and also makes them quite difficult to dissect. Such tiny ovaries are often easier to examine after staining nuclei with DAPI. This allows the various cell types to be identified based on nuclear size and morphology (Figs. 4B and 6A). In Fig. 4A, both mutant ovaries look similar; however, DAPI staining reveals very different defects. Tumorous ovaries contain very abnormal egg chambers filled with many small, presumably germline-derived nuclei (Fig. 4C). Agametic ovaries are characterized by a failure of the germline tissue to develop, and no nurse cell nuclei are visible (Fig. 4D). In such cases, the somatic tissue is often unaffected and forms ovariole-like structures.

Another common mutant phenotype is that of early arrest. In such ovaries, egg chambers develop to an intermediate stage such as Stage 6, 7, or 8 and then begin to degenerate. This leads to a characteristic beads-on-a-string phenotype shown in Fig. 5B. Early arrest ovaries are generally half the size of wild type. Another mutant class is one in which the transport of cytoplasm from the

---

Fig. 4 Early oogenesis defects. Wild-type (wt) and mutant ovaries were dissected and photographed under a dissecting scope (A). Tumorous (tu) and agametic (ag) types of mutations cause defects early in oogenesis and result in very small ovaries that are difficult to characterize morphologically (A). Staining nuclear material with the dye DAPI (B–D) allows nuclei to be visualized and defects in either their number or size to be determined. B–D were photographed from a fluorescence microscope using a 20× objective. The arrow in B shows a wild-type ovariole beginning at the germarium. In C, a tumorous phenotype is clearly seen. Egg chambers contain many small presumably germline nuclei (arrow), indicating abnormal cell divisions. Agametic ovaries (D), on the other hand, contain no germline tissue as seen by the absence of any nurse cell nuclei. The follicle cells often behave normally forming empty ovariole-like structures that appear as stacks of cells (arrow). The somatically derived oviduct (ovi) is indicated in the two mutants as a point of reference.

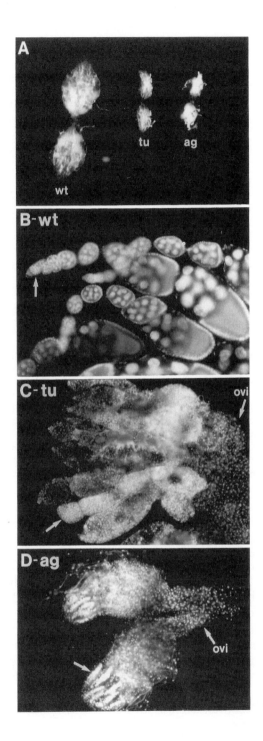

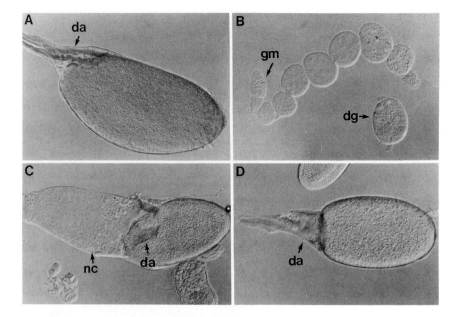

Fig. 5 Mid to late oogenesis defects. Mutations may disrupt mid to late stages of oogenesis. Fixed egg chambers were examined using a 20× objective. A mature wild-type Stage 14 egg chamber (A) is given for size comparison. In B, a typical early arrest ("beads-on-a-string") ovariole is illustrated. The germarium (gm) and early stages appear normal, but fail to develop beyond approximately Stage 6 and then degenerate (dg). Degenerating egg chambers often have a granular or vacuolar appearance. In C, an egg chamber in which cytoplasm transport was incomplete is shown. The small oocyte has been eclosed by a chorion and is attached to a large nurse cell cluster. The dorsal appendages are abnormal, probably due to the physical interference of the nurse cells with the follicle cells that secrete the egg shell. A dorsalized egg chamber is seen in D. The dorsal appendages have expanded ventrally and are much broader than wild type.

nurse cells to the oocyte is disrupted. Such egg chambers have a characteristic phenotype (Fig. 5C) in which a small oocyte, enclosed by a chorion, is attached to a persistent nurse cell cluster. Often the dorsal appendages in this class are abnormal, either fused or shorter and broader. This is probably a secondary effect of the nurse cell remnants interfering with the normal migration of the follicle cells between the oocyte and the nurse cells. Finally, mutations disrupting oogenesis can affect the polarity of the egg shell. An example of a dorsalized egg chamber in which the dorsal appendages have expanded ventrally is seen in Fig. 5D. In ventralized egg chambers, dorsal appendages are often fused, appear as a single appendage, or are absent entirely (not shown). Polarity defects can fall into the maternal effect lethal class if they can be fertilized.

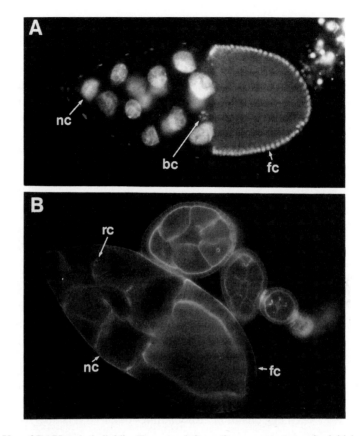

Fig. 6 Use of DAPI and phalloidin. Two very informative reagents to use for initial analysis of oogenesis defects are DAPI and rhodamine-conjugated phalloidin. DAPI stains nuclear material and can be used to identify the specific cell types. In A, a Stage 10 egg chamber, the large polyploid nurse cell nuclei (nc), and the small follicle cell nuclei (fc) are clearly visible. In addition, the nuclei of the migratory border cells (bc) are seen at the anterior border of the oocyte and the nurse cells. DAPI staining will reveal any abnormalities in number of cells, states of polyploidy, or general organization of the egg chamber. Notice that the oocyte cytoplasm autofluoresces in the UV channel. Rhodamine-conjugated phalloidin stains filamentous actin structures in all cells of the egg chamber (B). This is useful for identification of the state of cell membranes. The cells can be counted, the status of actin structures can be determined, and ring canals (rc) can be clearly seen and counted.

III. More Detailed Examination

Initial phenotypic analysis of live or fixed tissue will provide clues to the defect in a specific mutant. To test or confirm such hypotheses, there are a number of reagents available including DAPI (Fig. 6A) and rhodamine–phalloidin (Fig. 6B) for examination of specific cell types or processes during oogenesis. Some of these are commercially available while others may be obtained from individual researchers. Table I lists specific events or cell types, useful reagents for examining them, and references. This list is by no means complete, but will provide a useful starting point for analysis of mutant phenotypes.

If you suspect that oocyte determination or differentiation is affected, there are several markers that can be tested. Many mRNAs specifically accumulate in the oocyte starting very early in the germarium. Examples are *oskar* (Kim-Ha *et al.*, 1991), *orb* (Lantz *et al.*, 1992), and *hu-li tai shao* (Yue and Spradling, 1992). *In situ* hybridization with probes for any of these mRNAs (see Chapter 30) will show whether each egg chamber has an oocyte and whether it is in the correct position. In addition, normal accumulation in the very young oocyte indicates that cytoplasm transport in the germarial stages is normal. Possible defects in later cytoplasm transport can be examined using probes for maternal mRNA or protein or by using germline-specific enhancer trap lines. A probe for *bicoid* mRNA is particularly useful since it begins to accumulate in the oocyte in mid-oogenesis (around Stage 5) and is specifically located at the oocyte anterior later in Stage 10 (Berleth *et al.*, 1988). Similarly, antibodies to *vasa* (Lasko and Ashburner, 1990) or a probe for *nanos* mRNA (Gavis and Lehmann, 1992) can be informative both for general transport and for oocyte polarity since they accumulate in the posterior.

The cytoskeletal integrity of egg chambers can be assessed using various reagents. Ring canals connecting the germline cells can be visualized using either rhodamine–phalloidin (Fig. 6B) (Warn *et al.*, 1985) or *kelch* antibody (Xue and Cooley, 1993). You should be able to see 15 rings by focusing through each egg chamber. Rhodamine–phalloidin is also useful for examining the sub-

Table I
Useful Reagents for Examining Oogenesis

Subject	Reagent	Reference
Oocyte determination	*oskar* mRNA probe	Kim-Ha *et al.*, 1991
	orb mRNA probe	Lantz *et al.*, 1992
	hu-li tai shao mRNA probe	Yue and Spradling, 1992
Cytoplasm transport	*bicoid* mRNA probe	Berleth *et al.*, 1988
	nanos mRNA probe	Gavis and Lehmann, 1992
	Germline enhancer trap	Cooley *et al.*, 1992
	Vasa antibody	Lasko and Ashburner, 1990
Ring Canals	Rhodamine–phalloidin	Warn *et al.*, 1985;
	kelch antibody	Xue and Cooley, 1993
Cytoskeleton	Rhodamine–phalloidin	Fig. 6; Gutzeit, 1986; Cooley *et al.*, 1992
	tubulin antibody	Theurkauf *et al.*, 1992,1993
Border cell migration	DAPI	Fig. 6
	enhancer trap markers,	Montell *et al.*, 1992
	fasciclin III antibody	Brower *et al.*, 1981

cortical actin cytoskeleton and cytoplasmic actin bundles that form only in Stage 11 nurse cells (Gutzeit, 1986; Cooley *et al.*, 1992). Microtubules can be examined with a tubulin antibody (Theurkauf *et al.*, 1992,1993). Follicle cell migrations can be seen easily using DAPI to stain nuclei (Fig. 6A). In addition, there are antibodies or enhancer trap lines specific for migrating border cells (Montell *et al.*, 1992; Brower *et al.*, 1981).

IV. Protocols

A. Fixation of Whole Ovaries or Egg Chambers

1. Dissect out ovaries in cold EBR as described under Section IIB and transfer into an Eppendorf tube containing cold 1X EBR, on ice.
2. Pipet off EBR with long-stem Pasteur pipet. Add 100 μl devitellinizing buffer and 600 μl heptane. Agitate vigorously to ensure that the buffer is saturated with heptane and then gently for 10 min at room temperature.
3. Remove solution with a Pasteur pipet and rinse with 1X PBS.

B. Staining for β-Galactosidase Expression in *Drosophila* Ovaries

Spradling Lab Modifications of the Gehring Lab Bluebook Procedure.

1. Fix tissue as described under Section IVA and rinse in PBS.
2. Remove PBS and add 500 μl staining solution. Incubate at 37°C. Some strains require only minutes of incubation, while others can take up to 24 hr. Rinse stained ovaries with PBS and then put them in 50% glycerol, PBS. Stained ovaries can be stored in this manner at 4°C for at least 1 month.
3. To mount ovaries, remove 1–2 ovaries with 100 μl of glycerol solution from tube and place on a clean microscope slide. Dissect out egg chambers, completely removing all ovariolar tissue and separating egg chambers from each other. Too many egg chambers produces a crowded field that is hard to photograph. Place glass shards (from broken coverslips) under the coverslip to prevent flattening the egg chambers. Remove excess moisture with a Kimwipe. Examine under the microscope. For long-term storage, seal the slide with nail polish and store at 4°C.

C. DAPI Staining

1. Fix tissue as described under Section IVA and rinse in PBS.
2. Stain DNA with DAPI (1 μg/ml in PBS) for 5 min and then rinse.
3. Remove PBS and add about 50 μl glycerol:PBS (50% glycerol).

4. Place ovaries on a slide, dissect into individual egg chambers or ovarioles, coverslip, and apply fingernail polish to seal.
5. Examine in the UV channel of a fluorescence microscope.

D. Rhodamine–Phalloidin Staining of Actin in Ovarian Tissue

1. Dissect ovaries to individual egg chambers in EBR (removing them from the muscle sheath). Three or four ovaries-worth of egg chambers is a good amount for one microscope slide.
2. Put tissue into microfuge tube with EBR and store on ice until ready to stain.
3. Remove EBR from tissue with drawn out Pasteur pipet.
4. Fix as described under Section IVA.
5. Tap sides of tube so that tissue settles to bottom and then remove fixative with drawn out pipet.
6. Add about 500 μl EBR, let sit for 10 min (tissue will settle), and then repeat EBR rinse.
7. Take 10 μl previously aliquoted Rhodamine–phalloidin (Molecular Probes No. R-415) from $-20°C$ freezer and dry in Speed-Vac about 5 min, and resuspend in 100–200 μl EBR. Do this step as quickly as possible to minimize time exposed to light.

NOTE: PHALLOIDIN IS EXTREMELY TOXIC—WEAR GLOVES AND DO NOT INGEST.

8. Remove EBR from fixed tissue and add 50–100 μl Rhodamine–phalloidin.
9. Let sit in dark for 20 min.
10. Remove phalloidin and rinse twice with PBS.

NOTE: YOU CAN STAIN DNA WITH DAPI (1 μG/ML IN PBS) FOR 5 MIN AND THEN RINSE ONCE.

11. Remove PBS and add about 50 μl glycerol:PBS (50% glycerol) or 50 μl antifade.
12. Mix gently and then pipet onto clean microscope slide with Pasteur pipet.
13. Place glass shards under the coverslip to prevent flattening the egg chambers and fingernail polish to seal.
14. Examine in fluorescence microscope. Slides can be stored at $-20°C$.

E. Antibody Staining in Ovaries

1. Dissect ovaries in EBR and fix as in Section IVA.
2. Remove solution and rinse ovaries with PBS three times and then wash three times for 10 min each.
3. It will help antibody penetration to dissect apart the ovarioles.

4. Pretreat the ovaries by incubating in PBT for 10 min.

5. Add monoclonal supernatant (either undiluted or 1:1 with PBT) or diluted antiserum and incubate at 4°C overnight.

6. Rinse with PBT three times and then wash four times for 15 min each.

7. Add secondary antibody (usually 1:200 in PBT) and incubate 2 hr at room temperature.

8. Rinse with PBT, then wash four times for 15 min each, and rinse twice with PBS.

9. Add PBS:glycerol (1:1) and wait 20 min or so for ovaries to equilibrate.

10. Dissect egg chambers in PBS:glycerol or antifade. Place glass shards under the coverslip to prevent flattening the egg chambers and look.

F. Solutions

1× EBR
(Ringer's Solution)

130 mM NaCl
4.7 mM KCl
1.9 mM CaCl$_2$
10 mM Hepes, pH 6.9

Make a 10× EBR solution, filter sterilize, and freeze in 25-ml aliquots.

Devitellinizing Buffer
Make fresh with each use.

1 vol   Buffer B
1 vol   formaldehyde 36%
4 vol   H$_2$O

Buffer B

100 mM KH$_2$PO$_4$/K$_2$HPO$_4$ (pH 6.8)
450 mM KCl
150 mM NaCl
 20 mM MgCl$_2$ · 6H$_2$O

1× PBS
For 1 liter:

20.0 g   NaCl
 0.5 g   KCl

0.5 g KH_2PO_4
1.47 g Na_2HPO_4 anhydrous
(**or** 2.78 g $Na_2HPO_4 \cdot 2H_2O$)

Make a 10× PBS solution and autoclave.

β-Galactosidase Staining Solution

10.0 mM $NaH_2PO_4 : H_2O/Na_2HPO_4 : 2\ H_2O$ (pH 7.2)
150.0 mM NaCl
1.0 mM $MgCl_2 \cdot 6H_2O$
3.1 mM $K_4[Fe^{II}(CN)_6]$
3.1 mM $K_3[Fe^{III}(CN)_6]$
0.3%   Triton X-100

Just before use, preincubate at 37°C and then add 25 μl 8% X-Gal in DMSO (made fresh) for each 1 ml staining solution.

PBT

1× PBS
0.3% Triton X-100
0.5% BSA (Fraction V from Sigma)

Antifade
For 10 ml stock (keep at 4°C covered in foil):

0.233 g DABCO (1,4-diazabicyclo(2.2.2.) octane) Sigma D2522
800 μl ddH_2O
200 μl 1 M Tris–HCl, pH 8.0
9 ml glycerol

Acknowledgments

We acknowledge the support of Public Health Service Grant GM43301 (L.C.), the Pew Charitable Trusts (L.C.), and NIH Institutional Training Grant T32 HD07149 (E.V.).

References

Berleth, T., Burri, M., Thoma, G., Bopp, D., Richstein, S., Frigerio, G., Noll, M., and Nüsslein-Volhard, C. (1988). The role of localization of *bicoid* in organizing the anterior pattern of the Drosophila embryo. *EMBO J.* **7,** 1749–1756.
Brower, D., Smith, R., and Wilcox, M. (1981). Differentiation within the gonads of *Drosophila* revealed by immunofluorescence. *J. Embryol. Exp. Morphol.* **63,** 233–242.

Cant, K., Knowles, B., Mooseker, M., and Cooley, L. (1994). *Drosophila* singed, a fascin homolog, is required for actin bundle formation during oogenesis and bristle extension. *J. Cell Biol.* **125,** 369–380.

Cooley, L., Verheyen, E., and Ayers, K. (1992). *chickadee* encodes a profilin required for intercellular cytoplasm transport during *Drosophila* oogenesis. *Cell* **69,** 173–184.

Gavis, E., and Lehmann, R. (1992). Localization of *nanos* RNA controls embryonic polarity. *Cell* **71,** 301–313.

Gutzeit, H. O. (1986). The role of microfilaments in cytoplasmic streaming in *Drosophila* follicles. *J. Cell Sci.* **80,** 159–169.

Kim-Ha, J., Smith, J. L., and Macdonald, P. M. (1991). *oskar* mRNA is localized to the posterior pole of the *Drosophila* oocyte. *Cell* **66,** 23–35.

King, R. C. (1970). "Ovarian Development in *Drosophila melanogaster*." New York: Academic Press.

Lantz, V., Ambrosio, L., and Schedl, P. (1992). The *Drosophila orb* gene is predicted to encode sex-specific germline RNA-binding proteins and has localized transcripts in ovaries and early embryos. *Development* **115,** 75–88.

Lasko, P. F., and Ashburner, M. (1990). Posterior localization of vasa protein correlates with, but is not sufficient for, pole cell development. *Genes Dev.* **4,** 905–921.

Mahowald, A. P., and Kambysellis, M. P. (1980). Oogenesis. *In* "The Genetics and Biology of *Drosophila*," ed.), Vol. 2D, pp. 141–224. New York: Academic Press.

Montell, D. J., Rorth, P., and Spradling, A. C. (1992). *slow border cells,* a locus required for a developmentally regulated cell migration during oogenesis, encodes *Drosophila* C/EBP. *Cell* **71,** 51–62.

Price, J. V., Clifford, R. J., and Schüpbach, T. (1989). The maternal ventralizing locus *torpedo* is allelic to *faint little ball,* an embryonic lethal, and encodes the *Drosophila* EGF receptor homolog. *Cell* **56,** 1085–1092.

Schüpbach, T. (1987). Germ line and soma cooperate during oogenesis to establish the dorsoventral pattern of egg shell and embryo in *Drosophila melanogaster*. *Cell* **49,** 699–707.

Spradling, A. (1993). Developmental genetics of oogenesis. *In* "*Drosophila* Development" (M. Bate and A. Martinez-Arias, eds.), pp. 1–70. Cold Spring Harbor, NY: Cold Spring Harbor Laboratory Press.

Theurkauf, W., Alberts, B., Jan, Y., and Jongens, T. (1993). A central role for microtubules in the differentiation of *Drosophila* oocytes. *Development* **118,** 1169–1180.

Theurkauf, W., Smiley, S., Wong, M., and Alberts, B. (1992). Reorganization of the cytoskeleton during *Drosophila* oogenesis: Implications for axis specification and intercellular transport. *Development* **115,** 923–936.

Warn, R., Gutzeit, H., Smith, L., and Warn, A. (1985). F-actin rings are associated with the ring canals of the *Drosophila* egg chamber. *Exp. Cell Res.* **157,** 355–363.

Xue, F., and Cooley, L. (1993). *kelch* encodes a component of intercellular bridges in *Drosophila* egg chambers. *Cell* **72,** 681–693.

Yue, L., and Spradling, A. (1992). *hu-li tai shao,* a gene required for ring canal formation during *Drosophila* oogenesis, encodes a homolog of adducin. *Genes Dev.* **6,** 2443–2454.

PART V

Analysis of Gene Expression *in Situ*

Much current research is focused on the characterization of transcription and splicing factors, the nucleotide motifs that they recognize, and the macromolecular complexes of which they are part. The bulk of such work is carried out by isolating the relevant molecules and describing their structure and function outside of the nuclear context. Alternative approaches enable RNA transcription, splicing, and accumulation to be studied *in situ*. To gauge the value and potential of these latter methods, one need only recall the enormous contributions of hormone- or heat-treated polytene chromosomes to our understanding of coordinate gene regulation and acknowledge the equally noteworthy advances in cell and developmental biology made using *in situ* RNA localization. The four articles in this section illustrate very clearly how one can use *in situ* methods to study transcription, processing, and localization of RNA. These investigations oftentimes reveal aspects of gene regulation overlooked using more biochemical approaches; hence, the two distinct lines of inquiry blend together well.

Andres and Thummel begin the section with a chapter describing methods for studying gene activity during early metamorphosis (Chapter 29). Theirs might be viewed as a ''hybrid'' approach in that it involves rigorously staging larvae and prepupae, culturing hand-dissected imaginal discs (or larval organs), and isolating RNA from whole animals or organs. Included are methods for recognizing larval stages, prevention of larval anoxia, and state of the art RNA preparation. This paper sets the standard for analysis of transcription in intact animals and cultured tissue.

Lehmann and Tautz describe state of the art methods for *in situ* hybridization of RNA (Chapter 30). It is remarkable how far the original nonradioactive method of Tautz and Pfeifle has evolved and how many aspects of developmental biology that it has revealed. Particularly useful in the chapter of Lehmann and Tautz are the methods for double labeling RNA and protein and for simultaneously detecting digoxygenin- and biotin-labeled probes in the same preparation.

Zachar *et al.* next provide instructions for investigating mRNA splicing and transport *in situ* (Chapter 31). They are interested in the relationship between specific genes and subnuclear structures, and toward that end they exploit the very high concentrations of pre-mRNA characteristic of giant polytene nuclei, as well as the availability of particular gene constructs that are abundantly expressed in salivary glands. Zachar *et al.* describe how to prepare nuclear tissue by sectioning or smearing, and also the conditions for maximizing hybridization of single stranded RNA probes. Finally, they illustrate how one can detect either single or double probes hybidized to nuclear preparations. These approaches have the potential to revolutionize our understanding of nuclear compartments and the role they have in nucleic acid synthesis, processing, and transport.

Beyer finally provides protocols for viewing transcription in disrupted nuclei using Miller chromatin spreading methods (Chapter 32). The ability to view nascent transcripts of known structure at high resolution allows the patient observer to deduce the kinetics of processing in either normal or mutant strains. This approach has revealed that in many cases RNA splicing occurs concurrently with transcription and continues to refine our understanding of intron removal. It will be very interesting to use this method to investigate flies that over- or underexpress particular splicing factors.

CHAPTER 29

Methods for Quantitative Analysis of Transcription in Larvae and Prepupae

Andrew J. Andres and Carl S. Thummel

Howard Hughes Medical Institute
University of Utah
Salt Lake City, Utah 84112

I. Introduction

Although *Drosophila melanogaster* has provided a model system for studying the development of higher organisms, the inherent asynchrony of its development poses a significant problem to the investigator, particularly when one is studying the developmental events that occur in rapid succession during the onset of metamorphosis. It is at this time that pulses of the steroid hormone 20-hydroxyecdysone (20E) dramatically reprogram gene expression as they direct the animal through a series of complex behavioral and morphological changes. Initial studies of the effects of 20E on gene expression have depended largely on our ability to observe puffs in the polytene chromosomes of cultured larval salivary glands. Because the endogenous ecdysteroids are washed out during the culturing procedure, the glands are synchronized at a stage in which they are competent to respond fully to the addition of exogenous hormone. The pattern of puffs induced under these conditions closely parallels the puffing response *in vivo* (Ashburner, 1972). Furthermore, cultured salivary glands secrete a polypeptide glue into their lumen, mimicking the glue secretion that normally occurs *in vivo* (Boyd and Ashburner, 1977). Cultured imaginal discs can also be stimulated to undergo the early steps of metamorphosis by exposing them to the proper regime of hormone concentration changes (Fristrom *et al.,* 1986). Under ideal culture conditions, these discs will continue to differentiate into recognizable adult cuticular structures (Martin and Schneider, 1978).

In recent years, the isolation of a number of 20E-regulated genes has allowed their study in different tissues and at different times during development. In this chapter, we discuss the methodologies used to study gene activity during the early stages of metamorphosis. We first describe the methods currently available for staging third instar larvae and prepupae. Gene activity in these animals can then be assessed by isolating RNA from intact animals or dissected organs. Alternatively, extracted organs can be maintained in culture and treated with hormone *in vitro* to induce specific responses, after which RNA can be isolated and analyzed.

II. Staging of Third Instar Larvae and Prepupae

Although first, second, and third instar *Drosophila* larvae can be distinguished based on the number of teeth in their mouth parts (see Ashburner, 1989, pp. 174–175), the inherent asynchrony of larval development makes staging difficult. In our hands, first instar larvae synchronized at hatching vary by as much as 8 hr in the time at which they pupariate, 4 days later. More accurately staged third instar larvae can be obtained by picking animals as they molt from the second instar, but this labor-intensive technique will still result in an asynchronous population of animals at pupariation. A simple means of circumventing the tediousness of this approach is to propagate larvae on food containing 0.05% bromophenol blue. The dye is clearly visible inside the gut of the animal and has no apparent effects on development. As the larvae wander they stop feeding and the blue dye gradually clears from their intestine (Maroni and Stamey, 1983). Three types of wandering third instar larvae can be easily distinguished using this technique: those with dark blue guts, those with lightly stained guts (representing partial clearing of the intestine), and those with an almost completely clear gut (Fig. 1). As shown in Fig. 2, the majority of larvae with dark blue intestines will pupariate in 12–24 hr. Upon dissecting salivary glands from 25 of these animals and examining their puffing pattern, all were found to be at puff stage (PS) 1 (see Ashburner, 1967, for a description of puff staging). The majority of larvae with partially cleared intestines will pupariate in 5–12 hr (Fig. 2) and most are at PS2-6. Although a few animals were found at earlier puff stages, none were later than PS6 (Andres, 1990). Most clear gut larvae will pupariate in 1–6 hr (Fig. 2) and are at PS7-9. Again, no animals were at later puff stages, but a few were at puff stages earlier than PS7 (Andres, 1990). The developmental times will vary depending on the stock and culture

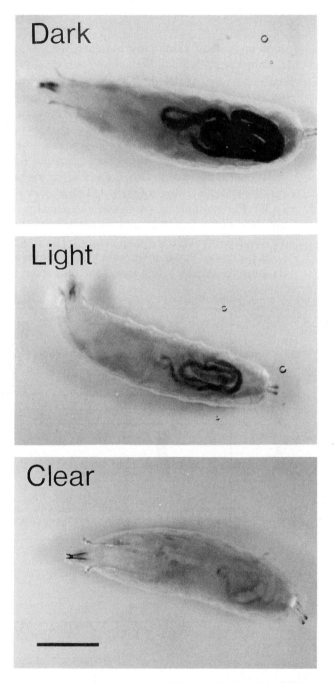

Fig. 1 Gut staging of late third instar larvae following the protocol of Maroni and Stamey (1983). Larvae are propagated under noncrowded conditions on standard medium containing 0.05% bromophenol blue. Examples are shown of a dark blue gut animal, light gut animal, and clear gut animal.

conditions. Best results are obtained when animals are reared at a density of 25 animals/ml medium. We achieve this density by adding 25 females and 10–25 males to a half-pint milk bottle and transferring them to a fresh bottle daily. As is obvious from Fig. 2, this staging technique is not precise in terms of predicting when animals will pupariate. Nevertheless, this approach allows one to easily obtain approximately staged animals from large unstaged populations. Furthermore, one can be sure that animals with dark blue guts are at PS1. These larvae have not yet experienced the high-titer late larval ecdysteroid pulse but are competent to respond fully to the hormone. To obtain reproducible results with later stage animals it is necessary to analyze them in larger groups (>30). For example, six of seven independent isolations of dark blue and clear gut third instar larvae generated similar time courses of *E74A* and *E74B* transcriptional induction (Boyd *et al.*, 1991).

At the end of larval development, the animal stops crawling, shortens, everts its anterior spiracles, and forms the characteristic barrel shape of a pupa. For

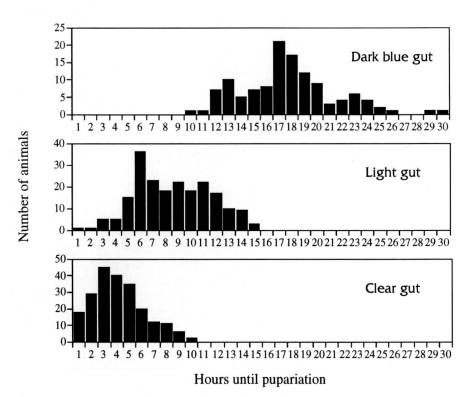

Fig. 2 Timing of pupariation of individual late third instar larvae selected on the basis of gut color. Crawling Canton S third instar larvae maintained at 25°C under noncrowded conditions were selected based on their gut color and observed at 1- to 2-hr intervals to determine when they pupariated. Numbers of animals examined were 121 animals with dark blue guts, 203 animals with light guts, and 218 animals with clear guts.

the first 10–12 hr, until head eversion, this animal is referred to as a prepupa (Ashburner, 1989, pg. 180). From head eversion until eclosion, approximately $3\frac{1}{2}$ days later, the animal is referred to as a pupa. Bainbridge and Bownes (1981) have published a detailed account of *Drosophila* prepupal and pupal development that allows one to stage living animals as they develop.

Prepupae and early pupae can be quite accurately staged by synchronizing animals at the moment of puparium formation. One way to achieve this is to select animals that have not yet tanned. Since tanning of the cuticle is first evident 15–30 min after puparium formation, animals selected in this manner will be relatively close to one another in developmental time. A more precise means of synchronizing prepupae can be achieved by picking animals as they pupariate. Clear gut third instar larvae can be easily observed by placing them on a damp piece of black filter paper in a petri dish. Newly pupariated animals can be identified by examining these animals every 15 min under a dissecting microscope. These animals are transferred with a small paintbrush to another petri dish containing a piece of white filter paper, a circle is drawn around the pupariated animal, and the time noted on the cover of the dish. Animals that have not fully pupariated will crawl out of the marked region and can be returned to the original plate (although this rarely occurs). The selected prepupae can then be allowed to develop for the appropriate time.

III. Organ Culture

In Chapter 6, J. Natzle describes a protocol for mass isolating and culturing imaginal discs from late third instar larvae. Although mass isolation provides a means of obtaining large quantities of mixed or purified larval organs, these materials are not ideal for small-scale studies. For example, RNA isolated from small samples of tissues that are individually selected out of a mixture of cultured larval organs is of relatively low quality (A. A. and C. S. T., unpublished results). This is probably due to the damage inherent in the mass-isolation procedure. Consequently, when smaller amounts of material are sufficient, or when organs are isolated from stages in which mass isolation is not possible (such as prepupal stages), hand dissection remains the preferred approach.

Crawling larvae with dark blue intestines are collected, washed briefly with water, and dissected in oxygenated Robb's saline at 25°C (Robb, 1969). Robb's saline can be efficiently oxygenated by exposing it to a stream of oxygen bubbles using an aquarium airstone or other diffuser. This step appears to be critical for avoiding a stress response due to anoxia, which can diminish both RNA and protein synthesis (F. Karim, unpublished results). The larvae are dissected in a Pyrex plate (Corning 7220) by grasping them on both sides of the pharynx with fine forceps and peeling off the cuticle. If necessary, organs can be teased away from the cuticle using finely sharpened tungsten needles. The larval organs and attached cuticle from at least six animals are then transferred to a fresh well or small petri dish that has been pretreated with 1% BSA. We have found the 12- or 24-well flat bottom assay plates (Falcon 3043, 3047) to be convenient for small-scale organ culture. If individual dissected organs are being examined, they can be easily incubated in 1.5-ml microcentrifuge tubes using 50–100 μl of Robb's saline. The use of microcentrifuge tubes at this point precludes the need to transfer the organs later for RNA extraction. These tubes should, however, be centrifuged at low speeds (<1000 rpm) to avoid tissue damage. The organs are washed several times with an excess of oxygenated Robb's saline. Oxygenated Robb's saline is then added to each culture and the organs are placed in a small styrofoam box at 25°C, into which a constant slow flow of oxygen is pumped. Sufficient Robb's saline should be used for all incubations to completely cover the organs while still maximizing the surface-to-volume ratio for optimum gas exchange. After incubating for 1 hr, to allow the organs to recover from any previous hormone exposure, the medium is replaced with oxygenated Robb's supplemented with 20E (Sigma H5142, dissolved in ethanol; 5×10^{-6} M is used for a maximal response). Cycloheximide can also be used to block protein synthesis, if desired. Addition of cycloheximide to 7×10^{-5} M is sufficient to block 98% of protein synthesis in cultured larval salivary glands while still allowing good recovery of protein synthesis upon washout (Ashburner, 1974). After the appropriate incubation, the organs are resuspended in Robb's saline and transferred to microfuge tubes which are centrifuged briefly at full speed to pellet the organs. The medium is then removed and RNA is extracted as described below.

IV. RNA Isolation

We use two methods to extract RNA from organs or intact animals. Both procedures yield samples that contain both DNA and RNA. Since DNA represents only ~10% of the total nucleic acids, however, it is of no consequence for subsequent Northern blot hybridization analysis. The DNA does not appear to transfer well by blotting, remains at the top of the blot, and is usually not detected by hybridization. If desired, DNA can be removed from these samples by poly(A) selection or digestion with RNase-free DNase. Both RNA extraction procedures are rapid and simple and allow one to routinely isolate intact RNA of high quality. In our hands, they have been more reliable than other procedures such as solubilization with guanidinium thiocyanate and pelleting through CsCl. The first protocol relies on incubation with proteases in the presence of SDS to remove the proteins in the extract prior to phenol extraction. This procedure can be easily scaled up to allow the extraction of larger quantities of material. The second protocol is more difficult to adapt for large-scale extractions, but is ideally suited for small-scale rapid RNA extraction.

A. RNA Isolation Using SDS Lysis Buffer

This protocol was modified from that described as Method II in Schulz *et al.* (1986).

1. Prepare a 2.5 mg/ml stock of Proteinase K (Boehringer) or Protease Type XI (Sigma) in water. Incubate at 37°C for 30 min to digest any contaminating nucleases. Filter-sterilize and store at 4°C for up to 1 month.

2. Homogenize up to 25 animals in 0.5 ml of SDS lysis buffer (20 mM Tris, pH 7.5, 0.2 M NaCl, 20 mM EDTA, 2% SDS). One can use either a small glass Dounce with a B pestle or a homogenizer that fits into the bottom of a 1.5-ml microcentrifuge tube. The blue disposable pestles available from Baxter (T4001-6) will fit the bottoms of most pop-top (but not screw-cap) microcentrifuge tubes. At this point the extract can be stored for up to 1 month at −20°C.

3. Transfer to a screw-capped microcentrifuge tube (to prevent leaking during vortexing in the next step) and add 50 μl of Proteinase K (a final concentration of 250 μg/ml). Incubate at 50°C for 1 hr.

4. Add 250 μl phenol (TE saturated, with an aqueous phase that is approximately neutral pH) and mix using a multiple tube vortexer at high speed for at least 5 min. Add 250 μl of chloroform and vortex again for ≥5 min. If a multiple tube vortexer is not available, the tubes can be strapped onto a standard vortexer.

5. Centrifuge for 5 min to separate phases.

6. Transfer the aqueous phase to a 1.5-ml microcentrifuge tube, add 0.5 ml chloroform, and vortex for ≥5 min.

7. Transfer the aqueous phase to a 1.5-ml microcentrifuge tube, add 1 ml RNase-free ethanol, and precipitate the nucleic acids by chilling at −20°C for >2 hr or −80°C for 30 min. Centrifuge for 5 min at 5°C to pellet the RNA.

8. Remove the supernatant, dry the pellet, and resuspend in RNase-free water to give a final concentration of 2–4 μg/μl. To determine the exact concentration, dilute 1 μl into 1 ml water and read the OD_{260}. Multiply by 40 to get the concentration in μg/μl. Store at −80°C or (for long-term storage) under liquid nitrogen. We have successfully retrieved intact RNA from tubes stored in liquid nitrogen for 5 years. Table 1 shows the micrograms of total nucleic acid that can be extracted from different larval tissues, as an average amount obtained per larva.

B. RNA Isolation by Direct Phenol Extraction

This procedure was originally described by K. Burtis (1985) and can be used for as little as a single imaginal disc to as many as 25 animals.

Table I
Estimated Yields (in μg) of Total Nucleic Acids
Extracted from a Single Larva Using SDS Lysis Buffer

Intact larva	10	Carcass*	2.3
Fat body	1.8	Hindgut	0.13
Hemolymph	0.2	Midgut	1.6
Malpighian tubules	0.18	Proventriculus	0.11
Pharynx	0.13	Trachae	0.13
Brain/ventral ganglion	0.9	Imaginal discs	0.6
Salivary glands	0.7		

Note. These amounts were derived by isolating tissues from a large number of animals. They are represented here as micrograms of nucleic acid per single animal.

* The carcass consists of tissues not easily dissected from the animal, including body muscles, peripheral nervous system, hemocytes, and epidermis.

1. The material to be extracted is transferred, with forceps, to either a 7-ml glass Dounce or a 1.5-ml microcentrifuge tube, and 400 μl of RNA extraction buffer is added (0.1 M Tris base, 0.1 M NaCl, 20 mM EDTA, 1% n-lauryl sarcosine). The mixture is then immediately homogenized with either a B pestle for the glass Dounce or a pestle that fits into the bottom of the microcentrifuge tube. The blue disposable pestles available from Baxter (T4001-6) will fit the bottoms of most pop-top (but not screw-cap) microcentrifuge tubes. Immediately add 400 μl phenol (TE saturated, with an aqueous phase that is approximately neutral pH) and either continue homogenization for at least 10 strokes or vortex thoroughly.

2. Centrifuge to separate phases. Repeat the phenol extraction. At this point, there is usually no detectable interphase. If there is, continue to phenol extract until the interphase is gone.

3. Extract with ether three times.

4. Add 1 ml cold RNase-free ethanol and let sit at room temperature for ≥30 min. Do not chill or other contaminants will precipitate out of solution. Centrifuge for 5 min to pellet the RNA. Dry the pellet, resuspend in water, and store at −80°C or under liquid nitrogen. It is best to resuspend the RNA samples in water to give a final concentration of 2–4 μg/μl. To determine the exact concentration, dilute 1 μl into 1 ml water and read the OD_{260}. Multiply by 40 to get the concentration in μg/μl.

Acknowledgments

We thank the members of the Thummel lab, past and present, for their contributions to these protocols. In particular, we thank Felix Karim and Lynn Boyd for their contributions to staging larvae, prepupae, F. Karim for his protocol on culturing hand-dissected larval organs, and Pam Reid for her help with the data presented in Fig. 2.

References

Andres, A. J. (1990). An analysis of the temporal and spatial patterns of expression of the ecdysone-inducible genes *EIP28/29* and *EIP40* during development of *Drosophila melanogaster*. Ph.D. thesis, Indiana University.

Ashburner, M. (1967). Patterns of puffing activity in the salivary gland chromosomes of *Drosophila*. I. Autosomal puffing patterns in a laboratory stock of *Drosophila melanogaster*. *Chromosoma* **21**, 398–428.

Ashburner, M. (1972). Patterns of puffing activity in the salivary gland chromosomes of *Drosophila*. VI. Induction by ecdysone in salivary glands of *D. melanogaster* cultured in vitro. *Chromosoma* **38**, 255–281.

Ashburner, M. (1974). Sequential gene activation by ecdysone in polytene chromosomes of *Drosophila melanogaster*. II. The effects of inhibitors of protein synthesis. *Dev. Biol.* **39**, 141–157.

Ashburner, M. (1989). "*Drosophila: A Laboratory Handbook*." Cold Spring Harbor, NY: Cold Spring Harbor Laboratory Press.

Bainbridge, S. P., and Bownes, M. (1981). Staging the metamorphosis of *Drosophila melanogaster*. *J. Embryol. Exp. Morphol.* **66**, 57–80.

Boyd, L., O'Toole, E., and Thummel, C. S. (1991). Patterns of *E74A* RNA and protein expression at the onset of metamorphosis in *Drosophila*. *Development* **112**, 981–995.

Boyd, M., and Ashburner, M. (1977). The hormonal control of salivary gland secretion in *Drosophila melanogaster*: Studies *in vitro*. *J. Insect Physiol.* **23**, 517–523.

Burtis, K. C. (1985). Isolation and characterization of an ecdysone-inducible gene from *Drosophila melanogaster*. Ph.D. thesis, Stanford University, Stanford, CA.

Fristrom, J. W., Alexander, S., Brown, E., Doctor, J., Fechtel, K., Fristrom, D., Kimbrell, D., King, D., and Wolfgang, W. (1986). Ecdysone regulation of cuticle protein gene expression in *Drosophila*. *Arch. Insect Biochem. Phys.* (**Suppl.**) **1**, 119–132.

Maroni, G., and Stamey, S. C. (1983). Use of blue food to select synchronous late third instar larvae. *Drosophila Information Service* **59**, 142–143.

Martin, P., and Schneider, I. (1978). *Drosophila* organ culture. *In* "The Genetics and Biology of *Drosophila*" (M. Ashburner and T. R. F. Wright, eds.), Vol. 2a, pp. 219–264. New York: Academic Press.

Robb, J. A. (1969). Maintenance of imaginal discs of *Drosophila melanogaster* in chemically defined media. *J. Cell Biol.* **41**, 876–885.

Schulz, R. A., Cherbas, L., and Cherbas, P. (1986). Alternative splicing generates two distinct *Eip28/29* gene transcripts in *Drosophila* Kc cells. *Proc. Natl. Acad. Sci. U.S.A.* **83**, 9428–9432.

CHAPTER 30

In Situ Hybridization to RNA

Ruth Lehmann* and Diethard Tautz†

* Whitehead Institute for Biomedical Research
Howard Hughes Medical Institute
Massachusetts Institute of Technology
Cambridge, Massachusetts 02142

† Zoologisches Institut
Universität München
D-80333 München
Germany

I. Introduction

The pattern of RNA distribution in tissues and embryos can be studied directly by *in situ* hybridization. Original techniques relied on radioactively labeled (^{35}S and ^{3}H) probes and hybridization to tissue sections. This technique, although applicable to any tissue, has many disadvantages: the exposure time of the hybridized material using photographic film emulsion covering the material can be quite long (depending on the abundance of the detected RNA and isotope used, from several days to weeks and months) and the analysis of the distribution of RNA within a whole animal or organ is difficult since it requires three-dimensional reconstruction of the sectioned material.

There is hardly a developmental study in *Drosophila* that does not apply whole-mount *in situ* hybridization techniques using nonradioactively labeled probes. This hybridization technique is extremely sensitive—able to detect even minute amounts of RNA—and its resolution allows single-cell analysis. The hybridization protocol is short (2 days), and observation of whole-mount embryos and tissue allows a direct three-dimensional image of the distribution pattern. In addition, for closer observation the tissue can be sectioned after the staining reaction is complete.

The technique of *in situ* hybridization using nonradioactively labeled probes was originally described by Tautz and Pfeifle (1989). This original protocol has been modified by several investigators, specifically regarding the production of more sensitive probes such as RNA probes and probes derived from small amounts of template using PCR. These modifications have been summarized in recent publications by Tautz (1992), Tautz *et al.* (1992), Patel and Goodman (1992), and Artero *et al.* (1992). As more developmentally important gene products are analyzed on the molecular level in *Drosophila*, it is often desirable to analyze more than one gene product in parallel. Cohen and Cohen (1992) have described methods to visualize RNA and protein in the same embryo using nonradioactive *in situ* hybridization techniques combined with enzyme-linked immunodetection of protein.

In the following, we will first describe basic protocols that deal with the principles of RNA hybridization. We will discuss various probe preparation methods and specific applications for detection of multiple products.

II. Embryo Collection and Fixation

The sensitivity and quality of RNA *in situ* hybridization critically depends on the condition used for fixation and pretreatment of embryos. Before fixation, the chorion and the vitelline membrane are removed from embryos. Dechorionation is achieved by treating embryos with commercial bleach. Devitellinization is achieved by osmotic shock by transferring embryos from a heptane/formaldehyde interface to methanol. The efficiency of devitellinization depends on the age (younger embryos and unfertilized eggs in general are more difficult) and sometimes even on the maternal genotype.

Optimal conservation of morphology by fixation has to be balanced by allowing the probe to properly penetrate the tissue and efficiently hybridize to the RNA. Fixation is achieved by treating the embryos with formaldehyde freshly prepared from paraformaldehyde. Thirty-seven percent formaldehyde in solution can also be used; however, longer fixation times may be required once formaldehyde oxidizes in an opened bottle. After fixation, tissue penetration is achieved by proteinase K treatment. The concentration and timing of this treatment are critical and should be determined for each batch of proteinase K and may differ for different tissues and different target RNA species. A treatment that is too short may result in high background and poor hybridization, while a treatment that is too long may result in poor morphology or even disintegration of the embryos.

A. Protocol 1A: Removal of Chorion and Vitelline Membrane, Fixation, and Hybridization Pretreatment of Embryos

1. Removal of Chorion and Vitelline Membrane

   1. Collect embryos from apple-juice agar plate (Wieschaus and Nüsslein-Volhard, 1986) with paintbrush and transfer into basket (made by melting the top 6–8 mm of a microcentrifuge tube to a fine metal mesh). All reactions and washes are done with 20–50 μl of settled embryos in microcentrifuge tubes.
   2. Rinse embryos in H_2O.

3. Transfer basket to 50% sodium hypochlorite (commercial bleach) for about 2–3 min.

4. Wash thoroughly with dH_2O and then rinse in 0.7% NaCl, 0.02% Triton X-100 and blot to remove excess liquid. Check under dissection scope that chorion is removed; dechorionated embryos appear shiny and are hydrophobic.

5. Fix embryos in a glass scintillation vial for 20 min by gently shaking in formaldehyde/heptane fixative.

6. Remove lower formaldehyde phase completely with a drawn-out pasteur pipet.

7. Remove most of the heptane (upper phase) and replace with 8 ml of fresh heptane; repeat this step once.

8. Add 8–10 ml of MeOH; shake vigorously for 30–60 sec.

9. Remove heptane phase (upper phase) and interphase (contains vitelline membranes).

10. Transfer embryos that have fallen to the bottom of the vial (devitellinized) with pasteur pipet or blue pipet tip with cut-off tip to a 1.5-ml microcentrifuge tube.

11. Wash in MeOH, 2–3 times over a 10-min period.

Although best sensitivity is obtained using freshly fixed embryos, embryos can be stored at −20°C at this step.

2. Prehybridization Treatment

All following steps are carried out in a 1.5-ml microcentrifuge tube with a solution volume of 1 ml on a rotating platform or on a rotator unless otherwise indicated.

12. Rehydrate into PBT, 3–5 min each step:

    7:3 MeOH:PBT
    5:5 MeOH:PBT
    3:7 MeOH:PBT
    100% PBT

(Alternatively aspirate MeOH such that 500 μl are left, add 500 μl PBT, rock vial 5–10 min, settle down aspirate, add PBT.)

13. Post-fix in 1 ml PBT + 4% formaldehyde for 15 min on rotating shaker. (This step is usually omitted when using DNA probes, but is advantageous for RNA probes, since the treatment of the embryos is a little more harsh.)

14. Wash in PBT for 5 × 5 min.

15. Incubate 2–5 min in 1 ml PBT with 4–30 μg proteinase K. Let embryos settle down for 1 min. The length of treatment and concentration of protease should be determined for each batch of proteinase K, since its activity can vary. Embryos that have been freshly fixed may need shorter proteinase K digestion times.

16. Stop by rinsing embryo 2 × 5 min with PBT (Optional: include 2 mg/ml glycine to ensure inhibition of proteinase K activity).

17. Refix in 4% formaldehyde in PBT for 20 min while rocking.

18. Wash in PBT for 5 × 5 min.

B. Protocol 1B: Mechanical Devitellinization Methods (after Tautz)

Devitellinization of *Drosophila* eggs and embryos with the above method is usually 50–70% effective. However, the vitelline membranes of other insects can be more difficult. Alternative methods are therefore required. For *Tribolium,* the following mechanical method has been used successfully (Sommer and Tautz, 1993): After fixation and methanol treatment the embryos are passed several times through a hypodermic needle (1.1 mm diameter). After devitellinization continue with Step 11.

C. Protocol 1C: Cold Shock Devitellinization Methods (after Sander)

This method was originally developed in the laboratory of Klaus Sander and was found to be useful for a variety of different insect species, including lower dipterans, *Tribolium,* and crickets.

1. Dechorionate and fix embryos as described above (fixation time should be extended to up to 40 min for larger embryos).

2. Remove fixation buffer and heptane.

3. Add 100% EtOH.

4. Wash twice for 10 min in EtOH.

5. Transfer into a microcentrifuge tube with screw cap.

6. Leave the embryos for 1–2 days at −80°C. Keep the tube in a horizontal position to avoid clumping of the embryos.

7. Throw the tube directly into a 65°C waterbath and shake it vigorously. The temperature of the waterbath or heat block may be up to 100°C. (If devitellinization was not successful, freeze the tube again, either in dry ice/ethanol or in liquid nitrogen, and repeat the heat shock.)

8. Transfer embryos into a larger tube.

9. Add PBT and EtOH to obtain a solution of 50% EtOH.

10. Add equal volume of heptane and shake vigorously.

11. Collect the embryos that sank to the bottom. Embryos that are at the interface may also be collected but need to be treated again.

After devitellinization, continue with Step 11.

D. Solutions

4% Formaldehyde from Paraformaldehyde

Add 1 g of paraformaldehyde to 12.5 ml of dH$_2$O.
Cover beaker with aluminum foil and stir on hot plate in fume hood to 60°C.
Add 1–2 drops of 1 *N* NaOH; the solution should clear at pH 9.4.
Add 12.5 ml 2× PBS and adjust pH if necessary to 7.2–7.4.
Fresh formaldehyde can be used for several days.

Formaldehyde/Heptane Fixative

2 ml 4% formaldehyde in PBS (freshly prepared from paraformaldehyde)
0.2 ml DMSO (optional for better penetration of dense material, ovaries, older embryos, imaginal discs)
8 ml heptane

PBS

130 m*M* NaCl
7 m*M* Na$_2$HPO$_4$
3 m*M* KH$_2$PO$_4$, pH 7.2

PBT

PBS with 0.1% (V/V) Tween 20

Proteinase K

Lyophilized from Boehringer Mannheim: Prepare 10-μl aliquots (4–30 mg/ml) and keep at −80°C; use fresh dilution of 1000× aliquot each time.
Alternatively use 20 mg/ml solution from Boehringer Mannheim at a final concentration of 15–30 μg/ml. This solution is stable at 4°C.

E. Trouble Shooting

1. Inefficient Devitellinization

Do not expose the embryos to bleach more than just necessary; if bleach is fresh, dilute 1:4 or more; stop treatment in time and wash very well afterward.

Make sure to remove formaldehyde phase completely and to replace heptane phase with fresh heptane before adding methanol.

Try longer fixation period if compatible with *in situ* hybridization signal.

III. Hybridization

Hybridization conditions are chosen to preserve tissue morphology and allow sensitive annealing of the probe to the target RNA. Formamide is included in the hybridization solution to allow for lower hybridization temperatures while maintaining high stringency of hybridization. The final formamide concentration of 50% is achieved stepwise to avoid disintegration of the embryos, which can be quite brittle after proteinase K treatment. To maintain good morphology while using high (65°C) hybridization temperature, the SSC/formamide solution is adjusted to pH 5. To reduce nonspecific hybridization of the probe ("background"), heparin and sonicated salmon sperm DNA are added to the hybridization solution. Salmon sperm DNA is denatured with the probe to compete for repetitive sequences often found in *Drosophila* genes. Washes can be carried out either in the hybridization buffer (Hb-A) or in a simplified washing/hybridization buffer (Hb-B) as indicated.

The hybridization Protocol 2 should be used as a guideline. Factors such as temperature and time of hybridization, probe concentration, and number and duration of washes should be optimized individually.

A. Protocol 2: Hybridization

1. Wash embryos in 1:1 hybridization buffer (Hb-B) and PBT for 5 min at 500 μl each.
2. Replace with 250 μl Hb-B.
3. Prehybridize at least 1 hr at 45°C (RNA probes higher T: 56–65°C) in 250 μl Hb-A.
4. Heat probe 5 min at 100°C (DNA) and 80°C (RNA) and then chill on ice.
5. Remove as much Hb-A as possible from embryos and add probe in Hb-A (see Protocols 5–8). Mix well by pipetting or flicking tube.
6. Hybridize 12 hr overnight. For DNA probes, hybridize at 45°C and RNA probes, hybridize at 56°C, or for higher stringency at 65°C, pH 5.
7. Remove probe and wash with 500 μl of fresh Hb-B at least 1 hr at 45°C (RNA probes higher T: 56–65°C.). Increasing the number of changes of Hb-B and length of washing can decrease background significantly (e.g., for 4–5 hr with 4–6 changes of Hb-B).
8. Remove Hb-B, in steps 3–5 min each, at RT:

    8:2 Hb-B:PBT

5:5 HB-B:PBT

2:8 Hb-B:PBT

Wash in PBT for 5 × 5 min.

B. Solutions

Hybridization Buffer (Hb-A)

50% deionized formamide

5× SSC (Sambrook *et al.* 1989)

100 μg/ml sonicated salmon sperm DNA (DNA aliquots (10 mg/ml) at −20°C)

50 μg/ml heparin (Sigma H3125, 100 mg/ml heparin stock (−20°C))

0.1% Tween 20

Store at −20°C and prewarm to hybridization temperature before use.

Options

1. For RNA probes, hybridization can be carried out at 65°C; if pH of hybridization solution is low, prepare 50% formamide and 5× SSC and adjust to pH 5 with HCl (use pH paper) and then add salmon sperm and heparin.
2. If probe preparation does not contain tRNA, add 10 μg/ml tRNA (20 mg/ml aliquots at −20°C).

Washing/Hybridization Buffer (Hb-B)

50% formamide, 5× SSC

C. Trouble Shooting

Good staining and good morphology are almost mutual exclusive. Hybridization conditions (duration, pH, temperature) and protease treatment have to be balanced such that the embryos just do not fall apart. To ease keeping the balance, prepare large batches of PBT, fixative, Hb-A, probes, embryos, and protease aliquots in advance.

1. Embryos Dissolve during Hybridization

Make sure hybridization solution is pH 5 if temperature is 65°C. Embryos fall apart if pH is 7.5. Otherwise, use lower temperature (56°C).

Vary proteinase K treatment: too low concentration causes weak signal and high background. Too high concentration causes loss of signal and bad morphology.

Variable results may be due to variations in room temperature, protease-lot, or time of previous storage of embryos in methanol (try half as much protease for fresh embryos).

Keep hybridization time at 65°C below 12 hr.

2. Embryos Clump Together during Washes after Hybridization or during Staining

If clumping becomes apparent during washes after hybridization, swirl embryos gently by moving the vials by hand from time to time; do not rock on shaking platform. Make sure embryos are not deformed by surface tension when trapped between air bubble and plastic.

IV. Detection

In the following we will describe only detection of digoxigenin-labeled probes. Detection of biotin-labeled probes as well as alternate substrates are described later (Protocol 11). Digoxigenin-labeled probes can be detected by antibodies coupled to alkaline phosphatase, peroxidase, or fluorescent dyes (see Tautz *et al.*, 1992 and Digoxigenin Application Manual, Boehringer Mannheim). Alkaline phosphatase seems most sensitive and different color substrates can be used. The most sensitive is 4-nitroblue tetrazolium chloride (NBT) plus X-phosphate, which produces a blue color after dehydration.

Color development usually occurs in a few minutes and should be observed under a stereomicroscope. If no significant background develops, the reaction can be developed for hours or even overnight at 4°C. Prior to use, the antibody can be preabsorbed; this may not be strictly necessary and may depend on the tissue or the particular antibody used.

A. Protocol 3: Preabsorption of Antibody and Staining

1. Before use, dilute anti-digoxigenin antibody (Ab) 1:200 in PBT and preabsorb against nonhybridized, fixed embryos (>200 μl embryos/ml of Ab solution, 4°C overnight). Preabsorbed Ab may be stored for several months at 4°C and diluted 1:10 for use. Antibody diluted 1:2000 may be reused at least twice.

2. Incubate in 0.5 ml of anti-digoxigenin antibody conjugate, 1:2000 in PBT, for 1 hr at RT with rocking.

3. Wash in PBT for 5 × 5 min.

4. Wash in staining solution for 2 × 5 min.

5. Transfer embryos to a crystallizing dish in 1 ml of developing solution. Add:

4.5 μl NBT (75 mg/ml in 70% dimethylformamide (DMF))

3.5 μl X-phosphate (50 mg/ml in DMF)

6. Stop reaction by washing well with PBT. Store embryos in PBT at 4°C until ready to embed.

B. Solutions

Staining Solution

0.1 M NaCl
0.1 M Tris–NCl, pH 9.0
0.05 M MgCl$_2$
0.1% Tween 20
1 mM Levamisol (inhibitor of endogenous phosphatases (optional))

C. Trouble Shooting

The staining reaction is very efficient. It is important, however, to prepare fresh staining buffer and to mix solution thoroughly each time after another component has been added. Further, once the reaction is complete, the embryos should be washed thoroughly.

1. Precipitates Form during Staining Reaction

Colorless, amorphous precipitates form if staining buffer is contaminated with PBT.

Yellow/brown needles sometimes form when staining goes overnight.

Little blue crystals in embryos: thoroughly wash embryos after staining. Store stained embryos before embedding in 70% EtOH, at RT, or in water rather than in PBS.

2. High Background/Low Signal

Negative controls: Hybridize embryos with deletion for the gene in question to control for nonspecific hybridization. Stain embryos that have not been hybridized to detect nonspecific binding of anti-digoxigenin antibody. If antibody binds nonspecifically, thoroughly preabsorb the antibody (two times if necessary; overdoing it results in reduced activity).

Reduce amount of probe. Check hydrolysis of probe. Check that probe recognizes insert of vector only.

Increase protease concentration and duration of hybridization, or vary pH of Hb-A (pH 5–6.5). Increase temperature during hybridization.

For low-abundant mRNAs DEPC treatment of PBT and sterile working might be important; gloves are not absolutely necessary if you take care not to touch inside of vials.

RNase treatment after the hybridization does not help: it reduces signal and increases background (even at 5 µg/ml).

V. Mounting Techniques

After detection, embryos can be embedded permanently in mounting media that will harden and clear the embryo. These procedures usually require dehydration of the embryos. Alternatively, embryos can be mounted in glycerol for immediate observation.

Each of the three protocols (A–C) described has its advantages and disadvantages. Embryos embedded in glycerol or GMM will not harden in the medium. This is somewhat cumbersome for storage of preparations but has advantages for photography, since position of embryos can be carefully adjusted after embedding. The cellular morphology usually looks better when embryos are embedded in glycerol. The dehydration necessary for embedding in GMM, araldite, Permount, or Euparal will change the brown color of the NBT-X phosphate staining to blue. Embryos embedded in GMM clear after mounting and internal structures are more visible. In Permount or Euparal, embryos are slightly less transparent, but Nomarski optics is much better. Embryos embedded in araldite can be sectioned for more detailed observation (see, for example, Leptin and Grunewald, 1990).

A. Protocol 4A: Embedding in Glycerol

1. Transfer embryos into microcentrifuge tube with 70% glycerol.
2. Equilibrate overnight.
3. Transfer embryos (with cut-off blue pipet tip) onto microscope slide.
4. Place coverslip over embryos.

B. Protocol 4B: Embedding in Gary's Magic Mounting Medium (GMM), Permount, or Euparal

1. Dehydrate embryos for 10 min each in 70, 90, and 100% ethanol.
2. For Euparal and Permount (Fisher scientific), replace EtOH with xylene for 2 × 5 min.
3. Transfer with as little ethanol (xylene) as possible to microscope slide into a drop of approximately 300–500 µl of GMM, Euparal, or Permount.
4. Mix embryos into embedding medium.
5. Let ethanol (xylene) evaporate (2–5 min) and place coverslip (22 × 40 mm) on embryos.

C. Protocol 4C: Embedding and Sectioning Using Plastic Resin

1. Dehydrate embryos in EtOH, 5 min each step: 30, 50, 70, 90, 95, and 100%.
2. Wash in 100% EtOH for 2 × 5 min.
3. Wash two times in acetone for 30 sec.
4. Remove most acetone and replace with ~200 μl araldite.
5. Remove acetone floating on top of araldite (otherwise embryos may destain).
6. Transfer to microscope slide with a cut-off blue pipet tip, keeping embryos moving to prevent sticking. Before placing coverslip onto embryo, make sure that all acetone is evaporated.
7. To prevent crushing of the embryos by coverslip, cut coverslip (22 × 50 mm) into 22 × 5-mm strips and place a strip on either side of the araldite drop and then place a coverslip (22 × 40 mm) on top so that it spans the two spacers.
8. Polymerize at 65°C overnight.
9. For sectioning, transfer embryos from ethanol (step C2) to acetone 2 × 2 min. Make sure that acetone and 100% EtOH are dry.
10. Transfer into 1:1 mixture of dry acetone and araldite in watchglass and let acetone evaporate overnight in fume hood. Replace with 100% araldite.
11. Embryos are placed into araldite in embedding forms and polymerized at 60°C for 2 days. Sections mounted in araldite can be viewed using Nomarski optics without further staining.

D. Solutions

GMM

1.5–2 g Canada balsam per milliliter of methylsalicylate (Lawrence *et al.*, 1986)

Araldite

LX112 resin from Ladd Research Industries
Mix components A and B at a 6:4 ratio for medium hardness.
Store mixture in 20-cc syringes at −20°C.

E. Trouble Shooting

Embryos shrink, bend, and darken upon contact with medium or air during embedding. Go from EtOH to medium in small steps (10, 20, 40, 80, of medium in EtOH).

Embedding: First spread embryos on slide and then add embedding medium around them and gently mix (and distribute) by swirling slide. If the embryos shrink before embedding medium is added, there is still too much ethanol in the medium.

Pink embryos: Wash embryos thoroughly after staining and store embedded preparation in the dark.

Bleaching of precipitate after embedding in araldite: Avoid remnants of acetone. For good penetration, use small steps from acetone to araldite (50:50, 20:80, 0:100). Let acetone evaporate overnight.

VI. Probe Preparation

Four different methods are used to label probes with digoxigenin modified nucleotides: (1) Synthesis of labeled dsDNA by random oligonucleotide priming (Protocol 5), (B) Synthesis of antisense RNA by *in vitro* transcription (Protocol 6) (C) Synthesis of antisense ssDNA by PCR (Protocol 7), and (D) End-labeling of specific oligonucleotides (Protocol 8).

A. Protocol 5: Probe Labeling by Random Priming

The random priming reaction is very reproducible and reliable (Feinberg and Vogelstein, 1983). It can be applied to any linearized DNA. Although purified fragments are the best templates, linearized DNA from whole plasmid minipreparations and λ-phage preparations can be used. The size of the template can vary between 100 nucleotides and 40 kb, although best results are obtained with templates that contain at least 1 kb of shared sequences with the target RNA. The principle is to anneal random hexamer primers to the denatured DNA template and to extend strand synthesis by incorporating dUTP modified with digoxigenin (dig-dUTP). The concentration of primer will determine the length of the fragments generated. The concentration of primers is generally somewhat higher for *in situ* hybridization than in other protocols.

1. Use 200–400 ng of purified DNA fragment for each reaction.
2. Denature DNA at 100°C for 5 min and then chill rapidly in an ice water bath.
3. Labeling reaction:

   13 μl denatured DNA

   2 μl 10× LB (0.5 M Tris–HCl, pH 7.2, 0.1 M MgCl$_2$, 1 mM DTT, 2 mg/ml BSA)

   2 μl dig-11-dUTP/dNTP labeling mix (Boehringer)

   2 μl 6 mg/ml hexamers (Pharmacia)

   1 μl Klenow (2–5 U/μl)

4. 0°C for 10 min
   15°C for 1 hr
   RT for 3 hr

5. Add EDTA to 10 mM, 10 μg yeast tRNA, and EtOH precipitate. Wash pellet with 70% EtOH and resuspend in 100 μl HB. Store at -20°C.

6. Add 10 μl of probe to 30–40 μl of embryos.

B. Protocol 6: Riboprobes (after Jiang *et al.*, 1991, and Gavis and Lehmann, 1992)

Labeled RNA probes are prepared by transcribing DNA *in vitro*. cDNA sequences are inserted into transcription vectors, e.g., Bluescript (Stratagene) or pGEM (Promega) such that the fragment is flanked by the SP6, T3, and T7 promoters. Depending on the type of RNA probe desired, sense and antisense probes can be generated from the same vector in a run-off transcription reaction using T7, T3, or SP6 polymerase (Melton *et al.*, 1984). (Dig-UTP) is used as a substrate for RNA synthesis. Single-stranded probes produced by *in vitro* transcription are more sensitive than those produced by random-primed labeled DNA. Another advantage of the riboprobe method is the possibility of using the sense strand as a control for nonspecific versus uniform hybridization.

1. The Boehringer Genius 4 kit may be used to synthesize the dig-11-UTP riboprobe, following the instructions (Section 11A, Steps 1–4) provided with the kit. Use about 1 μg of template, which should yield 5–10 μg of labeled RNA.

2. To determine how much RNA was synthesized, use 1 μl of a 20-μl reaction and dilute in 5× SSC. Heat and chill. Spot 1 μl of each dilution on nylon membrane. UV cross-link, block, and stain the membrane according to the instructions included in the detection kit. For reference, a dilution series of dig-labeled control RNA provided with the kit can be used.

3. Since the RNA is synthesized as a run-off transcript, the probe size has to be decreased after transcription. After synthesis of the RNA in a 20 μl reaction, add 30 μl dH$_2$O and 50 μl 2× carbonate buffer.

4. Incubate 40 min at 65°C.

5. Add 100 μl 0.2 M NaOAC and 1% acetic acid, pH 6.0, to stop reaction.

6. Precipitate RNA by adding

   20 μl 4M LiCl
   10 μl 20 mg/ml tRNA (phenol/CHCl$_3$ extracted)
   600 μl EtOH

   at -20°C for at least 15 min.

7. After precipitation, wash pellet with 70% EtOH and resuspend in 150 μl HB. Store at -20°C.

For use, add 1 μl of probe to 100 μl HB, heat to 80°C, chill for 5 min and then add to embryos.

Note on RNase precautions: Avoid potential RNase sources, especially for those steps in which the RNA probe should remain intact, i.e., probe production (*in vitro* transcription), storage, and hybridization. Reagents for the solutions at these steps should be DEPC-treated when possible and linearized plasmid template, tRNA solutions, and salmon sperm DNA solutions should be phenol/chloroform-extracted. Such precautions are unnecessary for the PBT solution (before or after hybridization) or for the hybridization buffer used for posthybridization washes.

C. Protocol 7: Probe Labeling by Polymerase Chain Reaction (PCR) (after Tautz *et al.*, 1992, and Patel and Goodman, 1992)

The labeling of probes by PCR generates specific single-stranded DNA probes. This protocol is based on the use of multiple rounds of DNA synthesis using *Taq* polymerase. In contrast to the standard PCR amplification that yields double-stranded products in this modified reaction, only one primer is used. This method shares many of the advantages of the riboprobe outlined above: (1) The single-stranded probes are very sensitive since the probe strands will not hybridize to each other. (2) Depending on the choice of primer, antisense and sense probes can be generated. (3) Despite the fact that the newly synthesized strand will not be used as a template, the yields of the PCR method are very good (from 200 ng of template, almost 1 μg of probe in a 25-μl reaction; Patel and Goodman, 1992). A potential disadvantage to riboprobes and oligonucleotide labeling is the fact that with the PCR reaction the probe size cannot be controlled. Further, although the probes can be produced from any template, this may require synthesis of specific primers. Gel-purified, linearized plasmid DNA (from a mini-preparation) or PCR-amplified, gel-purified DNA can be used as templates for the labeling reaction.

1. Amplify the desired probe in a standard PCR reaction (see following for the considerations about the size). Purify the fragment from an agarose gel.
2. Use about 200 ng of the DNA fragment in 16 μl of water.
3. Add 2.5 μl 10× PCR buffer (500 mM KCl, 100 mM Tris–HCl, pH 8.2, 15 mM MgCl$_2$).
4. Add 5 μl nucleotide solution (1 mM each of dATP, dCTP, and dGTP as well as 0.65 mM dTTP and 0.35 mM digoxigenin-dUTP (Boehringer Mannheim).
5. Add 1 μl PCR primer (20 μM stock) from the complementary strand of the RNA transcript.
6. Add 0.5 U *Taq* polymerase.

7. Carry out PCR under the following conditions: 94°C for 45 sec, 55°C for 30 sec, and 72°C for 60 sec for a total of 25 cycles.

8. Stop the reaction by freezing at −20°C. Purification of the probe is usually not necessary. The labeled probe can be stored at −20°C for at least a year.

9. Use 2 μl of probe in 100 μl of hybridization buffer.

10. The size of the probe should be about 200 bp. If longer fragments are used, they should be boiled at 100°C for 30 min to cause fragmentation.

D. Protocol 8: End-Labeling of Oligonucleotides (after Artero *et al.*, 1992)

This method is specifically useful for the detection of splice site variants and is described in detail by Artero *et al.* (1992). Antisense oligonucleotides that span the splice site are synthesized and labeled with digoxigenin-dUTP by terminal transferase tailing.

1. 30-mer antisense oligonucleotides are synthesized.

2. Labeling and tailing reaction:

   35 pmole oligonucleotides
   4 μl 5× tailing buffer (Boehringer Mannheim)
   6 μl 5 mM CoCl$_2$
   2.5 μl dig-11-dUTP (1 mM stock solution, Boehringer Mannheim)
   2 μl 50 μM dATP
   1 μl (25 U) terminal transferase
   ddH$_2$O to 20 μl

   Incubate for 1 hr at 37°C.

3. Ethanol precipitate in the presence of LiCl (see protocol 6.6).

4. Probe concentration for hybridization is 0.1–0.3 μg/ml.

E. Solutions

2× Carbonate Buffer

120 mM Na$_2$CO$_3$, 80 mM NaHCO$_3$
Adjust pH to 10.2, store in aliquots at −20°C.

VII. Double Labeling: RNA–Protein

In situ hybridization can also be combined with antibody staining or β-galactosidase staining. A detailed protocol is provided by Cohen and Cohen, 1992 (see also Aspiazu and Frasch, 1993). The protocol follows basically the standard procedure for whole-mount *in situ* hybridization. Antibody staining is done after the hybridization step and after the color of the *in situ* signal is fully developed. A standard protocol for antibody staining is utilized with a peroxidase color assay at the end. Applications have been described by Cohen (1990).

A. Protocol 9: *In Situ* Hybridization Followed by Antibody Staining (after Cohen and Cohen, 1992)

1. RNA Detection

1. Follow the entire fixation and *in situ* hybridization protocols (Protocols 1–3) until the detection reaction is completed.
2. Wash embryos 3 × 5 min in PBT.
3. Dehydrate embryos by washing them for 5 min each in 30, 50, 2 × 70% ethanol at RT.
4. Store embryos overnight at 4°C in 70% ethanol (without rotation).

2. Protein Detection

5. Wash embryos 3 × 5 min in BBT 250.
6. Add appropriate dilution of antibody into 500 μl BBT 250 and incubate at 4°C overnight. The optimal dilution will vary and need to be individually determined. [Most antibodies should be preabsorbed against fixed embryos prior to use. Preferentially, the embryos used for preabsorption should be of a stage that does not express the protein to be detected or should be deficient for the gene product. For fixation, follow Protocol 1, Steps 1.1–1.11 and then rehydrate and block embryos following Protocol 9, Steps 7 and 8. Add antibody to BBT in a 10-fold concentration of the final concentration. Use 100–200 μl of embryos per 500 μl of BBT. Roll overnight at 4°C. Store preabsorbed antibody at 4°C in fresh microcentrifuge tube and discard used embryos. Preabsorbed antibodies can be kept at 4°C for several weeks; for longer storage add azide (0.02%).]
7. Wash 4 × 10 min in BBT 250.
8. Wash 2 × 20 min in 500 μl BBT 250 with 2% normal serum (from the species in which the secondary antibody was raised).
9. Incubate for 2 hr at room temperature in 500 μl BBT-250 containing 2% normal serum and appropriate dilution of secondary antibody (secondary antibodies are preabsorbed overnight against fixed embryos at 1:10 and used at 1:500, as described in Protocol 3, Step 1.
10. Wash 6 × 10 min in PBT. Avoid BSA, azide, or animal sera because they may interfere with the avidin–biotin complex formation.
11. Incubate for 1 hr in 500 μl ABX mix (components and concentration of reagents in PBT depend on the Vectastain ABC kit (Vector labs) used).

12. Wash 3–5 × 10 min in PBT.

13. Transfer embryos in 1 ml PBT into watch glass or multiple depression slide.

14. Replace PBT with DAB staining solution.

15. Mix the solution and monitor the developing reaction under the microscope. Be prepared to stop the reaction quickly, as the staining often comes up within seconds. The components are reactive only for approximately 20 min; prolonged incubations will only increase background.

16. Stop by washing 3 × 5 min with PBT. Embryos can now be embedded following one of the methods outlined in Section V.

B. Protocol 10: Double Labeling for β-Galactosidase Expression and *in Situ* Hybridization (after Cohen and Cohen, 1992)

Transgenic lines expressing *lacZ* can also be stained for β-galactosidase (β-gal) activity first. For this purpose, embryos are fixed as in the standard protocol (Protocol 1), but are not devitellinized. The embryos are washed and stained in a standard β-gal assay. Devitellinization is carried out after the full development of the color precipitate. This is followed by a standard *in situ* hybridization protocol (Protocols 1–4). There are some problems with this approach: First, the embryos are very sticky as long as they have their vitelline membranes. It is therefore necessary to include high concentrations of a detergent (0.3% Triton X-100) in the respective buffers. The second problem is that β-gal staining is somewhat weaker than the staining that can be obtained by antibodies against β-gal. It might therefore be useful to follow Protocol 9 using an anti-β-gal antibody even in those cases in which an active β-gal would be produced in the embryo.

Applications of this technique have been published by Phillips *et al.* (1990) for imaginal discs and Cohen *et al.* (1991), and Cohen and Cohen (1992) for embryos.

1. Collect, dechorionate, and fix embryos as described in Protocol 1, Steps 1.1–1.5.

2. Stop shaking and let phases separate.

3. Transfer embryos with a cut blue pipet tip into a depression slide or staining dish. (The embryos will be sticky and have to be handled slowly and carefully; it can be helpful to wash the pipet tip and staining dish with PBX or spittle (works best!) prior to transfer).

4. Remove supernatant, add 0.5–1 ml PBX, and gently pipet solution up and down to expose embryos to the detergent.

5. Wash 3 × 5 min with PBX (all washes are done in 0.5–1 ml solution with mixing unless otherwise specified).

6. Exchange PBX with 500 μl of the X-Gal staining solution (this solution should be preincubated for 5 min at 37°C) and incubate at room tempera-

ture until the X-Gal staining reaction is sufficiently intense. The reaction can be monitored under the dissection microscope.

7. Stop reaction by washing embryos 3 × 2 min with PBX.

8. If the X-Gal staining reaction is strong, full color will be developed in less than 1 hr whereas weaker staining can take up to 24 hr. Minimal time should be used for the staining reaction, since prolonged incubation will reduce the intensity of the *in situ* hybridization signals in the subsequent reactions, presumably due to RNA degradation.

9. Transfer embryos into a scintillation vial and remove as much PBX as possible.

10. Add 5 ml of heptane and then 10 ml of methanol.

11. Immediately shake vigorously by hand or vortex for 15 sec.

12. Remove most of the heptane and add another 5 ml of methanol (most embryos devitellinize and sink to the bottom).

13. Transfer embryos into microcentrifuge tubes and proceed with RNA detection (Protocol 1, Step 1–Protocol 3).

C. Solutions

BBT

PBS
0.1% Tween 20
0.1% BSA
Filter sterilize

BBT 250

BBT with 250 mM NaCl

DAB Staining Solution

0.5 mg diaminobenzidine (DAB)/ml PBT, 0.006% H_2O_2

Prepare by adding 25 μl of a 20-mg/ml stock solution of DAB (stored in aliquots at −20°C in 50 mM Tris, pH 7.5; the stock may be frozen and reused repeatedly) to PBT. Add 2 μl of a 1/10 dilution of a 30% stock of H_2O_2 per ml of DAB/PBT solution.

Caution: DAB is carcinogenic and should be handled carefully and disposed of properly.

PBX

PBS with 0.3% Triton X-100

X-Gal Staining Solution

3.1 mM K$_4$[Fe^{II}(CN)$_6$]
3.1 mM K$_3$[Fe^{III}(CN)$_6$]
0.3% (v/v) Triton X-100
0.2% (w/v) X-Gal in PBS

VIII. Double Labeling: RNA–RNA

This protocol allows the simultaneous detection of two different RNAs using digoxigenin- and biotin-labeled probes. It is based on the original whole-mount *in situ* protocol by Tautz and Pfeifle (1989) for dig-labeled probes and on the modifications for the use of RNA probes (Jiang *et al.*, 1991; Gavis and Lehmann, 1992). The probes are labeled with dig-UTP and biotin-UTP (bio-UTP), respectively and are hybridized to embryos in the same hybridization reaction. The biotinylated probe is detected first, followed by detection of digoxigenin-labeled probe. Different color combinations can be achieved through choice of substrate. Useful combinations are (1) Digoxigenin-labeled probe with NBT/X phosphate (blue precipitate) and biotin-labeled probe with DAB/H$_2$O$_2$ (reddish brown precipitate) as substrates. (2) Digoxigenin-labeled probe with Vector red (red precipitate, autofluorescent) and biotin labeled probe with DAB/H$_2$O$_2$/Ni (purple precipitate) as substrates (see Fig. 1). The more abundant transcript should be hybridized with the biotin probe, since biotin-labeled probes are less sensitive than digoxigenin-labeled probes.

A. Protocol 11: Whole-Mount *in Situ* Hybridization with Two RNA Probes (after F. Pelegri)

This protocol works well for up to 20 μl of settled embryos. Larger amounts of embryos may lead to decreased signals. Removal of chorion and vitelline membrane and prehybridization treatments follow Protocol 1, Steps 1–18. Try to keep proteinase K concentration as low as possible (3–5 μg/ml).

1. Hybridization

For preparation of embryos and hybridization, follow Protocols 1 and 2. Riboprobes are prepared as described here.

1. Add 1 μl of each probe to 100 μl of Hb-A (for probe preparation, see Steps 14–20 in this section).
2. Hybridize at 56°C overnight.
3. After hybridization, follow washes as described in Protocol 2, Steps 7 and 8.

4. During washes, prepare AB complex (Elite PK kit, Vector labs): add 10 μl of solution A and 10 μl of solution B per ml of PBT. Mix by vortexing and let sit for 30 min exactly.

5. Add 0.5 ml of AB complex/PBT per tube and 50 μl (1:10 dilution) of anti-digoxigenin antibody (preabsorbed at a 1:200 dilution, see Protocol 3, Step 1). Roll 70' min at RT exactly.

6. Wash in PBT for 3 × 10 min.

7. Remove embryos from microcentrifuge tube and place in a watch glass or staining dish.

2. HRP Reaction

8. Remove as much PBT as possible from embryos and add DAB for 2 min.

9. Replace DAB with DAB/H_2O_2. Colored precipitate should appear within minutes.

10. Stop the reaction by washing 5× in PBT.

3. Phosphatase Reaction

11. Wash in alkaline solution 3 × 2 min. The HRP staining will change color and appear to bleach, but this is fine. Depending on the desired final color, different pH and substrates are used: For a blue precipitate, use alkaline solution at pH 9.0. For a red (also autofluorescent) precipitate, use alkaline solution at pH 8.3. Optional: add 10 μl of Levamisol (24 mg/ml) per ml of alkaline solution to inhibit endogenous phosphatases.

12. Add substrate solution and allow color to develop. Staining should appear in 2–10 min. Vector red staining may take considerably longer time; fortunately, the background for this substrate develops very slowly.

13. Rinse 5× with PBT. Store at 4°C or embed following Protocols 4A–4C.

4. Synthesis of Digoxigenin-Labeled RNA

14. Digoxigenin-labeled RNA probes are produced by *in vitro* transcription using SP6, T3, or T7 polymerase following the instruction in the Boehringer Mannheim RNA labeling kit (SP6/T7); see Protocol 6. Use 1 μl of probe for each hybridization.

5. Synthesis of Biotin-Labeled RNA

15. Biotin-labeled RNA probes are prepared as described in the protocol accompanying the biotion-21-UTP from Clontech (follow Clontech Protocol 4, Steps 1–3). After the 2-hr *in vitro* reaction, add dH_2O to a final volume of 50 μl (if necessary).

16. Add 50 μl of 2× carbonate buffer (see Protocol 6.3).
17. Incubate at 65°C for 40 min.

    Add 100 μl 0.2 M NaOAC, pH 6.0.

18. Precipitate the RNA by adding

    16 μl 5 M LiCl
    10 μl 20 mg/ml tRNA (phenol/chloroform extracted)
    600 μl ethanol

19. After precipitation, wash with 70% ethanol.
20. Resuspend in 150 μl hybridization buffer and store at −20°C. Use 1 μl of probe for each hybridization.

B. Solutions

Hb

See Hb-A Protocol 2; Store at −20 C and prewarm to 56°C before use.

DAB

For preparation of DAB and DAB/H_2O_2 solution, see Section VIIC. This substrate will yield a reddish-brown color. If a purplish color is desired, add NiCl to 0.03% (14 μl of NiCl solution (DAB substrate kit, Vector Labs) per milliliter of PBT). The presence of Ni also increases the sensitivity of detection.

Alkaline Solution (100 ml)

83 ml dH_2O
 2 ml 5 M NaCl
 5 ml 1 M MgCl$_2$
10 ml 1 M Tris (pH 8.3 or 9.0, accordingly)
10 μl Tween 20

Substrate Solution (for Blue Precipitate from Nonradioactive DNA Labeling and Detection Kit; Boehringer Mannheim)

4.5 μl NBT (vial 9)
3.5 μl X-phosphate solution (vial 10) per ml of Alkaline solution, pH 9.0

Substrate Solution (for Red (Autofluorescent) Precipitate from Vector Red Kit, Vector Labs)

14 μl each of solution A, solution B, and solution C per ml of Alkaline solution, pH 8.3

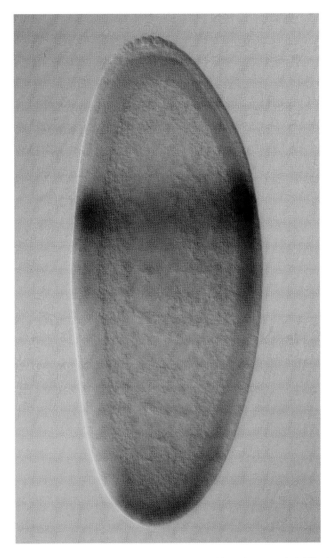

Fig. 1 RNA double labeling of blastoderm embryo. This embryo was hybridized with a dig-UTP-labeled probe complementary to the RNA of the gap gene *Krüppel* using Vector Red as a substrate to yield a red precipitate. In parallel, the embryo was also hybridized with a bio-UTP-labeled probe complementary to the RNA of the gap gene *knirps* using DAB as substrate intensified with Ni, which yields a purple-brown color (see Protocol 11). Preparation and photograph by F. Pelegri.

IX. Conclusions

The methods described here have been used mainly for visualizing gene products in *Drosophila* embryos, but applications for imaginal discs have also been described (e.g., Phillips *et al.,* 1990). Detection of RNA using nonradioactively labeled probes has been used very successfully in many organisms, such as mouse, chicken, frog, and zebrafish (e.g., Rosen and Beddington, 1993; Sasaki and Hogan, 1993; Parr *et al.,* 1993, Riddle *et al.,* 1993; Hemmati-Brivanlou *et al.,* 1990; Schulte-Merker *et al.,* 1992). Future improvements of these methods may yield probes of higher sensitivity and hopefully fixation techniques that would allow better preservation of tissue morphology. This is specifically important if the techniques described are to be applied at the ultrastructural level.

Acknowledgments

We thank the members of our labs, specifically L. Gavis, F. Pelegri, D. Curtis, L. Dickinson, R. Sommer, M. Hülskamp, R. Schröder, and M. Klingler for their contributions to the protocols described. R. L. is a HHMI associate investigator.

References

Artero, R. D., Akam, M., and Pérez-Alonso, M. (1992). Oligonucleotide probes detect splicing variants *in situ* in *Drosophila* embryos. *Nucleic Acids Res.* **20,** 5687–5690.

Aspiazu, N., and Frasch, M. (1993). *tinman* and *bagpipe:* Two homeobox genes that determine cell fate in the dorsal mesoderm of *Drosophila. Genes Dev.* **7,** 1325–1340.

Cohen, S. M. (1990). Specification of limb development in the *Drosophila* embryo by positional cues from segmentation genes. *Nature* **343,** 173–177.

Cohen, B., Wimmer E. A., and Cohen, S. M. (1991). Early development of leg and wing primordia in the *Drosophila* embryo. *Mech. Dev.* **33,** 229–240.

Cohen, B., and Cohen, S. M. (1992). Double labeling of mRNA and proteins in *Drosophila* embryos. *In* ''Nonradioactive Labeling and Detection of Biomolecules'' (C. Kessler, ed.), pp. 382–392. Berlin/Heidelberg: Springer-Verlag.

Feinberg, A. P., and Vogelstein, B. (1983). A technique for radiolabelling DNA restriction endonuclease fragments to high specific activity. *Anal. Biochem.* **132,** 120–125.

Gavis, E. R., and Lehmann, R. (1992). Localization of *nanos* RNA controls embryonic polarity. *Cell* **71,** 301–313.

Hemmati-Brivanlou, A., Frank, D., Bolce, M. E., Brown, B. D., Sive, L. H., and Harland, R. M. (1990). Localization of specific mRNAs in *Xenopus* embryos by whole-mount *in situ* hybridization. *Development* **110,** 325–330.

Jiang, J., Kosman, D., Ip, Y. T., and Levine, M. (1991). The *dorsal* morphogen gradient regulates the mesoderm determinant *twist* in early *Drosophila* embryos. *Genes Dev.* **5,** 1881–1891.

Lawrence, P. A., Johnston P., and Morata G. (1986). Methods of marking cells. *In* ''*Drosophila:* A Practical Approach'' (D. B. Roberts, ed.), pp. 229–242. Oxford: IRL Press.

Leptin, M., and Grunewald, B. (1990). Cell shape changes during gastrulation in *Drosophila. Development* **110,** 73–84.

Melton, D. A., Krieg, P. A., Rebagliati, M. R., Maniatis, T., Zinn, K., and Green, M. R. (1984). Efficient *in vitro* synthesis of biologically active RNA and RNA hybridization probes from plasmids containing a bacteriophage SP6 promoter. *Nucleic Acids Res.* **12,** 7035–7056.

Parr, B. A., Shea, M. J., Vassileva, G., and McMahon, A. P. (1993). Mouse *Wnt* genes exhibits discrete domains of expression in the early embryonic CNS and limb buds. *Development* **119**, 247–261.

Patel, N. P., and Goodman, C. S. (1992). Dig-labeled single stranded DNA probes for in situ hybridization. *In* "Nonradioactive Labeling and Detection of Biomolecules" (C. Kessler, ed.), pp. 377–381. Berlin/Heidelberg: Springer-Verlag.

Phillips, R. G., Roberts, I. J. H., Ingham, P. I., and Whittle, J. R. S. (1990). The *Drosophila* segmentation gene *patched* is involved in a position-signalling mechanism in imaginal discs. *Development* **110**, 105–114.

Riddle, D. R., Johnson, R. L., Laufer, E., and Tabin, C. (1993). *Sonic hedgehog* mediates the polarizing activity of the ZPA. *Cell* **75**, 1401–1416.

Rosen, B., and Beddington, R. S. P. (1993). Whole-mount *in situ* hybridization in the mouse embryo: Gene expression in three dimensions. *Trends Genet.* **9**, 162–167.

Sambrook, J., Fritsch, E. F., and Maniatis, T. (1989). "Molecular Cloning: A Laboratory Manual." Cold Spring Harbor, NY: Cold Spring Harbor Laboratory Press.

Sasaki, H., and Hogan, B. L. M. (1993). Differential expression of multiple fork head related genes during gastrulation and axial pattern formation in the mouse embryo. *Development* **118**, 47–59.

Schulte-Merker, S., Ho, R. K., Herrmann, B. G., and Nüsslein-Volhard, C. (1992). The protein product of the zebrafish homologue of the mouse *T* gene is expressed in nuclei of the germ ring and the notochord of the early embryo. *Development* **116**, 1021–1032.

Sommer, R. J., and Tautz, D. (1993). Involvement of an orthologue of the *Drosophila* pair-rule gene *hairy* in segment formation of the short germ band embryo of *Tribolium* (Coleoptera). *Nature* **361**, 448–450.

Tautz, D., and Pfeifle, C. (1989). A non-radioactive in situ hybridization method for the localization of specific RNAs in *Drosophila* embryos reveals translational control of the segmentation gene *hunchback*. *Chromosoma* **98**, 81–85.

Tautz, D. (1992). Whole mount in situ hybridization for detection of mRNA in *Drosophila* embryos. *In* "Nonradioactive Labeling and Detection of Biomolecules" (C. Kessler, ed.), pp. 373–377. Berlin/Heidelberg: Springer-Verlag.

Tautz, D., Hülskamp M., and Sommer, R. J. (1992). Whole mount in situ hybridization in *Drosophila*. *In* "In Situ Hybridization: A Practical Approach" (D. B. Roberts, ed.), pp. 61–73. Oxford: IRL Press.

Wieschaus, E., and Nüsslein-Volhard, C. (1986). Looking at embryos. *In* "*Drosophila*: A Practical Approach" (D. B. Roberts, ed.), pp. 199–227. Oxford: IRL Press.

CHAPTER 31

Looking at mRNA Splicing and Transport *in Situ*

Zuzana Zachar, Joseph Kramer, and Paul M. Bingham

Department of Biochemistry and Cell Biology
State University of New York at Stony Brook
Stony Brook, New York 11794

I. Introduction

Advances in experimental systems, in image acquisition/analysis, and in development of antibody and hybridization probes have opened subnuclear organization and compartmentalization to a newly penetrating level of analysis. This rapidly expanding body of information is continuing to produce fundamentally new insights (see Rosbash and Singer, 1993; Spector, 1993; Zachar *et al.*, 1993; and Kramer *et al.*, 1994 for discussions).

One of the most important of these advances has been exploitation of the unique properties of the giant salivary gland polytene nucleus of *Drosophila* larvae (Zachar *et al.*, 1993 and references therein). The *Drosophila* genome is small and is organized in five major chromosome arms arrayed in a monolayer around a single central nucleolus in the diploid nucleus. These simple organizational features are preserved in giant polytene nuclei. Individual polytene chromosomes result from association in precise register of ca. 500 copies of an interphase chromatid. Thus, the polytene nucleus represents an "exploded" version of the diploid interphase nucleus and allows cytogenetic analysis with extremely high resolution. As a result, in polytene nuclei—in contrast to diploid nuclei—the limited resolving power of the otherwise robust and convenient light microscope is sufficient to visualize relationships between specific, individual genes and associated subnuclear structures. Moreover, as a functional salivary gland polytene genetic locus consists of the ca. 500 tightly synapsed copies of the gene, very high local levels of pre-mRNA are produced. This permits easy detection of nascent transcripts and their processing products in spite of the modest sensitivity of current *in situ* detection systems.

In addition to their unique properties as a cytogenetic tool, salivary gland nuclei exist in the context of the highly mature *Drosophila* genetic system providing numerous useful tools. Among these are well-defined promoter segments allowing high level salivary gland expression. This permits the easy construction of reporter genes with any desired attribute (for example, a variant intron) and accessible to analysis in the opportune environment of the polytene nucleus.

We describe here protocols allowing *in situ* analysis of the organization of intact salivary gland polytene nuclei with particular attention to pre-mRNA metabolism.

II. Protocols

A. Construction of Chimeric Genes with High Expression Levels in Polytene Salivary Gland Nuclei

A crucial feature of this experimental system is the capability to engineer and reintroduce genes heavily transcribed in salivary gland nuclei and specifically designed for analysis. Detailed protocols for this purpose are beyond the scope of this chapter; however, discussion of this approach and description of necessary techniques can be found in Zachar *et al.* (1993).

B. Generation of Salivary Gland Nuclear Preparations for *in Situ* Detection of Pre-mRNAs and Splicing-Associated Proteins

The *Drosophila* salivary gland is enclosed in a basement membrane-like structure that must be breached to permit efficient access of nucleic acid and antibody probes. We routinely do this in two ways. The first is to generate "thick" cryosections (Section B1). These cut a salivary gland lobe in roughly five serial sections with complete and nearly complete nuclei in essentially every section.

The second strategy is to generate "smear" preparations (Section B2). The basement membrane is ruptured by gentle squashing that leaves most nuclear envelopes intact. The resulting nuclei are somewhat "flattened" (axial ratio of ca. 3:1 relative to the 1:1 ratio *in situ*); however, many nuclear structures are preserved and much information can be gained from such preparations.

1. Protocol—Larval Cryosections

1. Place 10–15 etherized, late third instar larvae into a depression slide containing a few drops of OCT cryosectioning compound (Baxter, Catalog No. M7148-4). Cover larvae with additional OCT and mix gently so that larvae become thoroughly coated with OCT. Place OCT-coated larvae into the bottom of a "peel-away" embedding mold (VWR, Catalog No. 15160-099) in two rows, with 5–7 larvae per row. Align the two rows of larvae head to head so that salivary glands are in the middle of the section. [The two salivary gland lobes occupy a large portion of anterior third of the larva (see Fig. 1 and Protocol B)]. Draw off extra OCT so that all larvae are in the same plane in a small residual volume of OCT.

2. Place mold into a shallow slurry of crushed dry ice and 2-methylbutane. As the larval layer begins to freeze (OCT goes from clear to opalescent white upon freezing and this requires only a few seconds), fill the mold with OCT almost to the top. Insert a cryostat specimen holder into the top layer of molten OCT and allow whole block to freeze (this will take about 3–5 min). (IMPORTANT: If the larva-containing layer freezes completely before the mold is filled with additional OCT, it will break off during sectioning.) Remove the frozen block from mold, immediately place it into cryostat set at −19°C, and

Salivary glands   Phase - - - - - - Fluorescence

Fig. 1 (Left) Dissected salivary gland lobes connected to mouth parts and associated structures. The gland lobes are nearly transparent. Note the opaque, elongated fat bodies associated with the lobes. (Center) Phase-contrast image of larval section. The larva extends from the anterior tip at left center off the right-hand border of the image. The very large salivary gland cells are visible in tissue beginning at center and extending horizontally to right-hand edge of image. (Right) Fluorescent image of DNA-specific stain. Note the double row of very large salivary gland nuclei and the numerous, much smaller diploid nuclei.

allow it to equilibrate to cryostat temperature (about 20 min). [Blocks can be stored for weeks or longer at −80°C; such blocks should be carefully reequilibriated to −20°C for a few hours before cutting sections.]

3. Cut sections of desired thickness. We use a Bright Instruments cryostat (Model No. OTF/AS/MR, 83608/1489). [Instruction in use of cryostats is available in most cell biology, histology, pathology, and neurobiology departments and from microtomy/microscopy texts.] Thirty-micrometer sections are appropriate for studies requiring largely intact nuclei. Thinner sections are useful for many purposes as well. Salivary gland tissue is easily recognized in sections—cells and nuclei are extremely large and cytoplasm is highly "textured" in contrast to other larval tissues. Permanent crytostat knives (Thomas, Catalog No. 6740-A20) used with the anti-roll guide plate work much better in our hands than disposable blades. Pick up individual sections with polylysine-coated slides [dip washed slides into a solution of 0.01% (w/v) polylysine (Sigma, Catalog No. P 8920) for 5 min and dry in a dust-free environment] and immediately place in formaldehyde/PBS fixative (3.7% formaldehyde from commercial formalin in PBS is suitable for virtually all purposes; PBS is 137 mM NaCl, 8.5 mM Na$_2$HPO$_4$, 4.4 mM KH$_2$PO$_4$, pH 7.0) for 10 min. Some earlier tissue *in situ* hybridization protocols call for air drying samples before fixation; however, this results in very large reductions in the salivary gland nuclear pre-mRNA hybridization signals.

4. Wash slides twice for 10 min in PBS. We avoid storage of slides for more than a few hours before proceeding to immunolocalizations or *in situ* hybridizations (following).

When the entire section curls up, the cryostat temperature is too cold. When individual larvae curl up while the OCT portion of section stays flat, it is because larvae were moist when placed into OCT. Remake the block, "washing" larvae more extensively in OCT to replace surface moisture.

2. Protocol—Smear Preparations

1. Dissect salivary glands from late third instar larvae in a roughly isotonic saline (e.g., PBS) as follows. With sharp, fine dissecting forceps (Inox No. 3 from Ernest Fullam, Inc., Latham, NY) the back third of the larva is held with the nondominant hand and the mouth hooks at anterior tip of the larva are grasped firmly in the dominant hand. (The mouth hooks are darkly pigmented and typically are in almost constant in-and-out motion. They are easy to spot.) A sharp pull on the mouth parts brings the associated structures—including salivary glands and various other tissues—out. Separate the gland lobes from the other structures including most of the fat body. (See left-hand panel in Fig. 1 for photograph of partially completed dissection.) With a little practice this should take about 15–30 sec.

2. Transfer glands to 20 μl of formaldehyde/PBS (as before) and incubate for 1 min. (We use 18-mm^2 Corning coverslips directly from the box and cleaned with a compressed air duster.)

3. Lift coverslip with a polylysine-treated slide (Section B1). Invert slide and gently move the coverslip back and forth by pushing on its side (not down on the top). The appropriate motion will break the basement membrane while leaving most of the nuclei intact. This can be mastered with brief practice monitored under the phase-contrast microscope.

4. Flip off coverslip without lateral movement. This is done by placing a razor blade snugly along one edge of the coverslip and then flipping the coverslip with a second blade forced under the opposite edge.

5. Immediately fix the slide for 10 min in formaldehyde/PBS. Wash as in Section B1.

C. Immunolocalization of Epitopes Associated with Pre-mRNA Metabolism

The major problems encountered in immunolocalization in salivary gland preparations are accessibility and preservation of epitopes and controlling non-specific binding of antibodies. Epitopes can be occluded for a variety of reasons. Light pronase digestion or treatment with 4 M urea (for 5 min followed by 2×10-min PBS washes immediately before blocking for immunolocalization) can sometimes reveal occluded epitopes. For example, we find that the me$_3$G snRNA cap is very poorly detected with available antibody probes in salivary gland preparations without prior pronase digestion.

Nonspecific Ab binding is traditionally reduced by preincubating in a blocking solution. We have tried a variety of these and find the block described here to give consistently superior results.

We apply block and antibody solutions in small volumes (30–50 μl) directly on slides and incubate these in a humidified chamber. To permit this, the section or smear sample is circled with a "Pap pen" (RPI, Mount Prospect, IL, Catalog No. 195500) producing a hydrophobic boundary that will restrain the solutions. We use 24×24-cm Nunc plates for humidified chambers. Slides are placed spanning two strips of plastic ca. 5 mm thick. The remaining surface of the plate is covered with 3-mm-thick soaked blotting pads. Wash steps are done in large volumes in coplin jars or their equivalent.

1. Protocol—Application of Antibodies

1. Shake excess PBS off the section or smear preparation, immediately circle sample with Pap Pen (above), and apply about 50 μl of block solution. [block is solution 5% (w/v) powdered nonfat dry milk in PBS containing 10% serum. We use standard generic milk purchased at the local grocery store and reconstituted, lyophilized "Normal Serum" (all immunochemicals are purchased from Jack-

son Immunoresearch Laboratories, West Grove, PA unless otherwise indicated). When possible, the serum should be from the same species as the secondary antibody (typically goat, sheep, or donkey) and never from the same species as the primary antibody.] Incubate for about 30 min.

2. Shake off blocking solution and add 30–50 μl of primary antibody (for example, a mouse anti-snRNP) in blocking solution. Incubate for 30–120 min at room temperature. [Optimal antibody inputs vary greatly. As a rule of thumb, ascites fluid and high titer polyclonal sera are generally used at 100- to 1000-fold dilutions and tissue culture fluid supernatants from monoclonals at 10- to 100-fold dilutions. Extending treatment times to 120 min (or overnight at 4°C) can improve signals in some cases.]

3. Wash slides in three changes of PBS for a total of 30 min.

4. Shake off PBS, add 50 μl of blocking solution, and incubate for ca. 5 min.

5. Shake off block and replace with 30–50 μl of secondary antibody (for example, a goat anti-mouse) in block solution. Incubate for 30 min. [For fluorochrome-conjugated secondaries, minimize light exposure until protective mounting medium is applied (see following). Optimal secondary antibody inputs vary greatly (anywhere from 30- to 2000-fold dilutions of standard commercial preparations) and must be calibrated.]

6. Wash in three changes of PBS for a total of 30 min.

2. Protocol—Detection of Fluorochrome-Conjugated Secondary Antibody Signals

1. Shake off PBS from washes above and place a small drop of mounting medium over the sample. [We use 0.2 M Tris, pH 8.6, 60% glycerol, 1 mg/ml p-phenylenediamine (Sigma, Catalog No. P6001; this mixture can be stored at −20°C in the dark for a couple of months), or a commercial preparation (Vectashield, Catalog No. H-1000, Vector Laboratories, Burlingame, CA).]

2. Place four small dots of nail polish around the outside of the section or smear and allow to dry briefly (30–60 sec).

3. Place a 22-mm^2 coverslip over the sample propped up on the hardened nail polish drops.

4. Anchor coverslips with a few drops of nail polish at edges or corners.

5. Signals are detected either by conventional epifluorescent or by confocal microscopy. The confocal microscope or equivalent is a powerful tool for exploiting the rich three-dimensional information in thick-section and smear salivary gland preparations. As well, in practice, the high-effective sensitivity and operating resolution of the confocal microscope often justifies its use even when three dimensional information is not a primary objective.

3. Protocol—Detection of Peroxidase-Conjugated Secondary Antibodies

For detection of peroxidase-conjugated secondary antibody, we find Vector-VIP peroxidase substrate (Vector Laboratories, Catalog No. SK-4600) superior to standard DAB-based protocols. Apply Vector-VIP as per manufacturer's directions. Terminate reaction by incubating slides in water for 5 min. Dehydrate slides through ethanol series (30, 60, 95, and 100%, for 2 min each) and incubate in Histoclear (National Diagnostics, Catalog No. HS-200) for 10 min. Drain off excess Histoclear and mount with Permount (Fisher Scientific, Catalog No. SP15-100).

Peroxidase-conjugated secondaries are superior to alkaline phosphatase (AP) conjugates (Section E5) for immunodetection of proteins in salivary preparations due to high backgrounds presumably resulting from endogenous phosphatase activity. However, the procedures used for *in situ* hybridization eliminate this background, and AP conjugates are preferable over peroxidase conjugates for *in situ* hybridization.

Peroxidase is inactivated by azide compounds often used as preservatives in antibody solutions. Thus, block solution for application of peroxidase-conjugated secondaries should generally be 5% milk only, without the serum addition.

Choice of secondary antibody is influenced by the following properties of each. Peroxidase-based detection is somewhat more sensitive than fluorochrome-based detection and thus is often better for low-abundance epitopes. The permanent slides resulting from peroxidase-based detection are also often desirable. Fluorochrome-based detection allows application of the confocal microscope. Moreover, fluorescent signal strength is generally linearly related to the amount of secondary antibody whereas, in general, peroxidase precipitate levels are not. Thus, fluorescence permits much better quantitation.

D. Generation of Labeled Riboprobes

Conventional denatured, double-stranded DNA probes (produced by random priming or nick-translation) are also useful in the procedures described here. However, we much prefer the versatility, reliability, and specificity provided by single-stranded RNA probes. When possible, all reagents used in this procedure should be pretreated with diethylpyrocarbonate (DEPC; Sigma) and autoclaved (or incubated at room temperature for 24 hr) before use. Nucleotides, nucleic acids, and enzymes cannot be DEPC treated.

1. Mix in a microfuge tube 2 μl transcription buffer (400 mM Tris, pH 8.0, 60 mM MgCl$_2$, 100 mM DTT, 20 mM Spermidine, 100 mM NaCl); 2 μl digoxygenin or biotin labeling mix (3.3 mM digoxygenin-UTP or biotin-UTP, 6.7 mM UTP, 10 mM each ATP, CTP, GTP; digoxygenin labeling mix is available from Boehringer Mannheim, Catalog No. 1277 065; biotin-UTP is available from Sigma); 300–500 ng of DNA template (linearized immediately "downstream" of probe segment); RNasin (Promega, Catalog No. N2513) to 1 unit per microliter of reaction; 2 μl RNA polymerase (SP6, T7 or T3); H$_2$O to 20 μl. Incubate at 37°C for 2 hr.

2. Terminate by adding EDTA to 20 mM. (Do *not* phenol extract.) To precipi-

tate, add 20 μg of glycogen (Boehringer Mannheim, Catalog No. 901 393) as carrier, 1/10 vol of 4 M LiCl followed by 3 vol of cold 95% EtOH. Incubate for 2 hr $-20°C$ or 30 min at $-70°C$.

3. Collect precipitate by centrifugation, wash with cold 70% EtOH, dry *in vacuo*. Dissolve pellet in 80 μl of water for 10–30 min at room temperature. Yields should be 10–20 μg per 20-μl reaction. We quantify yields by comparing ethidium fluorescence (after brief electrophoresis into conventional 1% submarine agarose gel) against a standard. We make a large batch of standard for this purpose by running a reaction with a trace amount of a ^{32}P NTP added and gel filtering the product before ethanol precipitation. This permits yield (percentage incorporation) and recovery to be precisely determined. This standard is then stored at $-80°C$ in single-use aliquots.

4. For probes longer than 300 nucleotides, it is important that they be degraded to a mean size of about 150–300 bases as follows: To 80 μl of RNA from Step 3, add 20 μl of 5× carbonate buffer (200 mM NaHCO$_3$, 300 mM Na$_2$CO$_3$; should be pH 10.4) and incubate at 60°C for 20 min. Terminate by adding $\frac{1}{10}$ vol of 3 M sodium acetate (pH 4.5), and 3 vol of cold 95% ethanol. Chill and collect precipitate as in Step 2.

5. Resuspend RNA at desired concentration in water with RNasin to 0.25 unit per microliter and store at $-80°C$. Probes are stable indefinitely (at least 2 years).

Assuming adequate care with DNA template, the most common source of poor yields in practice is inadequate polymerase activity. In our hands, commercial SP6 is particularly prone to lose activity on storage and T7 is particularly robust through storage.

E. *In Situ* Hybridization to Salivary Gland Pre-mRNAs

The major systematic problems in *in situ* hybridization are limited sensitivity and potentially significant background. Signal-to-noise ratios are improved substantially by controlling probe length and by preacetylation of tissue (following).

With the best available procedures, pre-mRNA detection is still relatively inefficient. In general, only relatively abundant RNA can be reliably detected. For example, dense nascent transcript arrays produced by strong salivary gland promoters are easily detected (Zachar *et al.*, 1993) but isolated, individual mRNA molecules are not.

Accessibility of a salivary gland RNA to probe is sometimes a problem. Light digestion with pronase prior to hybridization (following) allows detection of such RNAs. For example, U1snRNA is very poorly detected in salivary gland preparations without prior protease treatment. A side effect of proteolysis is the partial loss of some tissues. Fortunately, salivary tissue is highly resistant and well retained after these procedures.

1. Protocol—Pronase Digestion

Shake off excess PBS. Cover tissue with ca. 100 μl of 125–250 μg/ml pronase (Calbiochem, La Jolla, CA; Catalog No. 537088) in 50 mM Tris, pH 7.5, 5 mM EDTA (fine tune pronase to give limited but not extreme loss of nonsalivary tissue). Incubate for 10 min. Terminate digestion by placing slide into 2 mg/ml glycine in PBS for 1–15 min. Wash twice for 1 min in PBS. Refix in formaldehyde/PBS (above) for 10 min. Wash twice for 10 min each in PBS.

2. Protocol—Acetylation

Acetylation (slides from the preceding paragraph) requires continuous, vigorous stirring. We do this with a stir bar in a staining dish (VWR, Catalog No. 25445-009). The slide rack is raised above the level of the stir bar by placing it on two segments of glass tubing laid along the sides of the dish. Put slides into 500 ml 0.1 M triethanolamine (Sigma), pH 8.0 (titrate with HCl). Place staining dish in hood. While stirring vigorously, add 1.25 ml of acetic anhydride (Sigma). Reduce stir bar speed and incubate for 10 min. Wash twice for 2 min in PBS and incubate in 0.1 M Tris/0.1 M glycine buffer, pH 7.0, for 30 min. Wash briefly in PBS.

3. Protocol—Hybridization

Remove PBS from slides from the preceding paragraph by shaking vigorously and drying with kimwipe around the tissue to remove as much PBS as possible without allowing the tissue to dry. Quickly place a coverslip on each side of the tissue glued in place with a very small amount of rubber cement. Immediately add 40 μl of probe solution and cover with a third coverslip spanning the two surrounding the sample. Seal with rubber cement. Incubate slides upright in a slide rack in a humidified container (soaked paper towel in tupperware container works well) at 50°C overnight.

Hybridization solution is 36% formamide (deionized), 10% dextran sulfate, 0.02% Ficoll (type 400), 0.02% polyvinylpyrrolidone (mean MW = 360,000), 0.2 mg/ml BSA, 4× SSC (1× SSC is 150 mM NaCl, 15 mM Na$_3$ Citrate, 5 mM Na$_2$HPO$_4$, pH 7.0), 10 mM DTT, 1 mg/ml yeast tRNA (Sigma), and 1 mg/ml sonicated salmon sperm DNA (Sigma). Saturating inputs of single-stranded RNA probes (Section D) under these conditions are ca. 500 ng/ml/kb of sequence complexity. (Highly abundant targets such as ribosomal RNA will sometimes require higher probe inputs to achieve saturation due to probe depletion).

4. Protocol—Postwash

Remove rubber cement and all three coverslips (preceding paragraph) carefully. Wash slides in large volume of 2× SCC, 50% formamide at 50°C twice for 15 min and a third time for 1–3 hr. Rinse in PBS.

Many protocols call for RNasing as part of the postwash. If probes are partially degraded as recommended, this is generally unnecessary. When desired, this is done by washing slides in 2× SCC after the second 15-min wash, transfering slides into 20 μg per ml RNase A (Sigma) in 2× SSC (37°C for 1 hr), washing in 2× SCC (37°C for 30 min) and washing in 2× SCC, 50% formamide for 1–3 hr (50°C).

5. Protocol—Detection

Detection is accomplished with anti-digoxygenin antibodies conjugated either with a fluorochrome or alkaline phosphatase.

For fluorescent detection we use anti-digoxygenin Fab fragment (Boehringer Mannheim, Catalog No. 1207 741) typically at 1 : 30. Block, treat with antibody for 30 min, wash, and mount as described (Section C).

For alkaline phosphatase-based detection, we use the conjugate available from Boehringer-Mannheim (Catalog No. 1093 274) at 1 : 100 to 1 : 500. Block, treat with antibody, and wash as above (Section C). Transfer slides into AP buffer (100 mM Tris, 100 mM NaCl, 50 mM MgCl$_2$; pH 9.5) for 10 min. Shake off excess buffer and add ca. 100 μl per slide of phosphatase substrate [to make, add 45 μl of 75 mg/ml nitroblue tetrazolium salt and 35 μl of 50 mg/ml 5-bromo-4-chloro-3-indolyl phosphate (Boehringer-Mannheim, Catalog No. 1383 213 and 760 986), and 10 μl of 1 M levamisole (Sigma, Catalog No. L-9756) to 10 ml of AP buffer; make fresh before use]. Incubate in the dark in a humidified chamber for 1–24 hr depending on signal strength. Terminate reaction by incubating for 10 min in TE, pH 8.0. Mount with Gel/mount (Biomedica, Foster City, CA). Seal coverslip edges with nail polish if slides are to be kept in excess of a few days.

F. Double-Label *in Situ* Hybridizations

Double-label hybridization is quite useful for determining the relationships between different nucleic acid molecules in the same polytene nucleus. Recipes are described for denaturation of chromosomal DNA to permit hybridization and for double-label hybridization applicable to any combination of DNA and/ or RNA targets. (See Zachar *et al.*, 1993, for additional discussion.)

The strategy we use employs one probe labeled with biotin and a second with digoxygenin. A crucial difficulty is preventing artifactual cross-hybridization between the two probes. (This occurs when one region of one probe molecule hybridizes to material on the slide and a second region hybridizes to the second probe. This results in artifactual localization of the second probe at this site.) Shredding probes and avoiding excessive probe concentrations eliminates nonspecific interaction between probe RNAs and also greatly reduces the potential for artifactual cross-hybridization. However, it is advisable to also take steps to eliminate overt homology between the two probes. For example, current vector polylinkers are large and two probes made by transcription of the same vector in opposite orientations can cross-hybridize through polylinker sequences. Further, even low levels of incompletely digested template DNA leads to production of significant amounts of probe transcripts by extending through the cloned insert into contiguous vector sequences. When they are homologous, vector extensions of this sort can produce cross-hybridization between the two probes. These problems are most easily avoided by designing the two probe segments to be transcribed in the same orientation from the same vector. Some additional procedures can be found in Zachar *et al.* (1993).

1. Protocol—Denaturation of Chromosomal DNA for Hybridization

1. Cryosections (Section B1) are immediately fixed in 95% ethanol (rather than formaldehyde/PBS) for 30 min and rehydrated for 10 min in two changes of PBS.

2. Chromosomal DNA in sections is then denatured by treatment for 2 min in 100 mM NaCl, 10 mM Mops (pH 7.0), 70% formamide at 70°C followed by 2 min at 70°C in the same solvent containing 3.7% formaldehyde (from formalin). (This last solution produces highly noxious fumes—do this in a hood.)

3. Wash through three changes of PBS totaling 15 min.

2. Protocol—Double-Label Hybridization

1. Carry out hybridization (including proteolysis, acetylation, hybridization, and postwash) precisely as in Section E using each probe together at or near minimal saturating inputs (Section E).

2. Immunodetect hybrids by blocking (Section C), treating with antibodies in block for 30 min followed by three PBS washes totaling 30 min. Fluorochrome-conjugated antidigoxygenin antibodies are described above (Section E). We detect biotin with a mouse monoclonal antibody conjugated with the rhodamine-like fluor (Cy3) from Jackson ImmunoResearch Laboratories (Weston, PA) at around 1 : 100. We have had consistently poor results with avidin-based systems (very high backgrounds in salivary gland cytoplasm).

3. Mount as in Section C.

III. Concluding Technical Remarks

A powerful attribute of the strong salivary gland promoters (Section I) is that they generate sufficient amounts of RNA to allow S_1 protection analysis of relatively rare transcript forms, including nascent transcripts. Detailed discussion and protocols for this purpose are beyond the scope of this chapter and can be found in Zachar *et al.* (1993). However, we emphasize that this capability permits quantitative supplementation to the less quantitative assay of *in situ* hybridization. Many of the errors and weaknesses in early studies using *in situ* hybridization result from inadequate quantitation of this sort.

An enormously powerful supplement to investigation of nuclear components in intact nuclei using the procedures described here is analysis of the same components in squash or spread chromosome preparations. The first assay displays the entire nuclear distribution. The second (under appropriate conditions) can selectively display molecules directly bound to chromosomes and nascent transcripts once the "soluble" nuclear component is removed. Pairing these two assays allows powerful additional approaches to problems in nuclear organization and gene expression/regulation completely unattainable in diploid experimental systems.

In general, the detailed protocols for immunolocalization and *in situ* hybridization described here can also be applied to such squash preparations. Detailed protocols for making squashes are beyond the scope of this chapter. However, related protocols can be found in Chapters 19, 20, and 29. As well, we have had good success by applying detection protocols described here to squash preparations made according to protocol 30 in Ashburner (1989).

IV. Concluding General Remarks

Recent technical innovations discussed here and in other chapters in this volume open new, still substantially untapped, avenues to exploit the unique potential of the polytene genetic and cytogenetic experimental system. Application of these procedures, as is, to the problem of nuclear structure and function will provide substantial new insights.

In addition, crucial opportunities exist to further enhance these approaches—a number of important questions cannot be effectively addressed with available procedures. These include improving detection sensitivities (through probe strategies and microscopy). Increases in sensitivity of as little as 10-fold would allow a dramatic new round of progress. Further, procedures allowing electron microscopic examination of transcript arrays in chromatin spreads (see Chapter 32) from polytene nuclei would be powerful. Among many other things this would permit visualization of protein/RNA complexes on introns (including regulated introns) whose behavior has been previously defined using the unique properties of the polytene cytogenetic system.

Acknowledgment

The work described here was supported by NIH Grant GM32003.

References

Ashburner, M. (1989). "Drosophila, A Laboratory Manual." Cold Spring Harbor, NY: Cold Spring Harbor Laboratory Press.

Kramer, J., Zachar, Z., and Bingham, P. M. (1994). Nuclear pre-mRNA metabolism: Channels and tracks. Trends Cell Biol. **4**, 35–37.

Rosbash, M., and Singer, R. H. (1993). RNA travel: Tracks from DNA to cytoplasm. *Cell* **75**, 399–401.

Spector, D. L. (1993). Macromolecular domains with the cell nucleus. *Annu. Rev. Cell Biol.* **9**, 265–315.

Zachar, Z., Kramer, J., Mims, I. P., and Bingham, P. M. (1993). Evidence for channel diffusion of pre-mRNAs during nuclear RNA transport in metazoans. *J. Cell Biol.* **121**, 729–742.

CHAPTER 32

EM Methods for Visualization of Genetic Activity from Disrupted Nuclei

Ann Beyer, Martha Sikes, and Yvonne Osheim

Department of Microbiology
University of Virginia Health Sciences Center
Charlottesville, Virginia 22908

I. Introduction

A. The Miller Chromatin Spreading Method

This chapter will focus on use of the Miller chromatin spreading method as applied to studies of genetic activity in *Drosophila*. The Miller technique is a preparatory method for electron microscopy that yields gently dispersed interphase nuclear chromatin as the specimen. The method was developed by Oscar L. Miller, Jr. and his colleagues about 25 years ago (Miller and Beatty, 1969; Miller *et al.*, 1970) and was first applied to the visualization of the amplified ribosomal RNA genes of amphibian oocyte nuclei. The great values of the method are first, its ability to display chromosomal activities, such as DNA replication or RNA transcription, for analysis at the level of the individual genetic locus, while at the same time providing a view of the general population, and second, its conservation of the basic nucleoprotein structures on DNA and RNA, as opposed to most EM procedures for nucleic acids, which involve deproteination.

The preparation of a "Miller spread" involves hypotonic lysis of cells in a large excess of water at a pH of 8 to 9. Nuclear order is disrupted and nuclear contents are greatly dispersed. When this lysate is centrifuged at low speed onto an EM grid, the chromatin mass is deposited as a dispersed 2-D array. One loses the higher order structure of the nucleus as well as nuclear components that are not attached to chromosomes. One gains, however, a high-resolution view of such things as nucleosomal chromatin structure, replication forks, and unfolded active genes with their attached nascent transcripts (reviewed in Miller, 1981).

We have applied the method primarily to the study of premessenger RNA processing because we discovered that there is indeed a great deal of *cotranscriptional* RNA splicing activity, which is accessible to this ultrastructural approach (reviewed in Beyer and Osheim, 1991). For example, we can observe deposition of snRNP components at 5' and 3' splice sites with subsequent intron loop formation and intron removal (Fig. 1). An important aspect of this analysis is that multiple transcripts from the same gene can be analyzed, each one slightly older than the previous one, so that a temporal series of early RNA processing events is available within a single gene.

B. Advantages of *Drosophila* for Miller Spreading

Drosophila has proven to be the system of choice for these analyses. First, it is representative of eukaryotes in general in terms of RNA processing schemes, but its genome is relatively small (20 times smaller than a typical mammalian genome), simplifying the search for active genes in the largely inactive chromatin mass. It is not difficult to find active genes 10–40 kb long with up to 50 nascent transcripts, allowing analysis of up to a 30-min time frame with many "time points" (i.e., individual transcripts), for the study of early RNA processing events. Second, *Drosophila* provides a rich variety of developmental stages and tissue types that are easily dissected and when pulled apart in "pH 9 water" release their nuclear contents for analysis. Nucleoprotein structure is very well preserved because detergents usually are not necessary for Miller spreads with *Drosophila* as they are with many other cell types. Although many different cell types and organisms have been studied by us and other labs using the Miller chromatin spreading approach, we know of no other system better than *Drosophila* in providing the ability to visualize reasonable numbers of long, active, and minimally disrupted genes with relative ease. *Drosophila* also provides the obvious advantages of working with a system with well-established genetics and molecular and cellular biology.

C. Review of Chromatin Spreading Studies in *Drosophila*

The multiple, tandem genes encoding ribosomal RNA in *Drosophila* were first visualized by Hamkalo *et al.* (1973). In subsequent years, the spreading method was used to study the activation of these genes in early embryos, revealing that each promoter is activated individually and that transcription is not limited by Pol I availability (McKnight and Miller, 1976). Recently, Pol I elongation rate was measured on *Drosophila* rDNA genes by calculation of template distance traveled by elongating transcripts after inhibition of initiation, providing the first direct measurement of this activity (Mann and Miller, personal communication).

Aspects of replication in *Drosophila* have been elucidated by use of the Miller chromatin spreading method. Observations include determination of interorigin distance at the cellular blastoderm stage (McKnight and Miller, 1977), analysis of the timing of nucleosome disruption, reformation behind and in front of the replication fork (McKnight *et al.*, 1977), and confirmation that chorion gene amplification occurs by the onion-skin model, involving multiple rounds of chromosomal replication (Osheim and Miller, 1983; Osheim *et al.*, 1988).

Premessenger RNA transcription and processing have been extensively studied using this technique in *Drosophila*. The sister chromatid gene pairs that frequently are visualized in close proximity in embryo chromatin spreads (McKnight and Miller, 1979) have been very important to these studies because they allow analysis of two copies of the same active gene and thus confirmation of the reproducibility (and validity) of the parameters studied. Analyses of sister gene pairs have established the existence and discrete nature of transcription termination sites for Pol II (McKnight and Miller, 1979; Osheim *et al.*, submitted) as well as the existence of a characteristic and nonrandom ribonucleoprotein organization on the transcripts of a given gene (Beyer *et al.*, 1980,1981), which we have since shown can be correlated with splicing (Osheim *et al.*, 1985). This technique was the first to indicate that pre-mRNA splicing typically is cotranscriptional, precedes polyadenylation and is still the only approach to show the general nature of this phenomenon, although it has been confirmed by the study of specific genes (e.g., LeMaire and Thummel, 1990; Zachar *et al.*, 1993; Bauren and Wieslander, 1994).

II. Methods

A. Note on Experimental Protocols

The protocols presented here are very similar to methods previously published (Osheim and Beyer, 1989). The previous article is a more general consider-

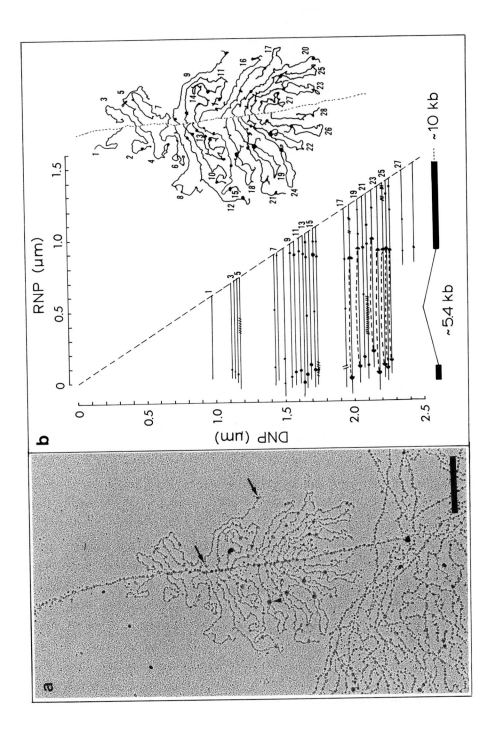

ation of the Miller chromatin spreading method, rather than specific applications to *Drosophila*.

B. Solutions

The two critical factors for optimal dispersal of chromatin are an alkaline pH (pH 8.5–9) and a very low ionic strength. The common dispersal solution, "pH 9 water," is deionized, glass-distilled water that has been adjusted just before use to this pH range by the dropwise addition of pH 10 buffer standard for pH meters. We currently are using Fisher brand buffer standard (Pittsburgh, PA) (Catalog No. SB116-500), which is a 0.05 *M* solution of potassium carbonate, potassium borate, and potassium hydroxide. All of the solutions are made with deionized, glass-distilled water, filtered through 0.2-μm-pore-size disposable Nalgene filter units and stored and used at room temperature, unless otherwise noted.

1. pH 9 water adjusted to pH 8.5–9. The pH adjustment should require between 1 and 3 drops of pH 10 buffer/100 ml of water. If more buffer than this is required, we suggest using specially pure water, such as Mallinckrodt high-performance liquid chromatography grade water (Catalog No. 6795). Make 100 ml of pH 9 water just prior to use.

2. Sucrose-formalin containing 10% (v/v) formalin (i.e., 10 ml of 37% (w/w) formaldehyde solution/100 ml; the brand and grade of formaldehyde are not important, we currently use Fisher brand but previously used Baker) and 0.1 *M* sucrose (RNase-free), adjusted to pH 8.5–9. Make the solution fresh weekly, but adjust the pH daily, if necessary.

3. Kodak Photo-Flo 200 [0.06% (v/v)] solution, pH 8.5–9. Make 100 ml of Photo-Flo fresh daily and store in a 100 ml beaker for convenient rinsing of grids.

4. Phosphotungstic acid [4% (w/v) PTA] stock stain solution. Microfilter and store in a brown glass bottle for up to 1 year.

5. Uranyl acetate [4% (w/v) UA] stock stain solution. Microfilter and store

---

Fig. 1   RNP fibril mapping analysis of an embryonic gene that displays cotranscriptional splicing. (a) Electron micrograph of an unidentified gene. The transcripts are not naked RNA; they are completely coated with (hnRNP?) protein to form a rather uniform fibril. RNP particles of ~20 nm (arrows) and ~40 nm (arrowhead) are deposited at nonrandom positions on this fibril. Bar, 0.5 μm. (b) RNP fibril map. Transcripts are linearized and aligned by abutting 3' termini to regression line. RNP particles are shown as closed circles, hairpin loops are shown as slashed regions, and the open intron loops are shown as dotted lines connecting the two "half-particles" at the points of intramolecular contact. The last two transcripts (Nos. 27 and 28) have been spliced. The proposed intron–exon structure of this ~10 kb transcript is shown at the bottom. (Inset in b) Interpretive tracing of electron micrograph. Dotted line is DNA; solid lines are RNA transcripts.

in a brown glass bottle for up to 1 year. (Caution: This solution contains a radioactive, toxic heavy metal salt.)

C. Materials

1. Film-coated electron microscope grids. We routinely use 300-mesh copper grids with one shiny side and one dull side (for orientation purposes). The grids are covered with a thin carbon film (30–50 nm) before use. The procedure we use for making this thin carbon film has been described in detail (Osheim and Beyer, 1989); it requires a vacuum evaporator and is somewhat demanding and unpredictable. As an alternative, one can purchase carbon film or plastic (Formvar) film coated with carbon from several EM suppliers, such as Ernest Fullam (Latham, NY), Ladd (Burlington, VT), Electron Microscopy Sciences (Fort Washington, PA), or Polysciences, Inc. (Warrington, PA). Carbon films are more fragile than plastic films and more subject to breaking under the EM beam, but they allow higher resolution analysis of the specimen because of their thinness and their purity.

2. Microcentrifugation chambers, used for centrifuging the specimen onto an EM grid. These chambers are made by slicing a ~25-mm-diameter plexiglass rod into discs 5–8 mm in height. A hole 3.5–4 mm in diameter, large enough to hold an EM grid, is drilled into the center of each disc, and a round coverglass is bonded to the bottom surface with epon. In bonding the cover glass, care must be taken to form a complete seal around the hole with epon, while avoiding epon extrusion into the central hole.

3. Dissecting microscope and light.

4. Microscope slides.

5. Double-stick tape.

6. Tweezers, Dumont No. 5, sharpened stainless steel (e.g., Ernest Fullam, Catalog No. 14340).

7. Tweezers, anticapillary, self-closing, Dumont pattern N4 (e.g., Ernest Fullam, Catalog No. 14140).

8. Long forceps, 10–12 inches.

9. Slide with one or two concavities ("depression slide") for chromatin dispersal (e.g., extra-thick hanging drop slide, Fisher Catalog No. 12-565B).

10. Pasteur pipets.

11. Ross optical lens tissue.

12. Bibulous paper.

13. Coverslips, 18 mm round.

14. Microtiter plate (96-well).

15. Small glass petri dish filled with 95% (v/v) ethanol.
16. Wash bottles, one with distilled water and the other with 95% ethanol.

D. Selection and Preparation of Experimental Sample

For experimental material, we have successfully used *Drosophila* oocytes, spermatocytes, embryos, larval fat body, salivary glands, imaginal discs, and adult brain. A very small amount of material is needed per grid—typically one oocyte, embryo, imaginal disc, or brain or a portion of a salivary gland, testis, or fat body. Presumably, other *Drosophila* tissues would also be amenable to chromatin spreading although we have no experience with them. Well-fed, uncrowded flies should be used for all experiments.

The general approach is to manually release the contents of the selected material into a small amount of pH 9 water (as described in Section IIE). For example, embryos of the desired stage are dechorionated by rolling on double-stick tape and then transferred to the pH 9 water, where the vitelline membrane is pulled apart and the embryonic contents are released into the hypotonic solution (Osheim and Beyer, 1989). Similarly, an oocyte of the desired stage is dissected from an ovariole. One can then spread the contents of the entire oocyte or select nurse cells or follicle cells by selecting the desired end of the oocyte (Osheim and Miller, 1983; Osheim and Beyer, 1989).

The experimental material is pulled apart under water using two tweezers while observing through the dissecting microscope. Maximal release of the water-soluble contents is desired at this stage; one need not be gentle. For example, after the embryonic vitelline membrane is pulled apart, the embryo pieces are tapped with the tines of the tweezers and/or shaken in the water to maximize release of the contents, which also is facilitated by the hypotonic nature of the spreading drop. One should not macerate any insoluble material such as the vitelline membrane, however, because it must be efficiently removed (with tweezers) so that it does not settle on the grid, obscuring the material of interest.

In dissecting the desired tissue for chromatin spreading, two things are import-ant—speed and maintenance of a low ionic strength. Since the object is to arrest ongoing *in vivo* genetic activity for EM visualization, it is important not to disturb the organism prior to chromatin dispersal. Heat or cold shock, or other trauma, will halt transcription initiation. Thus, one should start with an individual undisturbed living organism and do the dissection and release into pH 9 water as quickly as possible (<1 min). We avoid anesthetizing flies and instead try to knock them onto a food surface where one can be grabbed and transferred immediately to the dissection liquid. The dissection liquid should be pH 9 water rather than PBS or Ringer's. This ensures the lowest possible ionic strength in the spreading drop and thus the best chromatin dispersal. It also demands a very rapid dissection, which is desirable. If experimental material is exposed to physiological salt-containing solutions or Tris buffers, it becomes intractable to Miller spreading.

E. Basic Chromatin Spreading Procedure

1. Immediately before use, fill the well of a depression slide with 75 μl of pH 9 water.

2. Transfer the desired experimental material for chromatin spreading (Section IID) to the pH 9 water in the depression slide. After releasing the soluble contents of the sample into the water and removing any insoluble material, stir the solution continuously for at least 1 min using the tines of a tweezer and then pipet up and down several times (e.g., using a Pasteur pipet with a pulled-out tip), being careful not to create bubbles.

3. After dispersing for at least 5 min with occasional stirring (or up to 30 min), add 50 μl of the sucrose–formalin solution and mix, either by stirring or pipetting. While the chromatin is dispersing for an additional 10 min or so, prepare the microcentrifugation chamber (Steps 4–6).

4. Using a Pasteur pipet, overfill a clean, dry microcentrifugation chamber with the sucrose–formalin solution so that the solution rounds up above the chamber (creating a convex meniscus) but does not flow out. Make certain that no air bubbles are trapped in the chamber.

5. Using anticapillary tweezers, pick up a film-covered EM grid (Section IIC), holding the grid by the extreme outer edge. Swish the grid for 1 min in 95% ethanol in a small, open petri dish (which renders the grid hydrophilic) and immediately rinse the grid well with the sucrose–formalin solution, by releasing two or three refills of a Pasteur pipet over the grid. Without allowing the grid to dry, slide the grid under the rounded up meniscus on the microcentrifugation chamber with the film side up. Release the grid, which should sink gently to the bottom of the chamber.

6. Using a Pasteur pipet, remove most of the sucrose–formalin solution from the chamber, leaving the chamber $\frac{1}{4}$–$\frac{1}{3}$ full and with the grid flat on the bottom.

7. Gently mix the dispersing chromatin by pipetting and then carefully layer the chromatin mixture over the sucrose–formalin remaining in the chamber, completely filling the chamber. Place a round, 18-mm coverslip on top of the microcentrifuge chamber and press down firmly to seal the coverslip. Blot the excess liquid with bibulous paper. Typically, four chambers containing chromatin from four different samples are prepared during the same period, with the first samples dispersing for longer periods of time than the later samples.

8. Place the microcentrifugation chambers into the buckets of an appropriate swinging bucket rotor. We typically use a Sorvall GLC-1 tabletop centrifuge equipped with 100-ml swinging buckets, which have been fitted with rubber pads to form a flat surface. Long-handled forceps are used to place the chambers on the rubber pads. The chambers are then centrifuged at 2000–2500g for

6 min at room temperature without braking. Alternatively, a Sorvall centrifuge equipped with a HB-4 swinging bucket rotor can be used. Standard black plastic bottle caps with an outside diameter of ~2.8 cm are used as adaptors. A microcentifugation chamber is placed inside a bottle cap, and the cap is slowly lowered to the bottom of the bucket.

9. After centrifugation, remove the coverslip from the grid chamber and add sucrose–formalin to round up the meniscus over the well. Invert the chamber, hold it at eye level, and tap the side of the chamber gently with the back end of the tweezers. The grid should settle to the surface of the drop from which it is plucked with the anticapillary tweezers.

10. Holding the grid securely by the edge, gently swish the grid in a beaker of Photo-Flo solution for about 30 sec. Lift the grid from the Photo-Flo and blot the excess liquid with lens tissue, touching only the edge of the grid and the area between the tines of the tweezers. The grid should then air dry. This should be the first time the grid dries since it was dipped in ethanol (Step 5). The grid can now be stained or grids can be accumulated for later staining.

F. Contrast Enhancement

1. Fill two small beakers, one with 95% ethanol and the other with 0.06% Photo-Flo, for rinsing grids between staining steps.

2. Immediately before staining, prepare a working solution for each stain (PTA and UA) by adding three drops of the stock stain solution to nine drops of 95% ethanol in separate wells of a 96-well microtiter plate. Since the ethanol evaporates rapidly, make up fresh working solutions after staining every two grids.

3. Pick up a grid to be stained by the edge, using anticapillary, self-closing tweezers. Keep the grid in these tweezers until the staining process is completed and the grid is dry. Lower the grid into the PTA solution. Gently agitate the grid for 30 sec, being careful that it is completely submerged but not hitting the sides or bottom of the well.

4. Immediately immerse the grid in the beaker of 95% ethanol and swish the grid gently for 15–20 sec.

5. Lower the grid into the well containing the UA solution and agitate gently for 1 min.

6. Rinse the grid in the beaker of ethanol for about 30 sec.

7. Dip the grid into the Photo-Flo solution for a few seconds.

8. Blot the edge of the grid with lens tissue and air dry. The grids can now be viewed in the electron microscope, or to improve contrast, lightly shadowed with platinum.

9. If a vacuum evaporator is available, a thin metal shadow on the stained specimen is recommended to increase contrast. We use a platinum shadow

15–20 Å thick deposited at a 6–9° angle while the grids rotate at 30–60 rpm. For more details, see Osheim and Beyer (1989).

G. Electron Microscopy

We view grids in a JEOL 100CX transmission electron microscope operated at 80 kV with either a 40- or 60-μm objective aperture. One should use a rather low illumination when scanning grids to minimize film breakage and specimen contamination. Photographs should be taken as soon as a region of potential interest is spotted.

In viewing the grids, there are several considerations to keep in mind. First, the quality of the chromatin specimen can be quite variable, even for those grids made at the same time with the same solutions. If one makes eight grids of a certain type, there typically is at least one that is useful and informative and sometimes several. Second, chromatin distribution on the grid frequently is not uniform. Some regions of the grid may have none, some regions may have the desired concentration, and others may be overcrowded. In those regions with the desired concentration, the chromatin may or may not be dispersed sufficiently. [The chromatin region shown in Fig. 2 is an example of how chromatin will appear when additional dispersal is desired. Although genes can be detected in the chromatin mass, and the "polymerase backbones" representing the densely packed polymerases transcribing the genes are clearly visible, the individual RNA transcripts are in places difficult to trace. The particular example shown here, however, is tightly packed because the multiple DNA strands represent multiple replication bubbles from a chorion amplification control element (Osheim *et al.*, 1985).] Third, in scanning the grids to determine whether there is any useful material present, it may be necessary to view through the binoculars and to use a microscope magnification of 5000 to 10,000. This will certainly be true if one has not used a metal shadow to enhance the contrast of the chromatin. In general, be prepared to see more inactive nucleosomal chromatin than expected. Keep in mind Oscar Miller's counsel (as posted prominently in his lab): "To see genes in action, one must have patience, perseverance, and lots of luck."

H. Trouble Shooting and Adapting to New Tissues

The basic chromatin spreading procedure described here has been most frequently applied to the early *Drosophila* embryo (2.5–6 hr). It works well also for later embryos (at least up to 20 hr), and for adult brain, larval fat body, and imaginal discs. This approach, consisting of hypotonic shock in pH 9 water, should be the first approach attempted on a new tissue.

If one obtains chromatin spreads that are not as dispersed as desired, but in which one can detect individual chromatin strands and individual genes with well-extended transcripts (as in the example in Fig. 2), the first approach is to

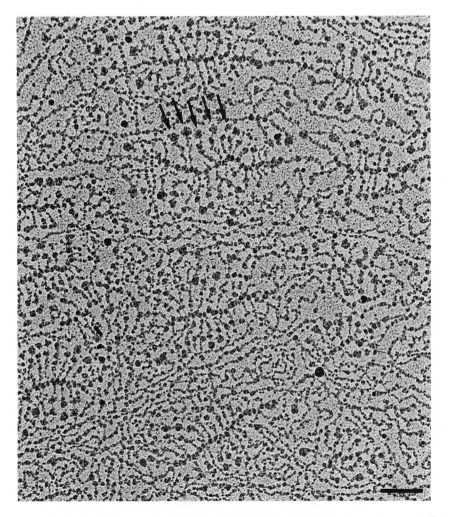

Fig. 2 Portion of a cluster of amplified s36 and s38 chorion genes from Stage 12 oocyte follicle cells. Many more gene copies (not shown) were discernible in the region surrounding the cluster shown. The five arrows point to the 5′ ends of five neighboring transcripts on an s38 gene. The three transcripts to the right (less mature end of the gene) display two ~20-nm particles at positions known to be the 5′ and 3′ splice sites. The two transcripts to the left display a single larger spliceosome particle due to the stable association of the two splice sites. (see text and Osheim *et al.*, 1985). Bar, 0.2 μm.

increase the time and the degree of mixing during dispersal (Section IIE Step 3). Increased pipetting through a small bore pipet to shear the chromatin may be most useful.

If one obtains inactive chromatin that is reasonably well spread, but genes in which the transcripts are not well spread, it is an indication that the pH of the spreading drop has fallen below pH 8 or so. An example of this phenomenon is shown in Fig. 3. To the practiced eye, this mass of entangled fibrils, including large electron dense particles with looped fibrils emanating from them, is obviously an active gene. The rosette structures on the transcripts (representing an intermediate in dispersal of splicing-associated structures) are diagnostic of a pH that is too low. Because the spreading solution is not buffered, the contents

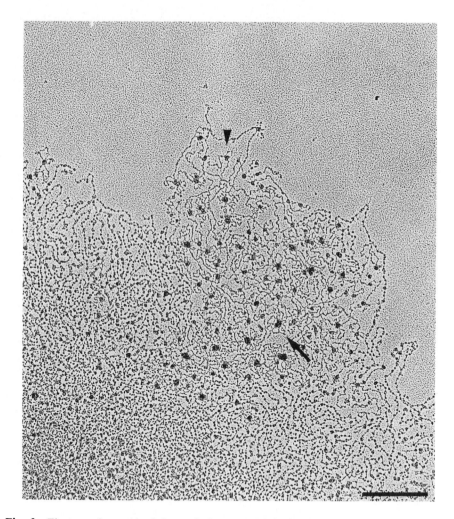

Fig. 3 Electron micrograph of chromatin from an adult brain with a poorly spread gene in the center. Genes with this "rosette" morphology (arrow) are diagnostic of a pH between 7 and 8 in the spreading solution, rather than the desired pH between 8.5 and 9. The rosettes are thought to be clusters of several intron loops (AB and YO, ms in preparation). Single intron loops also can be seen in the better dispersed regions (arrowhead). Bar, 0.5 μm.

of the specimen may contribute enough material to drop the pH. In this situation, one should experiment with (a) increasing the pH 9 water to specimen ratio, (b) buffering the spreading solution with a minimal amount of borate buffer (e.g., 0.5 mM), or (c) adding a detergent to the spreading solution (as discussed here), since this can increase the degree of dispersal as well as add buffering capacity to the spreading drop.

If there is poor release of chromatin from the sample, or very poor dispersal of the chromatin, one should then try the addition of a small amount of detergent to the spreading drop. Detergents will assist in the breakdown of cellular and nuclear membranes and may improve dispersal by partial removal of protein from the nuclear mass. The dishwashing detergent Joy (Proctor and Gamble) at 0.05% was originally found to be most useful for chromatin spreading (Miller and Bakken, 1972); it remains our most frequently used detergent. However, the composition of Joy detergent has changed somewhat over the years and we find that pre-1980 Joy gives the best results. Proctor and Gamble generously provided us with this vintage; we will be happy to provide a sample to anyone interested.

Examples of detergent used are the following: oocyte follicle cells were spread with 0.025% (w/v) sodium deoxycholate and 0.05% Joy at pH 8.5 (Osheim and Miller, 1983), and primary spermatocytes were spread with 0.05% Joy at pH 8.5 (unpublished). *Drosophila melanogaster* salivary glands were spread using 0.6% (w/v) Chaps at pH 8.5 (Hager and Miller, 1991). All of these detergent regimens allowed excellent retention of splicing components on nascent *Drosophila* transcripts. NP-40, sarkosyl, Triton X-100, and digitonin have also been used by various investigators in attempts to improve chromatin spreading in other organisms (reviewed in Osheim and Beyer, 1989). The detergent of choice and its concentration are determined empirically.

I. Analysis of RNA Processing Activity on Nascent Transcripts

1. Characteristics of Pol II Genes

Two kinds of genes typically can be seen in a chromatin spread—Pol I genes encoding preribosomal RNA and Pol II genes encoding premessenger RNA. For analysis of pre-mRNA processing, one must be able to recognize and exclude Pol I genes. Fortunately, this usually is quite easy because the multicopy Pol I genes occur in clusters, are tandemly linked, are highly transcribed, and are each about 2.6 μm in length (8 kb) (Miller and Beatty, 1969; McKnight and Miller, 1976). The third class of eukaryotic genes, those transcribed by Pol III, are very short (<300 nt) and generally go unnoticed in chromatin spreads.

Unlike Pol I genes, Pol II genes vary considerably in length and in transcript density and usually are not tandemly linked to another active gene. As mentioned earlier, we have taken advantage of the chromatin spreading technique to study early RNA processing events on Pol II genes. With the exception of unusual RNP morphologies seen on Y chromosome lampbrush loops in primary spermatocytes (de Loos *et al.*, 1984; our unpublished observations), we have found that there is very little variety in the basic ribonucleoprotein structures that form on nascent transcripts. These can be seen in Fig. 1a and consist of (a) a general coating of the transcripts with proteins assumed to be hnRNP proteins forming a rather uniform basic fibrillar structure of ~5 nm diameter, (b) RNP particles about 20 nm in diameter that form at specific sites on the transcripts of a given gene, (c) RNP particles about 40 nm in diameter that form by the stable association of two nearest-neighbor 20-nm particles, and (d) RNP loops, with a 40-nm particle at their base, also formed by the association of two 20-nm particles.

2. RNP Fibril Mapping

The positioning, time of formation, and temporal relationships between these structures have been analyzed by a technique we call RNP fibril mapping, as shown in Fig. 1b. These maps display schematized RNA transcripts from a single transcription unit, such that similar RNA sequences are aligned and RNP structures occurring on the transcripts are shown at the appropriate positions. First, the contour length of the transcripts and their position on the template is measured, using, for example, an electronic planimeter. Starting from the 5' end and moving into a gene, the transcripts typically increase in length as expected for RNA that is being transcriptionally elongated. The second step is to describe the slope of this initial length increase in an equation that is then used to draw a sloped line used for transcript alignment, as shown, for example, by the dashed line in Fig. 1b. (The least-squares method is used for the derivation of this equation using a series of points in which x is template position of the transcript and y is transcript length.) Third, each transcript is drawn as a straight line of the appropriate (measured) length with its 3' (template-attached) end abutted to the regression line at the appropriate position on the template. This alignment procedure ensures that like sequences are aligned on the various transcripts in the RNP fibril maps. Alignment is not perfect, due to variations in degree of RNP fibril decompaction that arise during the dispersal and centrifugation steps.

3. Splicing–Associated Structures on Pol II Transcripts

It was found that the alternative and simpler method of aligning transcripts at the left of the map by aligning the 5' ends is inadequate because the 5' ends of the transcripts very frequently are modified by cotranscriptional splicing events, which can remove large segments of the transcripts (Beyer *et al.*, 1981; Beyer and Osheim, 1988). This phenomenon is seen on the last two transcripts of the gene in Fig. 1. (Note that transcripts such as these, which are considerably shorter than expected for their position on the template, are not used for derivation of the regression line as described above.) Observation of the gene indicates that prior to this significant length reduction in the nascent RNA population, a loop formed on the transcripts (Transcript Nos. 18, 19, and 22–26) encompassing the length of RNA that is subsequently lost. The loop in turn had been formed by the stable association of two 20-nm RNP particles (seen on Transcript Nos. 9–16) and resulted in the formation of larger particles at the base of the loops. With our current knowledge of events in spliceosome formation and splicing, it is obvious that this is the expected series of ultrastructural events leading to intron removal (reviewed in Green, 1991). The surprising observation at the time was the degree to which the process was cotranscriptional, since splicing was thought to occur usually after transcript release and polyadenylation.

One major disadvantage of the Miller chromatin spreading method is that only in unusual circumstances can one identify the gene observed; thus, it usually is not possible to conclude that RNP particles and loops are occurring at positions expected for introns. One of the first confirmations that the process indeed represents splicing came from visualization of two specific genes with known intron lengths and positions. These were two of the chorion genes, which were identifiable by virtue of their multicopy nature in spreads of oocyte follicle cell chromatin (Osheim and Miller, 1983; Osheim et al., 1985), where the genes are specifically amplified (Spradling and Mahowald, 1980). Figure 2 shows a survey view including several copies of the amplified s36 and s38 chorion genes from Stage 12 follicle cells. Because the amplification is intrachromosomal, the multiple gene copies lie side by side or in close proximity. Details of the coincidence of intron position as related to particle position can be found in Osheim et al. (1985). One can see here, however, that these short genes, which have a single short intron near their 5' ends, exhibit two 20-nm particles at nonrandom positions near their 5' ends (which were shown to correspond to 5' and 3' splice sites (Osheim et al., 1985)), and that the two 20-nm particles are replaced by a single 40-nm particle on the more mature transcripts. (The intervening sequence is looped out but it is too short to be seen as a loop on most of these transcripts.) This is the same sequence of events seen on the gene in Fig. 1, albeit with much shorter introns. There is much additional evidence that the process visualized represents splicing, as discussed in Beyer and Osheim (1991). We also have found that both the spliceosomal particles and loops are absent on transcripts in *Xenopus* oocytes when U1 and/or U2 snRNA is depleted by antisense methods (Y.O. and A.L.B., unpublished).

This splicing-associated ultrastructure occurs on most Pol II genes seen in *Drosophila;* on longer genes, it is possible to observe the sequential formation and removal of multiple intron loops while the transcripts are still nascent (Beyer and Osheim, 1988). Although we have studied cotranscriptional splicing primarily in the early developmental stages of oogenesis and embryogenesis, it is not limited to these periods. For example, the poorly dispersed gene in Fig. 3, which includes several introns as judged by the presence of multiple loops, is from an adult brain. Other groups have observed cotranscriptional splicing in larval salivary glands, using either chromatin spreading (Hager and Miller, 1991) or other techniques (LeMaire and Thummel, 1990; Zachar *et al.*, 1993).

J. Heat Shock

Consequences of the heat-shock response and aspects of recovery from heat shock can be studied productively at the individual gene level by chromatin spreading. Immediately after subjecting the specimen to a 37°C heat shock for 30 min or longer, the nuclei become more difficult to disperse and the aesthetic quality of the spread is decreased due to the deposition on the grids of large electron-dense splotches of unknown origin. One can work around these obstacles, however, to study gene activity, with fairly frequent viewing of identifiable heat-shock genes (Y.O., M.S., and A.B., unpublished). There is also a considerable amount of transcriptional activity on genes that does not correspond to the typical heat-shock loci; the transcripts of these genes are unusual compared to their control counterparts in that they exhibit very little to no splicing activity on the nascent transcripts (cf. Yost and Lindquist, 1986).

K. Chromatin Spreads of Mutant Flies

The availability of mutant fly strains is another advantage to applying the chromatin spreading procedure to *Drosophila*. Although one cannot hope to visualize a specific mutant gene, one can use flies deficient in certain genes to one's advantage. For example, de Loos *et al.* (1984) used *Drosophila hydei* strains that were deleted for specific Y chromosome loops to correlate individual characteristic loop morphologies with specific lampbrush loops. *Bobbed* mutants (*bb8*), which were known to contain insertion sequences in 50% of their Pol I genes, were used to visualize these unusual rDNA genes (Jamrich and Miller, 1984).

We are using flies that overexpress individual hnRNP protein genes (Haynes *et al.*, 1990,1991) to visualize the general effects on splicing of a disruption of normal protein stoichiometries, with results indicating that while most genes are unaffected, others are dramatically affected and appear to be engaged in exon skipping.

III. Summary and Conclusions

This chapter has reviewed the technique of Miller chromatin spreading and its applications to the study of *Drosophila* RNA transcription and processing. The major drawback to the technique as applied to *Drosophila* is the inability to routinely visualize a specific gene of choice or to identify the genes visualized. Although individual genes on plasmid vectors can be visualized after microinjection into *Xenopus* oocytes, similar injection of genes into *Drosophila* embryos has been unsuccessful to date for the purpose of gene visualization (Y.O. and A.L.B., unpublished). Thus, only the chorion genes and certain heat-shock genes have been identified unambiguously. Likewise, techniques of *in situ* hybridization, which in theory could be useful for identifying specific genes, are incompatible with good-quality, high-resolution chromatin spreads (O'Reilly *et al.*, 1994).

However, a unique *in vivo* view is available using this approach, providing information that is lost or inaccessible by the *in vitro* or biochemical study of splicing. If the research question is chosen carefully, application of the chromatin spreading technique can be very valuable. We anticipate that study of splicing mutations (e.g., Kanaar *et al.*, 1993) by this approach will be very informative. In wild-type flies, it is straightforward to visualize four steps in the spliceosome assembly/splicing process on nascent transcripts, from 3' splice site recognition to intron removal. The order and timing of these steps is remarkably reproducible (Beyer and Osheim, 1988). We hope to be able to detect specific effects on these events due to various mutations in splicing proteins or due to their ectopic expression. Although effects on individual specific genes cannot be identified, the stage in *in vivo* splicing that is affected as well as the generality of the effect can be deduced.

Acknowledgments

Many, many thanks to Oscar L. Miller, Jr., who trained all of us in the chromatin spreading technique and who continues to provide advice and encouragement. Research in the authors' lab is supported by grants from the NSF and the NIH to A.L.B.

References

Bauren, G., and Wieslander, L. (1994). Splicing of Balbiani ring 1 gene pre-mRNA occurs simultaneously with transcription. *Cell* **76,** 1–20.

Beyer, A. L., Bouton, A. H., and Miller, O. L., Jr. (1981). Correlation of hnRNP structure and nascent transcript cleavage. *Cell* **26,** 155–165.

Beyer, A. L., Miller, O. L., Jr., and McKnight, S. L. (1980). Ribonucleoprotein structure in nascent hnRNA is nonrandom and sequence-dependent. *Cell* **20,** 75–84.

Beyer, A. L., and Osheim, Y. N. (1988). Splice site selection, rate of splicing and alternative splicing on nascent transcripts. *Genes Dev.* **2,** 754–765.

Beyer, A. L., and Osheim, Y. N. (1991). Visualization of RNA transcription and splicing. *In* "Seminars in Cell Biology" (R. H. Singer, ed.), Vol. 2, pp. 131–140. London: W. B. Saunders.

de Loos, F., Dijkhop, R., Grond, C. J., and Hennig, W. (1984). Lampbrush chromosome loop-specificity of transcript morphology in spermatocyte nuclei of *Drosophila hydei. EMBO J.* **3,** 2845–2849.

Green, M. R. (1991). Biochemical mechanisms of constitutive and regulated pre-mRNA splicing. *Annu. Rev. Cell Biol.* **7,** 559–599.

Hager, E. J., and Miller, O. L., Jr. (1991). Ultrastructural analysis of polytene chromatin of *Drosophila melanogaster* reveals clusters of tightly linked co-expressed genes. *Chromosoma* **100,** 173–186.

Hamkalo, B. A., Miller, O. L., Jr., and Bakken, A. H. (1973). Ultrastructure of active eukaryotic genomes. *Cold Spring Harbor Symp. Quant. Biol.* **38,** 915–919.

Haynes, S. R., Johnson, D., Raychaudhuri, G., and Beyer, A. L. (1991). The *Drosophila Hrb87F* gene encodes a new member of the A and B hnRNP protein group. *Nucleic Acids Res.* **19,** 25–31.

Haynes, S. R., Raychaudhuri, G., and Beyer, A. L. (1990). The *Drosophila Hrb98DE* locus encodes four protein isoforms homologous to the mammalian A1 hnRNP protein. *Mol. Cell. Biol.* **10,** 316–323.

Jamrich, M., and Miller, O. L., Jr. (1984). The rare transcripts of interrupted rRNA genes in *Drosophila melanogaster* are processed or degraded during synthesis. *EMBO J.* **3,** 1541–1545.

Kanaar, R., Roche, S. E., Beall, E. L., Green, M. R., and Rio, D. C. (1993). The conserved pre-mRNA splicing factor U2AF from *Drosophila:* Requirement for viability. *Science* **262,** 569–575.

LeMaire, M. F., and Thummel, C. S. (1990). Splicing precedes polyadenylation during *Drosophila* E74A transcription. *Mol. Cell. Biol.* **10,** 6059–6063.

McKnight, S. L., Bustin, M., and Miller, O. L., Jr. (1977). Electron microscopic analysis of chromosome metabolism in the *Drosophila melanogaster* embryo. *Cold Spring Harbor Symp. Quant. Biol.* **42,** 741–754.

McKnight, S. L., and Miller, O. L., Jr. (1976). Ultrastructural patterns of RNA synthesis during early embryogenesis of *Drosophila melanogaster. Cell* **8,** 305–319.

McKnight, S. L., and Miller, O. L., Jr. (1977). Electron microscopic analysis of chromatin replication in the cellular blastoderm *Drosophila melanogaster* embryo. *Cell* **12,** 795–804.

McKnight, S. L., and Miller, O. L., Jr. (1979). Post-replicative nonribosomal transcription units in *D. melanogaster* embryos. *Cell* **17,** 551–563.

Miller, O. L., Jr. (1981). The nucleolus, chromosomes, and visualization of genetic activity. *J. Cell Biol.* **91,** 15s–27s.

Miller, O. L., Jr., and Bakken, A. H. (1972). Morphological studies of transcription. *Acta Endocrinol. Suppl.* **168,** 155–177.

Miller, O. L., Jr., and Beatty, B. R. (1969). Visualization of nucleolar genes. *Science* **164,** 955–957.

Miller, O. L., Jr., Hamkalo, B. A., and Thomas, C. A., Jr. (1970). Visualization of bacterial genes in action. *Science* **169,** 392–395.

O'Reilly, M. M., French, S. L., Sikes, M. L., and Miller, O. L., Jr. (1994). Ultrastructural *in situ* hybridization to nascent transcripts in chromatin spreads. *Chromosoma,* **103,** 122–128.

Osheim, Y. N., and Beyer, A. L. (1989). Electron microscopy of RNP complexes on nascent RNA using the Miller chromatin spreading method. *In* "Methods in Enzymology" (J. N. Abelson and M. I. Simon, eds.), Vol. 180, pp. 481–509. Orlando: Academic Press.

Osheim, Y. N., and Miller, O. L., Jr. (1983). Novel amplification and transcriptional activity of chorion genes in *Drosophila melanogaster* follicle cells. *Cell* **33,** 543–553.

Osheim, Y. N., Miller, O. L., Jr., and Beyer, A. L. (1985). RNP particles at splice junction sequences on *Drosophila* chorion transcripts. *Cell* **43,** 143–151.

Osheim, Y. N., Miller, O. L., Jr., and Beyer, A. L. (1986). Two *Drosophila* chorion genes terminate transcription in discrete regions near their poly(A) sites. *EMBO J.* **5,** 3591–3596.

Osheim, Y. N., Miller, O. L., Jr., and Beyer, A. L. (1988). Visualization of *Drosophila melanogaster* chorion genes undergoing amplification. *Mol. Cell. Biol.* **8,** 2811–2821.

Spradling, A. C., and Mahowald, A. P. (1980). Amplification of genes for chorion proteins during oogenesis in *Drosophila melanogaster. Proc. Natl. Acad. Sci. U.S.A.* **77,** 1096–1100.

Yost, H. J., and Lindquist, S. (1986). RNA splicing is interrupted by heat shock and is rescued by heat shock protein synthesis. *Cell* **45,** 185–193.

Zachar, Z., Kramer, J., Mims, I. P., and Bingham, P. M. (1993). Evidence for channeled diffusion of pre-mRNAs during nuclear RNA transport in metazoans. *J. Cell Biol.* **121,** 729–742.

PART VI

Molecular and Classical Genetic Analysis

One of the great strengths of *Drosophila* research is that the community of workers has never been content with the advantages offered by the fly. Rather, they have continuously tinkered with the relevant genetic, cytological, and biochemical protocols to enhance the utility of the organism. From everyone's perspective, *Drosophila* research entered a new level of manipulability after the discovery of germ line transformation. One outcome of that advance is that it opened the way for the engineering of DNA constructs and special chromosomes that allow expression of a particular gene either ectopically or in a very limited manner (for example within only small subsets of fly tissues). In parallel, more biochemically or pharmacologically oriented workers have continued to improve methods for delivering substances into developing flies so as to inactivate specific systems or even particular proteins. Finally, cytologists continue to devise ever more precise assays for the functions of cells either over- or underexpressing particular genes and proteins.

Brand *et al.* begin the section with a review of techniques of ectopic expression (Chapter 33). They focus their discussion on two basic approaches, the first of which involves expressing the gene of interest using a heat-shock promoter and defined pulses of heat. The utility of this method is greatly extendable by using it in conjunction with the flp/FRT recombinase system, as clones of cells having stable ectopic expression patterns can be created during development and effects on a variety of tissues sampled rapidly. The authors also discuss advantages and disadvantages of the GAL4 system and in particular the flexibility offered by partitioning the activator sequence and target gene into distinct fly strains, ensuring survival of each. A second bonus of the GAL4 system is that the gene in question can be conveniently expressed in a variety of qualitatively and quantitatively distinct patterns, by utilizing various strains expressing GAL4 via tissue-specific promoters.

Xu and Harrison next present state of the art methods for creation of mosaics using the flp/FRT recombinase system (Chapter 34). They have constructed chromosomes having FRT sites to implement mitotic recombination and low perdurance markers to facilitate clone recognition. Using these chromosomes in conjunction with appropriate markers for either internal or external tissue, it is possible to create and recognize clones having virtually any mutation. One need only recombine said mutation onto a chromosome having a centromere-proximal FRT insertion. These advances have made mosaic analysis far more versatile than before and also eliminate the lethality commonly engendered by the necessity of irradiating larvae to enhance mitotic recombination in the course of conventional mosaic production.

Bieber discusses techniques for analysis of cellular adhesion in cultured *Drosophila* cells (Chapter 35). Much remains to be learned about the mechanisms by which cellular adhesion is regulated during embryogenesis, and assays must continue to be developed and refined to ensure progress. Schneider's line 2 (S2) cells have low intrinsic adhesion and hence are ideal recipients for genes encoding candidate proteins molecules. Bieber describes methods for characterizing these transfected cells using aggregation assays or antibody staining. He additionally recounts vital dye marking methods by which to monitor genotypically distinct (for example transfected vs untransfected) cells in the course of experiments.

Schubiger and Edgar have written a provocative chapter detailing a variety of methods for introducing drugs into embryos so as to perturb particular events (Chapter 36). The authors quite correctly point out that many objections to pharmacological experiments also apply to genetics and suggest that drug injection offers an alternative approach that can be used to verify or enhance a genetic experiment (and vice versa?). After reviewing appropriate injection methods, the authors summarize usage of drugs that inhibit DNA replication, transcription, protein synthesis, microtubules, and actin, as well as injection of a variety of labeling agents. As effects of more substances are characterized, the possibilities for using particular ones to implement a genetic screen or enhance a phenotype will continue to improve.

Finally, Beerman and Jay anchor the section (Chapter 37) with a description of a method that has the potential to become as important as conventional genetics, chromophore-assisted laser inactivation of proteins (CALI). The method involves conjugating particular antibodies with a dye, malachite green, binding the antibody to the target protein *in vivo*, and then ablating the antibody-antigen complex using laser irradiation. The chromophore efficiently targets the laser energy so that only the protein antigen is damaged. The method seems to work very well on 90% of proteins. Beerman and Jay provide instructions for antibody conjugation, irradiation, and phenotype analysis. In addition, they describe set up for the appropriate microlasers.

CHAPTER 33

Ectopic Expression in *Drosophila*

Andrea H. Brand,★ Armen S. Manoukian,† and Norbert Perrimon†‡

★ Wellcome/CRC Institute and
Department of Genetics
University of Cambridge
Cambridge CB2 1QR
United Kingdom

† Department of Genetics
‡ Howard Hughes Medical Institute
Harvard Medical School
Boston, Massachusetts 02115

I. Introduction

The identity of a cell is determined, in large part, by its characteristic pattern of gene expression. The ability to manipulate gene expression at will is thus a powerful tool for studying development. At least three possible consequences might result from activating a gene in a cell in which it is normally repressed: (1) ectopic expression will have no effect on development; (2) ectopic expression will cause the cell to undergo a change in fate; (3) ectopic expression forces neighboring cells to change fates. In this way, it is possible to test whether expression of a gene is necessary and sufficient to determine cell identity and whether its role is autonomous or nonautonomous. For example, misexpression of the ligand *boss* has been used to assay whether all of the cells that express the *sevenless* receptor tyrosine kinase are competent to respond to ligand-induced activation (Van Vactor *et al.,* 1991). It is also feasible to test whether a transcriptional activator is both necessary and sufficient for transcription of a target gene. Such an analysis can be particularly useful when a protein acts as an activator in one context and as a repressor in another. For example, at different stages of embryonic development, *even-skipped* acts as either a positive or a negative regulator of *fushi-tarazu* transcription (Manoukian and Krause, 1992).

There are now several different methods for ectopic expression in *Drosophila,* each with its merits and its limitations. The first technique is to drive expression of a gene using the transcriptional regulatory sequences from a defined promoter (Zuker *et al.,* 1988; Parkhurst *et al.,* 1990; Parkhurst and Ish Horowicz, 1991). Tissue-specific promoters allow transcription to be restricted to a defined subset of cells. Perhaps the best illustrated example of this technique is the use of the *sevenless* promoter to drive gene expression in a subset of cells in the eye imaginal disc. This approach has proven invaluable in analyzing signal transduction and cell fate determination in the eye (see, for example, Basler *et al.,* 1991). However, the method is limited by the availability of cloned and characterized promoters that can direct expression in a desired pattern. Furthermore, when the gene product to be expressed is toxic to the organism, it becomes impossible to establish stable transgenic lines carrying the chimeric gene.

The second method is to drive expression of a gene from a heat-shock promoter. A gene can then be turned on at a specific point in development by heat shocking the transgenic animal (Struhl, 1985; Schneuwly *et al.,* 1987; Ish-Horowicz and Pinchin, 1987; Ish-Horowicz *et al.,* 1989; Gonzales-Reyes and Morata, 1990; Blochlinger *et al.,* 1991; Steingrimsson *et al.,* 1991; Manoukian and Krause, 1992). An advantage of the heat-shock method is that it permits inducible expression throughout the organism and that the levels of ectopic expression can be easily regulated by altering the duration, or the temperature, of the heat shock (Manoukian and Krause, 1992). The disadvantages associated with the technique are that ectopic expression cannot be targeted to specific cells, that basal levels of expression are observed from heat-shock promoters, and that heat shock can induce phenocopies (Petersen and Mitchell, 1987; Petersen, 1990; Yost *et al.,* 1990).

A more recent inducible technique for ectopic expression relies on site-specific recombination catalyzed by the flp recombinase from *Saccharomyces cerevisiae,* which was first shown to function in flies by Golic and Lindquist

(1989). flp can promote recombination between two flp recombination targets, or FRTs. When the FRTs are inverted with respect to each other, flp recombination inverts, or "flips," the sequence between the two sites. When the FRTs are in the same orientation, flp recombination serves to excise the sequence between the FRTs. To generate an inducible transgene, Struhl and Basler (1993) and Buenzow and Holmgren (in preparation) cloned a flp cassette, consisting of a transcriptional termination sequence bounded by FRTs, between a constitutive promoter and the sequence they wished to express. When the cassette is in place, transcription terminates upstream of the coding sequence, and the gene is effectively silent. To excise the flp cassette, flp recombinase is expressed from a heat-shock promoter. After heat shock, flp recombinase is expressed and catalyzes the excision of the flp cassette, thus allowing the transgene to be transcribed.

Although the transgene is activated after a heat shock, the flp technique differs significantly from the heat-shock method described above. First, using the heat-shock method, a pulse of ectopic expression is induced. With the flp technique, once the transgene is activated, expression is maintained throughout development because it is driven by a constitutive promoter. Second, the heat-shock method induces ectopic expression throughout the animal. By comparison, since flp recombination is not 100% efficient, ectopic expression occurs in clones of cells, which vary in size according to the extent of the heat shock and the time in development when it is delivered. These two properties make the flp/FRT system ideal for mapping cell lineages: transcription can be activated within a clone of cells, and expression is maintained from that point forward. Specific cells and their progeny can thus be followed through development. Buenzow and Holmgren have used this technique successfully to map the lineages of the *gooseberry*-expressing neuroblasts in the embryonic central nervous system (in preparation).

Advantages of the flp/FRT system are that it is inducible and that, in principle, expression can be activated in any cell in the organism, although dividing cells may be favored. Struhl and Basler (1993) have used the system to ectopically express *wingless* in the leg and wing imaginal discs. A disadvantage of the method is that, because the clones are generated randomly, ectopic expression varies from animal to animal.

The GAL4 system (Fig. 1; Brand and Perrimon, 1993) is a technique that overcomes some of the drawbacks of other methods and offers great scope for future improvements. First, the system allows the rapid generation of individual strains in which ectopic expression of the gene of interest (the target gene) can be directed to different tissues or cell types. Second, the method separates the target gene from its transcriptional activator in two distinct transgenic lines. In one line the target gene remains silent in the absence of its activator; in the second line the activator protein is present but has no target gene to activate. This ensures that the parental lines are viable. Only when the two lines are crossed is the target gene turned on in the progeny, and can the phenotypic

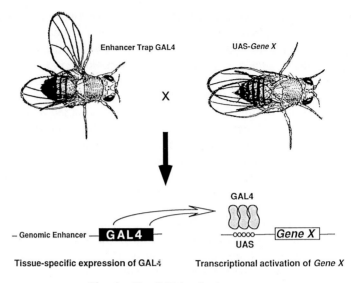

Fig. 1 The GAL4 activation system.

consequences of misexpression (including lethality) be conveniently studied. Finally, the method is designed to generate lines that express a transcriptional activator, rather than an individual target gene, in numerous patterns. The target can then be activated in different cell and tissue types merely by crossing a single line carrying the desired target to a library of activator-expressing lines. Thus, a library of different activator-expressing lines can direct each new target gene to be expressed in numerous distinct patterns.

Two approaches can be used to generate different patterns of GAL4 expression. The first is to drive *GAL4* transcription using characterized promoters (Fischer *et al.*, 1988; Brand and Perrimon, 1993). The second is based on "enhancer detection" (O'Kane and Gehring, 1987; Bier *et al.*, 1989; Bellen *et al.*, 1989; Wilson *et al.*, 1989). The GAL4 coding sequence is fused to the P-transposase promoter to construct the vector pGawB which, depending upon its genomic site of integration, can direct expression of GAL4 in a wide range of patterns in embryos, larvae, and adults (Brand and Perrimon, 1993). This eliminates the need to link numerous different promoters to the *GAL4* gene and allows expression in novel patterns from enhancers that have not yet been described. In addition, the enhancer detection/GAL4 vector can be mobilized to new genomic sites simply by P-transposition (Cooley *et al.*, 1988; Robertson *et al.*, 1988). In this way, a single transformant can be used to generate a large number of transgenic lines, each exhibiting a different GAL4 expression pattern.

GAL4-responsive target genes can be created using the vector pUAST (Brand and Perrimon, 1993), into which sequences can be subcloned behind a tandem array of five optimized GAL4-binding sites (hereafter referred to as the UAS, for *U*pstream *A*ctivation *S*equence) and upstream of the SV40 transcriptional terminator. It is possible, then (1) to subclone any sequence behind GAL4-binding sites; (2) to activate that target gene only within cells in which GAL4 is expressed; and (3) to observe the effect of this aberrant expression on development.

As with the other expression systems we have described, there are several drawbacks associated with the GAL4 system. To date, we have seen no GAL4-mediated misexpression in the female germline, in contrast with enhancer detection/*lacZ* screens in which one-third of the insertion lines express β-galactosidase in the germline (Fasano and Kerridge, 1988). During embryogenesis, GAL4-dependent activation has not been seen to occur prior to Stage 6 (about 3 hr of development). The GAL4 system is not yet generally inducible, although in specific cases it can be made inducible by activating a target gene that is temperature sensitive.

In the remainder of this chapter, we describe protocols for two of the techniques that are currently used to conduct ectopic expression experiments: the heat-shock method and the GAL4 system. Both of these techniques rely on conditional expression methods for the simple reason that attempts to drive ectopic expression of a gene by a heterologous promoter is generally unfruitful since misexpression of most genes whose expression is developmentally regulated is usually lethal to the organisms (Parkhurst and Ish-Horowicz, 1991).

II. The Heat–Shock Method

First, we describe some examples in which heat-shock-induced ectopic expression has been useful in studying the developmental role of a specific gene and then the protocols.

A. Applications of the Heat–Shock Technique

Complex regulatory interactions among the pair-rule class of *Drosophila* segmentation genes have been demonstrated (Ingham and Gergen, 1988). Models have been suggested to explain the mechanisms through which the pair-rule genes function to initiate segment polarity (DiNardo and O'Farrell, 1987; Lawrence *et al.*, 1987). One approach to directly test the validity of these models is to misexpress or overexpress a gene of interest and analyze its effect on the development of the animal. Such experiments allow one to test specific models of gene activities. For example, in loss-of-function *paired* mutants, the segment polarity gene *engrailed* is not activated in the odd-numbered parasegments of the embryo, suggesting that *paired* acts as an activator of *engrailed*. This hypothesis is substantiated by the result that ectopic *paired* expression using the heat-shock promoter can lead to the expansion of the *engrailed* stripes affected in *paired* mutant embryos (Morrissey *et al.*, 1991).

Misexpression of a specific gene can help unravel novel regulatory functions for the gene analyzed. For example, in *runt* mutant embryos all *engrailed* stripes are present, but are of unequal spacing. This result makes it rather difficult to understand whether *runt* acts as a negative or positive regulator of *engrailed* (Martinez-Arias and White, 1988). Using heat-shock *runt*, it has been possible to show that *runt* can act as a negative regulator of the odd parasegment *engrailed* stripes (Manoukian and Krause, 1993).

There are a number of cases in which the examination of mutant phenotype is not sufficient to fully understand the developmental function(s) of a specific gene. Since mutations reflect a cumulative requirement for a gene product, stage- or temporal-specific activity of a given gene may be difficult to understand. This is especially true if a specific gene has opposite functions at different times during development. For example, in *even-skipped* mutants, *fushi-tarazu* stripes expand, suggesting that *even-skipped* acts as a negative regulator of *fushi-tarazu*. However, by inducing ectopic *even-skipped* at different stages of embryogenesis, it has been discovered that *even-skipped* can act as both a positive and a negative regulator of *fushi-tarazu* transcription depending on the developmental stage of the embryo (Manoukian and Krause, 1992).

The kinetics of gene activity, i.e., the timing of the response of cells to a specific gene product, cannot be determined by simply looking at mutants. If mutations in two different genes result in similar alteration of the pattern of expression of a specific gene, the kinetic of response following ectopic expression can be used to classify the two genes relative to one another in a temporal pathway. For example, since ectopic *paired* expression results in the expansion of the domain of *engrailed* expression within a time frame during which ectopic *even-skipped* does not have an effect, it follows that *paired* can act faster in the regulation of *engrailed* than *even-skipped*. Therefore, one can assume that the *paired* transcription factor is more likely to be a direct regulator of *engrailed* transcription than the *even-skipped* homeobox gene.

Using heat-shock-inducible transgenes, one can attempt to disentangle complicated gene networks (segmentation gene hierarchy outlined previously). In this way, it has been possible to establish a temporal regulatory hierarchy of pair-rule genes acting to establish the spatial pattern of *engrailed* expression (Ish-Horowicz and Pinchin, 1987; Ish-Horowicz *et al.*, 1989; Manoukian and Krause, 1993).

The effects of changing concentrations of a gene product cannot be established using strict genetic mutant analysis. However, the use of heat-shock treatments allows one to assess the differential roles of different concentrations of a specific protein by varying the duration of heat shock. For example, it was found that the even-skipped protein may act as a local morphogen in the determination of cell fates in the embryo (Manoukian and Krause, 1992). Using this methodology, the properties of differing concentrations of most gene products can be analyzed *in vivo* (Manoukian and Krause, 1992; Lardelli and Ish-Horowicz, 1993).

B. Driving Expression from Heat–Shock Promoter

Using heat-inducible promoters (e.g., *hsp70* promoter), it is possible to induce the expression of specific genes in a majority of *Drosophila* cell types. Generally, a cDNA representing the gene of interest is cloned in P-element transformation vectors (see following). By creating a fusion between the cDNA and a heat-inducible promoter such as *hsp70,* it is possible to introduce an inducible transgene into the genome of *Drosophila*. This transgene will thus provide the luxury of expressing specific genes in all tissues following short heat pulses. There are a variety of P-element constructs available that utilize the *hsp70* (Struhl, 1985; Schnewly *et al.,* 1987; Krause *et al.,* 1988) and *hsp82* (Dorsett *et al.,* 1989) promoters. In most experiments, the heat-shock *fushi-tarazu* transgene made by Struhl (1985) can be induced more efficiently that those generated by Krause *et al.* (1988). Regardless of the type of construct, the efficiency of the heat-shock procedure is crucial and the general protocol discussed here has been proven successful for many different heat-inducible transgenes (Manoukian and Krause, 1992,1993; Fitzpatrick *et al.,* 1992).

1. Heat Treatment of Embryos

Protocol

1. Following collections on apple juice agar plates, embryos are collected on nylon sieves and washed. Excess liquid should be removed by wiping the bottom of the sieve with paper towels.

2. Dry embryos can easily be transferred to glass coverslips upon contact.

3. Embryos to be used for cuticle analysis are then covered with Halocarbon oil, while those to be used for histochemical analysis are covered with 80% glycerol solution. In both cases, just enough liquid should be used to cover only the embryos on the coverslip. Using a thin strip of plastic, the embryos should be gently spread to constitute a monolayer.

Note: While the embryos are covered with Halocarbon oil or glycerol, one can mount the coverslip on top of a slide and observe the embryos under a compound microscope. In this way, the embryos can be staged and inappropriate embryos can be discarded.

4. For efficient heat shocks, temperatures in excess of 37°C should be avoided. It is best to heat-treat embryos at 35–36°C. For this purpose, heat shocking in a water bath is recommended. Prewarmed water can be placed in a container and placed into the water bath. The embryos are simply heat shocked by floating the coverslip on the prewarmed water in the container. Using this methodology, heat-shock durations in excess of 10 min are excessive and can cause lethality (Manoukian and Krause, 1992). Usually heat shocks 5–7 min in duration are quite effective in the induction of expression throughout embryos.

Note: A less effective, yet simpler method involves the submersion of embryos in prewarmed water placed into Eppendorf tubes. Once the embryos are suspended in water, the tube is submerged into a standard water bath and heat shocked for the requisite time interval. Heat shocks in excess of 10 min may be necessary for the induction of certain transgenes.

5. Following heat shocks, embryos are transferred. For cuticle analysis, the embryos are transferred to apple juice agar plates and aged accordingly at the appropriate temperature. Embryos to be used for histochemical analysis are washed into sieves and the sieves are placed onto apple juice agar plates for aging and analysis. Excess water easily washes away the glycerol. For timed experiments, a standard approach is to keep the duration of heat shock constant and to age the embryos from the initial point of the heat shock.

The major advantage of this methodology is that heat pulses can be delivered with high efficiency within a short period of time. By delivering heat shocks of short duration, one can begin to analyze developmental events with a finer temporal accuracy (see following for examples).

2. Varying Levels of Ectopic Gene Expression

By varying the duration of the heat shock, one can vary the levels of ectopic gene expression (Manoukian and Krause, 1992). Heat shocks of 5–7 min can induce levels that are essentially two- to threefold wild-type levels in the embryo (Manoukian and Krause, 1992). Therefore, it is possible to analyze the effects of varying concentrations of a specific gene product within a certain time frame in the embryo. This methodology allows one to explore the possible morphogenetic and biological activities of genes *in vivo*.

3. Studying the Kinetics of Ectopic Gene Activity

By staging embryo collections and inducing ectopic gene expression at specific times during development, one can analyze the kinetics of the response of cells to the ectopic induction of a specific gene product. Embryos can simply be fixed at specific time intervals following the initiation of the heat shock and analyzed. In this way, the kinetics of response of cells following the induction of specific genes can be determined. If two or more gene products in the embryo affected the same developmental process in the same manner, one could use the kinetics of their activities as a method of classification. In this way, a temporal epistatic pathway can be established when a genetic epistatic analysis is impossible.

4. Heat Treatment of Larvae

It is also possible to deliver heat shocks at larval and postlarval stages of *Drosophila* development. Briefly, embryos are collected on apple plates and staged to hatching. Larvae are then staged by analyzing mouth parts and then transferred (washed onto a sieve) to glass food vials. Heat shocks are administered in glass vials in a water bath 10–15 min in duration (again temperatures of 35–36°C are best). These heat-shock conditions can efficiently induce extensive levels of ectopic protein (S. Scanga, A.S.M., and E. Larsen, in preparation).

5. Heat Treatment of Adults

Glass or plastic vials containing adults can be submerged in a 33–36°C water bath for 15–20 min. Several short heat pulses are delivered, with equivalent recovery times, at room temperature. This regimen seems to be preferable to extended periods at 37°C, both with respect to increasing the level of induction and to maintaining the viability of the flies. After the final heat shock, the flies are transferred to fresh food at room temperature.

C. Discussion

The heat-shock method fails when a gene is needed continuously: multiple heat shocks have to be administered, which can affect development. The GAL4 system is a better technique in this case. In addition, there are slight developmental delays that are associated with the heat-shock response in embryos and larvae. However, the methods described here are quite successful at minimizing developmental problems caused by severe heat shock. Developmental delays are kept at a minimum and phenocopies due to heat shock are greatly reduced. Some stages of *Drosophila* development may be inaccessible to the heat-shock method as successful methods to induce high levels of expression in the nurse cells and oocyte have not been reported. For sustained expression and misexpression in discrete patterns, the GAL4 system is a better approach.

III. The GAL4 System

We first describe examples in which the GAL4 system has proven useful in answering some developmental questions and then the vectors available for directing GAL4 expression in cell- and tissue-specific patterns, followed by a description of the vector, pUAST, for constructing GAL4-responsive target genes. Some of the strains currently available are listed in Table I. Next, we describe the generation of enhancer detection/GAL4 insertion lines and the method we use to screen for embryonic expression patterns. Finally, we outline the procedure for ectopic expression of target genes. In the discussion, we review the drawbacks of the technique and discuss the potential for improvements to the system.

Table I
GAL4 and UAS Insertion Lines

GAL4 insertion line (pGawB)	Chromosome	Expression	Reference
hsp70-GAL4[7-1]	II		A.B., unpublished
hsp70-GAL4[2-1]	III (viable)		A.B., unpublished
24B	III (viable)	Mesoderm	1 (Figs. 3B and 3C)
31-1	III	Nervous system	1 (Figs. 3D–3G)
Prd-GAL4	III	Stripes	1 (L. Fasano and C. Desplan, unpublished)
1J3	III (viable; hairy)	Stripes/discs	1 (Fig. 3A)
32B	III (viable)	Stripes/discs	1 (Figs. 9E and 9F)
69B	III (viable)	Stripes/discs	1 (Figs. 9G and 9H)
71B	III (viable)	Discs	1 (Figs. 9C and 9D)
30A	II (viable)	Discs	1 (Figs. 9A and 9B)
Rh2-GAL4	II (viable)	Ocelli	1 (Fig. 2B)
55B	III (viable)	Follicle cells	2 (Fig. 2A)
UAS Line (pUAST)			
UAS-lacZ[4-12]	II (viable)		1, 2
UAS-lacZ[4-8-2]	III (viable)		A.B., unpublished
UAS-Dras2[Val14]	II (viable)		1
UAS-Draf[gof F179]	III (viable)		A.B., X. Lu. N.P., in preparation
UAS-humraf[gof ra2]	X (viable)		2

Note. References: 1, Brand and Perrimon, 1993; 2, Brand and Perrimon, 1994.

A. Applications of the GAL4 System

The GAL4 system has been used effectively to induce cell fate changes in embryos. This has been achieved by directed misexpression of *even-skipped* (Brand and Perrimon, 1993), *abdominal-A* (Greig and Akam, 1993), *wingless* (K. Yoffe, E. Wilder, unpublished), and *engrailed* (A.B., unpublished). When a protein is required in a number of developmental processes or acts at several times in development, its separate roles can be conveniently studied by restricting ectopic expression to specific cells or tissues or to a particular stage of development.

Dominant mutations recovered by classical genetic techniques have been invaluable in identifying and ordering the components of signal transduction pathways. Dominant phenotypes generated with the GAL4 system can be used in a similar fashion. GAL4-directed expression of activated Raf within follicle cells is sufficient to dorsalize both the egg shell and the embryo (Brand and Perrimon, 1994). The dorsalized phenotype has been used in epistasis tests to show that Raf functions downstream of the EGF receptor in establishing the dorsoventral polarity of the egg.

The GAL4 system can facilitate structure/function analysis. First, the promoter of the gene to be analyzed (*Gene X*) is used to drive transcription of GAL4. The *Gene X* coding sequence is then subcloned behind the GAL4 UAS. In an ideal situation, expression of GAL4 from the *Gene X* promoter, which drives transcription of the *Gene X* coding sequence, is sufficient to rescue the *Gene X* mutant phenotype. Mutations can then be introduced into *UAS-Gene X,* and the GAL4 system can be used to express mutated forms of the protein in either a wild-type or a mutant background. In this way, gain-of-function, loss-of-function, hypomorphic, and dominant negative mutations can be identified.

Cells, and subcellular structures, can be tagged by GAL4-directed expression of markers such as β-galactosidase. β-Galactosidase can be either cytoplasmic or targeted to the nucleus by addition of a nuclear localization signal. The cytoplasmic β-galactosidase expressed from *UAS-lacZ*[4-1-2] (Brand and Perrimon, 1993) can fill axons in the adult nervous system, a property that is proving useful in mapping neurons and their projections and in generating three-dimensional reconstructions of the adult central nervous system (K. Kaiser and D. Shepherd, unpublished). The cell surface marker, CD2, has been placed under GAL4 control (O. Dunin-Borkowski and N. Brown, in preparation) and is being used to follow the development of the embryonic mesoderm.

Finally, the GAL4 system has proven effective in the conditional delivery of toxins, such as diphtheria toxin, as a means of directed cell killing (A.B., J. Haseloff, H. Goodman, and N.P., unpublished). Cell ablation can be used to study the role of cell–cell contact in establishing cell identity.

B. GAL4 Expression Vectors

1. hsp70–GAL4

To make an hsp70-GAL4 gene fusion, the GAL4 coding sequence was first excised from vector pLKC15 (a gift from L. Keegan) as a *Hind*III fragment. This fragment extends from a synthetic *Hind*III site inserted approximately 15 nucleotides upstream of the initiator methionine and includes the complete GAL4 coding sequence and its transcriptional terminator (L. Keegan, personal communication). The *Hind*III fragment was subcloned downstream of the hsp70 promoter in the vector pHSREM (Knipple and Marsella-Herrick, 1988) to give plasmid pF18-13. A *Not*I fragment carrying the hsp70 promoter/GAL4 coding sequence/hsp70 transcriptional terminator was removed from plasmid pF18-13 and subcloned into the *Not*I site of pCaSpeR3 to give pF89.

2. pGaTB (Fig. 2)

To create a vector for subcloning promoters upstream of GAL4, the heat-shock consensus sequences and the hsp70 TATA box and transcriptional start site were removed from pF18-13 by digestion with *Bgl*II and *Esp*I. A unique *Bam*HI site was inserted in their place, giving plasmid pGaTB. To make promoter/GAL4 gene fusions, we carry out a three-way ligation between (1) the

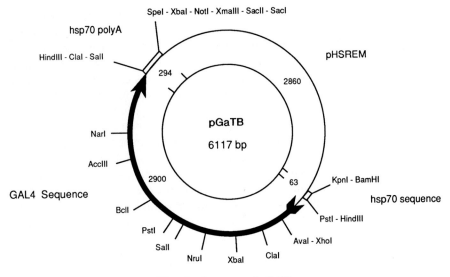

Fig. 2   Structure of pGaTB.

promoter fragment, (2) the *Bam*HI–*Not*I fragment containing the GAL4/hsp70 terminator fusion, and (3) a P-element vector (such as pCaSpeR2, 3, or 4). Alternatively, the GAL4/hsp70 fragment can first be subcloned into a P-vector, such as pCaSpeR2, leaving the *Bam*HI site as a unique site for subcloning promoters. In choosing the length of the promoter fragment to drive GAL4 expression, we use as much sequence as possible up to, but not including, the AUG.

3. pGawB

As a first step in creating an enhancer detection GAL4 vector, we modified the enhancer detection *lacZ* vector, plwB (Wilson *et al.*, 1989). To remove the *Not*I site in the vector, plwB was digested with *Not*I and the 5′ overhanging ends were filled using T4 polymerase. The resultant blunt ends were then ligated to make plasmid p41-4. To remove the P-transposase-*lacZ* fusion gene, p41-4 was digested with *Hin*dIII, the *Hin*dIII fragment was removed, and the plasmid was religated, forming p41-4-H3-1. A short linker oligonucleotide, formed by annealing the sequences 5′-AGCTTGGTTAACGCGGCCGC-3′ and 3′-ACCAATTGCGCCGGCGTCGA-5′, was then subcloned into the *Hin*dIII site of p41-4-H3-1. In the resultant plasmid, p41-4Hpa, the *Hin*dIII site is maintained and a unique *Hpa*I site is introduced.

To reconstitute the 5′ end of the P-element and the P-transposase promoter, we synthesized an oligonucleotide that extends from the *Hin*dIII site in the 5′ end of the P-element to nucleotide 140, followed by the sequence CGGCCGC, to create a *Not*I site. The oligonucleotide was subcloned as a *Hin*dIII-blunt-ended fragment into p41-4Hpa cut with *Hin*dIII and *Hpa*I, to create p41-4Hpa14. This plasmid is an enhancer trap vector into which any coding sequence can be cloned as a *Not*I fragment.

As a final step, the GAL4 coding sequence/hsp70 terminator was isolated from pGaTN by digestion with *Not*I. The *Not*I fragment was subcloned into the unique *Not*I site of p41-4Hpa14 to create pGawB.

C. GAL4-Responsive Genes

1. pUAST (Fig. 3)

We constructed a vector into which genes can be subcloned behind the GAL UAS (*U*pstream *A*ctivation *S*equence). A fragment containing five optimized GAL4-binding sites (the ''ScaI site'' 17-mer; Webster *et al.*, 1988) and a synthetic TATA box (Lillie and Green, 1989) separated by a unique *Not*I site from the SV40 terminator was excised from pF40X2-1 by digestion with *Nsi*I and *Spe*I. The fragment was subcloned into the P-element vector pCaSpeR3 (a gift

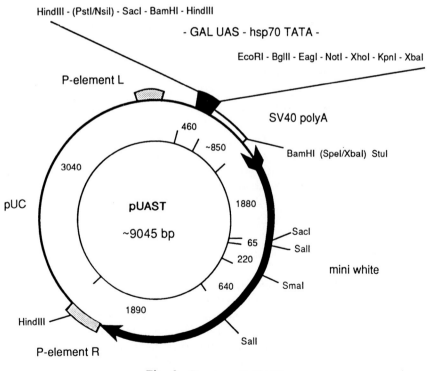

Fig. 3 Structure of pUAST.

from C. Thummel and V. Pirrotta) and cut with *Pst*I and *Xba*I, to give plasmid pF91. This vector was used to drive expression of *Dras2*^{Val14}, but appeared to be inefficient in promoting transcription and so was modified as follows.

pF91 was digested with *Xba*I, and a fragment spanning the TATA box was removed. This was replaced by a fragment containing the *hsp70* TATA box, generated as a PCR product using pCaSpeR-hs (a gift from C. Thummel and V. Pirrotta) as a template. The PCR product begins with an *Nhe*I restriction site, extends from 11 nucleotides upstream of the *hsp70* TATA box to the *Sac*II site in pCaSpeR3, and is followed by restriction sites for *Xho*I, *Kpn*I, and *Xba*I. The resulting vector, named pUAST, consists of five tandemly arrayed, optimized GAL4 binding sites, followed by the *hsp70* TATA box and transcriptional start, a polylinker with unique restriction sites for *Eco*RI, *Bgl*II, *Not*I, *Xho*I, *Kpn*I, and *Xba*I, and the SV40 small t intron and polyadenylation site. When subcloning sequences into pUAST, we try to minimize the amount of 5′ leader upstream of the coding sequence; ideally, the AUG is the most 5′ sequence.

2. *UAS-lacZ*

An *Adh-lacZ* fusion gene was removed from pCaSpeR-AUG-βgal (Thummel *et al.*, 1988) by digestion with *Eco*RI and *Xba*I and was subcloned in pUAST. The β-galactosidase expressed from this construct is cytoplasmic.

D. P-Element Transformation

Transgenic lines are generated by injection of CsCl-banded DNA, at a concentration of 600 μg/ml, into embryos of strain *yw; +/+; Sb,* P[*ry +*, *Δ2-3*]/TM6, *Ubx* (Robertson *et al.*, 1988) using standard procedures (Roberts, 1986). On average, we obtain three independent transformants per 100 embryos injected.

We find that the UAS transgenes are subject to position effects, such that integrants at different insertion sites are expressed at different levels. We therefore test and maintain between 5 and 6 independent insertion lines. This allows the generation of a phenotypic series: each GAL4 expressing line is crossed to several UAS lines, each of which expresses the target gene at a different level.

E. Enhancer Detection

Enhancer detection screens can be carried out to recover lines that express GAL4 in a cell- or tissue-specific manner. The P-element transposons are mobilized using the "jumpstarter" strain P[*ry +*; *Δ2-3*], which carries a defective P-element on the third chromosome at 99B (Laski *et al.*, 1986; Robertson *et al.*, 1988; Cooley *et al.*, 1988). This P-element expresses high levels of a constitutively active transposase, but cannot itself transpose.

The frequency with which new pGawB insertion lines are recovered is much lower than that previously reported for a similarly sized enhancer detection/ *lacZ* P-element (N. P., unpublished data). This might be attributed to the alterations made in the sequence of the 5′ end of the P-element pGawB that allow GAL4 to be expressed from its own AUG, rather than as a P-transposase-GAL4 fusion protein.

F. Screening for Embryonic Expression Patterns

Each GAL4 insertion line is crossed to a line carrying the *UAS-lacZ* reporter gene. Embryos from the cross are collected on agar/molasses plates and stained for β-galactosidase expression using X-Gal as a substrate. This is carried out in 25-well perspex plates sealed with Nitex mesh, as recommended by Nipam Patel. The embryos are dechorionated in 50% Clorox, rinsed in 1% Triton X-100, and washed into the wells through a funnel. The Triton is replaced with heptane, to which is added an equal volume of 3.7% formaldehyde in PBS, 0.1% Triton X-100. The embryos are fixed, with gentle swirling, for 7 min. The multiwell plate is moved into fresh heptane, and methanol is dropped into each well to crack the vitelline membranes. The plate is lifted up and down rapidly to mix the heptane and methanol. The plate is then moved into fresh methanol and then into PBS, 0.1% Triton X-100. The embryos must not remain in methanol for more than a few minutes in total, as longer exposure will inhibit β-galactosidase activity. The plate is then transferred to a staining solution of 10 mM PO$_4$ (pH 7.2), 150 mM NaCl, 1 mM MgCl$_2$, 3 mM K$_4$[FeII(CN)$_6$], 3 mM K$_3$[FeIII(CN)$_6$] containing a 1/50 dilution of X-Gal (25 mg/ml in dimethyl formamide). Staining takes from 30 min to overnight, at room temperature. After washing in PBS, 0.1% Triton X-100, embryos can be mounted in 70% glycerol.

After the initial screen, lines that express GAL4 in potentially interesting patterns are rescreened using anti-β-galactosidase antibodies. Similar screens can be carried out to look for expression in larvae or adults.

G. GAL4-Mediated Ectopic Expression

To ectopically express a gene, a line carrying the UAS-target gene is crossed to a line expressing GAL4 in the appropriate tissue or group of cells. GAL4-dependent phenotypes can then be examined in the progeny of the cross. As a first step in the analysis, ectopic expression is monitored directly by staining with antibodies against the target gene product or by *in situ* hybridization to the target gene transcript. The level of GAL4-directed expression naturally varies from line to line, but is generally robust. For example, ectopically expressed *even-skipped* is usually easier to detect than the endogenous protein. When a direct comparison is made between the hsp70 promoter driving a target gene directly and hsp70-GAL4 driving a UAS-target gene, higher levels of expression are seen using the GAL4 system (Brand and Perrimon, 1993). The level of GAL4-mediated ectopic expression also varies with temperature; more severe phenotypes are observed at 29°C than at 25 or 18°C.

When adult phenotypes are required, it is often easier to cross uncharacterized GAL4 lines to the UAS-target line and to screen for phenotypes first, prior to characterizing the GAL4 expression pattern.

H. Discussion

The GAL4 system offers a great deal of flexibility for targeting gene expression during development. However, there are several areas in which the technique can be improved. First, 50% of the lines that express GAL4 in a specific embryonic pattern also activate transcription in the salivary glands. It appears that the GAL4 vectors carry a position-dependent salivary gland enhancer that may have been fortuitously generated during their construction. Ideally, the sequence can be identified and removed from future vectors. Second, there is often variability in the level of GAL4-mediated transcription from cell to cell within an expression domain. Without effective anti-GAL4 antibodies, we cannot say whether the levels of GAL4 protein also vary from cell to cell or whether DNA binding or transcriptional activation is rate limiting. Third, although GAL4 can direct expression in a wide range of patterns in *Drosophila*, GAL4-mediated expression has not yet been seen in the female germline. GAL4 translation might be selectively repressed in the germline or the GAL4 transcript may be degraded. Since many translational regulatory sequences reside in the 3′ UTR of transcripts, altering the 3′ end of the GAL4 mRNA might overcome this problem. In pGawB and pGaTB, the GAL4 transcriptional terminator is followed by the hsp70 terminator. We have not determined where the GAL4 transcript from these vectors actually terminates. An alternative hypothesis is that GAL4 activates transcription in conjunction with another protein that is not present in germ cells.

A similar phenomenon might explain why, in embryonic development, the earliest that GAL4-mediated expression can be detected is after gastrulation. This holds true whether GAL4 is expressed using enhancer detection or using a defined promoter (J. Castelli-Gair and S. Greig, personal communication). Whereas GAL4 mRNA can be detected at the cellular blastoderm stage (K. Staehling-Hampton and F. M. Hoffmann, personal communication), transcription of the target gene is not seen until 3 to 4 hr of development. Once again, the GAL4 mRNA may not be translated or may be translated poorly. The availability of antibodies that recognize GAL4 in *Drosophila* embryos will help to resolve this issue. To date, the GAL4 antibodies that have been tested (including those that work on Western blots and in an ELISA) give high background, but no detectable signal, on embryos and imaginal discs.

The GAL4 system can in principle be used to express any gene of interest, including one that might be lethal to the organism. In the absence of GAL4, the toxic target gene should be silent and should be activated only in the progeny arising from a cross to a GAL4-expressing line. We have subcloned the diphtheria toxin A chain (DT-A) sequence and the *wingless* cDNA into pUAST, and in both cases have been unable to generate transformants, even in the absence of GAL4 (A.B., unpublished data). There may be a low level of transient expression from the injected DNA that kills the embryo. Once the UAS-gene construct is integrated in the genome, however, leaky expression is no longer a problem.

One way to prevent transient expression is to insert a flp/FRT cassette between the UAS and the target gene, forcing transcription to terminate prematurely. After the UAS/FRT cassette/target gene is integrated, the cassette can be removed by expression of flp recombinase from a heat-shock promoter. This technique has been used successfully to generate transformants carrying *UAS-wingless* (E. Wilder and N.P., in preparation). Taking a different approach, *UAS-DT-A* transformants were generated by engineering ribozymes to *trans*-splice the diphtheria toxin A chain coding sequence into the GAL4 mRNA (J. Haseloff and H. Goodman, unpublished). Only in the presence of the GAL4 mRNA should the ribozyme generate a fusion mRNA that can be translated. GAL4-dependent transcription of the ribozyme in *Drosophila* results in specific cell killing (A.B., J. Haseloff, H. Goodman, and N.P., unpublished).

The level of GAL4-induced expression can be modulated by increasing or decreasing the number of GAL4-binding sites upstream of the target gene or by using GAL4 derivatives (or activators fused to the GAL4 DNA-binding domain) that activate transcription to different degrees (Ma and Ptashne, 1987a,b; Gill and Ptashne, 1987; Johnston and Dover, 1988). Expression levels can also be increased by making either the GAL4 expression gene or the UAS-target gene homozygous or by introducing several copies of the target gene.

Finally, other components of the GAL regulatory pathway could be imported into flies to diversify further the GAL4 system. For example, a temperature-sensitive allele of GAL4 (Matsumoto *et al.,* 1978) has been described that, if expressed in flies, would allow ectopic expression to be restricted temporally as well as spatially. Alternatively, GAL4 can be used to drive expression of target genes encoding temperature-sensitive proteins, as has been done for *wingless* (E. Wilder and N.P., in preparation). Interestingly, the full-length GAL4 protein appears to be cold-sensitive when expressed in *Drosophila:* GAL4-mediated ectopic expression gives more severe phenotypes at 29°C > 25°C > 18°C.

To refine further the pattern of GAL4-dependent transcription, a negative regulator of GAL4, the GAL80 protein, might be introduced into *Drosophila* and expressed in a pattern that overlaps that of GAL4. A temperature-sensitive allele of GAL80 has also been described (Matsumoto *et al.,* 1978). Finally, because GAL4 can activate and maintain transcription at high levels, it may also be possible to use the GAL4 system to repress the expression of endogenous genes by using GAL4 to drive the transcription of antisense RNAs.

Acknowledgments

We are grateful to O. Dunin Borkowski, N. Brown, D. Sheperd, K. Kaiser, K. Staehling Hampton, M. Hoffmann, J. Castelli Gair, S. Greig, D. Buenzow, R. Holmgren, L. Keegan, K. Yoffe, E. Wilder, C. Thummel, V. Pirrotta, H. Goodman, and J. Haseloff for communication of unpublished results. A.B. is supported by The Wellcome Trust, A.M. is supported by National Cancer Institute of Canada, and N.P. is an Investigator with the Howard Hughes Medical Institute.

References

Basler, K., Christen, B., and Hafen, E. (1991). Ligand-independent activation of the sevenless receptor tyrosine kinase changes the fate of cells in the developing Drosophila eye. *Cell* **64**, 1069–1082.

Bellen, H. J., O'Kane, C., Wilson, C., Grossniklaus, U., Pearson, R. K., and Gehring, W. J. (1989). P-element-mediated enhancer detection: A versatile method to study development in *Drosophila. Genes Dev.* **3**, 1288–1300.

Bier, E., Vaessin, H., Shepherd, S., Lee, K., McCall, K., Barbel, S., Ackerman, L., Caretto, R., Uemura, T., Grell, E., Jan, L. Y., and Jan, Y. N. (1989). Searching for pattern and mutation in the *Drosophila* genome with a P-*lacZ* vector. *Genes Dev.* **3**, 1273–1287.

Blochlinger, K., Jan, L. Y., and Jan, Y. N. (1991). Transformation of sensory organ identity by ectopic expression of *Cut* in *Drosophila. Genes Dev.* **5**, 1124–1135.

Brand, A., and Perrimon, N. (1993). Targeted gene expression as a means of altering cell fates and generating dominant phenotypes. *Development* **118**, 401–415.

Brand, A., and Perrimon, N. (1994). Raf acts downstream of the EGF receptor to determine dorso-ventral polarity during *Drosophila* oogenesis. *Genes Dev.* **8**, 629–639.

Brand, M., Jarman, A. P., Jan, L. Y., and Jan, Y. N. (1993). asense is a Drosophila neural precursor gene and is capable of initiating sense organ formation. *Development* **119**, 1–17.

Cooley, L., Kelley, R., and Spradling, A. (1988). Insertional mutagenesis of the *Drosophila* genome with single P elements. *Science* **239**, 1121–1128.

DiNardo, S., and O'Farrell, P. (1987). Establishment and refinement of segmental pattern in the Drosophila embryo: Spatial control of engrailed expression by the pair-rule genes. *Genes Dev.* **1**, 1212–1225.

Dorsett, D., Viglianti, G. A., Rutledge, B. J., and Meselson, M. (1989). Alteration of *hsp82* gene expression by the gypsy transposon and suppressor genes in *Drosophila melanogaster. Genes Dev.* **3**, 454–468.

Fasano, L., and Kerridge, S. (1988). Monitoring positional information during oogenesis in adult *Drosophila. Development* **104**, 245–253.

Fischer, J. A., Giniger, E., Maniatis, T., and Ptashne, M. (1988). GAL4 activates transcription in *Drosophila. Nature* **332**, 853–865.

Fitzpatrick, V. D., Percival-Smith, A., Ingles, C. J., and Krause, H. (1992). Homeodomain-independent activities of the *fushi tarazu* protein in *Drosophila* embryos. *Nature* **356**, 610–612.

Gill, G., and Ptashne, M. (1987). Mutants of GAL4 protein altered in an activation function. *Cell* **51**, 121–126.

Golic, K., and Lindquist, S. (1989). The FLP recombinase of yeast catalyzes site-specific recombination in the *Drosophila* genome. *Cell* **59**, 499–509.

Gonzalez-Reyes, A., and Morata, G. (1990). The developmental effect of overexpressing a Ubx product in *Drosophila* embryos is dependent on its interactions with other homeotic products. *Cell* **61**, 515–522.

Greig, S., and Akam, M. (1993). Homeotic genes autonomously specify one aspect of pattern in the Drosophila mesoderm. *Nature* **362**, 630–632.

Ingham, P. W., and Gergen, J. P. (1988). Interactions between the pair-rule genes *runt, hairy, even skipped* and *fushi tarazu* and the establishment of periodic pattern in the *Drosophila* embryo. *Development Suppl.* **104**, 51–60.

Ish-Horowicz, D., and Pinchin, S. M. (1987). Pattern abnormalities induced by ectopic expression of the *Drosophila* gene *hairy* are associated with repression of *ftz* transcription. *Cell* **51**, 405–415.

Ish-Horowicz, D., Pinchin, S. M., Ingham, P. W., and Gyurkovics, H. G. (1989). Autocatalytic *ftz* activation and metameric instability induced by ectopic *ftz* expression. *Cell* **557**, 223–232.

Johnston, M., and Dover, J. (1988). Mutational analysis of the GAL4-encoded transcriptional activator protein of *Saccharomyces cerevisiae*. *Genetics* **120**, 63–74.

Knipple, D. C., and Marsella-Herrick, P. (1988). Versatile plasmids for the construction, analysis and heat-inducible expression of hybrid genes in eukaryotic cells. *Nucleic Acids Res.* **16**, 7748.

Krause, H. M., Klemenz, R., and Gehring, W. J. (1988). Expression, modification and localization of the *fushi tarazu* protein in *Drosophila* embryos. *Genes Dev.* **2**, 1021–1036.

Lardelli, M., and Ish-Horowicz, D. (1993). Drosophila *hairy* pair-rule gene regulates embryonic patterning outside its apparent stripe domains. *Development* **118**, 255–266.

Laski, F. A., Rio, D. C., and Rubin, G. M. (1986). Tissue specificity of *Drosophila* P element transposition is regulated at the level of mRNA splicing. *Cell* **44**, 7–19.

Lawrence, P. A., Johnston, P., MacDonald, P., and Struhl, G. (1987). Borders of parasegments in *Drosophila* embryos are delimited by the *fushi tarazu* and *even skipped* genes. *Nature* **328**, 440–442.

Lillie, J. W., and Green, M. R. (1989). Transcriptional activation by the adenovirus E1a protein. *Nature* **338**, 39–44.

Ma, J., and Ptashne, M. (1987a). Deletion analysis of GAL4 defines two transcriptional activating segments. *Cell* **48**, 847–853.

Ma, J., and Ptashne, M. (1987b). A new class of yeast transcriptional activators. *Cell* **51**, 113–119.

Manoukian, A. S., and Krause, H. M. (1992). Concentration-dependent activities of the *even-skipped* protein in *Drosophila* embryos. *Genes Dev.* **6**, 1740–1751.

Manoukian, A. S., and Krause, H. M. (1993). Control of segmental asymmetry in *Drosophila* embryos. *Development* **118**, 785–796.

Martinez-Arias, A., and White, R. A. H. (1988). *Ultrabithorax* and *engrailed* expression in *Drosophila* embryos mutant for segmentation genes of the pair-rule class. *Development* **102**, 325–338.

Matsumoto, K., Toh-e, A., and Oshima, Y. (1978). Genetic control of galactokinase synthesis in *Saccharomyces cerevisiae:* Evidence for constitutive expression of the positive regulatory gene *gal4*. *J. Bacteriol.* **134**, 446–457.

Morrissey, D., Askew, D., Raj, L., and Weir, M. (1991). Functional dissection of the *paired* segmentation gene in *Drosophila* embryos. *Genes Dev.* **5**, 1684–1696.

O'Kane, C. J., and Gehring, W. J. (1987). Detection in situ of genomic regulatory elements in *Drosophila*. *Proc. Natl. Acad. Sci. U.S.A.* **84**, 9123–9127.

Parkhurst, S. M., Bopp, D., and Ish-Horowicz, D. (1990). X : A ratio, the primary sex-determining signal in *Drosophila*, is transduced by helix-loop-helix proteins. *Cell* **63**, 1179–1191.

Parkhurst, S. M., and Ish-Horowicz, D. (1991). Mis-regulating segmentation gene expression in *Drosophila*. *Development* **111**, 1121–1135.

Petersen, N. S. (1990). Effects of heat and chemical stress on development. *Adv. Genet.* **28**, 275–296.

Petersen, N. S., and Mitchell, H. K. (1987). The induction of a multiple wing hair phenocopy by heat shock in mutant heterozygotes. *Dev. Biol.* **121**, 335–341.

Roberts, D. B. (1986). *In* "*Drosophila:* A Practical Approach" (D. B. Roberts, ed.), Oxford, England: IRL Press.

Robertson, H. M., Preston, C. R., Phillis, R. W., Johnson-Schlitz, D., Benz, W. K., and Engels, W. R. (1988). A stable source of P-element transposase in *Drosophila* melanogaster. *Genetics* **118**, 461–470.

Schneuwly, S., Klemenz, R., and Gehring, W. J. (1987). Redesigning the body plan of *Drosophila* by ectopic expression of the homeotic gene *Antennapedia*. *Nature* **325**, 816–818.

654   **Andrea H. Brand** *et al.*

Steingrimsson, E., Pignoni, F., Liaw, G-J., and Lengyel, J. A. (1991). Dual role of the *Drosophila* pattern gene *tailless* in embryonic termini. *Science* **254,** 418–421.

Struhl, G. (1985). Near-reciprocal phenotypes caused by inactivation or indiscriminate expression of the *Drosophila* segmentation gene *ftz*. *Nature* **318,** 677–680.

Struhl, G., and Basler, K. (1993). Organizing activity of wingless protein in Drosophila. *Cell* **72,** 527–540.

Thummel, C. S., Boulet, A. M., and Lipshitz, H. D. (1988). Vectors for *Drosophila* P-element-mediated transformation and tissue culture transfection. *Gene* **7,** 445–456.

Van Vactor, D. L., Cagan, R. L., Kramer, H., and Zipursky, S. L. (1991). Induction in the developing compound eye of *Drosophila:* Multiple mechanisms restrict R7 induction to a single retinal precursor cell. *Cell* **67,** 1145–1155.

Webster, N., Jin, J. R., Green, S., Hollis, M., and Chambon, P. (1988). The yeast UAS$_G$ is a transcriptional enhancer in human HeLa cells in the presence of the GAL4 *trans*-activator. *Cell* **52,** 169–178.

Wilson, C., Pearson, R. K., Bellen, H. J., O'Kane, C. J., Grossniklaus, U., and Gehring, W. J. (1989). P-element-mediated enhancer detection: An efficient method for isolating and characterizing developmentally regulated genes *Drosophila*. *Genes Dev.* **3,** 1301–1313.

Yost, H. J., Petersen, R. B., and Lindquist, S. (1990). RNA metabolism: Strategies for regulation in the heat shock response. *Trends Genet.* **6,** 223–227.

Zuker, C. S., Mismer, D., Hardy, R., and Rubin, G. M. (1988). Ectopic expression of a minor *Drosophila* opsin in the major photoreceptor cell class: Distinguishing the role of primary receptor and cellular context. *Cell* **53,** 475–482.

CHAPTER 34

Mosaic Analysis Using FLP Recombinase

Tian Xu* and Stephen D. Harrison†

* Boyer Center for Molecular Medicine
Department of Genetics
Yale University School of Medicine
New Haven, Connecticut 06536

† Department of Molecular and Cell Biology
University o California
Berkeley, California 94720

I. Introduction

In a mosaic animal, not all cells are genetically identical. In *Drosophila* it has been possible to introduce or generate cells of one genotype within an organism that is genetically distinct. These cells proliferate and generate a group of genotypically identical cells called a clone. By marking the cells of these clones it has been possible to answer many developmental questions (reviewed in Postlethwait, 1976; Ashburner, 1989). For example, by marking a developing cell it has been possible to establish the eventual fate of its progeny and the stage at which their fate becomes determined. In addition, the distribution and numbers of the marked progeny can tell us much about cellular division patterns of developing cells. In this chapter, we will discuss another use of clones of cells to study the phenotype of mutations in isolated patches of tissue, within an otherwise wild-type animal (reviewed by Hall *et al.*, 1976; Lawrence *et al.*, 1986; Ashburner, 1989). This application of mosaics is especially useful for the study of genes that are essential for the viability of the whole organism, because specifically removing their function in isolated clones does not necessarily impair overall survival. Furthermore, in these experiments the effects of mutations, lethal or otherwise, can be examined on both the mutant and surrounding wild-type tissue. We will describe the use of FLP recombinase and novel cell markers to greatly facilitate this type of analysis and, in addition, allow the rapid isolation of mutations, including lethals, which affect the development and function of the fly.

II. Mosaic Analysis

Clones of mutant tissue are usually generated by "mitotic recombination." A parental cell heterozygous for a mutant gene is induced to undergo exchange of homologous chromosome arms. In the following mitosis, segregation of the recombinant chromosomes often results in one mutant daughter cell and another wild-type daughter cell, which is referred to as a "twin spot cell" (Fig. 1A). If the parental mutant chromosome carries a marker mutation closely linked to the mutation, then the mutant clones will be distinguishable by the presence of the marker phenotype. Subsequent proliferation of the daughter cells results in a mutant and a wild-type twin-spot clone that often remain in close apposition. This technique has much wider applicability than alternative methods of generating mitotic clones that take advantage of the spontaneous loss of certain abnor-

---

Fig. 1 Producing and marking clones of cells with different doses of a given mutation (for example, *bib* on the second chromosome). The relevant chromosomes are illustrated with continuous or dashed lines with their centromere shown as a circle. (A) Generation of mitotic clones using X-rays. A strain, containing a chromosome carrying *bib*, is crossed to another carrying a closely linked marker (open triangle; P[w$^+$]). A cell in the heterozygous progeny is induced by X-ray irratiation to undergo exchange of homologous chromosome arms and in the following mitosis, segregation of the recombinant chromosomes often results in one mutant daughter cell and another wild-type, so called, "twin spot" cell. Clones of the *bib/bib* mutant cells in the adult eye can be identified as lacking the chosen marker (w$^-$ cells). (B) Generation of mitotic clones using FLP/ FRT and novel markers. A strain containing a chromosome carrying *bib* and a centromere-proximal FRT element (closed arrowheads; FRT) on the same arm is crossed to another carrying the same FRT element as well as a *hsFLP* element on a separate chromosome. Clones of cells homozygous for the *bib* mutation can be produced by inducing mitotic recombination between FRT sequences

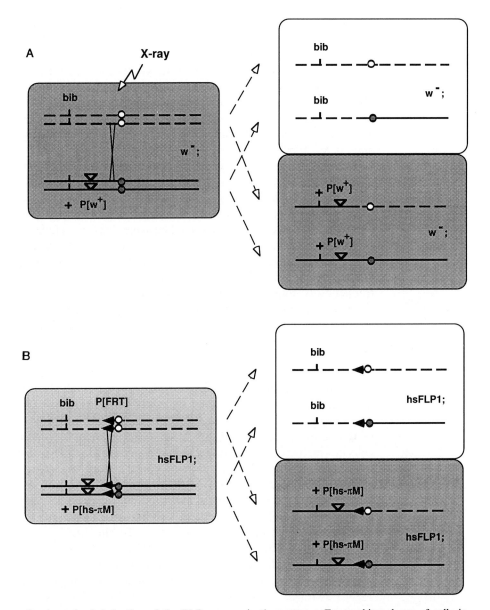

after heat-shock induction of the FLP enzyme in the progeny. For marking clones of cells in developing or adult tissues, the distal part of the FRT-carrying arm in the *hsFLP*-carrying strain also carries a P-transposon (open arrows) that contains an appropriate marker gene (for example, the πM marker; illustrated as P[hs-πM]). Cells in a *bib/bib* mutant clone can be identified as tissues lacking the given marker (πM^{-} *bib*/πM^{-} *bib*), while cells in the wild-type twin-spot clone (πM^{+} *bib*$^{+}$/πM^{+} *bib*$^{+}$) can also be recognized by the higher level of expression of the marker gene in these cells than the background heterozygous cells (πM^{+} *bib*$^{+}$/πM^{-} *bib*$^{-}$). For examples of actual clones see Fig. 3.

mal chromosomes and it is less technically demanding than the other alternatives of nuclear and cellular transplantation (reviewed by Ashburner, 1989). The agent commonly used to induce mitotic recombination is X-ray irradiation (Patterson, 1929; Friesen, 1936), although the doses required to give an acceptable frequency of clones often result in significant cell death, and developmental abnormalities that are unrelated to mosaicism can arise. A solution to these problems has been achieved by inducing mitotic recombination in *Drosophila* with the yeast 2-μm plasmid site-specific recombinase (FLP) (Fig 1B; Broach and Hicks, 1980; Jayaram, 1985). Golic and Lindquist (1989) have shown that expression of exogenous yeast FLP can lead to recombination between the FLP recombination targets, FRTs, in *Drosophila* cells. In this experiment, a transgenic pigment gene was removed from a chromosome by the induction of recombination between flanking FRT sites. Golic (1991) has also shown that the recombination event can be induced between FRT sites located on homologous chromosome arms. We have generated a series of *Drosophila* strains in which FRT sites are inserted at the bases of all the major chromosome arms (Fig. 2) and in this chapter we will describe how these may be used to generate a high frequency of mosaics for mutations in greater than 95% of the genes in the genome (Xu and Rubin, 1993). The high frequency afforded by these techniques also allows the study of clones in internal tissues: if the frequency of recombination is too low, then many animals must be painstakingly dissected before a clone is found.

To distinguish between mutant and twin-spot clones and the rest of the tissue, it is essential to clearly mark clones for mosaic analysis. The markers must obey the following criteria: they must have phenotypes that are easy to score; they must not affect the normal development or viability of the cells; they must be restricted to the cells of the clone; and they must be rapidly removed from cells in which they are no longer made. A variety of markers fitting these criteria have been traditionally used to label clones in adult animals (reviewed by Hall *et al.*, 1976; Lawrence *et al.*, 1986; Ashburner, 1989) and we have generated the chromosomes necessary to use a number of these in conjunction with the FLP/FRT system, including the *white* (*w*) gene in the eye and the *yellow* (*y*) gene in the adult cuticle (Fig. 2D; Xu and Rubin, 1993). Very few markers for internal and developing tissues exist and those that do can be used only in a restricted set of structures and can mark only mutant clones if the mutation is closely genetically linked to the marker gene (Janning, 1972; Kankel and Hall, 1976; Lawrence, 1981). We have constructed novel marker genes, which obey the criteria outlined above. The products of these genes are expressed in either the nuclei (hs-ΠM) or cell membranes (hs-NM) of most tissues (Fig. 3). Genes for each of these markers have been recombined onto all of the FRT-containing chromosomes described and consequently may be used to follow wild-type cell lineages, as well as mark the clones of any of the 95% of mutant genes that can be analyzed by FLP induced recombination. In the following protocols, we will explain in detail how to use the FRT-containing chromosomes, in combination with the described markers, to perform mosaic analysis in both external and internal tissues from a variety of developmental stages. We will also describe techniques that have been developed to use the FLP/FRT recombination system to positively mark mitotic clones and to generate mutant patches of cloned genes.

A. Recombination of Your Mutation onto an FRT–Containing Chromosome Arm

Since FLP-induced recombination occurs only at the FRT sites, the daughter cells will become homozygous for all the genes that are located distal to these sites (Fig. 1B). Thus, to use the FRT/FLP system to generate homozygous mutant clones, the mutation of interest must merely be recombined onto a chromosome arm that contains a centromere-proximal FRT insertion (Fig. 2A, *P[ry+, hs-neo, FRT]*, abbreviated as FRT; Fig. 4). The locations of these insertions are depicted in Fig. 2C, and some of their derivative strains are listed in Fig. 2D. Each strain has been designated by an abbreviation of its genotype, which starts with a number indicating the cytological location of the FRT insertion, followed by abbreviations for the cell markers. The letter F denotes the presence of the FLP gene. In these constructs, the FLP gene is expressed under the control of a heat-shock promoter (hsFLP; Golic and Lindquist, 1990). To identify recombinants that acquire your mutation, the FRT element has been dominantly marked with the *neomycin* gene (*hs-neo;* Steller and Pirrotta, 1985), whereas the distal end of the chromosome arm has been marked with a visible adult marker (either a P-element carrying the *y* or *w* genes on the autosomes or the *Bar* mutation on the X chromosome; Fig. 2D). Desired recombinants will be among those neomycin-resistant flies that lack the visible marker (Fig. 4).

To illustrate the recombination protocol, we will describe the generation of an FRT-containing chromosome arm that carries the *big brain* (*bib*) mutation.

1. Use Multiple Alleles

It is desirable to analyze at least three independent *bib* alleles, as chromosomes carrying lethal mutations often contain other mutations that have been previously induced or have arisen spontaneously over time. If there is such a contaminating mutation on the same arm as *bib*, mitotic clones will be homozygous for both mutations. Independent alleles of *bib* are unlikely to carry the same contaminating mutations.

2. Choice of Chromosome Arm

The *bib* gene is located at 30F and consequently there are four suitable chromosome arms that carry an FRT centromere-proximal to *bib*. These all carry an FRT at 40A, in addition to a visible marker (40-y+, marked with the *y* gene, and 40-w+, 40-ΠM, and 40-NM, marked with the white marker gene;

Fig. 2D). The 40-NM strain is the best choice as the white marker (which is in the same P-element as NM; see Section IIC1) is most closely linked to *bib*. In this case, about 7% of flies will recombine the white marker gene off the FRT-containing chromosome and these will be neomycin resistant (*neo^R*) and white-

---

Fig. 2 (A) Diagram of the *P[ry$^+$, hs-neo, FRT]* construct. The two tandem repeats containing the FRT sequences (closed arrowheads), two copies of the *hs-neo* gene (hatched bars), the 5′ and 3′ P-element ends (black boxes), the *rosy* gene (*ry$^+$*), and the pUC vector sequences (pUC) are indicated. A unique *Not*I site between the two FRT repeats is also indicated. (B) A representative example of the effect of heat-shock duration on the frequency of mosaicism induced at a *P[ry$^+$, hs-neo, FRT]* insertion. The experiment was performed on flies of the genotype shown. The frequencies of mosaicism are indicated by the percentage of eyes having at least one clone (continuous lines) and the percentages of flies having a clone(s) in one eye that also have a clone(s) in their other eye (dotted lines). Heat-shock treatments of first instar larvae (30–32 hr after egg laying) were carried out for the indicated number of minutes at 38°C. A total of ~4000 eyes were scored in five parallel experiments; standard deviations for each heat-shock period are indicated by vertical lines with end bars. (C) The locations of the functional centromere-proximal *P[ry$^+$, hs-neo, FRT]* insertions on the major *Drosophila* chromosome arms are illustrated. *P[ry$^+$, hs-neo, FRT]* elements located in the middle of chromosome arms are also indicated, although only the elements at 11A, 69A, and 72D have been tested for their ability to mediate mitotic recombination. Centromeres are indicated by the open circles. (D) List of strains constructed to facilitate mosaic analysis in developing and adult *Drosophila* tissues. The strains are listed according to the chromosome arm that carries the *P[ry$^+$, hs-neo, FRT]* element; each FRT-carrying arm is available with or without a *hsFLP* element on a separate chromosome. For the convenience of notation, each strain was designated by an abbreviation of its genotype. Each abbreviation starts with a number indicating the cytological location of the *P[ry$^+$, hs-neo, FRT]* element. Following this number, abbreviations for the cell markers located on that FRT-carrying arm are given. A letter F is written at the end of the abbreviation if a *hsFLP* element is present in the same strain. The *w* mutation in these strains is *w^{1118}* and the *ry* allele is *ry^{506}*. The *P[ry$^+$, y$^+$]* construct is described by Geyer and Corces (1987) and insertions of this element on the autosomes were provided by V. Corces. The genotype of the *hsFLP*-carrying strain is the same as the corresponding strain listed on the left except that it also carries the *hsFLP1* element (Golic and Lindquist; 1989) at 9F for autosomal *P[ry$^+$, hs-neo, FRT]* strains or a *hsFLP* element on the *MKRS* chromosome (Chou and Perrimon, 1992) at 86E for X-linked *P[ry$^+$, hs-neo, FRT]* strains; the allelism of the *ry* gene was not followed in the *hsFLP*-containing strains. To facilitate positively labeling mutant clones, two *P[mini-w$^+$, hs-πM]* elements that are not closely linked were placed on each of the FRT-carrying arms except 3L. (E) List of strains constructed to perform mosaic analysis in the adult cuticle and in the imaginal discs (Pascal Heitzler, personal communication). To use these strains, one will first cross an hsFLP-carrying strain with an FRT-containing strain and then cross the progeny from the first cross to a strain carrying the FRT-marker mutation of interest chromosome. For example, to obtain *sgg* mutant clones, one will first cross the *Dp(3; Y; 1)M2(emc$^+$, mwh$^+$), y, P[ry$^+$; hs-neo; FRT]19A / FM7, y^{31d}, B; emc^{FX119}, mwh, kar^2, ry^{506}* females to the *y; P[ry$^+$; hsFLP]38, Bc; Ki, kar^2, ry^{506}, Tb* males (see Fig. 2E) and then cross the *Dp(3; Y; 1)M2(emc$^+$, mwh$^+$), y, P[ry$^+$; hs-neo; FRT]19A; P[ry$^+$; hsFLP]38, Bc / +; mc^{FX119}, mwh, kar^2, ry^{506}/Ki, kar^2, ry^{506}, Tb* male progeny to the *sgg, f^{36a}, P[ry$^+$; hs-neo; FRT]19A / FM7, y^{31d}, B; mwh, kar^2, ry^{506}* females. After heat shock inducing mitotic recombination in the progeny from the last cross (Section B2), the *sgg, mwh, f^{36a}* mutant clones can be scored in the *y$^+$, B$^+$, Bc$^+$, Ki$^+$* female adults. The *neu$^{P[lacZ; ry+]A101}$* line is described by Bellen *et al.* (1989) and Huang *et al.* (1991). Other P-element transform lines are same as ones in Fig. 2D. The mutations, duplications, and balancers are described in Lindsley and Zimm (1992).

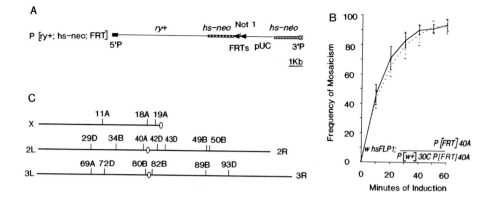

A P [ry+; hs-neo; FRT]

B Frequency of Mosaicism vs Minutes of Induction. w hsFLP1; P [FRT] 40A / P [w+] 30C P [FRT] 40A

C Chromosome map (X, 2L, 2R, 3L, 3R)

D

	Code	Genotype	FRT strains with hsFLP Code
		FRT strains without hsFLP	
X:	18-1	P[ry$^+$; hs-neo; FRT]18A; ry	18-1 F
	18-B	B, P[ry$^+$; hs-neo; FRT]18A; ry	
	18-w	w, P[ry$^+$; hs-neo; FRT]18A	18-w F
	18-πM	w, P[mini-w$^+$; hs-πM]10D, P[ry$^+$; hs-neo; FRT]18A	
	18-2πM	w, P[mini-w$^+$; hs-πM]5A, 10D, P[ry$^+$; hs-neo; FRT]18A	18-2πM F
	18-NM	w, P[mini-w$^+$; hs-NM]8A, P[ry$^+$; hs-neo; FRT]18A	
	19-1	P[ry$^+$; hs-neo; FRT]19A; ry	19-1 F
	19-y w	y, w, P[ry$^+$; hs-neo; FRT]19A	
	19-w sn	w, sn^3, P[ry$^+$; hs-neo; FRT]19A	
2L:	40-1	P[ry$^+$; hs-neo; FRT]40A; ry	40-1 F
	40-w$^+$	w; P[ry$^+$; w$^+$]30C, P[ry$^+$; hs-neo; FRT]40A	40-w$^+$ F
	40-y$^+$	y, w; P[ry$^+$; y$^+$]25F, P[ry$^+$; hs-neo; FRT]40A	40-y$^+$ F
	40-πM	w; P[mini-w$^+$; hs-πM]36F, P[ry$^+$; hs-neo; FRT]40A	40-πM F
	40-2πM	w; P[mini-w$^+$; hs-πM]21C, 36F, P[ry$^+$; hs-neo; FRT]40A	40-2πM F
	40-NM	w; P[mini-w$^+$; hs-NM]31E, P[ry$^+$; hs-neo; FRT]40A	40-NM F
2R:	42-1	P[ry$^+$; hs-neo; FRT]42D; ry	42-1 F
	42-w$^+$	w; P[ry$^+$; hs-neo; FRT]42D, P[ry$^+$; w$^+$]47A	42-w$^+$ F
	42-y$^+$	y, w; P[ry$^+$; hs-neo; FRT]42D, P[ry$^+$; y$^+$]44A	42-y$^+$ F
	42-πM	w; P[ry$^+$; hs-neo; FRT]42D, P[mini-w$^+$; hs-πM]45F	42-πM F
	42-NM	w; P[ry$^+$; hs-neo; FRT]42D, P[mini-w$^+$; hs-NM]46F	42-NM F
	43-1	P[ry$^+$; hs-neo; FRT]43D; ry	43-1 F
	43-w$^+$	w; P[ry$^+$; hs-neo; FRT]43D, P[ry$^+$; w$^+$]47A	43-w$^+$ F
	43-y$^+$	y, w; P[ry$^+$; hs-neo; FRT]43D, P[ry$^+$; y$^+$]44B	43-y$^+$ F
	43-πM	w; P[ry$^+$; hs-neo; FRT]43D, P[mini-w$^+$; hs-πM]45F	43-πM F
	43-NM	w; P[ry$^+$; hs-neo; FRT]43D, P[mini-w$^+$; hs-NM]46F	43-NM F
	43-2πM	w; P[ry$^+$; hs-neo; FRT]43D, P[mini-w$^+$; hs-πM]45F, 47F	43-2πM F
3L:	80-1	P[ry$^+$; hs-neo; FRT]80B; ry	80-1 F
	80-w$^+$	w; P[w$^+$]70C, P[ry$^+$; hs-neo; FRT]80B	80-w$^+$ F
	80-y$^+$	y, w; P[ry$^+$; y$^+$]66E, P[ry$^+$; hs-neo; FRT]80B	80-y$^+$ F
	80-πM	w; P[mini-w$^+$; hs-πM]75C, P[ry$^+$; hs-neo; FRT]80B	80-πM F
	80-NM	w; P[mini-w$^+$; hs-NM]67B, P[ry$^+$; hs-neo; FRT]80B	80-NM F
	80-mwh	w; mwh, jv, P[ry$^+$; hs-neo; FRT]80B	
3R:	82-1	P[ry$^+$; hs-neo; FRT]82B; ry	82-1 F
	82-w$^+$	w; P[ry$^+$; hs-neo; FRT]82B, P[ry$^+$; w$^+$]90E	82-w$^+$ F
	82-πM	w; P[ry$^+$; hs-neo; FRT]82B, P[mini-w$^+$; hs-πM]87E	82-πM F
	82-2πM	w; P[ry$^+$; hs-neo; FRT]82B, P[mini-w$^+$; hs-πM]87E, 97E	82-2πM F
	82-NM	w; P[ry$^+$; hs-neo; FRT]82B, P[mini-w$^+$; hs-NM]88C	82-NM F
	82-πM Sb y$^+$	y, w; P[ry$^+$; hs-neo; FRT]82B, P[mini-w$^+$; hs-πM]87E, Sb^{63b}, P[ry$^+$; y$^+$]96E	82-πM Sb y$^+$ F

E

Modified FLP/FRT-Marker Strains Constructed by Pascal Heitzler

X:	f^{36a}, P[ry+; hs-neo; FRT]19A; mwh, kar2, ry506
	N55e11, f^{36a}, P[ry+; hs-neo; FRT]19A / FM7, $y^{93}j12$, oc, ptg, B, P[ftz-lacZ; ry+]; mwh, kar2, ry506
	Dp(3; Y; 1)M2(emc+, mwh+), y, P[ry+; hs-neo; FRT]19A / FM7, y^{31d}, B; emcFX119, mwh, kar2, ry506
	Dp(3; Y; 1)M2(emc+, mwh+), y, M(1)oSp, P[ry+; hs-neo; FRT]19A / FM7, y^{31d}, B; emcFX119, mwh, kar2, ry506; Dp(1;4)r+/+
	y, w1118, P[mini-w+; hs-πM]5A, 10D, M(1)oSp, P[ry+; hs-neo; FRT]19A; kar2, ry506; Dp(1;4)r+/+
	y, w1118, P[mini-w+; hs-πM]5A, 10D, P[ry+; hs-neo; FRT]19A; Kgv, neuP[lacZ; ry+]A101, kar2, ry506/TM6C, ryCB
2L:	y, P[ry+; y+]25F, ckCH52, P[ry+; hs-neo; FRT]40A / CyO; kar2, ry506
	y, w1118, P[mini-w+; hs-πM]21C, M(2)mS6, P[ry+; hs-neo; FRT]40A / CyO; kar2, ry506
	y, w1118, P[mini-w+; hs-πM]21C, 36F, P[ry+; hs-neo; FRT]40A; Kgv, neuP[lacZ; ry+]A101, kar2, ry506/TM6C, ryCB
2R:	y; P[ry+; hs-neo; FRT]42D, pwn, P[ry+; y+]44B / CyO; kar2, ry506
	y, w1118, P[ry+; hs-neo; FRT]42D, P[mini-w+; hs-πM]45F, M(2)S7 / CyO; kar2, ry506
	y, w1118, P[ry+; hs-neo; FRT]42D, P[mini-w+; hs-πM]45F; Kgv, neuP[lacZ; ry+]A101, kar2, ry506/TM6C, ryCB
3L:	y, mwh, P[ry+; hs-neo; FRT]80B, kar2, ry506
	y, trc, P[ry+; hs-neo; FRT]80B, kar2, ry506/TM6C, ryCB, Sb, Tb
	y, w1118, jv, P[ry+; y+]66E, P[mini-w+; hs-πM]75C, P[ry+; hs-neo; FRT]80B, kar2, ry506/TM6C, ryCB, Sb, Tb
	y, w1118, P[ry+; y+]66E, M(3)i55, P[mini-w+; hs-πM]75C, P[ry+; hs-neo; FRT]80B, kar2, ry506/TM6C, ryCB, Tb
	w1118, M(3)i55, P[mini-w+; hs-πM]75C, P[ry+; hs-neo; FRT]80B, neuP[lacZ; ry+]A101, kar2, ry506/TM6C, ryCB, Tb
3R:	P[ry+; hs-neo; FRT]82B, kar2, ry506, Sb63b / TM6C, ryCB
	pr, pwn, P[ry+; hs-neo; FRT]82B, kar2, ry506, bx34e, Dp(2;3)P32(pwn+) / P[ry+; hs-neo; FRT]82B, kar2, ry506
	pr, pwn, P[ry+; hs-neo; FRT]82B, kar2, ry506, bx34e, Dp(2;3)P32(pwn+), M(3)w124 / TM6C, ryCB
	w1118, P[ry+; hs-neo; FRT]82B, P[mini-w+; hs-πM]87E, M(3)w124 / TM6C, ryCB, Tb
	y, w1118, P[ry+; hs-neo; FRT]82B, P[mini-w+; hs-πM]87E, P[ry+; y+]96E
FLP:	P[ry+; hsFLP]38, Bc; kar2, ry506
	C(1)DX,y,f; P[ry+; hsFLP]38, Bc; kar2, ry506
	y; P[ry+; hsFLP]1; Bc; kar2, ry506
	y; P[ry+; hsFLP]38, Bc; Ki, kar2, ry506, Tb
	pr, pwn, P[ry+; hsFLP]38 / CyO; Ki, kar2, ry506

Fig. 2 Continued

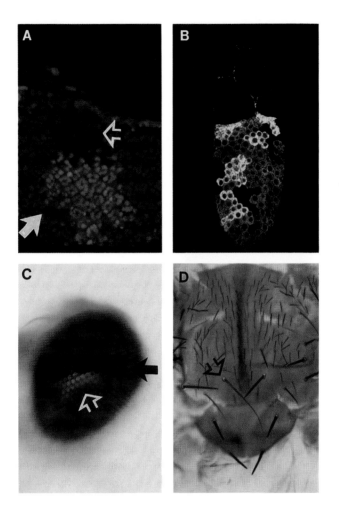

Fig. 3 (A) An apical optical section of a mosaic third instar larval eye disc in which recombination was induced in a *w⁻, hsFLP1; P[ry⁺, hs-neo, FRT]43D, P[mini-w⁺, hs-πM]45F/P[ry⁺, hs-neo, FRT]43D, +* animal. The disc was stained to visualize the πM marker. A clone of cells that lack the πM marker (open arrow) is accompanied by a twin-spot clone (closed arrow) that expresses the πM protein at a level higher than the heterozygous background cells. Clones of cells that carry two copies of a marker gene are always evident in the background of heterozygous cells carrying one copy of the same marker gene. However, cell-to-cell variations in the heat-shock response sometimes prevent the unambiguous determination of whether an individual cell carries one or two copies of the marker gene. The nuclei of polyploid cells in the peripodial membrane, seen at the top edge of the disc, are also brightly fluorescent. Posterior is up. (B) Twin-spot clones of ovarian follicle cells in a Stage 9 follicle were induced at the *P[ry⁺, hs-neo, FRT]40A* element and visualized using the NM marker inserted at 31E. Note that follicle cells containing zero, one, or two copies of the NM gene can be clearly distinguished. (C) A clone and its twin-spot clone in a mosaic adult eye of a *w⁻, hsFLP1;[mini-w⁺, hs-NM]31E, P[ry⁺, hs-neo, FRT]40A/+, P[ry⁺, hs-neo, FRT]40A* animal. The ommatidia in both twin-spot clones are recognizable since cells in the clone that do not carry the *P[mini-w⁺, hs-NM]* element lack any pigmentation (open arrow), whereas cells in the twin-spot clone carry two copies of the *P[mini-w⁺, hs-NM]* element and have a level of pigmentation (closed arrow) higher than the background cells, which carry only one copy of the same element. (D) Bristles of different clonal origins in the notum regions produced by inducing mitotic recombination in a *y⁻, w⁻hsFLP1; P[ry⁺, hs-neo, FRT]82B, Sb⁶³ᵇ, P[ry⁺,y⁺]96E/P[ry⁺, hs-neo, FRT]82B, +, +* animal; the genotypes of individual bristles can be identified according to their expression of the *y⁺* gene in the *P[ry⁺, y⁺]96E* element and by their expression of the *Sb⁶³ᵇ* mutation. Bristles made by cells that do not carry the *P[ry⁺, y⁺]* element and are homozygous for the *Sb⁺* gene are long and yellow (open arrows). Bristles that are homozygous for the *P[ry⁺, y⁺]* element and the *Sb⁶³ᵇ* mutation (closed arrow) are shorter than the heterozygous background bristles.

eyed (w^-). Because of the close linkage between *bib* and the marker, almost all of these flies will carry both the FRT element and the *bib* mutation on the same chromosome arm. In contrast, if the 40-y^+ strain is used, then only about 78% of the neo^R, y^- progeny will carry the *bib* mutation. The degree of linkage is given by the recombination distance and can be estimated from the cytological locations of the genes in the cytogenetic map prepared by Ashburner (1991; also see Lindsley and Zimm, 1992) and Fig. 2D. Strains used for recombination should not carry the hsFLP as it is not dominantly marked and cannot be followed through the crosses.

It is sometimes necessary to produce clones of cells that are only homozygous for the distal portion of a chromosome arm, and FRT elements have been inserted into the middle of the major chromosome arms for this purpose (Fig. 2C). For example, should you wish to mark cell identities of *bib* clones in the eye disc using enhancer trap line l(2)6433 (37B; Allan Spradling, personal communication), it would be necessary to homozygose only the *bib* mutation, as l(2)6433 has a lethal homozygous phenotype and would probably influence the clonal analysis. To accomplish this, FRT34D can be recombined onto both the *bib* and l(2)6433 chromosome arms and FLP induction will result only in the mitotic recombination of the sequences distal to 34D, which include *bib* but not the enhancer trap.

3. Genetic Selection of FRT-Containing Strains

Two copies of the *hs-neo* gene have been placed in the FRT construct (Fig. 2A). The *hs-neo* gene confers resistance to G418, a drug commonly used as a selective agent. Flies carrying the *hs-neo* element can usually be selected on G418-containing medium without heat-shock treatment, provided they are cultured at 25°C. However, not all the FRT insertions produce the same level of resistance; strains carrying insertions at 42D and 80B grow least well on G418 medium. Also, if any strains are cultured at a temperature below 25°C, they will grow poorly under selection. Under these circumstances, we recommend a 60-min incubation in a 38°C water bath, once or twice during early larval stages. The selection medium that we use is made by adding G418 solution to premade vials: a few holes are made in standard fly medium with a bundle of toothpicks, 0.2 to 0.3 ml of freshly made G418 solution (25 mg/ml, Geneticin, GIBCO) is added per 10 ml of standard fly medium, and the vials are allowed to air dry for several hours prior to use. G418 is stable in medium stored at 4–18°C for more than 2 weeks. If the neomycin selection does not work well, it may be necessary to adjust the G418 concentration or the number of heat-shock incubations. If necessary, the new conditions can be tested by setting up a cross that yields both FRT-containing and FRT-deficient progeny and then, by using appropriate markers, ascertain which classes of progeny survive.

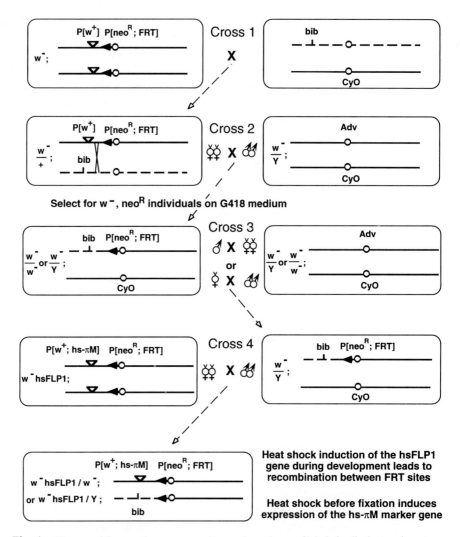

Fig. 4 Diagram of the genetic crosses used to produce clones of labeled cells that are homozygous for a previously identified mutation. The relevant chromosomes are illustrated with continuous or dashed lines with their centromere shown as open circles. Crosses 1–3: Recombining a previously identified mutation onto an FRT-carrying chromosome arm (also see Section A4). To induce clones of cells that are homozygous for a given mutation, for example, the *bib* mutation on 2L, the *bib* mutation must be genetically recombined onto the distal part of the FRT-carrying 2L arm. A strain carrying the *bib* mutation over a balancer chromosome (CyO) is mated to a strain that is homozygous for a centromere-proximal *P[ry+, hs-neo, FRT]* element (solid arrows; indicated as *P[neoR, FRT]*) on 2L (Cross 1). Non-Cy females heterozygous for both the *P[ry+, hs-neo, FRT]* element and the *bib* mutation are crossed to males from a balancer stock (Cross 2). Flies carrying the recombinant chromosome containing both the *P[ry+, hs-neo, FRT]* element and the *bib* mutation on the same arm *(P[ry+, hs-neo, FRT], bib)* are selected by virtue of their *neomycin* (G418) resistance whereas those that are also *w−* have recombined the *P[w+]* insertion away from the *P[ry+, hs-neo, FRT]*

4. The Experimental Procedure

Set up a vial containing about five of each *bib/CyO* male and 40-NM female flies on normal medium (Cross 1, Fig. 4). Collect 3 or 4 non-Cy virgins (*w/+*; P[mini-*w*⁺; hsNM]31E, P[ry+; hs-neo; FRT]40A / *bib*) and age them for 1 or 2 days on well-yeasted food (If your mutation is very close to the FRT, then you may wish to collect more virgins, as more offspring will be required to obtain the desired recombinant). Mate the virgins with about 10 *w; Adv-71/ CyO* males (Cross 2, Fig. 4) in a G418 vial containing enough dry-active yeast to feed the adults but not their larval offspring, which must be exposed to the antibiotic. Transfer the flies to a new vial every day for 3 to 4 days. From the surviving progeny, collect enough white-eyed individuals (it is easier to collect males as females must be collected as virgins) to ensure that at least one carries *bib* (this depends on the recombination distances, Section IIA2) and mate each individually to about 3 *w; Adv-71/CyO* flies on normal medium (Cross 3, Fig. 4) to generate balanced stocks. If the number of white-eyed flies emerging from the G418 vials is higher than expected, the selection is probably not working (Section IIA3). The balanced lines should be tested for the presence of *bib* by complementation crosses with other *bib* alleles. It is not sufficient to rely on the homozygous lethality of the balanced stock to determine the presence of *bib*, for the reasons outlined above (Section IIA1).

B. Analysis of Mutant Clones in Adult Cuticle

To analyze clones in a given adult external tissue, cross the FRT strain (usually males) that carries your mutation to another strain (usually females) carrying the same FRT, an appropriate marker and the hsFLP element (Cross 4, Fig. 4). Because the FLP gene is driven by a heat-shock promoter, it is induced the heterozygous progeny by a brief heat treatment at the appropriate developmental stage (Section IIB2).

---

element and are thus likely to have recombined the *bib* mutation onto the FRT-containing chromosome arm. Each w⁻, neoᴿ recombinant is mated to a balancer stock to balance the recombinant chromosome (Cross 3). Somatic clones homozygous for the *bib* mutation can be produced by crossing the P[ry⁺, hs-neo, FRT], *bib* recombinant to a strain that carries the same P[ry⁺, hs-neo, FRT] element and an appropriate cell marker (for example, P[hs-πM]) on 2L as well as a *hsFLP* element on a separate chromosome (Cross 4) and then inducing mitotic recombination between FRT sequences by heat-shock induction of the FLP enzyme at the desired developmental stage in the progeny. For marking clones of cells in developing tissues with the πM marker, the expression of the hs-πM gene is induced before fixation (see Section C3). Adv, Additional wing vein, a dominantly marked chromosome (Preiss and Artavanis-Tsakonas, 1991).

1. Adult Markers

a. Eye

To analyze clones in the adult eye, use the FRT-w^+F strains. The marker P-elements (P[w^+]) on these chromosomes carry the full-length w gene and produce pigment at, or close to, wild-type levels. Small clones of mutant ommatidia lacking the P[w^+] element will appear black if they lie in the center of the eye or white if at the periphery. Large w^- clones also appear white regardless of their position. In plastic sections, it is possible to identify individual photoreceptors that lack pigment. However, to achieve the best resolution, we recommend that clones are examined in males, which express higher levels of pigment as a result of dosage compensation of the w gene in the P[w^+] insertions. Some mutants affect cell viability and proliferation and consequently their clones are small or nonexistent. To distinguish this experimental outcome from an absence of mitotic recombination, it is useful to be able to visualize the wild-type twin-spot clones. This can be done by examining the eye of a female in plastic sections or by examining mosaic flies marked with the mini-*white* gene (P[mini-w^+; hs-II M] or P[mini-w^+; hs-NM]) under a dissection scope (Fig. 3C). In these cases, the w expression does not fully pigment the eye and twin-spot clones, which are homozygous for either the full-length or mini w gene, appear darker than the heterozygous background. The differential pigmentation is easier to see in younger flies. Small mutant clones can often be found because of their proximity to the wild-type twin-spot clones.

b. Adult Cuticle

To identify clones in the adult cuticle a number of markers exist. The y gene is a good multipurpose marker (FRT-y^+ F strains) and the y mutant bristles and cuticle are clearly distinguishable from wild-type tissue, especially in young animals, although it is often difficult to identify individual y cells along the border of a clone. Markers that can be used to give single-cell resolution have been recombined onto FRT chromosomes (Fig. 2D; the Sb^{63b} mutation on 3L and the $sn3$ mutation on X mark bristles and the jv and mwh mutations on 3L mark bristles and trichomes). A wonderful set of combined cuticular markers has been recombined onto every FRT-containing chromosome arm by Pascal Heitzler (Fig. 2E; personal communication). These markers allow one to perform cell by cell analysis of the borders between mutant and wild-type territories for both bristles and trichomes. For example, for a mutation on the X chromosome (such as sgg), cells in the mutant clone will be marked with both mwh and f^{36a}, whereas cells in the wild-type twin-spot clone will be marked with y (ck^{CH52} marked mutant clones and y marked wild-type clones for mutations on 2L; pwn marked mutant clones and y marked wild-type clones for mutations on 2R; mwh and y marked mutant clones and jv marked wild-type clones for mutations on 3L; pwn or Sb^+ marked mutant clones for mutations on 3R). The enhancer trap line $neu^{P[lacZ; ry+]A101}$, which expresses the β-galactosidase in the sensory mother cells (SMC) and their progeny, is also constructed in some of the stocks (Fig. 2E; Bellen *et al.*, 1989; Huang *et al.*, 1991). It is a very good marker for identifying SMC in clones in imaginal discs (see Section IIC2). In general, FRT chromosomes carrying multiple markers are particularly useful when screening more than one tissue for clonal phenotypes. The disadvantage of multiply marked stocks is their reduced viability.

c. Minutes

Minute mutations are haplo-insufficient gene disruptions that cause developmental delay and reduced bristle growth. Historically, *Minute* mutations have provided a way of increasing the size of X-ray clones. *Minute* mutations are recombined onto the parental chromosome that does not carry the mutation in which you are interested. In this case, the mutant clone will be wild type at the *Minute* locus and consequently will grow much faster than the surrounding heterozygous Minute tissue. This may also be useful in FLP-induced mosaic analysis (for strains, see Fig. 2E), particularly to study the long-range effects of wild-type tissue on the center of an otherwise small mutant clone. However, FLP recombinase will usually generate enough mutant tissue for analysis and, since *Minute* mutations affect cell proliferation and increase the relative size of the mutant clone, they complicate the interpretation of clonal phenotypes.

2. Generation of Clones

This protocol is designed to generate *bib* clones in the eye imaginal discs of first instar larvae. It is possible to modify this procedure for other developmental stages, as long as the cells of interest are still dividing, a prerequisite for clone induction.

Cross several *w; 40-bib/CyO* males with *w, hsFLP1; 40-w+* virgins at 25°C. Collect eggs for every 6–12 hr and age them for 30 to 40 hr. After aging, incubate the animals in a 38°C water bath for 60 min (the use of a water bath is essential, as air incubators will not provide adequate heat transfer). The size of clones can be increased or decreased by either earlier or later heat treatment, respectively. However, when generating large clones with early heat shock, a low frequency of small clones is also observed, which we presume to be the result of later mitotic recombination caused by persistence of the relatively stable FLP recombinase. In addition to adjusting the clone size, the clone frequency may also be modified by varying the length of heatshock (Fig. 2B).

3. Analysis of Clones

a. External Morphology

Clones in the eye or on the body can be directly examined and photographed under a dissection scope. Wings and legs can be pulled off with forceps, dipped in 90% ethanol for 30 sec, and in water for 30 sec, and then placed in a drop of Permount (Fisher Scientific) on a slide. These are examined under a coverslip with a compound microscope. Eyes and bodies can also be gently squashed and mounted in Permount. For a quick examination, any body parts can be placed under a coverslip with a drop of water and viewed with a compound microscope. For greater resolution, clonal phenotypes may be examined by scanning electron microscopy (SEM): dehydrate flies through 25, 50, and 75 and twice in 100% ethanol for 12 hr each. This is followed by an identical Freon 113/ethanol series, after which the samples are dried and stored under vacuum before being mounted onto stubs.

b. Plastic Sections of Eyes

Internal structures of the eye or other parts of the body can be examined in plastic sections. The following protocol is modified from Tomlinson and Ready (1987). Remove the head and gently cut away the eye that you do not need to expose the inside of the head. Place the head in 0.5 ml ice-cold 2% glutaraldehyde (Ted Pella Co.) in 0.1 M Na phosphate, pH 7.2. Spin the tube for 5 sec to allow the heads to settle to the bottom of the tube (this is not critical). Add 0.5 ml 2% OsO_4 (Ted Pella Co.) in 0.1 M Na phosphate (use gloves in a fume hood) and incubate the mixture for 30 min on ice. Remove the glutaraldehyde/OsO_4 mixture and wash the tissue with cold phosphate buffer by filling the tube completely. Remove the wash, add 0.5 ml 2% OsO_4 to the tissue, and incubate on ice for 1–2 hr. Discard the OsO_4 in a waste bottle in the hood and dehydrate the tissue by successive 5-min incubations on ice in the following ethanol solutions: 30, 50, 70, 90, and 100%. Repeat the final 100% EtOH incubation at room temperature. Replace the final alcohol treatment with propylene oxide and incubate for 10 min at room temperature. Repeat this step. Add 0.5 ml propylene oxide to 0.5 ml Durcapan resin (see following) and mix with the heads. (Handle resin with vinyl gloves as it is carcinogenic when unpolymerized). Incubate overnight at room temperature. Replace the resin/propylene oxide mixture with pure resin and incubate at least 4 hr. Using a needle, transfer a fixed head to a silicone rubber, flat embedding mold (Ted Pella Co.) that has been filled with resin. Place the head very close to the edge of the mold before orienting. Bake the resin at 70°C for 12 to 24 hr. Bake any waste resin before discarding. Trim the block and cut 1-μm sections. Transfer the sections to a puddle of water on a subbed slide (see following) with a stick. Dry the slide on a heating block at 80°C and then add a drop of DPX (Fluka) followed by a coverslip. [Durcapan resin: Mix epoxy resin (A), hardener (B), accelerator (C), and plasticiser (D) from Fluka at a ratio of 54 : 44.5 : 2.5 : 10, respectively. Store aliquots at −20°C. Subbed slides are prepared as following: heat 500 ml water to 80°C. Add 5 g gelatin (Sigma) and 0.5 g chromium potassium sulfate. Stir on a hot plate until dissolved (about 2 hr). Place new slides in racks, wash briefly in a warm hemosol solution, rinse with water. Dip slides in gelatin solution and dry, covered, overnight.]

C. Analyzing Clones in Developing or Internal Tissues

A given gene may be involved in multiple processes throughout development, giving rise to multiple phenotypes. It may, therefore, be misleading to draw inferences about the genes function based on the adult phenotype alone. Consequently, it is often prudent to characterize mutant phenotypes as a structure develops and this usually involves clonal analysis. The cuticle markers described (Section IIB1) cannot be used to mark internal and developing tissues and a general alternative approach is to use the IIM and NM markers described here (Section IIC1). In some cases specific cell markers are not required, particularly since FLP recombinase induces clones in most or all individuals. Two instances in which this might be the case are outlined below. If antibodies to the normal product of the mutant gene exist, then the clone may be recognized by the absence of staining, as has been done for *Notch⁻* clones (Rick Fehon, personal communication). Also, antibodies raised against normal cell components may reveal the abnormal phenotypes of the clone and thus define its boundaries (Heberlein *et al.*, 1993).

1. Novel Clonal Markers

The IIM and NM makers are fusions of the MYC epitope sequence with truncated sequences from the *Drosophila* P-transposase and Notch protein, respectively. IIM is localized in the nucleus and NM in the plasma membrane (Fig. 3). The choice of markers will depend on the tissue, the antibodies being used for double labeling (Section IIC2), and whether the marked cells are expected to display mutant phenotypes (for example, IIM might be useful to positively mark clones expected to contain multinucleate cells; see also Section IID). Further flexibility is afforded by two other markers: IIF carries the FLAG (Hopp *et al.*, 1988), rather than the MYC epitope, whereas SIIM contains the N-terminal amino acids of the *Drosophila* Src protein and is localized to both cell and nuclear membranes (Xu and Rubin, unpublished). Transcription of the marker genes is driven by the *hsp70* promoter, and, as it is induced only following heat shock (Lis *et al.*, 1983), the potential damage that could be caused by constitutive expression is limited. Even the heat pulse used to generate mitotic clones does not result in sustained high levels of marker due to their relative instability. To mark clones, a 1-hr heat shock at 38°C is sufficient; however, some chromosomes carry two marker elements, and in these cases the time can be reduced. Despite this, the expression is always low in a few tissues, including the germline cells of the ovary and the embryo during the first 13 divisions. The level of expression from the *hsp70* promoter is not saturating and tissues carrying one, two, or four copies can be distinguished (Fig. 3), making it possible to identify the wild-type twin-spot and the mutant clone simultaneously. This is particularly useful for studying dominant mutations, in which case the twin-spot clone is genotypically wild type, and in comparing the proliferation rates of wild-type and mutant clones of the same age (Section IID).

2. Double Labeling

Normally mutant clones will be generated so that they lack the marker gene expression (Fig. 1B); thus, it is relatively easy to label this mutant tissue with other cell identity or differentiation markers. However, in some experiments, in which the marker and mutation are on the same chromosome arm, the mutant patch is labeled and in this case it is best to chose a marker that will not obscure any double labeling. For example, if the cell identity or differentiation markers recognize the nucleus, the NM marker should be used. Alternatively, fluorescence microscopy with different fluorophores for each antibody can be used. As the marker proteins are recognized by anti-MYC or anti-FLAG antibodies that do not cross-react with β-galactosidase, mutant patches can be double-labeled with cell specifically expressed β-galactosidase, derived from enhancer-trap elements (O'Kane and Gehring, 1987; Blair et al., 1994).

3. Generating Clones in Developing Tissues

This protocol has been designed for imaginal discs, although it can be modified for use with other developing and internal tissues.

Induce clones in the first instar larvae (Section IIB) and then after several days at 25°C (larvae grown at 18°C often die during heat shock); place the vial containing wandering third instar larvae in a 38°C (no higher than 40°C) water bath for 60 to 90 min to induce the expression of the marker. Return the larvae to a 25°C incubator for 60 to 90 min prior to dissection. Fix the discs for 40 min in PLP (2% paraformaldehyde, 0.01 M Na$_3$H$_2$IO$_6$, 0.075 M lysine, 0.037 M sodium phosphate, pH 7.2) on ice. Wash the tissue four times for 10 min in PSN (0.037 M sodium phosphate, pH 7.2, 3% goat serum, 0.1% Saponin; it is important to remove the fix agent from the tissue before staining. The most effective way to wash the tissue well is to keep transferring it into fresh PSN) and then incubate overnight at 4°C with a 1 : 100 dilution (1 μg/ml) of anti-MYC ascites or 1 : 2 dilutions of anti-MYC supernatant (MAb MYC 1-9E10.2; Evan et al., 1985; American Type Culture Collection) or anti-FLAG (MAb M2, IBI) antibodies in PSN. After the antibody incubation, wash the discs four times in PSN and incubate with a 1 : 200 dilution of either FITC-, Texas Red-, Cy5-, or HRP-conjugated goat anti-mouse IgG (Jackson Laboratories) in PSN. After a 2- to 4-hr incubation at room temperature, rewash the discs four times in PSN and then mount them in 90% glycerol, 1X PBS, 0.5% n-propyl gallate. For double labeling experiments, in which you wish to label neuronal tissues, a 1 : 500 dilution of an FITC-conjugated goat anti-HRP IgG

(Organon Teknika Corporation) can also be added to the second staining solution.

Clones may be induced in other larval stages using essentially the same procedure. It is also possible to induce clones in embryos: embryogenesis consists of 13 synchronous nuclear divisions, followed by cellularization of the nuclei and 3 to 4 more divisions in local mitotic domains. While it is not possible to induce expression of FLP recombinase from the hsFLP construct during the first 13 divisions, small clones can be made within the mitotic domains by heat induction after cellularization (Dang and Perrimon, 1992; Harrison and Perrimon, 1993).

D. Positively Marked Clones

We have discussed the advantages of being able to recognize wild-type twin-spot clones with two copies of the *w* genes or the ΠM and NM cell markers (Sections IIB1 and IIC1). It can also be useful to positively mark mutant clones with two copies of a marker, especially if that marker will reveal aspects of the phenotype. To do this requires that the marker be recombined onto the mutant FRT-containing chromosome and such recombinants are usually obtained using an FRT-containing chromosome that already carries the mutation of interest. For example, to recombine ΠM onto a *bib* chromosome, cross the 40-2ΠM strain to *w; 40-bib/CyO*, essentially as described before (Section IIA4). Collect several non-Cy virgin females and balance the offspring that carry a single P[ΠM] on regular medium (these will have a lighter eye color than their parents). Test the balanced lines for *bib* by complementation.

In the approach described here, positively labeled clones express twice as much marker as the nonclonal tissue and so can be identified. In some cases, such as lineage analysis, it is preferable to generate positively labeled clones in a negative background. Two solutions to this problem use FRT sites and FLP recombinase to fuse a constitutive promoter to the coding sequences of the β-galactosidase gene. Harrison and Perrimon (1993) have developed FRT-containing strains in which the β-tubulin promoter and β-galactosidase sequences become fused together on the same chromosome after FLP-induced mitotic recombination at specially designed FRT sites. In the second approach, Struhl and Basler (1993) have generated P-elements in which the *Actin 5C* promoter is separated from the β-galactosidase coding region by a transcription termination sequence that is flanked by FRT sites. Expression of FLP recombinase causes the termination sequence to be recombined out of the P-element (Golic and Lindquist, 1989) and the consequent fusion of promoter and marker sequences initiates β-galactosidase expression. The β-galactosidase-labeled clones generated by both approaches will be especially suitable for lineage analysis in tissues composed of multiple layers of cells. Furthermore, the latter technique can also be used to generate patches of tissue that are mutant for cloned genes (Section IIE). However, more specialized FRT-containing chromosomes need to be constructed before the approach of Harrison and Perrimon (1993) can be used to generate mosaics for the majority of mutations.

E. Generating Mosaics for Cloned Genes

For mutations of cloned genes, there is an alternative way to make mosaic animals. Golic and Lindquist showed that if the *mini-white* gene is placed between two FRT sites oriented in the same direction, it can be removed from dividing cells by FLP-induced recombination ("flip-out"). This approach can be adapted for any gene for which the mutant phenotype can be rescued by cloned DNA. If the rescue fragment is flanked by FRT sites, then FLP recombinase will generate a small patch of mutant tissue and, if a marker gene is also included, the clone can be negatively labeled. A major advantage of this technique is that it removes only a single gene, unlike mitotic recombination between homologous chromosomes that homozygoses all the potentially mutant genes on a chromosome arm. Indeed, when cloning a novel gene, it should be possible to anticipate the generation of clones in this way and flank any rescue constructs with FRT sites. The P[*ry*+; *hs-neo*; FRT] element contains a unique *Not*I site between its two FRT sites and would be useful for this purpose (Fig. 2A; Xu and Rubin, 1993). The flip-out scheme can also be modified to generate positively labeled mutant clones, as suggested by Struhl and Basler (1993). In this case, the termination signal in their β-galactosidase containing P-element is replaced by the gene of interest.

These strategies have not yet been extensively applied and there are potential pitfalls. Unlike mitotic recombination, flip-out does not completely remove the cloned DNA from the cell and it may persist for a while as an extrachromosomal circle. If this DNA is transcribed, it could complicate the analysis of the phenotypes of small clones. Also, when positively labeling clones, β-galactosidase is constitutively expressed in mutant cells and could also potentially influence a mutant phenotype. If this is a concern, the *Actin 5C* promoter could be replaced with that from the *hsp70* gene.

Finally, Struhl and Basler (1993), have also shown that instead of the β-galactosidase gene, another gene of interest can be ectopically expressed in small clones using the same flip-out method. These authors have used this elegant approach to study the effect of ectopic *wingless* gene expression in imaginal discs (Struhl and Basler, 1993).

F. Germline Clones

The FLP-induced mitotic recombination has been shown to function in the male and female germline cells (Golic, 1991; Chou and Perrimon, 1992) as well as in the somatic cells (follicle cells) of the ovary (Harrison and Perrimon, 1993; Xu and Rubin, 1993). This allows the generation of germline clones.

1. Removing a Gene's Maternal Contribution

Embryos carry protein and RNA derived from their mother and it is often desirable to examine the effect of removing this contribution. As this often results in lethality, homozygous mothers cannot be used. One solution is to produce the embryos from a homozygous mutant clone in the germline cells of an heterozygous female and this can be done using mitotic recombination. To select only the progeny of the mutant clone, dominant female-sterile mutations (DFS) can be recombined onto the homologous, nonmutant chromosome. In this case, the heterozygous tissue and twin-spot clone will not produce eggs and only those from the mutant clone that does not carry the DFS mutation will survive (Wieshaus, 1980). This technique has now been fused with FRT/FLP technology: the DFS, Ovo^{D1} has been recombined onto an FRT-containing chromosome and can be used to generate a high frequency of mutant clones (Chou and Perrimon, 1992). The highest frequencies are obtained if FLP recombinase is induced in third instar larvae; indeed, there is an 80-fold preference for germline over somatic ovary clones at this larval stage, relative to that if FLP is induced in first instar larvae (Harrison and Perrimon, 1993).

2. Mosaic Analysis

Mosaic analysis of the germline and associated soma has proven very useful in addressing a variety of developmental issues, including the interaction between the germline and soma and the functional requirement in these tissues for genes that mutate to cause sterility. Clones can be induced in larvae and adults, provided that the induced flies are aged sufficiently to allow the clones to grow (ovary cells divide about once every 24 to 48 hr). This latter approach is particularly useful to generate clones in the nurse cells (Harrison and Perrimon, 1993). As stated before (Section IIC1) the heat-shock promoter does not induce high levels of ΠM and NM markers in the germline, so it may be necessary to use an enhancer-trap line (O'Kane and Gehring, 1987) expressed in the germline to mark clones or to replace the heat-shock promoter.

To produce mosaic ovaries, larvae or adult females from well-fed crosses are transferred into new vials, the cotton plugs are pushed down to restrict the movement of the animals, and they are then incubated in a 38°C water bath for 60 min. After aging, dissect the ovaries from the adults with a needle and tease them gently to separate the ovarioles prior to fixation. All tubes and pipets that are used to transfer ovaries should be rinsed with 1% BSA (or saliva; Tanya Wolff, personal communication) to prevent them sticking to the glass. Fixation and staining is as described previously (Section IIC3).

III. Genetic Screens

Screening for mutations that affect a given trait is one of the most powerful genetic approaches toward dissecting biological processes and has been applied to great effect in the study of development and function in *Drosophila*. However, conventional screens for mutations that affect the development or function of adult structures suffer from certain limitations. In particular, traditional screens require three generations and cannot identify homozygous mutations that kill the animal (Fig. 5A). These problems can be circumvented by screening for mutations that affect adult structures in mutant clones ("mosaic screens"). The two major advantages of this approach are that it does not necessarily select against lethal mutations and the screen can be accomplished in one generation. Although it has been appreciated that lethal mutations affecting adult structures could be identified in mosaics (Garcia-Bellido and Dapena, 1974), the low frequency of mosaicism induced by X-ray irradiation made such an approach impractical for systematic screens. Such screens are now practical with the increased frequency of mosaicism that can be induced for most of the genome using the FLP/FRT-containing strains. This system has made it possible to identify a new population of mutations that have escaped detection in previous screens.

There are two general approaches for screening for mutations that affect a particular structure in clones. First, one can recombine existing mutations, such as P-element-induced lethals, onto an FRT-containing chromosome arm and systematically assay their mutant clones for the desired phenotype (see Sections IIA and IIB). Alternatively, one can induce new mutations directly onto an FRT-containing chromosome arm. Flies carrying these mutations can then be crossed to nonmutagenized animals that carry the same FRT insertion, in addition to FLP recombinase, and mutant clones can be induced and assayed for phenotypes (F1 mosaic screens; Fig. 5B). Flies that show mutant clones would be kept and the mutant chromosome recovered from their germline cells. This approach is rapid as it requires only one *en masse* cross; however, because many clones are induced in each individual, there is theoretically a selection against mutations that kill flies that contain too many clones. In our experience, this has not been a problem. For example, in a screen that isolated 42 mutations with abnormal clones, 1 mutation caused lethality in more than 95% of mosaic offspring. Notwithstanding, if you believe that the bias of F1 screens will reduce your chances of isolating a desired mutation, an alternative is to cross the initial mutagenized FRT-containing animals to balancer stocks and subsequently examine the clones in the F2 progeny. Clearly, this involves much more work, as many individual crosses must be set up and, given the higher numbers of animals that can be screened, we recommend F1 screens.

A variety of mutagens can be used to disrupt genes, including X-rays, chemicals, and transposable elements. We have chosen to use X-rays, as these generate a variety of chromosomal aberrations ranging from point mutations to deletions and inversions. These latter classes of mutation are extremely useful in subsequent gene cloning and are not usually generated by chemical mutagens, such as EMS. In addition, X-rays usually induce double-stranded changes and consequently all the cells of a given F1 progeny will carry the identical mutagenized chromosomes. EMS, on the other hand, causes point mutations in only one strand of the DNA and thus F1 progeny are mosaic for the mutagenized chromosome and a wild-type copy. Consequently, not all mutant-bearing individuals will produce F2 offspring that carry the mutant chromosomes (see review by Ashburner, 1989). Finally, X-rays are easier to use safely.

Another mutagenesis strategy is to mutate FRT-containing chromosomes by hybrid dysgenesis. A potential problem in this approach would be the concomitant mobilization of the FRT sequences in the FRT elements with those used for mutagenesis. To overcome this, we have designed a strategy to immobilize the FRT elements. These P-elements carry both a *rosy* gene and a neomycin resistance gene and when exposed to transposase a fraction of the elements will undergo imprecise excision such that the *rosy* gene will be deleted, but *neo* will remain. These flies can readily be recovered and a significant proportion of them should retain the FRT sites, which are very tightly linked to *neo,* but which have lost the ability to transpose the FRT element because of the lack of functional 5' P-element sequences. Such elements are said to be "crippled." Despite this provision in the stocks, we have not yet used crippled (or normal) FRT sites in a P-element-based mutagenesis and cannot comment on its efficiency.

Many of the mutations isolated from our F1 mosaic screens are new and include both lethal and viable lesions. In addition, multiple types of mutations are recovered and their clonal phenotypes indicate that they affected genes in a number of biological pathways, such as neural development (remove, reduce, or increase photoreceptor cell and bristle numbers), pattern formation (ectopic bristle formation, mutiple bristles in a single position), morphogenesis (malformed bristles, malformed lenses, loss of pigmentation), cell differentiation (mutant clones form large scars), positive control of cell proliferation (small or absent mutant clones compared to the wild-type twin-spot clones), negative control of cell proliferation (overgrown and tumor-like mutant clones), and tissue polarity (bristles or photoreceptor cells with random polarities).

A. An F1 Mosaic Screen

Here, we present a detailed protocol for an F1 mosaic screen for mutations on the left arm of the second chromosome that affect eye morphology, although by appropriate choices of strains and the timing of FLP induction, it can be readily modified to analyze other chromosome arms and tissues.

Irradiate 50 males of the genotype *w, hsFLP1; 40-1* with 4000r of X-rays (this dosage gives approximately one lethal mutation per chromosome arm) and mate them to 150 virgin females of the genotype *w, hsFLP1; 40-w$^+$* in 15 vials at

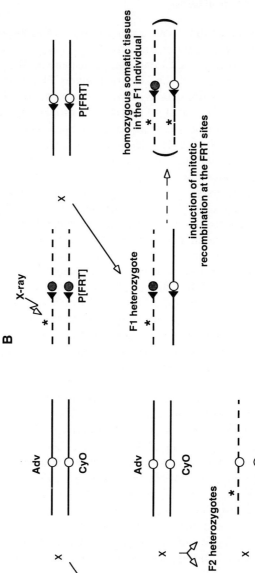

A

X-ray

*

○
○
⫶
⫶

X

F1 heterozygote

*

○

○
CyO
⫶

Adv

○

○
CyO

sibling F2 heterozygotes

X

Adv

○

○
CyO

*

○
CyO
⫶

F3 homozygotes

X

*

○
⫶

*

○
⫶

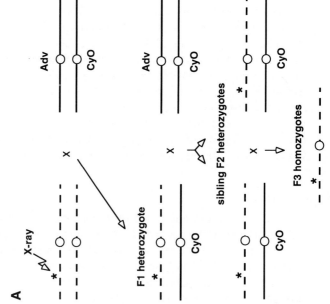

B

X-ray

*

●
▼

●
▼
P[FRT]
⫶
⫶

X

○
▼

○
▼
P[FRT]

F1 heterozygote

*

●
▼
⫶

○

homozygous somatic tissues
in the F1 individual

induction of mitotic
recombination at the FRT sites

*

▼
⫶

●
▼

*

○
▼

25°C. Over the course of the experiment, these crosses will produce 8000–10,000 progeny. Transfer the flies to new vials every 24 hr for 8 days, although the males should be removed after 4 days. The latter provision reduces the chances that chromosomes carrying the same X-ray-induced mutation are recovered more than once (Ashburner, 1989). Following transfer of the flies from a vial, it should be left a further 24 hr at 25°C and then placed in a water bath at 38°C for 60 min. This schedule is designed to induce clones in the eye of first instar larvae and to fit in with a normal work day; however, you may wish to collect eggs over a narrower time window so that you can obtain higher frequencies of mosaicism with more uniform clone sizes. Alternatively, earlier heat shocks will produce larger clones. After heat shock, return the vials to a 25°C incubator. Examine and collect F1 individuals (w, $hsFLP1$; 40-*/40-w^+) that contain mutant patches associated with w^- clones. A number of other classes of mutant clone may arise, including

1. Mutant phenotypes that are not associated with clones, which might be dominant effects of the mutagenized chromosome.
2. Small normal-colored mutant patches near w^- clones, which may be caused by spontaneous recessive mutations on the 40-w^+ chromosome.
3. Multiple clones in the same individual that do not always express the phenotype. These animals may be mosaic for the mutagenized chromosome (see before). It may be worth saving these flies in case they produce mutant offspring.

Individual mutant males should be mated to w, $hsFLP1$; 40-w^+/CyO virgins in separate vials and the clonal phenotypes reexamined in their w^+, non-Cy progeny. No more than 50% will show clonal phenotypes, although these should appear in multiple flies. If only w^+ cloneless progeny emerge and there are many dead pupae, it is possible that multiple clones of the mutation are killing the animal, as described previously. If the mutant clones are observed, generate a stock of the mutation by crossing w^-, Cy siblings (w, $hsFLP1$; 40-*/CyO) to each other. Mutant chromosomes can be recovered from F1 females, although this involves more work and it is usually simpler to just screen twice the number of males. However, if you wish to isolate a particularly interesting gene, F1 females can be crossed to 40-w^+F males and clones can be generated as before.

Fig. 5 Diagram of a standard F2 genetic screen and an F1 genetic screen. (A) A recessive mutation (*) induced in the germline of the parent is identified in homozygous flies of individual lines that are generated from three generations of crosses. CyO, balancer chromosome; Adv, dominantly marked chromosome. (B) Using a strain having a FRT element inserted near the centromere of a chromosome arm, induced recessive mutations on that arm can be identified in F1 heterozygous individuals by producing and examining somatic clones of cells that are homozygous for the mutagenized chromosome arm.

It is not important for F1 females to be virgins, since there is only a small chance that they were mated to a male that carried an independent mutation causing a similar clonal phenotype. F2 males that carry mutant clones can be balanced as before, but it is important to pick those that carry one w^+-marked chromosome, so that the w^- mutant chromosome can be followed.

IV. Concluding Remarks

The methods described here have greatly improved the efficacy of traditional mosaic analysis and make it feasible to use this technique to screen for novel mutations. However, in our use of these techniques, we and others have occasionally encountered minor difficulties and we will detail these below along with suggested solutions.

1. If certain tissues are made homozygous for some lethal mutations the flies will die. In this case, a high frequency of mosaicism may result in unacceptable mortality. Consequently, if survival following heatshock is low, we suggest that the duration of the heat treatment should be reduced (Fig. 2B).

2. Even at room temperature, the heat-shock promoter will drive low levels of FLP expression. This will result in somatic recombination in stocks carrying both the FRT sites and FLP recombinase. In our experience, this results, over time, in decreased viability and efficacy of the stocks. To reduce the severity of this deterioration we maintain the stocks at 18°C and regenerate them from their constituent chromosomes at least every year and before each major new experiment.

A. Limitations

The use of mitotic recombination for mosaic analysis has certain limitations. These include:

1. Following the initial mitotic recombination, daughter cells that become homozygous for a mutant gene may still contain some of the gene product derived from the parental cell. If this gene product is particularly stable, it may persist through several divisions and rescue aspects of the theoretical mutant phenotype.

2. As mitotic recombination can remove only a gene product following a mitotic division, it is not usually possible to generate clones in adult structures without also influencing the development of the clone. Thus, caution should be used in the interpretation of the abnormal function of mutant structures, as these abnormalities may also reflect an aberrant development.

3. The position of the FRT sites in our stocks mean that the approximately

5% of genes, those proximal to the FRTs or on the Y and fourth chromosome, cannot be analyzed.

4. The FLP enzyme is induced by heat stimulation of the *hsp70* heat-shock promoter. Activation of this promoter is ineffective at very early developmental stages, so it is difficult to generate clones during the early embryonic divisions and consequently the upper size of clones is limited. For most applications this is not a problem.

Some of these limitations may be overcome by modifications of the existing system. In addition, we envisage other developments that will broaden the applicability of these techniques. For example, we have placed the FLP enzyme under the control of a yeast UAS promoter (Xu and Rubin, unpublished). A number of fly strains that express the UAS-binding transcription factor GAL4 now exist in a variety of tissue-specific patterns (Brand and Perrimon, 1993; see also Chapter 33) and thus using a combination of these and the UAS-FLP strain, mitotic recombination could be restricted to particular target tissues or cells. As mentioned previously, the high frequency of clones generated using FLP recombinase enables one to perform screens for mutations that affect particular tissues in isolation. We believe that this will be particularly useful in the isolation of mutations that affect internal structures and those that affect behavior by disrupting only a defined portion of the nervous system. In the latter case, mosaic analysis could then be used to fine map the focus of the defect. Finally, the novel clonal markers that we have described make it possible to study the defects in individual mutant cells, and by double labeling with other antibodies that distinguish different subcellular compartments, it is possible to develop a detailed cell biological description of a mutant phenotype.

Acknowledgments

We thank Gerald Rubin, in whose laboratory most of these techniques were developed, and Drs. Pascal Heitzler, Norbert Perrimon, and Gary Struhl for unpublished information on the FLP-based mosaic analysis methods designed in their laboratories. Gratitude is also due to various members of the Rubin and Xu laboratories, in particular: Bruce Hay, Tanya Wolff, and Jenny Rooke for helpful discussions. This work was supported in part by a grant from the Lucille P. Markey Charitable Trust, a James S. McDonnell Foundation Molecular Medicine in Cancer Research Award (T. Xu). Stephen Harrison is a Howard Hughes Postdoctoral Fellow.

References

Ashburner, M. (1989). "*Drosophila:* A Laboratory Handbook." Cold Spring Harbor, NY: Cold Spring Harbor Laboratory Press.

Ashburner, M. (1991). *Drosophila* genetic maps. *Drosophila Information Service* **69,** 1–399.

Bellen, H. J., O'Kane, C. J., Wilson, C., Grossniklaus, U., Pearson, R. K., and Gehring, W. J. (1989). P-element-mediated enhancer detection: A versatile method to study development in Drosophila. *Genes Dev.* **3,** 1288–1300.

Blair, S. S., Brower, D. L., Thomas, J. B., and Zavortin, K. M. (1994). The role of *apterous* in the control of dorsoventral compartmentalization and PS integrin gene expression in the developing wing of *Drosophila*. *Development,* **120,** 1805–1815.

Brand, A. H., and Perrimon, N. (1993). Targeted gene expression as a means of altering cell fates and generating dominant phenotypes. *Development* **118,** 401–415.

Broach, J. R., and Hicks, J. B. (1980). Replication and recombination functions associated with the yeast plasmid, 2μ circle. *Cell* **21,** 501–508.

Chou, T.-B., and Perrimon, N. (1992). Use of a yeast site-specific recombinase to produce female germline chimeras in *Drosophila*. *Genetics* **131,** 643–653.

Dang, D. T., and Perrimon, N. (1992). Use of a yeast site-specific recombinase to generate embryonic mosaics in *Drosophila*. *Dev. Gen.* **13,** 367–375.

Evan, G. I., Lewis, G. K., Ramsay, G., and Bishop, J. M. (1985). Isolation of monoclonal antibodies specific for human c-myc proto-oncogene product. *Mol. Cell. Biol.* **5,** 3610–3616.

Friesen, H. (1936). Spermatogoniales crossing-over bei Drosophila. *Z. Indukt. Abstammungs. Vererbungsl.* **71,** 501–526.

Garcia-Bellido, A., and Dapena, J. (1974). Induction, detection and characterization of cell differentiation mutants in *Drosophila*. *Mol. Gen. Genet.* **128,** 117–130.

Geyer, P. K., and Corces, V. G. (1987). Separate regulatory elements are responsible for the complex pattern of tissue-specific and developmental transcription of the yellow locus in Drosophila melanogaster. *Genes Dev.* **1,** 996–1004.

Golic, K. G. (1991). Site-specific recombination between homologous chromosomes in *Drosophila*. *Science* **252,** 958–961.

Golic, K. G., and Lindquist, S. (1989). The FLP recombinase of yeast catalyzes site-specific recombination in the *Drosophila* genome. *Cell* **59,** 499–509.

Hall, J. C., Gelbart, W. M., and Kankel, D. R. (1976). Mosaic system. *In* "The Genetics and Biology of Drosophila" (M. Ashburner and Novitski, E., eds.), Vol. 21a, pp. 265–308. New York: Academic Press.

Harrison, D. A., and Perrimon, N. (1993). Simple and efficient generation of marked clones in Drosophila. *Curr. Biol.* **3,** 424–433.

Heberlein, U., Hariharan, I. K., and Rubin, G. M. (1993). *Star* is required for neuronal differentiation in the *Drosophila* retina and displays dosage-sensitive interactions with *Ras1*. *Dev. Biol.* **160,** 51–63.

Hopp, T. P., Prickett, K. S., Price, V. L., Libby, R. T., March, C. J., Cerretti, D. P., Urdal, D. L., and Conlon, P. J. (1988). A short polypeptide marker sequence useful for recombinant protein identification and purification. *Biol. Tech.* **6,** 1204–1210.

Huang, F., Dambly-Chaudiere, C., and Ghysen, A. (1991). The emergence of sense organs in the wing disc of *Drosophila*. *Development* **111,** 1087–1095.

Janning, W. (1972). Aldehyde-oxidase as a cell marker for internal organs in Drosophila melanogaster. *Naturwissenschaften* **59,** 516–517.

Jayaram, M. (1985). Two-micrometer circle site-specific recombination: The minimal substrate and the possible role of flanking sequences. *Proc. Natl. Acad. Sci. U.S.A.* **82,** 5875–5879.

Kankel, D. R., and Hall, J. C. (1976). Fate mapping of nervous system and other internal tissues in genetic mosaics of Drosophila melanogaster. *Dev. Biol.* **48,** 1–24.

Lawrence, P. A. (1981). A general cell marker for clonal analysis of *Drosophila* development. *J. Embryol. Exp. Morphol.* **64,** 321–332.

Lawrence, P. A., Johnston, P., and Morata, G. (1986). Methods of marking cells. *In* "Drosophila: A Practical Approach" (D. B. Roberts, ed.), pp. 229–242. Oxford: IRL Press.

Lindsley, D. L., and Zimm, G. G. (1992). "The Genome of *Drosophila melanogaster*." San Diego: Academic Press.

Lis, J. T., Simon, J. A., and Sutton, C. A. (1983). New heat shock puffs and β-galactosidase activity resulting from transformation of *Drosophila* with an hsp70-lacZ hybrid gene. *Cell* **35,** 403–410.

O'Kane, C., and Gehring, W. J. (1987). Detection *in situ* of genomic regulatory elements in *Drosophila*. *Proc. Natl. Acad. Sci. U.S.A.* **84,** 9123–9127.

Patterson, J. T. (1929). The production of mutations in somatic cells of *Drosophila melanogaster* by means of X-rays. *J. Exp. Zool.* **53,** 327–372.

Postlethwait, J. H. (1976). Clonal analysis of Drosophila cuticular patterns. *In* "The Genetics and Biology of Drosophila" (M. Ashburner and T. R. F. Wright, eds.), Vol. 2c, pp. 359–441. New York: Academic Press.

Preiss, A., and Artavanis-Tsakonas, S. (1991). Two mutants discovered in an X-ray mutagenesis. *Drosophila Information Service* **70,** 275.

Steller, H., and Pirrotta, V. (1985). A transposable P vector that confers selectable G418 resistance to *Drosophila* larvae. *EMBO J.* **4,** 167–171.

Struhl, G., and Basler, K. (1993). Organizing activity of wingless protein in Drosophila. *Cell* **72,** 527–540.

Tomlinson, A., and Ready, D. F. (1987). Cell fate in the *Drosophila* ommatidium. *Dev. Biol.* **123,** 264–275.

Wieshaus, E. (1980). "A Combined Genetics and Mosaic Approach to the Study of Oogenesis in Development and Neurobiology of Drosophila." New York/London: Plenum.

Xu, T., and Rubin, G. M. (1993). Analysis of genetic mosaics in developing and adult *Drosophila* tissues. *Development* **117,** 1223–1237.

CHAPTER 35

Analysis of Cellular Adhesion in Cultured Cells

Allan J. Bieber

Department of Biological Sciences
Purdue University
West Lafayette, Indiana 47907

I. Introduction

Many of the cellular interactions that occur during development reflect changes in the surface adhesive properties of individual cells. The powerful classical and molecular genetic approaches that are available with *Drosophila* offer an attractive system for the study of the cellular and molecular mechanisms underlying cell recognition and adhesion during development. In this chapter, we describe a cell culture assay that adds a new tool for the analysis of cellular adhesion in *Drosophila*.

Many cell adhesion molecules have been identified and characterized from *Drosophila* (for reviews, see Hortsch and Goodman, 1991; Bieber, 1991; Bunch and Brower, 1993). These molecules were often identified as putative cell adhesion molecules based on the presence of certain structural motifs that they shared with previously identified cell adhesion molecules from vertebrates, and the functions of many of the vertebrate cell adhesion molecules were demonstrated in cell culture assays that were pioneered in the laboratories of Takeichi and Edelman (Nagafuchi *et al.*, 1987; Edelman *et al.*, 1987). A cell transformation system using cultured *Drosophila* Schneider's line 2 (S2) cells has been established for the analysis of *Drosophila* cell adhesion molecules (for review, see Hortsch and Bieber, 1991). The general approach for testing a cDNA that encodes a putative cell adhesion molecule is presented in Fig. 1. cDNA clones are placed under the control of an inducible promoter and transfected into S2 cells. The protein is induced and the ability of the cells to form aggregates is assessed. Cell lines which express functional cell adhesion molecules will form large aggregates containing hundreds to thousands of cells.

Cell adhesion interactions can be classified as either homophilic, that is, an adhesive interaction between two molecules of the same species, or heterophilic, that is, an interaction between different protein species. By their nature, homophilic interactions have been the easiest to demonstrate but several heterophilic interactions have been demonstrated as well. Table I presents a list of the various adhesion molecules whose functions have been demonstrated using this approach.

In this chapter, we report the procedures that we have developed for the analysis of cellular adhesion in cultured *Drosophila* cells. The procedures are described as if homophilic adhesion molecules were being assayed. Obviously, assays for heterophilic interactions will require separate cell lines expressing both molecular components of the interaction but if clones for both molecules are available, the procedures described below should be readily adaptable.

II. Cell Transformation and Expression of Cell Adhesion Molecules

A. *Drosophila* Cell Lines and Cell Culture

A large number of *Drosophila* cell lines have been established and compared to many vertebrate cell lines they are quite easy to maintain. They can be grown in commercially available media supplemented with fetal calf serum and antibiotics. Most lines grow well at room temperature with air as the gas phase so no specialized incubation equipment is required. For a complete description of the available cell lines, their specific properties, and their maintenance requirements, see Chapter 9.

For the functional analysis of cell adhesion molecules in culture, we chose *Drosophila* Schneider's line 2 (S2) cells (Schneider, 1972). S2 cells were chosen for these assays because they show little tendency to adhere either to one another or to tissue culture plastic. Most of the cells grow in suspension as single, unattached cells with little in terms of processes or other morphology that might interfere with the assays.

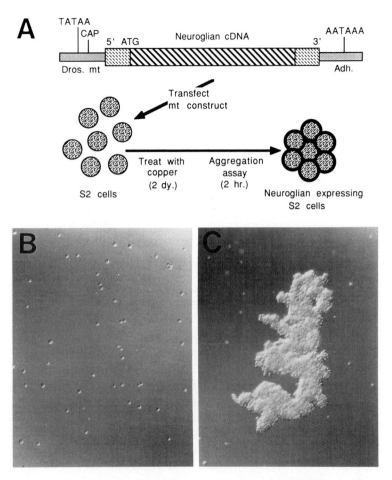

Fig. 1   A functional assay for homophilic cell adhesion molecules. (A) A schematic representation of the adhesion assay. A cDNA encoding a putative cell adhesion molecule, such as *Drosophila* neuroganglian, is placed behind the *Drosophila* metallothionein (mt) promoter. After transfection into *Drosophila* Schneider's 2 (S2) cells, the expression of the cDNA is induced by incubation for 2 days with 0.7 m*M* copper sulfate. The cells are then mechanically dissociated, shaken at 100 rpm for 2 hr and assessed for aggregation. (B) Cells that have been transformed with a construct in which the neuroglian cDNA is in the backward orientation to the promoter do not express the protein and do not undergo aggregation. (C) Cells which are induced to express neuroglian undergo a rapid and robust aggregation response. (Reprinted with permission from *Semin. Neurosci.* (1991) **3**, 309–320.

B. Expression Vectors

There are currently a variety of well-characterized promoters that can be used for the expression of cloned sequences in cultured *Drosophila* cells. For a list of the promoters and expression vectors that are known to work, see Chapter 9.

Table I
Cell Adhesion Molecules Whose Function
Has Been Demonstrated Using
Aggregation Assays in Cultured
***Drosophila* Cells**

Adhesion molecule(s)	Reference
Fasciclin III	Snow *et al.* (1989)
Chaoptin	Krantz *et al.* (1990)
Fasciclin I	Elkins *et al.* (1990)
Fasciclin II	Grenningloh *et al.* (1990)
Toll	Keith and Gay (1990)
Neurotactin	Barthalay *et al.* (1990)
Notch/Delta	Fehon *et al.* (1990)
	Rebay *et al.* (1991)
Neuroglian	Grenningloh *et al.* (1990);
	Bieber (1991)
Sevenless/Boss	Krämer *et al.* (1991)
Dtrk	Pulido *et al.* (1992)
Connectin	Nose *et al.* (1992)
Neuromusculin	Kania *et al.* (1993)

The expression vectors that we have used most commonly are pHT4 (Schneuwly *et al.,* 1987), pCaSpeR-hs (a derivative of Carnegie 4; Rubin and Spradling, 1983), and pRmHa-3 (Bunch *et al.,* 1988). The pHT4 and pCaSpeR-hs vectors express cloned sequences off the hsp70 promoter. These vectors have the added advantage that they are both suitable for P-element-mediated germline transformation of *Drosophila* embryos. The pCaSpeR-hs vector contains a multiple cloning site with several unique restriction sites, whereas the pHT4 vector contains a single unique *Kpn*I site for the insertion of cloned sequences.

The pRmHa-3 vector expresses cloned sequences off the *Drosophila* metallothionein promoter. All three of these expression vectors have worked well in our hands but our experience has shown that the highest steady state levels of transcript and protein accumulation are generally obtained by continuous induction off the metallothionein promoter as opposed to the short pulses of heat-induced expression off the hsp70 promoter. Like pCaSpeR-hs, the pRmHa-3 vector contains a convenient multiple cloning site but is not suitable for P-element transformation. More specific details on the use of these expression vectors are given here.

C. Cell Transformation, Selection, and Cloning

Several strategies have been reported for the transfection of *Drosophila* cells and for the selection of transformants. Calcium phosphate-DNA coprecipitation, lipofectin, and electroporation procedures have all been reported for the transfection of *Drosophila* cells. Several different selection systems are also available. Bourouis and Jarry (1983) used a vector containing the prokaryotic dihydrofolate reductase gene to confer methotrexate-resistance on transformed cells. Rio and Rubin (1985) reported the use of the bacterial neomycin resistance gene and resistance to G-418, a gentamicin derivative, as a selection system. For a more complete summary of the options available for transfection and selection, see Chapter 9).

Although several different transfection and selection systems are widely used, this laboratory currently uses calcium phosphate-DNA coprecipitation transfection to α-amanitin resistance almost exclusively. The plasmid pPC4 encodes the α-amanitin resistant RNA polymerase II gene from *Drosophila* (Jokerst *et al.*, 1989) and can be used to transform cultured *Drosophila* cells to α-amanitin resistance. Cotransfection with pPC4 and a nonselected plasmid, such as an expression vector, yields transformed cells that contain both the selected (pPC4) and nonselected DNA sequences. Although sensitivity to the drug varies with different *Drosophila* cell lines, we have found that treatment with modest doses of α-amanitin (5μg/ml) renders a very strong and efficient selection against untransformed S2 cells.

Described below is our procedure for calcium phosphate-DNA coprecipitation transfection to α-amanitin resistance. We have found this to be a consistent and reliable procedure. No special equipment or reagents are required. All culture procedures for α-amanitin selection can be carried out in commercially available Complete Schneider's medium with 12.5% fetal calf serum, 100 U/ml penicillin, and 100 mcg/ml streptomycin. The cells are cultured in bacterial grade or tissue culture plates or flasks in tightly sealed plastic containers to prevent dessication, at 25°C, with air as the gas phase.

Cell transfection by calcium phosphate-DNA coprecipitation with α-amanitin selection.

1. Plate 5 ml of cells at $1\text{-}2 \times 10^6$ cell/ml in 60-mm tissue culture plates. We usually allow the cells to sit overnight at this point, especially if they have been diluted from a culture that is near stationary phase growth.

2. Ethanol precipitate 4 μg of pPC4 with 16 μg of the nonselected plasmid and resuspend in 0.5 ml of sterile water. To prepare a 1-ml calcium phosphate-DNA precipitate, add the 0.5-ml DNA solution to a sterile test tube and then add 62 μl of 2 M $CaCl_2$ (filter sterilized). Gently agitate the solution by bubbling air through a sterile cotton-plugged Pasteur pipet that is resting in the tube. While the aeration continues, add 0.5 ml of 2× HBS (2× Hepes-buffered saline: 280 mM NaCl, 1.5 mM Na_2HPO_4, 50 mM Hepes) dropwise (1–2 drops/sec). Remove the pipet, cap the tube, and let it sit for about 30 min. After a few minutes the solution will begin to take on a bluish tinge but no visible granular precipitate should form. If the precipitate is too coarse, due to violent agitation during preparation or from adding the 2× HBS too rapidly, the transformation efficiency will be low. Add the 1 ml of precipitate dropwise over the surface of one plate of cells and allow the cells to sit for 15–hr. The S2 cell line does

not do well if incubated under high calcium conditions for much longer than 15–18 hr. The extensive washing in Step 3 is primarily designed to thoroughly wash out the calcium. (Note: Chen and Okayama (1987) have reported that replacing the Hepes with BES may give increased transformation efficiencies.)

3. Swirl the plates gently to dislodge any free precipitate and then decant the supernatant from each plate into separate 15-ml disposable centrifuge tubes. Add 4 ml of fresh media to the plate and swirl gently. Pellet any cells in the supernatant by a brief centrifugation (30 sec at about 3/4 speed on a clinical centrifuge), discard the supernatant, and then resuspend the cells with the 4 ml of medium from the parent plate. Add 4 ml of fresh media back to the plate. Repellet the cells, discard the supernatant, resuspend the cell pellet with the media from the plate, and add 5 ml of media to the plate. Repellet the cells, discard the supernatant, resuspend the cell pellet with the media from the plate, and then return the cells to the parent plate. Incubate the cells for 48 hr to allow expression of the α-amanitin resistance.

4. After 48 hr, dislodge the cells from the plate by blowing them off the plate with the medium. Pellet the cells and resuspend in 5 ml of fresh medium containing α-amanitin at 5 μg/ml (25 μl of 1 mg/ml α-amanitin in 100% ethanol).

5. About 7–10 days after addition of the drug, many of the cells may begin to lyse. The extent of cell lysis depends on the transformation efficiency. Poor transfections are followed by extensive cell lysis and it may take several weeks for the resistant cells to grow out. If the cell density gets too high or if the cellular debris on the plate gets too bad, wash the cells by pelleting and then resuspending at an appropriate cell density in fresh medium with α-amanitin. Always perform one transformation plate with 20 μg of the nonselected plasmid and no pPC4. No resistant cells should grow out on this plate and this control plate will give you some idea of how well the selection is working. As the resistant cells grow out, continue to passage the cells with 5 μg/ml α-amanitin. Once the cells have been passaged through several generations, you can discontinue the α-amanitin treatment. It is a good idea to freeze several aliquots of the cells at this time.

6. At this point, the mixed population of transformed cells can be analyzed for the expression of the nonselected plasmid or the cells can be cloned in soft agar to try to recover lines that express the nonselected plasmid at high levels. Antibody staining of mixed populations of transformed cells usually shows a wide range in the levels of expression between individual cells and we have found that aggregation assays often do not work well until more homogeneous, high-expressing lines are obtained.

Mixed populations of transformed cells can be cloned in 0.3% agar in the presence of an irradiated feeder layer of pPC4 transformed, α-amanitin-resistant S2 cells (S2:pPC4).

1. Pellet the S2:pPC4 cells and replate in exactly 4 ml, in Schneider's medium

without serum, on a 60-mm petri dish; tissue culture plates do not work as well here because the cells adhere more tightly to the surface of the plates. The volume of the cell suspension is important, but the concentration of cells can vary from 10^7 to 3×10^8 cells/ml without affecting the results. Expose the cells to 30,000 rads of gamma irradiation from a cesium or X-ray source. We usually irradiate the cells at 5 rads/s for about 6000 sec. The temperature of the plate should not increase noticeably during the irradiation. After irradiation, add 0.5 ml of fetal calf serum and allow the cells to sit for 10 min to loosen the cells from the plate. It should now be possible to blow the cells off the plate with a stream of media. Dilute the feeder cells to 1.25×10^6 cells/ml in Schneider's medium with 12.5% FCS, penicillin/streptomycin, and α-amanitin added to 5 μg/ml.

2. Prepare a 1.5% stock solution of agar in distilled water. Use a cell cloning grade agar such as Noble agar. Autoclave for 30 min to dissolve and sterilize and then hold the agar at 45–50°C until used. As long as sterility is maintained, this stock can be remelted by microwaving at least two or three times without losing the ability to gel.

3. Pipet 8 ml of the feeder layer cells onto one side of a 100-mm bacteriological grade petri dish, leaving most of the surface of the plate uncovered; again, tissue culture plates do not work as well here because the media tends to spread out on the plate and will not stay on one side. Add to this pool of media the viable transformed cells that you wish to clone. S2 cells clone with an efficiency of less than 50%, so to get a plate with 50–100 clones, you need to add a few hundred viable cells; we usually add these cells to the pool of feeder layer cells in a volume of 50–200 μl. Pipet 2 ml of the molten agar solution (45–50°C) onto the opposite side of the petri plate. Thoroughly mix the contents of the plate by swirling and allow the plates to solidify. The solidified agar is the consistency of very soft jello and the plates must be handled very carefully. Culture the plates at room temperature in tightly sealed plastic containers to minimize dessication.

4. Within 2 weeks, the clones will become visible. For each cloning experiment, one control plate that contains no viable cells should be prepared. On this plate there may be a few small clumps of cells but no clones will appear. On the plates that received the viable cells, round clones of viable cells will appear. These clones can get quite large (1–3 mm) but it is best to pick the clones when they are smaller to avoid cross contamination of cells between different clones. Also, to avoid cross contamination, it is best to keep the number of clones to 50–100 per plate.

5. To pick the clones, put the plate on an inverted microscope and while watching through the microscope, suck up a clone with an adjustable micropipeter sent at 10 μl. Expel each clone into a separate well of a 24-well microtiter plate containing 400 μl of medium that has been conditioned for 2–3 days over exponentially growing S2 cells, followed by filter sterilization. Amanitin should

be added to the conditioned medium prior to its use in growing up the clones. Once the clones have grown out to cover the bottom of the microtiter plate, they are ready to be scaled up (conditioned medium is no longer necessary) and tested for high-level expression of the transfected gene. We usually pick 24–48 clones from each transfection, but often end up testing only a dozen or so before we get a clone that satisfies our requirements for expression of the construct.

III. Analysis of Cellular Adhesion

A. Aggregation Assays

Once a mixed population of transformed cells has been obtained or clonal lines have been established, you are ready to test the cells for aggregation and to assay the cells for the proper expression of the protein.

When setting up cell aggregation assays, we usually begin with 5 ml of exponentially growing cells at $0.5–2 \times 10^6$ cells/ml in a 60-mm petri plate. Petri plates work better than tissue culture plates because the cells adhere less to the plastic. At cell concentrations greater than 2×10^6 cells/ml, even untransformed S2 cells form a few small aggregates; lower cell concentrations avoid this source of ambiguity.

Induction of constructs on the hsp70 promoter is accomplished by heat pulsing the cells for 15–30 min at 37°C. We usually heat shock the cells by placing a flask or plate of cells in a 37°C water bath. Since the media in a culture flask or plate is very shallow, it heats up quickly and you get an even heat shock. Cells can also be heat shocked in a test tube with good results. The heat-shock is followed by a recovery period of 1 hr. If the cells are in a tube or a flask, they should be transferred to a bacterial grade petri plate for the recovery period. After recovery, the cells are shaken gently at 100 rpm on a rotating platform shaker. For some cell adhesion molecules, aggregates will be clearly visible within 15–30 min after the start of shaking; for other molecules, or when mixed populations of transformed cells are used, it may take longer. Although aggregation may be observable within a few hours after heat shock, we have found that the accumulation of protein on the cell surface may not reach its maximum for 24 hr or more after heat shock.

Expression vectors that use the metallothionein promoter are induced by incubating the cells with 0.7 mM CuSO$_4$ (from a filter-sterilized 700 mM stock diluted 1/1000). We usually add the metal and then begin shaking the cells 24 hr later. As with the heat-shock-induced constructs, we have found that significant aggregation usually begins to occur long before maximal levels of protein accumulation are attained. When assaying a specific protein for the first time, it is a good idea to examine the kinetics of protein accumulation along with the aggregation properties.

Figure 1 shows the results of an aggregation assay with cells expressing *Drosophila* neuroglian, a member of the immunoglobulin superfamily of cell adhesion molecules. Although the results pictured in Fig. 1 are typical, the size, shape, and compactness of the cell aggregates vary somewhat with different molecules. Table I presents a summary of cell adhesion molecules whose function has been demonstrated using this system and a survey of these papers should provide a better feel for the variability of the assay.

Calcium is required for cell adhesion interactions that are mediated by the cadherin family of vertebrate cell adhesion molecules and a calcium requirement has been demonstrated for the interaction between the Notch and Delta proteins from *Drosophila* (Fehon *et al.*, 1990). As new molecules are identified and tested for their function in cell adhesion, it seems reasonable to test the requirements of these interactions for calcium and perhaps for other divalent cations as well. To determine whether any particular cell adhesion interaction is calcium dependent, the cells can be induced and then washed and aggregated in balance saline solution (BSS: 10 mM Hepes, 55 mM NaCl, 40 mM KCl, 15 mM MgSO$_4$, 5 mM CaCl$_2$, 20 mM dextrose, 50 mM sucrose, pH 6.95) from which Ca^{2+} has been omitted and EGTA added to 10 mM.

B. Analysis of Gene Expression

Immunoblotting of cell protein and antibody staining of the cells are the two most useful approaches for the analysis of gene expression in the transformed cells. When cell adhesion is being assayed in clonal cell lines, we usually try to assess the gene expression of each cell line first, followed by aggregation assays. High levels of antigen expression should correlate with the strongest aggregation results. Checking aggregation first and concentrating on the lines that work the best should be avoided, as it can easily lead to misinterpretation of the data.

When immunoblotting whole cell protein, we induce a small culture of cells (5 ml at 1×10^6 cells/ml), as previously described depending upon your choice of expression vector, and the cells are then pelleted and solubilized in 4 vol of 1.25× SDS–polyacrylamide gel sample buffer. The composition of this sample buffer will depend upon which SDS gel system you routinely use. Aliquots of 10–50 μl are separated electrophoretically, immunoblotted, and probed with an appropriate combination of antibodies. With the expression vectors described above, the levels of protein expression are quite high and sensitivity is usually not a problem. If sensitivity does become a problem, if you want to perform a biochemical characterization of the expressed protein, or if you want to isolate large amounts of the antigen, preparation of a membrane fraction might prove useful as a partial purification step. Chapter 17 provides protocols for the isolation and characterization of cell membranes and membrane proteins.

Direct antibody staining of transformed cells provides information that cannot be obtained from immunoblotting experiments. Antibody staining allows an assessment of the level of gene expression at the single cell level. Very often when aggregation assays are not working well or when mixed populations of transformed cells show that only a small fraction of the cells are aggregating, antibody staining reveals that the cells that are involved in aggregates are those cells with the highest levels of expression; obtaining new clonal lines that have uniform levels of high expression often solves the problem. Staining transformed cells with antibodies is a simple procedure and the following protocol has worked well for us.

Staining transformed cells with antibodies.

1. Cells can be fixed by adding an equal volume of 4% paraformaldehyde to the medium in which the cells have aggregated. Fix for 10–15 min at room temperature. Large cell aggregates will quickly settle to the bottom of a culture plate or test tube and the various reagents for antibody staining can be changed by simply removing and adding solutions over the top of the settled aggregates. For cells that do not aggregate, a brief centrifugation step may be necessary to pellet the cells between solution changes. To prepare a 4% paraformaldehyde fixative, make a NaOH solution of 2.52 g/100 ml, a $NaH_2PO_4 \cdot H_2O$ solution with 11.3 g/500 ml, and a $CaCl_2 \cdot 2H_2O$ solution with 0.5 g/50 ml. Dissolve 20 g of paraformaldehyde in 85 ml of the NaOH solution with vigorous stirring and gentle heating. Add 415 ml of the $NaH_2PO_4 \cdot H_2O$ solution. Add 6 g dextrose and stir until it dissolves. Add 2.5 ml of the $CaCl_2 \cdot 2H_2O$ solution. The solution should be clear and close to pH 7.4; adjust the pH if necessary. Filter the fixative and store refrigerated.

2. After fixation, wash the cells twice in phosphate-buffered saline (PBS:2 mM NaH_2PO_4, 8 mM Na_2HPO_4, 170 mM NaCl, pH 7.4) with 0.1% saponin. Add normal goat serum to 5% and incubate for 15–30 min at room temperature. Add an appropriate dilution of primary antibody and incubate for 1 hr at room temperature.

3. Wash the cells three times with PBS/saponin. Add normal goat serum to 5% and incubate for 15–30 min. Add an appropriate dilution of secondary antibody and incubate for 1 hr. We usually use fluorochrome conjugated secondary antibodies for this application, as opposed to enzyme conjugated antibodies, because the fluoresence signal makes it easier to discriminate between cells with different levels of antigen expression. The paraformaldehye fixative described above gives very low levels of autofluorescent background.

4. Wash the cells twice with PBS/saponin and once in PBS with 0.15% ascorbic acid to prevent fluorescence bleaching. The cells can now be mounted and examined.

One valuable application of cell aggregation assays is the structural/functional analysis of cell adhesion molecules by testing deletion constructs that have specific domains removed. It would not be surprising if certain deletion constructs affected protein folding so that the protein was not properly inserted into the membrane and antibody staining is one way to determine whether an expressed protein is in the membrane. For such experiments, it is often important that the membrane remain intact so that antibody cannot get into the cell. In this case, it is probably best to carry out the staining procedure on unfixed living cells. Stain the cells as described, leaving out the fixation step and the saponin. A balanced saline solution (BSS as described above) works best for the washes and incubations when dealing with living cells.

Some people find it easier if the cells are attached to a slide while carrying out the staining process. If this is your preference, the cells can be washed in BSS and then allowed to settle for 30 min on a poly-lysine-coated miroscope slide or coverslip. The cells stick quite firmly to the poly-lysine and the slide can be processed as described above by carefully flooding the slide with the various reagents and wash solutions. Both fixed and living cells can be stained in this manner.

C. Vital Staining of Cells with Fluorescent Dyes for Cell Mixing Experiments

Some aggregation experiments require that two different types of cells be mixed to determine the extent of the interaction between their expressed surface molecules. For example, when trying to determine whether a specific protein is a homophilic cell adhesion molecule, it is important to mix the transformed cells with ultratransformed S2 cells to determine whether the S2 cells express an endogenous heterophilic receptor for the cell adhesion molecule being tested. If no endogenous receptor is expressed, the cells that express the cell adhesion molecule will form homogenous cell aggregates that exclude the untransformed S2 cells. Mixing experiments are also necessary for the study of heterophilic cell adhesion interactions; separate cell lines that express the two interacting adhesion molecules are mixed and tested for their ability to form heterogenous cell aggregates that include both types of cells.

Mixing experiments therefore require a method for labeling the cells so that the different input cell lines can be distinguished from each other during the course of the assay. This is conveniently accomplished using vital fluorescent dyes. Mixing experiments can be performed either by labeling both cell lines with dyes that fluoresce at different wavelengths or by labeling only one cell line and leaving the other unlabeled. In either case, it is important to choose a dye that will not leach out of the cells once they have been labeled as this may lead to staining all cells during the assay and difficulties in cleanly interpreting the data.

DiI (1,1′-dioctadecyl-3,3,3′,3′-tetramethylindocarbocyanine perchlorate) is a cationic membrane dye that is widely used for the vital staining of cells. DiI stains very intensely and has absorption and fluorescence spectra compatible with rhodamine optical filter sets. The staining is very stable and will not wash out. To stain cells with DiI, we add 1 μl of a 2.5 mg/ml solution of DiI in ethanol per 1 ml of cell culture. We usually let the cells label overnight (several hours at least) at room temperature and then wash the cells twice in BSS or culture media before using them in a mixing experiment. DiO (3,3'-dioctadecyloxacarbocyanine perchlorate) is similar to DiI except that it has absorption and fluorescence spectra compatible with fluorescein optical filter sets. In our experience, DiO does not label as intensely as DiI and is generally more difficult to work with. We have found that the 16 carbon (dihexadecyl-) forms of DiI and DiO seem to label *Drosophila* cells more readily than the more commonly used 18 carbon (dioctadecyl-) forms.

CFSE (5-(and 6)-carboxyfluorescein diacetate, succinimidyl ester) is another useful dye. CFSE staining is cytoplasmic, with absorption and fluorescence spectra compatible with fluorescein optical filter sets. To stain cells with CFSE, we add 1 μl of a 5 mg/ml solution of CFSE in dimethylformamide per 1 ml of cell culture. The cells are intensely labeled within 15–30 min. Wash the cells twice in BSS or culture media before using them in a mixing experiment. Although staining with CFSE is rapid, the staining is not quite as stable as DiI and tends to leach out of the cells over a period of several hours. Nevertheless, it is very useful for some applications. The dichlorofluorescein derivatives of CFSE (such as 5-(and 6)-carboxy-2',7'-dichlorofluorescein diacetate, succinimidyl ester) seem to work the best on *Drosophila* cells.

When setting up mixing experiments, we usually label the cells with the fluorescent dyes first and then induce and aggregate. For each combination of cell lines to be tested, we set up several assays with different ratios (from 1 : 1 to 1 : 10) of the two different cell types. The total cell concentration in all assays should be constant at about $0.5–2 \times 10^6$ cells/ml.

Regardless of the type of experiment that is being performed, it is sometimes useful to have all cells in an assay uniformly labeled for fluorescein visualization. Hoechst dyes are most useful for this purpose. Hoechst is a bisbenzimide dye that binds to DNA. Its absorption and fluorescence spectra are compatible with DAPI optical filter sets. The dye can be added directly to the aggregating cells at a concentration of 250 μg/ml (from a 10 mg/ml stock in H_2O). Cell nuclei will stain intensely in 15–30 min and the cells should be examined with the dye present in the culture medium because the staining is easily washed out.

IV. Concluding Remarks

Expression of cell adhesion molecules in cultured cells has been a valuable approach for defining the functions of these molecules. By their nature, homophilic interactions have been the easiest to demonstrate but several heterophilic interactions have been demonstrated as well (see Table I). As other putative cell adhesion molecules are identified, it seems certain that further interactions of both types will be defined.

A distinction should be drawn between cell adhesion and the possible secondary effects of cell adhesion. Cell adhesion is a simple strengthening of the physical association between two cells. Some studies suggest that the binding of cell adhesion molecules may be accompanied by the generation of intracellular signals that can initiate changes in the participant cells (Schuch *et al.*, 1989). It should be remembered that although a given molecule may appear to function in cell adhesion, many of the cell adhesion molecules are structurally very complex and it is likely that they are involved in other cellular processes as well. Our understanding of the relative importance of adhesive interactions versus other intracellular interactions that are initiated by the binding of cell adhesion molecules is still rudimentary.

The cell aggregation system also offers a convenient assay for the functional analysis of specific domains within a cell adhesion molecule. Using recombinant DNA techniques, specific domains that might play a role in the adhesion process can be deleted or altered and the altered molecules can be expressed in cultured cells and assayed for their ability to promote cell aggregation. The mapping of specific repeats within the Notch protein that mediate its interaction with Delta and Serrate, is a nice example of this type of study (Rebay *et al.*, 1991).

Impressive progress has been made in the identification and functional analysis of cell adhesion molecules that are expressed during *Drosophila* development. Given the variety of molecular, genetic, and cell biological techniques that are now available with *Drosophila*, we are certain to see this system continue to yield exciting new insights into the molecular role of cell adhesion during development.

Acknowledgments

I thank Dr. Peter Cherbas and Dr. Corey S. Goodman in whose laboratories these procedures were developed. Work in my laboratory is supported by grants from the American Cancer Society and the National Science Foundation.

References

Barthalay, Y., Hipeau-Jacquotte, R., de la Escalera, S., Jimenez, F., and Piovant, M. (1990). *Drosophila* neurotactin mediates heterophilic cell adhesion. *EMBO J.* **9**, 3603–3609.

Bieber, A. J. (1991). Cell adhesion molecules in the development of the Drosophila nervous system. *Semin. Neurosci.* **3**, 309–320.

Bourouis, M., and Jarry, B. (1983). Vectors containing a prokaryotic dihydrofolate reductase gene transform *Drosophila* cells to methotrexate-resistance. *EMBO J.* **2**, 1099–1104.

Bunch, T. A., Grinblat, Y., and Goldstein, L. S. B. (1988). Characterization and use of the *Drosophila* metallothionein promoter in cultured *Drosophila* melanogaster cells. *Nucleic Acids Res.* **16**, 1043–1061.

Bunch, T. A., and Brower, D. L. (1993). *Drosophila* cell adhesion molecules. *Curr. Top. Dev. Biol.* **28**, 81–123.

Chen, C., and Okayama, H. (1987). High efficiency transformation of mammalian cells by plasmid DNA. *Mol. Cell. Biol.* **7**, 2745–2752.

Edelman, G. M., Murray, B. A., Mege, R-M., Cunningham, B. A., and Gallin, W. J. (1987).

Cellular expression of liver and neural cell adhesion molecules after transfection with their cDNAs results in specific cell–cell binding. *Proc. Natl. Acad. Sci. U.S.A.* **84,** 8502–8506.

Elkins, T., Hortsch, M., Bieber, A. J., Snow, P., and Goodman, C. S. (1990). *Drosophila* fasciclin I is a homophilic adhesion molecule that along with fasciclin III can mediate cell sorting. *J. Cell Biol.* **110,** 1825–1832.

Fehon, R. G., Kooh, P. J., Rebay, I., Regan, C. L., Xu, T., Muskavitch, M. A. T., and Artavanis-Tsakonas, S. (1990). Molecular interactions between the protein products of the neurogenic loci Notch and Delta, two EGF-homologous genes in *Drosophila. Cell* **61,** 523–534.

Grenningloh, G., Bieber, A. J., Rehm, E. J., Snow, P. M., Traquina, Z. R., Hortsch, M., Patel, N. H., and Goodman, C. S. (1990). Molecular genetics of neuronal recognition in *Drosophila:* Evolution and function of immunoglobulin superfamily cell adhesion molecules. *Cold Spring Harbor Symp. Quant. Biol.* **55,** 327–340.

Hortsch, M., and Bieber, A. (1991). Sticky molecules in not-so-sticky cells. *Trends Biochem. Sci.* **16,** 283–287.

Hortsch, M., and Goodman, C. S. (1991). Cell and substrate adhesion molecules in *Drosophila. Annu. Rev. Cell Biol.* **7,** 505–557.

Jokerst, R. S., Weeks, J. R., Zehring, W. A., and Greenleaf, A. L. (1989). Analysis of the gene encoding the largest subunit of RNA polymerase II in *Drosophila. Mol. Gen. Genet.* **215,** 266–275.

Kania, A., Han, P-L., Kim, Y-T., and Bellen, H. (1993). *Neuromusculin,* a *Drosophila* gene expressed in peripheral neuronal precursors and muscles, encodes a cell adhesion molecule. *Neuron* **11,** 673–687.

Keith, F. J., and Gay, N. J. (1990). The *Drosophila* membrane receptor Toll can function to promote cellular adhesion. *EMBO J.* **9,** 4299–4306.

Krämer, H., Cagan, R. L., and Zipursky, S. L. (1991). Interaction of *bride of sevenless* membrane-bound ligand and the *sevenless* tyrosine-kinase receptor. *Nature* **352,** 207–212.

Krantz, D. E., and Zipursky, L. S. (1990). *Drosophila* chaoptin, a member of the leucine-rich repeat family, is a photoreceptor cell-specific adhesion molecule. *EMBO J.* **9,** 1969–1977.

Nagafuchi, A., Shirayoshi, Y., Okazaki, K., Yasuda, K., and Takeichi, M. (1987). Transformation of cell adhesion properties by exogenously introduced E-cadherin cDNA. *Nature* **329,** 341–343.

Nose, A., Mahajan, V. V., and Goodman, C. S. (1992). *Connectin:* A homophilic adhesion molecule expressed on a subset of muscles and the motoneurons that innervate them in *Drosophila. Cell* **70,** 553–567.

Pulido, D., Campuzano, S., Koda, T., Modolell, J., and Baracid, M. (1992). *Dtrk,* a *Drosophila* gene related to the *trk* family of neurotrophin receptors, encodes a novel class of neural cell adhesion molecule. *EMBO J.* **11,** 391–404.

Rebay, I., Fleming, R. J., Fehon, R. G., Cherbas, L., Cherbas, P., and Artavanis-Tsakonas, S. (1991). Specific EGF repeats of Notch mediate interactions with Delta and Serrate: Implications for Notch as a multifunctional receptor. *Cell* **67,** 687–699.

Rio, D. C., and Rubin, G. M. (1985). Transformation of cultured *Drosophila melanogaster* cells with a dominant selectable marker. *Mol. Cell. Biol.* **5,** 1833–1838.

Rubin, G. M., and Spradling, A. C. (1983). Vectors for P-element mediated gene transfer in *Drosophila. Nucleic Acids Res.* **11,** 6341–6351.

Schneider, I. (1972). Cell lines derived from the late embryonic stages of *Drosophila melanogaster. J. Embryol. Exp. Morphol.* **27,** 353–365.

Schneuwly, S., Klemenz, R., and Gehring, W. J. (1987). Redesigning the body plan of *Drosophila* by ectopic expression of the homeotic gene *Antennapedia. Nature* **325,** 816–818.

Schuch, U., Lohse, M. J., and Schachner, M. (1989). Neural cell adhesion molecules influence second messenger systems. *Neuron* **3,** 13–20.

Snow, P. M., Bieber, A. J., and Goodman, C. S. (1989). Fasciclin III: A novel homophilic adhesion molecule in Drosophila. *Cell* **59,** 313–323.

CHAPTER 36

Using Inhibitors to Study Embryogenesis

Gerold Schubiger* and Bruce Edgar†

* Department of Zoology
University of Washington
Seattle, Washington 98195

† Fred Hutchinson Cancer Research Center
Seattle, Washington 98104

I. Introduction: Using Chemical Inhibitors to Study Development

A common approach for studying developmental processes is to physiologically perturb the system, at precise developmental stages, and identify the earliest developmental deviation from controls. Such methodology can define the function of the target affected by the insult. The obvious choice for such perturbances are genetic mutations, because they block the function of a single gene product. However, some developmental problems are not amenable to mutational analysis. Alternate approaches include the introduction of pharmacological inhibitors, specific antibodies, or antisense nucleic acids into the embryo at specific times. There are strong opinions for or against pharmacological approaches, but interestingly, some of the objections raised against drug studies also apply to genetic analysis. For example, the analysis of developmental stages long after initial perturbation is usually meaningless for both pharmacological and genetic perturbations. Second, the analysis of a phenotype based on one mutant allele is as tenuous as the use of one dose of a drug. Third, the reversal of the perturbation is a robust control for both genetic and pharmacological approaches. Finally, opposite effects of mutants as well as drugs help to assess the function of the targets. For example, the comparison of hypermorphic and hypomorphic alleles is a very powerful analysis of function. Pharmacologically, such comparisons can be made using drugs with opposite effects or by titrations of drug dose (Table I).

II. Introduction of Substances into the Embryo

A. Introduction

Two methods are used to introduce foreign substances into *Drosophila* embryos: microinjection (Santamaria, 1986; Spradling, 1986) and permeabilization (Limbourg and Zalokar, 1973). The manipulations required for permeabilization are relatively easy; thus, this technique is used if large numbers of embryos must be treated simultaneously. Standard mass fixation and devitellinization methods can also be used. Since the permeabilization media can be changed during an experiment, this technique is used if several different substances are administered at different times. These advantages must be weighed against certain drawbacks. Only small molecules (MW <400) can be introduced by permeabilization. Thus, this method is useful for introduction of many drugs and isotopic labels, but not for proteins and nucleic acids. Since permeabilization involves equilibration of the embryonic cytoplasm with the surrounding media, introduced substances have their effects rather gradually. It is inconvenient to stage individual embryos during permeabilization; thus, postfixation staging is used. Finally, much care is required to avoid traumatizing permeabilized embryos by overdesiccation or anoxia.

Table I
Summary of Drug Targets and Effects

Drug	Concentration	Method of introduction	Cross cell membranes	Target molecule/process	Effect	Reversibility	Reference
Aphidicolin[a]	400 µg/ml	I/P[b]	+	DNA pol α	Blocks DNA replication	+(P)	Sheaff et al., 1991
Hydroxyurea	>50 mM	I/P	+	Nucleotide reductase	Blocks DNA replication	–	Broadie and Bate, 1991
ARA-CTP	30 mM	I	–	DNA replication	Blocks DNA replication	–	Edgar and Schubiger, 1986
Cytochalasin D[a]	500 µg/ml	I	+	Filamentous actin	Blocks polymerization	–	Cooper, 1987
Phalloidin	1000 µg/ml	I	?	Actin	Stabilizes filaments	–	Cooper, 1987
Colcemid	200 µg/ml	I/P	+	Microtubules	Blocks polymerization	+; by UV	Rieder and Palazzo, 1992
Taxol[a]	0.5 mM	I	+	Microtubules	Stabilizes microtubules	–	Forry-Schaudies, et al., 1986
D_2O	>50%	P	+	Microtubules	Stabilizes microtubules	?	Dustin, 1979
Cycloheximide	1000 µg/ml	I/P	+	Protein synthesis	Blocks translation	+(P)	Vazquez, 1979
Pactymycin	1000 µg/ml	I	?	Protein synthesis	Blocks translation	–	Vazquez, 1979
Puromycin	10,000 µg/ml	I/P	+	Protein synthesis	Blocks translation	–	Zalokar and Erk, 1976
α-Amanitin	100 µg/ml	I	?	RNA pol II	Blocks pol II and pol III transcription	–	Edgar and Schubiger, 1986
[35S]methionine	>1000 Ci/mM	I	+	Protein label	n/a[c]	–	Zalokar, 1976
[32P]NTP	>3000 Ci/mM	I	?	RNA label	n/a	–	Edgar and Schubiger, 1986
[32P]dNTP	>3000 Ci/mM	I	?	DNA label	n/a	–	Edgar and Schubiger, 1986
BrdU	50 mM/1 mg/ml	I/P	+	DNA label	n/a	–	Bodmer et al., 1989
WGA	2.7 mg/ml	I	?	Binds nuclear pores	Immediate cell cycle arrest; variable phase	–	Finlay et al., 1987
Antisense RNA	2000 µg/ml	I	–	Specific mRNA	Degrades mRNA or blocks translation	–	Schuh and Jäckle, 1989
Antisense DNA	1000 µg/ml	I	–	Specific mRNA	Degrades mRNA or blocks translation	–	–

[a] Stock solution in DMSO; DMSO must not be injected at >1%.
[b] I, Injection; P, permeabilization.
[c] n/a, not applicable.

Microinjection is advantageous because known quantities of a substance can be introduced into the embryo virtually instantaneously, at a time chosen precisely by visual staging. Responses are more rapid and more uniform than with permeabilization, and live real-time analysis of the effects by observation or video microscopy is possible. Artifacts introduced by desiccation and anoxia are also more easily controlled. Molecules of any size, provided they are soluble in water, oil, or <1% DMSO, can be introduced. Higher concentrations of DMSO are toxic. Substances that are poorly soluble in water are problematic. For example, limited amounts of cytochalasin can be dissolved in water after heating and vortexing, but often crystallize in the injection needle, so the final concentration cannot be determined. Experiments in which two substances are injected at separate times are tricky, because the needle has to be inserted twice into the same opening. The disadvantage of microinjection is that each embryo must be handled manually, limiting the number that can be processed. We usually inject 200 or 300 embryos over the course of 2–3 hr. Because embryos are sensitive to abuse of all types—particularly physical insult, desiccation, and anoxia—parallel controls with buffer-injected embryos, or embryos permeabilized with permeabilization media alone, are mandatory.

B. Protocol 1: Egg Collection and Staging

Often it is important to have synchronously developing embryos. For such collections, start with parents of the same age. From a $\frac{1}{2}$-hr egg collection, parents are raised under noncrowded conditions. After eclosion, the adults are transferred to bottles with fresh food and aged for 3 days. Optimal density is important, and removal of half of the males is advised. About 180 females per bottle aged between 3 and 10 days after eclosion are ideal for egg collection. Egg laying is highest at sunrise and sunset. Since development can proceed in the uterus, eliminate overaged embryos from a collection by treating females with CO_2 or making several precollections. First, transfer adult flies to a fresh food bottle for 30–60 min. Then make at least three precollections in 10-min intervals on 1.5% agar plates (small petri dishes that fit into the openings of empty *Drosophila* bottles). We make furrows into the agar, sprinkle the plate with a couple of yeast granulates, and add a drop of acetic acid. After precollections, collections for experiments are made. We make timed collections as short as 8 min. We use five collection bottles and collect up to 200 embryos per plate within 8 min. Humidity in the bottle, and the selective loss of females, can be reduced by increasing the percentage of agar or using fewer flies. Alternatively, plastic collection bottles with breathable stoppers may be used.

C. Protocol 2: Permeabilization of Embryos

Permeabilization is performed essentially as described by Limbourg and Zalokar (1973) and Bodmer *et al.* (1989). Rinse and brush embryos from collection plates into a Nitex screen basket (150 μm mesh), using 0.7% NaCl, 0.05%

Triton X-100 (embryo bath). To dechorionate them, immerse the basket in 50% bleach until the dorsal appendages disappear (2–3 min). Rinse extensively in embryo bath and then rinse briefly with water. Blot off excess liquid with a Kimwipe and air dry the embryos until water droplets have evaporated (3–5 min). Swirl the embryos, in the basket, in octane (Aldrich gold label) for 5 min. Next, swirl the embryos into a monolayer on the Nitex, blot the basket with absorbant paper, and air dry just long enough to allow evaporation of excess octane (1–2 min), while observing under a dissecting microscope. Then put the basket in a dish and gently cover the embryos with Grace's insect medium or Schneider's medium (Gibco) containing the substance to be introduced. Incubate at least 10 min or up to several hours. The embryos should adhere to the Nitex underneath the medium. A minimum volume of medium is suggested to avoid drowning. For prolonged incubations after treatment, remove the basket from the media, blot dry, and cover the embryos with halocarbon oil (HC700; Halocarbon Products, N. Augusta, SC). For fixation, the basket with embryos is removed from the medium, blotted and swirled in heptane, and the embryos are pipetted into a two-phase heptane/formaldehyde fixative as described by Karr and Alberts (1986). If they adhere to the basket, spritz fixative into the basket to remove them. Permeabilized embryos may be devitellinized after fixation by shaking in a two-phase mixture of methanol/heptane. Efficient devitellinization fails if the embryos are overdessicated. Some protocols use an isopropanol wash to remove residual water and increase embryo permeability (see Ashburner, 1989): after bleach treatment and rinsing, incubate the embryos in isopropanol for 30 sec with gentle agitation. Remove the basket, blot it dry, and then rinse briefly with octane to remove residual isopropanol. Then incubate in octane and proceed as described above.

D. Protocol 3: Microinjection

Injection methods and equipment were originally developed for nuclear transplantation by Geyer-Duszynska (1967). The method was then modified by Zalokar (1971) and Illmensee (1972) and in the Schneiderman lab (for example see Okada et al., 1974). For detailed description, see Ashburner (1989) and Santamaria (1986). The method we describe here is a simplification of that described by Okada et al. (1974). For our purpose, we use needles smaller than those for nuclear transplantation (0.5–2 μm diameter vs 10–13 μm) and thus we damage the embryo less. We also work at room temperature. Our method allows us to inject about 30 pl, or 2–5% of egg volume.

1. Preparation of Injection Slides

Embryos are affixed to a microscope slide. Glue may be prepared by soaking 10 ft of half-inch double-stick adhesive tape (Scotch 3M) in 10 ml heptane on a rocker overnight. Certain batches of tape are reported to be toxic to embryos. The glue/heptane solution is pipetted onto coverslips and dried. Gluey coverslips are affixed to microscope slides with 2–5 μl of glycerol or water. Alternatively, glue or double-stick tape can be applied directly to microscope slides.

2. Hand Preparation of Embryos for Injection

Hand dechorionization and alignment is preferred for injecting very young embryos [0–20 min after egg deposition (AED)]. Embryos are loaded on a watchmaker forcep by moving it along a furrow in the agar where eggs are laid. They are transferred to double-stick tape and spread out. The embryos are dried for 3 min. Two prongs of the forceps are then squeezed together and pushed against the embryo so that the chorion breaks. First, we crack all of the embryos on the tape. Then we peel about 30 embryos out of the chorion by nudging with the forceps from the side. Once an embryo is exposed, transfer it to the injection slide. Rather than holding embryos between the two prongs of the forcep, embryos are transferred using a single prong. This is less damaging and quicker. To avoid anoxia, leave space equal to at least three embryo widths between embryos. Limit the process to 10 min per slide; otherwise the embryos will desiccate. Low microscope illumination reduces desiccation. If not used for injection immediately, store the slide with the embryos in a humid chamber. For this, we use petri dishes (diameter 16 cm) lined with soaking wet filter paper.

3. The Agar Block Method of Preparing Embryos for Injection

First, dechorionate embryos with 50% bleach, as described. Then, cut a small block of agar from a plate with a razor blade and transfer it to a microscope slide. Pipet 50–100 dechorinated embryos, in embryo bath, onto the agar block using a cut-off yellow tip (Pipetman). Wick off excess embryo bath from around the embryos with a tissue paper. Then, with one prong of a blunt forcep, slide individual embryos to the cut edge of the block and align them perpendicular to the edge. The poles of the embryos should protrude slightly beyond the edge of the agar. Twenty to thirty embryos may be lined up in this manner. Again, leave at least three embryo widths open between embryos. When a row of 20–30 embryos has been made, apply a glued coverslip (as before) or slide on top of the embryos. They adhere to the glue in the same arrangement as on the agar block. Then put the slide into a humid chamber for storage or a desiccation chamber in preparation for injection.

4. Desiccation of Embryos

Transfer the slide of embryos to a chamber containing desiccant or leave the slide in the open air. Desiccation times are 5–8 min for hand-dechorionated embryos and 8–12 min for bleach-dechorionated embryos, but vary according to ambient humidity and temperature. If the embryos develop wrinkles during desiccation, discard them. When desiccation is complete, apply a thin line of HC-700 halocarbon oil (Halocarbon Product Corp. North Augusta, SC) to the embryos. Heavy oils like this are user-friendly because embryos adhere to the slide better and the oil does not run. However, the embryos survive better in lighter oils, which also give a better optical quality. After desiccation, inject the embryos immediately. This restores them to normal turgor.

5. Injection Equipment

The embryo injection setup we use was originally developed by Zalokar (1971) and summarized by Santamaria (1986). Many combinations of micromanipulators and microscopes may be used. We prefer an upright compound microscope with fixed stage and a focusing head (Nikon), with long working distance $5\times$, $10\times$, or $20\times$ objectives. Traditionally, Leitz micromanipulators have been used, but the Newport (NPMX100L) is less expensive and more convenient. Positive pressure or vacuum can be applied to the needle by a large air-filled syringe, or by an oil-filled syringe and tubing. Syringe systems generate a constant positive pressure, which is released when the needle is inserted into the embryo. Alternatively, pressurized air from a lab hose line, with an open thumb valve, can be used to spritz fluid into the embryo after needle insertion (Fig. 1).

6. Injection

Needles are pulled from Drummond microcaps (25, 50, or 100 μl) using a standard needle puller (Kopf, Sutter, or equivalent) and broken by collision with a microscope slide to give a beveled tip with a diameter of 0.5–2 μm. Needles may be filled from the tip or backloaded with a drawn-out micropipet that fits into the capillary needle. For tip filling, apply a small drop of injection solution to a piece of parafilm affixed to a microscope slide, place the needle in the micromanipulator, and fill by suction. Injection solutions must be clean; debris can be removed by centrifugation. Filled needles may be stored for some time, provided the tips are dipped in oil. Insert the needle and needle holder into a micromanipulator at an angle 6° to the horizontal. For injection, insert the needle near one pole of the embryo and move the tip near to the far pole, then draw the needle back out, leaving a thin stream of fluid down the midline of the embryo. Remove the needle when the embryo has been restored to normal turgor. Leaving the needle fixed and moving the microscope stage and embryos is faster than manipulating the needle. Following injection, remove blebs of cytoplasm leaking out of the embryos by suction, using a needle with tip diameter >10 μm. Injected embryos should be kept in humid chambers.

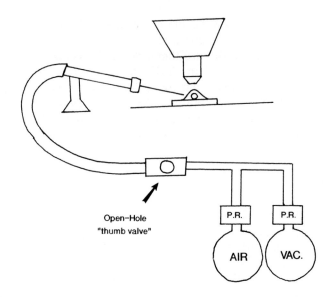

Fig. 1 Poor man's microinjection system. This system uses hose air and vacuum to deliver positive and negative pressure to the injection needle. Adjustable pressure regulators (P. R.) are attached to the hose lines, and an open hole in a plastic insert in the line is closed with the thumb, transiently applying pressure or vacuum (VAC.) to the needle. Thick-walled tubing should be used to assure even pressure transfer. The injection needle is mounted on a micromanipulator and positioned under a long working distance objective, using a fixed stage, upright microscope.

7. Fixation of Injected Embryos

Injected embryos must be devitellinized manually. Transfer the coverslip with injected embryos into a small dish and submerge it in heptane. Spritz heptane over the embryos until the embryos float off the coverslip. Then quickly remove the coverslip and the heptane and add fresh heptane. This rinse removes dissolved glue and oil. Finally, swirl the embryos to the center of the dish and pipet them into a glass vial with fixative, using a Pasteur pipet. We use a fixative of 1:1 heptane and 10% formaldehyde in PBS in the aqueous phase. Fix by gentle rocking for 20–30 min.

8. Devitellinization of Injected Embryos

After fixation, pipet the embryos using a cut-off yellow tip (Pipetman) from the fixative interface onto the outside of the Nitex screen of a Nitex basket that has been turned upside-down. Wick excess liquid away from the embryos and apply a glued or sticky-taped slide to them. Embryos will stick to the slide. Residual liquid left on the embryos before transfer from the Nitex prevents them from sticking. Quickly, to avoid desiccation, cover embryos with a drop of PBS. Now remove embryos from their vitelline membranes by pushing and teasing with a tungsten needle under a dissecting microscope. Tungsten needles are sharpened by dipping the tip of the tungsten wire into a solution of 1 *M* NaOH and running a 6 V current through. We devitellinize 50–100 embryos per hour by this method. Overfixed embryos adhere to the vitelline membrane, making devitellinization impossible, whereas underfixation causes the embryo to tear during dissection. When embryos are freed from their membranes, add some PBS containing 0.05% Triton X-100 and pipet them into an Eppendorf tube with a cut-off yellow tip. They can then be stained with DNA dyes or antibodies or processed for *in situ* hybridization (see Chapters 25 and 30).

9. Video Imaging

There are several problems with the video imaging of preblastoderm-staged embryos. Embryos develop best under oil without a coverslip, which is optically not ideal. *Drosophila* eggs are rich in yolk and thus opaque. However, the cytoplasm surrounding each nucleus prior to cycle 9 is optically different than the yolk. This is best seen under low-viscosity oil (3S voltalef Ugine-Kuhlmann,98,bd Victor Hugo, 92-CLICHY, France), but this oil runs easily and does not protect the embryo for very long. To overcome this problem, we use a "video chamber" (Fig. 2).

We use a Hamamatsu C2400 video camera with contrast enhancement. Recordings are done on a Gyyr or Panasonic time-lapse VCR. Optical disc recorders (e.g., Panasonic) offer higher resolution, but are expensive

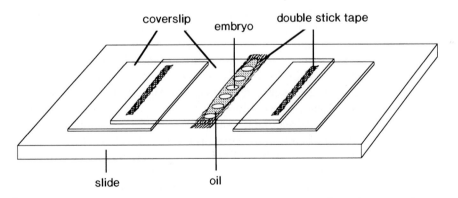

Fig. 2 Video chamber. On a slide, we glued two coverslips (22 mm^2) leaving a space between of about 10 mm. A small strip of double-stick tape (about 2- to 3-mm wide) is placed between the coverslips, leaving tapeless space on each side. Eggs are then positioned on the sticky tape and covered with low-viscosity oil (oil only on the tape). Care must be taken not to crowd embryos to prevent anoxia. The preparation is then covered with a coverslip contacting the oil drop and held in place by a small strip of tape on top of the glued coverslip and viewed with DIC optics and an immersion objective if necessary.

($10,000–$15,000). Another type of recording system uses Hi-8-mm video tape. Hi-8-mm tapes are reusable (video discs are not rewriteable) and much cheaper than video discs. The disadvantages of the Hi-8-mm tapes are that the time-lapse rate is limited (1 frame every 15 sec) and for time-lapse recording a computer interface is necessary. We predict that major improvements will be achieved using the high-definition TV broadcast system, which will provide 1000 lines of resolution. Video imaging after injection of fluorescently labeled molecules such as histones, actin, tubulin, or dextrans is an excellent way to track the cytological effects of specific inhibitors in real time in the syncytial blastoderm (Kellogg et al., 1988). However, observation of live embryos with either confocal or normal fluorescent microscopes is not possible for visualization of internal structures, such as nuclei before nuclear migration.

E. Protocol 4: Immunoblotting Staged Embryos

Embryos fixed in methanol can be staged postfixation and then analyzed by Western blotting (Edgar et al., 1994). This is useful for tracking levels and modification of proteins. Dechorionate embryos in 50% bleach for 2 min, rinse well, and rapidly devitellinize and fix in a two-phase mixture of methanol and heptane for 5 min with gentle mixing. Rinse embryos three times in ice-cold methanol and twice in PBS containing 0.05% Tween 20 (PBT) and protease inhibitors. Other compounds, such as phosphatase inhibitors, can also be added to this buffer. Stain embryos for DNA with 4 μg/ml bisbenzamid (Hoechst 33258) for 5 min, rinse twice in PBT, transfer to 40% PBT/60% glycerol, and ice for 2 hr to clear. Fixed embryos can be stored in this solution at −80°C without deterioration. Stage and select embryos visually using UV illumination on a compound microscope with 6.3 and 20× objectives. Use a tungsten needle and Pipetman for manipulation. Hand-picked embryos (1–10 per sample) are dissolved in 50% PBT, 50% 4× SDS sample buffer, mixed well, boiled for 5 min, and subjected to SDS–PAGE followed by immunoblotting (Harlow and Lane, 1988). Blots are probed using 3% milk as a blocking agent, HRP-conjugated secondary antibodies, and ECL chemiluminescent detection (Amersham). Filters may be stripped for reprobing by incubation in 50 mM Tris, pH 7.5, 2% SDS, 100 mM mercaptoethanol at 50°C for 30 min. This method may be used for permeabilized embryos. For Western blotting of injected embryos, rinse them, after injection, from the coverslip as described above (e.g., fixation of injected embryos) and pipet them into a 0.75-ml Eppendorf tube. Pipet off excess heptane and then air dry residual heptane with a stream of air. Homogenize the embryos briefly with a pestle made from a Pasteur pipet that has been heat-fused at the tip. Add a small volume of SDS sample buffer, homogenize again, boil for 5 min, and run on SDS–PAGE.

III. Summary of Drug Uses and Effects

A. DNA Replication Inhibitors

Aphidicolin binds and inactivates DNA polymerase alpha (Sheaff *et al.*, 1991). It acts rapidly and can be used either before or after cellularization. It can be delivered by either injection or permeabilization. When injected into syncytial-stage embryos, aphidicolin does not block all aspects of the cell cycle. Although DNA replication is arrested, Raff and Glover (1988,1989) reported that cycles of chromosome condensation and decondensation, centrosome replication, and spindle formation continue. Pole cells and cortical buds are formed without nuclei. The continuation of cytoskeletal cycling without DNA replication produces chromosome bridges and micronuclei. In contrast, injections of aphidicolin to block S phases of cycle 14 or later cause a coordinate arrest of all aspects of the cell cycle (Foe *et al.*, 1993). Hydroxyurea, which inhibits nucleotide reductase, may also be used to inhibit DNA replication in the embryo. Its effects, however, are slower than aphidicolin in early embryos.

B. Transcription Inhibitors

α-Amanitin is the drug of choice. We have not experimented with actinomycin D (Zalokar and Erk, 1976; Arking and Parente, 1980) as it is reported to affect DNA synthesis as well as transcription. Although α-amanitin is reported to block RNA polymerase II in a dose-specific manner, in *Drosophila* embryos it blocks both pol I and pol II with the same dose specificity (Edgar, unpublished). RNA pol III is several hundredfold less sensitive to α-amanitin. For a control, use the α-amanitin-resistant mutant C4, which has a mutation in a subunit of polymerase II and is about 100-fold less sensitive to α-amanitin than wild type (Greenleaf and Weeks, 1980). Blocking transcription allows identification of maternally controlled processes in the early embryo. For example, the first 13 cell cycles proceed normally without transcription, but cellularization is blocked. α-Amanitin is also useful for determining mRNA stabilities *in vivo*. After injection, embryos can be frozen at different time points and then processed for detection of a specific mRNA by Northern blotting, primer extension, or RNase protection (Edgar *et al.*, 1986,1989; Weir *et al.*, 1988).

C. Protein Synthesis Inhibitors

Puromycin, pactymycin, and cycloheximide affect the *Drosophila* embryo similarly. We use cycloheximide, because after injection, polyribosomes are rapidly stabilized, arresting protein synthesis (Vasquez, 1979). It can be used both before and after cellularization. Cycloheximide blocks nuclear division within one or two cycles of injection and blocks DNA replication (Zalokar and

Erk, 1976; Edgar and Schubiger, 1986). Many movements of the preblastoderm embryo are blocked, for example axial expansion of the nuclei and movements of the cortical cytoplasm. However, nuclear migration to the cortex continues (Baker *et al.*, 1993). Low concentrations of cycloheximide can slow the cell cycle in early embryos without arrest (Boring *et al.*, 1989). Cycloheximide is useful in assessing the stability of specific proteins *in vivo*. For such experiments, the drug is applied and embryos are frozen or methanol fixed at various time points and then processed for Western blotting. Because cycloheximide causes interphase arrest, and transcription continues, it can be used to distinguish primary responses in cascades of gene expression (Ashburner *et al.*, 1974). Cycloheximide also stabilizes mRNAs that are normally labile and in some cases causes ectopic transcription (Edgar *et al.*, 1986,1987,1989). To control for nonspecific effects, one can use a different inhibitor (such as pactymycin or puromycin) to demonstrate similar effects (Edgar *et al.*, 1989).

D. Microtubule Inhibitors

Although many agents destabilize microtubules, we have the most experience with colcemid (Rieder and Palazzo, 1992). It is stable, water-soluble, and penetrates cell membranes. At all stages of development, colcemid causes cell cycle arrest in the first metaphase following treatment. Colcemid also blocks nuclear migration and cellularization (Zalokar and Erk, 1976; Edgar *et al.*, 1987). A good control is lumicolchicine, an inactive form of colchicine. The effects of colcemid can be reversed with UV light (366 nm). We find that 40% of embryos injected with colcemid ($10^{-4} M$) at cycle 6 can be rescued by a 30-sec exposure to UV light 20 min after injection. Heavy water (D_2O) and taxol stabilize microtubules. Injection of D_2O does not have a noticeable effect on early development. However, dechorionated embryos incubated in 100 or 50% D_2O (without permeabilization) have abnormal microtubule networks that which are almost twice as large as in controls. Development after cycles 9 and 10 is arrested and nuclear and spindle morphology are abnormal. Chromosomal bridges, tripartite spindles, and micronuclei are some of the obvious abnormalities. Taxol is dissolved in DMSO; thus, it is only of limited value. When injected into early embryos, taxol causes elongation of spindle and astral microtubles.

E. Actin Inhibitors

Cytochalasins bind microfilaments, inhibiting both polymerization and depolymerization (Cooper, 1987). Cytochalasin D acts more quickly than cytochalasin B. The biggest drawback of cytochalasins is that they must be dissolved in DMSO. Both Zalokar and Erk (1976) and Hatanaka and Okada (1991) permeabilized embryos and incubated them in high concentrations of cytochalasin. They report that cytochalasin inhibits axial expansion of nuclei in preblastoderm stage embryos, but cortical migration is observed. However, no specific details are reported. We injected >100 embryos at or before cycle 4 with 500 μg/ml Cytochalasin D and found that only one-third of the injected embryos had the expected abnormalities. One third showed abnormal nuclear morphology and distribution, indicating that the drug has additional affects, and one-third developed normally. Thus, caution is necessary in interpreting data obtained with this compound. Cytochalasins have also been reported to block yolk contractions (Foe and Alberts, 1983) and cellularization of the syncytial embryo (Edgar et al., 1987). Because gastrulation depends on microfilaments, cytochalasin can be used to arrest embryos in cellular blastoderm configuration without blocking gene expression (Baker et al., personal communication). If an estimated concentration of 250 μg/ml of cytochalasin B is injected 20 min before gastrulation movement begins, 25% of the injected embryos maintain a "blastoderm configuration" hours after injection. Most of the embryos are partially arrested. Phalloidin has the opposite effect of cytochalasin: it stabilizes microfilaments, causing the cytoplasm to gel (Planques et al., 1991). Unfortunately, phalloidin also has toxic side effects. Blastoderm arrest is observed only if embryos are injected with 1 mg/ml phalloidin after cellular blastoderm is completed, just as the ventral furrow forms. However, blastoderm arrest is achieved in only 5% of the injected embryos; 60% continue to develop, most of them abnormally, and 35% die shortly after injection.

F. Labeling Agents

We have in vivo labeled RNA, DNA, and protein efficiently enough for detection of discrete species on gels, from single embryos, by injection of [^{32}P]UTP, [^{32}P]dTTP, and [^{35}S]methionine, respectively, at the highest specific activities available (Edgar and Schubiger, 1986). Phosphate labeling of phosphoproteins by injection of [γ-^{32}P]ATP also works, but in our hands is not efficient enough for phosphopeptide mapping experiments. For injections of radioisotopes, a lucite shield should be rigged for the microinjection apparatus. DNA may be labeled very efficiently with BrdU using either permeabilization (Bodmer et al., 1989) or injection (Edgar and O'Farrell, 1989). This allows the detection, in situ, of replicating nuclei by immunostaining using an anti-BrdU antibody (Becton-Dickinson). Chromatin, mitotic spindles, and the actin cytoskeleton may be tagged by injection of fluorescently labeled proteins such as histones, tubulin, or actin (Kellogg et al., 1988).

G. Antisense Nucleic Acids

Although antisense RNA and single-stranded DNA oligonucleotides have been used to degrade or block translation of specific mRNAs in other systems, their utility in Drosophila embryos is limited. Phenocopies of maternal-effect mutants such as cactus (Geisler et al., 1992) and pecanex (LaBonne et al., 1989) and early-acting zygotic mutants like Krüppel (Schuh and Jäckle, 1989;

Schmucker *et al.*, 1992) have been obtained by injection of size-reduced antisense mRNA. However, such phenocopies are often weak. Injection of concentration series, and control injections of sense RNA or sense DNA oligos (in cases in which antisense DNA oligos are used), are mandatory, since high concentrations are toxic. Antisense inhibition of abundant transcripts, such as *string* and *cyclin B,* and others, has been unsuccessful (Edgar, unpublished).

H. Serine Protease Inhibitors

Active extracellular serine proteases are involved in dorso-ventral embryonic axis and imaginal disc development. The function of these proteases can be studied by trypsin treatment, which activates proteases, or by the use of specific protease inhibitors (e.g., aprotinin, soybean trypsin inhibitor, or -1-antitrypsin (Pino-Heiss and Schubiger, 1989; Stein and Nüsslein-Volhard, 1992).

I. Other Inhibitors

Many inhibitors have been tried with various inconclusive results. Zalokar and Erk (1976) used mitomycin, cordycepin, and deoxyadenosine, which interfere with DNA replication, but report rather irregular effects. Low concentrations of 2,4dinitrophenol were used to inhibit metabolism and seemed to specifically elongate spindle length at anaphase. Arking and Parente (1980) showed that actinomycin D, rifampin, and rifamycin SV blocked development. Wheat germ agglutinin (WGA) blocks protein transport into the nucleus and RNA to the cytoplasm (Finlay *et al.*, 1987). We have found that injection of 2.7 μg/μl WGA into early embryos arrests the cell cycle at different phases, and halts development, although cortical migration of nuclei can still occur. WGA block can be reversed by addition of *N*-acetylglucosamine in other systems, but we have not tried this in *Drosophila*. Leptin *et al.* (1992) could not confirm the observation of Naidet *et al.* (1987) that injected RGD peptides arrested gastrulation. The inhibition by RGD injection cannot be due to PS integrin function, since embryos lacking both maternal and zygotic -PS subunit function gastrulate normally (Wieschaus and Noll, 1986; Leptin *et al.*, 1989).

IV. Concluding Remarks

We see three areas in which inhibitors will continue to be useful in *Drosophila* cell biological studies. These are (1) blocking specific enzymes or cytoskeletal components; (2) determining turnover rates of specific mRNAs and proteins; and (3) assessing the significance of generic functions in the progression of development. Although we have discussed points 1 and 2 above, the third point merits explanation. Because many inhibitors affect general cell functions, they can be used to test whether there are critical periods in development that are particularly sensitive to perturbations in these functions. For example, transcriptional rate is highest at cellular blastoderm (Anderson and Lengyel, 1979,1981), and such high rates of transcription may be necessary for the progression of development after cellular blastoderm. We can assume that specific gene products are necessary for these rate controls, but we have no idea of phenotypes that would be caused by loss of these gene functions. We foresee the possibility that lowering the rate of transcription using inhibitors might generate a phenotype that could be used as a diagnostic tool in a mutant screen.

Acknowledgments

We thank Jayne and Robert Baker for providing unpublished data and making helpful suggestions. This work was supported by NIH Grant GM 33656 to G. S. and a grant from the Markey Charitable Trust to B. A. E.

References

Anderson, K. V., and Lengyel, J. A. (1979). Rates of synthesis of major classes of RNA in Drosophila embryos. *Dev. Biol.* **70**, 217–231.

Anderson, K. V., and Lengyel, J. A. (1981). Changing rates of DNA and RNA synthesis in *Drosophila* embryos. *Dev. Biol.* **82**, 127–138.

Arking, R., and Parente, A. (1980). Effects of RNA inhibitors on the development of *Drosophila* embryos permeabilized by a new technique. *J. Exp. Zool.* **212**, 183–194.

Ashburner, M. (1989). "Drosophila, A Laboratory Handbook." Cold Spring Harbor, NY: Cold Spring Harbor Laboratory Press.

Ashburner, M., Chihara, C., Meltzer, P., and Richards G. (1974). Temporal control of puffing activity in polytene chromosomes. *Cold Spring Harbor Symp. Quant. Biol.* **38**, 655–662.

Baker, J., Theurkauf, W. E., and Schubiger, G. (1993). Dynamic changes in microtubule configuration correlate with nuclear migration in the preblastoderm Drosophila embryo. *J. Cell Biol.* **122**, 113–121.

Bodmer, R., Carretto, R., and Jan, Y. N. (1989). Neurogenesis of the peripheral nervous system in *Drosophila* embryos: DNA replication patterns and cell lineages. *Neuron* **3**, 21–32.

Boring, L. B., Sinervo, B., and Schubiger, G. (1989). Experimental phenocopy of a *Minute* maternal-effect mutation alters blastoderm determination in embryos of *Drosophila melanogaster*. *Dev. Biol.* **132**, 343–354.

Broadie, K., and Bate, M. (1991). The development of adult muscles in *Drosophila:* Ablation of identified muscle precursor cells. *Development* **113**, 103–118.

Cooper, J. (1987). Effects of cytochalasin and Phalloidin on actin. *J. Cell Biol.* **105**, 1473–1478.

Dustin, P. (1979). "Microtubules." Berlin: Springer-Verlag.

Edgar, B. A., and Schubiger, G. (1986). Parameters controlling transcriptional activation during early *Drosophila* development. *Cell* **44**, 871–877.

Edgar, B. A., Kiehle, C. P., and Schubiger, G. (1986). Cell cycle control by the nucleo-cytoplasmic ratio in early *Drosophila* development. *Cell* **44**, 365–372.

Edgar, B. A., Odell, G. M., and Schubiger, G. (1989). A genetic switch, based on negative regulation, sharpens stripes in *Drosophila* embryos. *Dev. Genet.* **10**, 124–142.

Edgar, B. A., Sprenger, F., Robert, J., Duronio, R., Leopold, P., and O'Farrell, P. (1994). Distinct mechanisms regulate the cell cycle at four successive stages of *Drosophila* embryogenesis. *Genes Dev.,* **8**, 440–452.

Edgar, B. A., and O'Farrell, P. H. (1989). Genetic control of cell division patterns in the *Drosophila* embryo. *Cell* **57**, 177–187.

Edgar, B. A., Odell, G. M., and Schubiger, G. (1987). Cytoarchitecture and the patterning of fushi tarazu expression in the *Drosophila* blastoderm. *Genes Dev.* **1,** 1126–1247.

Finlay, D. R., Newmeyer, D., Price, T., and Forbes, D. J. (1987). Inhibition of *in vitro* nuclear transport by a lectin that binds to nuclear pores. *J. Cell Biol.* **104,** 189–200.

Foe, V. A., and Alberts, B. M. (1983). Studies of nuclear and cytoplasmic behaviour during the five mitotic cycles that precede gastrulation in *Drosophila* embryogenesis. *J. Cell Sci.* **61,** 31–70.

Foe, V. E., Odell, G. M., and Edgar, B. A. (1993). Mitosis and morphogenesis in the Drosophila embryo: Point and counterpoint. *In* "The Development of *Drosophila*" (M. B. and A. Martinez-Arias, eds.) pp. 149–300. Cold Spring Harbor Laboratory Press.

Forry-Schaudies, S., Murray, J. M., Toyama, Y., and Holtzer, H. (1986). Effects of colcemid and taxol on microtubules and intermediate filaments in chick embryo fibroblasts. *Cell Motil Cytoskel.* **6,** 324–338.

Geisler, R., Bergmann, A., Hiromi, Y., and Nüsslein-Volhard, C. (1992). *cactus,* a gene involved in dorsoventral pattern formation of *Drosophila* is related to the I-k-B gene family of vertebrates. *Cell* **71,** 613–621.

Geyer-Duszynska, I. (1967). Experiments on nuclear transplantation in *Drosophila melanogaster.* *Rev. Suisse Zool.* **74,** 614–615.

Greenleaf, A. L., and Weeks, J. R. (1980). Genetic and biochemical characterization of mutants at an RNA polymerase II locus in *D. melanogaster.* *Cell* **21,** 785–792.

Harlow, E., and Lane, D. (1988). "Antibodies, A Laboratory Manual." Cold Spring Harbor, NY: Cold Spring Harbor Laboratory Press.

Hatanaka, K., and Okada, M. (1991). Retarded nuclear migration in *Drosophila* embryos with abberant F-actin reorganization caused by maternal mutations and by cytochalasin treatment. *Development* **111,** 909–920.

Illmensee, K. (1972). Developmental potencies of nuclei from cleavage, preblastoderm, and syncytial blastoderm transplanted into unfertilized eggs of *Drosophila melanogaster.* *Roux's Arch. Dev. Biol.* **170,** 267–298.

Karr, T. L., and Alberts, B. M. (1986). Organization of the cytoskeleton in early *Drosophila* embryos. *J. Cell Biol.* **102,** 1494–1509.

Kellogg, D. R., Mitchison, T. J., and Alberts, B. M. (1988). Behavior of microtubules and actin filaments in living *Drosophila* embryos. *Development* **103,** 675–686.

LaBonne, S. G., Sunitha, I., and Mahowald, A. P. (1989). Molecular genetics of *pecanex,* a maternal-effect neurogenic locus of *Drosophila melanogaster* that potentially encodes a large transmembrane protein. *Dev. Biol.* **136,** 1–16.

Leptin, M., Grunewald, B., and Stein, D. (1992). No effect of RGDS peptides. *Nature* **355,** 777.

Leptin, M., Bogaert T., Lehmann, R., and Wilcox, M. (1989). The function of PS integrins during *Drosophila* embryogenesis. *Cell* **56,** 401–408.

Limbourg, B., and Zalokar, M. (1973). Permeabilization of *Drosophila* eggs. *Dev. Biol.* **35,** 382–387.

Naidet, C., Semeriva, M., Yamada, K., and Thiery, J. P. (1987). Peptides containing the cell-attachment recognition signal Arg Gly Asp prevent gastrulation in *Drosophila* embryo. *Nature* **325,** 348–350.

Okada, M., Kleinman, I. A., and Schneiderman, H. A. (1974). Restoration of fertility in sterilized *Drosophila* eggs by transplantation of polar cytoplasm. *Dev. Biol.* **37,** 43–54.

Pino-Heiss, S., and Schubiger, G. (1989). Extracellular protease production by *Drosophila* imaginal discs. *Dev. Biol.* **132,** 282–291.

Planques, V., Warn, A., and Warn, R. M. (1991). The effects of microinjection of rhodamine–phalloidin on mitosis and cytokinesis in early *Drosophila* embryos. *Exp. Cell Res.* **192,** 557–566.

Raff, J. W., and Glover, D. M. (1988). Nuclear and cytoplasmic mitotic cycles continue in *Drosophila* embryos in which DNA synthesis is inhibited with Aphidicolin. *J. Cell Biol.* **107,** 2009–2019.

Raff, J. W., and Glover, D. M. (1989). Centrosomes, not nuclei, initiate pole cell formation in *Drosophila* embryos. *Cell* **57,** 611–619.

Rieder, C. L., and Palazzo, R. E. (1992). Colcemid and the mitotic cycle. *J. Cell Sci.* **102,** 387–392.

Santamaria, P. (1986). Injecting eggs. *In* "*Drosophila,* A Practical Approach." (D. B. Roberts, ed.), pp. 159–174. Oxford: IRL Press.

Schmucker, D., Taubert, H., and Jäckle, H. (1992). Formation of the *Drosophila* larval photoreceptor organ and its neuronal differentiation require continuous *Krüppel* gene activity. *Neuron* **9,** 1025–1039.

Schuh, R., and Jäckle, H. (1989). Probing *Drosophila* gene function by antisense RNA. *Genome* **31,** 422–425.

Sheaff, R., Ilsley, D., and Kuchta, R. (1991). Mechanism of DNA polymerase alpha inhibition by aphidicolin. *Biochemistry* **30,** 5065–5072.

Spradling, A. (1986). P-element-mediated transformation. *In* "*Drosophila,* A Practical Approach" (D. B. Roberts, ed.), pp. 175–197. Oxford: IRL Press.

Stein, D., and Nüsslein-Volhard, C. (1992). Multiple extracellular activities in *Drosophila* egg perivitelline fluid are required for establishment of embryonic dorsal ventral polarity. *Cell* **68,** 429–440.

Vasquez, D. (1979). "Inhibitors of Protein Synthesis." Berlin: Springer-Verlag.

Weir, M. P., Edgar, B. A., Kornberg, T., and Schubiger, G. (1988). Spatial regulation of *engrailed* expression in the *Drosophila* embryo. *Genes Dev.* **2,** 1194–1203.

Wieschaus, E., and Noll, E. (1986). Specificity of embryonic lethal mutations in *Drosophila* analyzed in germline clones. *Roux's Arch. Dev. Biol.* **195,** 63–73.

Zalokar, M. (1971). Transplantation of nuclei in *Drosophila melanogaster. Proc. Natl. Acad. Sci. U.S.A.* **68,** 1539–1541.

Zalokar, M. (1976). Autoradiographic study of protein and RNA formation during early development of *Drosophila* eggs. *Dev. Biol.* **49,** 425–437.

Zalokar, M., and Erk, I. (1976). Division and migration of nuclei during early embryogenesis of *Drosophila melanogaster. J. Microbiol. Cell* **25,** 97–106.

CHAPTER 37

Chromophore-Assisted Laser Inactivation of Cellular Proteins

Anke E. L. Beermann and Daniel G. Jay

Department of Molecular and Cellular Biology
Harvard University
Cambridge, Massachusetts 02138

METHODS IN CELL BIOLOGY, VOL. 44

I. Introduction

Drosophila is one of the best established genetic systems for the study of biological questions. However, it is not always possible to obtain a mutation in a gene of interest and for the determination of the stage specificity and site of action, genetic and molecular methods are limited.

As an alternative means of addressing functional interactions in biology with a high degree of spatial and temporal resolution, we developed chromophore-assisted laser inactivation (CALI). CALI is analogous to cellular laser ablation (Miller and Selverston, 1979) but the damage is localized to a protein of interest by covalently attaching the chromophore, malachite green, to an antibody that specifically binds to that protein (Jay, 1988). The laser energy is targeted to effectively inactivate the bound protein while unbound proteins are unaffected. Moreover, this perturbation of function occurs at the time of laser irradiation and only within the diameter of the laser spot. This opens up the possibility of perturbing the function of a given protein in a spatially and temporally controlled manner.

In our studies on the mechanism of CALI, we have shown that CALI is spatially restricted to within 34 Å of the dye molecules (Linden *et al.*, 1992; Liao *et al.*, 1994) and that the mechanism of CALI is mediated by photogenerated hydroxyl radicals. The high reactivity and very short lifetime of this radical species is responsible for the spatial specificity of CALI (Liao *et al.*, 1994). Spatial specificity has been demonstrated for the enzymes β-galactosidase and acetylcholinesterase *in vitro* (Linden *et al.*, 1992) and for the T-cell receptor *in vivo* (Liao *et al.*, submitted).

CALI has been successfully applied to a broad range of proteins (see Table I). Specific inactivation has been achieved for 18 of 20 proteins tested. For two *Drosophila* proteins, Patched and fasciclin I, the application of CALI results in phenocopies that resemble the corresponding hypomorphic mutations (Schmucker *et al.*, 1994; T. N. Chang *et al.*, unpublished results). The method has also allowed us to determine the *in vivo* functions of three proteins during insect neural development. In the grasshopper limb bud, the cell adhesion molecules fasciclin I and II function in axon adhesion and axonogenesis, respectively (Jay and Keshishian, 1990; Diamond *et al.*, 1993). The *Drosophila* segment polarity gene product Patched plays a role in neurogenesis that is independent of its segmentation function (Schmucker *et al.*, 1994). These experiments demonstrate the potential of CALI in addressing a broad array of biological questions.

A further refinement of CALI is micro-CALI. Micro-CALI employs microscope optics to focus a laser beam down to 10-μm diameters such that proteins can be inactivated within parts of cells. While this method has not yet been used on *Drosophila* embryos, it has been used on both grasshopper embryos and chick dorsal root ganglion neurons in culture (Diamond *et al.*, 1993; Sydor and Jay, 1993). In addition to the spatial and temporal resolution achieved by this technique, a further benefit has been a higher efficacy such that greater

Table I
Proteins Inactivated by CALI

Enzymes	Signal transduction molecules
β-galactosidase (1)	Cyclophilin A (*in vitro*) (7)
Alkaline phosphatase (1)	Calcineurin (*in vitro* and *in vivo*) (7)
Acetylcholinesterase (1)	pp60src (*in vivo* and *in vitro*) (7)
Horseradish peroxidase (7)	

Membrane proteins	Intracellular proteins
Grasshopper fasciclin I (*in vivo* (2)	Actin (*in vitro*) (7)
Drosophila fasciclin I (*in vitro* and *in vivo*, mimics hypomorphic mutation) (7)	Talin (*in vivo*) (6)
Grasshopper-fasciclin II (*in vivo*) (3)	Tau (*in vivo*) (7)
α, β, and ε chains of the T cell receptor (*in vivo*) (4)	
Drosophila Patched protein (mimics hypomorphic mutation) (5)	

Proteins not inactivated by CALI

Hexokinase (1)
Glyceraldehyde-3-phosphate dehydrogenase (1)

Note. (1) Jay, 1988. (2) Jay Keshishian, 1990. (3) Diamond *et al.* 1993. (4) J. C. Liao, L. Berg, and D. G. Jay, submitted. (5) Schmucker *et al.,* 1994. (6) Sydor and Jay, 1993. (7) Unpublished results.

than 80% of the cells subjected to micro-CALI show a result compared to approximately 40% observed for large scale CALI (Diamond *et al.*, 1993). This higher efficacy and the presence of unirradiated contralateral controls within the same embryo has allowed us to define the *in vivo* function of proteins of interest with a high level of confidence.

In the following chapter, we will first outline the methods for malachite green labeling of antibodies and setting up the laser. We will than describe how to perform CALI on *Drosophila* embryos and prepare the resulting embryos for immunohistochemistry and cuticle preparation. We will conclude with a description of the methods for micro-CALI and a discussion of the limitations and advantages of the technology presented.

II. Malachite Green Labeling of Antibodies

Antibodies and other reagents are labeled with malachite green isothiocyanate (MGITC) according to Jay (1988). A solution of malachite green isothiocyanate (Molecular Probes Inc., Eugene, OR) is freshly prepared by dissolving the dry reagent with dimethylsulfoxide at a concentration of 10 mg ml^{-1}. MGITC is stored desiccated at −20°C and is stable for at least a year under these conditions but is readily destroyed by moisture. The dye was added to a solution of 400 μg of antibody in 500 μl of 0.5 M NaHCO$_3$, pH 9.5, in four 2-μl aliquots once every 5 min with continuous rocking. The mixture is incubated for an additional 15 min. The labeled reagent is separated from free label by gel filtration with a prepacked PD-10 column (Pharmacia) eluted with 10 mM Tris–HCl, 150 mM NaCl, pH 7.4. The final concentration of the antibody was 200 μg ml^{-1}. The ratio of labeling was obtained by measuring the optical density of the labeled solution at 620 nm (molar absorptivity × 150,000 M^{-1} cm^{-1}) to determine the concentration of the dye and dividing that value by the reagent concentration.

The labeling method is similar to other isothiocyanate reactions with free amino groups (Mann and Fish, 1972). Two competing reactions occur when malachite green isothiocyanate is added to protein solution: reaction with the amino groups to form a stable thioester and hydrolysis of the isothiocyanate. A high concentration of the isothiocyanate is added due to its loss by the competing hydrolysis reaction. High pH favors the reaction with amino groups so the labeling is done at pH 9.5 but higher pH does not seem to enhance the reaction and risks antibody denaturation. It is essential that the antibody solution does not contain free amino groups other than the antibody (e.g., buffers such as Tris and glycine). These will react readily with the isothiocyanate and lead to poor antibody labeling.

The efficacy of CALI is dependent on the labeling ratio (Linden $et al.$, 1992) and having approximately 6 to 10 dyes per antibody is optimal. The labeling ratio is dependent on the relative amount of MGITC to protein and usually an equal weight of MGITC to protein will provide optimal labeling. The hydrolysis product of MGITC slowly forms an insoluble purple precipitate. The gel filtration separates this product but it may clog the column. Precentrifugation on a microfuge of the labeling solution circumvents this problem (full speed for 30 sec).

Much of the label on the protein is not covalently attached but associated hydrophobically with the surface. Slowly, this adhering dye will dissociate but the halftime is approximately weeks at −20°C and days at room temperature. This limits the shelf life of the labeled antibody and we aliquot, quick freeze, and generally use a batch for less than 3 months. When the protein concentration is too low, the hydrolysis reaction predominates and labeling efficiency is poor. We generally aim for a protein concentration of 0.4 mg/ml or greater. This concentration is usually limited by the short supply of the antibody (i.e., commercially obtained antibodies are costly and come in small quantities).

III. Laser Instrumentation and Setup

Nd:YAG (Neodynium:Yttrium/Aluminum/Garnet)-driven dye laser (GCR-11 with HG-2 doubling crystal, PDL-2; Spectra Physics Corp.)

DCM laser dye in methanol (Exciton Corp.) oscillator concentration, 175 mg/liter; amplifier concentration, 39 mg/liter

Right angle prism, holder, and mounting rods (Newport Corp.)

Convex lens and lens holder (same)

Laser parameters: Peak power, 56Mwatts/cm^2; spot size, 2 mm; pulse width, 8.5 nsec; frequency, 10 Hz. Energy/pulse = 15 mJ.

Whenever pulsed lasers are used, extreme caution should be exercised, especially concerning eye protection. Due to the short pulse width of the Nd:YAG laser, the peak power is very high (Mwatts) although for very short durations and a single stray beam could cause blindness. Protective goggles are always worn, unless specified and beam blockers are placed to prevent stray reflections. Contact with skin should also be avoided. Investigators are advised to carefully follow laser safety protocols provided with the laser system.

When aligning the laser beam, use the lowest lamp energy possible such that a red spot is just visible. At this power, goggles are not necessary, although precaution should be taken to avoid a direct beam. One must wear goggles that block 620-nm light whenever the lamp energy is high. An overview of the setup is shown in Fig. 1. At the exit port, a glass microscope slide is interjected at a 45° angle that reflects 1/7 of the light by 90° to a light meter allowing us to measure the laser energy. The main laser beam is directed down vertically using total internal reflection of a right angle prism held in the beam path by a rod and prism holder that is mounted approximately 50 cm from the exit port. An interjected planar-convex lens (approximate focal length, 70 mm) is placed below the prism on the same supporting rod such that the vertically directed beam goes through the center of the lens. The relative distance between the lens and prism determines the spot size and the lens is mounted such that the spot size is 2 mm.

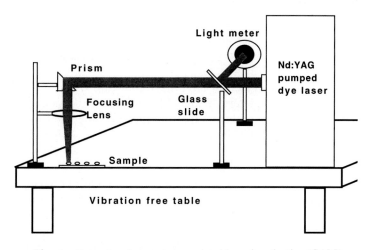

Fig. 1   Setup for chromophore-assisted laser inactivation (CALI).

To determine the spot size, a black and white polaroid photographic print is placed in the vertical beam path at the table surface. The lamp energy is raised to approximately 15 mJoules per pulse (remember to wear goggles), a single pulse is emitted to photobleach a spot on the film, and the diameter of the spot is measured. The sample is placed in the center of the photobleached spot, the alignment is confirmed by a single laser pulse, and the laser is switched to repeated pulse mode for the duration of the desired irradiation time. The laser beam has a Gaussian profile, so placement of the sample in the center of the spot, where power is highest, is recommended. Irradiation is stopped by switching the laser back to single pulse mode, the next sample is moved into the path of the beam, and the laser is returned to continuous mode for the prescribed time.

The important parameters for the laser beam are wavelength ($\sim$620 nm) and peak power density ($\sim$50 Mwatts/cm^2). We employ a Nd:YAG-driven dye laser because it provides these parameters for a spot size of 2 mm and a nitrogen-driven dye laser, which is adequate for a spot size of 10 μm. Other lasers will work as long as these parameters are met.

IV. CALI of *Drosophila* Embryos

A. Embryo Preparation

Synchronous embryos are dechorionated, positioned on a glue-covered coverslip, desiccated, and layered with halocarbon oil in preparation for injection of labeled antibodies and laser irradiation. This is done basically according to Wieschaus and Nüsslein-Volhard (1986) and Spradling (1986).

1. Optimizing of Egg Laying

Young (4–6 days old) flies are placed in an egg laying cage (see Section VI) on a fresh agar plate with a big drop of yeast. Estimate the number of flies per egg cage and do not add more flies than can cover the agar plate. Let them feed for 1–2 days.

On the injection day, start changing the plates (prewarmed, with small a yeast drop) every 30 min–1 hr. Initiate this process about 2 hr before you want to start injections so that the flies become used to constant interruptions. Do not disturb the flies during the egg-laying periods and cover the cage if necessary. In a regular day–night rhythm, flies tend to lay eggs in the early to late afternoon and early evening.

It is important to feed the flies well and to change the plate every day (fresh plate, fresh yeast) when you are not injecting. *Drosophila* females need yeast as a stimulant to lay eggs; without it, they will stop laying eggs at once and tend to retain eggs unless they have optimal conditions. After 10 days at 25°C, discard the flies. When they become old, they lay many unfertilized eggs.

2. Preparation for Injection

The embryos are first washed and freed of their chorion before they are aligned and transferred to a glue-covered coverslip. They are then desiccated and the coverslip is mounted on the stage for injection.

Pick the embryos off the plate with a moist brush into a little basket filled with 0.1% Triton X-100 to wash the embryos and dip the basket dry on a Kleenex. Then place the basket with the embryos in 50% Clorox for 2 min to remove the chorion. Quickly dip dry, wash thoroughly 2 times in 0.1% Triton X-100 and dip dry again. It is important to maintain a high survival rate of the experimental embryos. This is best done by handling the embryos as gently as possible and treating them with Clorox for no longer than 2 min.

With the moist brush you can now transfer the embryos onto a cut-out agar block (about 1.5 × 5 cm, from a fly agar plate). Once the chorion is removed, the embryos are surrounded by the vitelline membrane only and are very delicate and sensitive to desiccation. The agar keeps the embryos uniformly moist during the alignment. It helps to cut the agar with a straight edge. Under the dissecting microscope, orient embryos along this edge with a fine needle or a brush. For injections into the posterior, align the embryos with their anterior end to the right edge of the agar block (posterior to the injection needle later). A marker for the anterior end is the micropyle (the site of sperm entry). Figure 2A shows the embryo alignment. For the laser treatment, the *Drosophila* embryos should be aligned in pairs, with a distance of about 4 mm in between each pair. Control embryos that are not going to be irradiated can be aligned in groups of five embryos so that more embryos will fit on a coverslip.

Under the dissecting microscope, pick up the aligned embryos with the previously prepared glue-covered coverslips (see Section VI for glue preparation recipe). Use a thin layer of glue on the coverslip; the glue should be put on only shortly before injections. Desiccate the embryos by placing the coverslip in a drying chamber (see Section VI) for 15–20 min. The exact time of desiccation must be determined empirically and depends on several parameters: the relative humidity and temperature of the room and the genotype of the flies. The embryos have to be desiccated properly, because they have a high turgor. If not, the cytoplasm will leak out as soon as the injection needle is inserted, resulting in a lower survival rate. The ideal embryo just fails to leak cytoplasm when injected. After desiccation, cover the embryos with Halocarbon oil (HC-400). Discard the remaining embryos and start a new round with a fresh collection.

B. Injection

The injection procedure for *Drosophila* embryos has been described by other authors in detail (Spradling, 1986; Santamaria, 1986; Ashburner, 1989).

There are two choices for the injection system: an air pressure system and an oil-based system. A good, basic injection apparatus will suffice. It consists

A

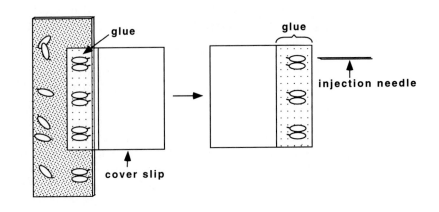

agar block with aligned embryos   embryos under injection microscope

B

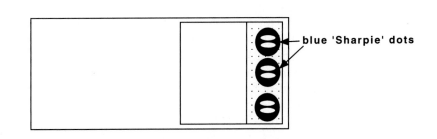

embryos before laser treatment

Fig. 2 Application of CALI to *Drosophila* embryos. A) Alignment of embryos. B) Embryos before laser treatment.

of a micromanipulator (Narishige, Leitz) connected to a microinjector composed of a syringe (a plastic syringe for the air system, a glass microinjector (Narishige Model IM-5A/B) for the oil system) connected with plastic tubing to the needle holder.

The air pressure system works with air pressure applied with a 10- to 20-ml plastic syringe and is relatively simple to set up. If you connect it to a hydraulic/piezoelectric system, it is possible to exactly determine the volume you are injecting. Refer to Spradling (1986) for further information.

In the oil system, all the parts of the micromanipulator and the microinjector are filled with desiccated Paraffin oil. This system allows much finer control; tiny manipulations of the microinjector are sufficient to eject fluid out of the needle or suck fluid into the needle. Because of the oil, fluids or cells in the

needle do not dry out easily, so transplantation of cells is also possible. It is necessary to keep this system clean (using 70% ethanol) and to avoid air bubbles in the system.

The microscope for injection (we use Zeiss or Olympus) should have a 10× phase objective lens (bright field is also acceptable) and 10× wide-field ocular lenses. The needle is oriented by using the manipulator, but injection is carried out by moving the microscope stage that carries the embryos. A vibration-free table is helpful, but a good, solid bench in a quiet room will work.

Needles are pulled from Borosil capillary tubing (see Section VI) by a needle puller (Vertical pipet puller, David Kopf Instruments, Model 700C). Determine an appropriate setting of the magnet and the heater. Break the needles under the microscope in a oil-free extra needle holder on the edge of a dust-free, clean double-slide stage or under the dissecting scope with a fine tungsten forceps. After breaking, the ideal injection needle has a tapered tip with a sharp, beveled opening of 7–10 μm.

Centrifuge the injection solution (labeled antibody) briefly before loading to prevent clogging of the needle. Labeled antibodies are generally used at a concentration of 0.2 mg/ml but one should determine the optimal concentration empirically. Needles in the oil system have to be filled from the front. First, remove the air out of the system: hold the needle holder upward (air bubbles rise to the top) and carefully press some oil out via the syringe until you are sure you have expelled all the air bubbles in the needle holder (they should be the only ones in the system if you handle it well). Insert the needle, fill it to the tip with oil, and mount the needle holder with the needle to the micromanipulator. Now you can load the needle by directly sucking the injection solution into the tip, from a small drop (1.5 μl) covered with oil, on a siliconized coverslip.

Mount the coverslip with the aligned and desiccated embryos onto a triple slide stage with an oil drop and put it under the microscope. The triple slide stage is constructed from two slides glued with an oil drop on top of each other, with a third slide that has been cut in half, on top of the other two. Move the embryos into view slowly. The needle should be above the embryos (otherwise you break the needle easily). Lower it slowly into the same plane of focus as the embryos. Move only the microscope stage with the embryos and once positioned, do not move the needle; instead, drive the single embryo into the needle via the stage. Release a small drop of fluid into the embryo and retract the needle by reversing the direction of the stage as soon as the embryo is filled but before any leakage occurs.

Drosophila embryos have a unique feature of insect embryos in that the early embryo is a syncytium; they go through their first 13 nuclear divisions without cytokinesis. Therefore, antibodies that are injected before the formation of the cell membranes (blastoderm-early gastrulation) diffuse throughout the embryo. For CALI of extracellular proteins, the antibody is injected into the yolk at the cellular blastoderm stage in a fashion similar to that described (also see Schmucker *et al.*, 1994).

The time between the injection of the labeled antibody and the laser treatment depends on your experiment. For early acting genes, laser irradiate at once after injection; for later acting genes (*fasciclin*) you have to laser irradiate later. Let the injected embryos develop in a moist chamber until the desired stage.

C. Laser Irradiation

Before the first experiment, you need to establish the parameters for irradiation. The time of laser irradiation and the energy of the laser beam need to be determined to optimize for conditions for CALI that minimize nonspecific embryo death. We have found that 1 min of laser irradiation at a pulse energy of 15 mJ results in effective inactivation of Patched without affecting embryo viability (Schmucker *et al.*, 1994).

Choose your controls carefully. They should include:

1. A set of embryos injected with the unlabeled antibody alone, subjected to laser irradiation.
2. A set of embryos injected with the malachite green-labeled antibody, but not laser irradiated.
3. A set of embryos injected with a malachite green-labeled protein not related to the experiment and subjected to laser irradiation. We use malachite green-labeled BSA or a nonspecific antibody at a concentration of 0.2 mg/ml.

For injections and laser treatment, the embryos should be aligned in pairs, with a distance of about 4 mm in between each pair. Because the diameter of the laser beam is 2 mm, only one pair of embryos fits under the laser beam at a time. Glue the coverslip with the injected embryos on a slide with a little drop of oil and mark each pair with a blue magic marker dot underneath: laser irradiation photobleaches the blue ink so that you know that the embryos have been laser-treated (see Fig. 2B).

Where the laser beam hits the table, there should be a piece of exposed photographic paper (Polaroid) with a bleached spot marking the area of the beam. Position the first pair of embryos on this spot. Now set the timer. *Drosophila* embryos are usually laser irradiated at 15 mJ for 1 min. Switch on the laser beam and irradiate the embryos pair after pair. One should note that irradiating 100 embryos (50 pairs) takes 50 min. Due to the rapid development of *Drosophila* embryos, a fresh round of embryos is generally prepared for each laser session.

After CALI, place the coverslips with the embryos on an agar plate, make sure the oil will not run off, and put the agar plate in a moist chamber. Depending on the defect you are looking for, allow the embryos to develop to the desired stage.

≡ V. Preparation for Phenotype Analysis

The preparation of the embryos after CALI should be chosen based on the phenotype that you wish to observe. We present two methods that allow for immunohistochemistry and cuticle preparation, respectively.

A. Immunohistochemistry

These methods are done according to Siegfried Roth (Stein *et al.*, 1991). After CALI and incubation until the embryos develop to the desired stage, the halocarbon oil is removed and the embryos are fixed. Gentle agitation of the coverslip in a small petri dish filled with heptane will wash off the oil. Put one end of the coverslip into a slit cut into a rubber silicone stopper and put the stopper into the glass vial that contains the heptane fix (5 ml heptane, 4 ml Hepes or Pems buffer, 0.5 ml 37% formaldehyde, as used in Macdonald and Struhl, 1986). The vial should be filled nearly to the top. Wrap parafilm around the stopper to avoid leaking. Fix by rotating the glass vial gently on the rotator for 20 min (put the glass vial into a box). Ignore embryos which do not stick to the coverslip.

After the fixation, put the coverslip into a small petri dish with methanol for a maximum of 2 min. This step will improve the morphology of the tissue. Then transfer the coverslip into a tray with distilled water. Due to the hypotonic conditions, the embryos will regain turgor. Now the embryos can be devitellinized with a needle. Transfer the embryos with a micropipet or a brush (with just 5 or 6 hairs) into an Eppendorf tube containing the methanol. Proceed with the antibody staining as usual (Macdonald and Struhl, 1986).

B. Cuticle Preparation

Development of the cuticle takes 2 or 3 days at room temperature (20°C). If the embryos have not developed long enough, the cuticle will not be fully developed. On the other hand, the embryos rot after more than 3 days at room temperature, resulting in poor preparations. Unaffected embryos will have hatched and crawled away by this time.

It is necessary to remove the halocarbon oil covering the embryos because it will mess up the later steps of cuticle preparation. With a drop of oil, mount the coverslip with the injected embryos onto a slide, which is easier to handle. Under the dissecting scope with a white background field, roll the embryos off the glue very carefully using a fine needle and lift them out of the oil. Transfer the embryos one by one onto a clean slide in tiny little oil drops. Rinse the slide with the transferred embryos gently with heptane using a Pasteur pipet over a petri dish. Be careful not to rinse the embryos off the slide and have Hoyers/lactid acid at hand. Let the heptane evaporate but be careful: the embryos will easily dry out. Just before they dry out, add one small drop of

Hoyers/lactic acid 1:1 and cover with a clean coverslip (avoid air bubbles). Put just enough mounting medium on top of the embryos such that it distributes evenly under the coverslip to sufficiently flatten the embryos. Clear the embryos for microscopy in a 60°C oven overnight (Van der Meer, 1977; Wieschaus and Nüsslein-Volhard, 1986).

VI. Materials

A. Equipment for Embryo Injections

Egg-laying cage, manmade (machine shop)—0.5×8.8-cm plexiglass tubing, cut in 10 cm height, one end covered with steel mesh, the other end carved 1 cm at the rim to fit the size of the bottom of the 100×15-mm standard agar plate (Fisher Scientific) filled 0.5 cm with agar.

Fine brush

Little baskets—Eppendorf tubes, cut off about 1 cm under the rim, carefully melt onto a fine steel mesh.

Microscope slides

Coverslips (22×22 cm)

Razor blade

Spatula

Fine needle

Brown mailing tape

Glue—The glue must be transparent, harmless to the embryos, soluble in heptane and sufficiently adhesive. Brown mailing tape (2 inches by 1 ft of Scotch 3MM, preferably TESA from Beiersdorf AG, Germany) is cut into pieces and stuffed into a glass vial containing 10–15 ml heptane. Shake the vial well for 5 min, remove the pieces of tape, and centrifuge remaining glue for 10 min at maximum speed. The supernatant is the glue. Keep the glue in glass screw-cap vials. The glue can be redissolved in heptane if it dries out.

Glue-coated coverslip—Apply a single layer of the glue with a pipet to 22×22-mm coverslips along a strip of approximately one-third of the width and let dry.

Drying chamber—"Drierite" (anhydrous calcium sulfate, W. A. Hammond Drierite Company, U.S.A.) in a covered glass petri dish, with Vaseline along the edge to form an air-tight seal.

Borosil capillary tubing, with dot fiber for rapid filling (1.0 mm OD $\times$ 0.75 mm ID/Fiber, FHC Brunswick, ME).

Injection setup (see Section IVB).

B. Reagents for Embryo Injections

Grape or apple juice agar plates (Wieschaus and Nüsslein-Volhard, 1986) fresh + prewarmed (room temperature), in 100×15-mm standard plastic petri dishes.

0.1% Triton X-100

50% Clorox (commercial bleach)

Halocarbon oil (HC-400; Halocarbon Products Corporation, N. Augusta, SC)

C. Equipment for Cuticle Preparations

Dissecting scope

Pasteur pipets

Fine needle

Slides

Small petri dishes

Clean coverslips: 12×12mm, 18×18 mm

60°C oven

D. Reagents for Cuticle Preparations

Heptane

Hoyer's mounting medium and lactic acid. Mix fresh 1:1 in an Eppendorf tube and centrifuge for 10 min (Wieschaus and Nüsslein-Volhard, 1986).

VII. Micro-CALI Setup and Application

The recent advent of micro-CALI has provided a high level of spatial and temporal resolution in the functional disruption of specific proteins during cellular processes (Diamond *et al.*, 1993). It has allowed for the generation of spatial discontinuities of protein function within a cell. Furthermore, the small laser that is employed for micro-CALI is much less expensive than the Nd:YAG laser and thus more accessible to other laboratories. Many laboratories already have this instrument to perform cell ablations. We present here the methods for setting up and applying micro-CALI (Fig. 3).

A. Equipment

Dye laser and microscope interface:

Nitrogen-driven dye laser (VSL-337 or VSL-337ND and DLM-110) (Laser Science Co., Newton, MA).

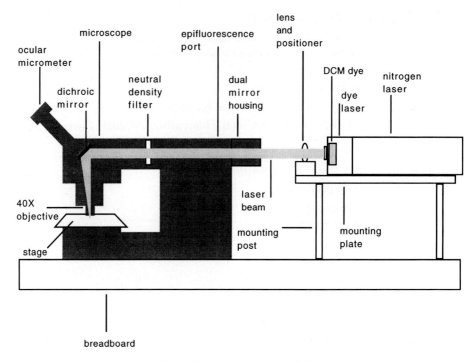

Fig. 3   Setup for micro-CALI.

DCM laser dye 35 mg/20 ml of DMSO (Exciton Corp. Dayton, OH).

Dual mirror lamp housing attachment (Catalog No. 447230. Carl Zeiss Inc., Thornwood, NY) NB—This is for a Zeiss microscope; consult your microscope dealer for an equivalent adaptor.

Dichroic mirror (reflecting light through objective at wavelengths greater than 600 nm) (Chroma Inc.)

Hard Mounting (all hard mounting equipment is available from Newport Corp., Fountain Valley, CA).

LE series breadboard, 2 ft/3 ft/2 inches (1 inch distance between screw holes).

Two mounting plates (Catalog No. 290-TP).

LP, five-axis positioner, and a series of planar-convex lens for steering and focusing the beam.

Seven Posts and 1/4-20 set screws as needed (We make our own aluminum posts at appropriate height).

B. Setup and Alignment of Micro–CALI

1. Mount dual mirror lamp housing with one opening directed to the fluorescence lamp and the other directed toward the laser.

2. Hard mount laser onto the breadboard using platforms and posts such that the beam is directed approximately to the center of the mirror. The combined height of the platform, the posts, and the distance from the bottom the laser to the aperture where the beam is located should be equal to the height of the center of the mirror.

3. Install the dichroic mirror in a vacant fluorescence filter set. You may use the position employed for bright-field microscopy (the image will be tinted slightly blue by reflecting wavelengths > 600 nm).

4. Set the fluorescent filter so that the dichroic mirror is in the optical path and turn the lever on the dual mirror mount such that the laser beam will be reflected into the microscope. Place a piece of white paper on the mirror housing. Turn on the laser and observe the red spot on the white paper. Although birefringence bands are visible, the majority of the light should be in one spot. Factory specifications for this laser with DCM is around 30 μJ per pulse. Measure this if you have a power meter. Move the microscope such that the beam is approximately at the center of the mirror. For safety reasons, minimize the time that the laser is on while manipulating the microscope and by all means, turn it off between steps.

5. Remove an objective lens from the nosepiece and coarsely adjust the alignment of the beam by maximizing the red light that goes through the microscope. This is easily visualized by laying a piece of lens paper or a kimwipe on the stage with the illuminator aperture closed down to a minimum (make sure that the illuminator is centered). The red beam and the illuminator spot should coincide. Adjustment is done in two ways; moving the microscope on the breadboard and slight rotation of the dual mirror housing.

6. Place a low-power objective in the microscope and repeat the alignment procedure. The beam should be parfocal with the stage. If this is not the case, you may need to interject a focusing lens in a five-axis positioner in between the laser and the microscope. This seems necessary for the smaller laser because there is a significant divergence of the beam. The slightly larger laser (VSL-337ND) has almost no divergence, so an interjected focusing lens is not required. Choose a focusing lens with focal length to tighten the beam somewhat before it reaches the dual mirror. This is trial and error and one should be careful not to focus onto the face of the objective lens. Perform this using neutral density filters to attenuate the power.

7. The power should now be sufficient for a single pulse of laser light through the 40× objective to photobleach a spot on a field of blue Sharpie ink

drawn on a slide. This provides sufficient power for micro-CALI. The size of the spot and location in the field of view are determined using an ocular micrometer. It should be noted that the spot is not circular but elongated due to the laser beam profile of the nitrogen-drive dye laser. The ocular micrometer is aligned with the location of the laser spot in the microscopic field. The coverslip is removed and replaced by the slide or dish containing the cell of interest. The cell is placed into position by aligning it with the ocular micrometer. The laser is turned on and continuous illumination is applied (usually 2 to 5 min).

VIII. Limitations

While CALI has been effective in 90% of the cases examined (see Table I). CALI may not work for the all proteins of interest. This is likely due to the heterogeneity of protein sensitivity to hydroxyl radical damage as well as the distance between the antibody binding site and a domain that is critical for the protein's function. However, CALI does dramatically enhance the probability of a binding antibody being a blocking reagent.

Thus far, CALI has resulted in hypomorphic and not amorphic phenotypes for the cases studied *in vivo*. For *in vitro* studies using CALI, there is generally 5–10% of the initial activity that is not affected. This residual amount of activity after CALI may be sufficient for normal function for some cellular processes. In addition, recovery of function may occur by *de novo* protein synthesis in some cases, although the rate of turnover varies from protein to protein (generally in the range of hours to days; e.g., 10 to 20 hr for the TCR complex; Klausner *et al.*, 1990). For example, T-cell activation was markedly perturbed when measured at 15 hr after CALI of the T-cell receptor but not at 36 hr (Liao *et al.*, submitted). The effects of CALI and micro-CALI were observed as long as 36 hours for fasciclin I and II in Ti1 neuron development (Diamond *et al.*, 1993).

An additional concern is the possibility that CALI is inactivating other proteins within a multisubunit complex. Although the effects of CALI drop fourfold for each intervening protein in a complex (Linden *et al.*, 1992; Liao *et al.*, submitted), there is a slight effect on neighboring subunits and one must take this into account when interpreting data.

The success of CALI is dependent on the specificity and quality (titer) of the antibodies employed. One should check if antibody binding alone has effects. This is generally not so; most antibodies do not block function. There are also possible concerns with antibody accessibility and retention. It is necessary to ensure that the antibody is present in sufficient quantity to bind all of the protein to be studied that is present at the site of interest. This can be tested by fixing injected embryos at the proposed time of CALI and performing immunohistochemistry using secondary antibody alone.

IX. Concluding Remarks

CALI and micro-CALI provide a high level of spatial and temporal resolution that is not offered by antibody blocking or antisense technology but only with difficulty by genetic techniques. For micro-CALI, the laser is aimed at only one cell at a time in the embryo or culture dish, and other similar cells in the system can act as nonirradiated controls.

CALI also provides a means of functional perturbation in systems that lack genetic techniques. It has been successfully applied to grasshopper embryos (Jay and Keshishian, 1990: Diamond et al., 1993), mouse T-cells (Liao et al., submitted), and chick neurons in culture (Sydor and Jay, 1993).

Compensatory mechanisms seem to be problematic in genetic approaches to studying complex developmental processes. The acute nature of the inactivation caused by CALI may not be subject to these effects so that one can observe the effects of an instantaneous loss of function that does not allow time for compensatory mechanisms. In addition, many proteins that are required early in development also play roles in later developmental processes. These later processes are inaccessible to knock-out mutations in these genes since the embryo dies before this stage. CALI may prove useful in studying these later processes.

CALI is a powerful new approach for studying a wide array of biological questions. Its ability to disrupt function in a temporally and spatially specific manner make it particularly useful in addressing molecular mechanisms of development.

Acknowledgments

The authors thank Martin Hülskamp and Madeline Crosby for critical reading of the manuscript. This work is funded by the Human Frontiers of Science, the Lucille P. Markey Charitable Trust, and the National Institutes of Health. DGJ is a Lucille P. Markey scholar and the John L. Loeb Associate Professor of the Natural Sciences.

References

Ashburner, M. (1989). Injecting eggs and embryos. In "Drosophila—A Laboratory Handbook," pp. 223–228. Cold Spring Harbor, NY: Cold Spring Harbor Laboratory Press.

Diamond, P., Mallavarapu, A., Schnipper, J., Booth, J. W., Park, L., O'Connor, T. P. and Jay, D. G. (1993). Fasciclin I and II have distinct roles in the development of grasshopper pioneer neurons. Neuron 11, 409–421.

Jay, D. G. (1988). Selective destruction of protein function by chromophore-assisted laser inactivation. Proc. Natl. Acad. Sci. U.S.A. 85, 5454–5458.

Jay, D. G., and Keshishian, H. (1990). Laser inactivation of fasciclin I disrupts axon adhesion of grasshopper pioneer neurons. Nature 348, 548–550.

Klausner, R. D., Lipincott-Schwartz, J., and Bonifacino, J. S. (1990). The T cell antigen receptor: Insights into organelle biology. Annu. Rev. Cell Biol. 6, 403–431.

Liao, J. C., Roider, J., and Jay, D. G. (1994). Chromophore-assisted laser inactivation of proteins is mediated by the photogeneration of free radicals. *Proc. Natl. Acad. Sci. U.S.A.* **91,** 2659–2663.

Linden, K. G., Liao, J. C., and Jay, D. G. (1992). The spatial specificity of chromophore-assisted laser inactivation of protein function. *Biophys. J.* **61,** 956–962.

Macdonald, P. M., and Struhl, G. (1986). A molecular gradient in early *Drosophila* embryos and its role in specifiying the body pattern. *Nature* **324,** 537–545.

Mann, K. G., and Fish, W. W. (1972). Protein polypeptide chain molecular weights by gel chromatography in guanidinium chloride. *Methods Enzymol.* **26,** 28–42.

Miller, J. P., and Selverston, A. I. (1979). Rapid killing of single neurons by irradiation of intracellularly injected dye. *Science* **206,** 702–704.

Santamaria, P. (1986). Injecting eggs. *In* "*Drosophila*—A Practical Approach" (D. B. Roberts, ed.), Oxford: IRL Press.

Schmucker, D., Su, A., Beermann, A., Jäckle, H., and Jay, D. B. (1994). Chromophore-assisted laser inactivation of patched protein switches cell fate in the larval visual system of *Drosophila*. *Proc. Natl. Acad. Sci. U.S.A.* **91,** 2664–2668.

Spradling, A. (1986). P element-mediated transformation. *In* "*Drosophila*—A Practical Approach" (D. B. Roberts, ed.), Oxford: IRL Press.

Stein, D., Roth, S., Vogelsang, E., and Nüsslein-Volhard, C. (1991). The polarity of the dorsoventral axis in the *Drosophila* embryo is defined by an extracellular signal *Cell* **65,** 725–735.

Sydor, A. M., and Jay, D. G. (1993). Regional laser inactivation of talin in the neuronal growth cone results in a transient loss of filopodial extension and retraction. A.S.C.B. meeting abstract supplement.

Van der Meer, J. M. (1977). Optical clean and permanent whole mount preparations for phase-contrast microscopy of cuticular structures of insect larvae. *Drosophila Information Service* **52,** 160.

Wieschaus, E., and Nüsslein-Volhard, C. (1986). Looking at embryos. *In* "*Drosophila*—A Practical Approach" (D. B. Roberts, ed.). Oxford: IRL Press.

INDEX

Page numbers in italics indicate figures; *t* after a page number indicates tables.

Formaldehyde/methanol fixation, for
  embryonic cytoskeletons, 495–498
Founder cells, identifying, 243
Fractionation
  additional, in isolating motor proteins,
    283
  isolation of mature oocytes by, 139–140
Freeze-substitution, in electron microscopy
    immunolabeling, 433
Freezing cells, 166
Fungal contamination, of culture, 23
  of imaginal discs, *in vitro,* minimizing, 119,
    121

G

GAL4 system of inducing ectopic gene
    expression, 637–639, 643–651
  applications, 644–645
  discussion, 650–651
  enhancer detection, 648–649
  expression vectors, 645–647
  GAL4-responsive genes, 647–648
  P-element transformation, 648
  screening for embryonic expression patterns
    in, 649
GAL4-upstream activating sequence (UAS),
    in following muscle differentiation *in situ,*
    245
Gamma rays, mutation induction by, 36
Gastrula-stage cells, elutriation, 136–137
Gel electrophoresis, flight muscle protein
    preparation for, 252
"Gene first" strategy, 43–46
  mutation selection in, 43–46
Generating local transposition line protocols,
    in targeted mutagenesis, 88–89
Genes
  actions, analysis, 57–64
    genetic mosaics in, 58–64
    molecular probes in, 58
  activity, electron microscopy visualization,
    613–629; *see also* Electron microscopy
    for visualization of genetic activity
  ectopic expression, 635–652; *see also*
    Ectopic gene expression
  extracellular matrix, temporal and spatial
    expression, 320–323
  GAL4–responsive, 647–648
  interactions, exploring, double mutants in,
    67–69

interesting, identified by mutation in
    "mutation first" strategy, cloning,
    41–43
  nautilus, in muscle precursor, 243
  nomenclature, 26
  reporter, in following muscle differentiation
    in situ, 244–245
  in situ expression, analysis, 563–630
    electron microscopy methods for
      visualization of genetic activity,
      613–629; *see also* Electron
      microscopy in visualization of
      genetic activity
    mRNA splicing and transport *in situ,*
      599–610; *see also* mRNA splicing
      and transport *in situ*
    quantitative, for transcription in larvae
      and prepupae, 565–572; *see also*
      Transcription in larvae and prepupae,
      quantitative analysis
  in situ hybridization to RNA, 575–597
  twist, in muscle precursor, 243
Genetic analysis, molecular and classical,
    631–732; *see also* Molecular and classical
    genetic analysis
Genetic mosaics, making, 60–64
  by chromosome loss, 60–61
  by injection of cells or nuclei, 63
  by mitotic recombination, 61–62
Genetics, 33–71
Genetics and Biology of Drosophila, The, 6
Genome of Drosophila melanogaster, The,
    4–5
Genomic Southern blot hybridization, for P-
    element insert detection, 83–84
Germline cell markers, 448, *449*
Germline clones, making, 63–64
Glycerol clearing, of embryos for neuronal
    subset imaging, 472
Glycerol solutions, 483
Gooseberry mutant, 477
Grape plates recipe, 107–108

H

Hatching failure, muscle dysfunction and, 250
Heat-shock-induced ectopic gene expression,
    636, 639–643
  applications, 639–640
  driving, 641–643
  heat treatment of adults, 643
  heat treatment of embryos, 641–642

VOLUMES IN SERIES

Founding Series Editor
DAVID M. PRESCOTT

Volume 1 (1964)
Methods in Cell Physiology
Edited by David M. Prescott

Volume 2 (1966)
Methods in Cell Physiology
Edited by David M. Prescott

Volume 3 (1968)
Methods in Cell Physiology
Edited by David M. Prescott

Volume 4 (1970)
Methods in Cell Physiology
Edited by David M. Prescott

Volume 5 (1972)
Methods in Cell Physiology
Edited by David M. Prescott

Volume 6 (1973)
Methods in Cell Physiology
Edited by David M. Prescott

Volume 7 (1973)
Methods in Cell Biology
Edited by David M. Prescott

Volume 8 (1974)
Methods in Cell Biology
Edited by David M. Prescott

Volume 9 (1975)
Methods in Cell Biology
Edited by David M. Prescott

Volume 10 (1975)
Methods in Cell Biology
Edited by David M. Prescott

Volume 11 (1975)
Yeast Cells
Edited by David M. Prescott

Volume 12 (1975)
Yeast Cells
Edited by David M. Prescott

Volume 13 (1976)
Methods in Cell Biology
Edited by David M. Prescott

Volume 14 (1976)
Methods in Cell Biology
Edited by David M. Prescott

Volume 15 (1977)
Methods in Cell Biology
Edited by David M. Prescott

Volume 16 (1977)
Chromatin and Chromosomal Protein Research I
Edited by Gary Stein, Janet Stein, and Lewis J. Kleinsmith

Volume 17 (1978)
Chromatin and Chromosomal Protein Research II
Edited by Gary Stein, Janet Stein, and Lewis J. Kleinsmith

Volume 18 (1978)
Chromatin and Chromosomal Protein Research III
Edited by Gary Stein, Janet Stein, and Lewis J. Kleinsmith

Volume 19 (1978)
Chromatin and Chromosomal Protein Research IV
Edited by Gary Stein, Janet Stein, and Lewis J. Kleinsmith

Volume 20 (1978)
Methods in Cell Biology
Edited by David M. Prescott

Advisory Board Chairman
KEITH R. PORTER

Volume 21A (1980)
**Normal Human Tissue and Cell Culture, Part A: Respiratory,
    Cardiovascular, and Integumentary Systems**
Edited by Curtis C. Harris, Benjamin F. Trump, and Gary D. Stoner

Volume 21B (1980)
Normal Human Tissue and Cell Culture, Part B: Endocrine, Urogenital, and Gastrointestinal Systems
Edited by Curtis C. Harris, Benjamin F. Trump, and Gary D. Stoner

Volume 22 (1981)
Three-Dimensional Ultrastructure in Biology
Edited by James N. Turner

Volume 23 (1981)
Basic Mechanisms of Cellular Secretion
Edited by Arthur R. Hand and Constance Oliver

Volume 24 (1982)
The Cytoskeleton, Part A: Cytoskeletal Proteins, Isolation and Characterization
Edited by Leslie Wilson

Volume 25 (1982)
The Cytoskeleton, Part B: Biological Systems and *in Vitro* Models
Edited by Leslie Wilson

Volume 26 (1982)
Prenatal Diagnosis: Cell Biological Approaches
Edited by Samuel A. Latt and Gretchen J. Darlington

Series Editor
LESLIE WILSON

Volume 27 (1986)
Echinoderm Gametes and Embryos
Edited by Thomas E. Schroeder

Volume 28 (1987)
***Dictyostelium discoideum:* Molecular Approaches to Cell Biology**
Edited by James A. Spudich

Volume 29 (1989)
Fluorescence Microscopy of Living Cells in Culture, Part A: Fluorescent Analogs, Labeling Cells, and Basic Microscopy
Edited by Yu-Li Wang and D. Lansing Taylor

Volume 30 (1989)
Fluorescence Microscopy of Living Cells in Culture, Part B: Quantitative Fluorescence Microscopy—Imaging and Spectroscopy
Edited by D. Lansing Taylor and Yu-Li Wang

Series Editors
LESLIE WILSON AND PAUL MATSUDAIRA

ISBN 0-12-564145-1

90018

9 780125 641456